Enabling the Internet of Things

Massimo Alioto
Editor

Enabling the Internet of Things

From Integrated Circuits
to Integrated Systems

Springer

Editor
Massimo Alioto
Department of Electrical & Computer Engineering
National University of Singapore
Singapore, Singapore

ISBN 978-3-319-51480-2   ISBN 978-3-319-51482-6 (eBook)
DOI 10.1007/978-3-319-51482-6

Library of Congress Control Number: 2016963262

Printed on acid-free paper

This Springer imprint is published by Springer Nature
The registered company is Springer International Publishing AG
The registered company address is: Gewerbestrasse 11, 6330 Cham, Switzerland

To Maria Daniela and Marco, and now also Marina.
My deepest merriment begins with M. And them, indeed.

To my family, including my beloved nephews and niece:
Rachele, Davide, Gaetano and Francesco.

And to everyone who has inspired my curiosity and love for life,
including Giusi, Rodolfo, Alfio, Dora, Gaetano, Annamaria
and Santina. And thankfully many, many others.

Preface

Decades of exponential improvements in integrated circuit manufacturing and design have been spurred by (and given rise to, in a virtuous circle) relentless reduction in the cost per transistor, as well as many other interesting consequences. The cost reduction keeps following a well-proven learning curve, which includes Moore's law as a corollary, and will continue in spite of the end of this law. According to the Bell's law, such technology trend has led to inexorable shrinking of electronic systems, which are now approaching the sub-centimeter scale and below. At the same time, the Koomey's and Gene's laws promise the reduction in the energy consumption by another two orders of magnitude. Those trends promise the possibility of integrated electronic systems that are very inexpensive, small, and extremely low power. In other words, we will increasingly see systems that are pervasive in space and long-lived in time. At the same time, Metcalfe's law (or Sarnoff's law for the least optimistic) traces the fast-growing value of connectivity, thanks to the rapidly increasing number of connected users, and, more in general, connected objects.

At the same time, several megatrends are demanding more pervasive and continuous sensing, as well as sensemaking and transfer of physical data. Accelerated urbanization and increasing worldwide population requires sustainable usage and sharing of resources, as well as more livable and smarter environments at all scales (from home to city). Pervasive sensing and sensemaking are also being required by assistive and proactive technologies (e.g., robotics, decision support) that increasingly relieve humans from routine tasks, repetitive labor, and recently data-driven decision-making. The sharing economy is demanding the ability to spatially track, to physically monitor and manage objects, to encourage responsible usage, and to charge users by the actual usage. Well-being and other human factors are being modeled and monitored to create healthy environments where humans can be happy and productive. Geosocialization and participatory sensing are progressively involving objects other than individuals or as support to human activities. Three-dimensional remote physical interaction with reality provides sensory feedback, thus demanding ubiquitous sensing to enable this ability on a wider scale and on a finer granularity.

The push and the pull effect of the above technological trends and applications is converging on and creating a virtuous circle that we now call the "Internet of Things" (IoT). The IoT can evidently create a huge value and bring unprecedented benefits to the society. To set this on a trend perspective, we can extrapolate Hick's law to artificial intelligence and cloud computing: more physical data will enable us to take more automated decisions with an effort that is only logarithmic in the space of decision choice. The IoT is ultimately a powerful enabler to share on a larger scale, make technology more human centric and real time, and decouple socioeconomic progress from intensive use of resources. And, interestingly, IoT silicon technology becomes so small that the user is immersed in it (there is no more "user experience," in a sense), with interesting implications in terms of market and perceived value.

In spite of the daily IoT-related claims in the chip design community, the tiny sensing nodes of the IoT at its edge (the "IoT nodes") are still in their technological infancy. Several challenges need to be tackled, such as energy efficiency and related lifetime, cost, security, and interoperability, among others. Such challenges need to be tackled in a holistic manner, developing both an understanding of the different parts of IoT nodes and an insight into the big picture and the strong linkage to applications and related requirements.

To the best of our knowledge, this is the first book on integrated circuit and system design for the Internet of Things. This book develops in both the "vertical" and the "horizontal" dimension. Vertically, it provides a comprehensive view on the challenges and the solutions to successfully design chips for IoT nodes as systems (from circuits to packages), a broad analysis of how chip design needs to evolve to meet those challenges, and a fresh perspective grounded on historical and recent trends. Horizontally, the book covers in one place the very diverse domain-specific expertise of the subareas involved in the design of IoT nodes, which was previously scattered across a large number of talks, journals, and conferences.

This book provides a design-centric perspective, providing an understanding of what the IoT really means from a design point of view. Typical specifications of commercial IoT nodes are discussed, and constraints imposed by IoT applications are translated into design constraints that chip designers are used to deal with. Design guidelines to meet them are systematically discussed in every chapter.

This book started in the form of talks at various venues, such as VLSI Symposium, HotChips, and ISCAS, where I had very interesting conversations with several other speakers. Those talks were motivated by the lack of a cohesive and detailed source of accessible knowledge on the design of IoT nodes. The idea to write this book came exactly from those conversations, which later continued throughout the interaction with chapter authors. They really made this book possible, providing their deep insights and invaluable expertise. I deeply thank all outstanding researchers and designers who contributed to the chapters of this book, sharing their expertise in an accessible and concise manner for the benefit of our community.

This book is structured as follows. Chapter 1 describes the big picture in view of technological trends, an overview of the challenges ahead and the possibilities that research has recently opened, and some link to the economics of the IoT and social megatrends. Chapter 2 provides a system-level perspective of IoT nodes. Then, Chaps. 3–7 cover the design of digital subsystems of IoT nodes, from architectures to circuits, and memories in CMOS and other emerging technologies. Chapter 8 is about hardware-level security techniques, whereas Chap. 9 focuses on System-on-Chip design methodologies. Power management and energy harvesting are covered in Chaps. 10 and 11. Analog interfaces and analog–digital converters are discussed in Chaps. 12 and 13. Short-range radios are discussed in Chap. 14. Batteries as further essential component of IoT nodes are the focus of Chap. 15. Packaging is the topic of Chap. 16. Finally, Chaps. 17 and 18 describe two system integration examples, exemplifying the design techniques introduced in the previous chapters. As a common thread, all chapters include a final section on perspectives and trends, which provides a glance into the future, and a good starting point for further research and advances.

There are many ways to use this book. In particular, it can serve as a reference to practicing engineers working in the broad area of integrated circuit/system design of IoT nodes, in view of the wide and detailed coverage of state-of-the-art solutions for IoT and the fresh perspective on the future of such technologies. The book is also very well suited for undergraduate, graduate, and postgraduate students, thanks to the rigorous and lean coverage of topics and selected references.

Singapore   Massimo Alioto
December 2016

Contents

About the Editor

Massimo Alioto (M'01–SM'07-F'16) was born in Brescia, Italy, in 1972. He received the Laurea (M.Sc.) degree in Electronics Engineering and the Ph.D. degree in Electrical Engineering from the University of Catania (Italy) in 1997 and 2001, and a Bachelor of Music in Jazz Studies from the Conservatory of Music of Bologna in 2007.

He is currently an Associate Professor at the Department of Electrical and Computer Engineering, National University of Singapore, where he leads the Green IC group and is the Director of the Integrated Circuits and Embedded Systems area. Previously, he was Associate Professor at the Department of Information Engineering of the University of Siena. In 2013 he was also Visiting Scientist at Intel Labs—CRL (Oregon) to work on ultra-scalable microarchitectures. In 2011–2012, he was Visiting Professor at the University of Michigan, Ann Arbor, investigating on active techniques for resiliency in near-threshold processors, energy-quality scalable VLSI design, and self-powered circuits. In 2009–2011, he was Visiting Professor at BWRC—University of California, Berkeley, investigating on next-generation ultra-low power circuits and wireless nodes. In the summer of 2007, he was a Visiting Professor at EPFL—Lausanne (Switzerland).

He has authored or co-authored more than 220 publications in journals (80+, mostly IEEE Transactions) and conference proceedings. One of them is the second most downloaded TCAS-I paper in 2013. He is co-author of three books, *Enabling the Internet of Things—From Integrated Circuits to Integrated System* (Springer, 2017), *Flip-Flop Design in Nanometer CMOS—From High Speed to Low Energy* (Springer, 2015) and *Model and Design of Bipolar and MOS Current-Mode Logic: CML, ECL and SCL Digital Circuits* (Springer, 2005). His primary research interests include ultra-low power VLSI circuits, self-powered and wireless nodes, near-threshold circuits for green computing, energy-quality scalable VLSI circuits, hardware-level security, circuits for on-chip learning, and circuit techniques for emerging technologies.

Prof. Alioto was a member of the HiPEAC Network of Excellence (EU) and the MuSyC FCRP Center (US). In 2010–2012 he was the Chair of the "VLSI Systems and Applications" Technical Committee of the IEEE Circuits and Systems Society, for which he was also Distinguished Lecturer in 2009–2010 and member of the DLP Coordinating Committee in 2011–2012. He is also member of the Board of Governors of the IEEE Circuits and Systems Society (2015–2017). In the last 5 years, he has given 50+ invited talks in top universities and leading semiconductor companies. He currently serves as Associate Editor-in-Chief of the *IEEE Transactions on VLSI Systems*, and served as Guest Editor of various journal special issues (e.g., IEEE TCAS-I issue on Internet of Things in 2017, IEEE TCAS-II issue on green computing in 2012). He also serves or has served as Associate Editor of a number of journals, such as *IEEE Transactions on VLSI Systems*, *ACM Transactions on Design Automation of Electronic Systems*, *IEEE Transactions on CAS—part I and part II*, *Microelectronics Journal*, and others. He serves or has served as panelist for several funding agencies and research programs in the USA and Europe. He was Technical Program Chair (*ICECS*, *PRIME*, *VARI*, *NEWCAS*, *ICM*, *SOCC*) and Track Chair in a number of conferences (*ICCD*, *ISCAS*, *ICECS*, *VLSI-SoC*, *APCCAS*, *ICM*). Prof. Alioto is an IEEE Fellow.

IoT: Bird's Eye View, Megatrends and Perspectives

Massimo Alioto

This chapter opens the book and provides a summary of the challenges and the opportunities that are offered by the Internet of Things (IoT), with emphasis on the aspects that are relevant to integrated circuit and system design from circuits to packaging for IoT nodes. The chapter is organized along a chronological perspective, first reviewing technology historical trends beyond mere Moore's law, and summarizing recent past achievements and capabilities that are making the IoT possible. Then, present challenges are described, as pathway to up-coming advances and developments in the design of IoT nodes. Finally, mega-trends are examined to unearth clues on longer-term evolution of the IoT and the implications on integrated system design.

1.1 The Internet of Things: Context and Overview

The concept of the IoT seems to first appear in Kevin Ashton in a presentation delivered at Procter & Gamble in 1999 (Ashton 2009), which was then described as a large-scale network of smart RFIDs. On a broad perspective, the IoT lies at the intersection of the Internet realm with

M. Alioto (✉)
National University of Singapore, Singapore 117583, Singapore
e-mail: malioto@ieee.org

its pervasive networking, cloud and related advances (e.g., big data), the physical world through distributed sensing and people's activities, in the unprecedented form of mostly real-time fine-grain and aggregated data from the knowledge coming from environments, goods, resources, tools, infrastructures, among the others.

So far, the IoT has been defined in several different ways, and its meaning has become so broad that it oftentimes includes any object on earth that is connected to the Internet, such as connected cars, drones, smartphones, smart appliances, industrial tools, and so on. Under such generic definition based on pure Internet connectivity, the IoT has been already realized as the number of computing devices connected to the Internet surpassed the worldwide population back in 2008–2009 (Evans et al. 2011).

This book focuses on the IoT as pervasive, unobtrusive, systematic and coordinated introduction of sense-, compute-, communication-ability and sensemaking of physical data in a very large number of objects on earth. This is enabled by the introduction of extremely miniaturized integrated systems ("IoT nodes") with very long lifetime (e.g., decades) that are autonomous in many respects, from functionality, to energy, to the way they interact with the physical world and the network infrastructure. From this perspective, the IoT pushes such capabilities beyond personal devices (e.g., smartphones), embedding them in everyday objects and living environments. This book addresses the challenges involved in the creation of IoT nodes in the form of integrated circuits, covering the different areas involved in this process including architecture, circuit building blocks, design methodologies, packaging and system demonstrations. Being the IoT an extensive topic, the scope of this book purposely excludes the challenges related to the integration of IoT nodes into a cohesive and scalable network comprising inter-operable and heterogeneous nodes, and related communication protocol and software layers.

A commonly agreed target of the IoT is to expand the number of connected devices per person to the order of a thousand, thus reaching an unprecedented scale of trillions of connected devices (Gaudin 2015). The number of connected devices is expected to grow to 30–50 billion devices by 2020, with an expected market CAGR growth of 15–35% (Markets and Markets; https://newsroom.cisco.com/press-release-content?type=press-release&articleId=1771211; Ericsson Mobility Report; http://www.gartner.com/newsroom/id/3165317; Greenough and Camhi 2015; Worldwide Internet of Things Forecast 2015; TechNavio 2015; Machina Research 2015; Bauer et al. 2014; Jankowski et al. 2014; Dobbs et al. 2015; Digital Universe of Opportunities 2014), [IoT Analytics, Oct. 2014], (http://www.postscapes.com/internet-of-things-market-size/). Some forecasts question such fast growth and

predict a somewhat slower growth (Nordrum 2016). The IoT market size is expected to have a global economic impact of 2.5–11.1 T\$ by 2022–2025 (Dobbs et al. 2015; Jankowski et al. 2014).

As shown in the simplified architecture in Fig. 1.1, the IoT is structured into three tiers of devices. At the bottom, IoT nodes perform sensing and interact with the physical world. To assure scalability and ubiquitous network access, gateways and concentrators collect, protect (under users' control) and route data from several and physically proximal IoT nodes, and route it to servers. The latter perform data aggregation and knowledge extraction, and deliver physically-enhanced cloud services. Some additional intermediate levels of aggregation might be needed, depending on the amount of data generated, the area covered by a sub-network, and the density of IoT nodes, among the others. For example, concentrators might actually be a sub-set of the network below an Internet hub/gateway, which is here omitted as this would be simply part of the existing Internet infrastructure.

The hardware requirements of the devices in the three tiers in Fig. 1.1 are very different, by virtue of their significantly different number and level of pervasiveness. The number of IoT nodes is expected to be approximately two orders of magnitude larger than the number of concentrators, which in turn is plausibly higher than the number of server blades by another two orders

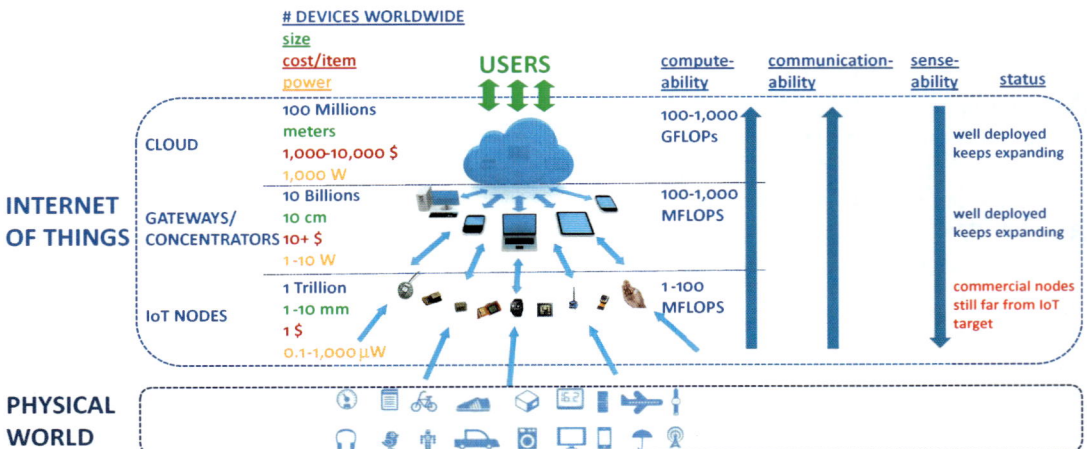

Fig. 1.1 A simplified architecture of the IoT

of magnitude. To be embeddable in objects and the living environment, the form factor of IoT nodes is expected to be in the scale of millimeters, which is at least an order of magnitude smaller than concentrators, whose size is expectedly in the same order as today's wireless routers (10-cm range). The form factor of server blades is another order of magnitude larger. The cost target for IoT node is widely accepted to be in the 1-dollar range (Ricker et al. 2016), as might be expected by observing that an average customer in the consumer electronics market would likely spend as much as a top-of-the-line smartphone to populate their home and objects with 1000 IoT nodes. Concentrators are allowed to have a larger cost in view of their lower number, expectedly by an order of magnitude at least, considering their non-trivial computational and wireless bandwidth requirements. In turn, cloud servers clearly entail a larger cost by at least two orders of magnitude. Similarly, concentrators are expected to deliver at least two orders of magnitude more compute power compared to IoT nodes, which can typically have very limited (e.g., sub-Mega Operations per Second—MOPS) or moderate computational capability (100 MOPS). Cloud servers are certainly required to have a much larger computational power compared to concentrators, by at least two orders of magnitude.

Due to their large number and ubiquity, IoT nodes need to be untethered and hence their power budget is very small, and is as low as sub-μW for miniaturized systems powered by energy harvesters. Due to their larger size and lower density, concentrators are expected to be mostly tethered, and hence their power can be much larger (e.g., in the order of Watts). A server blade dissipates a power that is two orders of magnitude larger.

IoT nodes have design requirements that are markedly different from existing Internet-connected devices (e.g., networked computers and smartphones), as they aim at facilitating convergence of several tasks onto a single platform (Jankowski et al. 2014). Instead, IoT nodes need to pursue hardware specialization and application specificity, mostly for the very stringent power requirements, as discussed in the following sections. As a result, IoT technologies currently tend to substantial fragmentation, posing a fundamental challenge in terms of economy of scale and interoperability, which adds to the expected fragmentation due to lack of standardization in this early phase of its development.

Internet cloud services and wireless networks will be greatly affected by the expansion of the IoT, due to the large number of connected nodes. The IoT is indeed being responsible for data deluge issues that impact the network traffic, and the power associated with wireless communications. Regarding the data deluge, currently only 1% of enterprise data is being used to generate valuable knowledge, and is mostly utilized for alarms or real-time control (Dobbs et al. 2015). Such poor data utilization for useful purposes and value creation will further worsen due to the volume increase determined by the IoT, as the worldwide data is expected to grow by $4\times$ in 2015–2020 (Digital Universe of Opportunities 2014; Jankowski et al. 2014; https://newsroom.cisco.com/press-release-content?type=press-release&articleId=1771211). By 2020, data generated by IoT devices will account for 10% of the world's data (Digital Universe of Opportunities 2014) (i.e., approximately 44 zettabytes). Hence, the IoT will demand better data utilization as well as pre-selection and filtering of valuable data to be processed and stored in the cloud. Regarding the volume of wirelessly transmitted data, in 2020 the IoT is expected to generate $1000\times$ more data than in 2015 (Digital Universe of Opportunities 2014), with an overall power consumption that would become comparable to expected total worldwide energy production of 25 PWh (Callewaert 2016). Accordingly, the wireless power consumption in the IoT needs to be substantially reduced for sustainability reasons, which adds to the issues raised by the tight power limitations of IoT nodes (see later). In addition, the large number of IoT nodes requires an acceleration in the transition from the 32-bit IPv4 Internet protocol suitable for 4×10^9 different addresses, to the 128-bit IPv6 protocol that can handle up to 10^{38} addresses, with some challenge imposed by the different 64–96 bit length of RFID identifiers (Atzori et al. 2010). Finally, the Internet as we know it today was mostly designed for non-real-time sharing of documents and data, with resiliency being the main concern (Greenemeier et al. 2013). Due to the generation of large amounts of real-time data, the IoT pushes the Internet towards its limit and hence needs to be structured in a more decentralized manner to assure sustainable scalability.

The above issues related to the IoT data deluge are drastically mitigated by moving intelligence from the cloud to the concentrators and most importantly to the IoT nodes in Fig. 1.1, i.e. making the IoT nodes "smarter" (or "cognitive", if intelligence means ability to detect and classify patterns) than they are today. Indeed, pre-processing in the IoT nodes and more distributed intelligence reduce the data volume, as only partially aggregated data needs to be sent over the network, as opposed to raw data.

1.2 Brief Review of IoT Applications

1.2.1 Considerations on the IoT Market Volume

The IoT as a whole is inherently a general-purpose technology, similarly to computers and mobile devices in the past decades. Like any other general-purpose technology, it can boost true productivity and create a value that is substantially higher than its market size, as it can serve as catalyst for bigger change (Brynjolfsson and Hitt 1998). Indeed, the IoT can further improve efficiency, economy of scale, ability to react to and predict demand in capex, labor and energy. Also, the IoT is expected to enable better coordination and usage monitoring of buildings, machinery, manufacturing processes, factories, supply chain and resources. The IoT will impact a very wide diversity of applications, from agriculture to consumer products, automotive, healthcare, retail, manufacturing and supply chains (e.g., Industry 4.0), telecommunications, logistics, public sector, financial, transportation and shipping, smart environments from homes to

buildings and nations, toys, worksites, smart infrastructures, energy, lifestyle/entertainment, among the others.

As opposed to previous technological waves in the semiconductor history, the IoT is the first one that is so pervasive that it becomes invisible to the users, with several implications on the value capturing in the semiconductor industry. For example, only 5–10% of the IoT technology spending is expected to fuel the semiconductor industry market (Dobbs et al. 2015), whereas more value (15–20% each) will be captured by software and integration services. Plausibly, most of the value of the IoT will come from the data aggregation and the real-time response (or actuation) of cloud services, as well as the demand prediction for new proactive services and tasks that no longer need us to "push a button" (or click a mouse) to be executed. To capture more value from the large market volume and by delivering integration services (e.g., from IoT nodes to software for data aggregation and sensemaking), semiconductor companies will likely become more vertically integrated through acquisitions, close partnerships and industrial consortia. As further benefit, this trend will also favor IoT node inter-operability and standardization.

1.2.2 Summary of Current and Prospective Applications of the IoT

The IoT is a very fragmented application scenario (Vermesan and Friess 2014), and encompasses a wide range of applications, some of which are summarized in the following.

In the agriculture sector, the IoT infrastructure can monitor the quality, the actual usage and the availability of resources, for better and predictive management (e.g., irrigation) and storage (e.g., avoid waste of feed and fertilizing). Monitoring the environmental conditions permits to support the growth of animals and plants (e.g., aquaculture), optimally time the next course of action, and ultimately assure quality (e.g., wine) and raise the efficiency in the production process.

In automotive, the IoT enables the monitoring of the state of a vehicle down to its critical components, from initial shipping to usage, to assess their correct utilization (e.g., detecting bumps, vibration) and maintenance (e.g., opening of containers, wearing parts). Based on actual usage, predictive maintenance can be performed to lengthen the vehicle lifetime, and lower the upkeep. Such capabilities enabled by the IoT are also very useful in fleet management and car sharing services. Also, distributed sensing and global sensemaking enables traffic control through differentiated and personalized road pricing to encourage virtuous behavior and prioritize tasks for commercial (e.g., car pooling with multiple passengers sharing cost) and private vehicles (e.g., fast delivery for critical goods), through virtual/dynamic city area boundaries.

In public transportation, the occupancy and utilization can be monitored to assure an adequate quality of service, detect potential danger (e.g., potential collision between vehicles and pedestrians), and predict short term demand based on crowd monitoring in strategic locations. On the road side, excessive congestion and pollution can be managed with real-time demand-response schemes where the road pricing is dynamically adjusted through real-time observations and utilization prediction, based on previous history and real-time data in strategic locations. Also, the transportation of dangerous goods and the circulation of slow (or frequently stationary) vehicles can be optimally coordinated with the ordinary traffic to minimize their negative impact. Again, the IoT offers unprecedented opportunities to share resources efficiently, while preserving their running condition.

Consumer electronics substantially benefits from the IoT, as its pervasiveness permits to track smart goods (e.g., positioning systems for object retrieval) and detect their exposure to anomaly conditions (e.g., overheating, physical shocks). IoT sensing can signal spatial co-presence of objects and specific people (e.g., kids) to signal potential danger, or to recommend activities to complete when all necessary objects tools are available in the same space. Smart objects for personal care and hygiene can be used to remind of regular but infrequent care activities based on dentist's suggestions shared in the cloud, and motivate positive behavior in children. Smart clothing can remind of periodic cleaning based on actual usage. Smart toys can be selectively enabled only upon the occurrence of desired conditions to create positive habits (e.g., only at certain times or lighting conditions), and prevent danger by disabling them under the presence of others (e.g., toddlers). Smart jewelry can be used to unobtrusively track activity, measure exposition to solar light and other environmental conditions, and make emergency calls.

Energy management at different scales can be made more effective by the IoT. At the city scale, the smart grid offers several opportunities to leverage the sensing and sensemaking capabilities of the IoT to optimize the energy usage across many users, a better coordinated usage and planning of alternative energy sources, ultimately reducing the overall energy and the currently large gap between the peak and the average consumption.

Health care is another important application area in which the IoT promises to fundamentally contribute to. As few examples, the miniaturization and long lifetime of IoT nodes provides an unobtrusive mean to constantly monitor vital signs and other related parameters (e.g., behavioral) and develop deeper understanding of the patient's health evolution. In addition, the availability of big data from a large number of patients offers an unprecedented opportunity to explore correlations, build models and tools for predictive diagnosis, early treatment and make drug discovery more efficient and effective. Similar considerations hold for the elderly and the disabled, as constant non-obtrusive monitoring allows for better and highly responsive/predictive care, while preserving individual's independency and offloading hospitals. Remote supervision also enhances the ability to share professionals across a larger number of individuals and patients, thus driving the care cost down.

Industrial processes and logistics can also highly benefit from the IoT, as it can enable ubiquitous sensing of operating conditions, real-time tracking of semifinished products, detection of events that slow down the process throughput and potential safety issues. The data generated in the production line can be intelligently shared with the quality assurance process and across different sites, to raise the yield and reduce cost. On the warehouse side, product location and storage conditions can be tracked for more efficient product delivery and distribution. Similarly, sharing real-time and historical data on parts with the procurement process makes restocking more efficient, and reduces the inventory cost through more strategic purchasing strategies. The presence of IoT within machines can enable early and self-diagnosis, predictive (rather than reactive or pre-scheduled) maintenance, pre-emptive vendor support to prevent known failures. Again, the IoT enables better economy of scale, efficiency and makes processes leaner.

In the area of retail, smart malls can provide real-time shopping recommendations, matching available offers with individual customers, discard products for potential customers with allergy issues and provide other personalized services (e.g., for customer fidelization). The tight coupling between the individual and collective customers' behavior, the store setting and the warehouse permits to streamline the inventory management, offer better shopping experience, dynamically adjust in-store display based on the predicted demand, and cut inventory costs.

Through the IoT, smart homes can manage utilities more efficiently by controlling individual appliances based on actual utilization and needs, and purchasing electricity when cost is lowest within the day in demand-response energy pricing schemes. Unprecedented levels of security (e.g., perimeter access control) are achievable thanks to the pervasiveness of IoT nodes and sensemaking ability. Occupant recognition permits to adjust lighting, sound, air conditioning/heating based on individual preferences. This can be done in a predictive manner, so that occupants do not need to "push any button", leveraging the fine-grain knowledge of occupants' habits and the ability of the cloud to generalize and extract trends and predictions.

Smart homes promise to automatically order supplies before exhaustion and upon the availability of online offers, as a major step forward compared to the today's Amazon Dash wireless button that simply orders goods online when the button is pushed (Amazon Dash). At the same time, waste management will be made more efficient by sensing the actual demand, and pricing based on actual consumption habits, thus encouraging virtuous behavior. The IoT can also make residential compost recycling easier and automated, through the monitoring of humidity and temperature trends.

Smart buildings can leverage the IoT to be more adaptive to the actual demand and needs of the occupants, while ensuring the highest safety and comfort standards. Indeed, air quality and thermal/acoustic/visual comfort can be monitored and controlled for the first time with a granularity that goes down to the single room, with obvious advantages in terms of comfort assurance and energy cost. Beyond normal building operation, the real-time capability of the IoT enables the ability to respond to critical events (e.g., fire) quickly, minimizing the human and material losses in case of emergencies.

Through the IoT, smart cities can manage resources more efficiently, be made much more resilient to temporary malfunctions and disasters, and encourage virtuous behavior. Smart and weather-adapting lighting, water/gas leakage monitoring, smart parking with dynamic pricing and area allocation, no physical boundaries and automated parking advice are just a few examples of how to use the IoT to solve today's urban challenges. Ubiquitous vision can enable an unprecedented level of safety and security, detecting potential danger and provide crucial information on crowd behavior and citizens' needs (e.g., for adaptive and predictive transportation management, real-time digital signage recommendations to prevent immediate danger). Other than enabling ubiquitous and augmented surveillance, vision in IoT offers physical augmentation to social media and recommendation systems (e.g., venue recommendation based on crowdedness, and crow sentiment), and human activity monitoring to achieve better match between demand and supply of services dynamically. Similarly, distributed audition permits to develop situational awareness, build real-time noise urban maps to mitigate noise pollution at critical times, and localize noise events for safety assurance. Smart irrigation of green spaces and parks is another sub-area where the IoT has potential to make an impact. Smart tourism promises to give tourists the ability to have an immediate understanding of the city, such as availability, crowdedness or quietness of different places to receive dynamic recommendations on tours that adapt to their disposition, other than already available factual information on places. Waste management can be made more efficient and priced fairly as discussed before, while detecting potentially dangerous and inappropriate waste that would need to be disposed with different procedure (again, encouraging virtuous behavior).

Smart infrastructures will also benefit from the IoT in terms of safety (e.g., structural monitoring of bridges) and security (e.g., automated identification of unattended bags and suspicious behavior). Distributed IoT nodes will enable gesture-based natural human-infrastructure interface, where users do express their preferences anywhere and any time, being constantly observed, instead of pushing buttons on an electronic booth or controllers (e.g., thermostat). This also introduces the new capability to average out requests and preferences from multiple users, thanks to the distributed nature of such human-infrastructure interface.

In the areas of wildlife and nature preservation, the IoT can monitor both the activity and the living conditions of wildlife, as well as the quality of available natural resources (e.g., water), their level of pollution, forest fire detection, earthquake early detection, counteraction of illegal activities against wildlife.

1.3 Requirements of IoT Nodes

The distinctive features of IoT nodes nodes are defined by the requirements imposed by IoT applications in terms of physical constraints, type of interaction with the external world,

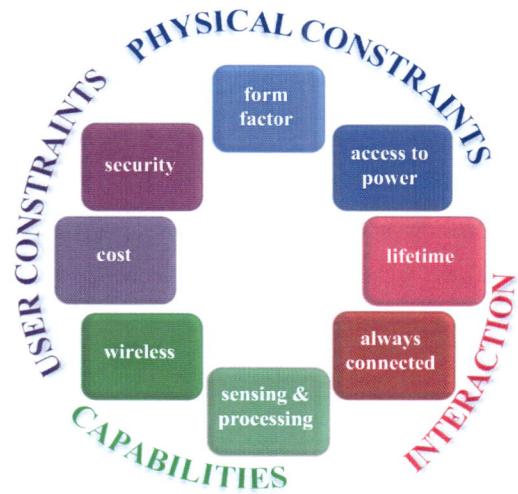

Fig. 1.2 Summary of IoT node requirements

required capabilities of IoT nodes and user requirements, as summarized in Fig. 1.2 and discussed in the following.

1.3.1 Physical Constraints

Physical constraints of IoT are dictated by size considerations and the necessity to untether IoT nodes and avoid any maintenance (e.g., battery replacement), as dictated by their large number. Regarding the form factor, IoT nodes need to be sufficiently small to make the deployment of IoT nodes non-intrusive, with a typical volume ranging from cubic millimeters to hundreds of mm^3. Being untethered, IoT nodes need to be energy autonomous and rely on a battery and/or an energy harvester as energy source (Alioto 2012).

In purely battery-powered nodes, the average IoT node power P_{avg} needs to be small enough to achieve the desired lifetime $t_{lifetime} = E_{battery}/P_{avg}$, for a given battery energy capacity $E_{battery}$. Figure 1.3 shows the lifetime of an IoT system versus its average power consumption, assuming optimistically that the battery self-leakage and ageing are negligible. From this figure, smartphone or button cell batteries assure a reasonably long lifetime of a decade (or more) for P_{avg} in the order of few hundreds of nWs.

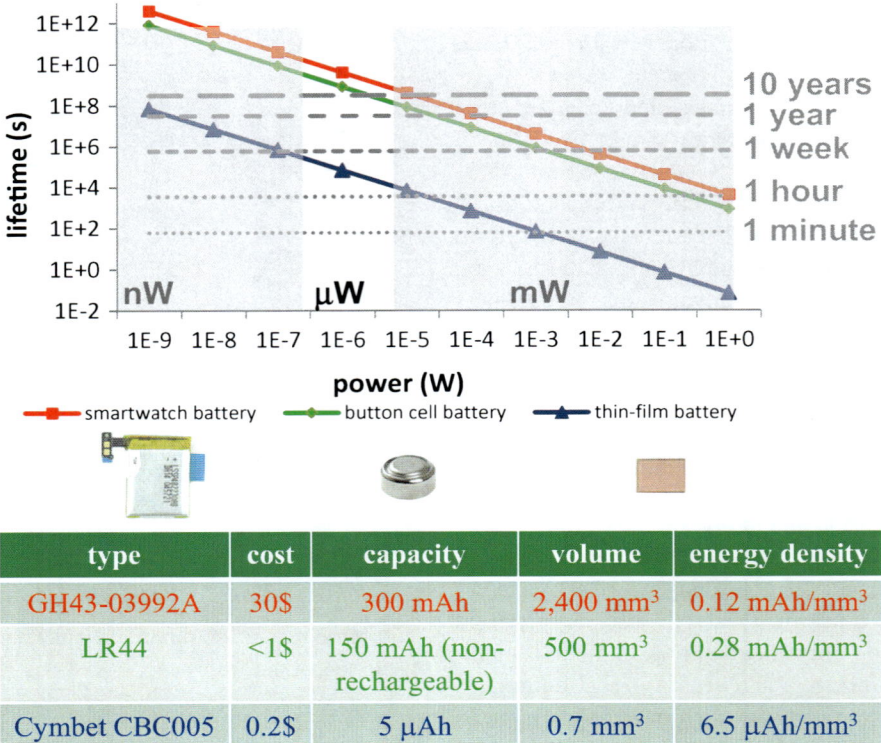

Fig. 1.3 Lifetime vs. average power consumption for different batteries

Larger P_{avg} mandate the addition of an energy harvester, whose size is generally proportional to P_{avg}. Figure 1.4 shows the harvester size required for a given P_{avg} for various energy sources. From this figure, millimeter-sized photovoltaic (indoor), thermo-electric (on-body patch) and airflow (indoor) harvesters can indefinitely sustain P_{avg} in the order of µWs (Alioto 2015). Tens of µWs are sustainable under more abundant energy sources, such as photovoltaic (outdoor), thermo-electric (industrial machines) and body vibration (e.g., walking) harvesting (Alioto 2015). GSM radio-frequency energy harvesting can instead sustain only tens to very few hundreds of nWs. From Fig. 1.3, printed and solid-state batteries (see Chap. 15) enable aggressive miniaturization at the cost of much shorter lifetime, whose extension requires the addition of an energy harvester in all practical cases. As a third energy source option, the battery can be suppressed altogether by pairing the energy harvester with a small energy source (e.g., off-chip supercapacitor, on-chip capacitor)

to deliver the peak power, if the former does not have adequate instantaneous power capability, as dictated by its size.

As opposed to purely battery-powered systems, energy harvested IoT nodes can operate nearly-perpetually, as long as the harvester power exceeds P_{avg} (i.e., the harvester size is large enough), and can hence indefinitely sustain the power required by the IoT node. On the other hand, an increase in the targeted lifetime $t_{lifetime} = E_{battery}/P_{avg}$ for a given P_{avg} requires the adoption of proportionally larger batteries. Hence, energy harvesters are invariably more compact than batteries for long enough lifetime targets. Figure 1.4 shows the breakeven lifetime at which harvester and battery have the same size, assuming a battery with energy density equal to typical alkaline button cell batteries (e.g., LR44). From this figure, harvesting is always more compact for all practical lifetime targets under abundant energy sources, such as photovoltaic (outdoor), thermo-electric (industrial machines) and body vibration (e.g., walking) (Alioto 2015). On the

Fig. 1.4 Lifetime vs. average power consumption for different batteries

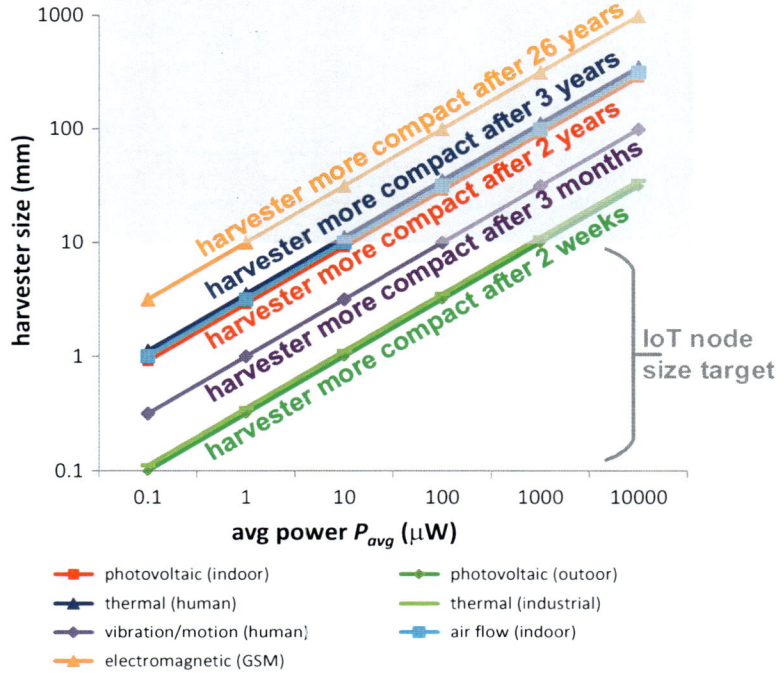

other hand, harvesters become more compact than batteries for targeted lifetimes of 2–3 years and longer, and hence in most of IoT applications. GSM radio-frequency harvesters are instead always larger sized than the battery counterpart.

Regardless of the specific lifetime target and energy source architecture, the volume of IoT nodes is certainly dominated by off-chip components, and in particular by the energy source, as the antenna can be made very thin. In other words, the size of IoT nodes is essentially set by their power consumption, which hence become a very stringent and crucial requirement in any IoT node design.

1.3.2   Interaction with the External World

With reference to Fig. 1.2, the interaction of IoT nodes with the external world needs to last at least the lifespan of the object/environment they are embedded in, as battery replacement is not an option due to the large number or the inaccessibility of nodes. When deployed in buildings or

other living environments and infrastructures, this translates into a lifetime of several decades. Industrial applications, transportation and shipping might require a shorter lifetime, although still in the order of a decade. The lifetime requirement can be further relaxed in other applications such as retail, worksites, lifestyle/entertainment. Hence, the above considerations on the power budget of IoT nodes apply almost unmodified in a very wide range of applications.

Meeting power budgets of few µWs or below is feasible only if the IoT node actively performs tasks (e.g., sensing, processing) only infrequently. In other words, power needs to be aggressively reduced by duty cycling the IoT node operation, alternating active tasks and long sleep periods as depicted in Fig. 1.5, with periodicity set by the wake-up cycle T_{wkup}. From an architectural standpoint, this means that IoT nodes are organized into an always-on (ALWON) sub-system that manages the periodicity of the wake-up cycle and stores information across active tasks, and a duty-cycled (DCYC) sub-system that periodically performs the active task (Alioto 2012). Hence, the average IoT node power can be written as the sum of the

Fig. 1.5 Lifetime vs. average power consumption for different batteries

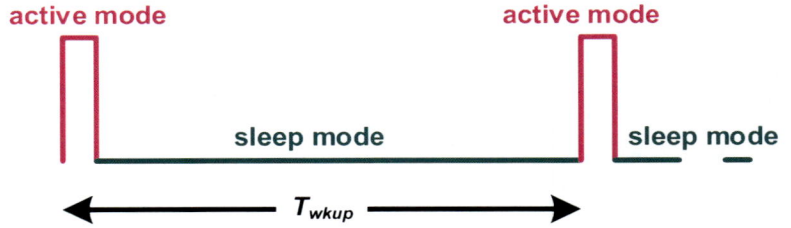

ALWON power P_{ALWON} and the DCYC energy E_{DCYC} (Alioto 2012):

$$P_{avg} = P_{ALWON} + \frac{E_{DCYC}}{T_{wkup}}. \qquad (1.1)$$

From Eq. (1.1), the power reduction can be reduced by reducing the power (energy) of the always-on (duty-cycled) sub-system. In other words, nearly-minimum power design needs to be pursued in the always-on sub-system, while nearly-minimum energy design is the objective in the duty-cycled one (see Chap. 4). Of course, larger T_{wkup} and hence more infrequent active operation mitigates power, although T_{wkup} is upper bounded by the application, depending on how frequently data needs to be updated. Such system-level tradeoffs are discussed in Chap. 2, whereas approaches to further reduce the leakage power cost of storing information across tasks is discussed in Chaps. 5–7.

1.3.3 On-Board Capabilities of IoT Nodes

IoT nodes need to have sensing, computation, and wireless communication capabilities. In IoT nodes design for a specific purpose, sensing can be typically made more inexpensive by tailoring the MEMS design and the analog interface around the specific application. This permits to substantially reduce the complexity that is experienced by general-purpose platforms, and hence the cost. As simple example, Fig. 1.6 shows that sensors for IoT applications cover a wide range of resolutions, hence using the appropriate ADC resolution (see Chap. 13) is necessary to avoid using general-purpose platforms with over-

designed resolution (and hence higher cost and power). From Fig. 1.6, most of IoT applications require a minimum resolution that is below 12 bits, and 8 bits are sufficient for a rather wide range of practical cases. On the secondary y axis, the figure also reports the energy per conversion, assuming an energy per conversion step of 30 fJ, which is relatively optimistic especially for larger resolutions (Murmann 1997).

The datarate range of the above sensors is plotted in Fig. 1.7. This figure shows that most of the sensors require only thousands of bits per second when operating continuously, whereas tasks related to vision and audio processing need orders of magnitude higher datarates (up to 10 Mbps in the case of compressed VGA video streaming). From the above considerations, the specifications of IoT node sensing interfaces are actually quite relaxed, thus cost and power consumption are far more important aspects than pure performance. Both challenges are well addressed by tailoring such circuits around the specific application. The power consumption of the ADC is proportional to the datarate in Fig. 1.7 and the energy per conversion in Fig. 1.6, and is plotted in Fig. 1.8. From this figure, the power consumption of ADCs for IoT nodes spans a very wide range, mostly because of the wide energy per conversion range in Fig. 1.6, as dictated by the exponential relationship between resolution and energy (Freyman et al. 2014). This confirms that tailoring the ADC to the specific application is crucial in IoT, and the same consideration applies to most of the other building blocks and sub-systems.

Let us now consider the case where the raw sensor data is transferred directly to concentrators and cloud. Assuming a best-in-class radio consuming 5 nJ/bit (ISSCC 2016), the resulting power to wirelessly transmit such

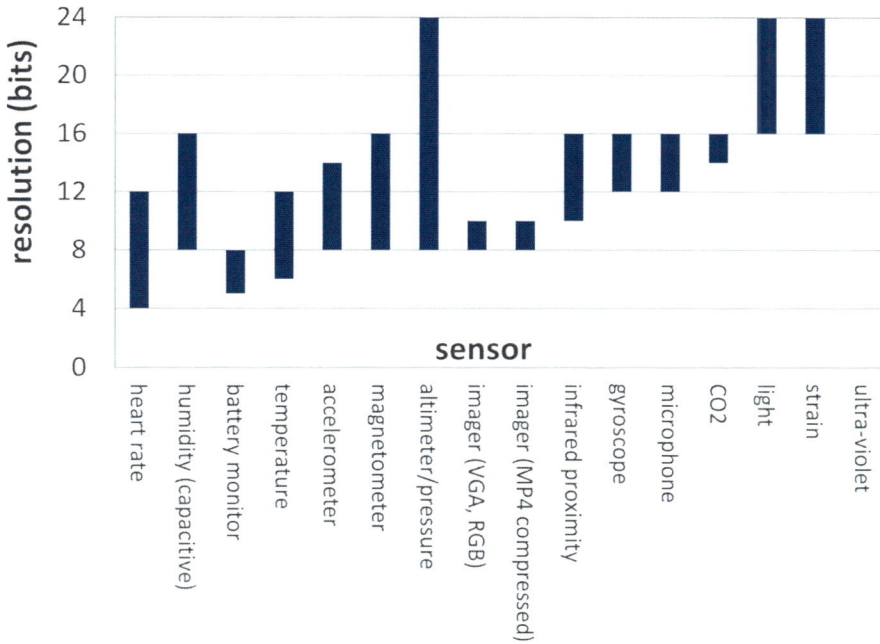

Fig. 1.6 Resolution range required by various sensors

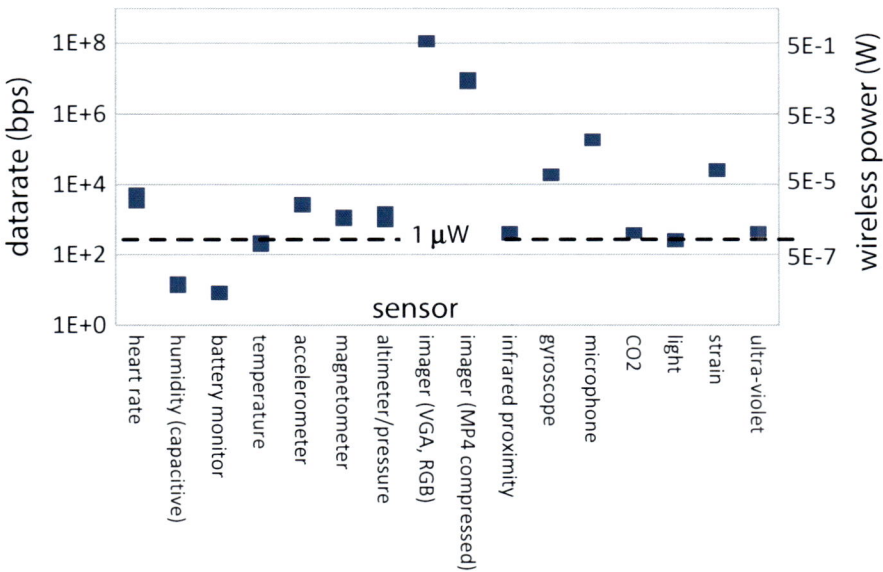

Fig. 1.7 Datarate range required by various sensors, and wireless power required to continuously transmit data (energy/bit assumed to be 5 nJ/bit (ISSCC 2016))

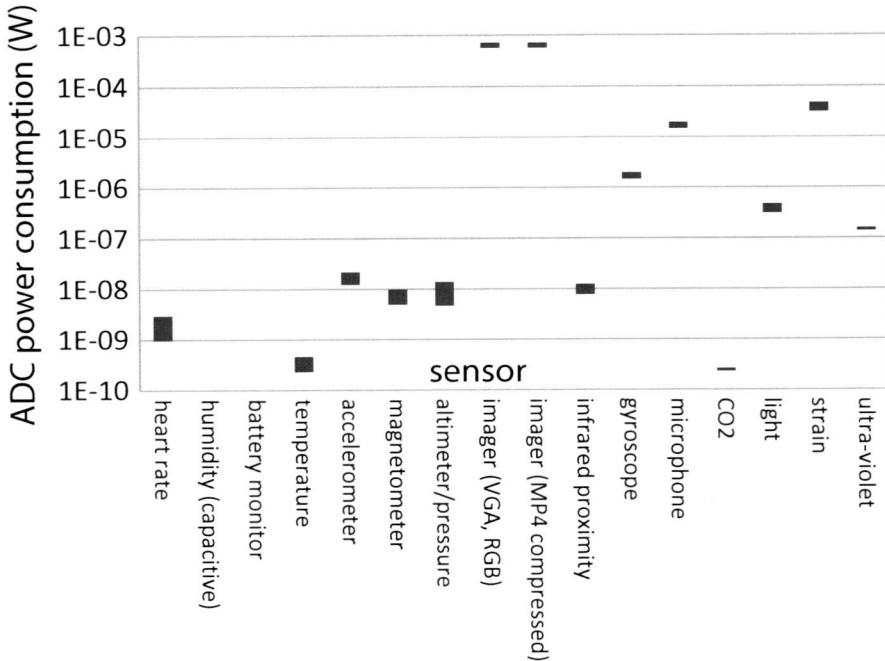

Fig. 1.8 ADC power consumption under sampling rate associated with the datarate in Fig. 1.7, assuming an energy/conversion step of 30 fJ

raw data is reported on the secondary y axis in Fig. 1.7. From this figure, most of sensors certainly exceed 1 μW and hence the range of practical IoT node power targets mentioned above. Hence, mere computation offloading to concentrators and cloud through raw data transmission is not an option for IoT nodes operating continuously.

For some environmental sensors (e.g., temperature, CO_2, light, UV), duty cycling discussed in the previous subsection is applicable since measurements do not need to be taken continuously, as the related phenomena exhibit slower time constants. For such sensors, a duty cycle of percentage points reduces the average datarate down to hundreds of tens of bits/second, and the power down to tens of nWs. Often times, the other sensors cannot be duty cycled as the dynamics of the related phenomenon does not really allow it (e.g., accelerometers, gyroscopes, imaging, audio). In these cases, further power reduction can be achieved by leveraging the well-known computation-communication tradeoff (Min et al. 2001), moving computation onto the IoT node to reduce the volume of wireless data

communication. For example, the IoT node can be proactive and monitor for critical or important events (e.g., the crossing of a threshold, or an increase rate larger than a pre-set value), and transmit data only upon their occurrence. From Fig. 1.7, this is particularly crucial in applications involving large datarates, such as continuous vision and audio sensing. In such applications, more intelligence needs to be embedded in the IoT node, such as the ability to perform pattern recognition and classification. Other options to trade off computation and communication are in the choice of the data representation and sampling approach (e.g., compressive sensing, including computation in the compressive sensing domain (Shoaib et al. 2015)), as well as signal dimensionality reduction (e.g., in-node feature extraction, which is equivalent to compression, with the further advantage that it is often a necessary task to be performed anyway in many algorithms).

From the above considerations, the wireless power is always an issue in IoT nodes, and hence requires the choice of appropriate communication standard for the intended range and datarate, as will be discussed in the next section.

1.3.4 User Constraints

Other important requirements of IoT nodes come from the user, and are mainly related to cost and security. Regarding the cost, consumer applications dictate a target of approximately 1 $/node as was discussed in Sect. 1.1 (which limits the die cost to a fraction of it). This clearly puts pressure on the financials of the semiconductor industry due to the limited room for profit margin, and can be addressed through large sales volumes (say, at least several tens of millions per year) and specialized hardware for reasons related to cost, power and form factor (see previous subsections). On the other hand, achieving such volumes is difficult even deploying an IoT with a trillion devices. Indeed, the IoT space is highly fragmented, and only few applications are so pervasive that they require such large volumes (https://www.mckinsey.de/files/mckinsey-gsa-internt-of-things-exec-summary.pdf). Similarly, it is hard to justify the non-recurring engineering cost of a new chip design for applications that do not require more than tens of million pieces. This will require the development of an ecosystem that favors design reuse, and platform-based design approaches. Regarding the recurring costs, IoT node cost reduction certainly requires more aggressive on-die integration to limit the cost of off-chip component assembly and testing. For example, circuits for power delivery and harvesting need to avoid off-chip passive components (see Chaps. 10 and 11), and innovative integration techniques and packaging becomes crucial to assemble heterogeneous components in an inexpensive and ultra-compact manner (Heterogeneous Integration Roadmap) (see Chap. 16).

As an additional challenge, IoT nodes are required to have a long lifetime (e.g., decades), which translates into a missed opportunity to replace the nodes for a very long time. In the long run, this will expectedly make the IoT market very different from the consumer market, which typically relies on periodic new waves of demand stimulated by incoming generations of products with improved features (and predictable release timeline, which allows planning).

The above challenges need to be addressed through platform-based design and moderate reconfigure-ability to reduce the design cost and widen the range of targetable applications, especially in consumer electronics. On the other hand, such challenges are mitigated in applications where the ability of directly receiving information on large numbers of objects is particularly valuable. For example, this is the case of manufacturing, logistics and smart cities, whereas it is not for single users in a smart home.

Security is another important requirement coming from the user, as the IoT offers a very large number of backdoors to attackers, in view of its large scale. In addition, traditional solutions to counteract cyber-attacks (e.g., firewall, cryptography) are not applicable to IoT nodes, due to their very limited power budget and cost. As further challenge, the dispersed deployment of IoT nodes makes it hard to keep track of individual nodes, thus exposing them to physical attacks. Such challenges require novel security approaches that embrace the hardware level rather than being confined at the network or software level, in order to reduce the cost and energy, and assure chip-level authentication (see Chap. 8).

1.4 Looking at the Past: IoT as Natural Outcome of Technological Trends

The IoT can be shown to be a natural consequence of historical trends that are relevant to its distinctive features, such as size, energy, sales volume, cost, with other software implications discussed at the end of the section. Other considerations on the evolution of the communication infrastructure will be made in Sect. 1.7.

The Bell's law observes that a new computer class has appeared every 10 years, thus bringing exponential improvements in computer size ($100\times$ smaller every 10 years) and cost (Bell et al. 1972; Bell 2008; Fojtik et al. 2013), as summarized in Fig. 1.9. This has driven the computer market expansion in computer units by a factor of 10–$20\times$ every 10 years (Tsai 2014), as in Fig. 1.9. Based on the current dominant wave

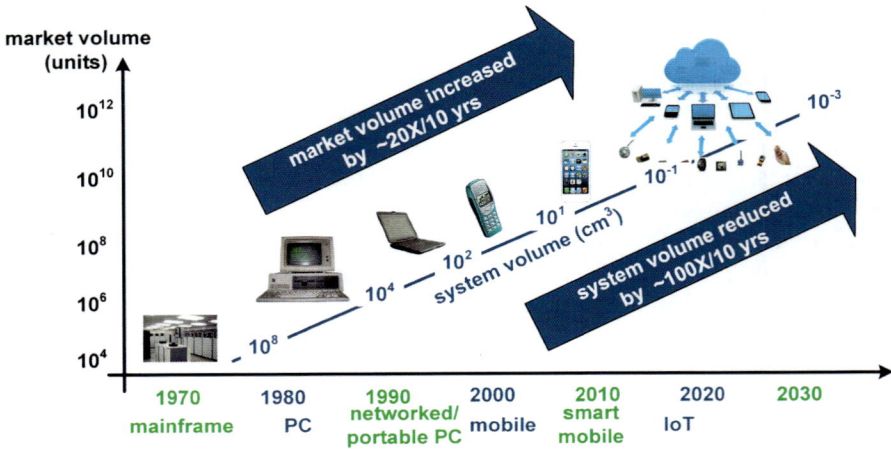

Fig. 1.9 Scaling laws of computer market size and system volume

of mobile computing, the next technological wave is expectedly composed of systems with a volume of hundreds of mm^3 or lower, i.e. the IoT. Essentially, the last few decades have seen computers and integrated systems move from desks to pockets, and then anywhere.

The energy efficiency trend for computers and Digital Signal Processors is well captured by the Koomey's (Koomey et al. 2011) and Gene's law (Frantz et al. 2000; Karam et al. 2009), which are representative of control-heavy and data-heavy architectures. As summarized in Fig. 1.10a–b, both classes of computing have equally benefited from technology advances, which can be expected to hold true in the future as well. Historically, most of 100× energy reduction achieved every 10 years has been obtained through technology scaling (90% according to (International Technology Roadmap for Semiconductors 2015)). In the IoT domain, energy efficiency improvements will not come from technology scaling, as advanced CMOS technologies are simply too expensive for the very low cost target of IoT nodes, as discussed below. Hence, keeping the same pace in the energy reduction requires major innovation at system level (Chap. 2), in key building blocks (Chaps. 3–7, 10–14), through innovative design methodologies (Chap. 9), and suitable batteries (Chap. 15).

Regarding the cost of IoT nodes, and in particular the cost per transistor, it is useful to recall the concept of "learning curve" for the silicon manufacturing, as plotted in Fig. 1.11. In general, the learning curve comes from the observation that the doubling in the large-scale cumulative production typically leads to a fixed reduced cost per unit at any point in time (Jaber 2011). The semiconductor industry is only one of the many examples of steady (exponential) cost reduction that comes from the accumulated knowledge and improvements in the overall design/production process. In particular, Fig. 1.11 shows that the transistor manufacturing cost fits a 55% learning curve, as doubling the overall number of manufactured transistors reduces the cost to 55% of the original cost. Such relentless cost reduction is at the basis of the Moore's law (Jovanovic and Rousseau 2002), which has driven the learning curve of transistor cost down relying on a steady (exponential) doubling of transistors per die at each CMOS generation, assuming the cost reduction comes from shrinking (Hutcheson 2009) (which is no longer true at 20 nm and below due to higher lithography cost, thus making those CMOS generations unsuitable for low-cost IoT nodes, as discussed in Sect. 1.8). The cost per transistor has been constantly reduced through shrinking in the past decades, although this is clearly not the only way to do it. In other words, the learning curve in Fig. 1.11 actually transcends Moore's law, and is expected to continue for a long time in spite of the end of the latter (Rhines

Fig. 1.10 Scaling laws of energy in (**a**) computers (Koomey's law from (Koomey et al. 2011)), (**b**) Digital Signal Processors (Gene's law)

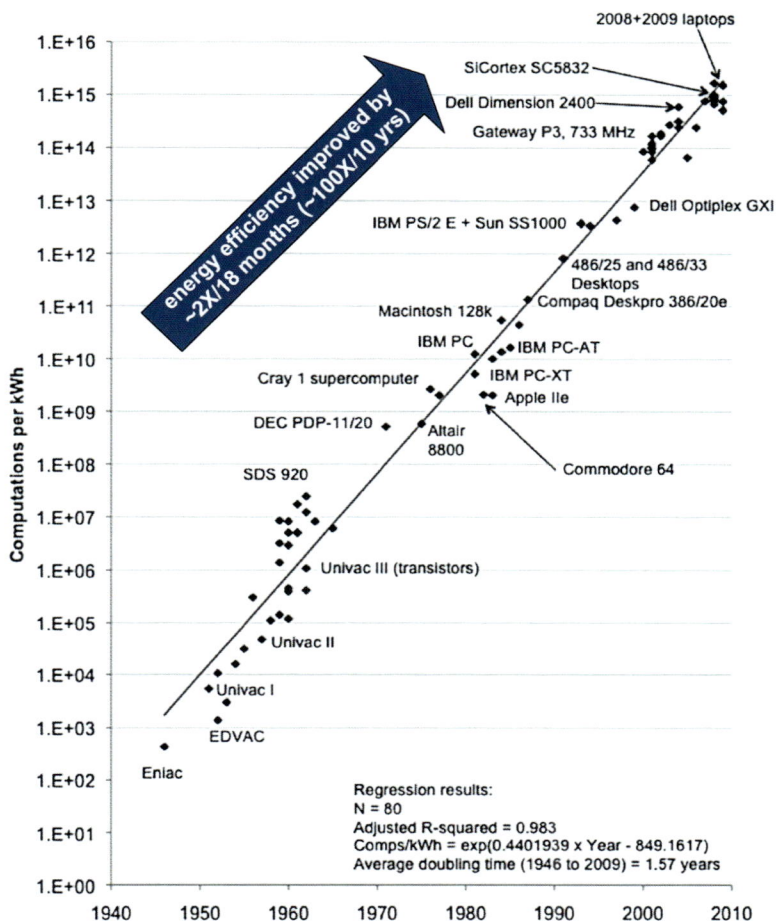

(a)

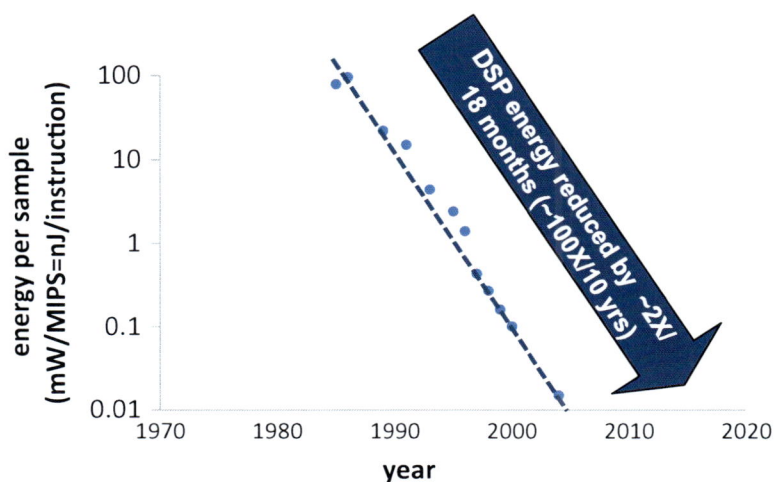

(b)

Fig. 1.11 Learning curve of transistor manufacturing

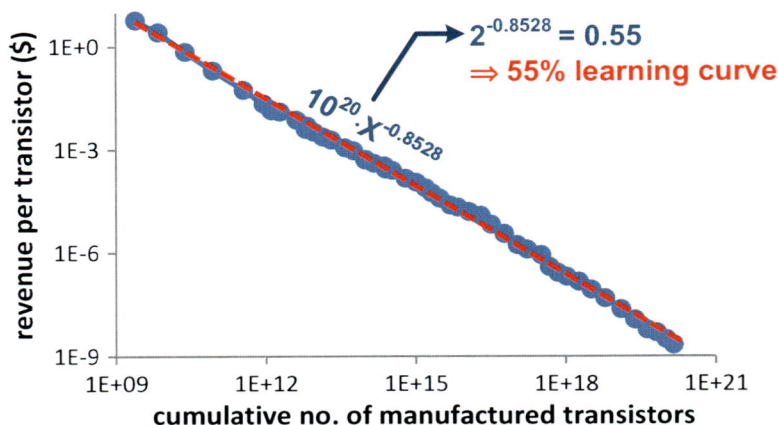

2015), leveraging on the accumulated knowledge of semiconductor industry, and various other approaches described in Sect. 1.8.

More considerations on trends in software and communication infrastructure will be made in Sect. 1.8.

1.5 Looking at the Present: Typical Specifications of IoT Nodes

1.5.1 Architecture of IoT Nodes

A relatively general architecture of IoT nodes is depicted in Fig. 1.12 (covered in Chap. 2), which details its main sub-systems, including processing and security assurance (Chaps. 3–9), power conversion and delivery (Chaps. 10 and 11), analog interfaces (Chaps. 12 and 13), radios (Chap. 14), energy sources (Chap. 15), system integration and assembly (Chap. 16). Two examples of detailed architectures of IoT nodes are provided in Chaps. 17 and 18. In this section, a review of the main features of IoT node architectures is provided.

As in Fig. 1.12, sensors are connected to analog interfaces that include an amplifier with programmable gain (and sometimes analog filters), and are multiplexed to share a single ADC for all analog channels. Analog voltages are generated by a DAC for actuation. Energy interfaces involve an energy storage element (e.g., battery, supercapacitor), and energy harvesters for energy

replenishment. Circuits to monitor the state of the battery are included to adapt the power management strategy to the actual energy availability (e.g., brown-out detector). Specialized circuitry is embedded for the generation of proper multiple voltages from the single (and not perfectly stable) voltage coming from the battery. Voltage booster and battery recharger are inserted between the energy harvesters and the battery. A Power Management Unit (PMU) coordinates the interaction between the energy source, the harvester and the power mode of the node.

The task performed in active mode (see Sect. 1.3.2) is started by a periodic trigger (time-driven) or specific events (event-driven), depending on the power mode and the specific application. Time-driven control is determined by deriving a proper clock from the system clock, whose frequency is typically 32,768 Hz to serve as Real-Time Clock and hence perform accurate timestamping and inter-node synchronization. Event-driven control is achieved by constantly monitoring digital and analog signals of interest, respectively through digital transition detectors and analog comparators that detect the crossing of a meaningful threshold to signal an event occurrence. When more sophisticated comparison is needed for signals coming from sensors, the ADC and processing are kept on to acquire samples and process them until the event of interest is detected.

In Fig. 1.12, memories are an important part of IoT node processing. RAM is needed for the

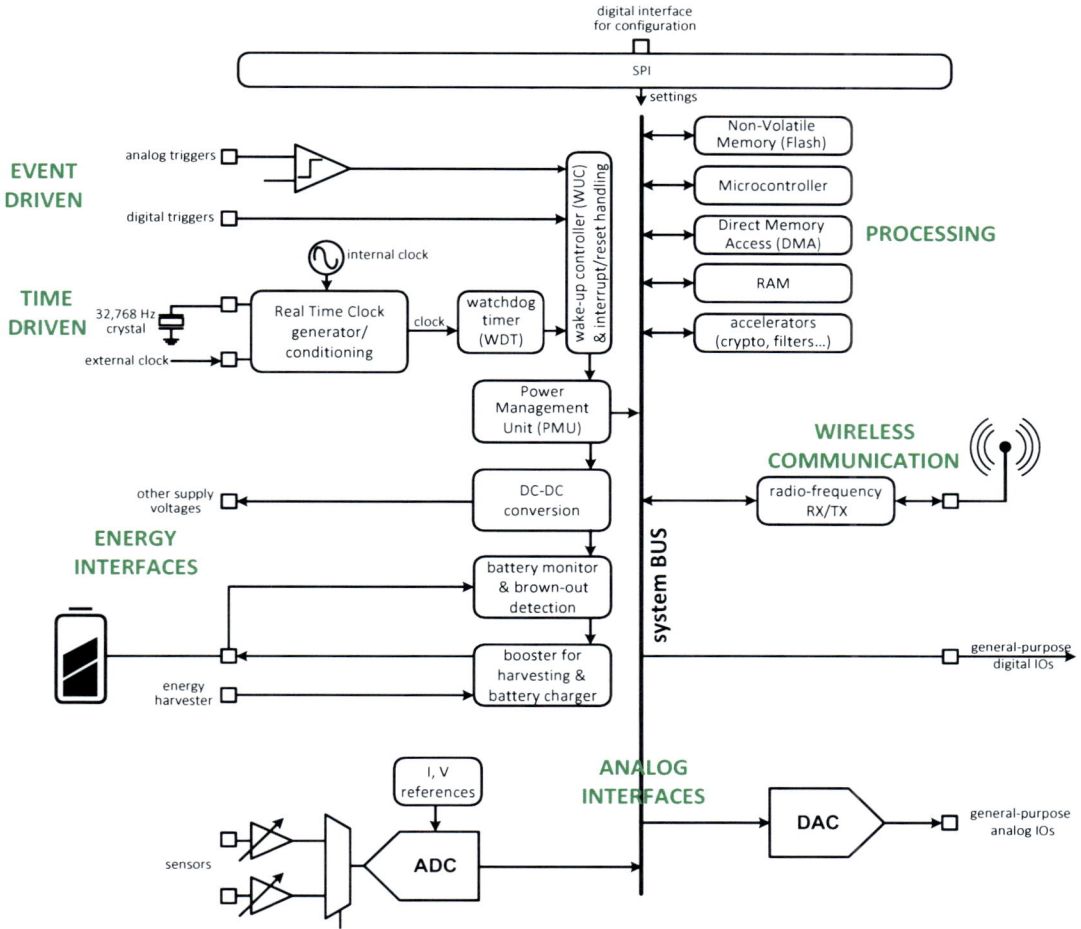

Fig. 1.12 Relatively general architecture of IoT nodes with detailed sub-systems

microcontroller/microprocessor execution, as well as to store data in sleep mode. The Non-Volatile Memory (NVM) contains the instructions, settings and also data that needs to be retained for a long time, thus suppressing the leakage power consumption associated with retention.

1.5.2 Typical Specifications of Commercial IoT Nodes

Today's IoT nodes are implemented according to various system integration approaches:

- Systems on Board (SoBs), using off-the-shelf components that are assembled on a printed circuit board.
- Systems on Chip (SoCs), usually in the form of MicroController Units (MCUs) with additional peripherals such as analog interfaces and radios.

To develop some quantitative understanding of existing IoT nodes, we performed a survey of more than 90 commercially available "motes" (i.e., IoT nodes in the form of SoB), more than 30 MCUs and more than 30 sensor hubs (i.e., a sub-system collecting, fusing and processing

Table 1.1 Survey of motes, MCUs and sensor hubs

	Motes (SoB)			MCUs (SoC)			Sensor hubs (SoC)		
	Min.	Avg.	Max.	Min.	Avg.	Max.	Min.	Avg.	Max.
Cost ($)	10	295	1000	0.8	12	29	2	7	25
RAM (kB)	2	64–128	512	8	30	32	4	–	16
Flash (kB)	32	32,000–64,000	64,000	2	16	32	–	–	–
Lifetime (months)	3	9	11	–	–	–	–	–	–
Volume (mm^3)	300	73,000	200,000	–	–	–	–	–	–
ADC resolution (bits)	8	11	16	10	11	12	12	14	16
DAC resolution (bits)	10	11	12	4	6	12	–	–	–
RF receiver sensitivity (dBm)	−110	–	−90	–	–	–	–	–	–
Min. operating voltage (V)	1.8	3.3	15	1.6	1.6	1.8	1	1.8	3.2

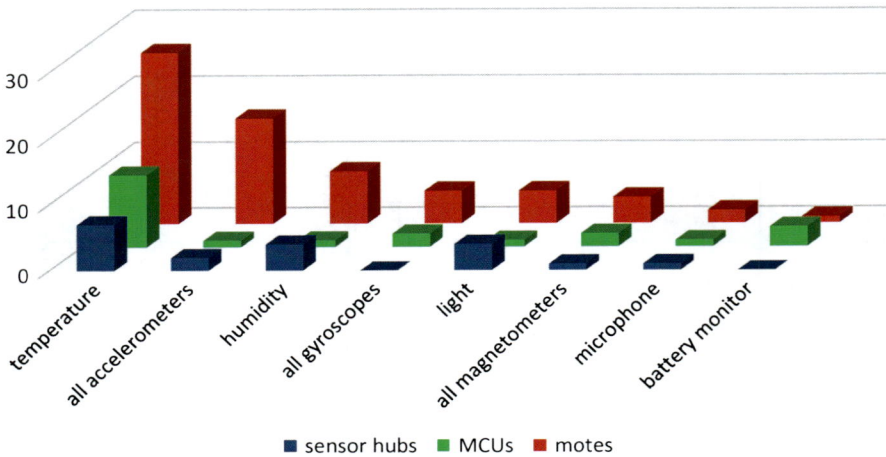

Fig. 1.13 Survey on sensors available on board in commercial sensor hubs, MCUs (SoCs) and motes (SoBs)

data from multiple sensors to offload energy-hungry SoCs, typically encountered on a smartphone). The results are summarized in Table 1.1 and are commented below.

Regarding the cost, SoBs represents a design approach with very low capital entry barrier, being the prototyping and development cost relatively small. However, the recurring cost is high and the specifications (e.g., power) are drastically worse than SoCs, as separate general-purpose components are used. From Table 1.1, the typical cost of motes is in the order of hundreds of dollars, and only in a few cases it is as low as tens of dollars, largely exceeding the cost requirements of IoT nodes (see Sect. 1.1). SoCs are an order of magnitude more inexpensive, with a typical cost of approximately 10$.

This means that even SoCs are still far from the IoT node cost target, which poses a challenge for the development of the IoT, as will be further discussed in Sect. 1.8.

Regarding the availability of on-board sensors, Fig. 1.13 shows that about 30% of available SoB Printed Circuit Boards (PCBs) include a temperature sensor, some also include accelerometers (mostly 3-axis), humidity sensors. Less than 5% of the SoBs include on-board light sensors, magnetometers, microphones and battery monitors. Other sensors are available only externally. Expectedly, SoCs have a much narrower choice of on-board sensors, mostly for environmental monitoring (e.g., temperature, humidity, light), and motion sensors (mostly 3-axis accelerometers and gyroscopes). Some MCUs have an embedded battery monitor for power management purposes. Sensor hubs do not include a battery monitor as they are employed in smartphone platforms, where battery management is performed by the Application Processor.

Further observations can be made on typical features of existing IoT nodes, from Table 1.1:

- RAM capacity: in motes, it ranges from KBs to 512 KB, and is typically around 64–128 KB, as allowed by inexpensive stand-alone (general-purpose) RAM chips. SoCs have much lower capacity in the order of 8–32 KB, as enabled by fairly expensive on-chip SRAM.
- Flash capacity: motes can have up to 64 MB of Flash, again as allowed by inexpensive stand-alone memory chips. MCUs have embedded Flash on chip, which is expectedly much lower capacity (up to 32 KB) due to their larger cost. Sensor hubs typically do not have any non-volatile memory, as they can rely on the Application Processor.
- Battery lifetime: the battery lifetime claimed by mote vendors is in the order of months, which is well below the typical IoT node targets of decades (see Sect. 1.1).
- Volume: the volume of SoBs is several orders of magnitude larger than IoT targets, due to the unavoidably large size of PCBs. Full on-chip integration enables substantial system shrinking, with packaging playing a fundamental role (see Sect. 1.3.4 and Chap. 16).
- ADC resolution: as required by the wide range of sensors in Fig. 1.6, it ranges from 8 to 16 bits, although MCUs have a somewhat narrower range (10–12 bits).
- DAC resolution: it is in the order of 10–12 bits in SoBs, and goes down to much lower values in MCUs, which are better tailored to the lousy requirements of many practical applications. Sensor hubs do not have a DAC, as this is not required in sensor data gathering.
- Wireless receiver: it has typically a sensitivity of −90 to −100 dBm. The radios of most of motes and MCUs operate in the 2.4 GHz band, with very few exceptions working at 433 MHz

(which offers lower path loss), with a datarate ranging from a few kbps to 1 Mbps.

Operating voltage: it typically ranges from 1.8 to 3.3 V in motes and MCUs, as respectively required by off-the-shelf chips and compatibility with commercial peripherals for MCUs. Sensor hubs can operate down to 1 V, as dictated by the low power consumption requirements of smartphone platforms and the adoption of such voltages in the related chip ecosystem.

To build a tighter link between the above specifications and the applications in Sect. 1.2.2, Table 1.2 shows the sensor-application matrix, which lists typical sensors employed in each application area.

As shown in Fig. 1.14, the power consumption of current IoT nodes is dominated by the wireless communication power, considering short-range communication standards with low average and peak current, as required by IoT nodes (e.g., ZigBee and Bluetooth Low Energy—BLE). This is particularly true for nodes employing ZigBee radios (see, e.g., Fig. 1.14a), whereas the impact of the wireless communication power becomes around 50% of the power budget with BLE radios, as they offer better energy per bit (Siekkinen et al. 2012). On the other hand, BLE handles only short-range point-to-point communications and few nodes, as opposed to ZigBee or Z-Wave that have true networking capabilities (e.g., meshes). For IoT nodes that are connected directly to the Internet, their number certainly exceeds the available number of addresses in the IPv4 Internet protocol (32-bit address, i.e. billions of unique addresses). Techniques to reuse and mask addresses such as network address translation and private network addressing cannot deal with the prospective explosion in the number of connected devices. The IPv6 protocol solves this issue, as it comprises 128 bits for the address (i.e., virtually unlimited unique addresses), and can maintain compatibility with IPv4 by using the last 32 bits for traditional IPv4 addressing. Since early 2010s, the IPv6 protocol has been rapidly adopted by new users, and is expected to be

Table 1.2 Sensor-application matrix

Application	Accelerometer	Altimeter/pressure	Battery monitor	Gas	Galvanic skin response	Gyroscope	Heart rate	Humidity	Imager	Light	Microphone	Magnetometer	Proximity (IR)	Strain	Temperature	Ultra-violet
Agriculture	✓	✓	✓	✓		✓		✓	✓	✓			✓	✓	✓	✓
Automotive	✓	✓	✓	✓	✓	✓	✓	✓	✓	✓	✓	✓	✓	✓	✓	✓
Consumer	✓	✓	✓	✓	✓	✓	✓	✓	✓	✓	✓	✓	✓	✓	✓	✓
Energy	✓		✓	✓		✓				✓						✓
Healthcare	✓		✓	✓	✓	✓	✓		✓	✓	✓	✓	✓	✓	✓	✓
Industrial	✓	✓	✓	✓	✓	✓			✓	✓	✓	✓	✓	✓	✓	✓
Lifestyle/entertainment	✓	✓	✓	✓	✓	✓	✓	✓	✓	✓	✓	✓	✓		✓	✓
Logistics	✓	✓	✓	✓		✓			✓	✓	✓	✓		✓	✓	✓
Manufacturing	✓		✓	✓		✓			✓	✓	✓		✓	✓	✓	✓
Public sector	✓	✓	✓	✓	✓				✓	✓	✓		✓		✓	
Retail	✓	✓	✓	✓	✓	✓		✓	✓	✓	✓	✓	✓		✓	✓
Shipping	✓	✓	✓	✓		✓		✓	✓	✓			✓	✓	✓	
Smart buildings	✓	✓	✓	✓		✓		✓	✓	✓	✓		✓		✓	✓
Smart cities		✓	✓	✓				✓	✓	✓	✓		✓		✓	
Smart homes		✓	✓	✓				✓	✓	✓	✓		✓		✓	
Smart infrastructures	✓	✓		✓		✓				✓	✓	✓	✓	✓	✓	✓
Transportation	✓		✓	✓	✓	✓		✓	✓	✓	✓	✓	✓	✓	✓	

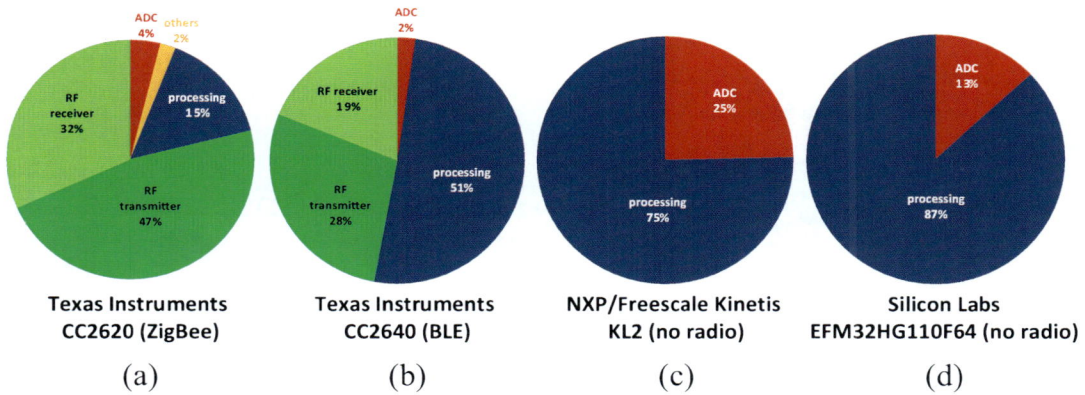

Fig. 1.14 Power breakdown of commercial motes (*on the left*) and MCUs (*on the right*)

very widespread by the time the IoT reaches its inflection point. From the IoT node point of view, the adoption of IPv6 entails some communication overhead, due to the increase in the length of the header by few tens of bytes (Siekkinen et al. 2012). This suggests that IoT nodes carrying little information per measurement (e.g., temperature, humidity) should temporarily store and cluster measurements, and transmit multiple measurements in a burst fashion.

When lower power radios are employed (see, e.g., Fig. 1.14b), the power consumed by the processing becomes a major contribution, and it can reduced through substantial voltage reduction and other techniques, which will be discussed in Chaps. 3–5, 9 and 12–13 from the perspective of power utilization, and Chaps. 10 and 11 for the power delivery. As shown in Fig. 1.14c, d, the analog interface consumes a power that is a fraction of the processing contribution. Chapters 17 and 18 will present two examples of IoT node architectures and designs based in SoB and SoC implementations, and will discuss the underlying design tradeoff.

1.6 Present and Future Challenges in Chips for IoT Nodes: Energy Efficiency

The scarce and often intermittent power availability in IoT nodes tightly constrains their power budget, and demands for energy efficiency improvements by orders of magnitude compared to today's commercial IoT nodes in Sect. 1.5.2. In practice, excessive power consumption leads to discontinuous operation depending on the residual (instantaneous) energy availability in battery-powered (purely energy-harvested) IoT nodes. As an analogy to the "dark silicon" issue in high-performance SoCs, where a spatial portion of the chip needs to be kept off to meet the power budget (Esmaeilzadeh et al. 2012) (see Fig. 1.15a), IoT nodes become "dark" (i.e., entirely off) when their power consumption cannot be sustained by their energy source. In other words, inadequate energy efficiency of IoT nodes again leads to dark silicon as occurs in high-performance SoCs, although on a temporal dimension rather than spatially (see Fig. 1.15b).

1.6.1 The Wireless Power Issue and the Communication-Computation Tradeoff

As discussed in the previous section, the wireless communication power contribution needs to be substantially reduced to create room for energy efficiency improvements. Best-in-class commercial radios consume an energy in the order of tens of nJ/bit, and several academic prototypes with energy around 1 nJ/bit have been demonstrated for both receivers and transmitters with a range in

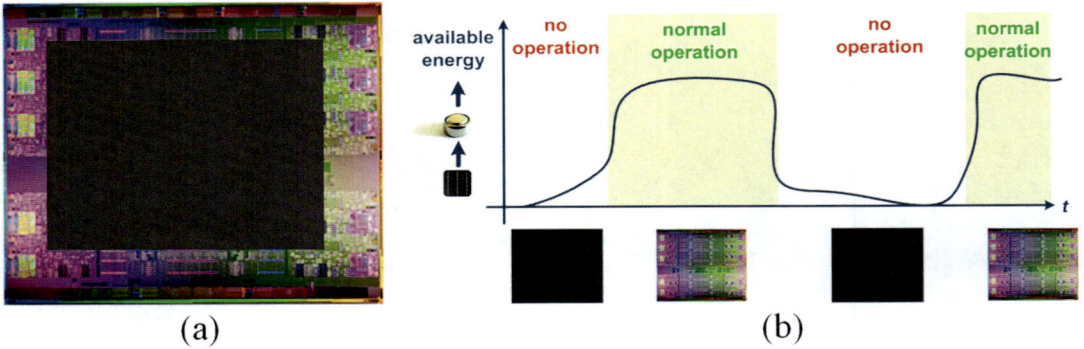

(a)   (b)

Fig. 1.15 Dark silicon in (a) high-performance SoCs (spatial dimension), (b) IoT nodes (temporal dimension)

the order of 10 m (ISSCC 2016). Unfortunately, limited savings in the energy per bit are expected from circuit and modulation techniques in the decade ahead, compared to today's state of the art. Indeed, from the Shannon-Hartley theorem on the channel capacity (Shannon 1949), the energy per bit during transmission cannot be really reduced through the choice of the modulation scheme and spectral efficiency improvements, as existing schemes with reasonable low complexity (e.g., BPSK, BFSK, OOK) are already very close to best-in-class highly complex modulations (unaffordable in IoT nodes, such as M-ary FSK with large M), and only 10 dB higher than the minimum theoretical limit (Otis and Rabaey 2007). Also, the adoption of low-complexity modulation schemes keeps the energy overhead for modulation/demodulation low, which in the foreseeable future can be expected to be moderately reduced (e.g., by units, not orders of magnitude) through energy-efficient circuit techniques. Channel coding will also bring some further energy reduction in the energy per bit, using error correcting codes (e.g., Cyclic Redundancy Check—CRC) and packet retransmission schemes, as customary in most of low-power wireless network standards, and recently in cellular standards for IoT (e.g., EC-GSM (Extended Coverage-GSM), EC-GSM, NB-IoT (Narrow Band IoT)). For example, Hamming (Golay) codes are able to reduce the energy per bit by $1.4\times$ ($2.5\times$) at the additional computation energy cost of 10 pJ/bit (30 pJ/bit) (Desset et al. 2003), as scaled down to 40 nm. Other codes may potentially reduce the energy per bit further, but this would come at the

cost of an unaffordable energy cost associated with encoding/coding (e.g., 250 pJ/bit in convolutional codes, 800 pJ/bit in Reed-Muller and Turbo codes, and more than 1 nJ/bit in Reed-Solomon codes).

Overall, the above advances in ultra-low power radios will only moderately reduce the energy per bit, beyond the great achievements of research in the field in the last 15 years. Such energy per bit reduction will translate into an even smaller saving in the consumption of IoT nodes, due to the additional fixed energy cost of radio start-up (e.g., waiting oscillators to settle to the correct frequency), which becomes a major energy contributor in typical IoT nodes sending short packets.

Another way to substantially reduce the wireless power is to reduce the communication range R, as the transmitted power can be reduced quadratically in free space (or faster, in indoor or obstructed environments) when R is reduced (Friis 1946). Pragmatically, a $10\times$ wireless power reduction can be achieved by shortening the range from the 10-m range down to a few meters. Such range reduction requires a denser distribution of concentrators, whose number increases quadratically when reducing the range, as shown in Fig. 1.16. Although the range reduction and increase in concentrator density is technologically feasible even today, this entails an unacceptable cost increase due to the substantially larger number of concentrators, and considering that their cost is easily tens of times the cost of a single IoT node (the cost of a concentrator is probably similar to a wireless

Fig. 1.16 Quadratic increase in the number of concentrators per IoT system under range reduction from (**a**) to (**b**)

● **IoT node**

■ **concentrator**

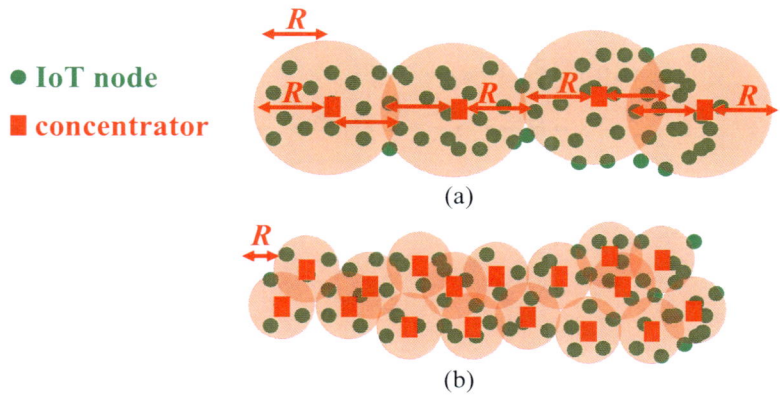

(a)

(b)

router). In other words, reducing the range of IoT nodes is not an option in real applications due to economic reasons, rather than technological ones. Accordingly, the range of IoT nodes is expected to remain in the 10-m range for most of the prospective applications.

The above trend in the IoT node range is somewhat in contrast with cellular communications, in which the distance between phones and the base station keeps shrinking, due to the more pervasive adoption of small cells (Keene 2015) (e.g., pico and femto cells), and the integration with non-licensed bands offering more capillary distribution of access points (e.g., WiFi as in LTE-U networks, which clearly pose other challenges due to the expansion of the cellular traffic into existing bands (Ngo et al. 2016)). This difference is explained by the different economy of scale. Indeed, the cost of cells in the cellular network infrastructure is shared by a very large number of users, whereas IoT sub-networks are spatially fragmented and tend to be proprietary, thus their cost is born by a smaller number of users per device. As further fundamental difference with respect to cellular and broadband networks, the preferential traffic direction is expected to be upstream, as massive quantities of physical data are used to receive aggregate information and decision outcomes (Zhang et al. 2015).

The above considerations again confirm that the most effective way to reduce the wireless

power is to make IoT nodes smarter ("cognitive", "attentive") than they are today, as was observed in Sect. 1.1. In other words, further reductions in IoT node consumption are subject to improvements in the energy cost associated with processing. As aggressive technology scaling is not an option in IoT nodes (see Sect. 1.8), various tradeoffs need to be explored to further reduce energy, as discussed in the next subsection.

To express the communication-computation tradeoff more quantitatively, we will refer to "simple" IoT nodes if the wireless power while transmitting raw sensed data $P_{wireless,raw}$ is well below the targeted power budget P_{budget} (i.e., $P_{wireless,raw}/P_{budget} \ll 1$). In other words, the wireless power for raw data is not a major bottleneck to the overall power consumption, and likely intelligence can be entirely kept in the cloud (i.e., there is no fundamental need for on-board intelligence to bring down the power). On the other hand, we will refer to "smart" nodes when the wireless power associated with raw data would simply exceed the power budget (i.e., $P_{wireless,raw}/P_{budget} \gg 1$), hence on-board processing has been necessarily added to aggregate information in a more compact form that can be transmitted more inexpensively. Assuming a radio with state-of-the-art energy per bit E_{bit}, the notion of "simple" and "smart" nodes essentially depends on P_{budget} set by the energy source, and $P_{wireless,raw} = E_{bit} \cdot N_{bit,measure} \cdot f_{measurements}$, being

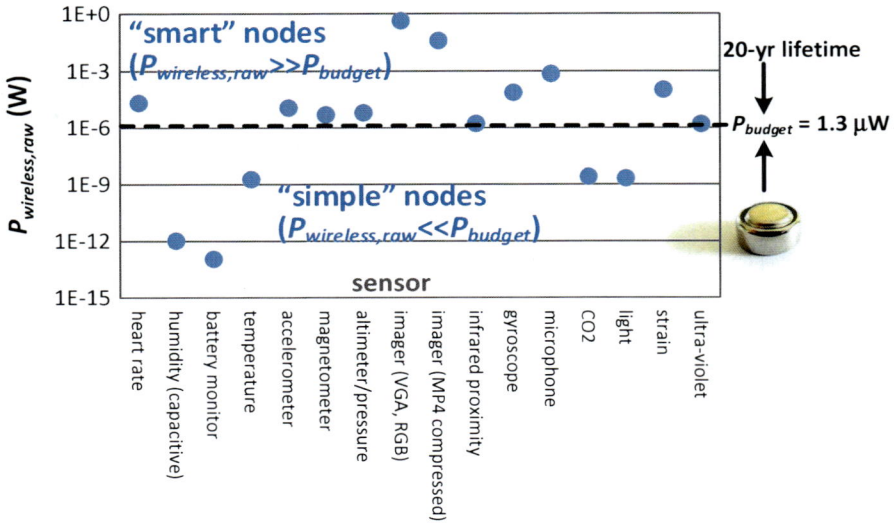

Fig. 1.17 Average wireless power $P_{wireless,raw}$ required by sensors in Fig. 1.7, assuming best-in-class 5-nJ/bit radio and comparison with power budget imposed by LR44 battery under 20-year lifetime target. IoT nodes above the black dashed line need on-board intelligence to reduce the amount of wireless data communication

$N_{bit,measure}$ the number of bits transmitted per measurement, and $f_{measurements}$ the number of measurements per unit of time.

Numerical examples are presented in Fig. 1.17, for various sensors with datarate taken from Fig. 1.7, assuming an LR44 button cell battery as only source of energy with capacity of 150 mAh, a 5-nJ/bit radio, and a targeted lifetime of 20 years. From this figure and for LR44 battery-powered nodes, environmental sensors (e.g., humidity, temperature) and battery monitors are "simple" nodes, i.e. there is no point in adding intelligence on board, as the wireless power to transmit raw data is much smaller than the power budget anyway. This is mainly due to the aggressive duty cycling enabled by the slow dynamics of the physical quantities monitored. All other types of IoT need to be "smart", in the sense that some on-board processing is needed to reduce the amount of transferred data. Clearly, smart IoT nodes with a larger $P_{wireless,raw}$ can benefit more from the insertion of on-board intelligence. Hence, IoT nodes with larger $P_{wireless,raw}/P_{budget}$ should internally process the sensed data at a higher level of their semantic understanding, so that the output to be transmitted becomes more succinct while still retaining the core knowledge that needs to be extracted from the data (i.e., some of the cloud computation is moved into the IoT node). The above considerations are all relative to the available power budget, hence relaxing (tightening) P_{budget} has an immediate effect on which sensors need to be smart (essentially, the black dashed line in Fig. 1.17 moves and hence splits the sensors into the two categories accordingly).

As few representative examples of levels of semantic understanding encountered in IoT applications, Fig. 1.18 ordered by data succinctness. The simplest type of processing (i.e., lowest level of semantic understanding) above raw data is compression, which does not really extract any knowledge, or translation into a more efficient representation (e.g., compressive sensing (Baraniuk 2007; Candès et al. 2006; Donoho 2006)). Similarly, a very simple type of processing encountered in IoT applications is represented by the evaluation of aggregate metrics to enable simple critical event detection (e.g., comparison of average/peak value or slope of recent data with a threshold (WebSphere Process Server)). Higher and relevance. Salient and relevant features can then be analyzed in terms of semantic understanding is achieved through

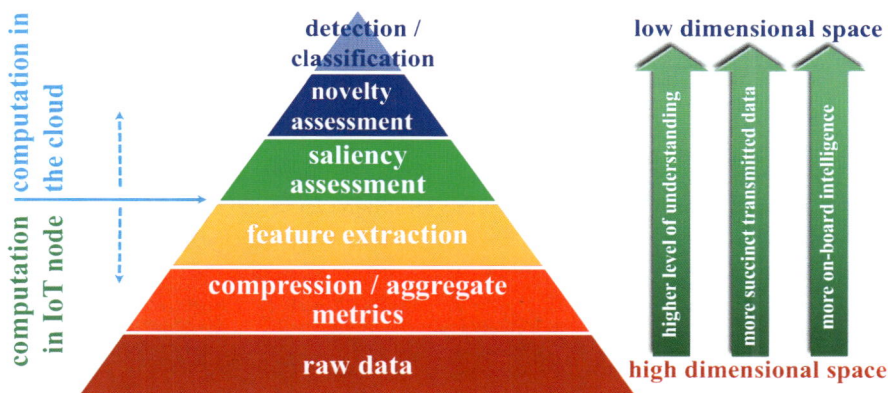

Fig. 1.18 Levels of semantic understanding ordered from low (raw data) to high (detection/classification) and allocation of related computation between IoT node and cloud. Smarter IoT nodes embrace more levels of understanding to reduce the amount of wirelessly transmitted data

feature extraction, which compresses information and also identifies potentially interesting patterns, as preliminary step for higher levels of understanding. On a higher level of understanding, extracted features can be analyzed in terms of their saliency novelty, as they should be further processed only if they bring novel information. Relevant and novel features are then processed through the more computationally expensive detection/classification engine. As in Fig. 1.18, lower and higher levels of understanding are properly allocated between the IoT node and the cloud, with smarter nodes embracing a larger number of levels of understanding.

In summary, some computational tasks that are traditionally performed in the cloud are expected to be moved into IoT nodes, whenever this achieves a more favorable balance between the reduced wireless datarate (thanks to the reduced data dimensionality at higher levels of understanding) and the increased computational effort performed in the IoT node. As a result, the IoT is expected to require an even more distributed processing model compared to the conventional cloud computing based on virtualization. Indeed, in the IoT computation can be dynamically allocated between IoT nodes, concentrators and the cloud, depending on the energy availability of the former ones. Then, conventional datacenter-scale distributed

processing model applies to computation performed in the cloud.

1.6.2   Opportunities to Achieve Highly Energy-Efficient Processing

Trading off performance and energy is an obvious option in the processing sub-system of IoT nodes. Such dynamic tradeoff is primarily enabled by dynamic voltage scaling, as was briefly mentioned in Sect. 1.5.2. From a circuit point of view, operation at nearly-minimum energy is achieved through the techniques and design guidelines described in Chaps. 4 and 5, and the automated design methodologies in Chap. 9. Architecturally, energy efficiency in the processing sub-system is improved through parallelism and the adoption of energy-efficient accelerators (i.e., trading off energy efficiency for area efficiency), as discussed in Chap. 3.

Another general approach to mitigate the consumption in IoT nodes is to resort to event-driven sensing and processing paradigms, rather than more conventional time-driven schemes where data is continually sampled and processed while waiting for specific conditions that need to be detected. Event-driven sensing and processing is inherently asynchronous and triggers processing and communication only when really needed. Indeed, event-driven operation permits to stop the clock and limit the overall power to few always-on analog blocks (e.g., the comparator in Fig. 1.12) and the leakage of the processing sub-system. This approach is clearly advantageous when the active task in Fig. 1.5 is triggered by specific events whose occurrence has a wide temporal distribution, rather than a strict temporal periodicity.

In non-real time systems whose dynamics of actuation and/or decision is much slower than the data acquisition, latency is not an issue to a certain extent. In other words, the value of sensed data comes from its collection and aggregation, rather than the most recent datum. In such cases, the IoT node senses and stores data in a time-driven fashion, occasionally processes data in a batch fashion, and then transmits only if the stored data is representative of a condition of interest that should be sent to the cloud (or examined by the nearby concentrator to develop a more global perspective). In other words, the IoT node behaves like a smart data logger, as it normally stores without transmitting and uses on-board intelligence to decide when wireless communication is truly needed. Such approach is exemplified in the SoC design discussed Chap. 18, where NVM is tightly integrated with logic to enable such capability. From a design perspective, such approach involves a tradeoff between the wireless power reduction and the increased power cost due to the data storage in a memory (all other power contributions are unaffected, and can be ignored in terms of power optimization). The latter contribution becomes important when meaningful events occur infrequently, as the leakage energy $P_{lkg,MEM} = V_{DD} \cdot I_{off,MEM} \cdot T_{MEM,avg}$ for storage is proportional to the average storage time $T_{MEM,avg}$, and hence becomes large for slow sampling and long observation times. This tradeoff can be made much more favorable by embedding a non-volatile memory for acquired data, which motivates Chaps. 6 and 7. In these chapters, the consideration of emerging non-volatile memories also aims to make the above tradeoff

more favorable by mitigating the energy per write, compared to state-of-the-art embedded Flash memories. Indeed, the write energy per bit in Flash memories is in the order of nJ/bit (Aitken et al. 2014), which is a few tens of times larger than low-power DRAM (e.g., LPDDR2 (Malladi et al. 2012)), and tens of thousand times larger than SRAM (ISSCC 2016). The lower write energy per bit in emerging NVMs makes data storing more advantageous at smaller $T_{MEM,avg}$.

Another important design dimension that can be traded off to reduce the energy per computation is the quality of processing, leveraging the fact that algorithms, architectures and circuits for processing are substantially overdesigned in terms of their accuracy in most of the applications (Alioto 2016; Frustaci et al. 2015). Accuracy is a very general term that can involve arithmetic precision (e.g., approximate computing), circuit resiliency to occasional faults (e.g., voltage overscaling (Hegde and Shanbhag 2001)), classification accuracy or number of classes in a classifier. The fundamental idea is that higher quality requires more energy per computation, or equivalently the energy can be reduced whenever lower quality can be targeted. Energy-quality scalable systems allow dynamic tradeoff between energy and quality of computation, minimizing the former for a given quality target. As an illustrative example of energy-quality scalable system, Fig. 1.19 shows the energy-quality tradeoff in the processing of a video stream on a smartphone. As in Fig. 1.19a, the quality of the video stream processing (e.g., video decoding) can be degraded under poorer lighting conditions, when the user is paying less attention, when the battery is close to discharged, or when the user is being physically active. As in Fig. 1.19b, lighting, attention (e.g., camera), battery monitor and motion sensors can detect the usage context and generate an appropriate quality target, which is achieved by letting the operating system (OS, or a hardware controller) adjust the energy-quality (EQ) knobs that are supported by the hardware architecture. Then, the hardware provides the OS with the actual quality measurement for feedback loop closure.

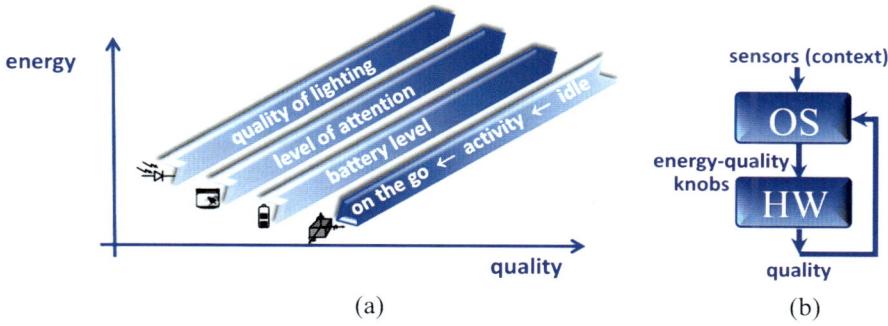

Fig. 1.19 Example of energy-quality scaling allowed during video stream processing in a smartphone: (**a**) energy can be reduced when lower quality is acceptable, based on the usage context, through (**b**) a feedback control loop that senses such context, generates the corresponding quality target and adjusts the related energy-quality knobs, and measures the actual quality for further tuning

Fig. 1.20 Slack between the quality available from the processing platform and the actual quality required by the specific chip, application, usage scenario and dataset

As summarized in Fig. 1.20, digital platforms are designed with a given level of processing quality, as constrained by the architecture for general-purpose designs, or by the quality imposed by the most stringent task in application-specific designs. Hence, a "quality slack" exists and can be reduced to improve the energy efficiency. Such slack is imposed by the requirement of adequate circuit resiliency, i.e. no hardware faults are allowed, although actually the application might occasionally allow them. The slack is also due to the fact that the design is sized for the most demanding application or task, whereas the one at hand could run at lower quality. Also, the lack of knowledge of the usage context forces the quality to be set at its maximum level, although this could actually be relaxed (see above example in Fig. 1.19). Finally, the required quality of processing actually depends on the specific incoming data, although the quality is usually set the worst-case dataset (e.g., object classification might become difficult under cluttered scene).

IoT nodes can greatly benefit from the concept of energy-quality scaling, as their on-board processing deals with physical signals, which are noisy in nature and hence can be processed with a precision that adapts to the level of noise. Similarly, in certain cases smart IoT nodes might execute statistical and soft (e.g., machine learning) algorithms, which are very resilient to occasional faults or approximations. Also, IoT nodes might be involved in sensing physical quantities that are related to human perception (e.g., vision, hearing), which have intrinsic redundancy and are inherently resilient against simplifications and uncertainty. Finally, "swarm intelligence" and intelligent global behavior (Beni et al. 1989) arises due to the massively distributed nature of IoT nodes, and it can be exploited to reduce the processing ability of individual nodes, thus offering additional opportunities for energy-quality scaling.

Other opportunities to reduce energy come from context-awareness and tight coupling with task under execution in power management.

Indeed, the distributed nature of the IoT gives puts certain nodes (e.g., some concentrators) in a position to have a global or semi-global perspective on the context, and hence they can help setup the power management strategies in individual IoT nodes. In other words, power management in IoT has the potential to become distributed, and leverage the global understanding to mitigate the power consumption in individual IoT nodes. On a smaller scale, tight coupling between power delivery and processing offers additional opportunities for energy efficiency improvements (Ramadass and Chandrakasan 2008). For example, adjusting the power delivery to the instantaneous performance requirement of the processing sub-system permits to improve both the energy per computation of the latter (Ramadass and Chandrakasan 2008; Kwong et al. 2009), and the power efficiency of the former (Zimmer et al. 2016). Furthermore, reconfigurable on-chip DC-DC converters offer the ability to adapt to the system active-sleep pattern, maximizing the energy efficiency via optimal sleep/wakeup power balancing (Lueders et al. 2014; Alioto et al. 2013), while avoiding the area cost of several different regulators to cover the available power modes (Kristjansson 2006).

1.7 Present and Future Challenges in Chips for IoT Nodes: Security

1.7.1 Security Challenges in IoT Nodes

Security represents a very broad challenge in the IoT, due to several factors:

- Traditional security schemes such as public-key cryptography are not feasible in most of IoT nodes, due to their stringent power and cost requirements
- The large number of IoT nodes and the resulting scale of the IoT network create an unprecedented very large number of "backdoors" that can be exploited by attackers to perform physical and network attacks
- The always-connected feature of IoT nodes makes them rather vulnerable to eavesdropping and software attacks, data stealing and device cloning (Narendra Mahalle and Railkar 2015)
- In our view, the personal data stored in the cloud will be likely shared across different service providers, which will deliver their service through "cloud apps" that will run on a datacenter-scale software platform. The user will likely give access to the service providers as we do today when we install apps on a smartphone. This data sharing with multiple service providers poses new security challenges, which have to be solved mostly on the server side.
- The limited energy availability of IoT nodes makes them vulnerable to resource enervation and Denial of Service (DoS) attacks (Narendra Mahalle and Railkar 2015; Aitken et al. 2014).
- The long lifetime of IoT nodes makes software and firmware updates unavoidable, and hence mandates strong authentication mechanisms to assess the authenticity and the integrity of the updates and patches, under the tight power budget of IoT nodes.

The above security challenges are clearly perceived to be crucial by users, and a credible solution is well known to be a fundamental prerequisite for the future expansion of the IoT (Hofschen et al. 2015). This is already being reflected in the growth of IoT security spending, with an expected CAGR close to 25% until 2018 (http://www.gartner.com/newsroom/id/3291817). In line with the scope of this book, security will be mostly discussed from the IoT node standpoint, and introducing new concepts that assure truly distributed and ubiquitous security, as discussed in Chap. 8.

Regarding the first of the above challenges, public-key cryptography cannot be embedded in most of IoT nodes, although it is a fundamental building block in today's information security architectures (e.g., for crypto-key exchange over an insecure channel, authentication, and others). Indeed, its computational effort is unaffordable under typical power budgets of IoT node (see Sects. 1.1 and 1.6). For example, software implementations on an MCU of the popular encryption RSA algorithm at typical 2048-bit key size entails an energy consumption per bit in the order of mJ/bit (Nehru et al. 2014; Hu et al. 2009), i.e. 10^9 times larger than state-of-the-art energy per bit in private cryptography (e.g., AES (Zhao et al. 2015), see Chap. 8), and 10^6 times larger than energy/bit of best-in-class radios (see Sect. 1.3.3). RSA decryption entails a $10\times$ larger computational effort and energy (Hu et al. 2009). This means that public-key cryptography is affordable only if it is used roughly once every 10^9 transactions, as in this case its impact on the average power becomes comparable to private-key cryptography. Exchanging a key every 10^9 transactions translates into one RSA execution every 192 years (!), assuming that one measurement (and transmission) every 10 min is performed by the IoT node. In other words, public-key cryptography is not really usable in power-constrained IoT nodes, and novel methods to exchange keys over the insecure wireless channel are needed, as discussed in Chap. 8.

On the other hand, application-specific on-chip accelerators for RSA encryption consume an energy in the order of uJ/bit (Hu et al. 2009), i.e. 10^6 times larger than private cryptography. This comes at the cost of additional complexity in the order of 40 kgates, which is comparable to the gate count of an entire MCU (e.g., MSP430 (Kwong et al. 2009)), three times a low-end processor for IoT (ARM Cortex M0), 5–7× more complex than AES crypto-cores, and 40× more complex than recently proposed crypto-cores for IoT (e.g., Simon (Beaulieu et al. 2015)). Hence, public-key cryptographic accelerators might be affordable in terms of power for rather infrequent key exchange (once every few months), but at significant hardware cost. Some research effort is being performed to devise new public-key crypto-algorithms that are affordable in IoT nodes (Arbit et al. 2015). Also, crypto-accelerators for hardware-software

security integration are being made available (Atmel), where keys, certificates and other sensitive security data used for authentication are stored in secure hardware and protected against software, hardware and back-door attacks.

1.7.2 Opportunities to Address Security Challenges on the IoT Node Side: Physically Unclonable Functions (PUFs) and PUF-Enhanced Cryptography

Due to the pervasive nature of the IoT, security measures need to be distributed and ubiquitous. Distributed security is necessary to preserve scalability in the IoT, introducing a centralized mechanism to preliminarily build the trust among IoT nodes once and for all, so that inter-node communication does not need any involvement of other devices to manage their mutual trust, as summarized in Fig. 1.21. Regarding the need for ubiquitous security, in the IoT a hardware root of trust is needed in every single piece of hardware to assure its authenticity (and possibly integrity), as well as the integrity and confidentiality of sensitive data used to establish trust (Atmel). Such hardware root of trust also drastically simplifies difficult tasks such as the key exchange over an insecure channel (see previous subsection), as will be discussed in Chap. 8.

A hardware root of trust is a secret shared between each chip and a secure database server, and is typically based on chip IDs that uniquely identifies the node and are stored in the chip and the server. The (non-volatile) chip IDs can be classified as extrinsic and intrinsic. Extrinsic keys are created outside the chip and shared with it by writing them onto it before deployment. Conversely, intrinsic keys are directly created and available on the chip once it is fabricated, and must be securely read before deployment to share the secret with the database managing trust.

As mainstream approach to hardware-level security (e.g., RFIDs, MCUs), extrinsic keys can be stored in eFuse-based, One-Time

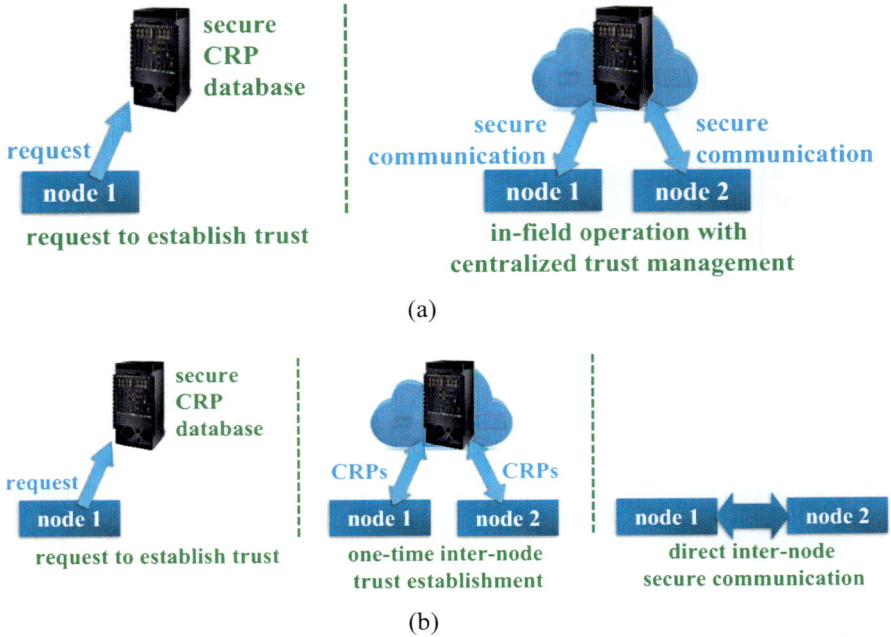

Fig. 1.21 (**a**) Centralized security architectures require other devices to constantly maintain the trust between nodes, (**b**) distributed architectures preliminarily establish trust between nodes, which then communicate securely and directly afterwards

Programmable Read Only Memory (OTPROM) or Flash memories (Aitken et al. 2014; Rosenblatt et al. 2013; Uhlmann et al. 2008). In IoT nodes, the Flash memory is certainly needed to store the program and intermediate data (see Sect. 1.5.2), hence it is the natural mainstream solution to store keys on IoT nodes. Unfortunately, extrinsic keys can be recovered with little or moderate effort through several types of attacks (van Tilborg and Jajodia 2005). For example, eFuses that disable the testing and programming interface can be restored with invasive methods (e.g., FIB) or bypassed through fault attacks (Helfmeier et al. 2013; Tehranipoor and Wang 2012). OTPROM or Flash key storage is prone to cloning through de-layering or manipulation/read-out through a number of well-established techniques to read out on-chip memories, and (Kommerling et al. 1999; Nedospasov et al. 2013; Skorobogatov et al. 2010).

Physically Unclonable Functions (PUFs) are able to generate intrinsic keys, and are far less vulnerable against physical attacks, compared to extrinsic key storage. Indeed, keys not really stored on chip, but are recreated on demand through circuit techniques that are sensitive to random variations, and insensitive to systematic or fully-correlate variations (see Chap. 8). Also, they also provide tamper evidence, at least to a certain extent (Helfmeier et al. 2014). Clearly, PUFs are still somewhat vulnerable to sophisticated physical attacks, and their resiliency against such attacks is currently under investigation (Nedospasov et al. 2013; BlackHat; Helfmeier et al. 2014).

Due to their ability to inexpensively introduce hardware-level security into individual silicon dice, PUFs have drawn substantial attention, and have already been adopted in a few commercial products (ICTK, Co. Ltd. 2014; Intrinsic-ID 2016; Invia PUF IP; QuantumTrace and Product 2013; Verayo Inc. 2013). For the sake of simplicity, a PUF can be thought of as a digital sub-system that provides chip-specific, unpredictable and perfectly repeatable responses to

Fig. 1.22 Summary of PUF operation, with chip-specific responses being stored in a secure server for authentication via comparison. Responses are unique to the die, and can be regarded as "silicon fingerprint"

incoming challenges (i.e., inputs), as summarized in Fig. 1.22. The chip-specific responses permit to identify and authenticate the die in an unambiguous manner. Authentication is largely performed by storing challenge-response pairs (CRPs) in a secure server, each pair being used to perform authentication strictly in a single transaction (being transmitted over an insecure channel). In Chap. 8, this model is questioned since the need for a fresh CRP for each transaction fundamentally conflicts with the long lifetime requirement of IoT nodes (i.e., a large number of fresh CRPs during the lifespan of the PUF). In particular, it is shown that such conventional CRP-based security scheme requires a PUF capacity that translates to a silicon cost of more than a dollar only for the PUF itself. This is clearly unaffordable and requires the adoption of a more efficient usage of CRPs.

A much more inexpensive approach for authentication will be discussed in Chap. 8, which introduces the new concept of "PUF-enhanced cryptography", which merges Physically Unclonable Functions (PUFs, or "silicon fingerprint") for hardware-level security and traditional cryptography. Interestingly, as will be discussed in Chap. 8, PUF-enhanced cryptography drastically reduces the PUF capacity requirement, solving the above mentioned issue. In addition, PUF-enhanced cryptography enables novel approaches to exchange crypto-keys over an insecure channel without resorting to conventional energy/area-hungry public cryptography schemes (see previous subsection). As discussed in Chap. 8, this concept permits to combine the best of the two worlds: (1) chip-specific operation, excellent energy/area efficiency of PUFs, (2) while hiding CRPs behind a crypto-core,

thus reducing the number of CRPs during the IoT node lifespan, and lowering the power by strengthening the crypto-key with the PUF key. Also, in Chap. 8 it is shown that recent advances in energy-efficient crypto-cores now allows full and continuous encryption of any packet under very small power budgets (Zhao et al. 2015) (e.g., down to UHF RFIDs), as AES encryption/decryption can be performed at a small fraction of a μW at typical IoT throughputs (e.g., hundreds of kbps).

In reality, the responses of PUFs are not perfectly unpredictable and repeatable (i.e., random and stable), and significant research effort is being spent to improve such features in an inexpensive manner. To gain deep understanding of the state of the art and fairly compare the different available options, Chap. 8 of this book introduces for the first time the public PUF database (http://www.green-ic.org/pufdb), named "PUFdb". This database summarizes the state of the art in PUFs, and reports up-to-date technology trends and advances in the field of PUFs, through the comparative evaluation of relevant metrics that are discussed in Chap. 8.

In perspective, the long lifetime of IoT nodes poses additional challenges, in addition to their limited power budget and hence capabilities. Indeed, security standards will change over time and will impose higher standards. On one end, this means that IoT node will need to be margined in terms of hardware-level security and cryptography to exceed current requirements, with obvious cost implications. At the same time, security primitives in IoT nodes need to be flexible enough to adapt to future standards and requirements, as well as to allow for introducing hardware patches for unanticipated types of attacks and vulnerabilities.

A possible approach can be borrowed from today's accelerators for cryptography, which combine highly efficient primitives for operations common to most of crypto-algorithms, while allowing flexibility to reuse them in a wide range of cryptographic standards (see, e.g., (Silicon Labs 32-bit Microcontroller Application Notes; NXP Cryptographic Acceleration Technology; STMicroelectronics Nescrypt cryptoprocessor datasheet; Texas Instruments Cryptoaccelerators)). Future accelerators will need to incorporate PUFs as additional primitives.

As additional challenge, there is a fundamental conflict between chip debug-ability and hardware-level security (Bhadra et al. 2016; Ray et al. 2015). Indeed, the former requirement demands architectural support for internal state observability through the testing port, and some controllability through firmware patching. In turn, such architectural support weakens the level of hardware security since it offers opportunities and backdoors to discover and leverage vulnerabilities, having access to internal signals and states (e.g., through assertion checker and coverage monitor circuits), and expose them outside the chip. This can enable the attacker to recover sensitive information (e.g., digital rights management keys), settings of programmable fuses, defeature bits, among the others (Bhadra et al. 2016). This tradeoff is difficult to manage, and requires deeper integration of design methodologies for verification and security, which in turn demands novel tools with such ability. A possible solution to cleanly manage such tradeoff is to introduce an explicit "hardware firewall", which manages independently the access to such architectural support based on the context and software under execution. For example, such hardware firewall can preventing such access under fault injection (as detected by timing, voltage, temperature sensors), physical intrusion (as detected by on-chip sensors), or unsuccessful authentication. The same firewall could also manage the testing port and forbid access to critical primitives (e.g., PUF) and related signals transferred to other sub-systems.

On the larger scale of an IoT node cluster, the distributed nature of the IoT also offers new possibilities to address issues caused by malfunctioning or hacking in a single node, thanks to the partial redundancy that is available across nodes placed in the same vicinity. For example, the measurements of multiple IoT nodes sensing the same physical quantity have some degree of correlation, if not too far from each other. This offers the opportunity to identify anomalies (due to faults or attacks) in individual nodes, based on the past observations. Distributed sensor fusion offers additional opportunities as the deviation of a given physical quantity can affect other quantities, hence the correlation between different sensors in different IoT nodes can be leveraged similarly.

1.8 Present and Future Challenges in IoT Nodes: Cost

The challenges related to the tight IoT cost requirement need to be looked at from the perspective of semiconductor economics trend (Rhines 2015), as exemplified by the learning curve in Fig. 1.11. To this aim, the cost per transistor and the cumulative number of transistors are separately plotted in Fig. 1.23 versus time. Since the inception of integrated circuit manufacturing, the overall number of manufactured transistors has doubled every 18 months, i.e. expectedly faster than the transistor density as per Moore's law, which has increased by 1.6–1.7× every 18 months. The economy of scale and the efficiency improvements from the accumulated design/manufacturing knowledge has translated into a halved cost per transistor every 22 months, being a 55% learning curve (see Sect. 1.4).

Figure 1.23 is useful to develop a sense of how semiconductor manufacturing will help drive down the cost per transistor in the IoT, assuming we will continue to innovate to maintain the same cost reduction rate (quite reasonable, based on the past experience). On the other hand, the cost reduction in IoT nodes is expected to be driven by very different factors, compared to the past. Indeed, cost reductions has been mainly pushed by the technology scaling and Moore's law, with

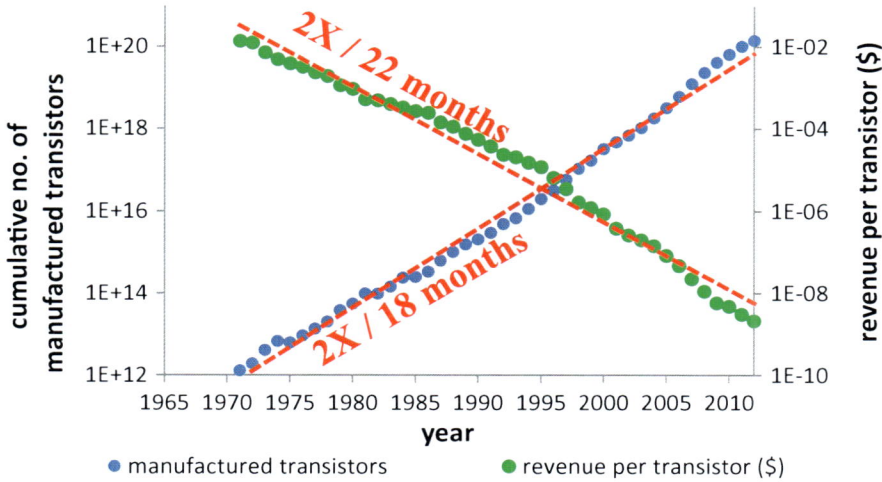

Fig. 1.23 Plot of cumulative manufactured transistors and cost per transistor (adjusted for inflation) versus year, and related scaling rules

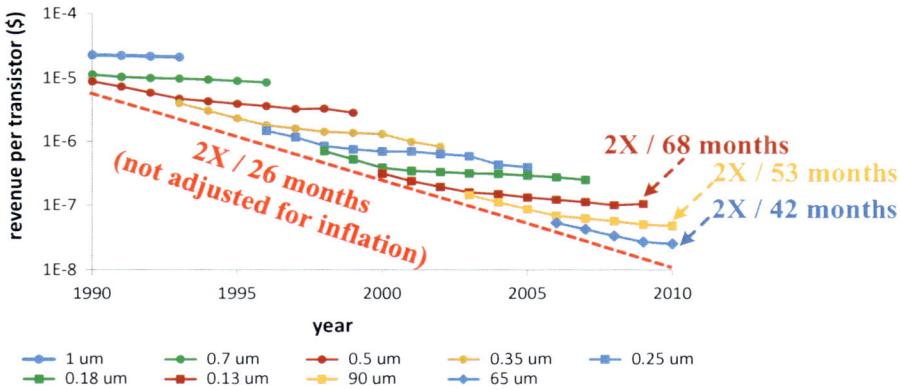

Fig. 1.24 Plot of cost per transistor (not adjusted for inflation) for several CMOS generations vs. time (data from (Kurzweil 2005; Goodall et al. 2002))

steady reduction enabled at each new CMOS generation as in Fig. 1.24 (the cost reduction is somewhat slower than in Fig. 1.23 since the former does not account for inflation). Within the same generation, some cost reduction has also been achieved over time, thanks to the amortization of the initial investments across a larger production volume, and the depreciation compared to new generations. The cost reduction rate within the same generation at 90–65 nm technologies is approximately twice as slow as

the envelope across multiple generations (Kurzweil 2005; Goodall et al. 2002).

Unfortunately, it will be very difficult to leverage aggressive technology scaling (e.g., below 28 nm) for the implementation of IoT nodes in the foreseeable future, due to various reasons. Indeed, embedded Flash is a fundamental requirement in IoT nodes (see Sect. 1.5.2), hence the technology of choice is severely restricted by the availability of such process option. As of 2016, 40 and 55 nm are the most

Fig. 1.25 Plot of cost per
million gates versus CMOS
technology generation, for
both prototyping and mass
production (data from
(Jones 2014))

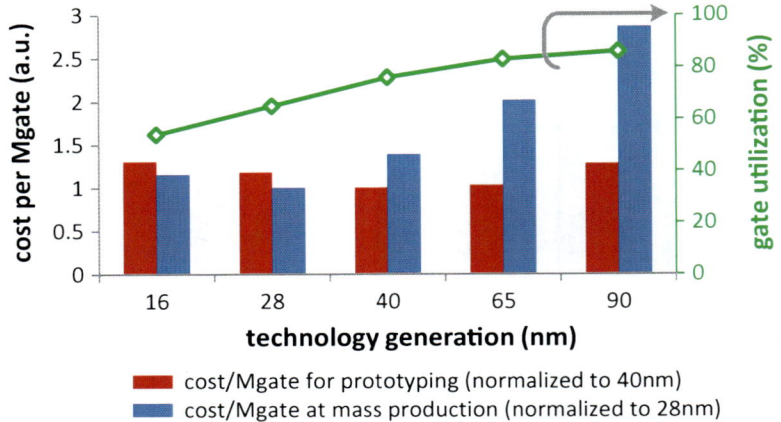

cost/Mgate for prototyping (normalized to 40nm)
cost/Mgate at mass production (normalized to 28nm)

advanced technologies offering eFlash, whereas 28 nm is being developed for mass production and makes sense only for highly digital-intensive IoT nodes (Taito et al. 2015). 2T-NOR bitcells are generally a preferable option in IoT nodes, in view of their low power consumption and relatively small capacity (Strenz et al. 2012). No further downscaling of eFlash (e.g., at 20 nm) is expected, being technologically very challenging and excessively expensive (Jurczak et al. 2015). Other emerging NVM technologies are being considered as a replacement of successive generations (e.g., ReRAM and STT-MRAM, as discussed in Chaps. 6 and 7), although with uncertain roadmap for the market introduction.

As more fundamental reason why IoT nodes will hardly scale below 28 nm in the foreseeable future, the cost per transistor in 28 nm has reached a minimum, as shown in Fig. 1.25 (Jones 2014). The cost per transistor increases again at 20 nm and finer technologies due to higher cost associated with double patterning and poorer gate utilization per unit area, sanctioning the end of Moore's law. From Fig. 1.25, 40–65 nm generations are very reasonable options for cost-effective IoT nodes, with 28-nm becoming preferable only upon availability of eFlash and for digital-dominated SoCs. Then, we envision that the range of feasible technologies for IoT nodes will remain the same for several years, dictating a radical shift in the business model of chip design (see next

section). Some further considerations on the technology selection will be made in Chap. 18.

From the above considerations, chips for IoT nodes will not be able to benefit from further technology scaling, and will only enjoy half of the cost reduction that we have observed (see Fig. 1.24). The remaining half will need to come from other strategies that enable further cost reduction while not shrinking the process (i.e., extend the lifespan of mature technology generations such as 65, 40, and ultimately 28 nm). For example, we envision that 450-mm wafers should actually be used at such well-established processes, rather than being limited to up-coming cutting-edge generations (Global 450 Consortium (G450C) Program). Indeed, IoT offers an unprecedented opportunity to scale up the manufacturing volume at such well-established generations, and hence truly benefit from the cost reduction allowed by larger wafer size. Furthermore, such cost benefit is expected to be even more pronounced compared to most advanced CMOS generations, as lithography takes a smaller fraction of the wafer processing time, as no double patterning is needed (the cost associated with does not scale down at larger wafers). Another very interesting option to extract value from well-established technologies at low cost is to introduce the FDSOI option at 40 and 65 nm (Esteve et al. 2016), leveraging its low cost and benefits in terms of voltage scalability and leakage control.

As another important recurring engineering cost, testing of chips for IoT nodes needs to be kept to a minimum, in spite of their low voltage operation and hence large sensitivity to process, voltage and temperature variations (see Chap. 4). For example, their sensors are generally needed to be calibrated in a post-silicon manner, between testing and run time (i.e., at additional testing cost or higher energy in in-field operation). In perspective, such cost can be driven down by leveraging the distributed nature of IoT systems, and the intrinsic spatial redundancy of IoT sensors (Lee et al. 2014; Miluzzo et al. 2008; Stojmenovic 2005). The research work in this field has been mainly algorithmic and dealt with sensors that have not been conceived to calibrate each other. In perspective, mutual calibration should be enabled by incorporating it in IoT nodes in first place, offering architectural support to the network for faster, continuous and accurate mutual calibration. We envision that such approach will help create "supersensors", i.e. sensors exceeding their intrinsic capabilities through synergy with the neighboring IoT nodes. This can be achieved through conventional sensor fusion (i.e., combining information from different physical dimensions), as well as cross-calibration by measuring (1) their operating conditions (e.g., using an IoT node sensing temperature) to suppress the margin that is traditionally added to the sensor static characteristics, (2) their usage intensity for aging/drift estimation and compensation (as needed by their long lifetime).

To mitigate their Non-Recurring Engineering cost (NRE), IoT nodes need to be designed efficiently through an application-driven combination of platform-based design, design composability (both intra-die and inter-die, i.e. reuse and System-in-Package design), reconfigurability and low-cost adaptation. As an example, Application-Specific Instruction set Processors (ASIP) with processor extensions (i.e., accelerators) offer flexibility to be tailored around an application and achieve both area (i.e., cost) and energy efficiency, while mitigating the design effort through automated generation of RTL, Software Development Kit, and verification support.

As discussed in Sect. 1.1, this book focuses on the design of integrated circuits for IoT nodes, and hence does not explicitly address challenges at higher levels of abstraction, such as interoperability, communication protocols, data management, detailed IoT architecture with distributed processing and software allocation, among the others. Such challenges are crucial to the IoT, and mandate the creation of proper abstraction layers that can abstract the specific behavior of IoT nodes and concentrators, to create a scalable, secure, IoT infrastructure with predictable latency and availability as a whole (Zhang et al. 2015). Ultimately, the challenge is to create a true ecosystem that seamlessly collects pushes data to the cloud on the enterprise side (communication with sensors), create repositories and structure data, perform storage and security, analyze and visualize data for sensemaking. The reader is encouraged to explore the related topics in the vast existing literature.

1.9 Looking at the Future: IoT Market towards the End of Moore's Law and Related Trends

1.9.1 Perspectives on the Growth of the IoT

The IoT is currently at its very early stage of adoption and development, and more precisely it corresponds to the "innovators" adoption stage, according to the well-known S-curve adoption model of innovation (Rogers 2003). The S-curve is the cumulative distribution of the time of adoption, and is usually Gaussian distributed (Rogers 2003). The adoption stages are defined by the distance from the mean μ of the time of adoption, expressed in number of standard deviations σ. For example, from Fig. 1.26 innovators are very early adopters, as they adopt a specific innovation 2σ before μ, thus accounting for 2.5% of the total adopters at maturity stage in a Gaussian S-curve. Fast growth is observed until the average/median adopter is included, then growth slows down. After $\mu + \sigma$, the adoption tends to saturate, leading the market to a maturity stage. This is for example the case of the smartphone market around 2016–2017 (IDC 2016), which is

Fig. 1.26 The S-curve
describes the stages of
adoption of innovation in
the form of cumulative
distribution of the time of
adoption

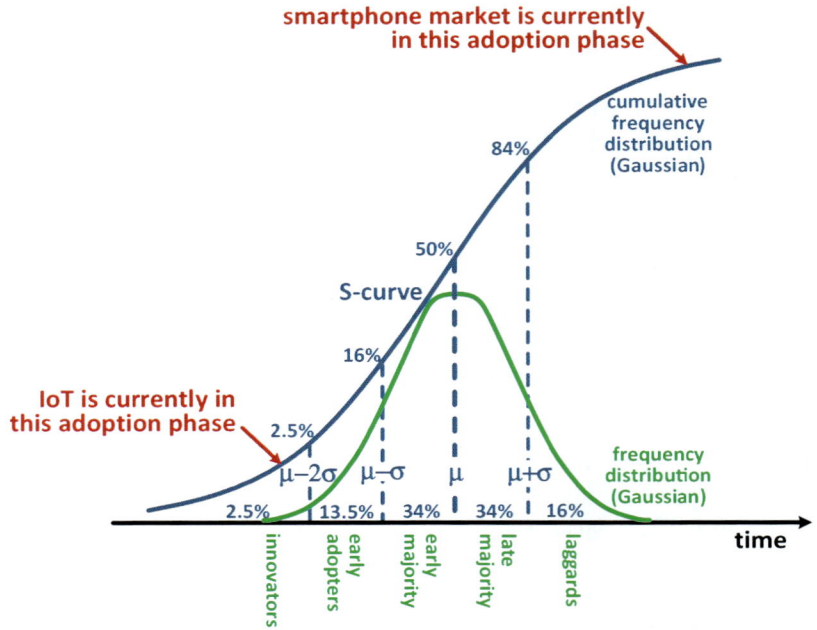

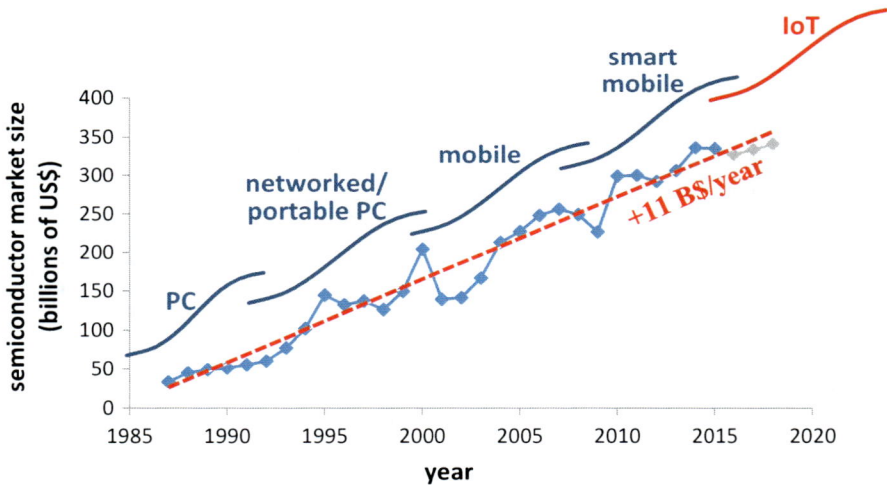

Fig. 1.27 Semiconductor annual sales over time, and technology waves leading to growth

only a few percentage points, after the impetuous
growth in the previous 10 years.

The above considerations show that the
smartphone market wave is now in the
"laggards" phase, as shown in Fig. 1.26. Accordingly, the growing IoT wave is expected to trigger the next growth in the semiconductor market
size as shown in Fig. 1.27, as steadily occurred
through previous technology waves (Rhines

2015), which have grown the market size by
approximately 11 B$ per year in the last three
decades (this size is expected to grow as the
logarithm of the IoT market volume increase,
comparing the trends in Figs. 1.9 and 1.27 and
assuming they will not change drastically).

The push towards the next phase of adoption
in IoT in Fig. 1.26 will likely come from very
few key applications that are less impacted by the

open challenges in IoT described in Sect. 1.3 (e.g., cost/profitability, inter-operability), and benefit more from the IoT infrastructure (e.g., applications where having large numbers of coordinated sensing points is perceived as a strong need). In terms of profitability, in Sect. 1.3.4 it was pointed out that profitability in the IoT area is not trivial in spite of the large production volume, due to the extreme diversity of IoT applications, which in turn limits the volume in each specific application. Design approaches mitigating this profitability challenge have been briefly discussed in Sect. 1.3.4, and will require better coordination within the semiconductor value chain. Meanwhile, the applications that inherently have larger profit margin will develop faster than others.

Similarly, applications where a proprietary/closed system is acceptable will develop faster than those requiring pervasive standardization and open hardware/software platforms. On both respects, industrial applications are particularly well suited for initiating the next stage of adoption of the IoT. Indeed, industrial applications put less pressure on the profit margin per IoT node, as a disproportionately larger value comes from the savings and the efficiencies that are enabled by the distributed real-time monitoring capability. At the same time, industrial environments do not currently need to be pervasively connected to the outside world, thus making the adoption of proprietary/closed systems (i.e., non-interoperable with the outside world) acceptable in the foreseeable future. And, overall, the ability to coordinate large numbers of IoT nodes is of great value to industry, as efficiencies come from economy of scale and global coordination (from machines, to processes, to inventory and sales).

After fostering Industry 4.0 (i.e., a business-to-business commercialization model), we envision that the IoT will penetrate smart homes as second large-scale application over time (and first as business-to-customer). Indeed, the cost of individual IoT nodes is not a key factor in smart homes, as they can be progressively upgraded (as opposed to an industrial plant) and their value is well perceived by users, as the IoT truly becomes part of their life. In smart homes, new capabilities will be introduced by the IoT in terms of conditioning/heating system management, security, lighting, entertainment, appliances, assisted living, safety and security (i.e., well beyond the existing simple thermostats or surveillance cameras.). The need for deeper integration with the external world (e.g., cloud services, smartphone apps) will make the start of the smart homes technology wave slower than Industry 4.0. We believe that the third wave of IoT technologies will be in smart buildings and then smart cities, as the experience and standardization efforts developed in smaller settings will be valuable to grow to the next scale. Other applications such as retail, transportation, environment, security and public safety, will develop later, and will augment the capabilities of the IoT in the above application spaces. Several other views have been expressed on the evolution of the IoT (see, e.g., (Beica 2015; Jankowski et al. 2014; Dobbs et al. 2015; https://www.mckinsey.de/files/mckinsey-gsa-internt-of-things-exec-summary.pdf; Lueth 2014)).

1.9.2 Perspectives on the Impact of the IoT on the Semiconductor Industry Structure

The value chain of IoT includes silicon integration (i.e., sensors, chips with processing and communication abilities, actuators), system integration (software, network/server infrastructure), and applications (cloud services and end-user applications) (https://www.mckinsey.de/files/mckinsey-gsa-internt-of-things-exec-summary.pdf). The silicon integration is of crucial importance to the realization of the IoT, and semiconductor manufacturing will be boosted by More-than-Moore integration focusing on integrated MEMS/ASICs, actuators (power devices, LEDs), microfluidics, integration with printed electronics and flexible substrates, among the others. However, the value perceived by the end users will mostly come from the services that will be delivered through it. Indeed, such services ultimately define the user experience, being the infrastructure and the IoT nodes physically invisible or inaccessible. This has implications on the viable business models for IoT, both for single users, and small/large-sized enterprises.

The above mentioned value shift towards the service will expectedly favor bottom-up small business approaches that deliver high-value services that stem from a strong expertise in a specific application area (domain knowledge). In other words, we expect that the IoT will develop in a similar way as the Internet did, with many small enterprises leveraging their deep expertise in a specific application domain, and learning about the technologies that the IoT is made up of. Similarly, once the Internet infrastructure was available in the 1990s, innovation (and its growth) was created "on the fringes", by people and small organizations focused on applications, instead of being specialized companies with strong technological expertise (e.g., chip design) in search of promising applications. This explains why 50% of the IoT solutions are expected to be provided by companies which are less than 3 years old in 2017 (Gartner 2014).

The above trend of application-focused companies driving the IoT will need to be supported by an ecosystem that makes chip design more accessible, in terms of design effort and financial entry bar. In the last decade, technological advances (e.g., cloud computing) have drastically lowered the financial bar for starting new companies (Horowitz et al. 2014), and especially for software startups (Wertz et al. 2012). We envision that this will extend to hardware startups as well, through an ecosystem built on consortia of silicon service providers that are linked to foundries with well-established technologies (e.g., 65–40 nm, or 28 nm) congregating to average out demand volatility and have predictable/high wafer throughput, create economy of scale through collaborative cloud design environments, with a new breed of Electronic Automation Design (EDA) tools that lower the expertise bar to build integrated systems, and platform-based design methodologies (e.g., based on reconfigurable architectures) that favor design reuse and rapid design closure.

To create such economy of scale and enable the above mentioned growth of IoT through bottom-up business and under mature CMOS technologies, the EDA industry will need to provide value in a way that deviates from the traditional business fueled by technology scaling. Indeed, we envision that user-friendliness of EDA tools for IoT node design will become a major differentiator in the future, considering that the value will not come from the development of tools for new technology generations. User-friendly tools are indeed expected to bring substantial value to customers who are focused on applications rather than chip design, as they will give them the capability to build chips with reduced effort, team size and related capital expenditures. On the EDA tool vendor side, the reduced profit margin will be (over) compensated by the much wider market volume, thanks to the larger demand of a wider basis of small application-focused enterprises performing chip design. To continue create value, EDA tool vendors will likely offer vertically integrated software packages and pay-per-use emulation services, where IoT nodes (and maybe complete systems) can be simulated embracing their multiphysics nature, from electrical to microfluidics, chemical and mechanics, including packaging aspects at their interface.

Other developments in the software development are expected to become part of chip design environments. Among the others, they will need to become more collaborative for rapid development through distributed and virtual/elastic teams, similarly to the raise of Collaborative Development Environments (CDE) introduced in software design since early 2000s (Q&A with Grady Booch: Collaborative Development Environments 2006), and CAD/CAM environments (Wang and Zhang 2002). In turn, the adoption of such collaborative environments will favor rapid communication between designers, application engineers, manufacturing, sales and customers, and more frequent pivoting towards the objectives. As a consequence, the principles of agile design will ultimately permeate chip design, as an approach to further reduce time to market, improve responsiveness to customers' needs, and tight coupling between design and applications (Beck et al. 2001; Patterson et al. 2015; Beck and Andres 2005; Agile SoC). Development in short sprints will be enabled by the reduced design effort in individual tasks, as enabled by reuse and platform-based design.

Cloud-based chip design platform are emerging and will likely become mainstream in design under well-established CMOS generations (see, e.g., (Silicon Cloud)). These cloud-based design platforms will primarily create further economy of scale in design team of application-focused companies, as they will enable pay-per-use cost models, thus avoiding the large cost of entry related to the purchase of EDA tools, and related system management costs (Brayton et al. 2015). Also, cloud-based design platforms will favor collaboration within enterprises, and will create new opportunities for synergy between companies (e.g., rapid and unceasing exchange of designs for strategic partnerships). Also, cloud-based design platforms will help build trust between entities generating Intellectual Properties (IP), foundries and system integrators, as access to resources of each party is strictly disciplined in the cloud. The coexistence of several designs and related aggregate data (without details on design execution, for obvious confidentiality reasons) on their performance will create unprecedented opportunities to develop across-design understanding of the underlying design tradeoffs in a data-driven manner. This will enable new capabilities that go beyond the individual designer's experience, and enable rapid architectural exploration, IP and process selection, and verification/debugging, NRE cost prediction and manufacturing cost model creation through data mining and machine learning-based models. Linking this new type of collective knowledge to a collaborative design environment will further accelerate the capabilities of design teams. For example, new online (private or public) forums can be created to discuss about design issues while linking them to specific parts of the design or simulation outcomes, thus simplifying troubleshooting, favoring communication and knowledge building. Such more collaborative and shared environment can rely on a relatively stable core set of available IPs thanks to the longer lifetime of IPs, as enabled by the adoption of mature (and stable) technology generations. As side benefit, the longer lifetime of IPs will amortize their NRE design cost across a larger number of systems, thus reducing the cost. As a side benefit of cloud-based design environments, workflow management (e.g., for agile design) will be integrated directly with the design tools for more predictable design schedule, and nearly-real time management of related resources (e.g., cashflow, inventory, sales).

The above envisioned bottom-up model of IoT companies will not put an end to the traditional top-down technology push from Integrated Device Manufacturers (IDMs), of course. In the future, IDMs will need to innovate by developing truly vertical design capabilities, ranging from chip design to software. This will be financially viable in applications with large market size (e.g., wearables). As another important role of IDMs, they will be able to build complete security/trust chains that go from design to manufacturing and die-specific key and certificate management, as needed by hardware-level primitives such as PUFs (NXP).

Finally, the growth in the market segment of IoT nodes will certainly boost (with a lever effect) several other segments, being responsible for the production of massive amounts of data. In other words, the IoT market expansion will actually lead to the expansion of markets related to servers, storage and communications. For this reason, we envision that traditional players in such spaces might actually subsidize the creation of the IoT infrastructure, with the goal of expanding their core business.

1.10   Looking at the Future: Convergence of IoT and Other Social Megatrends

The growth of the IoT is expected to be fostered by the convergence with other social megatrends. Accelerated urbanization and increased human population are posing fundamental challenges in terms of sustainability, livability, and demands more efficient resource management (even space and time, other than water, electricity, transportation, etc.) and opportunities for more pervasive sharing (Ericsson report on "City Index"). An examples of opportunity to share services and goods is the e-mobility (e.g., bikes, smart city cars, self-driving cars and drones for good delivery). All these services need to track goods, detect anomalies, and keep track of their maintenance status.

Geo socialization is also well recognized as an up-coming trend that will supersede today's social networking, adding the geographical dimension as additional component to facilitate social interaction and communication for leisure and business purposes (Frost and Sullivan Report). The IoT will play a key role in geo socialization, as it will provide the necessary physical information to augment social media (e.g., auditive awareness, cognitive cameras).

Pervasive assistive or proactive robot technology for are likely to be part of our life in one or two decades. Such robots will provide domestic help, perform repetitive labor in factories, and assist the elderly. Such robots will need to be aware of the physical context in which they are immersed, beyond mere Internet connectivity. Again, the IoT will be instrumental to augment their capabilities with physical understanding.

Constant and data-driven product upgrade will become a norm, following today's examples from the automotive industry (Tesla). Such upgrades are pushed from the vendor to continuously improve the product, based on the data generated by the sensors embedded in all similar products. Accordingly, the IoT will play a fundamental role in understanding the in-field usage and performance of products, and will be instrumental in closing the loop for regular product improvements.

Three-dimensional remote physical interaction is another megatrend that has clear convergence with the IoT. 3D remote interaction permits to manipulate objects and receive sensory feedback (e.g., visual, tactile), and has applications to immersive conferencing, medical training, remote surgery, among the others.

Distributed sensing is again the key to recreate a remote environment.

Wide adoption of participatory sensing is another megatrend that is expected to converge with IoT. In participatory sensing, users deliberately agree to share their sensor information (currently on a smartphone) with a cloud platform that aggregates data and provides valuable information and recommendations to the participants (e.g., traffic congestion). We envision that an extended form of participatory sensing will take place where users choose to share the data sensed by their IoT nodes for the benefit of all the participants (e.g., fire, house break-in and loud events detection).

As further interesting megatrend, the sharing economy is proving that aggregate advice from individual recommendations is perceived as trustworthy, in a time when public institutions are losing people's trust (Botsman 2016). The IoT might strengthen this trend, augmenting shared data with its physical dimension. Recently, the human factor is playing a more important role in IoT product planning and design, with happiness being modeled, sensed and kept in the actuation loop to make the built environment closer to the actual needs of human beings (Yano et al. 2015).

In the end, the IoT will enable new capabilities that can truly improve the quality of life of a large number of people on earth. Indeed, it will allow more efficient and real-time management of common resources, based on actual and predicted demand. It will make technology more human centric, as it will sense and learn the behavioral patterns and the users' needs, to overcome the need for "pushing buttons" and enable proactive adaptation to such needs. It will also take the sharing economy to the next level, as embedding sensors in all shared objects permits to measure actual utilization of goods, resources, tools, and monitor their operating conditions. In turn, this encourages responsible usage, incentivizes virtuous behavior that favors sustainability and resource share-ability, leveraging the capability to charge users by the effective usage (instead of merely charging by time, as is customarily done in equipment leasing, car sharing, and others).

In summary, the pervasive nature of the IoT and its ability to distribute intelligence across objects has a very strong potential impact on our ability to share on a larger scale, to ultimately improve our lives and decouple socioeconomic progress from intensive use of resources (Ericsson Mobility Report).

Acknowledgments The author acknowledges the kind support by the MOE2014-T2-1-161 grant from the Singaporean Ministry of Education, and Mohsen Shaghasemi for gathering data on commercial IoT nodes. Special thanks go to Vivek De, Giovanni De Micheli, Gerhard Fettweis, Niraj Jha, and Samuel Naffziger for the interesting discussion and their valuable feedback.

References

A1006 Secure Authenticator for anti-counterfeit applications, NXP website on http://www.nxp.com/documents/leaflet/A1006-LF-SECURE.pdf?fasp=1&WT_TYPE=Brochures&WT_VENDOR=FREESCALE&WT_FILE_FORMAT=pdf&WT_ASSET=Documentation&fileExt=.pdf

Agile SoC website, http://www.agilesoc.com/about/

R. Aitken, V. Chandra, J. Myers, B. Sandhu, L. Shifren, G. Yeric, Device and technology implications of the Internet of Things, in *Proceedings of IEEE Symposium VLSI Circuits (VLSI-Symposium)* (2014)

M. Alioto, Ultra-low power VLSI circuit design demystified and explained: a tutorial. IEEE Trans. Circ. Syst. **59**(1), 3–29 (2012)

M. Alioto, Designing (relatively) reliable systems with (highly) unreliable components,—Keynote at *IEEE NEWCAS 2016*, Vancouver (CA), 24–26 June (2016)

M. Alioto, Energy harvesters for IoT: applications and key aspects—Short course at *VLSI Symposium 2015*, Kyoto, 15 June (2015)

M. Alioto, E. Consoli, J. Rabaey, "EChO" reconfigurable power management unit for energy reduction in sleep-active transitions. IEEE J. Solid State Circuits **48**(8), 1921–1932 (2013)

Amazon Dash, https://www.amazon.com/Dash-Buttons/b?ie=UTF8&node=10667898011

A. Arbit, Y. Livne, Y. Oren, A. Wool, Implementing public-key cryptography on passive RFID tags is practical. Int. J. Inform. Secur. **14**(1), 85–99 (2015)

ARM Cortex M0 datasheet, https://www.arm.com/products/processors/cortex-m/cortex-m0.php

K. Ashton, That 'Internet of Things' Thing. RFID J. http://www.rfidjournal.com/articles/view?4986. Accessed 22 June 2009

Atmel website, The first hardware interface library for TLS stacks used in IoT edge node apps, http://blog.atmel.com/2016/02/22/atmel-launches-the-industrys-first-hardware-interface-library-for-tls-stacks-used-in-iot-edge-node-apps/

L. Atzori, A. Iera, G. Morabito, Internet of things—a survey. Comput. Netw. **54**(15), 2787–2805 (2010)

R.G. Baraniuk, Compressive sensing [Lecture Notes]. IEEE Signal Process. Mag. **24**(4), 118–124 (2007)

R. Beaulieu, S. Treatman-Clark, D. Shors, B. Weeks, J. Smith, L. Wingers, The SIMON and SPECK lightweight block ciphers, in *Proceedings of the DAC 2015*, San Francisco, USA (2015)

K. Beck, C. Andres, *Extreme programming explained*, 2nd edn. (Addison-Wesley, Boston, 2005)

K. Beck, J. Grenning, R. C. Martin, M. Beedle, J. Highsmith, S. Mellor, A. van Bennekum, A. Hunt, K. Schwaber, A. Cockburn, R. Jeffries, J. Sutherland, W. Cunningham, J. Kern, D. Thomas, M. Fowler, B. Marick, Manifesto for Agile Software Development (2001). http://agilemanifesto.org/

R. Beica, MEMS and sensors: applications and key aspects—Short course at *VLSI Symposium 2015*, Kyoto, 15 June (2015)

G. Bell, Bell's law for rise and death of computer classes. Commun. ACM **51**(1), 86–94 (2008)

C.G. Bell, R. Chen, S. Rege, Effect of technology on near term computer structures. IEEE Comput. **5**(2), 29–38 (1972)

G. Beni, J. Wang, Swarm intelligence in cellular robotic systems, in *Proceedings of the NATO advanced workshop on robots & biological systems*, Italy (1989)

J. Bhadra, S. Ray, Security challenges in mobile and IoT systems, in *Proceedings of the SOCC 2016*, Seattle, USA (2016)

BlackHat conference website. https://www.blackhat.com/

R. Brayton, L.P. Carloni, A. Sangiovanni-Vincentelli, T. Villa, Design automation of electronic systems: past accomplishments and challenges ahead. Proc. IEEE **103**(11), 1952–1957 (2015)

E. Brynjolfsson, L.M. Hitt, Beyond the productivity paradox. Commun. ACM **41**(8), 49–55 (1998)

S. Callewaert, Communication infrastructure disruptions caused by the forthcoming IoT data deluge: the FD-SOI solution, in *Proceedings of the SOI Consortium Symposium*, San Jose (2016). http://www.soiconsortium.org/fully-depleted-soi/presentations/SOI-Consortium-FD-SOI-Symposium-Sanjose-2016/

E. Candès, J. Romberg, T. Tao, Robust uncertainty principles: exact signal reconstruction from highly incomplete frequency information. IEEE Trans. Inform. Theory **52**(2), 489–509 (2006)

Cisco Visual Networking Index Predicts Near-Tripling of IP Traffic by 2020, white paper, https://newsroom.cisco.com/press-release-content?type=press-release&articleId=1771211

C. Desset and A. Fort, Selection of channel coding for low-power wireless systems, in *Proceedings of the IEEE Vehicular Technology Conference* (2003), pp. 22–25

R. Dobbs, J. Manyika, J. Woetzel, The internet of things: mapping the value beyond the hype, McKinsey Global Institute report (2015). http://www.mckinsey.com/business-functions/business-technology/our-insights/

the-internet-of-things-the-value-of-digitizing-the-physical-world

D. Donoho, Compressed sensing. IEEE Trans. Inform. Theory **52**(4), 1289–1306 (2006)

Ericsson Mobility Report, https://www.ericsson.com/res/docs/2016/ericsson-mobility-report-2016.pdf

Ericsson report on "City Index". https://www.ericsson.com/networked-society/trends-and-insights/city-index?utm_source=programmatic&utm_medium=dbm-1stpartydata&utm_campaign=cityindex

H. Esmaeilzadeh, E. Blem, T.S. Amant, K. Sankaralingam, D. Burger, Dark silicon and the end of multicore scaling. IEEE Micro **32**(3), 122–134 (2012)

E. Esteve, No reason for FD-SOI Roadmap to follow Moore's law!, SemiWiki forum, 26 April (2016). https://www.semiwiki.com/forum/content/5720-no-reason-fd-soi-roadmap-follow-moores-law.html#

D. Evans, The internet of things—How the next evolution of the internet is changing everything, in *CISCO white paper* (2011). http://www.cisco.com/c/dam/en_us/about/ac79/docs/innov/IoT_IBSG_0411FINAL.pdf

Extended Coverage-GSM—Internet of Things standard. http://www.gsma.com/connectedliving/extended-coverage-gsm-internet-of-things-ec-gsm-iot/

M. Fojtik, D. Kim, G. Chen, Y.-S. Lin, D. Fick, J. Park, M. Seok, M.-T. Chen, Z. Foo, D. Blaauw, D. Sylvester, A millimeter-scale energy-autonomous sensor system with stacked battery and solar cells. IEEE J. Solid State Circuits **48**(3), 801–813 (2013)

G. Frantz, Digital Signal Processor Trends, *IEEE Micro* (2000)

L. Freyman, D. Fick, M. Alioto, D. Blaauw, D. Sylvester, A 346 μm² VCO-based, reference-free, self-timed sensor interface for cubic-millimeter sensor nodes in 28 nm CMOS. IEEE J. Solid State Circuits **49**(11), 2462–2473 (2014)

H.T. Friis, A note on a simple transmission formula. Proc. IRE **34**(5), 254–256 (1946)

Frost & Sullivan report, Six degrees apart: geo socialization: the next trend in social networking, http://www.growthconsulting.frost.com/web/images.nsf/0/8b2a3d36c83d296c652577ee00270b79/$FILE/Megatrends_GeoSocialization.pdf

F. Frustaci, M. Khayatzadeh, D. Blaauw, D. Sylvester, M. Alioto, SRAM for error-tolerant applications with dynamic energy-quality management in 28 nm CMOS. IEEE J. Solid State Circuits **50**(3), 1310–1323 (2015)

Gartner Says By 2017, 50 Percent of Internet of Things Solutions Will Originate in Startups That Are Less Than Three Years Old. http://www.gartner.com/newsroom/id/2869521. Accessed 9 Oct 2014

Gartner Says 6.4 Billion Connected "Things" Will Be in Use in 2016, Up 30 Percent From 2015. http://www.gartner.com/newsroom/id/3165317. Accessed 10 Nov 2015

Gartner Says Worldwide IoT Security Spending to Reach $348 Million in 2016. http://www.gartner.com/newsroom/id/3291817. Accessed 25 April 2016

S. Gaudin, Get ready to live in a trillion-device world, in *ComputerWorld*, Sept 11 (2015). http://www.computerworld.com/article/2983155/internet-of-things/get-ready-to-live-in-a-trillion-device-world.html

Global 450 Consortium (G450C) Program website. http://www.f450c.org/

Global Internet of Things (IoT) Market 2015–2019, TechNavio report, Feb (2015), http://www.technavio.com/report/global-internet-of-things-iot-market-2015-2019

R. Goodall, D. Fandel, H. Huffet, Long-Term Productivity Mechanisms of the Semiconductor Industry, in *Proceedings of the Ninth International Symposium on Silicon Materials Science and Technology*, Philadelphia, USA (2002)

Q&A with Grady Booch: Collaborative Development Environments, https://web.archive.org/web/20081011045609/http://www.alphaworks.ibm.com/contentnr/cdeintro. Accessed 7 Dec 2006

L. Greenemeier, When Will the Internet Reach Its Limit (and How Do We Stop That from Happening)? *Scientific American*, Feb 12th (2013). https://www.scientificamerican.com/article/when-will-the-internet-reach-its-limit/

J. Greenough, J. Camhi, The internet of things: examining how the IoT will affect the world, Business Intelligence report (2015)

H. Jones, Why Migration to 20 nm Bulk CMOS and 16/14 nm FinFETs Is Not Best Approach for Semiconductor Industry (white paper) (2014). http://www.soitec.com/pdf/WP_handel-jones.pdf

R. Hegde, N.R. Shanbhag, Soft digital signal processing. IEEE TVLSI **9**(6), 813–823 (2001)

C. Helfmeier, C. Boit, D. Nedospasov, J.-P. Seifert, Cloning Physically Unclonable Functions, in *Proceeding of the IEEE International Symposium on Hardware-Oriented Security and Trust (HOST)* (2013), pp. 1–6.

C. Helfmeier, C. Boit, D. Nedospasov, S. Tajik, J.-P. Seifert, Physical Vulnerabilities of Physically Unclonable Functions, in *Proceedings of the DATE 2014*, Grenoble (France) (2014)

Heterogeneous Integration Roadmap, http://cpmt.ieee.org/technology/heterogeneous-integration-roadmap.html

S. Hofschen, Internet of things—Secure Authentication and Communication Are Vital for Success, *Silicon Trust website*, https://silicontrust.wordpress.com/2015/01/20/internet-of-things-secure-authentication-and-communication-are-prerequisites-for-business-success/. Accessed 20 Jan 2015

B. Horowitz, The hard thing about hard things: building a business when there are no easy answers, HarperBusiness (2014)

H. Bauer, M. Patel, J.Veira, The Internet of Things: Sizing up the opportunity (2014) http://www.mckinsey.com/industries/high-tech/our-insights/the-internet-of-things-sizing-up-the-opportunity
http://www.postscapes.com/internet-of-things-market-size/
W. Hu, P. Corke, W. C. Shih, L. Overs, secFleck: a public key technology platform for wireless sensor networks, in *Proceedings of the EWSN'09*, Cork (Ireland) (2009), pp. 296–311.
G.D. Hutcheson, The economic implications of Moore's law, in *Into the nano era*, ed. by H. Huff (Springer, Berlin, 2009)
ICTK, Co. Ltd., (2014). http://www.ictk.com/servicenproduct/puf
International Technology Roadmap for Semiconductors: 2015 edition. http://www.semiconductors.org/main/2015_international_technology_roadmap_for_semiconductors_itrs (2015)
Intrinsic-ID, SRAM PUF: the secure silicon fingerprint, White Paper (2016)
Invia PUF IP. http://invia.fr/infrastructure/physical-unclonable-function-PUF.aspx (2016)
ISSCC 2016 Trends. http://isscc.org/doc/2016/ISSCC2016_TechTrends.pdf
M.Y. Jaber, *Learning curves: theory, models, and applications* (CRC Press, Boca Raton, 2011)
S. Jankowski, J. Covello, H. Bellini, H. Ritchie, D. Costa, IoT primer the internet of things: making sense of the next mega-trend, *Global Investment Research, The Goldman Sachs Group* (2014)
B. Jovanovic, P.L. Rousseau, Moore's Law and Learning by Doing. Rev. Econ. Dyn. **5**(2), 346–375 (2002). http://www.nyu.edu/econ/user/jovanovi/MooreLawRed.pdf
G. Jurczak, Advances and Trends of RRAM technology, *SemiconTaiwan 2015, Taipei (Taiwan)* (2015). http://www.semicontaiwan.org/en/sites/semicontaiwan.org/files/data15/docs/2_5_advances_and_trends_in_rram_technology_semicon_taiwan_2015_final.pdf
K. L. Lueth, IoT market segments—Biggest opportunities in industrial manufacturing, IOT Analytics report (2014). http://www.icinsights.com/news/bulletins/Internet-Of-Things-Boosts-Embedded-Systems-Growth/
L. Karam, I. Alkamal, A. Gatherer, G.A. Frantz, D.V. Anderson, B.L. Evans, Trends in multicore DSP platforms. IEEE Signal Process. Mag. **26**(6), 38–49 (2009)
I. Keene, Market trends: small cell infrastructure, femtocell, picocell and carrier wi-fi hot spot deployment plans start to solidify, Gartner report, 19 Mar (2015)
O. Kommerling, M. Kuhn, Design principles for tamper-resistant security processors, *USENIX Workshop on Smartcard Technology*, Chicago, IL, 10–11 May 1999. http://www.cl.cam.ac.uk/Research/Security/tamper (1999)
J. Koomey, et al., Implications of historical trends in the electrical efficiency of computing. IEEE Ann. History Comput. (2011)

E. Kristjansson, Low Power ARM7 Design (2006) [Online]. http://www.microcontroller.com/ARM7_Low_Power_Design_White_Paper.htm
R. Kurzweil, The Singularity Is Near (Viking Press, 2005). http://www.singularity.com/charts/page60.html
J. Kwong, Y.K. Ramadass, N. Verma, A.P. Chandrakasan, A 65 nm Sub-Vt microcontroller with integrated SRAM and switched capacitor DC-DC converter. IEEE J. Solid State Circuits **44**(1), 115–126 (2009)
B.-T. Lee, S.-C. Son, K. Kang, A blind calibration scheme exploiting mutual calibration relationships for a dense mobile sensor network. IEEE Sensors J. **14**(5), 1518–1526 (2014)
M. Lueders, B. Eversmann, J. Gerber, K. Huber, R. Kuhn, M. Zwerg, D. Schmitt-Landsiedel, R. Brederlow, Architectural and circuit design techniques for power management of ultra-low-power mcu systems. IEEE Trans. VLSI Syst. **22**(11), 2287–2296 (2014)
M. Alioto, A. Alvarez, Physically unclonable function database, [Online]. http://www.green-ic.org/pufdb
K. T. Malladi, F. A. Nothaft, K. Periyathambi, B. C. Lee, C. Kozyrakis, M. Horowitz, Towards Energy-Proportional Datacenter Memory with Mobile DRAM, in *Proceedings of the ISCA 2012*, Portland, USA (2012).
Markets and Markets report—"Internet of Things Technology Market by Hardware, Platform, Software Solutions, and Services, Application, and Geography—Forecast to 2022". http://www.marketsandmarkets.com/Market-Reports/iot-application-technology-market-258239167.html
E. Miluzzo, N. D. Lane, A. T. Campbell1, R. Olfati-Saber, CaliBree: A Self-calibration System for Mobile Sensor Networks, in *Proceedings of the 4th IEEE international conference on Distributed Computing in Sensor Systems*, Santorini (Greece) (2008), pp. 314–331
R. Min, M. Bhardwaj, S.-H. Cho, E. Shih, A. Sinha, A. Wang, A. Chandrakasan, Low-Power Wireless Sensor Networks, in *Proceedings of the 14th International Conference on VLSI Design* (2001), pp. 205–210
B. Murmann, ADC Performance Survey 1997–2016. http://web.stanford.edu/~murmann/adcsurvey.html
P. Narendra Mahalle, P.N. Railkar, *Identity Management for Internet of Things* (Rivers Publishers, Netherlands, 2015)
D. Nedospasov, J.-P. Seifert, C. Helfmeier, C. Boit, Invasive PUF Analysis, in *Proceedings of the 2013 Workshop on Fault Diagnosis and Tolerance in Cryptography* (2013), pp. 30–38
V. Nehru, H. S. Jattana, Efficient ASIC Architecture of RSA Cryptosystem, in *Procedings of the Fourth International Conference on Advances in Computing and Information Technology (ACITY 2014)*, Delhi, India (2014)
T. Ngo, Why Wi-Fi Stinks—and How to Fix It, *IEEE Spectrum*, http://spectrum.ieee.org/telecom/wireless/why-wifi-stinksand-how-to-fix-it. Accessed 28 Jun 2016

A. Nordrum, Popular internet of things forecast of 50 billion devices by 2020 is outdated, *IEEE Spectrum*, http://spectrum.ieee.org/tech-talk/telecom/internet/popular-internet-of-things-forecast-of-50-billion-devices-by-2020-is-outdated, Accessed 18 Aug 2016

NXP Cryptographic Acceleration Technology, http://www.nxp.com/products/identification-and-security/network-security-technology/cryptographic-acceleration-technology:NETWORK_SECURITY_CRYPTOG

B. Otis, J. Rabaey, *Ultra-Low Power Wireless Technologies for Sensor Networks* (Springer, Berlin, 2007)

D. Patterson, B. Nikolic, Agile Design for Hardware, Part I, EETimes, July 27 (2015), http://www.eetimes.com/author.asp?doc_id=1327239

QuantumTrace, LLC PUF IP Product (2013), http://www.quantumtrace.com/Products/IP/PUF%20IP/

R. Botsman, We've stopped trusting institutions and started trusting strangers, TED talk (2016), http://www.ted.com/talks/rachel_botsman_we_ve_stopped_trusting_institutions_and_started_trusting_strangers

Y.K. Ramadass, A.P. Chandrakasan, Minimum energy tracking loop with embedded DC-DC converter enabling ultra-low-voltage operation down to 250 mV in 65 nm CMOS. IEEE J. Solid State Circuits **43**(1), 256–265 (2008)

S. Ray, J. Yang, A. Basak, and S. Bhunia, Correctness and Security at Odds: Post-silicon Validation of Modern SoC Designs, in *Proceedings of the 52nd Annual Design Automation Conference* (2015)

W. C. Rhines, Cost Challenges on the Way to the Internet of Things—Keynote at *IEEE ICECS 2015*, Cairo, Egypt (2015)

T. Ricker, First Click: This $1 chip will connect your things to the city for free, The Verge, http://www.theverge.com/2016/2/17/11030692/Lorawan-internet-of-things-network-amsterdam. Accessed 17 Feb 2016

E.M. Rogers, *Diffusion of Innovations*, 5th edn. (Simon and Schuster, New York, 2003)

S. Rosenblatt, D. Fainstein, A. Cestero, J. Safran, N. Robson, T. Kirihata, S.S. Iyer, Field tolerant dynamic intrinsic chip ID using 32 nm high-k/metal gate SOI embedded dram. IEEE J. Solid State Circuits **48**(4), 940–947 (2013)

C.E. Shannon, Communication in the presence of noise. Proc. IRE **37**(1), 10–21 (1949)

M. Shoaib, N.K. Jha, N. Verma, Signal processing with direct computations on compressively sensed data. IEEE Trans. VLSI Syst. **23**(1), 30–43 (2015)

M. Siekkinen, M. Hiienkari, J. K. Nurminen, J. Nieminen, "How Low Energy is Bluetooth Low Energy? Comparative Measurements with ZigBee/802.15.4," in *Proceedings of the WCNC 2012* (2012), pp. 232–237

Silicon Cloud website, https://siliconcloudinternational.com/white-paper/

Silicon Labs 32-bit Microcontroller Application Notes, http://www.silabs.com/products/mcu/Pages/32-bit-mcu-application-notes.aspx

S. Skorobogatov, Optical Fault Masking Attacks, in *Proceedings of the 2010 Workshop on Fault Diagnosis and Tolerance in Cryptography* (2010), pp. 23–29

STMicroelectronics Nescrypt cryptoprocessor datasheet, http://www.st.com/en/secure-mcus/sr31z052.html

I. Stojmenovic (ed.), *Handbook of Sensor Networks: Algorithms and Architectures* (Wiley, Hoboken, 2005)

R. Strenz, Embedded Flash Technologies: Enabler for Automotive μCs & Smartcards, in *Workshop on Innovative Memory Technologies MINATEC*, Grenoble, France, 21 June (2012), http://leti.congres-scientifique.com/workshopmemories2012/11_Strenz_Leti_InnovativeMemoryWorkshop_2012.pdf

Y. Taito, M. Nakano, H. Okimoto, D. Okada, T. Ito, T. Kono, K. Noguchi, H. Hidaka, T. Yamauchi, A 28 nm Embedded SG-MONOS Flash Macro for Automotive Achieving 200 MHz Read Operation and 2.0 MB/s Write Throughput at Tj of 170 °C, in *IEEE ISSCC Dig. Tech. Papers* (2015), pp. 132–133

M. Tehranipoor, C. Wang, *Introduction to Hardware Security and Trust* (Springer, Berlin, 2012)

Tesla, https://www.tesla.com/

Texas Instruments Cryptoaccelerators, http://processors.wiki.ti.com/index.php/Cryptography_Users_Guide#Devices_Supported

The Digital Universe of Opportunities: Rich Data and the Increasing Value of the Internet of Things, IDC report, April (2014). http://www.emc.com/leadership/digital-universe/2014iview/index.htm

The Global IoT Market Opportunity Will Reach USD4.3 Trillion by 2024, Machina Research report, (2015). https://machinaresearch.com/news/the-global-iot-market-opportunity-will-reach-usd43-trillion-by-2024/

The Internet of Things—Opportunities and challenges for semiconductor companies.pdf, McKinsey Global Institute and Global Semiconductor Alliance report, May (2015); Executive summary available at https://www.mckinsey.de/files/mckinsey-gsa-internt-of-things-exec-summary.pdf

M.-K. Tsai, Cloud 2.0 Clients and Connectivity—Technology and Challenges, keynote at IEEE ISSCC (2014). http://isscc.org/videos/2014_plenary.html

G. Uhlmann, T. Aipperspach, T. Kirihata, C. Kothandaraman, Y. Li, C. Paone, B. Reed, N. Robson, J. Safran, D. Schmitt, and S. Iyer, A commercial field-programmable dense eFUSE array memory with 99.999% sense yield for 45 nm SOI CMOS, in *IEEE ISSCC Dig. Tech. Papers* (2008), pp. 406–407

H.C.A. van Tilborg, S. Jajodia (eds.), *Encyclopedia of Cryptography and Security* (Springer, New York, 2005)

Verayo Inc. (2013), http://www.verayo.com/tech.php

O. Vermesan, P. Friess, *Internet of Things—From Research and Innovation to Market Deployment* (River Publishers, Netherlands, 2014)

H.F. Wang, Y.L. Zhang, CAD/CAM Integrated System in Collaborative Development Environment. Robot Comput. Integrated Manuf. **18**, 135–145 (2002)

WebSphere Process Server documentation. https://www. ibm.com/support/knowledgecenter/SSQH9M_7.0.0/ com.ibm.btools.modeler.advanced.deploy.doc/ measures/aggregates.html

B. Wertz, "The One Barrier to Entry Startups Should Focus on," on VersionOne website, http://versionone. vc/the-only-barrier-to-entry-you-should-care-about/. Accessed 5 Nov 2012

Worldwide Internet of Things Forecast, 2015–2020, IDC report, May (2015). http://www.idc.com/getdoc.jsp? containerId=256397

Worldwide Smartphone Growth Forecast to Slow to 3.1% in 2016 as Focus Shifts to Device Lifecycles, IDC report, June (2016). http://www.idc.com/getdoc.jsp? containerId=prUS41425416

K. Yano, T. Akitomi, K. Ara, J. Watanabe, S. Tsuji, N. Sato, M. Hayakawa, N. Moriwaki, Profiting From IoT: The Key Is Very-Large-Scale Happiness Integration, keynote at *VLSI Symposium 2015* (2015)

B. Zhang, N. Mor, J. Kolb, D. S. Chan, N. Goyal, K. Lutz, E. Allman, J. Wawrzynek, E. Lee, J. Kubiatowicz, The cloud is not enough: saving iot from the cloud, in *HotCloud'15 Proceedings of the 7th USENIX Conference on Hot Topics in Cloud Computing*, Santa Clara, USA (2015)

W. Zhao, Y. Ha, M. Alioto, Novel self-body-biasing and statistical design for near-threshold circuits with ultra energy-efficient aes as case study. IEEE Trans. VLSI Syst. **23**(8), 1390–1401 (2015)

B. Zimmer, Y. Lee, A. Puggelli, J. Kwak, R. Jevtic, B. Keller, S. Bailey, M. Blagojevic, P.-F. Chiu, H.-P. Le, P.-H. Chen, N. Sutardja, R. Avizienis, A. Waterman, B. Richards, P. Flatresse, E. Alon, K. Asanovic, B. Nikolic, A RISC-V Vector Processor with Simultaneous-Switching Switched-Capacitor DC–DC Converters in 28 nm FDSOI. IEEE J. Solid State Circuits **51**(4), 930–942 (2016)

IoT Nodes: System-Level View

2

Pascal Urard and Mališa Vučinić

In this chapter, we detail the key elements of the wireless sensor network nodes architectures. We also review the most important tradeoffs to make in order to maximize the system energy efficiency, keeping the cost of the solution under control. We also review the pairing and registration operations and detail the security requirements as well as the impact of security on the energy efficiency and the cost of the solution.

2.1 Architecture of IoT Nodes

There are multiple possible architectures of IoT nodes. Depending on the mission profile, the use-case conditions, the pairing conditions to setup the network, topology of the network, and for sure the application use-cases, then the architecture shall be optimized in a direction or another. However, some elements are common to the possible architectures:

- One or several processors, typically MCUs (microcontroller units) like STM32, based on ARM CortexM solutions
- a communication unit (i.e.,: radio for wireless sensor networks),
- one or several sensors or actuators

- a battery: many possibilities, from alkaline to long-life Lithium based

The current market trend is to enlarge battery life by increasing the global solution energy efficiency. One of the research directions is to get rid of the disposable batteries and create an energy-autonomous solution at a reasonable price thanks to small energy harvesters. In the case of such an autonomous node the system would also embed:

- An energy harvester enabling energy harvesting from the environment: photovoltaic cell or vibration harvester, or even a Seebeck effect harvester enabling to create some energy from a delta of temperature between two materials, or between a hot surface and a radiator.
- an energy storage, usually a rechargeable battery or a supercap
- a power management unit, typically taking care of adapting harvester voltage value to the battery, and managing battery charge/discharge, plus eventually energy distribution to the system (Fig. 2.1).

In the following, we take as an example a wireless sensors network end-device node. This node is collecting data from the environment (sensor) or act on the environment (actuator). The so-called 'environment' is a general term that can represent multiple domains. The sensing

P. Urard (✉) • M. Vučinić
STMicroelectronics, Crolles, France
e-mail: urard.pascal@gmail.com

© Springer International Publishing AG 2017
M. Alioto (ed.), *Enabling the Internet of Things*, DOI 10.1007/978-3-319-51482-6_2

Fig. 2.1 Basic blocks and functions

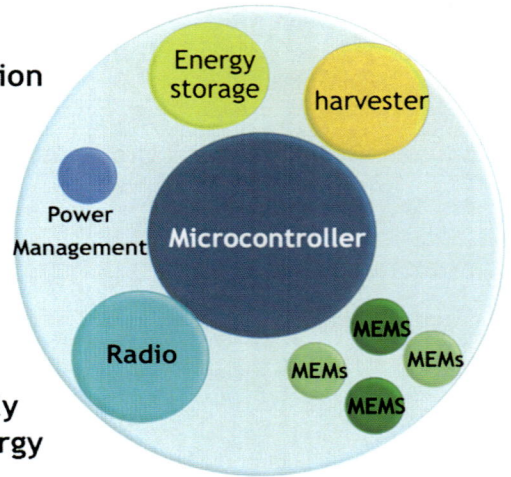

Collecting data
Treating information
Communicating…

…and possibly
harvesting energy

function is for sure adapted to each environment. In the case of:

- a motor: sensing T° or vibration, …
- a room: sensing T°, pressure, humidity, light, gas (CO or CO_2 or any chemicals), noise, presence of roommate, …
- a forest: sensing humidity, T°, fire, light, gases, chemicals, noise, and even the presence of bugs,…

The first task is to precisely define the use cases and the targeted cost of the solution. This enables to select what kind of sensors have to be used, as well as the amount of local computing in the node (by opposition to raw-data sending through the network) and the kind of MCU needed. Then the global architecture often respects the following tradeoffs.

- In order to keep a low-cost solution, the application processor can manage directly sensors and actuators:
- Sensor power supply: in the vast majority of the low-power sensors, sensor power supply can be directly connected to one of the MCU digital General Purpose Input/Output (GPIO) pins.
- Actuator power supply: in some other case, when the load need in terms of current is higher than the MCU capability (tens of mA), the actuator can be connected to the power

source through a switch or a relay, managed by a GPIO. This is typically the case of an actuator like a motor, or in the case of some high-current consumption sensors.

- The digital sensors interfaces can be directly connected to the I2C or SPI buses of the application processor
- The analog sensors can be connected directly to analog-to-digital IOs. Some MCUs like STM32 can share this ADC between several analog inputs, enabling to reuse this important component to manage several analog sensors.
- In the case a sensor is used as a wake-up function, then the usually available "wake-up" output of the sensor should be connected to the processor interrupt controller to turn on the MCU
- In the case a sensor needs some specific analog voltage values, MCUs also embed a digital-to-analog converter (DAC).

Concerning MCU choice, modern low-power MCU solutions like STM32L family embed a built-in DCDC converter enabling to drastically reduce the power consumption of the processor itself. In the case such a processor is used, the sensors and eventually the radios powered by this processor will also benefit of this power reduction. Let's take the example of a 3 V battery-operated sensing system embedding an MCU and a low-power sensor. In the case the MCU has an eDCDC (i.e.,: embedded DCDC, which is the case of the STM32L151), then a 1.2 V sensor that consumes ~5 mA under 1.2 V will only draw 2 mA from the 3 V battery if powered through the MCU.

In many cases, the radio or the radio subsystem has its own DCDC or can also take advantage of the eDCDC of the main processor.

Radio subsystem can rely on a module or on separated components. Module solution is easier to integrate but also more expensive. In recent generations of MCUs (e.g., NXP/Freescale), the radio subsystem covering Bluetooth-low-energy and 802.15.4 is integrated in the main processor.

One of the first choices to be done is the system partitioning at processor level. Several possibilities exist at SoC level as shown in Fig. 2.2.

Depending on the use case, the chosen architecture of the MCU should be more or less powerful. One of the key requirements of a low-power solution is to not wake-up the processor(s) for nothing. Some system analysis with the different tradeoffs shall be done, taking into account the different wake-up of the MCU(s), the power-consumption during these active periods, and also the cost of waking-up the processors. Choosing a powerful MCU to manage at the same time application and communication may be lower cost than two processors, but also require to turn-on regularly a large power-consuming processor. On the other hand,

keeping an application processor for system and application management and an additional network co-processor for the radio and protocol management enables to have a quite robust solution where both processors can be turned on or off independently. It also enables to set the optimization of each subsystem independently from the other (i.e.,: choose for each function the most adequate solution in terms of computing capacity, without affecting the other functions).

Having only one processor is more complex because depending on radio protocol, radio actions may be higher priority than application, interrupting an on-going sensing in some cases (so we have to define priorities and possibly require software-level concurrency contol).

The battery voltage choice is also a key parameter of the system line-up, as the battery voltage is heavily affecting the global architecture of the node. If the battery voltage is higher than the energy harvester output voltage, then a DCDC boost is needed to charge de battery. This component will enable to target high charging voltage batteries like LiPO. Another important parameter in the choice of the battery is the internal resistance value. Minimizing this resistance enables to increase system power efficiency. A high internal resistance (10s or 100s of ohms) will lead to an important voltage drop when the system is running, and may require raising the input voltage value when the harvester is charging the battery. Some batteries

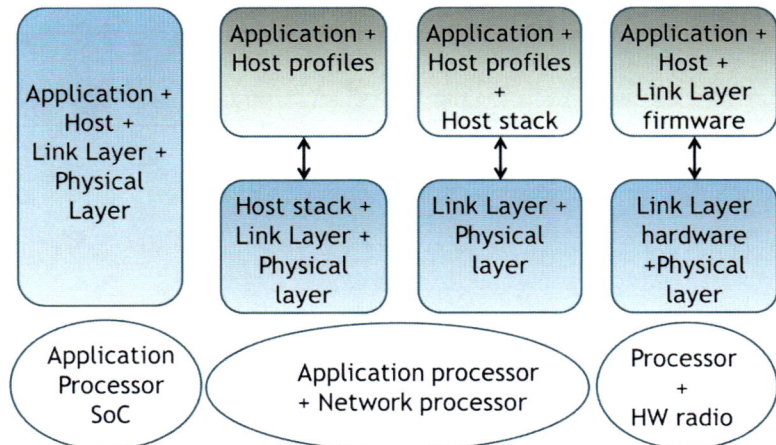

Fig. 2.2 System level radio partitioning

must be charged at a fixed given voltage. In this case, the battery current charge decreases when the battery charge reaches its maximum capacity. In other case, the battery must be charged in current, always keeping the current bellow a maximum given in order to avoid any damage to the battery. In this case, the PMU shall adapt the impedance of the harvester to the battery, and battery voltage (V_{batt}) increases with the battery charge, up to a certain point where the charge has to be stopped to preserve the battery. Continuing the charge may damage it.

There are some key advices that could save time to design a performant solution:

- The battery voltage range should be chosen taking into account the energy harvester output voltage. Minimizing the number of voltage conversions enables better system efficiency.
- The harvester has to be chosen to meet the requirements of your use cases: typical/minimal light and output charging voltage to minimize the need for complex PMU. A PV cell can easily and at low cost be designed on purpose in many places worldwide.
- In the case of unknown harvesting conditions or risk of non-respected use-cases, an MPPT (maximum power point tracking) or pseudo-MPPT power management unit can be chosen to adapt to any situation. This option has a higher cost.

- A quick estimation of the battery efficiency is given by the following ratio: V_{batt} when using the system/V_{batt} when charging. The higher the internal resistance of the battery, the lower is this ratio.

2.2 Requirements for IoT Nodes

Depending on the usage and the targeted market one radio protocol will have to be chosen to enable interoperability. For industrial market purpose, it is admitted that sub-gigahertz radios are the ones to target. For home-automation and building automation, it can be both. The market seems to push 2.4 GHz at home level: NEST-labs, acquired by Google and promoting Thread, seem to rely primarily on Bluetooth low-energy and 802.15.4 @ 2.4GHz (web:), but Zwave and at a lower scale the energy-harvested solution Enocean still offer sub-gigahertz solutions. At local or regional area, we see upcoming solutions like Lora or Sigfox enabling long (resp. very long) distance of transmission at the cost of a low or super-low data rate. There are plenty of applications that could take advantage of such solutions in the coming years.

We can divide these solutions in two topologies (see Fig. 2.3): the solutions enabling star network topology such as Wifi, Lora, Sigfox, or even GSM or LTE, and the ones enabling

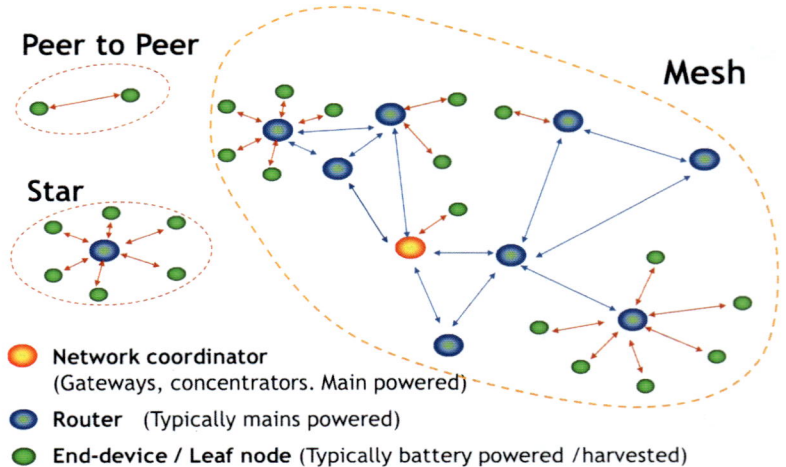

Fig. 2.3 Network topologies and definitions

Mesh networks such as 802.15.4x including Zigbee & Thread, Zwave, Enocean,...

We discuss in the next section the impact of topology on the power consumption.

2.2.1   Power

In order to target long battery life duration or energy harvesting compatibility, there should be a ratio of at least x500 between the system sleeping time and its functional time.

- Processor in terms of current is in the range of mA (e.g., 5–10 mA for CortexM3 running at 12 MHz under a 3 V supply, which results in 15 to 30 mW power) when in activity, but only around 1uA in deep sleep mode (3 μW). It should logically evolve in the coming years, thanks to Moore's law and additional low power techniques to achieve a division by 3 of this power consumption, reaching 5 to 10 mW for the same function.
- Energy cost of wake-up and go-to-sleep has to be taken into account in the tradeoff, as it is not negligible. Processors usually offer several kinds of sleep modes (light, medium, deep). Each of them can have some interest, depending on your application and context. Consider that you have to empty a large on-board capacitance to make the system sleep; the price-of-wake-up may be higher than putting the system in an intermediate sleep mode where this capacitance remains charged. So it is important to take this parameter into account, and not put the system in some deeper sleep mode when the power budget is finally larger than staying in some lighter sleep mode for a given time.
- Sensors have to be turned off as much as possible (i.e.,: between two sensing phases), or at least forced to low-power mode if they have some, especially when they have the mission to wake-up the system.
- Radio has to be mainly off.

We can consider two families of systems:

- System with regular wake-up

- System with wake-up on purpose (e.g., alarm).

In the vast majority of modern sensors, this function is natively embedded in the sensors, providing a mode with activity watch. So the design can directly use this feature, usually called *low-power mode*.

In order to successfully design an ultra-low-power solution, one should think energy harvesting and duty-cycling. See Fig. 2.4.

For sure, every component of the node has to be carefully chosen to minimize power. However, in the first place, the radio choice and radio protocol are one of the major contributors in power consumption.

Minimizing power by minimizing protocol overhead is an excellent starting point. However as we will see in the next sections this may go against interoperability. The first adopted systems in the past used to propose proprietary optimized solutions for radios and non-IP protocols. However, the future seems to be standards-based and IP-based. The Moore's law helps to gain in power efficiency: the former generations of radios used to consume 20 mA @ 3 V to send a frame at 0dBm (1 mW in the air). In 2015, the best solutions on the market consume bellow 6 mA for the same service, and the trend is still to reduce further, towards 3 mA in the 2 to 4 coming years.

Once the radio power is under control, the leakage has to be addressed. Let's imagine a node working during 20 ms every 2 min (mean schedule in real operating mode) with a mean power of 30 mW (10 mA under 3 V). It would consume around 600 μJ per active cycle. If during the inactive period, the total current (board level leakage + low-power-watch-dogs + counters) is around 2 μA (6 μW), then the total leakage would be 720 μJ per inactive cycle. The Pareto of the power consumption would start by the leakage! So the leakage can really become your number one issue to meet your total power budget. In the case your objective is to build an energy-harvested platform, the global requirements should lead you to consider every leakage above 30 nA in stop mode, and any way to save 5 μJ or more in active mode.

Duty Cycle = Active/Inactive down to 0.1%

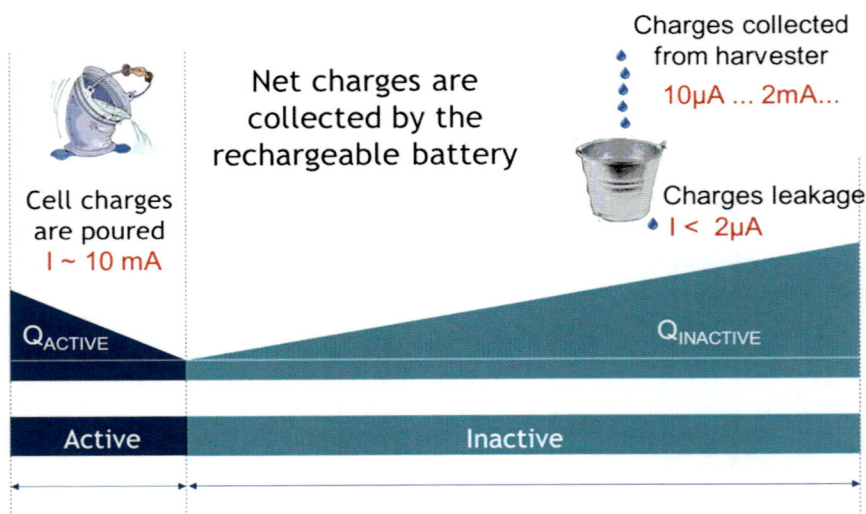

Net charges are
collected by the
rechargeable battery

Charges collected
from harvester
10µA ... 2mA...

Cell charges
are poured
I ~ 10 mA

Charges leakage
I < 2µA

Q_{ACTIVE}

$Q_{INACTIVE}$

Active

Inactive

Fig. 2.4 Heavy duty cycling enables Ultra-Low-Power solutions

2.2.2 Cost

Radio protocol has a direct impact on the global cost, because of its huge impact on the power efficiency. Also, factors like whether the upper protocol is IPv6 or not-IP, if security is enabled or not will directly impact useful data rate, and so the global energy per useful bit transmitted, leading to expend the battery size or reduce the lifetime of the node operating in harsh conditions where the eventual energy harvester would not be able to harvest anything (i.e.,: dark for a PV cell).

The battery choice, as seen in the previous section will heavily impact the cost of the node.

Choosing components with high power consumption would lead to choose a large battery, like the ones of the cellphones. Even if the price in volume may seem reasonable, it shall impose to add a specific power-management unit, enabling to charge this battery up to 4.2 V. Such a high voltage will prevent to connect directly the MCU to the battery: MCU maximum operating voltage is usually around 2.6 V.

The cost of not-optimizing the global power consumption would lead to a poor battery life for some or even all the nodes of the network. This

issue was not a major problem few years ago, but it is more and more reported by final customers that the battery budget per year can be a show-stopper after a first trial. Imagine you have to change every year heavy-duty long-duration lithium-AA batteries in 10 devices. The cost after 5y shall be higher than the system itself. The pain to change the batteries every year in a more-than-20 devices network very quickly becomes upsetting or too expensive if a specialist has to do it. We can easily understand why it is requested to optimize the solutions, still maintaining the cost low:

- by choosing an optimal system partitioning
- by optimizing global power efficiency
- by minimizing additional components and voltage conversions.

Let's take as an example the final cost of an energy-harvested node in volume, the GreenNet node V2.1 as shown in Fig. 2.5 (Urard et al. 2015). It was targeting a Bill-Of-Material (BOM) in high volume below 12 USD. This R&D project has been an ST-internal demonstrator to understand the WSN challenges and

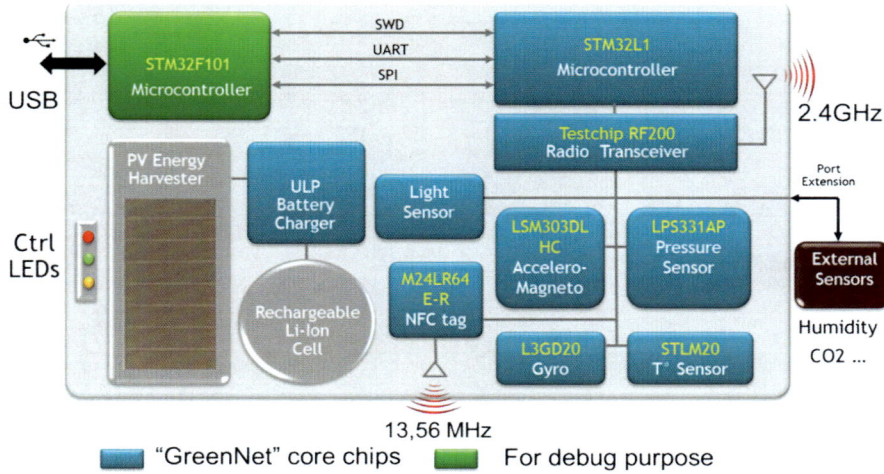

Fig. 2.5 GreenNet V2.1: Energy Harvester secured IPv6 Node content

requirements, regarding system energy efficiency on the hardware side, and secured-IPv6 power-optimized solution compatible with energy harvesting on the protocol side, in cooperation with the Laboratoire d'Informatique de Grenoble (LIG). GreenNet is a successful demonstrator but has not been commercialized. Some of the key elements are detailed in Chap. 17.

2.2.3   Interoperability

Wireless Sensor Networks have suffered for years from the fragmentation of the radio offer: too many different standards, not enough interoperability. Zigbee is the typical example where 4 main "application profiles" were offered as defensive layers, one for each of the markets: Smart Energy, Home Automation, Building Automation, and Lighting. Thread, pushed by Google through Nest-labs, has pushed further by proposing a single profile for all the use-cases. IPv6 is a definitely a trend: ensuring interoperability by offering an IPv6 solution. Nowadays, it seems also as one of the preferred ways to adopt a secured solution. IPv6 security is constantly evolving, so progressing at low cost can only come from solutions that can be shared among the standards.

Figure 2.6. shows the key elements of various 802.15.4 solutions and highlights in red some of the most interesting added values to increase the quality of service or the energy efficiency of the network. Please note the similarity between those solutions. Most of them are using 6LoWPAN for IPv6 interoperability. The two on the right (circled in red) are the latest ones, with a larger adoption of 802.15.4e, now part of 802.15.4-2015 specification. However, in some particular cases detailed in the coming chapters, there is a room for improvement and some part of the 802.15.4-beaconed may in the future be reused to further improve the energy efficiency of 802.15.4e standard.

802.15.4x solutions aren't the only WSN protocols to adopt IPv6: as an example, 802.15.1 (Bluetooth) compliancy to IPv6 has been developed in 2015 and should be available during 2016.

2.2.4   Security

Few years ago (2013), while the author was demonstrating GreenNet in Paris, audience was not convinced, at the time, about the need of secured connections for WSN. In only 2 years, the number of hacks worldwide, the need for more secured solutions at all levels, especially

	802.15.4	**Wireless HART**	**802.15.4e**	**802.15.4-beaconed**
Appli, L5-L7	HTTP/CoAP	App WHART	CoAP	CoAP
Transport, L4	TCP/UDP	Transport WHART	UDP (TCP)	UDP (TCP)
Network, L3	IPv6/6LowPAN	Routing WHART	IPv6/6LowPAN	IPv6/6LowPAN
Data Link, L2	MAC	MAC	MAC	MAC
Physical, L1	PHY	PHY	PHY	PHY

| Always ON nodes
No Synchronization | • TDMA
• Frequency Hopping
• **Fixed**
  •SuperFrame duration
  •slot duration
• All proprietary L2-L5 | • TDMA
• Frequency Hopping
• **Programmable**
  • SuperFrame duration
  • slot duration
• Standard transport RPL routing | • TDMA
• **Programmable**
  • Beacons Interval
  • Active Part
  • Up/Down Latency
• Light Routing Protocol L2-L3 for energy savings |

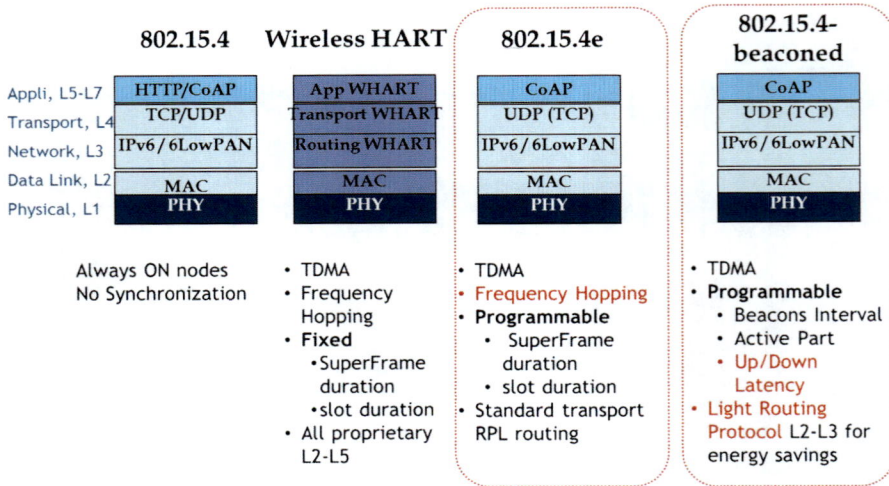

Fig. 2.6 2.4 GHz 802.15.4 enlarged protocols family

private data, and the fact that the interesting information for hackers is maybe not the one you think about, made people change their mind. Nowadays, no question any more: security is a must.

We present these hereafter an overview of typical threats and attacks in WSNs. Some more details can be found in (Vučinić et al. 2015). Security of a system can be studied within a given model. Internet protocols typically consider the traditional Dolev-Yao model (Dolev and Yao 1982) where the attacker has full control over the network.

More precisely, the attacker can:

- Intercept messages,
- Modify messages,
- Block messages,
- Generate and insert new messages

It is important to understand that cryptographic algorithms are considered "perfect" and the attacker can decrypt/forge a message only if he possesses the corresponding key. In the networking context, "message" corresponds to a Protocol Data Unit (PDU) of an abstraction layer under study. For instance, if we consider security solution at the link layer (radio protocol), message corresponds to a radio frame.

Traditionally, there are two typical classes of attacks:

- Passive attacks: Such as eavesdropping and traffic analysis, where the attacker gains knowledge on ongoing communication by passive means. For instance, if messages are sent in clear, attacker is able to read full message content. If network messages are encrypted, attacker may still be able to infer some information by studying communication patterns, timing, or message length.

- Active attacks: Attacker actively participates in the communication by re-playing old messages (replay attack), modifying messages and playing Man in the Middle (MITM), pretending to be another entity in order to gain unauthorized access to a resource and similar. A particular class of active attacks are Denial of Service (DoS) attacks, where the attacker's ultimate goal is to disrupt the availability of a network service, such as the alarm notification, typically by exhausting physical resources (memory, energy, bandwidth) on the target node.

An important point to note is that the Dolev-Yao model typically considered in protocol design is a formal model that does not take into account physical compromise of a node. Therefore, research around WSNs (Atakli et al. 2008; Chan et al. 2003; Karlof and Wagner 2002;

Rezvani et al. 2015; Vempaty et al. 2012) has often taken into account a more powerful, Byzantine attacker (Awerbuch et al. 2004). In such scenarios, attacker has access to local cryptographic material on the node and we cannot rely on cryptographic techniques to prevent attacks (Rezvani et al. 2015). Indeed, Byzantine attacker can compromise a set of nodes and through them inject false data that passes all cryptographic checks.

Becher et al. (2006) conclude that physical compromise of a device in order to extract keying material and obtain full control over it, as assumed by the Byzantine model, is not as easy as often perceived in WSN literature. It requires costly equipment, expert knowledge on hardware and hard determination of the attacker. An interesting observation of this study is that such attacks often require that a node be removed from the network for a non-trivial amount of time making detection of unusual activity via neighbor discovery protocols a simpler approach than specialized Byzantine-tolerant schemes. Common sense practices, such as disabling JTAG port or Bootstrap Loader (BSL) once the product is deployed, go a long way towards making attacks in the field more difficult.

We do recognize that in many IoT deployments, devices will be physically available to the general public and as such, system designers should take into account the threat of a physical compromise and extraction of the keying material. We emphasize that final IoT products should either have hardware- level or software-level protection against physical tampering, i.e., tamper-resistant packages or schemes to detect unauthorized access to the hardware (Becher et al. 2006; Liu et al. 2007).

2.2.4.1 Wireless Network Threats

- Physical Jamming. The most basic attempt to disrupt the network service is the attack on physical resources—the radio channel. Attacker can generate high-power signal that will interfere at different receivers in the network and increase the error rate, possibly completely disrupting wireless operation (Law et al. 2005; Li et al. 2007; Raymond

and Midkiff 2008). This DoS attack is often called jamming and is mostly a concern in military scenarios. Common defense is channel hopping that increases the bandwidth attacker has to jam, which can require a substantial power supply and thus make the attack less practical. Also, network-level redundancy can help in order to route around the jammed area.

- Traffic Injection. Injecting false traffic in the network can have multiple consequences. Firstly, it is possible to affect network applications, e.g., by introducing a bogus temperature reading to trigger the Heating, Ventilation, and Air Conditioning (HVAC) system, or even to directly control an actuator, such as the pressure regulating valve in the industrial automation system. Similarly, one can obtain full control over the network by forging network maintenance packets (Karlof and Wagner 2002), e.g., beacons, and corrupting neighbor tables of the nodes. Secondly, attacker can launch DoS attacks by generating significant traffic loads that can cause network collapse in terms of depleted energy due to, e.g., multihop forwarding. First-level protection against such attacks is link-layer security—network nodes should not accept any radio frame other than those secured with link-level keys they possess locally. At the application level, access rights should be properly configured in order to limit the damage if one of the nodes in the network is compromised. For instance, node measuring the temperature should not be allowed to issue pressure valve regulating commands. Second-level defense is common sense programming—if some of the network nodes gets compromised and starts injecting cryptographically-valid traffic, one should locally check the rate at which it is forwarding packets or performing local operations instead of blindly following the protocol.

- Attacks on Join Protocol. Link-layer security protects the wireless network in "steady" state, when all the nodes have joined and have been provisioned with necessary keying material. Before we admit a new node in the network, it is necessary to perform some checks. Join protocols are technology-specific but some common points exist:

- The joining node may initiate the join protocol multiple hops away from the gateway
- Several messages may need to be exchanged between the joining node and the gateway before the "admittance" decision can be made. This necessitates that intermediate nodes in the mesh forward the messages that may come from a rogue joining node (attacker), which opens up the possibility of DoS attacks. Although this threat can never be fully neutralized, a common strategy is to minimize the damage a potential attacker can do. As such, one may ensure that joining messages do not instill state information in the network and can control the rate at which intermediate nodes forward join protocol messages.

- Attacks on Routing Protocol. Due to their distributed nature, WSNs are prone to attacks that involve an attacker that can for example:
  - Selectively forward messages if it is within the network, or jam radio transmissions and cause collisions from the outside.
  - Advertise false routes in order to attract the surrounding traffic and create a sinkhole.
  - Present multiple false identities to other nodes in order to reduce effectiveness of fault-tolerant schemes.
  - Create radio "tunnels", so called "wormholes", between two distant parts of the network in order to appear closer to the gateway and create a sinkhole at the other end of the tunnel. Such attacks can only partly be neutralized by using link-layer security in order to reject radio frames coming from the outside. When an attacker is inside the network, i.e., a compromised node, defense requires careful design of the routing protocol that takes security into account from the beginning (Karlof and Wagner 2002).

- Privacy issues. Sensor and actuator networks that make part of our daily life bring along various privacy issues. While management of data collected by these networks in itself represents a privacy concern, we focus on information that may leak to an outsider. Obviously, data confidentiality at the link layer (protected radio frames) is the first step to improve user's privacy. In many IoT scenarios, however, radio communication alone suffices to reveal some information about the user. For example, a presence sensor may initiate radio communication when a person enters a building (Tschofenig et al. 2015) or a light switch may indicate that the state has been toggled by emitting a radio frame. Typical defense would involve injecting dummy traffic in the network but that may not always be feasible due to the local energy constraints.

2.2.4.2 Countermeasures

Threats described in above section are typically fought using security mechanisms at 2 levels: Level 2 (L2, security between radio neighbors) for hop-to-hop security or link-layer security and Level 5 for end-to-end security (L5, security between an IoT node and e.g., smartphone). Each level of protection needs to meet 4 fundamental security goals (Fig. 2.7):

1. Confidentiality: ensured by using encryption.
2. Integrity: we want the message to arrive safely, and not loose parts of it. Ensured by appending a checksum at the end of a message.
3. Authentication: Am I talking to the right node and is the message I received coming from the right node? Ensured by an authentication protocol and by using secure checksums, computed using a secret key.
4. Availability: Achieving the guaranteed level of operational performance even in presence of Denial of Service attacks (DoS) is a non-trivial task for a system designer. In terms of link-layer security, message filters and access control lists implemented in hardware allow certain degree of confidence but are alone not enough due to possible jamming attacks. Radio technologies that use frequency hopping (802.15.4e and BLE) help in preventing networks to be stuck on a single

Security is a MUST

Message confidentiality
Ensured by encryption
Prevent attacker from understanding the message

Message integrity
Ensured by Hash[1] function
Prevent message alteration during the transit phase

Message authentication
Ensured by digital certificate
Prevent from pretending to be node from you network

Availability
Ensured by firewall at gateway
Make sure that service is ready for use when
expected. Avoids DoS[2]

99.999 %

(1) : Algorithm mapping variable length data to a fixed length. (2) : Deny of Service

Fig. 2.7 The four fundamental security goals of a secured solution

radio channel and make it more difficult to the attacker to disrupt the service. However, availability is a system-level goal and thus must not be treated only at the node level, but also at the gateway level thanks to:

(a) The gateway firewall
(b) The decoupling of fast Internet world (HTTPS) and the slow one (CoAP over DTLS).

Attacks on the routing protocol are, from the point of view of confidentiality and authentication, defended using end-to-end security mechanisms, where even the on-path attacker is not able to modify or read the data, as it is not in possession of the end-to-end keys.

2.3 Power-Related Challenges and Design Tradeoffs

Chapter 2 of Varga's PhD thesis (Varga 2015) presents an overview of the latest technologies used in IoT as well a presentation of the synchronized and unsynchronized operation mode. These are presented hereafter.

2.3.1 Node Availability and Duty-Cycled Operation

As for the system, the radio shall sleep as long as it makes sense. This is different than "as long as possible" for several reasons. First reason: the need for synchronization in radio protocols. In this trade-off, the topology plays a major role.

- Case of star network topologies: we can consider the central node (concentrator) as always-on like on Wi-Fi or Sigfox or Lora.
- Case of Mesh or extended Star topology (extended star enables each node to have one additional peer attached to it). In this case there are multiple solutions:
  - Either the routers are always or mainly listening (e.g., Enocean, Zwave, Zigbee, 802.15.4-by-default) in order to receive information from the other nodes of the mesh and transmit them to the next node
  - Either both emitter and receiver are mainly off. This is the trend of the new radio standards. Bluetooth-low-Energy (BLE), 802.15.4e, 802.15.4-beaconned-option standards are proposing mainly off

solutions with synchronized wake-ups. This enables better energy efficiency versus older standards. We can say the system has slots: an active period where the radio transmits or listens to the RF activity (in case of message for the node), and an inactive period where the radio is turned off. By extension we say the system is slotted.

Slotted systems have however a constraint to solve: the need for synchronization between emitters and receivers. As there is no always-listening node over the air, a de-synchronized network would have a very poor quality of service, unless the active slots represent a very high percentage of the time, reducing by this way the efficiency and the interest of such solutions regarding previously existing ones.

As a conclusion: in order to maximize energy efficiency, new systems have the requirement to be mainly-off so they are synchronized in order to transmit the data in the shortest time for both nodes, produce the relevant acknowledge (emitter get the feedback that the data has been received) and go back to sleep mode as soon as possible.

There are two ways to maintain the synchronization of the network.

- Either there is enough and regular communications: the transmitted data or the first-level acknowledge can carry the synchronization information: this is the case of the 802.15.4e: when two nodes communicate together, the receiver sends to the emitter a low-level acknowledge embedding the synchronization data (e.g., timing corrections for the next wake-up). This solution requires a one-to-one communication: a specific rendez-vous between the two nodes. This is possible in the 802.15.4e standard in which a specific time slot and frequency channel is attributed by the router (or the master node) to each of the nodes that needs to communicate with him.
- Either the network synchronization is maintained globally thanks to the usage of beacons: the master node sends a beacon in broadcast, all the surrounding nodes needing to

communicate with him can get by decoding this beacon, the time of the next available beacon, the time slot when to emit if they need and even the fact that they have some information to request to the master node in the case they have: Bluetooth low-energy, 802.15.4 beaconed-option are working this way.

All these possibilities enable to have more or less energy efficient protocols see for instance Romaniello (2015). In Fig. 2.8, we can consider that green bubble could enable energy-harvested nodes in some specific conditions.

- In the first category, we can fit mains-powered devices. We can also consider part of this category the networks where all the devices have been declared as routers. This is typically the case when all regular devices offer routing capability to enable mesh network: Zwave & Zigbee.
- ZigbeePro-GreenPower, Enocean, would fit in the second category, enabling end-device to be super-low-power and energy-harvested at the cost of a mains-powered, always-listening router device. However, bi-directionality is not granted, preventing low-cost battery-operated actuators or IPv6 secured link.
- BLE and 802.15.4e and 802.15.4-beaconned would be categorized on the third category (row), enabling bidirectional IPv6 and secured networks, at the cost of a need for synchronization through beacons or regular data exchanges.
- 802.15.4-beaconned could also be categorized in the fourth category in some specific conditions (low-latency network), enabling energy-harvested routers, as demonstrated in the GreenNet project.

2.3.2 Activity Profile and Power Modes

One of the goals of an embedded code software designer in charge of providing a solution for an autonomous wireless sensor network is to minimize the amount of energy spent in the nodes.

A beacon = a sync signal sent by a parent node to all his children.

- This signal is propagated along the network from parents to children
- Need such signals periodically to maintain network synchronous, even if there is not data to transmit

Few data to transmit E.g. every 20s or more	Fair amount of data transmitted E.g. every 20s or less
Sync signals aim to be short.	Make sense for the payload to enable synchronization
→ **Beaconing advised**	→ **802.15.4e advised**

(a)

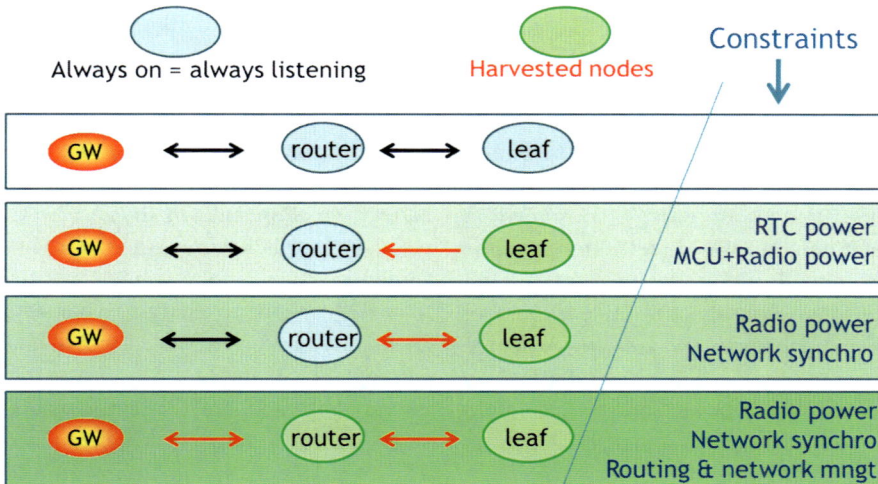

(b)

Fig. 2.8 (a) 802.15.4-beaconned option versus 802.15.4e; (b) network topology impacting the power (*red arrows* mean ultra-low-power protocols, *bidirectional arrows* mean bidirectional communication)

In the case of a sensor node operating regularly, there are few main ways to achieve this goal:

- Maximizing the deep-sleep periods of time. In deep-sleep mode, the application processor is programmed at the lowest level of power, still maintaining its stack in retention mode (or equivalent) in the System RAM and the logic glue, and turning off the NVM. In some case having the local low-speed oscillator running for the next RF rendezvous, or having some sensors in low-power mode, able to wake up the node in the case the environment parameter (e.g., temperature, vibration) goes beyond a programmed limit. In this case, the node would register this kind of alarm and send it to the gateway as soon as the next communication slot is available. So on one hand, the amount of time spent in deep-sleep mode shall be as long as possible. However, waking-up from deep-sleep requires some energy, so on the other hand, the number of time we put the system in deep-sleep has to be minimized. One unique long period between each sensor measurement is an ideal case and answers to both constraints.

- When the duration of the sleeping period is not sufficient (e.g., less than 2 ms in the case of the GreenNet node) some intermediate stop mode may be used: reducing the power by a fair amount, still consuming more than the deep-sleep, but at a lower penalty cost for wake-up.
- Turn the radio off as often and as soon as possible.
- The sensors have to be turned to their lowest power mode when in run mode and to be switched off asap.

In the case of an alarm node able to operate anytime, like a light-switch, the amount of energy required by the node is directly proportional to the latency you would accept to wait in order to operate the action. Same principle for an actuator: the amount of energy consumed to enable actuation control is directly linked to the actuator latency. E.g.: in the case of a light switch, we would consider only the amount of energy to be able to command the switch, not the energy consumed by the light. The total latency would then be: switch latency to send the command + network latency + actuator latency to receive and execute the order.

In the case the alarm node is operating without any battery but with a capacitor coupled to a pulsed-energy harvester generating 100 uJ or less, then the only way to operate is to have an always-on router, quite-always listening, that will immediately take the message and relay it inside the network.

2.4 Cost-Related Challenges and Design Tradeoffs

Is it better to design a single node with multiple programmable sensors you can activate over the air, or it is worse putting only one sensor per node? The answer has a direct impact on the cost of the solution. It depends on the cost of maintenance versus the cost of fabrication. Only a detailed cost study taking into account the forecasted volumes, price per node in each case, and the cost of SW maintenance of several applications versus 1 unique application can give you a precise answer.

2.4.1 Impact of Power on Cost of IoT Node and Concentrator

The most expensive part of a sensor node can easily become the battery. Having a low-cost coin battery is a plus as its price in volume is around 1USD (2032 LiMn). However, it is limited to 200 mAh (announced) and the usable part would only be around 40mAh (absolute limit) in the case you want to preserve the battery from short life duration.

What would be the maximum activity rate of a node using such a battery? We first consider it is acceptable to cycle by 0.5% per day on such a battery which means 1mAh (i.e., 2.6 Coulombs) per day. In other words, the mean current consumed by such a node would have to be lower than 41.6 µA if no energy is harvested during 24 h. Any energy harvested would enable more activity. For example a 6 h harvesting duration per day with a 20 cm^2 PV cell would enable the same system to have a mean current of 55 µA. This would correspond to an ultra-low-power system operating every 20–30 s, which is not so bad. If the components are consuming more, or if the communication needs to occur more often, then the battery may have to be upgraded to a larger capacity, from another type, more expensive, with sometimes more complex charging protocols.

Regarding indoor routers: in the general case one doesn't have any choice: routers currently must have a powerful battery or be mains-powered. In few particular use-cases however, it is possible to specify an energy-harvested node that handles the router function. However due to the poor harvesting capabilities, this kind of router would need to be placed very close to an energy source, like a window.

2.4.2 Impact of Protocol on Cost

We have seen in the previous chapters that a trend for interoperability is the use of IPv6 protocol. However, it defers from the usual "fast" internet world (TCP-IP based) in several points. Figure 2.9 shows the equivalence between the WSN "slow" internet world on one hand (low latencies in a WSN are hard to achieve, and are not granted in energy-constrained networks), and "fast" internet world (i.e.,: usual internet). The HTTP protocol is replaced by CoAP (Constrained Application Protocol). The main role of CoAP is to reduce the number and overhead of the exchanged packets. From end device node point of view, CoAP is also interfacing the two internet worlds, protecting the WSN from repeated high-speed requests.

In the IPv6 frame in Fig. 2.9, there is a fairly good reduction of the number of bytes exchanged in a CoAP-based case: from 681 to 111, but over a short IP packet as it is defined in 802.15.4 (127 bytes), the payload is not representing more than 20 to 27 bytes, depending on the options (long or short addresses, security level, ...). This gives a useful bit efficiency of only 15–20%.

This efficiency would be better with long Internet packet (2047 bytes) like in 802.15.4 g, where these long packets are supported. The IPv6 overhead would remain the same in absolute value, but would be more acceptable with a much longer payload. So there is still room improvement regarding protocol overhead.

In summary, IPv6 heavily impacts the useful data rate. But in order to enable interoperability, as well as implement regularly latest security updates from IETF, it is worth implementing an IPv6 solution, compared to proprietary ones. In another chapter, we will show through the GreenNet demonstrator how efficient such a network can be.

2.5 Pairing and Security

2.5.1 Pairing, Registration, and Installation of IoT Nodes

Pairing is a mandatory phase of a Wireless Sensor Network. The area of pre-paired objects is quite over. Pairing by the final users enables to complete an existing network with new nodes, and customize each network with various set of nodes, even coming from multiple vendors. Technically speaking, pairing enables the node to enter into the network by providing the ID of the network and some network-type specific data like the discovery RF channel to be able to register at each level of the stack, plus some security parameters if the network is secured.

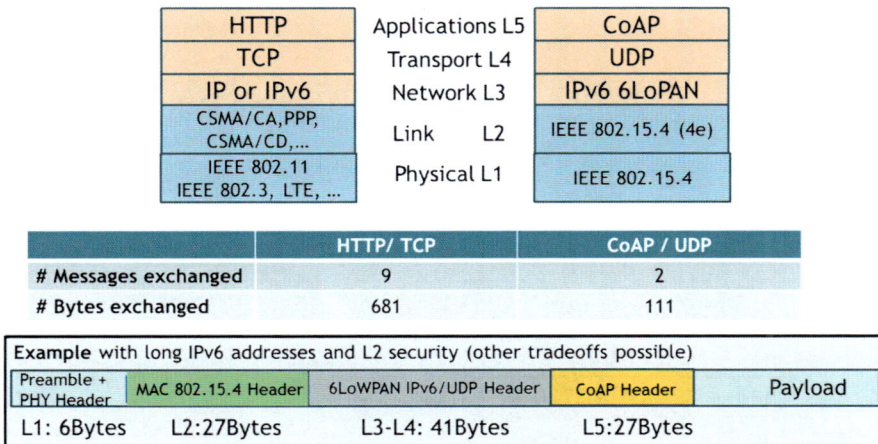

	Applications L5	
HTTP		CoAP
TCP	Transport L4	UDP
IP or IPv6	Network L3	IPv6 6LoPAN
CSMA/CA,PPP, CSMA/CD,...	Link L2	IEEE 802.15.4 (4e)
IEEE 802.11 IEEE 802.3, LTE, ...	Physical L1	IEEE 802.15.4

	HTTP/ TCP	CoAP / UDP
# Messages exchanged	9	2
# Bytes exchanged	681	111

Example with long IPv6 addresses and L2 security (other tradeoffs possible)				
Preamble + PHY Header	MAC 802.15.4 Header	6LoWPAN IPv6/UDP Header	CoAP Header	Payload
L1: 6Bytes	L2:27Bytes	L3-L4: 41Bytes	L5:27Bytes	

Fig. 2.9 Protocol suites typically used in traditional Internet, versus IP-based 802.15.4 Wireless Sensor Networks

A clear trend to ease secured pairing is to use NFC as an out-of-band (OOB) channel. NFC-forum (http://nfc-forum.org) specifies the way to enable secured pairing for BLE. The same principle could be used for 802.15.4 standard and its extensions. The principle is to exchange handover data using NFC. The negotiated handover protocol, introduced in January 2014 with the 1.3 release of the NFC-forum Connection Handover Specification [http://nfc-forum.org/our-work/specifications-and-application-documents/specifications/nfc-forum-technical-specifications/], enable to use a smartphone to pair securely 2 devices like an IoT node and a Concentrator or a Gateway.

The registration is a phase coming after the pairing, requiring a fair amount of exchanges between a node and the concentrator/gateway. Registration protocol is usually described in the application profile (e.g., Smart Energy Profile—SEP2.0 specified by Zigbee Alliance). An example of SEP2.0 registration for a Temperature sensor is given in Fig. 2.10.

We use the term installation for the physical installation of the node in its *final* working place. A node can always be moved, but regular moves may lead to frequently rebuild the routing tables, which can be energy hungry. This is why we consider ultra-low-power WSN as quasi-static networks.

2.5.2 Impact of Security on Power

Security of WSNs typically relies on Level 2 (MAC) and L5 (CoAP over DTLS) that both provide the four security goals, as seen in Sect. 2.2.4. The impact of L2-802.15.4 security on power is not as bad as often perceived, if we consider the nodes already paired. We measured the impact of link-layer security in terms of energy on the GreenNet demonstrator, using a fully autonomous scenario in harsh environment (e.g., sensor nodes communicating during 31 ms once every 4min20s). The overall cost of IEEE 802.15.4 security in our scenario ranges from 1.94 to 4.18%, depending on the security level. For energy-harvested platforms, such as GreenNet, this result directly corresponds to the requirement that 1.94–4.18% extra energy needs

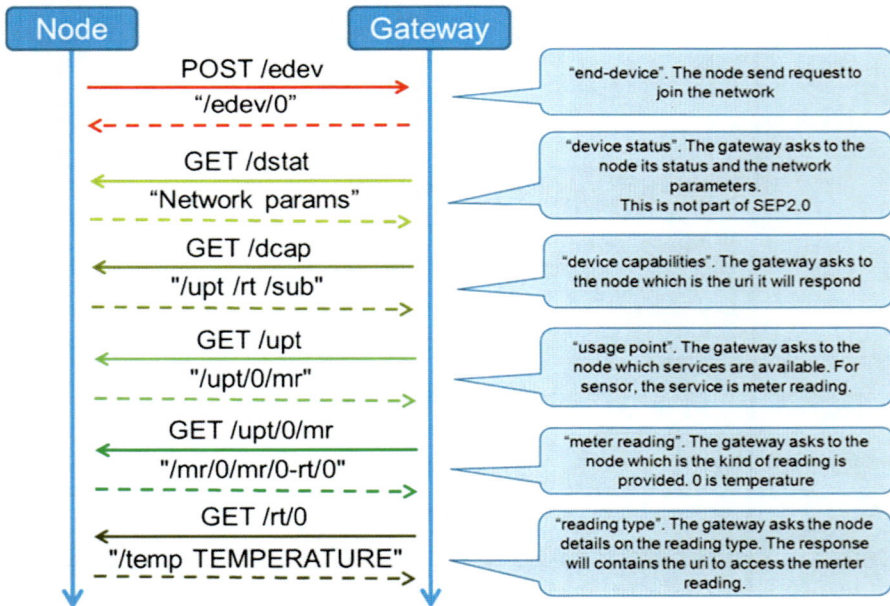

Fig. 2.10 SEP registration procedure example

to be harvested from the environment, in respect to the scenario without security (Vučinić et al. 2015).

L5 end-to-end security impact would be more important, but still smaller than the impact of using IPv6 transmissions. During pairing phase however, the Level 5 security ensured by DTLS implies to exchange a common key. Typically L5 messages shall be exchanged between the gateway and each node, penalizing the energy of the node, and the energy of all the router nodes on the way if the registration is done prior installation.

It is advised to define both pairing and registration protocols as well as installation use-cases early enough, to take their energy impact into account during the architectural definition phase of the nodes. In the case you intend to use energy-harvested or battery-operated routers, we would advise to perform pairing and registration phases close to the gateway, prior to physical installation, to avoid many messages exchanges through the full network.

This exchange of data can take a fair amount of time, up to several minutes in the case of very slow, ultra-low-power networks. This could indirectly raise some energy issues if the physical installation of a given node occurs before the registration is finalized. Effectively, if a node is paired in a place (e.g., nearby the gateway) and installed in another place far from the pairing place (i.e.,: lower in the network tree, at a 'distance' of several hops), the required exchanges to update the routing tables in order to follow a "moving node" inside the network would come on top of the registration exchanges. Some registration exchanges may be lost and would have to be sent several times, leading to burn a fair amount of the energy stored in the nodes.

In the case the network is very slow because very low-energy (e.g., one active mode per minute), it is possible to fix this issues by speeding-up the pairing and registration phases. Fast-exchanges can be used to quickly perform these phases in a few seconds. Then any move of the registered node to its final place inside the network becomes much more energy friendly. This technique has been developed for the GreenNet demonstrator. In this network pairing and registration are done at high rate: up to 32 frames per second if the routers enable it. Once the registration is performed, the node 'slows down' to its normal rate (e.g., 1 communication per minute or less).

2.5.3   Impact of Security, Pairing and Installation on Cost

The bill of material may be affected by the solutions chosen for Pairing and Installation protocols. As seen above, the need for security on top of IPv6 may induce to enlarge the battery capacity and to add some NFC specific chip. Some energy-independent low-cost solutions exist to perform NFC pairing (e.g., Dual-interface I2C-NFC EEPROM). An NFC-specific antenna must also be added on the IoT board. There are some electromagnetic rules to comply with, in order to avoid NFC (12.56 MHz) antenna to interact with the node radio antenna (2.4 GHz or Subgig). You can refer to (AN2866 Application Note) to design a 12.56 MHz customized tag antenna. In the case your IoT node is small enough and isn't compatible with printed NFC antenna size, some discrete coil antenna components are also available from distributors. In some case, the most expensive component may finally be the NFC antenna.

2.6   Battery Lifetime and Examples

Let define what we call battery lifetime: it is the time a node will operate in its normal/typical mode, without replacing the battery.

A first approximation of the battery lifetime (in seconds) can be done by dividing the battery capacity provided by the battery maker (in Coulombs), by the mean power consumption of the node (in Amperes). However battery capacity is usually provided in mAh. 1 mAh means 1 mA delivered during 1 h, which corresponds to 3.6 Coulombs.

Let take a 2000 mAh "AA" battery. It would provide 7200C. If the system consumes only

3.2 uA, then the theoretical battery lifetime would be 2.25E9 seconds which corresponds to about 71 years. This duration does not take into account battery aging or battery auto-discharge (that reaches up to 20% per year in the case of Alkaline batteries), as well as battery usage conditions (temperature, humidity, chemicals, non-continuous discharge, peak current). All these parameters affect the battery lifetime. Long-life non-rechargeable lithium batteries have a much better auto-discharge specification than alkaline ones. However, because of the other aging parameters, it is difficult to predict the exact battery duration. This is why consumer systems are claiming 10-year autonomy but very few makers (if any) really guarantee the battery lifetime.

In the case the node has to use an energy harvester, choosing the right battery capacity is not trivial. We describe hereafter a way to proceed.

The battery capacity provided by the battery maker is obtained when the battery is charged at its maximum capacity then totally discharged with a constant current value which is specified for each kind of battery. This current is usually much smaller than the peak current of a node, but

higher than the mean current. The actual battery capacity of your system will depend on the way you use this battery:

- To preserve the battery capacity, it is advised to never charge the battery to 100% of its capacity, but keep a 5–10% margin.
- To preserve battery life, it is required to never discharge completely the battery. Battery cycling (deep discharge) usually reduces the battery capacity; however the effective impact depends on the battery model. Deep cycles (60–80% discharge) have heavy impact compared to light cycles (10% or less). As the harvester is refueling regularly, it is possible to calculate the daily cycle and choose a battery capacity to maintain in typical conditions this cycle around 0.5% of the ideal capacity. In this case the impact on aging can be considered as negligible.
- When operating in a real environment, every day is different. For example during the weekend, there may be less light in the offices and the harvester may not be able to recharge the battery, leading to a weekly cycle, deeper than the daily one see (see Fig. 2.11). You should estimate this weekly cycle and make sure it

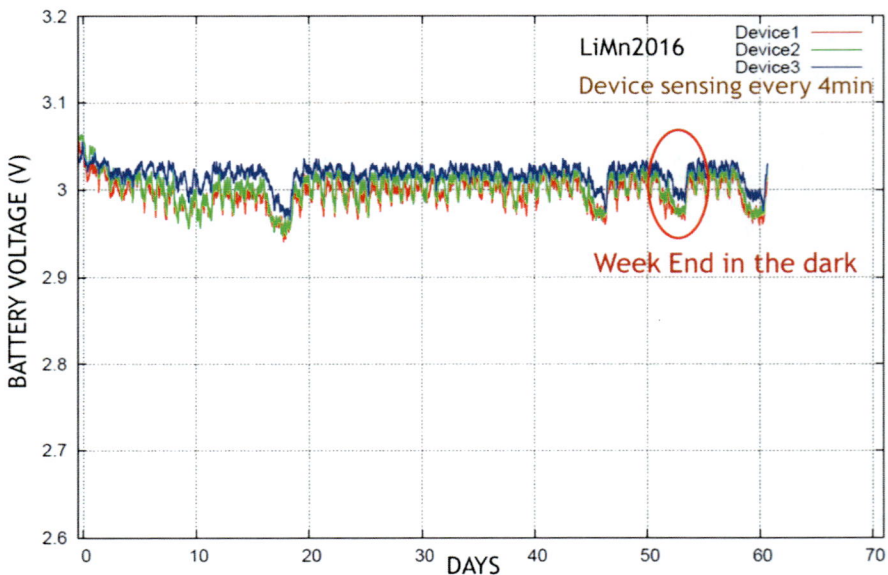

Fig. 2.11 Energy-Harvested GreenNet node rechargeable battery voltage

represents less than 2% of the ideal capacity. In some case you may enlarge the battery size.

- Every week is different, and every month is different. Using the same way, you should estimate monthly cycles be less than 15% and yearly cycles to be less sufficient to target a 10 year life. Values differ for each kind of battery, but a 40% cycle can be considered in many cases as a sufficient value.
- Tradeoff network loss and recovery algorithm to be lower than a monthly cycle in terms of battery discharge
- Take pairing and installation into account as a major cycle

2.7   Global System Power Optimization

The power optimization is of tremendous importance when designing an Ultra-Low-Power node and it is very often a never-ending process. We saw all along the chapter some critical points to keep high the energy efficiency. Here are few additional points to check out, in order to still be in your power budget:

- 15 to 30 nA over-leakage on each of the GPIO seems not so much, but given the number of GPIOs, it may lead to overpass your power budget. Programming GPIOs correctly in the lowest possible energy state for long sleep is not always simple, but very often profitable.
- Tradeoff quartz capacitances and precision to avoid wake-up in advance. You need to compensate the local oscillator error of both the emitter and the receiver by doubling the max possible error. The longer the sleep between two synchronizations, the larger the error.
- In the case of synchronized networks, a large temperature change may lead to the wake-up of the node for synchronization. Regarding energy efficiency in such networks, it is better to wake-up a node to keep synchronization than to recover a loss of synchronization.
- Minimize routing activity: routing protocols are very hungry in terms of transactions, limit

broadcast transmissions and network maintenance to its minimal activity

- Tradeoff deep-sleep to save more charges than the amount of charge you loose when you discharge external capacitors. In some cases, some higher leakage sleep mode needing less energy to recover, can be more profitable for the energy efficiency
- Tradeoff higher-level software. Usage of DTLS + CoAP seems a good compromise to ensure an IPv6 network with fairly good energy efficiency.
- Checkout the wires between non-powered debug-purpose on board material. Some pernicious leakage could be hidden there.
- When using multiple sensors, check that sensors are really off even if several sensors were running asynchronously.
- In some case, depending on your node architecture, grouping sensors wake-up and radio wake-up enables to reduce wake-up penalty.

Many of those points concern software impact on hardware. Software impact is effectively huge in term of power efficiency. Ultra-low-energy efficient software programming is not an easy task, and is a new domain for many software engineers.

Figure 2.12 gives an overview of the commercially available nodes in November 2015, and provides a comparison with the results with the GreenNet development. Node names depicted in italic are not commercially available.

2.8   Perspectives and Trends

The current generations of IoT nodes on the market are mainly AAA battery operated, consuming 50–100 mW of power to transmit a message at 0 dBm. Most of them are operating thanks to always-on routers. Few of them are IPv6 compliant, few of them offer a secured solution, few other are energy-harvested but none are able to provide all those features at the same time. In the future, the growing multiplicity of sensors, including image sensors, as well as the needs for lower cost, interoperable and secured

Node	MCU #bits	RAM [kB]	CPU ON [mA/MHz]	CPU sleep [µA]	Tx 0dBm [mA]	Rx [mA]	Har-vested	Batt. Size/type [mAh]
GreenNet	32	32	0.185	0.44	4.9	4.5	Y	25 LiMn
Hikob Azure [H14]	32	16	0.180	0.6	12.8	11.8	Y	2000
SmartMeshIP [WDS13]	32	72	0.176	0.8	5.4	4.5	OPT	2AA
M3OpenNode [FIT15]	32	64	1.138	25	11.6	10.3	N	650 LiPo
OpenMote [O15]	32	32	0.438	0.4	24	20	N	2AAA
WisMote [W15]	16	16	0.312	1.69	25.8	18.5	N	2AAA
TelosB [M04]	16	10	1.8	5.1	19.5	21.8	N	2AA
Waspmote15.4 [WD15]	8	8	1.07	7.2	45	50	OPT	N/A
MICAz [C08]	8	4	1.0	<15	17.4	19.7	N	2AA

Fig. 2.12 GreenNet versus commercially available nodes (Varga et al. 2015)

solutions seem to be paving the way towards the use of secured-IPv6 networking solutions. Easier to maintain than proprietary solutions, IETF compliant, enabling complete interoperability, these solutions seem to be the definitive trend of the WSN.

There are mainly three limitations to adopt secured IPv6: first, the need for interoperability was not strong enough during years in the WSN domain. This has evolved thanks to the Thread initiative. The other two reasons are:

- In some case the cost increase due to additional need for more SRAM in the processor to store and process IPv6 frames.
- In some other case the power consumption reached with the current solutions (MCU + radio + IPv6 secured protocols) which leads to either decrease the lifetime of the battery, or increase the battery capacity, impacting the price of the global solution, preventing the usage of energy harvesting as energy source. In order to use energy harvesting with a plain 802.15.4 secured IPv6, we would need an

MCU and digital computing part of the radio able to process 5–10 times more data keeping the same power budget. The new generation of MCUs, using CMOS 40 nm Ultra-Low-Power with embedded non-volatile memories technology should enable to offer 90 nm-like retention current with a dynamic power reduction from $x2$ to $x5$ on the digital part depending on the needed frequency, thanks to DVFS techniques. This technology will definitely help to fix the power consumption issue, on one hand by keeping power consumption at a fair level, compatible with low-cost batteries, and on the other hand by enabling much more computing capability than the previous technology nodes for the same power budget. One step further will be reached few years after thanks to 28FDSOI-ULP technology, achieving to save an additional $x2$ to $x10$ dynamic power versus CMOS40 (depending on the targeted frequency), keeping the leakage at a fair level. Once again, cost shall be the limiting factor for adoption and only the needs for additional features like extended signal processing capability, extended memory storage or even graphical display at node level may justify the need for adoption.

At node level, we start to see NFC-pairing adopted by a fair amount of the solutions reaching the market in 2016. Using NFC for pairing and registration should become a must in a near future.

Among the future evolutions of the radio protocol solutions, two major opportunities are foreseen at the time:

- Bluetooth-low-energy is one of them and seems evident when targeting Personal Area Network communication. BLE should move short term to IPv6 however its poor networking capability and its relatively short distance of communication may not enable to replace WSN-oriented radios (802.15.4 family). In order to solve these issues, BLE will eventually offer real mesh network capability and even some longer transmission range option. The success of these options will depend on how it compares to alternatives such as 802.15.4 family.
- The 802.15.4e evolution for industrial applications, part of the 802.15.4-2015 is enabling the IPv6 frequency hopping multi-channel mesh network in 2.4GHz networks. This should increase quality of service in crowded environment.

Both those standards may survive together, and we may see a generalization of the radio combos (e.g., BLE + 802.15.4 in 2.4GHz) each solution enabling to target a different network depending on the required service. As an example: a node could use BLE to transmit some information like an image to a smartphone when it is in the range, but would use 802.15.4 or 4e for infrastructure management. Multi-radio nodes sharing the same antenna for cost reason shall become the standard in a near future.

As an alternative to Mesh WSN enabling to repeat the messages from nodes to nodes to cover long distances, star topology networks could be used in so-called Wide Area Network (WAN).

Among the existing solutions, you can find Lora, SigFox and Weightless solutions. All of them operate in the Subgig ISM bands, with reduced data rates (100bit/s to 100 kb/s) compared to 2.4 GHz technologies. Business model should be different in this case, as it requires the usage of an existing infrastructure usually managed by operators, enabling some pay-per-use options. Those solutions will certainly survive in parallel to WSN-based solutions, targeting different needs.

References

AN2866 Application Note, How to design a 12.56 MHz customized tag antenna. http://www.st.com/st-web-ui/static/active/cn/resource/technical/document/application_note/CD00221490.pdf

I.M. Atakli, H. Hu, Y. Chen, W.S. Ku, Z. Su, Malicious node detection in wireless sensor networks using weighted trust evaluation, in *Proceedings of the 2008 Spring Simulation Multiconference*, SpringSim'08 (Society for Computer Simulation International, San Diego, 2008), pp. 836–843

B. Awerbuch, R. Curtmola, D. Holmer, C. Nita-Rotaru, H. Rubens, Mitigating byzantine attacks in ad hoc wireless networks. Department of Computer Science, Johns Hopkins University, Technical Report, Version, 1 (2004)

A. Becher, Z. Benenson, M. Dornseif, Tampering with motes: real-world physical attacks on wireless sensor networks, in *Proceedings of the Third International Conference on Security in Pervasive Computing*, SPC'06 (Springer-Verlag, Berlin, 2006), pp. 104–118

H. Chan, A. Perrig, D. Song, Random key predistribution schemes for sensor networks, in *Proceedings of 2003 Symposium on Security and Privacy 2002* (May 2002), pp. 197–213

D. Dolev, A.C. Yao, On the security of public key protocols. IEEE Trans. Inform. Theory **29**(2), 198–208 (1982)

HIKOB,   http://www.hikob.com/wp-content/uploads/2015/06/HIKOB_AZURE_LION_ProductSheet_EN.pdf

C. Karlof, D. Wagner, Secure routing in wireless sensor networks: attacks and countermeasures. Ad hoc Netw. **1**(2), 293–315 (2002)

Y.W. Law, L. van Hoesel, J. Doumen, P. Hartel, P. Havinga, Energy-efficient link-layer jamming attacks against wireless sensor network mac protocols, in *Proceedings of the 3rd ACM Workshop on Security of Ad Hoc and Sensor Networks*, SASN'05 (ACM, New York, 2005), pp. 76–88

M. Li, I. Koutsopoulos, R. Poovendran, Optimal jamming attacks and network defense policies in wireless sensor networks, in *26th IEEE International Conference* on *Computer Communications*, INFOCOM 2007 (IEEE, May 2007), pp. 1307–1315

F. Liu, X. Cheng, D. Chen, Insider attacker detection in wireless sensor networks, in *26th IEEE International Conference on Computer Communications*, INFOCOM 2007 (IEEE, May 2007), pp. 1937–1945

M3 Open Node motes, https://www.iot-lab.info/hardware/m3/

MICAz mote, http://www.openautomation.net/uploadsproductos/micaz_datasheet.pdf

OpenMote, http://www.openmote.com/hardware/openmote-cc2538-en.html

D.R. Raymond, S.F. Midkiff, Denial-of-service in wireless sensor networks: attacks and defenses. IEEE Pervasive Comput. **7**(1), 74–81 (2008)

M. Rezvani, A. Ignjatovic, E. Bertino, S. Jha, Secure data aggregation technique for wireless sensor networks in the presence of collusion attacks. IEEE Trans. Dependable Secure Comput. **12**(1), 98–110 (2015)

G. Romaniello, Energy efficient protocols for harvested wireless sensor networks. PhD thesis, Universite de Grenoble, March 2015

TelosB motes, http://www4.ncsu.edu/~kkolla/CSC714/datasheet.pdf

H. Tschofenig, T. Fossati, TLS/DTLS Profiles for the Internet of Things. draft-ietf-dice-profile-14. Work in progress, August 2015

P. Urard, G. Romaniello, A. Banciu, J.C. Grasset, V. Heinrich, M. Boulemnakher, F. Todeschni, L. Damon, R. Guizzetti, L Andre, A. Cathelin, A self-powered IPv6 bidirectional wireless sensor & actuator network for indoor conditions, in *2015 Symposium on VLSI Circuits (VLSI Circuits)* (IEEE, June 2015), pp. C100–C101

L.-O. Varga, G. Romaniello, M. Vucinic, M. Favre, A. Banciu, R. Guizzetti, C. Planat et al., GreenNet: an energy-harvesting IP-enabled wireless sensor network. IEEE Internet Things J **2**(5), 412–426 (2015a)

L.-O. Varga, Multi-hop energy harvesting wireless sensor networks: routing and low duty-cycle link layer. PhD thesis, Grenoble Alps University, December 2015

A. Vempaty, O. Ozdemir, K. Agrawal, H. Chen, P.K. Varshney, Localization in wireless sensor networks: byzantines and mitigation techniques. IEEE Trans. Signal Process. **61**(6), 1495–1508 (2012)

M. Vučinić, Architectures and protocols for secure and energy-efficient integration of wireless sensor networks with the internet of things. PhD thesis, Grenoble Alps University, November 2015

Waspmote, http://www.libelium.com/downloads/documentation/waspmote_datasheet.pdf and http://www.digi.com/pdf/ds_xbeemultipointmodules.pdf

T. Watteyne, L. Doherty, J. Simon, K. Pister, Technical overview of SmartMesh IP. in *Proceedings of IMIS'13* (Washington, DC, 2012)

WisMote, http://wismote.org

Ultra-Low-Power Digital Architectures for the Internet of Things

3

Davide Rossi, Igor Loi, Antonio Pullini, and Luca Benini

This chapter introduces the architectures implementing the digital processing platforms and control for Internet of things applications. It will provide a review of the state of the art Ultra-Low-Power (ULP) micro-controllers architecture, highlighting the main challenges and perspectives, and introducing the potential of exploiting parallelism in this field currently dominated by single issue processors.

3.1 Definitions and Motivations

The last years have seen an explosive growth of small, battery powered devices that sense the environment and communicate wirelessly the sensed data after some data processing, recognition, or classification. Collectively referred to as the Internet of Things (IoT), all these devices share the need for extreme energy efficiency and power envelopes of few milliwatts. From a system level perspective, in architectures targeting such applications, the data acquisition

part is implemented by a sensing subsystem, realized with low-power sensors such as visual imagers, microphone arrays, Micro Electro-Mechanical Systems (MEMS), or bioelectrical sensors. Sensed data is then digitally processed with low-power microcontrollers (MCUs), and transmitted through a wireless communication subsystem consisting of low-energy TX/RX radio transceivers implementing low-energy stacks. While these three subsystem are sometimes split over more than one chip, the market is trending toward fully-integrated single-chip solution. The entire IoT node is powered by harvesters or small form factor batteries (power supply/conversion subsystem). Despite of an almost 10x reduction of the transceiver power in just a few years, its share in the overall power budget of most wireless sensor nodes and wearables remains dominant (De Groot 2015). In this scenario, a high-potential approach to reduce the system energy is to increase the complexity of near-sensor data analysis and filtering by providing more computational power to the processing sub-system. This approach can dramatically reduce the amount of wireless data transmitted, that could be reduced to a class, a signature, or even just a simple event. For this reason, the availability of powerful, flexible and energy-efficient digital processing hardware in close proximity to sensors plays a key role in the internet of things revolution. The aim of this chapter is to review the state of the art of

D. Rossi (✉) • I. Loi
University of Bologna, Bologna, Italy
e-mail: davide.rossi@unibo.it

A. Pullini
ETH, Zurich, Switzerland

L. Benini
University of Bologna, Bologna, Italy

ETH, Zurich, Switzerland

© Springer International Publishing AG 2017
M. Alioto (ed.), *Enabling the Internet of Things*, DOI 10.1007/978-3-319-51482-6_3

ultra-low power computing platforms for near-sensor processing, and highlight the main challenges and perspectives related to these architectures.

Most of low-power commercial microcontrollers cannot provide the required performance levels for several applications within the power budgets offered by small form factor coin batteries and energy harvesters. A promising approach to achieve up to one order of magnitude of improvement in energy efficiency of integrated circuits with respect to "business as usual" CMOS in strong inversion is ultra-low voltage, near-threshold computing. The key idea is to lower the supply voltage of chips to a value only slightly higher than the threshold voltage. Aggressive voltage scaling has been extensively analyzed in the literature, including its limitations and disadvantages (Dreslinski et al. 2010). One of the main issues with low-voltage operation is performance degradation, which can limit the degree of use of voltage-scaling for a given processing requirement.

A commonly adopted approach to overcome the performance loss in ultra-low voltage devices on hardwired functions implemented in *Application Specific Integrated Circuits* (ASICs). By exploiting dedicated circuits, digital processing systems are able to match the performance requirements of applications, even at a very low operating voltage, with frequencies of tens to hundreds of KHz and power consumptions in the range of few μW to hundreds μW. In some cases, these dedicated systems, implemented as *System-On-Chip* (SoC) or *System-In-Package* (SiP), integrate digital signal processing circuits, analog front end, analog signal processing circuits, and power supply circuits (batteries, harvester or both) leading to extremely compact form factors.

The dedicated ASIC approach has been extensively used in traditional fields of applications of ultra-low power devices, such as wearable or implantable sensors for health monitoring (Zhang et al. 2012; Yoo et al. 2012; Yakovlev et al. 2012). Although these devices minimize power consumption, they are not flexible as

their function is limited to the specific purpose for which they are designed. Also, their performance is not scalable as they are designed with a specific use case in mind. While adoption of dedicated circuits is attractive to tackle the stringent constraints of implantable applications for health monitoring (tens of microwatts), these two important aspects limit the exploitation of dedicated circuits for the majority of IoT applications since the Non Recurrent Engineering (NRE) costs required for their development cannot be amortized over large volume products.

When the targeted algorithms are more generic, so that can they be re-utilized for more than a single application, the above described restrictions can be relaxed by providing some run-time configurability to the integrated circuits, and by increasing the operating range via voltage and frequency scaling. As this approach has mainly been applied to the signal processing field, this class of computing devices is usually referred as *Application (or domain) Specific Signal Processors* (ASSPs). Several examples of this class of devices apply to visual sensors, where several basic functions implemented with dedicated accelerators or specialized processors can be shared among different applications (Park et al. 2013; Hsu et al. 2012; Jeon et al. 2013).

The ultimate step toward flexibility is given by the exploitation of the software programmability of instruction processors. In last few years some instruction processors working in the near-threshold or sub-threshold operating region have been presented from both industry (Ambiq 2015) and research (Bol et al. 2013). Some of the proposed devices are also able to work on a wide range of operating points, as the low voltage operation might not be always suitable to match the requirements of the application targets (Gammie et al. 2011). Again, when operating at low voltage, sequential instruction execution coupled with very slow operating frequency may lead to insufficient performance for the application requirements. Explored solutions to improve performance of ultra-low power processors while maintain a high degree of flexibility also rely on the exploitation of software parallelism. The exploitation of multi core platforms can provide benefits with respect to single processor cores for high application workloads as it has been demonstrated that parallel computing at low voltage can be more energy efficient than sequential computing at a higher voltage under certain assumptions (Dogan et al. 2011). This concept has been extensively exploited for high-end embedded applications, where multi-core architecture has become the de-facto standard. On the other hand, when the application workload is low, or when the workloads are not easily parallelizable, the introduced multi-core platforms suffer from energy efficiency losses with respect to single core platforms, mainly caused by static (primarily leakage-induced) power consumption, due to the larger area and architectural overheads. Hence, while multi-core platforms are starting to be seen with interest in the world of ultra-low-power applications, a great effort still needs to be done to target the applications driving the IoT domain.

The rest of the chapter is organized as follow: Section 3.1 describes the architecture of off-the-shelf microcontrollers typically employed in IoT applications. Section 3.2 describes the main challenges of IoT computing platforms, mainly related to the exploitation of performance scalability exploiting parallel processing. Section 3.3 provides an example of research Parallel Ultra-Low-Power computing platform (PULP). Finally, Sect. 3.4 provides some concluding remarks, and highlights challenges and perspectives.

3.2 Ultra-Low-Power Microcontroller Architectures

The applications targeted by current off-the-shelf microcontroller (MCUs) architectures require to periodically fetch environmental information from a wide variety of sensors, which is then transmitted via wireless through low-power antennas after a limited amount of processing. In this scenario, to maximize the power efficiency of the system, the data processing hardware should be active only for the small amount of time required to read, process and transmit the information, while it is *idle* for the rest of the time. It is usual to refer to this mechanism as *duty cycling*. Figure 3.1 shows the typical power consumption pattern of a microcontroller employed for a generic IoT application. For most of the time, the MCU is in a *deep sleep* state where it is waiting for an event to wake up, triggered by an internal timer or by an external event generated by one of the sensors. Once the event is captured the device restores the power supply and restarts clocks (wake-up), restores the state of the CPU (stack, data etc.) and then can start fetching new data from I/Os, process and transmit it. Once done with the processing, it can save the state and go back to sleep.

Today microcontrollers (MCUs) feature several power modes to deal with different application scenarios where the various states may have a very different duration, duty cycles and performance requirements. For example if an application has to deal with frequent wake-ups, paying at each wake-up the price of saving and restoring

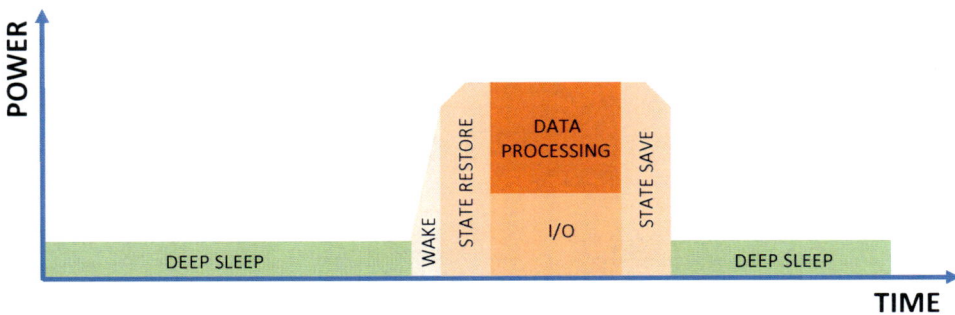

Fig. 3.1 Typical power profile of IoT applications

its full state is highly inefficient and a power
mode with higher off current but with state retention in this case is much more valuable.

3.2.1   Power Management

Having an efficient off and idle states is a key
requirements for micro controllers. In many IoT
applications sleep-mode current is the biggest
contributor to the overall power consumption.
When considering sleep-mode current we should
consider also the current required to reactivate
the circuit. Wakeup currents have a big impact on
the choice of the sleep mode. There are many
different techniques to reduce the power consumption when idle (Sect. 10).

The simplest approach is *clock gating*, which
only implies stopping the clock. In this mode
dynamic power is cut and only leakage current
is flowing. This mode is state retentive and
requires no time to go back to active if the
clock generator is still running. Clock gating
usually has a very fine grained granularity,
mostly all blocks in the system can be put in
idle mode independently since it does not have

any overhead at circuit level. Wakeup time after
an idle state is dependent only on the status of the
clock generator. If the clock generator is kept on,
the wakeup time is negligible; but the clock
generator has also been stopped for more aggressive power savings. The wake-up time is strongly
impacted by the type of clock generator. In
today's microcontrollers there are different
types of clock generation unit to deal with the
different operating corners that always guarantee
the maximum efficiency. A good example of
state of the art clocking offer in microcontrollers
is the latest STM32L4 family. In the micro of the
L4 family there are five available clock sources
(see Fig. 3.2). There are two low speed
generators that generate a 32 kHz frequency:
one oscillator using an external quartz (LSE)
and another one using an internal RC oscillator
(LSI), which is more power-efficient but less
precise. The same is done for the faster 48 MHz
clock, where there is an oscillator using an external clock (HSE) and an internal configurable RC
oscillator (MSI). The fifth clock source is a fixed
16 MHz internal RC oscillator. The L4 has also
three PLLs capable of multiplying the frequency
and reaching up to 180 MHz (Microelectronics

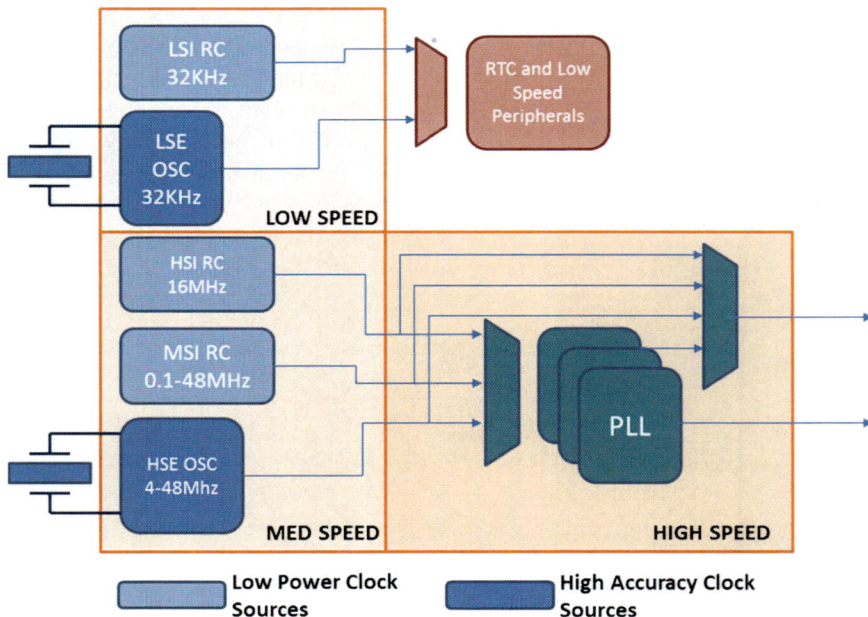

Fig. 3.2 Example of state of the art clock distribution

2016). Sleep mode turns the clocks to the core off, but the user has the option to leave on the peripherals' clocks. The power in this mode is not just leakage but dynamic current of the peripherals that are left on. In this mode data can be still received, and the core retains its state and continues operation when required. All clocks to the digital logic are turned off and the analog sub-systems can be controlled to have flexible wake up times depending on the application requirements. The lowest power mode is when all the analog clocking elements are turned off. Wake up time is determined by the selection of the wake up clock source. The fastest time is from the low-power oscillator and the slowest time is from the crystal oscillator and the PLL.

To reduce leakage in idle mode, clock gating is used in conjunction with voltage scaling. When the clock is stopped we can afford to lower the voltage as low as the retention voltage for the memory elements. The cost of this at circuit level is the required flexibility in the power supply. The effect on the application is that we require more time to exit the idle mode due to the settling time of the power supply. Further reduction can be obtained by not only lowering the voltage but also applying a reverse body biasing to increase the threshold voltage of the devices (Rossi et al. 2016a).

The lowest leakage is obtainable only with power gating, where the power supply is turned off. All microcontrollers available today feature a deep-sleep mode where the device power supply is disconnected and only a small always-on part of the chip controlling the wake up is kept alive. Always-on domain usually have an RTC to have the possibility to wake up the system after a certain amount of time and have a set of memory elements (flops or SRAMs) to keep track of the previous state. The amount of features active on the always-on domain is usually selectable by the user to give maximum flexibility depending on the application requirements. The startup behavior of the analog modules can have a major impact on the amount of time spent in active mode; voltage regulators or references utilizing external decoupling caps can take milliseconds to settle. Therefore, it is important for a systems designer to analyze the overall wake-up and settling time for both the digital and analog circuitry to factor in the true cost of this wasted energy. State of the art devices as the Ambiq Apollo have deep sleep currents as low as 100 nA with RTC on (Ambiq 2015).

3.2.2 IO Architecture

Peripheral subsystems in microcontrollers include an extensive set of peripherals needed to connect to the wide variety of sensors available on the market. Although MCUs traditionally feature low bandwidth interfaces like UART, I2C, I2S or standard SPI, they lately include also higher bandwidth peripherals like USB or camera and display interfaces. Low bandwidth peripherals are usually attached to a shared bus and high bandwidth peripherals are usually connected to the system bus. Traditionally, the I/O was constantly supervised by the CPU and the CPU was responsible of handling peripherals events and regulate data transfer to/from peripheral and memory. Recent MCUs have increased the complexity of the I/O subsystem to support more power modes to be able to selectively turn on peripherals only when needed (Figure 3.3).

Further reduction in power can be obtained by increasing the intelligence of the peripherals and have they run without CPU supervision while the CPU is in sleep mode. The "smart peripheral" approach is used more and more in the most advanced MCUs. It has many variants with different names, but the general idea is the same. For example ATMEL "SleepWalking" features in the latest SAM-L2 family enable events to wake up peripherals and put them back to idle when data transfer is done without any CPU intervention for instance using a DMA (Atmel 2015). Another example is the Renesas' RL78 which has a "snooze mode" where the analog-to-digital converter (ADC) operates while the processor is asleep. An I2C slave or CAN controller can watch for an address before it captures incoming data and then wakes up the CPU (Renesas 2014). Cypress Semiconductor's PSoC flexible family was one of the pioneers with their configurable digital and analog peripherals. Part of the configuration is the ability to link peripherals together without the CPU being involved. The company's latest PSoC 4 BLE (Bluetooth Low Energy) can operate the BLE radio while the CPU is idle (Cypress et al. 2016). Microchip's Configurable Logic Cell (CLC) highlights the more conventional approach to configurable peripherals. A microcontroller may have one or more CLC blocks. Each block has a selectable set of inputs and outputs with limited logic capability so the output of one peripheral can be fed to another. As with the PSoC, linked peripherals can often handle simple algorithms faster than the CPU. ST Microelectronics' STM32 has "autonomous peripherals" that use a "peripheral interconnect matrix" to link peripherals together. Timers, DMAs, ADCs, and DACs can be linked together. Or with their latest "Analog Chain" where comparators, references, DACs and ADCs can be interconnected to generate complex triggers (Microelectronics 2016). The trend towards more functionality and complexity is highlighted by Microchip's PIC16F18877 family. It has a 10-ADC tied to a computational unit that can do accumulation and averaging. It can even do low-pass filter calculations in hardware. The CPU can sleep until a filtered result exceeds programmed limits (Microchip 2016).

3.2.3 Data Processing

Reduction of energy consumption for the data processing part should be improved under different aspects, by improving the micro-architecture of the CPU increasing the amount of data that can be processed per cycle, or by optimizing the circuit design to reduce the power consumed per cycle.

The Architecture of CPUs used in MCUs has rapidly evolved in the last years, moving quickly from the 8 bit architectures to the now widespread 32 bit ARM Cortex-M architectures. The architectural evolutions are driven by the constantly increasing demand for performance for the always more complex applications. The vast majority of CPUs used in microcontrollers are single issue machine in which instructions are executed in order. The only exception today is the Cortex-M7 recently released by ARM which has a dual-issue pipeline. The choice of simple single issue micro-architecture is mainly dictated by energy efficiency for the target performance. We are not yet in the performance range to justify multiple-issue superscalar architectures. Engineers focused more on optimization to the micro architecture to improve as much as possible the IPC (instruction per cycle) and the data level parallelism.

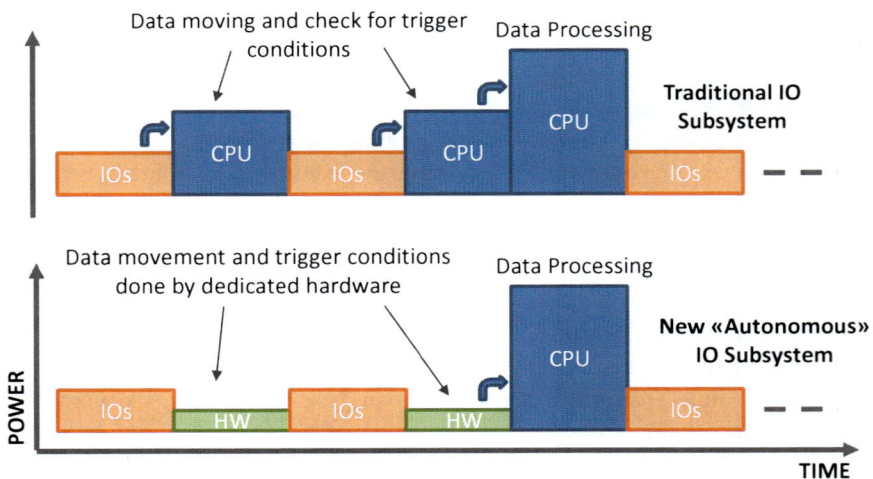

Fig. 3.3 Power profile of new "Autonomous" I/O subsystem

The most common improvements include for example hardware loops to speed up loops on tiny kernels typical of most DSP applications. It is usually done with one or more instructions that setup a dedicated logic which controls the number of interactions of a loop and the fetch stage, with the benefit of reducing the amount of cycles needed to compare and jump back at the end of a loop. The benefit is huge when small kernels are considered (Gautschi et al. 2015).

Other non DSP-specific improvements are loads and stores with pre- and post-increment whose main usage is to reduce the number of instructions needed to access consecutive memory locations.

More on the DSP optimizations we have for example the use of SIMD (Single instruction Multiple Data) where a single arithmetic instruction can operate on multiple data. On high-end CPUs this is often coupled with wider access to memory to handle multiple data at the data path width. On microcontrollers it is much more common to use their less power hungry version based on data size lower than the data path width. It is common to have the possibility to process 2 half-words or 4 chars with a single 32bit instruction. Good examples are the DSP extensions introduced in the Cortex-M4 (ARM 2010).

Architectural optimization to support single cycle multiplications and dedicated instructions for MACs, as well as hardware support for saturation arithmetic to better handle fixed point arithmetic, are other examples of how CPUs are extended to increase energy efficiency during the execution of DSP algorithms.

Those were the optimizations for the data and control logic but with the introduction of the ARM cores we also saw improvements on the instruction side. The ARM7TDMI ARM introduced the THUMB instructions, which allows some instructions to be coded using 16-bit instead of 32-bit, reducing the pressure on the instruction memory, the overall code size and instruction memory requirements. Table 3.1 summarizes the main features of few MCUs widely used for ultra-low-power applications, while Figure 3.4 shows their typical architecture.

Table 3.1 Summary of commercial low power MCUs

MCU	16 F1503	MSP430FR6x	SAML21x	Kinetis KL17	EMF32 Pearl Gecko	Apollo	STM32F745xx
Instruction set architecture	PIC	MSP430	ARM Cortex M0+	ARM Cortex M0+	ARM Cortex M4	ARM Cortex M4	ARM Cortex M7
Datapath	8-bit	16-bit	32-bit	32-bit	32-bit + FPU	32-bit + FPU	32-bit + FPU (dual issue)
Memory	I: 2 bB SRAM	128 kB FRAM	256 kB FLASH	256 kB FLASH	256 kB FLASH	512 kB FLASH	1 MB FLASH
	D: 128 B SRAM		64 kB SRAM	32 kB SRAM	32 kB SRAM	64 kB SRAM	340 kB SRAM
Max Freq. (MHz)	20	16	48	48	40	24	216
Current (µA/ MHz)	30	100	35	54	60	34	700
Deep sleep current (µA)	0.02	0.02	0.2	0.28	0.02	0.12	2
State-retentive current (µA)	NA	0.02	1.3	1.96	1.4	0.193	2.75
Retentive deep sleep	No	Yes	No	No	No	No	No

Fig. 3.4 Typical architecture of off-the-shelf MCUs for IoT

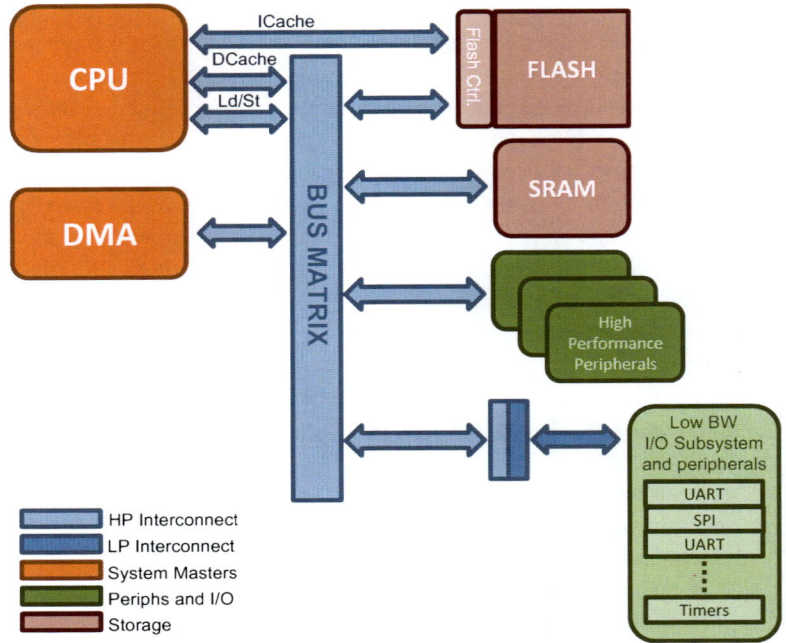

ICache
DCache
Ld/St
CPU
Flash Ctrl.
FLASH
SRAM
DMA
BUS MATRIX
High Performance Peripherals
Low BW I/O Subsystem and peripherals
UART
SPI
UART
Timers

HP Interconnect
LP Interconnect
System Masters
Periphs and I/O
Storage

3.2.4 Non-volatile Memories

As briefly mentioned, before energy spent to save and restore the state before and after a deep sleep state has a deep impact on how often and for how long a deep sleep state could be used. Such cost depends almost entirely on the type of memory used for the storage. Embedded flash memories are the most common type of non-volatile memory used but, due to their high write energy and low endurance, they are not suitable to be used for state retention during frequent power cycles (Chap. 6). In today microcontrollers, state retention is implemented by keeping active part of the circuit, with the consequent reduction of the effectiveness of the deep sleep state.

Recent developments in non-volatile memories show many possible solution to this problem. Texas Instruments has recently included in their MSP430 family some microcontrollers based on FeRAM (Texas 2011). FeRAM have a structure very similar to DRAM where a bit cell is made of one transistor

and one capacitor and they share with DRAM the high speed. The capacitor is implemented by using a ferroelectric material (usually PZT lead-zirconate-titanate) which is polarized in two possible states that are kept without requiring any refresh. Other types of memory currently under study and close to commercial applications are MRAM, STT-MRAM, and PCRAM. Magnetic RAM (MRAM) are based on memory cells constituted of two magnetic storage elements, one with fixed polarity and one with switchable polarity. Depending on the state of the switchable element the resistance of the overall cell changes and that is what is sensed by the read circuitry. STT-MRAM is a particular implementation of MRAMs more suitable for scaled technology. It uses spin-polarized currents enabling smaller and less power demanding bit cells. Currently, STT-RAM is being developed in various companies including Everspin, Grandis, Hynix, IBM, Samsung, TDK, and Toshiba. Due to its easy integration in the CMOS process as well as its performance, power and scalability properties, it is one of the most appealing solutions. Chapter 7 will present a detailed explanation. Another example of NVM currently investigated by the industry is the Phase Change RAM (PCRAM) in which the bit cell is made of materials that can exist in two phases (e.g., crystalline and amorphous) and result in different resistance. A more detailed description of NVM memories in the context of IoT architectures is presented in Chaps. 6 and 7.

3.2.5 A Step Forward: Near-Threshold MCU Architectures

Today's low-power MCU architectures operate in the super-threshold domain during active phases of computation, relying on duty cycling and heavily optimized deep sleep modes to improve energy consumption. For example, some commercial devices, such as Ambiq Apollo (Ambiq 2015) provides a sub-threshold RTC to minimize deep-sleep power (~300 nW), but it operates at 0.9 V, which is only 100 mV below the nominal operating voltage for the 90 nm process technology utilized for its implementation, providing active power efficiency similar to other commercial MCUs. Although this approach provides a very low power for applications requiring low computational workloads (e.g., a temperature sensor wakes up the microcontroller once per minute to transmit sensed data) the situation drastically changes when the applications require nearly always-active operation. In this scenario, the energy consumption of a microcontroller can increase by up to three orders of magnitude (i.e., from tens of μW to tens of mW, on average). Near threshold computing is emerging as a promising approach to achieve major energy efficiency improvements of ultra-low-power digital architectures (Dreslinski et al. 2010). However, this comes at the cost of performance degradation and increased sensitivity with respect to process, voltage and temperature (PVT) variations. While the performance degradation can be managed by adjusting the operating voltage of the device according to the required

performance target, compensation of PVT variations has to be achieved in a transparent way with respect to the end user, which cannot be aware of the operating conditions of the device.

3.2.6 Compensation of Process and Environmental Variations in Near-Threshold

The variability of ultra-low-power devices operating at low voltage has been extensively analyzed in the last few years from the research community (Alioto 2012). Some of the approaches leverage design-time techniques to improve resiliency of the circuits with respect to process and temperature variations, including standard-cell design, clock tree optimization, and automatic synthesis. All these techniques are extensively analyzed in Chap. 4. On top of design level techniques to mitigate the impact of variations, their compensation is usually addressed at the architectural level, integrating mechanisms able to probe the PVT conditions of the circuit and aging, and provide a feedback to knobs exposed at system level that allow to compensate the variations by adjusting the supply voltage of the circuit.

Widely explored approaches to implement the probing circuits to adapt the supply voltage are process monitoring blocks (PMBs), canary circuits, or razor flip-flops (Ernst et al. 2005). PMBs are generic structures implementing ring oscillators with different characteristics (e.g., PMOS only, NMOS only, inverters with long wires). Depending on the specific oscillator probed from a PMB it is possible to acquire specific information about the process condition of PMOS and NMOS, temperature, or voltage. Although this approach is generic, as it does not require design of specific hardware for each chip, some additional logic (e.g., a lookup table) or software processing is required to merge the data and provide a feedback to the actuator. A canary circuit is a replica of the critical path plus some delay element, often programmable, which makes the monitor super-critical (Calhoun and

Chandrakasan 2004). As canary circuits are designed to clone a specific critical path, they are not generic components, but they provide a more direct information on the criticality of the operating point. The global process variations as well the environmental conditions can be monitored with both PMBs and canary circuits, and the supply voltage can be adapted according to these conditions. However, there are still considerable margins to be assumed to ensure a reliable operation of the circuit. For example, local process variations and IR drop cannot be covered, as the monitor circuit is a copy of the critical path placed in a different location. A yet more direct approach is to use the speed monitor within the respective circuit block. The razor concept (Ernst et al. 2003; Blaauw et al. 2008) provides energy reduction guaranteeing reliable operation by lowering the supply voltage to the point of first failure. Timing failures are detected by shadow latches placed on the critical path of the design controlled using a delayed clock, eliminating all margins due to global and local PVT variations. By comparing the values latched by the flip-flop and the shadow latch, a delay error in the main flip-flop is detected. The value in the shadow latch, which is guaranteed to be correct, is then used to correct the delay failure. The main drawback of the razor approach is that is very design specific, requiring major manual adjustments to the microarchitecture of the SoC, as no automatic razor flops mechanism is available in commercial synthesis tools. This makes the adoption of this approach very challenging for a wide set of designs, and impossible when the IP cores are provided as hard macros or encrypted netlist by processors or IPs providers.

Although the most traditional way of compensation for PVT variations leads to adjust the supply voltage of the circuits, other approaches have been explored at the architectural level: clock gating the specific pipeline stage where a fault is detected (Ernst et al. 2003), using counter flow pipelining (Charles et al. 1994), through architectural replay (Blaauw et al. 2008), or through insertion of reconfigurable pipeline stages (Bortolotti et al. 2013). The main benefit of the architectural compensation or error recovery mechanism is that they can react in a single clock cycle to the detection of a fault or a critical situation. On the other hand, once again, they are very invasive from the micro-architectural viewpoint. Alternative approaches to compensate for PVT variations rely on the adoption of adaptive body biasing (Tschanz et al. 2002, 2007). This approach has several advantages with respect to modulation through supply voltage. First of all, dynamically adapting threshold voltage of transistors only increases leakage power of the circuit, while adaptation of supply voltage has an impact on both leakage and dynamic power. With body biasing it is possible to apply independent polarization to PMOS and NMOS. Hence, it is possible to optimize the leakage power consumption of a circuit by applying asymmetric body biasing. Moreover, body biasing is very effective when the device operates in near-threshold, as in this operating region small changes of the threshold voltage of transistor provide significant increase (or reduction) of the operating frequency (Rossi et al. 2016a). Finally, since the polarization of the PWELL and NWELL only requires transient currents, modulation of body biasing can be implemented with simple and energy efficient circuits (e.g., charge pumps), as opposed to supply voltage control, which requires the adoption of DC/DC or voltage regulators. As a drawback, the adoption of body biasing to address large PVT variations is somehow challenging due to the limited capabilities of bulk technology to provide extended body bias ranges. For this reason, joining the degrees of freedom provided by voltage scaling and adaptive body biasing appear as a suitable solution to compensate for PVT variations while tracking the best energy operation.

3.3   From Single Core to Multi Core

The trend to develop multicore-systems is a general tendency for several high-end embedded, desktop and server platforms. Energy efficiency requirements have forced processor developers to add multi-core capabilities instead of increasing the system clock frequency in single-core systems (Parkhurst et al. 2006). Cache coherency and memory bandwidth determine the architecture of such multi-core systems, and when private caches are involved, support for cache coherency across the entire multi-core system is desired, as it enables the software running on embedded processing system to balance load and allocate tasks seamlessly between cores.

Cache coherence protocols must react immediately to writes, and invalidate all cached read copies in the private cache banks, and this is source of significant complexity. Complexity translates into cost, both from silicon and energy perspective (Martin et al. 2012). Data synchronization in multi-core systems prevents data from being invalidated by parallel access whereas event synchronization coordinates concurrent execution. One common mechanism to achieve data synchronization is a lock. Event synchronization forces processes to join at a certain point of execution. Barriers can be used to separate distinct phases of computation and they are normally implemented without special hardware using locks and shared memory (Culler and Singh 1999). Efficient data transfer between cores in a multi-core system is critical for balanced system performance. A multi-core environment introduces high demands on the interconnect infrastructure and the interconnect needs to be able to handle multiple streams simultaneously. Complex interconnect are widely used to sustain bandwidth and QoS requirements (e.g., AXI ACE), and again complexity is translated into cost.

On the other side, micro-controller systems can take benefits of lean a simple architecture with respect to high-end single/multi-core systems. For this reason, they are much more attractive for a wide category of IoT devices. For instance basic micro-controllers typically have no data/instruction caches, while multicores needs complex memory hierarchies to sustain bandwidth and computational power requirements. But the current trend in IoT devices is that, modern use cases are demanding more and more computational power (eg. speech and image recognition, surveillance etc.). To achieve this target with a prefixed energy budget, standard micro-controllers are not sufficient to

provide this amount of computational power, and the trend is to migrate from a single-core to a multi-core architecture, while maintaining low the complexity of the memory hierarchy to target high energy efficiency.

For this reason, multi-core architectures are beginning to penetrate the microcontroller business segment: recently, a new class of heterogeneous dual-core MCU products appeared in the market. These devices have cores with different instruction set architectures (i.e., Cortex M0 and Cortex M4) (NXP Semiconductors 2015) and rely on the architectural heterogeneity to achieve energy efficiency with a principle similar to the mid-to-high-end big-little multi-cores (Peter 2011). In these architectures, the little core is mainly meant for control of peripherals and low workload tasks, while the big processor perform heavy data processing tasks. The main advantage of these architectures is that cores are completely decoupled, easing the partial shut-down of the platform when one of the cores is not used. On the other hand, this decoupled architecture, where each core has its own binary and process data on private memories, doesn't allow for easy and fair workload distribution among cores, and requires to keep private copies of data-buffers, which dramatically reduces the computational efficiency of the platform when true data level parallelism has to be exploited.

A more convenient approach to the design of parallel low-power architectures is a tightly coupled clusters sharing, similarly to what happens in GPGPUs, a multi-banked L1 memory. With respect to traditional architectures featuring per-core private data memory, where data buffers are processed on the local, low-latency L1 memory, and shared with the other cores through a higher high-latency L2 memory, the tightly coupled data memory approach significantly reduces the data-sharing overhead, as the memory banks can be accessed by all cores with a fixed latency (usually one cycle). This way allows to efficiently exploit both data and task parallelism, as opposed to more traditional private L1 memory scheme that allows to run efficiently only task parallel applications, enabling better performance scalability when the required workloads allow for true data-level parallelism.

3.3.1 Energy Benefits and Challenges for Parallel ULP Processors

Figure 3.5 shows the tradeoff for multi-core vs. - single-core systems, under different workload conditions. Each workload represents a specific voltage-frequency pair that exploits the required computation power. Under high workload requirements the single-core is less energy efficient when compared to multi-core due to the quadratic dependency of supply voltage with dynamic power (Dogan et al. 2011). Indeed, to achieve the same throughput the single-core needs to operate with a supply voltage much higher than that of the multi-core, where CMOS devices are far from their maximum energy efficiency point. However, when the voltage is close to the threshold, leakage become dominant and circuit slowdown increases, so from an energy perspective, single core is much more efficient, therefore, this region is not interesting for always-on parallel computing over multiple cores (Fig. 3.5).

On the other hand, when we introduce a deep-sleep mode typical of the MCU domain, under low workloads, the multicore solution is still attractive. In Fig. 3.6, in case of a single core execution, energy is consumed in the whole active period, then when computation is done, the system is put in deep-sleep. In the active period, the energy drained by the core is given by the sum of core and system energy. Assuming to run the same application on a multi-core platform, then the application runtime (active region) will decrease (in the ideal case, will be N times smaller, where N is the number of cores). In this region, we have multiple cores and the system draining energy from the power supply, and assuming that the system power is the same in both cases, and that core energy is the same, by reducing the active period, some amount of energy is saved, thus better energy efficiency is achieved.

This scenario highlights the importance of a power management strategy at the system-level to develop efficient shutdown policies of unused cores and workload consolidation policies for minimizing the dynamic and leakage power consumption for the idle cores (Rossi et al. 2016a). An effective strategy to eliminate dynamic power of unused processors during the execution of sequential portions of code is architectural clock gating. Indeed, joining hardware support for synchronization mechanism with architectural clock gating of cores provides significant energy savings during execution of sequential or not perfectly balance code, while minimizing the synchronization overhead, not present in single-core platforms. However, even at low speed levels a processor consumes a significant amount

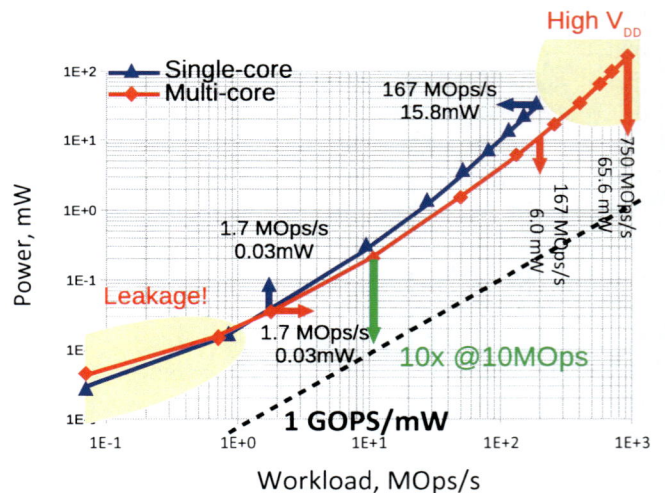

Fig. 3.5 Power efficiency of multi-core vs. single core under different operating points. Each operating point is function of both workload and frequency/voltage that enable such performance

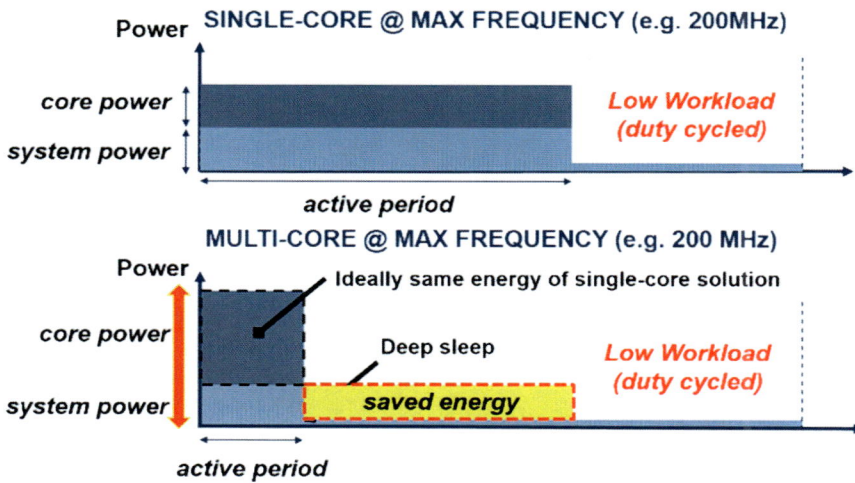

Fig. 3.6 Energy efficiency Multi-core vs Single core on parallel application

of static energy, caused, for example, by leakage current. A possible alternative to reduce both leakage and dynamic power consumption of multi-core platform is per-core DVFS. Exploiting a per-core DVFS scheme, each core is allowed to operate with and independent voltage and frequency, thus theoretically achieving the best possible energy efficiency for each given workload, and can be shut down when idle. However, this comes with level shifters and dual-clock FIFOs at the boundaries of every core introducing significant data sharing overhead, due to the handshaking required to implement the clock domain crossing between the processors and the data memory. Moreover, the small form factor of processors typically adopted in this domain might not justify the costs of a so complex solution (i.e., per-core DC/DC converters or LDOs are required). A simpler but effective approach to reduce leakage power of unused cores lies in per-core power gating. Exploiting a per-core power gating architecture all the cores belong to the same frequency domain, and operate at the same voltage when active, eliminating clock synchronization and level shifting overhead, but they are able to shut down independently when idle. Although this approach is effective as it can reduce leakage power by several orders of magnitude, it requires a ring of PMOS transistors around each core to

implement the power gating. This significantly increases the area of the cores, and most important significantly degrades the performance of the circuits when operating at low voltage. Indeed, the presence of one additional PMOS stacked over the pull-up network of standard cells has been shown to lead significant performance degradation in the near threshold operating region (Alioto 2012). Finally, power gating does not allow for state retention, as during the idle phases all the cells within a gated region are not supplied. In this context, implementing a state retention mechanism requires more complex state-retentive memories and flip-flops, with the associated overheads in terms of area, power, and additional routing required to bring to all these components the retention voltage. Another alternative to manage leakage power of idle cores is reverse body biasing (RBB). RBB can reduce leakage power by up to one order of magnitude (Rossi et al. 2016a). Although this is not enough for implementing deep-sleep modes typical of MCUs, it provides an effective and fast (less than 100 ns settling time) way to reduce the leakage power of idle cores in a ULP parallel architecture. Indeed, in contrast to multiple voltage domains and power gating approaches, this architecture has minimal overhead in term of isolation, as it usually requires a small ring of deep N-well around the region to isolate P-wells and N-wells in a typical triple well process. Moreover, it does not require level shifters and isolation with power gating transistors, since all the regions belong to the same voltage domain. Finally, employing leakage reduction through body biasing allows standard registers to maintain the state during the sleep phases, thus avoiding the usage of more complex state-retentive flip-flops and the related overhead in terms of power routing and area.

3.3.2 Memory Hierarchy for Parallel ULP Processors

IoT is a network of physical objects that can refer to a wide variety of devices. In such scenario it is not possible to derive a single architecture that fits all the possible cases, hence the device architecture is tailored upon the computation requirements of the use case, and energy constraints. The memory hierarchy of such architectures plays a dominant role in the performance/energy metrics: for example, in real time systems caches would never exist for the fact that different layers of memory hierarchy have such different access/speed characteristics, and the global access time, e.g., in case of miss, is not predictable, and may lead to deadline violations and system failures (safety issues). Moreover, it is hard for the software to explicitly control and optimize data locality and transfers (Kalokerinos

et al. 2009). Scalability issues are real in multi-core system, where the cache coherency complexity puts a limit to the number of cores that can be integrated in the chip.

Data Scratch-Pad Memories (SPMs on Fig. 3.7a) are a valid alternative to caches for on-chip data memories. SPM is small on-chip memory bank (e.g., SRAM) mapped into the processor's address space, and tightly coupled with the processor pipeline. SPMs are faster and consume less energy with respect to larger or off-chip memories, and their inherent predictability have made them popular in real-time systems for instance. SPM also offer better scalability by allowing explicit control and optimization of data placement and transfers. These systems, then employ specialized data memories hierarchies for better efficiency for targeted data. However these memory structures need a mechanism to move data from the different levels of the memory hierarchy to the SPM, and transferring data to/from this address space (explicit communication) may lead to inefficiencies that can vanish the benefits of specialization. Remote direct memory accesses (RDMA) is a wide used technique to move data from/to the SPM from the different levels of the memory hierarchy. This technique is efficient if the programming overhead to trigger the DMA transfer is minimized, and it is affordable in the cases when the producer knows who the consumers will be, or when the consumer knows its input data set ahead of time. Moreover,

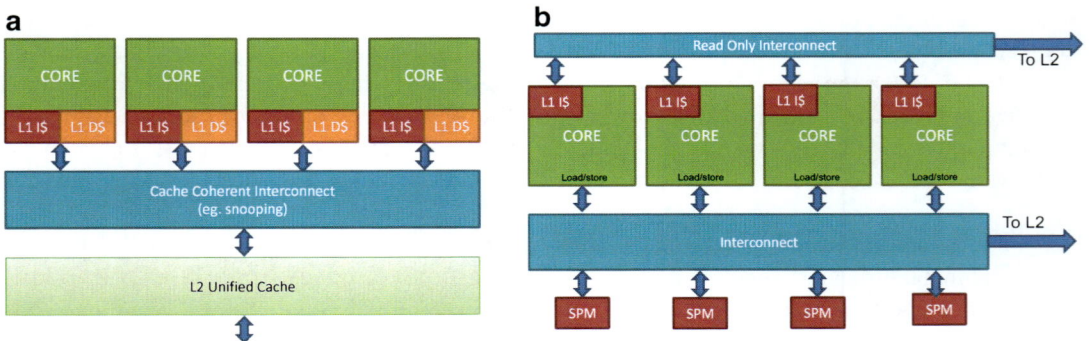

Fig. 3.7 Overview of cached (**a**) vs. cacheless (**b**) multicore system. Private caches pose a limit to scalability, increase complexity (e.g., snooping), and introduce

non uniform access time. Cacheless approach removes any coherency issues, and improve energy efficiency (simple architecture), and predictability

SPMs can replace data caches only if they are supported by an effective compiler. The mapping of memory elements of the application benchmarks to the SPM is done mostly for data, since for instructions there are no coherency issues, therefore having a cache is highly desirable on this side. However, the major benefit to adopt SPMs vs. on-chip caches is the better energy efficiency, better access timing, and better silicon area usage. Caches using static RAM consume power in the range of 25–45% of the total chip power, and energy can be reduced by 40% if replaced by SPM (Banakar et al. 2002). The Area-Time product metric can be reduced to 46% in favor of the SPM.

3.3.3   SPMs in the Near-Threshold Region

On the other side, with the introduction of near-threshold or even sub-threshold operation to digital circuits, memory design has gained renewed attention as being more susceptible to the side effects of low voltage operation.

A classic SRAM cell is a ratioed circuit relying on the relative drive strength of the transistors involved, and parametric variations of the individual devices can lead to functional failures of the cell. Low-voltage SRAMs use different supply voltages for the digital domain and memories. This approach entails additional complexity on system level (level shifters and power distribution). Other approaches for designing low-voltage SRAMs include design of 7 T, 8 T, or 10 T bit-cells, decoupled read/write operations, read and write assist techniques, and adaptive and resilient SRAMs design (Chap. 5).

One important factor is given by the leakage power, which is proportional to the chip area, which generally is dominated by memories. Scaling the voltage in the NTC area can leverage to 10X better static power consumption, which in IoT devices can lead to huge power improvement since those devices are most of the time in standby. If, from one side, scaling the voltage can lead to some power benefit, on the other side, SRAMs become slower and slower and more susceptible to process variation, leading to failures when the voltage is below 0.6 V in a 28 nm process (Teman et al. 2015).

Figure 3.8 shows the power breakdown for a multicore system, a cluster of four RISC processors, equipped with private instruction

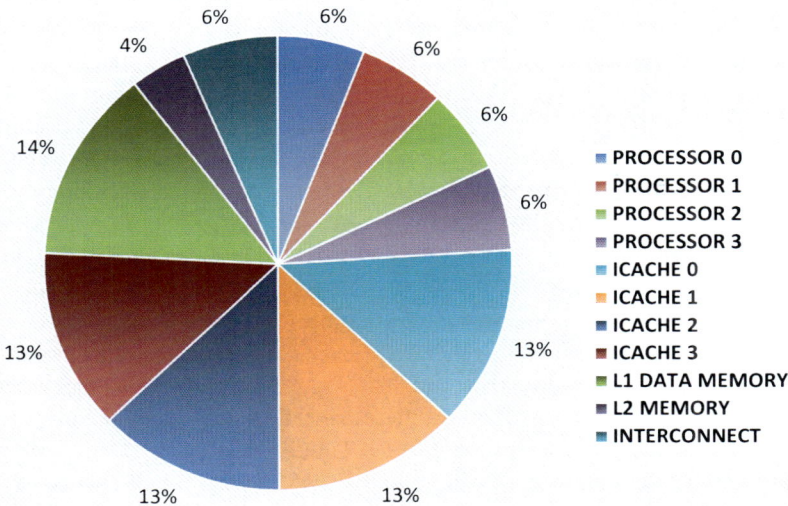

Fig. 3.8 Power Breakdown for a QuadCore with private instruction caches, based on monolithic RAM memories

caches, based on SRAM memories (TAG and DATA). As shown in the figure, private I\$ are responsible for more than 50% of the power budget. One of the alternatives that has been proposed in recent years to overcome some of these shortcomings is to synthesize memories from standard cells. Standard Cell Memories (SCMs) are soft-digital blocks, described in a hardware description language (HDL) at register-transfer level (RTL), and mapped to Standard Cell (SC) libraries. By using SCMs instead of SRAMs, designers can define each memory block according to the specific needs of each component and achieve the specifications required by their design as part of the standard digital design implementation flow. One of the primary advantages that SCMs have over traditional SRAM is their robustness, especially at low-voltages. Since SCMs are constructed exclusively from SCs, they scale along with the core digital logic, and continue to operate well below the limit of standard 6 T SRAM arrays (~600–700 mV). This advantage is accompanied by a loss of memory density, since the basic SCM storage cell is much larger than a standard 6 T SRAM bitcell. However, this trade-off is common to all low-voltage SRAM solutions (Teman et al. 2015). SCM offers better energy efficiency with respect to SRAM based SPMs In the application example provided in (Meinerzhagen et al. 2010) it was shown that the use of the considered SCM architecture reduces the power consumption 37% compared to the use of SRAM.

3.3.4 Architecture of Memory Subsystem for Parallel ULP Processors

The conventional memory hierarchy for low-power multi-core architectures is generally composed by a private L1 subsystem and a shared L2 level (as depicted in Fig. 3.7a). In the private L1 subsystem, local data keeps a copy of the accessed data, potentially replicating the same data in the different private local memory subsystem. If the L1 is based on data caches, and cores are working on the same dataset, then all those access patterns are cached locally, leading to data replication, which results in reduced effective memory capacity, and poor energy efficiency. Secondly, L1 private cache need to be coherent, and as highlighted in the previous section, cache coherency pose serious limit to multi-core scalability, and it brings a not negligible intrinsic cost (power).

Moreover, for each processor, the available memory capacity is limited by its local memory size, and each inter-core communication is expensive because is done though message-passing mechanism (mailbox etc.). The major benefit of the private caching scheme is a lower cache hit latency.

On the other hand, the shared caching scheme always maps data to a fixed location. Because there is no replication of data, this scheme achieves a lower on-chip miss rate than private caching (because of large aggregate cache capacity), and simple and efficient mechanism for inter-core communication, and finally no coherency issues (at L1 level). However, the average cache hit latency is larger, because cache blocks are simply distributed to all available cache slices. To mitigate this penalty, several architectural solutions have been proposed (e.g., victim cache).

However, both in case of shared or private scenarios, the data cache is not attractive for an energetic point of view. Secondly a wide range of applications from image and video processing domains have significant data storage requirements in addition to their computational requirements and recent and studies (Benini et al. 2000) have shown that regular data access patterns found in array-dominated applications can be better captured if SPM is employed, instead of a more traditional data cache.

Based on the assumption discussed before, in ultra-low-power multi-core systems, the data-cache is replaced by several SPM blocks (Fig. 3.7b) that are linked together in a multi-banked shared data memory, and referred as Tightly Coupled Data Memory (TCDM). TCDM is tightly integrated on the processors load/store unit interface, and shared through a lean, fast and thin distributed crossbar, that provides shared access to several SPM banks.

The TCDM therefore implements a multi-ported multi-banked shared data memory, where each processor load/store interface is plugged directly on one master port of the TCDM, while the SPM memory banks are connected on the slave side. The internal arbiter handles routing and flow control of request coming from different processors and directed to the same SPM bank. Memory banks are mapped on the global address space, and are accessible from each master port with a simple flow control protocol (req/grant). To reduce the collision probability, those banks are mapped with a word level interleaving scheme, meaning that adjacent addresses are mapped on adjacent banks, with a typical granularity of 32 bits (same as the processor architecture). To further decrease the pressure on memory side, the number of SPM banks are doubled with respect the number of processors. Since TCDM is a shared architecture, it loses the determinist latency feature of private SPM, but on the other side, the embedded crossbar ensures a maximum worst case latency (with round robin arbitration, the maximum latency for the worst case is equal to the number of cores).

To make this approach affordable, shared and private SPM schemes must behave in the same way in the case of best case (no collision on shared accesses). The processor load/store interface requires to be carefully optimized, since SPM are now shared, and there is the crossbar logic in between processor pipeline and SPM. Second, in case of conflicts (e.g., two or more processors making a request on the same shared memory bank) the processor pipeline must be frozen until the request is granted, therefore and additional stall must is added in the processor pipeline architecture.

Private instruction caches are usually able to achieve higher speed, due to their simpler design (deep integration with processor pipeline), but similarly to private data cache, the reduced capacity (vs. the aggregate capacity of shared instruction cache) lead to an increase in the miss ratio. Although large private instruction caches can significantly improve performance, they have the potential to increase power consumption, therefore are not affordable for ultra-low-power multicore systems.

The optimal solution from a point of view of the energy saving is the shared instruction cache

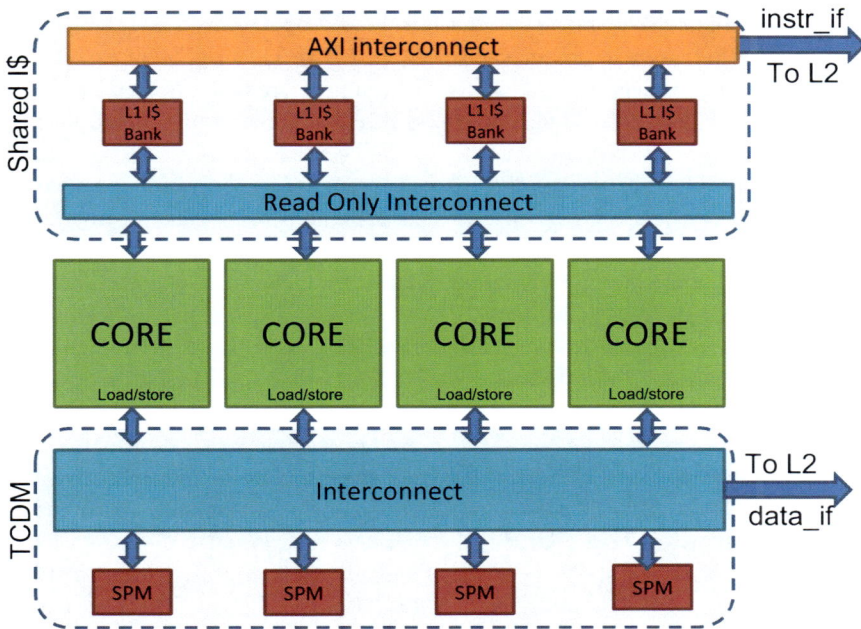

Fig. 3.9 Cluster of four processing elements, with tightly coupled data memory (SPM) and shared instruction cache

(Fig. 3.9). This approach can is an attractive solution to improve performance and energy efficiency while reducing the footprint area of the chip, since the total capacity seen by each processor is given by the aggregate capacity of the whole shared instruction cache (Loi et al. 2015). The shared instruction cache has no coherency issues (cache is read only), and it is implemented as multi-banked, multi ported read-only cache. Each cache bank is shared through a thin, fast read-only interconnect (which arbitrates the fetch request among the available cores), and mapped at cache line level address interleaving to better distribute accesses in all available cache banks (to reduce the collision rate). Similarly to the TCDM, the latency is not deterministic because both hot/miss and local collision may happen. Local contention may lower the instruction per cycle metric (IPC) in each core, so some additional features are required to reduce performance degradations.

Adding a pre-fetcher in between the shared instruction cache and processor instruction-fetch interface, with the capability to fetch one cache line in a single cycle (wide interface) further reduce the local collision rate at cache bank level, leading to reduce the IPC, very close to 1. With these optimizations the shared cache can be considered a valid alternative to private cache subsystem, with a negligible overhead due additional resources, but with the big advantage of an increased memory capacity which finally leads to improve hit ratio and less refills though the L2 level hierarchy (with a sensible improvement of energy/power consumption).

Both shared TCDM and Shared instruction cache, can be implemented using SCM, making these architecture suitable even in the Near Threshold region, where ultra-low-power requirements are mandatory, and where SRAM based memory are not operative.

3.4 Design Example: The PULP Platform

PULP is a multi-core platform achieving leading-edge energy efficiency and featuring tunable performance across a wide range of workloads (Rossi et al. 2016a; Rossi et al. 2016b). The aim of PULP is to satisfy the computational demands of IoT applications requiring flexible processing of data streams generated by multiple sensors, such as accelerometers, low-resolution cameras, microphone arrays, vital signs monitors. As opposed to single-core MCUs, a parallel ultra-low-power programmable accelerator allows to meet the computational requirements of these applications, without exceeding the power envelope of a few mW typical of miniaturized, battery-powered systems.

3.4.1 SoC Architecture

The compute engine is a cluster with a parametric number (2–16) of cores (Fig. 3.10). Cores are based on a power-optimized micro-architecture implementing the OpenRISC ISA. Pipeline depth is optimized at four balanced stages, enabling full forwarding with single stalls only on load-use and mis-predicted branches. Further pipelining would not improve near-threshold energy-efficiency because L1 memory access time dominates cycle time, while register and clocking overhead increase power (Gautschi et al. 2014). The original OpenRISC ISA and the core micro-architecture is extended for energy efficient digital signal processing, supporting zero-overhead hardware loops with L0 I-buffer, load and store operations embedding pointer arithmetic, SIMD vector instructions and power management instructions (Gautschi et al. 2015). The cluster can be configured with either private or shared instruction cache with instruction broadcasting support. By coupling the shared I-cache with L0 buffers, it is possible to greatly reduce cache pressure, resulting in a much higher energy efficiency (Loi et al. 2015). The cores do not have private data caches, avoiding memory coherency overhead and greatly increasing leakage and area efficiency for data memory. A L1 multi-banked Tightly Coupled Data Memory (TCDM) acts as a software-managed shared data scratchpad. The TCDM features a parametric number of word-level interleaved

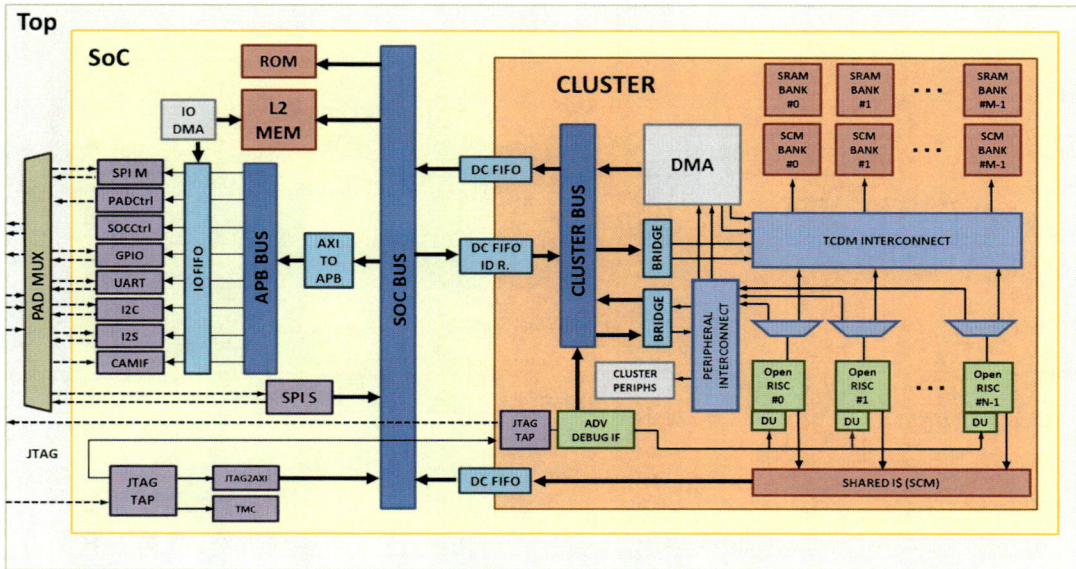

Fig. 3.10 PULP platform architecture

banks connected to the processors through a non-blocking interconnect to minimize banking conflicts (Rahimi et al. 2011). Each logical bank is implemented as a heterogeneous memory, composed of a mix of SRAMs and latch-based Standard Cell Memory (SCM) banks. Instruction caches can also be implemented using SCMs. While SRAMs achieve a higher density than SCMs (3x-4x), SCMs are able to work over the same voltage range as logic, extending the operating range of the whole cluster to the very limit of the technology; moreover, their energy/access is lower than that of SRAMs for the small cuts needed in L1 (Teman et al. 2015). Depending on the availability of low-voltage memories in the targeted implementation technology, different ratios of SCM and SRAM memory can be instantiated at design time. Reconfigurable pipeline stages, controlled by the processors through a memory mapped interface, can be optionally added to the TCDM interconnect to deal with the performance degradation and variability of SRAMs at low voltage.

The cluster has AXI4-compliant interfaces. Off-cluster (L2) memory and peripheral access latency is managed by a tightly coupled DMA

optimized for low power with just ten cycles programming latency, up to 16 outstanding transactions and a private physical channel for DMA control for each core. Various peripherals are available for PULP SoCs, including SPI interfaces with streaming support, I2C, I2S, a camera interface, GPIOs, a bootup ROM and a JTAG interface for debug and test purposes. SPI interfaces can be configured in master/slave single or quad mode. PULP SoCs can operate standalone or as a slave accelerator of a standard host processor (e.g., an ARM Cortex-M microcontroller). To reduce the overall number of pads, and make low-cost wire bonding packaging of the SoC suitable for IoT applications, the IO interfaces can be multiplexed.

3.4.2   Power Management Architecture

To provide high energy efficiency across a wide range of workloads, the PULP cluster and the rest of the SoC are in different power and clock domains. Fine-grained tuning of the SoC and cluster frequencies is achieved through two FLLs (Frequency-Locked Loops) (Miro-Panades

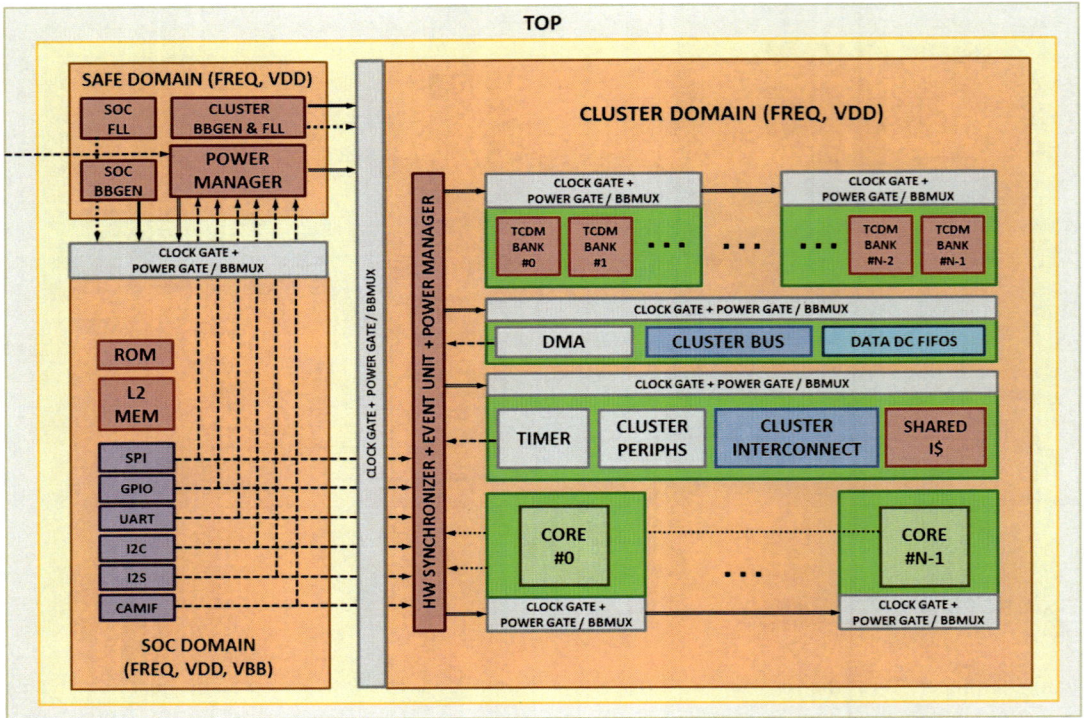

Fig. 3.11 Power management architecture

et al. 2014). Each processor within the cluster, as well as other data transfer, and memory resources can be separately disabled, clock-gated and reverse body biased when idle. The concept of fine-grain power management is shown in Fig. 3.11. In the PULP platform, the body-bias multiplexers allow to dynamically select the back-bias voltage of each processor, enabling ultra-fast transitions between the active mode and the sleep mode. Although the same concept applies also to leakage management with power gating, in the context of a near-threshold processor body biasing offers two key advantages for partial shut-down of small blocks. First it is intrinsically state-retentive, and does not require to increase the complexity of flip flops and memories with additional supply rails. Second, power gating requires a ring of transistors around the digital area to me managed, causing a relevant overhead for small blocks, and add a PMOS to the pull-up network of the digital logic, which degrades the performance in near-threshold. Depending on the required workload, the cluster

is able to keep active an arbitrary number of processing elements, while the others consume zero dynamic power and up to 10x less leakage power. An event unit automatically manages transitions of the cores between the active and idle states. Processors can be put in idle state with a write operation on a memory mapped control register. After going in sleep mode, a core remains idle until a configurable event is triggered. Events can be issued by all IO peripherals, the DMA, and timers. Emergency and general-purpose events are also supported. The same approach is replicated hierarchically at cluster and SoC level. A power manager based in a safe domain controls the state of the SoC and cluster domains, managing dynamic and leakage power during sleep states and generating wake-up sequence when a termination event is received, for example due to the completion of a transfer from an IO peripheral to the L2 memory. This mechanism allows to minimize the intrinsic overheads of a relatively complex parallel computing platform, guaranteeing that only the strictly necessary blocks are active during the phases of parallel or sequential computation, and data transfer. On the other hand, power gating is more effective to implement system-level deep-sleep modes of heavily duty-cycled applications, where the whole system needs to be managed for longer periods (tens to hundreds micro seconds) and leakage power needs to be reduced by several orders of magnitudes (100 nW–10 μW).

3.4.3 Programming PULP

OpenMP, OpenCL and OpenVX programming models are available for PULP, as they are well-known standards for shared memory programming, and widely adopted in embedded MPSoCs (Stotzer et al. 2013). GCC- and LLVM-based tool-chains and light-weight implementations of all these environments have been tailored to PULP's explicitly managed, scratchpad-based memory hierarchy. To achieve energy efficiency, however, it is necessary that the implementation of the different execution models efficiently exploits the ULP features of the hardware. The alternance of sequential and parallel code parts is inherent to all the programming paradigms based on the fork-join model. While minimizing the impact of Amdahl's law by extracting the maximum degree of concurrency in applications is paramount for every parallel system, for ULP parallel system the way idleness is implemented is key to achieve energy efficiency. Idle power of unused cores might be significant, causing huge energy efficiency drop. To this end, special hardware for accelerating key software patterns has been developed. The PULP software runtime integrates a clock-gating based thread docking scheme to eliminate dynamic power, coupled with Reverse Body Bias (RBB) to reduce leakage, when worker threads are idling (e.g., in sequential regions of the program). In addition, the key operations required in fork/join thread management are HW-accelerated, dramatically increasing the performance with respect to polling or event-based synchronization.

3.4.4 Extending PULP

When the application requirements are so strict that they cannot be matched by parallel execution on power-optimized processors, customization of the cluster may be required. It is possible to include hardware accelerators as an extension of the baseline PULP cluster sharing L1 memory (Dehyadegari et al. 2015). Integration of HW accelerators in the PULP cluster is fully modular; the IPs used to couple accelerators with the cluster are parametric and support manually or HLS-designed accelerators using either a streaming dataflow model or a memory-mapped one. In the context of high-performance computing, deep learning (Memisevic 2015) and more specifically Convolutional Neural Network (CNN) algorithms have rapidly grown since 2012 thanks to their outstanding capabilities in the recognition (Hannun 2014), and classification (Russakovsky 2014) fields. Being these algorithms approximation tolerant, a new key challenge is to exploit them as hardware accelerators for deeply embedded applications by replacing their native single- or double- precision implementations with integer or approximated floating point arithmetic to improve energy efficiency. Another advantage of CNNs is their affinity to be rapidly adapted to several application domains, such as embedded audio and video processing (Wu et al. 2015). In addition, the computational patterns typically implemented by CNNs are very similar to those of more traditional filters used in several deeply embedded classification applications such as FIR filtering or FFT. Hence, CNN hardware accelerators join the typical advantages of application specific computing (i.e., significant boost in performance and energy efficiency with respect to equivalent software implementations) with generality, matching two of the key requirements of IoT computing devices. Vision-PULP is an extension of the basic platform oriented to embedded vision that features a 2D convolutional accelerator able to perform two 16-bit 5×5 convolutions per cycle (Conti et al. 2015). This improves energy efficiency by a factor of 40x or more in convolutional workloads, reaching up to 80 GOPS of performance and 3000 GOPS/W.

3.5   Trends and Perspectives

Table 3.2 provides a summary of recent single-core and multi-core ultra-low-power and energy efficient digital computing architectures targeting the IoT application domain. Until last few years, most IC design has mainly been driven by performance and area to reduce the cost production, while power consumption has always been a secondary concern for optimization. With the coming of for near-sensor processing and IoT we are experiencing a shift from a computation-driven to a data-driven environment, requiring a radical change in the design of the hardware platform architectures. In this scenario data are generated by several sensors, often asynchronously (event-based computing), that activate the computational platform at

regular intervals, with long idle periods in between. Hence, the computational platforms have to be always-on, powered by harvester or small coin batteries, and their life-time can be as long as several years, as expected by their target applications. As a consequence, both sleep modes and active modes need to be carefully optimized to reach the goal of zero-energy sense, classify and transmit IoT systems.

In the next generation of IoT nodes energy saving will have to be considered as a system-wide concern. All the components within an IoT node need to be optimized for power, including sensor interfaces (Chaps. 12 and 13), digital platforms (Chap. 9) and RX/TX transceivers (Chap. 14), power supply and conversion subsystem (Chaps. 10, 11, 15). Optimized software also plays a crucial role system-wide to manage power state of all the components forming the IoT node. While DVFS joint with parallelism has been demonstrated to be extremely effective in reducing the energy of digital blocks, this technique can only be applied in a very limited way

Table 3.2 Summary of recent ultra-low-power and energy efficient processors and DSPs

	SLEEP WALKER (Bol et al. 2013)	REISC (Ickes et al. 2011)	DSP (Gammie et al. 2011)	FRISBEE (Wilson et al. 2014)	CENTR P3DE (Fick et al. 2012)	PULP (Rossi et al. 2016a)	PULP2 (Rossi et al. 2016b)
CPU	MSP430	ReiSC	TMS320C64x	FRISBEE	ARM CortexM3	OpenRISC	OpenRISC
Data format	16-bit	32-bit	32-bit VLIW	32-bit VLIW	32-bit	32-bit	32-bit
Number of cores	1	1	1	1	64	4	4
I\$/D\$/L2 (bytes)	16 k/2 k/n.a.	8 k/8 k/n. a.	32 k/32 k/ 128 k	4 k/4 k/n.a.	64 k/512 k/n. a.	16 k/4 k/ 16 k	16 k/4 k/ 16 k
Technology	CMOS	CMOS	CMOS	FD-SOI	CMOS	FD-SOI	FD-SOI
Node	65 nm GP	65 nm LP	28 nm LP	28 nm HP	130 nm	28 nm LP	28 nm HP
VDD range (mem) [V]	0.4 (1.0)	0.54–1.2 (0.4–1.2)	0.6–1.0	0.4–1.3	0.65–1.15 (0.8–1.65)	0.44–1.2 (0.54–1.2)	0.32–1.2 (0.45–1.2)
Max freq. (MHz)	25	82.5	331	2600	80	475	825
Power dens. (uW/MHz)	7.7	10.2	409	62	317	72	20.7
Best Perf. (MOPS)	25	57.5	662	2600	1600	1800	3300
Energy eff. (MOPS/ mW)	64.5	68.6	4.5	16	3.9	60	193

and with some restrictions to analog blocks. For example, memories are very critical IPs, and often form a bottleneck for energy efficiency of ultra-low-power platforms, since the bit-cells, as well as some of the analog IPs in the periphery requires higher voltages than the logic to be operational in a reliable way. Although several solutions has been proposed in research (Chap. 5), silicon vendors still strive to provide memory generators optimized for low-voltage operation in production design kits.

An emerging trend for ultra-low-power systems, supplied only with energy harvesters or small batteries is that of energy-driven computation. This computational paradigm, referred as *transient computing*, can be applied to applications with very low requirements in terms of real-time constraints. For example a sensor that collects and transmit environmental conditions (e.g., temperature, pressure). When an *energy burst* is captured and accumulated in some energy storage (e.g., super-caps) by the harvester the acquisition and computation can start. When the energy on the accumulator— monitored by the hardware platform—is going to finish, the hardware platform goes into a deep-sleep or even shut-down mode, after saving its internal state and temporary data on a non-volatile support. When a further energy burst occurs, the hardware platform restores the state from its non-volatile support and continues acquisition, processing and transmission. This is yet an example showing the new requirements in terms of tighter system-level integration between the MCU and the peripheral parts of the node (e.g., batteries, sensors, transceivers) whose power consumption need to be constantly monitored and predicted to adopt power-management and shut-down strategies, while guaranteeing to always be able to recover processing from a consistent state. In this scenario, a key role will be played by non-volatile memories, which has to be re-designed and optimized for their new scope (Chaps. 6 and 7).

Most of today's IoT systems are fairly low volume, and intended for a cost conscious market. Hence, computing devices for IoT are implemented with very cheap and not scaled

technology nodes, ranging from 180 to 65 nm. Advanced nodes such sub 20 nm, or FD-SOI, would have advantages at the transistor level but their mask set and wafer cost will only be justified only with the expected increase of IoT product volumes, that would reduce manufacturing costs as well. A big concern of scaled technology nodes is the availability of embedded Flash as today 40 nm is the most advanced node featuring embedded Flash. Another issue that has to be addressed deals with process and temperature variations. IoT systems are intrinsically subject to variations during their long lifetime due to degradation of batteries and power supply network lowering the supply voltage, or aging. Usage of scaled technology magnifies variations caused by process and temperature. More specifically, scaled transistors operating at low voltage are subject to thermal inversion phenomena, which causes a strong dependency between the ambient temperature and maximum operating frequency. This issues will have to be tackled in next generation of IoT commercial platforms with advanced techniques to tolerate or compensate these variations, such as in situ error detection and correction, adaptive voltage scaling, and adaptive body biasing.

References

M. Alioto, Ultra-low power VLSI circuit design demystified and explained: a tutorial. IEEE Trans. Circuits Syst.—Part I (Invited) **59**(1), 3–29 (2012)

Ambiq Micro, Ultra-low power MCU family. Apollo Datasheet rev 0.45 (Sep. 2015)

ARM, Cortex-M4, Technical Reference Manual r0p0 Issue A (Mar. 2010), pp. 3.8–3.10

Atmel, SMART ARM-based Microcontrollers. SAM L22x Rev. A datasheet (Aug. 2015)

R. Banakar, S. Steinke, B. Lee, M. Balakrishnan, P. Marwedel, Scratchpad memory: a design alternative for cache on-chip memory in embedded systems. in *Tenth International Symposium on Hardware/Software Codesign (CODES)* (Estes Park, 2002)

L. Benini, A. Macii, E. Macii, M. Poncino, Increasing energy efficiency of embedded systems by application-specific memory hierarchy generation. IEEE Design Test Comput. **17**(2), 74–85 (2000)

D. Blaauw, et al., Razor II: in situ error detection and correction for PVT and SER tolerance, in *IEEE*

International Solid-State Circuits Conference, 2008. ISSCC 2008. Digest of Technical Papers (IEEE, 2008)

D. Bol, J. De Vos, C. Hocquet, F. Botman, F. Durvaux, S. Boyd, D. Flandre, J. Legat, SleepWalker: a 25-MHz 0.4-V Sub-mm 2 7-µW/MHz microcontroller in 65-nm LP/GP CMOS for low-carbon wireless sensor nodes. IEEE J. Solid-State Circuits **48**(1), 20–32 (2013)

D. Bortolotti, D. Rossi, A. Bartolini, L. Benini, A variation tolerant architecture for ultra low power multi-processor cluster, in *2013 23rd International Workshop on Power and Timing Modeling, Optimization and Simulation (PATMOS)* (Karlsruhe, 2013), pp. 32–38

B.H. Calhoun, A.P. Chandrakasan, Standby power reduction using dynamic voltage scaling and canary flip-flop structures. IEEE J. Solid-State Circuits **39**(9), 1504–1511 (2004)

F. Conti, L. Benini, A ultra-low-energy convolution engine for fast brain-inspired vision in multicore clusters, in *Design, Automation & Test in Europe Conference & Exhibition* (9–13 Mar 2015), pp. 683–688

D.E. Culler, J.P. Singh, *Parallel Computer Architecture, a hw/sw Approach* (Morgan Kaufmann, San Francisco, 1999)

Cypress, Designing for low power and estimating battery life for BLE applications, AN92584 Rev. C Application Note (Mar. 2016)

C.E. Molnar, R.F. Sproull, I.E. Sutherland, The counterflow pipeline processor architecture. IEEE Des. Test Comput. **11**, 48–59 (1994)

H. De Groot, IoT and the cloud: a hacked personality and an empty battery head-ache or an intuitive environment to make our lives easier? S3S (2015)

M. Dehyadegari, A. Marongiu, M.R. Kakoee, S. Mohammadi, N. Yazdani, L. Benini, Architecture support for tightly-coupled multi-core clusters with shared-memory HW accelerators. IEEE Trans. Comput. **64**(8), 2132–2144 (2015)

A.Y. Dogan, D. Atienza, A. Burg, I. Loi, L. Benini, Power/performance exploration of single-core and multi-core processor approaches for biomedical signal processing, in *Proceedings of 21st International Workshop*, PATMOS 2011 (Madrid, 26–29 Sept. 2011), pp. 102–111

R.G. Dreslinski, M. Wieckowski, D. Blaauw, D. Sylvester, T. Mudge, Near-threshold computing: reclaiming Moore's law through energy efficient integrated circuits. Proc. IEEE **98**(2), 253–266 (2010)

D. Ernst et al., Razor: a low-power pipeline based on circuit-level timing speculation, in *Proceedings of 36th Annual IEEE/ACM International Symposium on Microarchitecture, 2003. MICRO-36* (2003), pp. 7–18

D. Ernst, S. Das, S. Lee, D. Blaauw, T. Austin, T. Mudge, N.S. Kim, Razor: circuit-level correction of timing errors for low-power operation. IEEE Micro **34**(6), 10–20 (2005)

D. Fick et al., Centip3De: a 3930DMIPS/W configurable near-threshold 3D stacked system with 64 ARM Cortex-M3 cores. in *2012 I.E. International Solid-State Circuits Conference* (San Francisco, 2012), pp. 190–192

G. Gammie, N. Ickes, M.E. Sinangil, R. Rithe, J. Gu, A. Wang, H. Mair, S. Datla, B. Rong, S. Honnavara-Prasad, L. Ho, G. Baldwin, D. Buss, A.P. Chandrakasan, Uming Ko, A 28 nm 0.6 V low-power DSP for mobile applications, in *2011 I.E. International Solid-State Circuits Conference Digest of Technical Papers (ISSCC)* (20–24 Feb. 2011), pp. 132–134

M. Gautschi, D. Rossi, L. Benini, Customizing an open source processor to fit in an ultra-low power cluster with a shared L1 memory, in *Proceedings of the 24th Edition of the Great Lakes Symposium on VLSI-GLSVLSI'14* (2014), pp. 87–88

M. Gautschi, A. Traber, A. Pullini, L. Benini, M. Scandale, A. Di Federico, M. Beretta, G. Agosta, Tailoring instruction-set extensions for an ultra-low power tightly-coupled cluster of OpenRISC cores, in *IFIP/IEEE International Conference on Very Large Scale Integration (VLSI-SoC)* (October 2015)

A. Hannun, Deep speech: scaling up end-to-end speech recognition, arXiv (2014)

S. Hsu, A. Agarwal, M. Anders, S. Mathew, H. Kaul, F. Sheikh, R. Krishnamurthy, A 280 mV-to-1.1 V 256b reconfigurable SIMD vector permutation engine with 2-dimensional shuffle in 22 nm CMOS, in *2012 I. E. International Solid-State Circuits Conference Digest of Technical Papers (ISSCC)* (19–23 Feb. 2012), pp. 178–180

N. Ickes, Y. Sinangil, F. Pappalardo, E. Guidetti, A. P. Chandrakasan, "A 10 pJ/cycle ultra-low-voltage 32-bit microprocessor system-on-chip, in *2011 Proceedings of the ESSCIRC (ESSCIRC)* (Helsinki, 2011), pp. 159–162

D. Jeon, Y. Kim, I. Lee, Z. Zhang, D. Blaauw, D. Sylvester, A 470 mV 2.7 mW feature extraction-accelerator for micro-autonomous vehicle navigation in 28 nm CMOS, in *2013 I.E. International Solid-State Circuits Conference Digest of Technical Papers (ISSCC)* (17–21 Feb. 2013), pp. 166–167

G. Kalokerinos, V. Papaefstathiou, G. Nikiforos, S. Kavadias, Ma. Katevenis, D. Pnevmatikatos, X. Yang, FPGA implementation of a configurable cache/scratchpad memory with virtualized user-level RDMA capability, in *International Symposium on Systems, Architectures, Modeling, and Simulation, 2009. SAMOS '09* (Samos, 2009), pp. 149–156

I. Loi, D. Rossi, G. Haugou, Exploring multi-banked shared-L1 program cache on ultra-low power, tightly coupled processor clusters, in *Proceedings of the 11th ACM Conference on Computing Frontiers-CF '15* (July 2015), pp. 1–10

M.M.K. Martin, M.D. Hill, D.J. Sorin, Why on-chip cache coherence is here to stay. Commun. ACM **55**(7), 78–89 (2012)

P. Meinerzhagen, C. Roth, A. Burg, Towards generic lowpower area-efficient standard cell based memory architectures, in *Proceedings of IEEE International Midwest Symposium on Circuits and Systems* (Aug. 2010), pp. 129–132

R. Memisevic, Deep learning architectures, algorithms and applications, in *Hot Chips: A Symposium on High Performance Chips* (Cupertino, 23–25 Aug. 2015)

Microchip, Analog-to-digital converter with computation technical brief. TB3146 Application Note (Aug. 2016)

ST Microelectronics, STM32L4x5 advanced ARM®-based 32-bit MCUs. RM0395 Reference Manual (Feb. 2016)

I. Miro-Panades, E. Beignè, Y. Thonnart, L. Alacoque, P. Vivet, S. Lesecq, D. Puschini, A. Molnos, F. Thabet, B. Tain, K.B. Chehida, S. Engels, R. Wilson, D. Fuin, A fine-grain variation-aware dynamic Vdd-hopping AVFS architecture on a 32 nm GALS MPSoC. IEEE J. Solid-State Circuits **49**(7), 1475–1486 (2014)

NXP Semiconductors. 2015. LPC5410X Product Data Sheet. NXP. Rev 2.2

J. Park, I. Hong, G. Kim, Y. Kim, K. Lee, S. Park, K. Bong, H.-J. Yoo, A 646GOPS/W multi-classifier many-core processor with cortex-like architecture for super-resolution recognition, in *2013 I.E. International Solid-State Circuits Conference Digest of Technical Papers (ISSCC)* (17–21 Feb. 2013), pp. 168–169

J. Parkhurst, J. Darringer, B. Grundmann, From single core to multi-core: preparing for a new exponential, in *2006 IEEE/ACM International Conference on Computer Aided Design* (San Jose, CA, 2006), pp. 67–72

Peter Greenhalgh, big.LITTLE processing with ARM Cortex-A15 & Cortex-A7. Technical Report (ARM Ltd., 2011)

A. Rahimi, I. Loi, M.R. Kakoee, L. Benini, A fully-synthesizable single-cycle interconnection network for Shared-L1 processor clusters, in *Design, Automation & Test in Europe Conference & Exhibition* (Mar. 2011), pp. 1–6

Renesas, RL78/G13 Rev. 3.20 User's Manual Hardware (July 2014)

D. Rossi, I. Loi, G. Haugou, L. Benini, Ultra-low-latency lightweight DMA for tightly coupled multi-core clusters, in *Proceedings of the 11th ACM Conference on Computing Frontiers—CF '14* (July 2014), pp. 1–10

D. Rossi, A. Pullini, I. Loi, M. Gautschi, F.K. Gurkaynak, A. Bartolini, P. Flatresse, L. Benini, A 60 GOPS/W, −1.8 V to 0.9 V body bias ULP cluster in 28 nm UTBB FD-SOI technology. Solid-State Electron. **117**, 170–184 (2016a)

D. Rossi, A. Pullini, I. Loi, M. Gautschi, F.K. Gurkaynak, A. Teman, J. Constantin, A. Burg, I.M. Panades, E. Beignè, F. Clermidy, F. Abouzeid, P. Flatresse, L. Benini, 193 MOPS/mW @ 162 MOPS, 0.32V to 1.15V voltage range multi-core accelerator for

energy-efficient parallel and sequential digital processing, Cool Chips (2016b)

O. Russakovsky, ImageNet large scale visual recognition challenge. Int. J. Comput. Vis (2014)

E. Stotzer, A. Jayaraj, M. Ali, A. Friedmann, G. Mitra, A. P. Rendell, I. Lintault, OpenMP on the low-power TI keystone II ARM/DSP system-on-chip, in *OpenMP in the Era of Low Power Devices and Accelerators* (2013)

A. Teman, D. Rossi, P. Meinerzhagen, L. Benini, A. Burg, Controlled placement of standard cell memory arrays for high density and low power in 28 nm FD-SOI, in *20th Asia and South Pacific Design Automation Conference (ASP-DAC), 19–22 January, 2015* (2015), pp. 81–86

Texas Instruments, Low-power FRAM microcontrollers and their applications, SLAA502 White Paper (June 2011)

J.W. Tschanz et al., Adaptive body bias for reducing impacts of die-to-die and within-die parameter variations on microprocessor frequency and leakage. IEEE J. Solid-State Circuits **37**(11), 1396–1402 (2002)

J. Tschanz et al., Adaptive frequency and biasing techniques for tolerance to dynamic temperature-voltage variations and aging, in *IEEE International Solid-State Circuits Conference, 2007. ISSCC 2007. Digest of Technical Papers* (11–15 Feb. 2007), pp. 292–604

R. Wilson et al., A 460MHz at 397mV, 2.6GHz at 1.3V, 32b VLIW DSP, embedding F_{MAX} tracking, in *2014 I. E. International Solid-State Circuits Conference Digest of Technical Papers (ISSCC)* (San Francisco, 2014), pp. 452–453

M. Wu, R. Iyer, Y. Hoskote, S. Zhang, B. Deadman, M. Bhartiya, Y. Satish, Design of an ultra-low Power SoC testchip for wearables & IOT, in *Hot Chips: A Symposium on High Performance Chips* (Cupertino, 23–25 Aug. 2015)

A. Yakovlev, D. Pivonka, T. Meng, A. Poon, A mm-sized wirelessly powered and remotely controlled locomotive implantable device, in *2012 I.E. International Solid-State Circuits Conference Digest of Technical Papers (ISSCC)* (19–23 Feb. 2012), pp. 302–304

J. Yoo, Y. Long, D. El-Damak, M. Bin Altaf, A. Shoeb, Y. Hoi-Jun, A. Chandrakasan, An 8-channel scalable EEG acquisition SoC with fully integrated patient-specific seizure classification and recording processor, in *2012 I.E. International Solid-State Circuits Conference Digest of Technical Papers (ISSCC)* (19–23 Feb. 2012), pp. 292–294

F. Zhang, Y. Zhang, J. Silver, Y. Shakhsheer, M. Nagaraju, A. Klinefelter, J. Pandey, J. Boley, E. Carlson, A. Shrivastava, B. Otis, B. Calhoun, A batteryless 19 μW MICS/ISM-band energy harvesting body area sensor node SoC, in *2012 I.E. International Solid-State Circuits Conference Digest of Technical Papers (ISSCC)*. (Feb. 2012), pp. 298–300

Near-Threshold Digital Circuits for Nearly-Minimum Energy Processing

4

Massimo Alioto

This chapter addresses the challenges and the opportunities to perform computation with nearly-minimum energy consumption through the adoption of logic circuits operating at near-threshold voltages. Simple models are provided to gain an insight into the fundamental design tradeoffs. A wide set of design techniques is presented to preserve the nearly-minimum energy feature in spite of the fundamental challenges in terms of performance, leakage and variations. Emphasis is given on debunking the incorrect assumptions that stem from traditional low-power common wisdom at above-threshold voltages.

In this analysis, the main emphasis is given on the energy consumption, as performance requirements in IoT nodes are easily achievable with near-threshold circuits in most cases, as discussed in Chap. 1 and in the following. Sustained higher levels of performance can always be achieved through architectural techniques (see Chap. 3), whereas occasional performance boosts can be obtained through circuit techniques (see below).

4.1 Preliminary Considerations on Near-Threshold Operation

4.1.1 Transistor Current vs. Supply Voltage and Transregional Model

Voltage scaling is well known to be a very effective knob to reduce the energy per computation at the cost of degraded performance (Burd et al. 2015). The performance degradation at supply voltages V_{DD} lower than the nominal voltage is determined by the reduction in the transistor on-current I_{on}, which in turn depends on the operating region (i.e., the voltage range). The transregional EKV model can be conveniently used to express such dependence in all regions (Enz and Vittoz 2006):

$$I_{on} = I_0 \cdot IC = I_0 \cdot [\ln(e^v + 1)]^2 \qquad (4.1)$$

where IC is the inversion coefficient (i.e., normalized current), I_0 is the specific current $2 \cdot n \cdot \mu \cdot C_{OX} \frac{W}{L} (kT/q)^2$, and v is the normalized gate overdrive $v = (V_{DD} - V_{TH})/[2 \cdot n \cdot (kT/q)]$. In the above equations, n is the transistor sub-threshold factor, μ is the carrier mobility, C_{OX} is the MOS capacitance per unit area, W/L is the aspect ratio, V_{TH} is the transistor threshold voltage, and kT/q is the thermal voltage.

In the EKV model in (4.1), a transistor operates in weak inversion when $IC < 0.1$

M. Alioto (✉)
National University of Singapore, Singapore, Singapore
e-mail: malioto@ieee.org

© Springer International Publishing AG 2017
M. Alioto (ed.), *Enabling the Internet of Things*, DOI 10.1007/978-3-319-51482-6_4

Fig. 4.1 Qualitative trend of I_{on} transistor current (log scale) versus the gate overdrive $V_{DD} - V_{TH}$

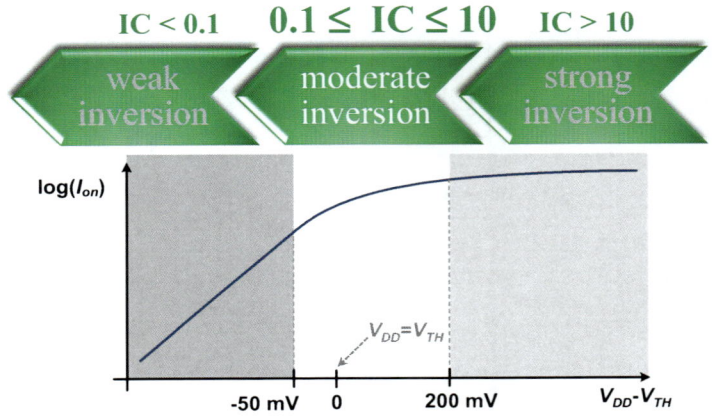

(i.e., for $v < -1$), which from (4.1) corresponds to voltages below $V_{TH} - 50\,mV$ for typical n (~1.3–1.5) and operating temperatures (Sansen 2006). On the other hand, a transistor operates in strong inversion for $IC > 10$ (i.e., for $v > 3.1$), and hence for voltages above $V_{TH} + 200\,mV$ (Sansen 2006). Near-threshold operation occurs for intermediate voltages, as summarized in Fig. 4.1.

The above traditional EKV model is very useful for quick estimates, but it oversimplifies the I–V characteristics at voltages above V_{TH}. Indeed, eq. (4.1) leads to $I_{on} \approx I_0 \cdot v^2$ in strong inversion, and its quadratic trend is far from the linear trend that is observed in actual nanometer CMOS technologies.[1]

Introducing voltage-dependent coefficients in (4.1) solves the issue, but leads to impractically complicated expressions for pencil-and-paper evaluations. To retain its simplicity while employing constant coefficients, (4.1) is here modified according to

$$I_{on} = I_0 \cdot \ln\left(e^{\frac{V_{DD}-V_{TH}}{n \cdot (kT/q)}} + 1\right) \quad (4.2)$$

which is plotted in Fig. 4.2 along with the actual I–V characteristics for 28-nm NMOS and PMOS transistors. The model is 10% (20%) within

circuit simulations on average (in the worst case), hence it is well suited for quick estimates and design purposes.

4.1.2 Transistor Current and Gate Delay in Different Regions

By the definition summarized in Fig. 4.2, sub-threshold voltages correspond to transistor operation in weak inversion, above-threshold are associated with strong inversion, and near-threshold voltages correspond to intermediate voltages between $V_{TH} - 50\,mV$ and $V_{TH} + 200\,mV$. For typical standard threshold voltages,[2] near-threshold voltages are in the range of 400–600 mV, approximately.

At above-threshold voltages such that $e^{\frac{V_{DD}-V_{TH}}{n \cdot (kT/q)}} \gg 1$, Eq. (4.2) is approximately a linear function of $V_{DD} - V_{TH}$ as expected

---

[1] Indeed, sub-100 nm CMOS technologies typically have an I–V characteristics that is proportional to $(V_{DD} - V_{TH})^{\alpha}$ with $\alpha \approx 1$ (Sakurai and Newton 1990).

[2] Operation at near-threshold voltages tends to increase V_{TH} compared to the value at nominal voltage, due to DIBL (see Sect. 5.2.2). For standard V_{TH} of 350–380 mV at nominal voltage, it is common to have V_{TH} in the order of 400–450 mV when operating at near-threshold voltages (see, e.g., Fig. 4.7). Observe that the "standard V_{TH}" nomenclature might be attributed to different threshold voltages in some processes.

Fig. 4.2 Plot of I_{on} transistor current (log scale) in (4.2) versus the magnitude of the gate-source voltage V_{GS} (in CMOS logic gates, V_{GS} = V_{DD})

VOLTAGE RANGE	sub threshold	near threshold	above threshold
TRANSISTOR REGION	weak inversion	moderate inversion	strong inversion
EKV MODEL $IC=I/I_0$ RANGE	IC < 0.1	0.1 < IC < 10	IC > 10
V_{DD} RANGE	<V_{TH}-50 mV	intermediate	>V_{TH}+200 mV

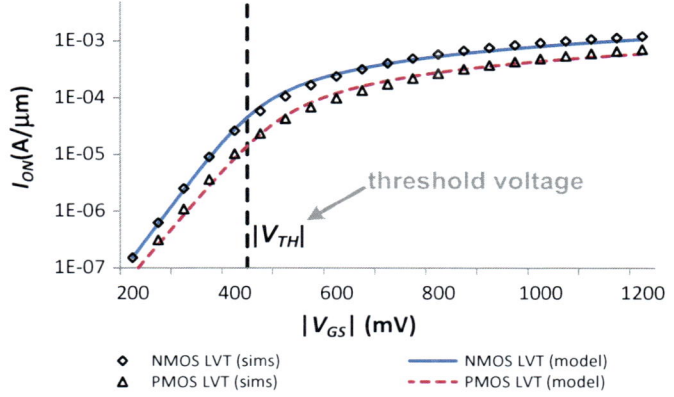

◇   NMOS LVT (sims)   —— NMOS LVT (model)
△   PMOS LVT (sims)   - - - PMOS LVT (model)

$$I_{above-threshold} \approx \left(I_0/n\frac{kT}{q}\right) \cdot (V_{DD} - V_{TH})$$

$$(4.3)$$

whereas at sub-threshold voltages it can be approximated as[3]

$$I_{sub-threshold} \approx I_0 \cdot e^{\frac{V_{DD}-V_{TH}}{n \cdot v_t}} \qquad (4.4)$$

which exponentially decreases when lowering the voltage. At near-threshold voltages, Eq. (4.2) can be approximated as

$$I_{near-threshold} \approx \frac{I_0}{2} \cdot \left[1.5 + \left(\frac{V_{DD} - V_{TH}}{n \cdot kT/q}\right)^{1.35}\right]$$

$$(4.5)$$

which is within 15% of the exact I–V characteristics in Fig. 4.2. From (4.5), the near-threshold I–V characteristics is a power law, and is steeper than in the above-threshold region.

Let us now consider a CMOS logic gate driving a capacitive load C, which includes the capacitive parasitics of the gate itself. As usual

---
[3] Indeed, $\ln(e^x + 1) \approx e^x$ for $x < 0$ (i.e., for $V_{DD} < V_{TH}$) in (5.2).

(Weste and Harris 2011), its propagation delay τ_{PD} can be expressed as $(C/I_{on}) \cdot (V_{DD}/2)$:

$$\tau_{PD} = \frac{C}{I_{on}} \cdot \frac{V_{DD}}{2} \approx \frac{C}{2 \cdot I_0} \cdot \frac{V_{DD}}{\ln\left(e^{\frac{V_{DD}-V_{TH}}{n \cdot (kT/q)}} + 1\right)} \cdot$$

$$(4.6)$$

As shown in Fig. 4.3, from (4.3) and (4.6), voltage downscaling leads to an approximately linear delay (i.e., performance) degradation, when operating above threshold. As discussed in Sect. 4.3, the energy is typically dominated by the dynamic contribution, hence a quadratic energy saving is observed above threshold. On the other hand, an exponential increase in the gate delay is observed in the sub-threshold region. Also, due to the heavier leakage contribution at low voltages (Sect. 4.3), the energy reaches a minimum energy point (MEP), and it tends to increase again when further lowering V_{DD}. Hence, near-threshold voltages are an ideal compromise between energy and performance in energy-centric VLSI designs. Indeed, the near-threshold gate delay is still reasonably small, and energy is close to its minimum value across all voltages. This motivates this chapter, and the adoption of near-threshold circuits for VLSI processing in the IoT domain.

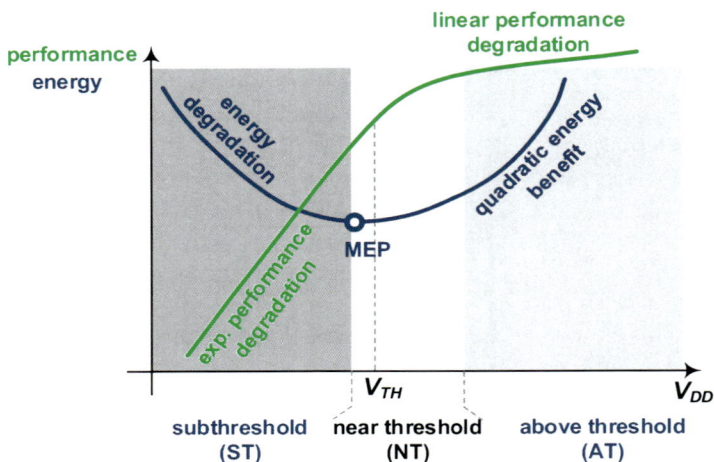

Fig. 4.3 Qualitative trend of performance (gate delay) and energy per operation versus supply voltage V_{DD}

4.2 Near-Threshold Transistor and Circuit Properties

In this section, properties of transistors at near threshold are discussed to provide general circuit design guidelines. Preliminary considerations on voltage scaling and threshold voltage dependence on sizing are respectively provided in Sects. 4.2.1 and 4.2.2. The impact of transistor stacking and PMOS/NMOS imbalance are discussed in Sect. 4.2.3 to guide the topology selection during circuit design. As second and equally fundamental aspect of circuit design, transistor strength adjustment is discussed in Sect. 4.2.4.

4.2.1 Impact of Aggressive Voltage Scaling on Transistor Current and Delay

The considerations on the delay degradation under voltage scaling in the previous section were based on the assumption that the gate load C is independent of V_{DD}. Observe that the load C comprises wire parasitics and transistor gate capacitances. The above assumption certainly holds in wire-dominated loads (as wire parasitics are voltage-independent), whereas it is somewhat pessimistic in gate-dominated loads. Indeed, as

shown by Fig. 4.4, the transistor gate capacitance tends to moderately decrease at voltages close to or below V_{TH}, and hence makes the delay degradation more graceful than discussed above, although to a minor extent.

According to the above observation, the above qualitative considerations on the delay at near- and sub-threshold voltages fully apply to any practical design. As an example, Fig. 4.5 shows the trend of the fan-out-of-4 delay $FO4$ (i.e., the delay of an inverter gate driving four equal inverters). This metric is widely used at process level to characterize the speed of the technology, at circuit level to abstract the circuit design from the process details, and at architectural level since the clock cycle normalized to $FO4$ is typically a constant that is defined by the architecture (Harris). In short, $FO4$ characterizes the system performance versus voltage for a given architecture. From Fig. 4.5, operation in the middle of the near-threshold region degrades the performance by approximately a factor of 10, compared to operation at nominal voltage. This is generally true regardless of the adopted technology (Dreslinski et al. 2010).

A very distinctive property of near-threshold operation is the stronger delay sensitivity to a given absolute change in the gate overdrive (i.e., both V_{DD} and V_{TH}), compared to above-threshold designs. This is partially explained by the steeper I–V characteristics (4.5) compared to (4.3) (the exponent of v is respectively 1.35 and 1). But the

Fig. 4.4 Gate capacitance normalized to value at nominal voltage versus supply voltage V_{DD} (28 nm, LVT and RVT transistors)

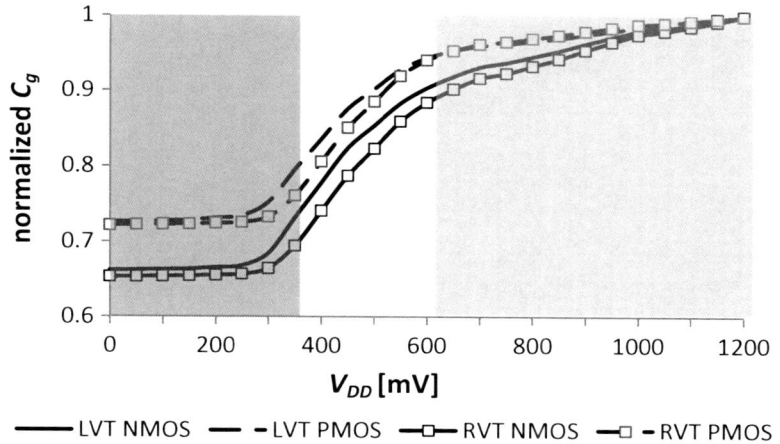

Fig. 4.5 Fanout-of-4 (*FO4*) delay normalized to value at nominal voltage versus supply voltage V_{DD} (28 nm, LVT and RVT transistors)

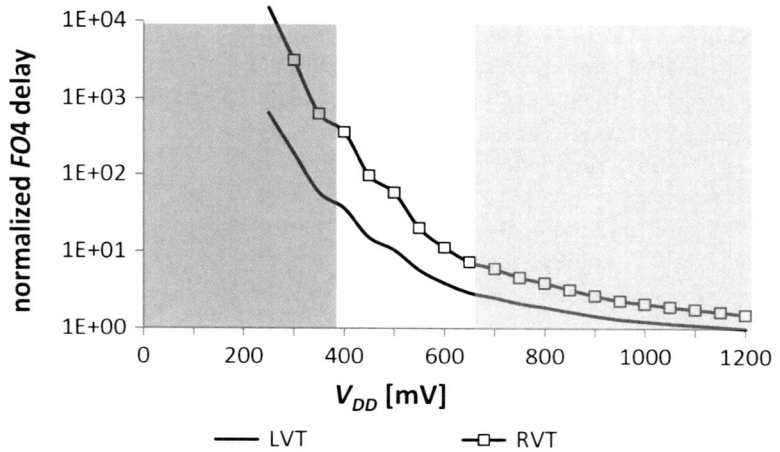

main reason is due to the very large sensitivity of $v = (V_{DD} - V_{TH})/[2 \cdot n \cdot (kT/q)]$ to a given change in V_{DD}, as V_{DD} is much closer to V_{TH} compared to above-threshold voltages. The relative I_{on} improvement due to a 100-mV supply voltage increase (i.e., boosting) for a 28-nm technology is shown in Table 4.1. As expected, the impact of voltage boosting at near-threshold voltages is substantially larger than above threshold, with improvements in I_{on} in the range of 2-4X. This unique feature permits to have significant speed adjustment capability with very limited amount of boosting, which needs to be thoroughly exploited in near-threshold-designs.

The above considerations equally apply to the threshold voltage, as I_{on} is a direct function of the

Table 4.1 I_{on} Improvement due to supply voltage boosting by 100 mV

V_{DD}	I_{on} improvement
400 mV	4.05X
500 mV	2.24X
600 mV	1.7X
800 mV	1.31X
1 V	1.17X

gate overdrive $V_{DD} - V_{TH}$. For example, the I_{on} and speed sensitivity to a 100-mV V_{DD} shift in Table 4.1 hold for the same change in V_{TH} (although with negative sign). In other words, increasing V_{TH} by 100 mV (i.e., the typical difference between a low and regular V_{TH}) at near-threshold supply voltages leads to a 2-4X

Fig. 4.6 Ratio between
FO4 of RVT and LVT
transistors versus supply
voltage V_{DD}

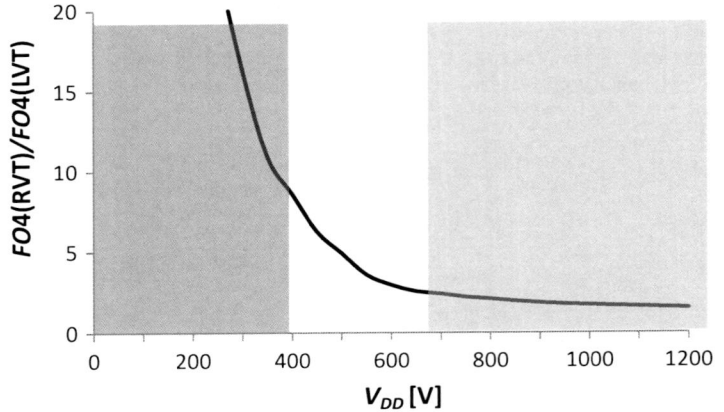

reduction in speed. This is shown in Fig. 4.6, which plots the inverter delay ratio under regular-V_{TH} (RVT) and low-V_{TH} (LVT). At nominal voltage, the different V_{TH} has a moderate impact on the performance, whereas such difference is much more pronounced at near-threshold voltages.

In summary, the high sensitivity of performance to V_{DD} and V_{TH} makes them very powerful knobs at near-threshold voltages, although the (same) sensitivity to their variations poses a challenge at the same time, as will be discussed in the following sections.

4.2.2 Impact of DIBL and Sizing on Threshold Voltage

In the previous subsection, V_{TH} was implicitly considered constant. In view of the large sensitivity of I_{on} to V_{TH}, the dependence of V_{TH} on transistor voltages and sizing needs to be explicitly considered at near-threshold voltages.

Regarding the dependence of the transistor voltages, V_{TH} tends to be quite sensitive to the drain-source voltage due to the Drain Induced Barrier Lowering (DIBL) effect (Tsividis 1999). Due to the DIBL effect, V_{TH} increases in an approximately linear fashion when the magnitude of the drain-source voltage is reduced. Due to the body effect, V_{TH} decreases (increases) under Forward FBB (Reverse, RBB) Body

Biasing, i.e. for positive[4] (negative) body bias voltages V_{BB} (Tsividis 1999). The approximately linear dependence in both effects is captured by the following equation

$$V_{TH} = V_{TH0} - \lambda_{DIBL} V_{DS} - \lambda_{BB} V_{BB} \qquad (4.7)$$

where V_{TH0} is the threshold voltage extrapolated for very low V_{DD} and $V_{BB} = 0\,V$, λ_{DIBL} is the DIBL coefficient and λ_{BB} is the body effect coefficient. The DIBL coefficient is in the order of 0.1 V/V or larger for technologies suited for IoT, and hence denotes a pronounced dependence of the threshold voltage on the drain-source voltage. As an example, Fig. 4.7 shows the change in V_{TH} versus the drain-source voltage (i.e., V_{DD}) in 28-nm transistors. When V_{DD} is reduced down to near-threshold voltages, V_{TH} typically increases by around 100 mV compared to operation at nominal voltage. This needs to be explicitly taken into account when choosing the type of threshold voltage at design time.

On the other hand, the threshold voltage dependence on the body voltage is well known to be rather weak in advanced technologies, although it is appreciable in 90-nm generations or older. Considering the strong sensitivity of performance and leakage on V_{TH}, body biasing

---

[4] These considerations hold for NMOS transistors. For PMOS transistors, change the sign in all voltages. Regarding the body effect, FBB (RBB) refers to body voltages V_{BB} below (above) V_{DD}.

Fig. 4.7 Threshold
voltage deviation vs. drain-
source voltage due to DIBL

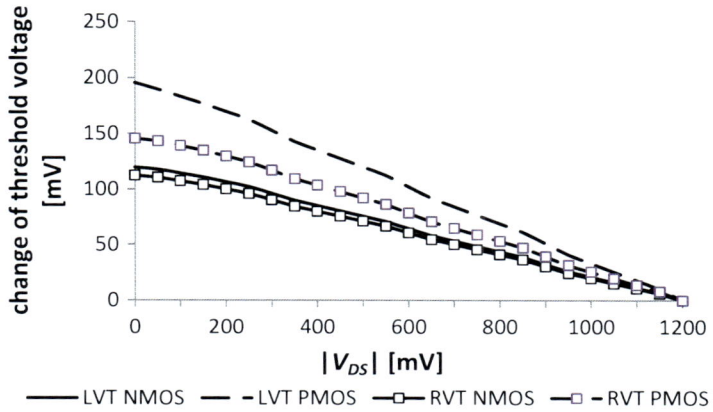

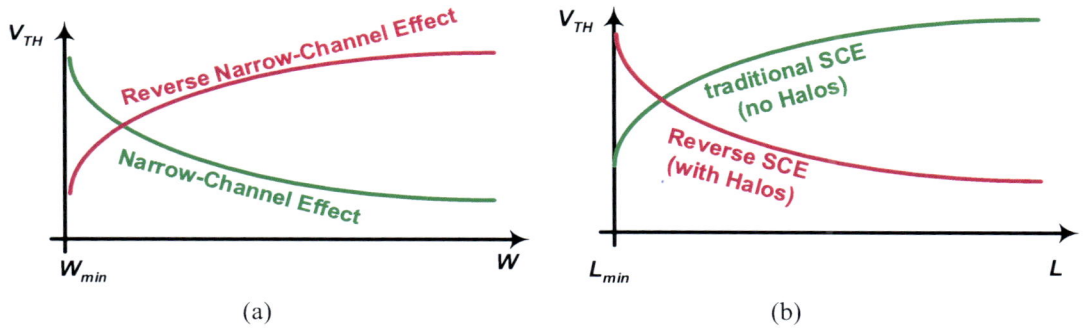

Fig. 4.8 Qualitative trend of threshold voltage vs. (**a**) transistor channel width, (**b**) transistor channel length (due to Short Channel Effect)

is still a viable option in near-threshold circuits in 90-nm technology bulk generations, or in more recent generations in FDSOI CMOS technology.

The transistor threshold voltage also depends on the size, especially when the latter is close to the minimum allowed by the process. The qualitative dependence on the channel width W (length L) is qualitatively depicted in Fig. 4.8a, b. From Fig. 4.8a, the reduction of W leads to a decrease (increase) in V_{TH} due to the Reverse (traditional) Narrow-Channel Effect RNCE (NCE) (Tsividis 1999). The dominance of one of the two effects mainly depends on the transistor isolation technology (e.g., Shallow Trench Isolation vs LOCOS), device structure (bulk, FinFET, FDSOI) and several parameters. On the other hand, from Fig. 4.8b, the reduction

of L leads to an increase (decrease) in V_{TH} due to the Reverse (traditional) Short-Channel Effect RSCE (SCE) (Tsividis 1999). The dominance of one of the two former mainly depends on whether the transistor body is lightly doped or includes halos to counteract short-channel effects (Tsividis 1999).

From the above considerations, transistor sizing can affect the performance in ways that are more complicated than the usual linear dependence of I_{on} on W/L, due to the additional (strong) dependence of V_{TH} on size at near-threshold voltages. As an example, Fig. 4.9 shows that the I_{on} current trend versus W deviates from the traditional linear dependence of $I_{on} \propto W$. For the specific considered technology, the current is increasing faster than

Fig. 4.9 I_{on} normalized to
current in minimum-sized
transistor vs. channel width
(normalized to minimum
width W_{min})

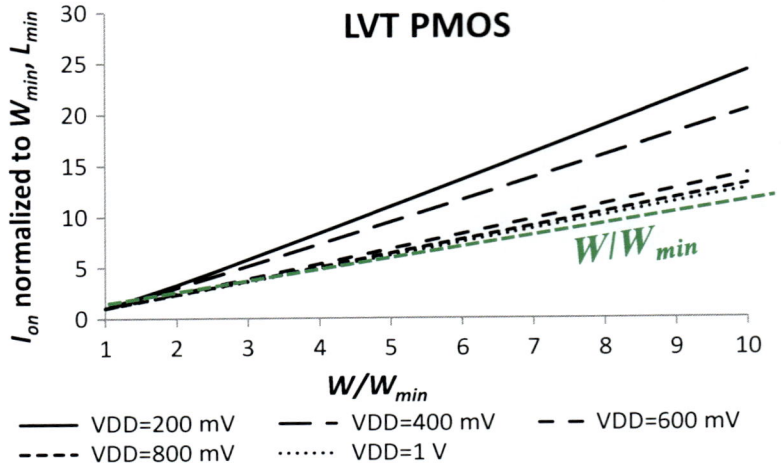

Fig. 4.9 I_{on} normalized to current in minimum-sized transistor vs. channel width (normalized to minimum width W_{min})

$I_{on} \propto W$, denoting that the NCE effect dominates (i.e., wider channels lead to large current than expected due to the simultaneous reduction in V_{TH}). This effect is clearly more pronounced at low voltages due to the stronger dependence of I_{on} on V_{TH}, whereas it is negligible at nominal voltage.

Other technologies might have opposite behavior due to dominant RNCE (i.e., I_{on} increases slower than W, due to the progressive increase in V_{TH} due to the increase in V_{TH}). On the other hand, Fig. 4.10 shows that I_{on} decreases faster than $1/L$ at near-threshold voltages, due to the dominance of SCE. Again, this dependence is 1.5–3X stronger than at nominal voltage due to the stronger dependence of I_{on} on V_{TH} at low voltages. Other technologies might have different behavior, due to the dominance of RSCE.

4.2.3 PMOS/NMOS Strength Ratio, Stacking and Wire Delay

Another important effect observed at near-threshold voltages is the deviation of the ratio of the PMOS and NMOS strength (i.e., I_{on}) at iso-size, compared to nominal voltage. This is due to the different dependence of PMOS and NMOS I_{on} across different voltages. Indeed, from (4.3)–(4.6) the transistor strength has a

mild dependence on V_{TH} and is mostly defined by the carrier mobility at nominal voltage. Hence, differences in V_{TH} between PMOS and NMOS do not significantly impact the strength. On the other hand, the strength has a strong dependence on V_{TH} at near-threshold (and lower) voltages, hence even moderate differences in V_{TH} between PMOS and NMOS substantially alter their strength ratio. The latter can be smaller or larger than the value at nominal voltage, depending on the V_{TH} differences between PMOS and NMOS (including DIBL), and hence the specific technology. Figure 4.11 shows the trend of the PMOS/NMOS strength ratio in a specific 28-nm technology, which at near-threshold voltages can be reduced by up to 2.5X compared to nominal voltages. At lower voltages, the impact is even larger, due to the exponential dependence of I_{on} on V_{TH} in (4.4). This deviation of the PMOS/NMOS strength ratio clearly threatens the noise margin of CMOS logic gates, thus degrading robustness and exposing logic gates to malfunctions due to variations. This also emphasizes the imbalance between the rise and fall delay, thus degrading performance.

Analogously, the strength of stacked transistors (i.e., connected in series) can heavily deviate from the strength of a single transistor, compared to operation at nominal voltage.

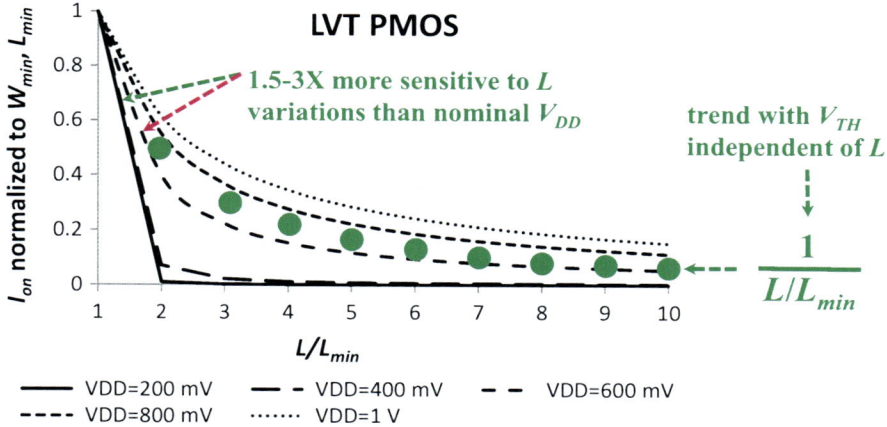

Fig. 4.10 I_{on} normalized to current in minimum-sized transistor vs. channel length (normalized to minimum length L_{min})

Fig. 4.11 Ratio between the strength of PMOS and NMOS versus supply voltage V_{DD} (LVT transistors)

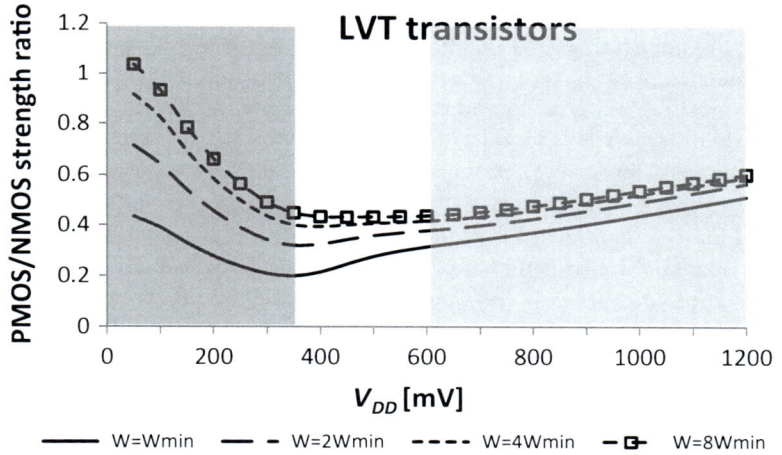

This can be shown by the I_{on} stacking factor X_{on}, defined as the factor by which the I_{on} current is reduced due to the transistor stacking, compared to a single transistor (assuming all transistors have the same size as the single one). The trend in Fig. 4.12 shows that the stacking factor tends to peak around near-threshold voltages, and the phenomenon is more evident under a larger number of stacked transistors. At lower (sub-threshold) voltages, the stacking factor goes back to smaller and threshold-voltage independent value (Alioto 2012).

The stacking factor peaking at near-threshold voltages can be observed in any CMOS technology, as the presence of stacked transistors reduces the drain-source voltage of each stacked transistor, and hence leads to a further increase in V_{TH} (and decrease in the strength) due to DIBL, compared to a single transistor. This explains why the near-threshold current delivered by four stacked transistors is up to 7X lower than a single transistor at iso-size, although this factor is about half of it at nominal voltage. Due to the same reason, the degradation

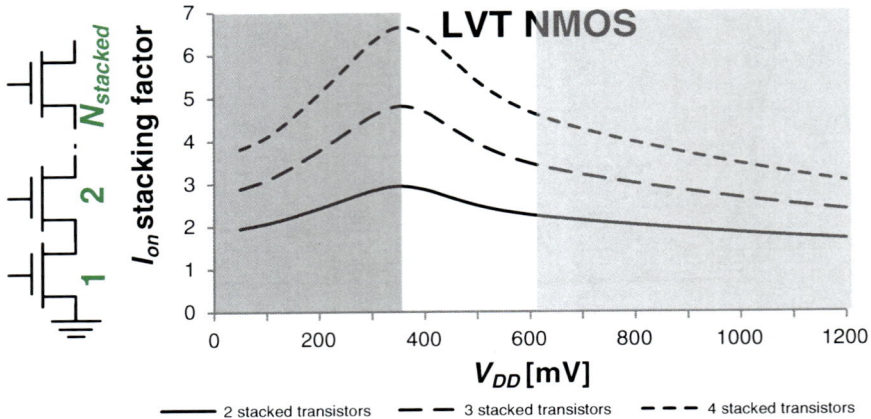

Fig. 4.12 Ratio between the strength of PMOS and NMOS versus supply voltage V_{DD} (LVT transistors)

for two and three stacked transistors is much less pronounced, and is acceptable from a performance point of view. Hence, as general circuit design guideline, the maximum number of stacked transistors in near-threshold designs needs to be lower (e.g., 3) than at nominal voltage (typically up to four).

Finally, another fundamental difference encountered in near-threshold designs is the deviation of the ratio between the gate and wire delay, compared to nominal voltage. Indeed, at lower voltages, the gate delay increases as in (4.6), whereas the wire delay remains constant. As an example, Fig. 4.13 considers a wire whose delay matches the delay of a single gate designed for high performance (i.e., its delay is about one $FO4$ (Sutherland et al. 1999)) at nominal V_{DD}. This corresponds to a global wire with a length in the order of a millimeter. From this figure, the wire delay at near-threshold voltages represents only a small fraction (in the order of 10X smaller) of the gate delay. This means that above-threshold designs and architectures that aims at mitigating the impact of wire delay are definitely overdesigned and performance/energy sub-optimal at near-threshold voltages. Hence, near-threshold circuits and architectures need to be very different from traditional above-threshold solutions, due to drastically smaller

impact of wire delay. This brings back circuit and architectural solutions that were abandoned in the late 90s, due to the then incumbent impact of wires on the system performance.

In summary, the large performance sensitivity on V_{DD} and V_{TH} represents a very interesting opportunity in near-threshold circuits, but also poses various challenges. Among those, it is not possible to maintain a fixed delay ratio between cells with different amount of stacking and threshold voltage, when scaling the voltage (even without variations). This, in addition to the substantial performance degradation due to stacked transistors, suggests that the maximum fan-in of near-threshold CMOS standard cell should be three. Within the same cell, it is not possible to maintain a stable PMOS/NMOS strength ratio across different voltages. For the same reasons, ratioed and dynamic logic styles are unfeasible at near-threshold voltages (not to mention the larger impact of leakage and variations, as discussed in Sects. 4.3 and 4.6). Similarly, topologies that are inherently based on current contention and positive feedback need to be definitely avoided (e.g., cross-coupled non-clocked inverters in flip-flops). Unfortunately, this cannot be avoided in SRAM bitcells and register files for reasons due to density, and other sophisticated techniques need to be deployed (see Chap. 5).

Fig. 4.13 Ratio of wire and gate delay normalized to value at nominal voltage versus supply voltage V_{DD} (28 nm, LVT transistors)

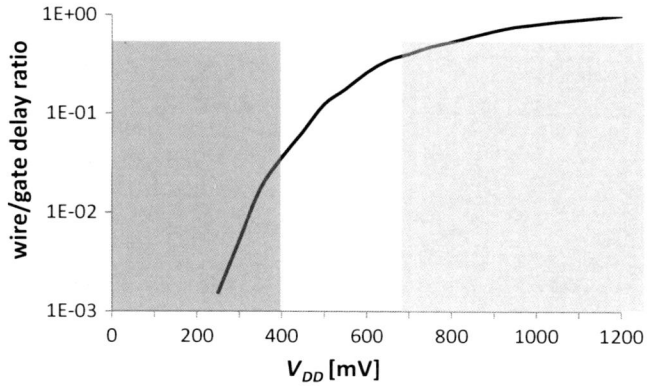

4.2.4 Knobs to Adjust Transistor Strength

From the previous subsection, the transistor strength can be adjusted with the following knobs:

- transistor size
- body biasing
- V_{TH} selection
- V_{DD} tuning and fine-grain boosting.

From the previous subsection, transistor sizing is relatively effective, and can be more or less effective than at nominal voltage, depending on the dominance of RNCE over NCE, and SCE over RSCE. Body biasing can significantly alter the transistor strength only in old technologies (e.g., 90 nm), or in recent FDSOI technologies, with a typical 30% range of adjustment at near-threshold voltages.

In view of the strong dependence of I_{on} on the gate overdrive discussed in Sect. 4.1, the transistor strength can be substantially modified through the proper selection of the threshold voltage, and the fine-grain boosting of V_{DD} to selectively increase I_{on} where required. Regarding the V_{TH} selection, a 2–4X I_{on} (and delay) change was previously shown to be feasible when changing V_{TH} from one type (e.g., RVT) to the next available one (e.g., LVT). However, V_{TH} selection at near-threshold voltages poses various additional challenges, compared to

operation at nominal voltage. Indeed, the sensitivity of I_{on} to V_{TH} translates into a strong sensitivity to its process variations. Also, the delay ratio of an RVT and LVT logic gate (see Fig. 4.14. for an inverter gate) strongly depends on the supply voltage. In other words, mixing standard cells with different V_{TH} poses the problem of having different delay scaling in different portions of the system. In turn, this makes timing closure certainly more difficult and might reduce the energy benefit of dynamic voltage scaling, as the critical path(s) depends on the voltage.

Let us now consider fine-grain voltage boosting, which consists in selectively over-driving appropriate transistors with a voltage above V_{DD}. As shown in the illustrative example in Fig. 4.15, this might be the case of a single large transistor M1 (e.g., sleep transistor, large buffer) that drives a sub-circuit containing several smaller transistors. Let us assume that the gate of M1 is overdriven at $V_{DD} + \Delta V_{DD}$ as opposed to all other transistors and logic gates, which are powered at V_{DD}. Due to the strong (super-linear) I_{on} increase in M1 due to the gate voltage boosting by ΔV_{DD}, the transistor can be significantly undersized while maintaining the same strength as the transistor that is driven by V_{DD}. In view of the strong dependence of I_{on} on the gate voltage in M1 at near-threshold voltages, a small amount of boosting ΔV_{DD} permits to substantially reduce the area occupied by M1. This is shown in the example in Fig. 4.15 in 28 nm, where the area of M1 can be reduced by

Fig. 4.14 Ratio of I_{on} of
LVT and RVT transistors
normalized to value at
nominal voltage versus
supply voltage V_{DD}

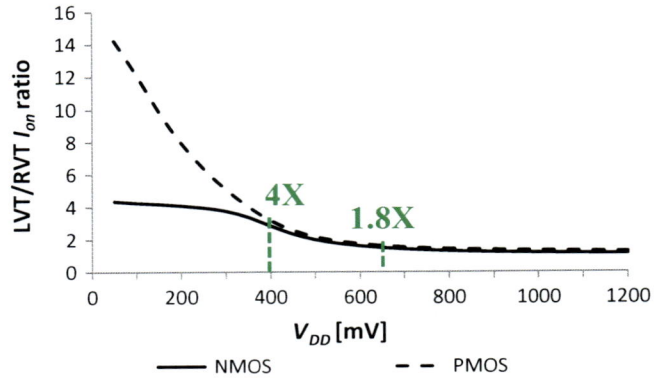

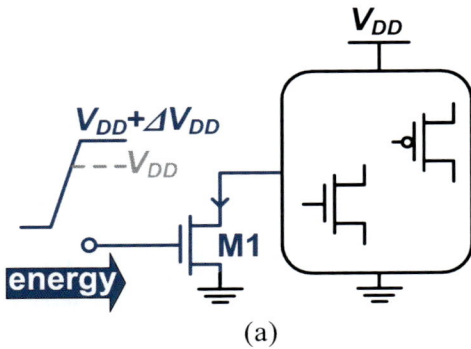

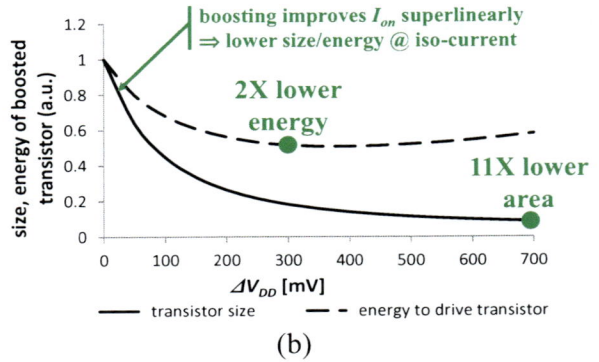

Fig. 4.15 (a) In-principle circuit with large transistor whose gate voltage is boosted by ΔV_{DD}, (b) area and energy
improvement vs. ΔV_{DD}

up to an order of magnitude while maintaining the same strength, through an amount of boosting in the order of a few hundreds of mVs. Similarly, such selective boosting permits to super-linearly reduce the leakage current of M1. At the same time, the gate capacitance $C_{g,M1}$ of M1 is reduced super-linearly as well, whereas the supply voltage is increased by the very limited amount ΔV_{DD}. This means that the dynamic energy $C_{g,M1} \cdot (V_{DD} + \Delta V_{DD})^2$ to switch M1 ON is reduced overall. In the example in Fig. 4.15, a 2X energy reduction can be achieved through selective boosting of M1, when adopting an adopting an optimal ΔV_{DD} of 300 mV (this voltage depends on the specific technology). Very similar energy saving is observed for ΔV_{DD} in the order of 100–200 mV. On the other hand, larger amount of boosting slightly increases the energy consumption to turn on M1, since the transistor starts

operating above threshold (i.e., I_{on} becomes less sensitive to ΔV_{DD}), and the energy cost $C_{g,M1} \cdot (V_{DD} + \Delta V_{DD})^2$ of boosting increases substantially due to the quadratic dependence.

From the above considerations, near-threshold circuits can be made more energy- area-efficient by selectively boosting portions of the circuit that contain large (and hence energy- and area-hungry) transistors. As opposed to traditional multi-V_{DD} approaches that are applied at the module level, in this case the supply is boosted with fine granularity (i.e., down to the single transistor). Such fine-grain voltage boosting also offers the opportunity to equalize imbalanced logic across pipestages. As fundamental challenge, fine-grain boosting entails significant area overhead, due to the additional level shifters to drive boosted-voltage domains, and to the slight additional cost of distributing multiple voltages at

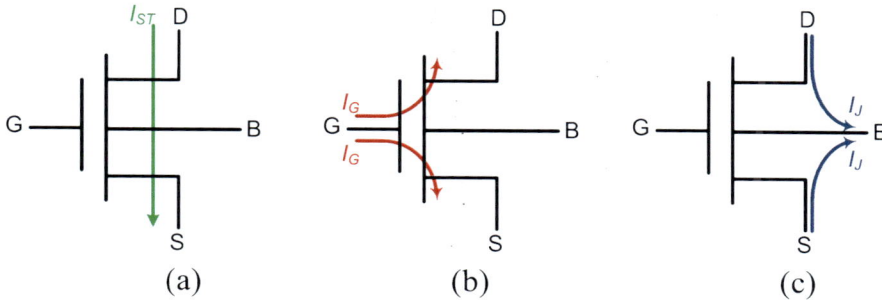

Fig. 4.16 Transistor leakage contributions: (**a**) sub-threshold, (**b**) gate, (**c**) substrate

the physical design level (Flynn et al. 2007). In other words, near-threshold circuits should certainly take advantage of fine-grain voltage boosting, but innovative techniques are needed to minimize the unavoidable overhead. Some recently proposed ideas to address this challenge will be presented in Sect. 4.1.

4.3 Energy Trends

In this section, the impact of voltage scaling on the energy is reviewed, providing models and design guidelines for minimum-energy operation.

4.3.1 Transistor Leakage Current at Near-Threshold Voltages

The MOS transistor leakage contributions are summarized in Fig. 4.16. The dominant contribution is due to the sub-threshold leakage in (4.4), which flows between drain and source and is due to the diffusion of minority carriers between the two terminals (Tsividis 1999). The gate leakage flows from gate to source/drain or vice versa, depending on the applied voltages, and tends to be exponentially smaller than the sub-threshold contribution when lowering V_{DD} (Narendra and Chandrakasan 2006). Similar considerations hold for the substrate leakage, which is mostly due to the Band-to-Band Tunneling (BTBT), and the inverse saturation current of the source-bulk and drain-bulk pn junctions (Narendra and Chandrakasan 2006). Hence, the transistor leakage current at near-

threshold voltages is well approximated by (4.4), where the gate-source voltage (assigned to V_{DD} in (4.4)) has to be set to zero. By substituting the dependence of V_{TH} in (4.7), the near-threshold leakage current of an NMOS transistor immediately results to

$$I_{lkg} = I_0 \cdot e^{-\frac{V_{TH0} - \lambda_{BB} V_{BB}}{n\,kT/q}} \cdot e^{\frac{\lambda_{DIBL} V_{DD}}{n\,kT/q}} \quad (4.8)$$

where the first exponential term is set by the threshold voltage (including body biasing, when applied), and the second expresses the DIBL effect and hence the leakage dependence on V_{DD}. The latter dependence is exponential, and typically operation at near-threshold voltages reduces the leakage current by about an order of magnitude for typical DIBL coefficients in the order of 0.1 V/V, compared to nominal voltage. The consistent exponential trend across voltages in (4.8) is shown in Fig. 4.17 for a 28-nm technology, along with a leakage current reduction at near-threshold voltages by 4–8.5X.

4.3.2 Energy Consumption of Digital Systems at Near-Threshold Voltages

The total energy per operation[5] in a near-threshold VLSI digital system is essentially equal to the sum of the dynamic and the leakage energy. Indeed, the short-circuit energy

---

[5] An operation is here defined as the basic task that the considered system is executing, e.g., an instruction in a CPU or GPU, a new output sample in a DSP, an arithmetic operation in an Arithmetic Logic Unit.

Fig. 4.17 Leakage current I_{off} of LVT transistor (normalized to value at nominal voltage) versus supply voltage V_{DD}

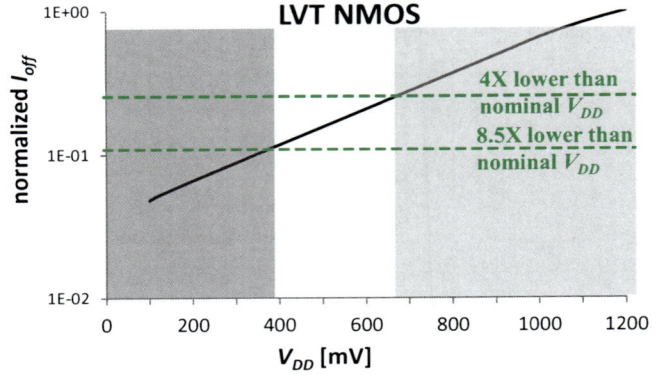

contribution (Weste and Harris 2011) is negligible at near-threshold voltages, as opposed to operation at nominal voltage. This is because the transistors for input voltages around $V_{DD}/2$ is a small sub-threshold current, which also rapidly vanishes when the input voltage deviates from $V_{DD}/2$ to settle to its stable value (Alioto 2012) (due to the exponential I–V characteristics in the sub-threshold region).

The dynamic energy per operation is given by

$$E_{dyn} = \alpha_{SW} \cdot C \cdot V_{DD}^2 \cdot CPO \qquad (4.9)$$

where $\alpha_{SW} \cdot C \cdot V_{DD}^2$ is the energy per cycle, being C the total capacitance within the circuit, α_{SW} is the activity factor (Weste and Harris 2011) (i.e., the fraction of C that is switched in a cycle, on average). In (4.9), it was considered that an operation in general takes an average number of cycles CPO (Cycles per Operation), which depends on the specific (micro)architecture, and the dataset to a minor extent (e.g., in microprocessors).

The leakage energy per operation can be expressed as the product of the average leakage power $V_{DD} \cdot I_{off}$ (being I_{off} the average leakage current), the clock cycle T_{CK} and CPO (Alioto 2012). T_{CK} can be expressed as $FO4 \cdot LD_{eff}$, where $LD_{eff} = T_{CK}/FO4$ is the number of the

number of $FO4$ delays (i.e., cascaded inverters with fan-out of 4) that can fit the cycle time. Hence, LD_{eff} represents the effective logic depth per pipestage, which is a constant defined by the (micro)architecture.[6] Hence the leakage energy per operation can be written as $E_{lkg} = V_{DD} \cdot I_{off} \cdot T_{CK} \cdot CPO$, or equivalently

$$E_{lkg} = V_{DD} \cdot I_{off} \cdot FO4 \cdot LD_{eff} \cdot CPO. \qquad (4.10)$$

In (4.10), the only parameters that depend on V_{DD} are V_{DD} itself, I_{off} and $FO4$. When downscaling the voltage, the first term decreases linearly and I_{off} decreases exponentially due to DIBL, although not very rapidly since V_{DD} is multiplied by $\lambda_{DIBL} \ll 1$ in (4.8). On the other hand, $FO4$ rapidly increases as in (4.6) when V_{DD} is reduced down to near-threshold voltages and below. The overall effect of the three factors leads to an increase in E_{lkg} at near- and sub-threshold voltages when V_{DD} is reduced, as opposed to the dynamic energy. The leakage energy tends to increase very rapidly when decreasing V_{DD} down to the transistor threshold

---

[6]$T_{CK}/FO4$ is essentially constant in gate-delay dominated critical paths when varying V_{DD}, as all gate delays generally scale like $FO4$ (Harris et al. n.d.; Weste and Harris 2011). At near-threshold voltages, this assumption is generally correct, as the wire delay is typically much smaller (see Sect. 5.2.3).

Fig. 4.18 Leakage energy vs. supply voltage V_{DD} for different logic depths LD_{eff} equal to $25FO4$ and $50FO4$ (28 nm, LVT and RVT transistors)

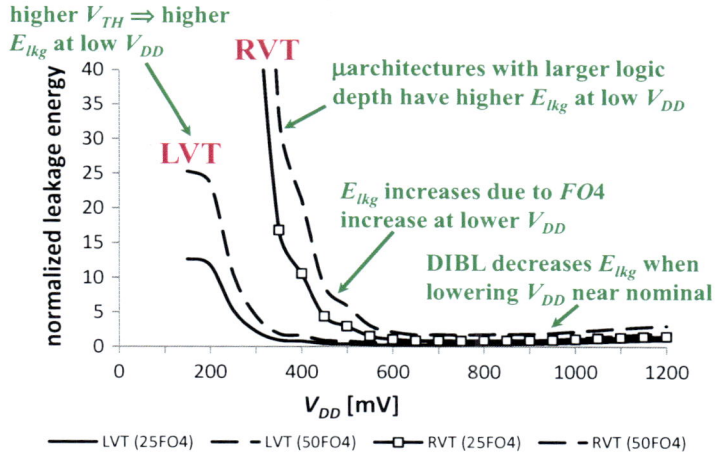

voltage, due to the resulting rapid increase in the gate delay in (4.6). This is shown in Fig. 4.18, where the leakage energy of the reference digital circuit in Sect. 4.4 is plotted versus V_{DD} under RVT and LVT transistor flavor. As expected, the leakage energy under RVT flavor rapidly increases at higher voltages compared to LVT, due to the higher threshold voltage. This figure also shows that E_{lkg} tends to shoot up at larger voltages, under microarchitectures with larger logic depth (e.g., $LD_{eff} = 50$ instead of 25). This is because such microarchitectures suffer from larger leakage energy from (4.10), and hence the rapid increase can be observed at larger voltages. On a side note, Fig. 4.18 also shows that E_{lkg} has an opposite behavior at above-threshold voltages (i.e., it decreases when decreasing V_{DD}), due to the dominance of the exponential effect of DIBL over the linear $FO4$ increase.

From (4.9)–(4.10), the total energy per operation E_{TOT} of a given VLSI system or sub-system results to

$$E_{TOT} = E_{dyn} + E_{lkg} = E_{cycle} \cdot CPO \quad (4.11)$$

where the energy per cycle $E_{cycle} = E_{TOT}/CPO$ is defined as

$$E_{cycle} = \alpha_{SW} \cdot C \cdot V_{DD}^2 + V_{DD} \cdot I_{off} \\ \cdot FO4 \cdot LD_{eff} \quad (4.12)$$

The qualitative trend of (4.11)–(4.12) versus V_{DD} in Fig. 4.19 shows that the voltage down-scaling reduces the dynamic energy, but increases the leakage energy. Hence, a minimum-energy point (MEP) is observed at a voltage $V_{DD,opt}$ that optimally balances the dynamic and leakage energy, thus leading to the minimum energy[7] E_{min}. The MEP voltage $V_{DD,opt}$ typically lies in the sub-threshold or near-threshold region (Hanson et al. 2006a; Hanson et al. 2006b), as discussed in the next section. Due to the flatness of the MEP, near-threshold operation permit true- or nearly-minimum energy operation, as fundamental design target of this chapter.

Figure 4.20a–d shows the energy trend and the presence of the MEP in various integrated prototypes, including an FFT core from MIT (Wang and Chandrakasan 2005), an 8-bit microprocessor from Umich (Hanson et al. 2008), an IA-32 processor from Intel (Jain et al. 2012), and an AES core from NUS (Zhao et al. 2015). From these figures, the energy curve is relatively flat around the MEP, hence the minimum- or nearly-minimum energy per operation does not require a stringent precision in the generation of the supply voltage. In practical designs, a change in V_{DD}

---

[7] Since the energy per operation E_{TOT} in (5.11) is proportional to E_{cycle}, in the following we will simply refer to the energy per cycle in (5.12), unless otherwise specified. All considerations are immediately extended to E_{TOT} by simply multiplying E_{cycle} by CPO.

Fig. 4.19 Qualitative trend of dynamic, leakage and total energy per cycle E_{cycle} (or equivalently total energy per operation E_{TOT}) vs. supply voltage V_{DD}

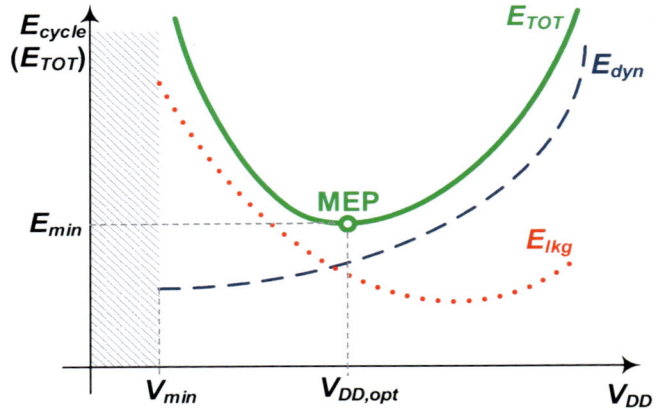

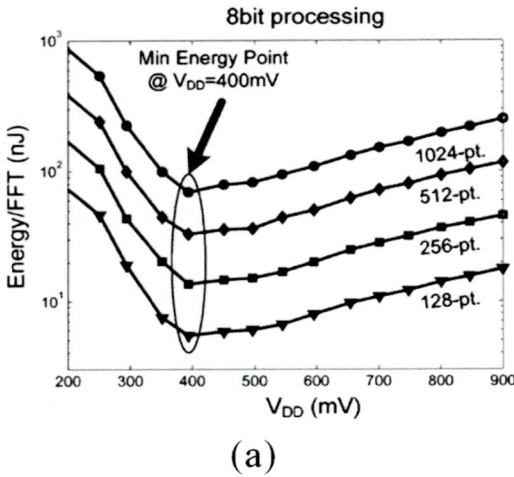

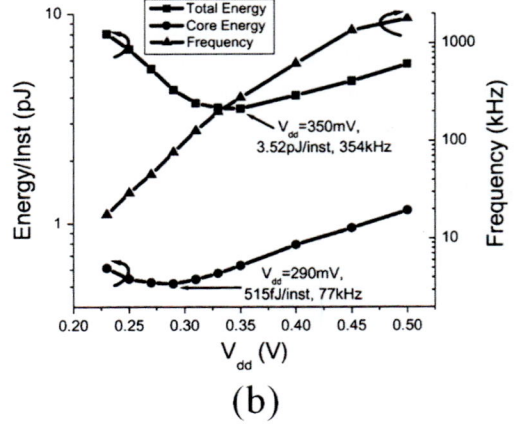

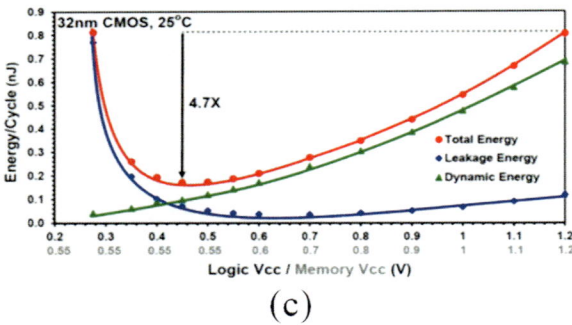

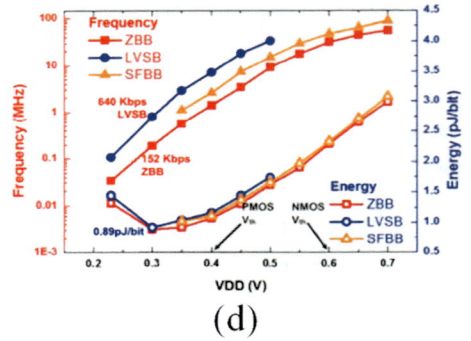

(a)

(b)

(c)

(d)

Fig. 4.20 Energy vs. V_{DD} and minimum-energy point in (a) FFT core ((Wang and Chandrakasan 2005) from MIT), (b) 8-bit microprocessor ((Hanson et al. 2008) from Umich), (c) IA-32 processor ((Jain et al. 2012) from Intel), (d) AES core ((Zhao et al. 2015) from NUS)

around the optimal voltage $V_{DD,opt}$ by various tens of mVs (e.g., 30-50 mV) keeps the energy very close to E_{min} (e.g., within a few percentage points). The MEP voltage $V_{DD,opt}$ in the above examples covers the typical range encountered in real designs (300–450 mV). The detailed analysis on the dependence of the MEP position on process and design parameters is presented in the next subsection.

Let us observe that the leakage energy increase at low voltages limits the energy reductions enabled by aggressive voltage scaling, compared to the quadratic reduction that would be achievable if the total energy were dominated by E_{dyn}. Indeed, in the latter case the minimum achievable energy would be given by (4.9) with V_{DD} equal to the minimum operating voltage V_{min} that ensure correct operation, as in Fig. 4.19. The related energy saving compared to nominal voltage is reported in Table 4.2, which represents an upper bound of the energy savings achievable for quick estimates. Observe that the potential energy savings in circuits with wire-dominated load are lower than the case of gate-dominated load. Indeed, in the former case the dynamic energy reduction is purely quadratic, whereas the latter also benefits from the simultaneous load reduction due to the reduction in the transistor gate capacitance at low V_{DD} (see Sect. 4.2.1 and Fig. 4.4). From this table, operation at near-threshold voltages potentially reduces the energy by up to an order of magnitude, compared to the nominal voltage. At the same time, the presence of leakage narrows down the range of voltages at which energy reduction is truly allowed.

As even more crucial observation, the leakage energy is a substantially larger fraction of the overall energy budget at near-threshold voltages, compared to nominal voltage. Indeed, E_{lkg} (E_{dyn}) at near-threshold voltages is larger (smaller) than at nominal V_{DD}. Table 4.3 shows the detailed energy breakdown measured in the microprocessor in (Jain et al. 2012), which includes a level-1 cache. Above threshold, the leakage energy is 14% and is well in line with the expectations at nominal voltage. In near-threshold region, the leakage energy raises to a much larger 42% as expected, and in sub-threshold region it completely dominates the overall energy.

For all the above reasons, mitigating the leakage energy is a crucial goal of near-threshold designs, and is far more important than traditional low-power above-threshold designs. In addition, Sect. 4.4 will show that traditional low-power techniques to mitigate leakage are rather ineffective when V_{DD} is pushed down to near threshold.

4.3.3 Trans-Regional Energy Model

From the previous subsection, the MEP is set by the optimal balance between dynamic and leakage energy. In other words, the MEP voltage $V_{DD,opt}$ and the resulting minimum energy E_{min} both depend on the ratio between leakage and total energy. This means that the MEP position in the energy-voltage plane changes according to this ratio, as discussed below.

When the leakage energy significantly increases for some reason, whereas the dynamic energy remains constant, the total energy clearly increases and $V_{DD,opt}$ increases as well (i.e., the MEP moves to the right, and upwards, as summarized in Fig. 4.21a). Indeed, in this case

Table 4.2 Dynamic energy reduction vs. VDD

V_{DD} (mV)	V_{DD}^2 energy saving (load = wire only)	$C_g \cdot V_{DD}^2$ energy saving (load = transistor only)
200 mV	36X	54X
400 mV	9X	11.6X
600 mV	4X	4.4X
800 mV	2.2X	2.4X
1 V	1.4X	1.4X
1.2 V	1X	1X

Table 4.3 Measured energy breakdown in Jain et al. (2012)

	V_{DD} (V)		E_{lkg}(%)			E_{dyn}(%)		
	Logic	L1C	Logic	L1C	Total	Logic	L1C	Total
Sub threshold (V_{min})	0.28	0.55	62%	33%	**95%**	4%	1%	5%
Near threshold (MEP, $V_{DD,opt}$)	0.45	0.55	27%	15%	**42%**	53%	5%	58%
Above threshold (nominal V_{DD})	1.2	1.2	11%	3%	**14%**	81%	5%	86%

logic = Core, L1C = L1 Cache

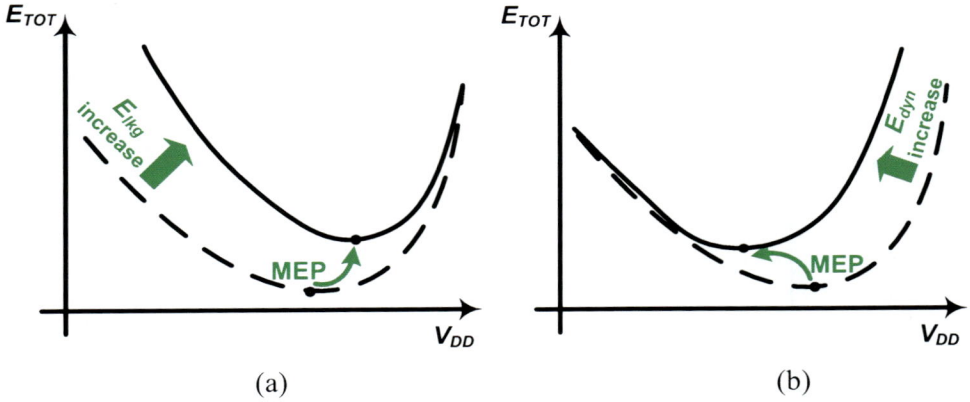

(a)   (b)

Fig. 4.21 Qualitative description of how the minimum-energy point (MEP) changes position when (**a**) E_{lkg} increases (E_{dyn} kept constant), (**b**) E_{dyn} increases (E_{lkg} kept constant)

the leakage energy tends to be a larger fraction of E_{TOT}, hence V_{DD} needs to be increased to reduce E_{lkg} (as explained by Fig. 4.19). On the other hand, when the dynamic energy increases at iso-leakage energy, E_{dyn} becomes a larger fraction of E_{TOT}, hence it becomes more important to reduce E_{dyn} and hence $V_{DD,opt}$ decreases (i.e., the MEP moves to the left, and upwards, as summarized in Fig. 4.21a). From the above considerations, the MEP tends to move to the right when the temperature is increased and/or the circuit activity is reduced, due to a different input data profile or power mode (i.e., different

modules are activated). Since the input dataset and the temperature are time varying and are unpredictable at design time, a feedback scheme that tracks the actual MEP through energy sensing (or estimation) and adjusts the supply voltage accordingly (Ramadass and Chandrakasan 2008).

To have a more quantitative understanding of the dependence of the MEP position on process, design and environmental parameters, let us consider an analytical model of the energy. In detail, eq. (4.12) can be written in the following more useful form[8]:

$$
\begin{aligned}
E_{cycle} &= \alpha_{SW} \cdot C_{TOT} \cdot V_{DD}^2 \left[1 + LD_{eff} \cdot \frac{I_{off,TOT}}{\alpha_{SW} \cdot C_{TOT} \cdot V_{DD}} FO4 \right] \\
&\approx \alpha_{SW} \cdot C_{TOT} \cdot V_{DD}^2 \left[1 + LD_{eff} \cdot \frac{\left(gatecount \cdot \overline{I_{off}} \right)}{\alpha_{SW} \cdot \left(gatecount \cdot \overline{C_{cell}} \right)} \cdot \frac{5C_{in,min}}{2I_{on,min}} \right] \\
&\approx \alpha_{SW} \cdot C_{TOT} \cdot V_{DD}^2 \left[1 + 2.5 \cdot LD_{eff} \cdot \frac{I_{0,min} \cdot e^{-\frac{V_{TH}}{n\frac{kT}{q}}} \cdot \frac{strength}{X_{stack,off}}}{\alpha_{SW} \cdot \frac{C_{cell}}{C_{in,min}}} \cdot \frac{1}{I_{0,min} \ln \left(e^{\frac{V_{DD}-V_{TH}}{n\cdot(kT/q)}} + 1 \right)} \right] \\
&= \alpha_{SW} \cdot C_{TOT} \cdot V_{DD}^2 \left[1 + ILDR \cdot e^{-\frac{V_{DD} \left(\frac{1+\alpha_{X_{off}}}{n\frac{kT}{q}} \right)}} \cdot f_{ILDR}(V_{DD}) \right] \quad (4.13)
\end{aligned}
$$

where (4.7) and (4.26) were used, and the intrinsic leakage-dynamic energy ratio $ILDR$ (i.e., the contribution of E_{lkg}/E_{dyn} that is independent of V_{DD}) was defined as

---

[8] The following analysis is inspired by Hanson et al. (2006a), Hanson et al. (2006b), Bo et al. (2004) and generalizes the results to arbitrary designs, instead of being valid only for simple cascaded inverters.

$$ILDR = 2.5 \cdot LD_{eff} \cdot \frac{\overline{strength}}{X_{stack,off}\big|_{V_{DD,nom}} \cdot e^{-\alpha_{X_{off}} \cdot \frac{V_{DD,nom}}{n \cdot kT/q}}} \cdot \frac{1}{\alpha_{SW} \cdot \frac{\overline{C_{cell}}}{C_{in,min}}} \tag{4.14a}$$

and $f_{ILDR}(V_{DD})$ was defined as

$$f_{ILDR}(V_{DD}) = \frac{e^{\frac{V_{DD}(1+\lambda_{DIBL})-V_{TH0}}{n\frac{kT}{q}}}}{\ln\left(e^{\frac{V_{DD}(1+\lambda_{DIBL})-V_{TH0}}{n\cdot(kT/q)}}+1\right)} \tag{4.14b}$$

Also, in (4.13) it was observed that

- the total leakage current $I_{off,TOT}$ of the design under consideration is equal to the gate count (*gatecount*) multiplied by the average leakage per standard cell $\overline{I_{off}}$
- $\overline{I_{off}}$ can be expressed as the leakage current of a minimum-sized inverter $I_{0,min} \cdot e^{-V_{TH}/n\frac{kT}{q}}$, multiplied by the average cell strength $\overline{strength}$ (where 1X refers to the minimum-sized inverter) and divided by the average off-stacking factor $X_{stack,off}$ (i.e., the factor by which the leakage current is reduced due to transistor stacking)
- the total capacitance C_{TOT} is equal to the gate count multiplied by the average switched capacitance per standard cell $\overline{C_{cell}}$
- FO4 can be thought of as the delay of a minimum-sized inverter driving four equal inverters, and hence its total load capacitance is $4C_{in,min}$ ($C_{in,min}$ is the input capacitance of a minimum-sized inverter) plus its parasitic capacitance, which is approximately equal to $C_{in,min}$ (Sutherland et al. 1999); $I_{on,min}$ is the current delivered by such minimum-sized inverter (see (4.4)), and $I_{0,min}$ is the I_0 parameter in (4.1) and (4.2) pertaining to the same inverter.

In (4.13), the average off-stacking factor $X_{stack,off}$ is evaluated at the nominal voltage $V_{DD,nom}$ of the adopted technology, and its downscaling at low voltages is accounted for by the technology-dependent parameter $\alpha_{X_{off}} \ll 1$ (see (4.26)). Eq. (4.13) is strongly affected by ILDR. From (4.13), the latter parameter represents the voltage-independent (i.e., intrinsic) contribution of the ratio between leakage and dynamic energy, and is defined by

- the architecture through LD_{eff} (i.e., faster designs with low LD_{eff} exhibit lower ILDR)
- the function implemented by the design under consideration, which in turn sets the standard cell usage statistics (i.e., the average off-stacking factor $X_{stack,off}$) and the average fan-out $\overline{C_{cell}}/C_{in,min}$ (i.e., the average equivalent number of minimum-sized inverters that load the cells); such dependence tends to be fairly weak, due to the averaging effect across cells in large designs
- the performance target in the automated cell sizing phase, as $\overline{strength}$ and the average fan-out $\overline{C_{cell}}/C_{in,min}$ both tend to be larger for tighter timing constraints (again, faster designs have lower ILDR); accordingly, such dependence tends to be fairly weak as well
- the input dataset, which in turn sets the circuit activity (i.e., α_{SW}).

In summary, parameter ILDR is mainly set by the architecture and the input statistics, and low values of ILDR are associated with more active and faster designs. In practical cases, ILDR ranges from a few hundreds for rather fast and active designs with heavy dynamic energy, to a several tens of thousands in very slow and inactive circuits. Higher values are observed only when an additional constant power contribution comes from external blocks.

Table 4.4 Numerical Examples for *ILDR* in 28 nm ($V_{DD,nom}$ = 1.2 V)

Design	LD_{eff}	α_{SW}	$\frac{\overline{C_{cell}}}{C_{in,min}}$	*strength*	*ILDR*	$V_{DD,opt}$	$\left.\frac{E_{lk}}{E_{dyn}}\right|_{MEP}$
High performance, active	20	15%	20	6	110	180 mV	0.62
Low performance, little active	100	3%	6	3	3100	320 mV	0.27

Observe that *ILDR* is defined at nominal voltage and all parameters not explicitly related to $V_{DD,nom}$ are essentially independent of the voltage,[9] and hence can be evaluated from the report of synthesis/place&route at such voltage without requiring the full characterization of the library at different voltages. A few numerical examples in 28 nm are reported in Table 4.4, assuming $\lambda_{DIBL} = 0.1$ (i.e., $\alpha_{X_{off}} = 0.098$ from (4.26)), $X_{stack,off}\big|_{V_{DD,nom}}$ equal to 20 (i.e., average of two stacked transistors in this technology) at nominal voltage. From the technology scaling viewpoint, k_0 tends to slightly decrease at finer technologies, due to stronger DIBL and hence larger $X_{stack,off}\big|_{V_{DD,nom}}$. As a simpler approach, *ILDR* can also be estimated as the value that makes the ratio E_{lkg}/E_{dyn} (i.e., $ILDR \cdot e^{-V_{DD}\left(1+\alpha_{X_{off}}\right)/n\frac{kT}{q}}$ $\cdot f_{ILDR}(V_{DD})$ in (4.13)) equal to the value that is obtained from power analysis at RTL level.

4.3.4 Considerations on the MEP Voltage

Typically, the MEP mostly lies in the deep sub-threshold region (Hanson et al. 2006a), and sometimes near-threshold (Hanson et al. 2006b). In the former case, $f(V_{DD}) \approx 1$ in (4.13) since $V_{DD} < V_{TH0}$ in (4.14b), hence the energy is independent of the transistor threshold voltage. This is because the latter affects both the leakage and the on-current in the same way (i.e., both are proportional to $\exp\left(-V_{TH}/n \cdot \frac{kT}{q}\right)$ in sub-threshold). In this case, V_{TH} is chosen

exclusively based on the performance requirement (i.e., targeted *FO4*), according to (4.4).

Let us now analyze the optimum voltage $V_{DD,opt}$ that minimizes the energy in (4.13) assuming the MEP to be in sub-threshold. Although a closed-form solution cannot be found, a good approximation for a single-V_{TH} design is logarithmic (similar to (Bo et al. 2004; Hanson et al. 2006a, b))

$$V_{DD,opt} \approx \frac{n}{1+\alpha_{X_{off}}} \frac{kT}{q} \left[1.25\ln(ILDR) - 0.5\right]$$

(4.15)

which has a maximum error of 4% for practical values of *ILDR*, as plotted in Fig. 4.22a under the above 28-nm parameters and $V_{TH0} = 0.35$ V. From this figure, the MEP voltage logarithmically increases with the constant slope in (4.15), which is independent of the adopted threshold voltage, as shown in Fig. 4.22b.

As expected from the above considerations and Fig. 4.21, the MEP moves to the right for slow and less active circuits, and to the left for circuits with dominating dynamic energy and low logic depth. Observe that operation at $V_{DD} < V_{min}$ severely degrades the die yield, hence minimum-energy designs need to adopt a supply voltage equal to the minimum between $V_{DD,opt}$ in (4.15) and V_{min}. From Fig. 4.22a, b, this means that fast (i.e., with low LD_{eff}) designs with *ILDR* lower than a few thousands cannot really achieve true-minimum energy, due to voltage scaling limitations imposed by robustness issues.

The above analysis assumed that the MEP lies in the deep sub-threshold region, which is correct as long as $V_{DD,opt}$ in (4.13) is lower than $V_{TH} - 50$ mV (see Fig. 4.1), i.e., when $ILDR < e^{\frac{V_{TH}-50\ mV}{1.5n\frac{kT}{q}}+1.6}$. The latter boundary value for *ILDR* in 28 nm is typically in the order of 1,000–2,000 for $V_{TH} = 350$ mV (including

---

[9] For example, the average fan-out $\overline{C_{cell}}/C_{in,min}$ is independent of V_{DD} since the wire capacitance is constant, and the transistor gate capacitance does not change substantially (see Fig. 5.4). Similarly, the logic depth, the activity and the average strength do not depend on V_{DD}.

Fig. 4.22 MEP voltage vs *ILDR* for 28-nm technology with (**a**) V_{TH0} = 0.35 *V* and detailed analytical model, (**b**) V_{TH0} = 0.35 *V*, 0.45 *V*, 0.55 *V*, 0.65 *V* (exact solution via numerical minimization of (4.13), approximate expression as in (4.15)–(4.17))

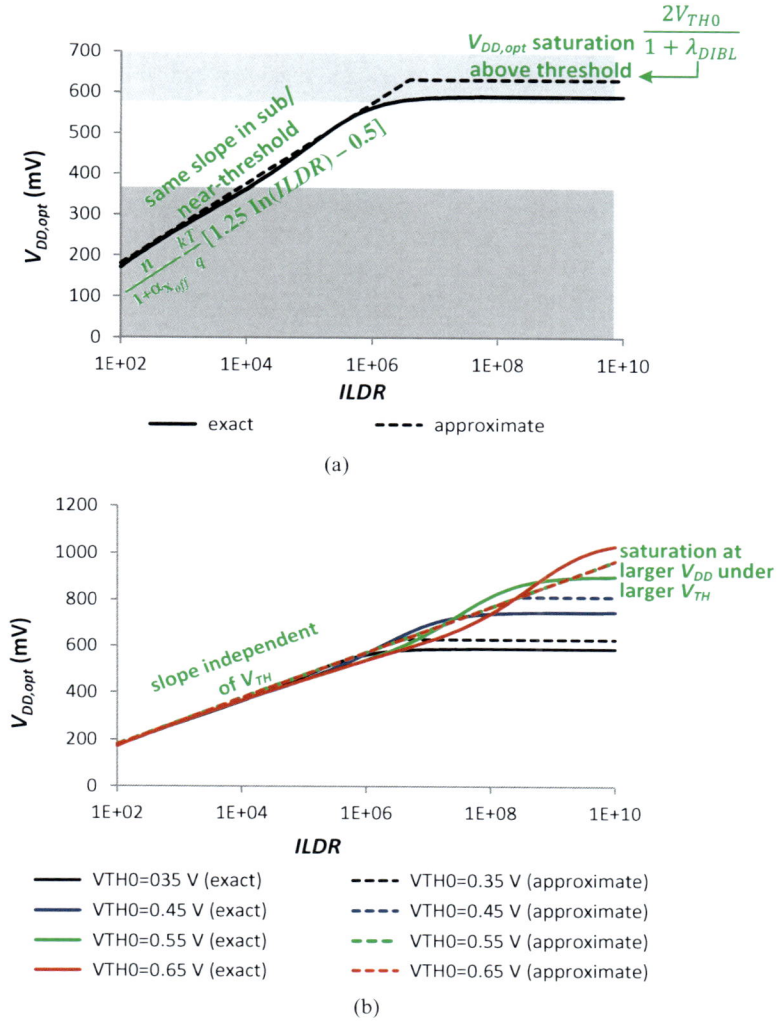

(a)

(b)

DIBL), 4000–5000 for $V_{TH} = 400$ mV, and 10,000 for $V_{TH} = 450$ mV. Slightly larger values are typically found in older technologies, due to the lower subthreshold factor n.

Interestingly, Fig. 4.22a, b show that (4.15) can be extended to the near-threshold region (i.e., larger *ILDR*), as it still predicts the MEP voltage with good accuracy. Hence, the MEP voltage is again independent of the threshold voltage, even in the near-threshold region. For even larger values of *ILDR* such that $V_{DD} > V_{TH0} + 200$ mV (see Fig. 4.1), the MEP moves to the above-threshold region and eventually saturates to a

value $V_{DD,opt}|_{ILDR \to \infty}$. Indeed, $f(V_{DD})$ in (4.14b) becomes approximately equal to $e^{\frac{V_{DD}(1+\lambda_{DIBL})-V_{TH0}}{n\frac{kT}{q}}} \cdot \left(\frac{V_{DD}(1+\lambda_{DIBL})-V_{TH0}}{n \cdot (kT/q)} \right)$, and the resulting $V_{DD,opt}$ is found by minimizing (4.13) for *ILDR* $\to \infty$:

$$V_{DD,opt}|_{ILDR \to \infty} = \frac{2V_{TH0}}{1 + \lambda_{DIBL}} \qquad (4.16)$$

which was found to be always within 12% of the exact solution that minimizes (4.13) in 28 nm (and typically within 5%). $V_{DD,opt}$ saturates because E_{lkg} increases when increasing V_{DD} in the above-threshold region, as opposed to sub- and near-threshold (see Fig. 4.18 and related discussion). In other words, it does not make sense to increase V_{DD} beyond (4.16) from an energy viewpoint, as this would surely increase both dynamic and leakage energy, and hence the total energy. Indeed, (4.16) represents the

voltage at which E_{lkg} is minimum (i.e., such that $V_{DD} \cdot I_{off} \cdot FO4$ is minimum, from (4.12)).

In summary, the above considerations suggest that $V_{DD,opt}$ can be simply modeled by extending (4.15) to the near-threshold and part of the above-threshold region, and limiting it to its asymptotic maximum value in (4.16):

$$V_{DD,opt} \approx \min\left(\frac{n}{1 + \alpha_{X_{off}}}\frac{kT}{q}[1.25\ln(ILDR) - 0.5], \ \frac{2V_{TH0}}{1 + \lambda_{DIBL}}\right) \qquad (4.17)$$

which has a typical (maximum) error of 4% (7%) across the very wide range of *ILDR* in Fig. 4.22a, b. Eq. (4.17) is a useful tool to estimate the MEP position by knowing the type of design (i.e., *ILDR*), and a few other technology-dependent parameters. From (4.17), the transistor threshold choice affects only the value of *ILDR* and the voltage at which the MEP saturates at. In particular, larger V_{TH0} moves saturation towards exponentially larger k_0 and proportionally larger $V_{DD,opt}|_{ILDR \to \infty}$.

4.3.5 Considerations on the MEP Energy

From (4.13), the resulting energy at the MEP in deep sub-threshold region (i.e., under (4.15)) can be written as

$$E_{min} = \alpha_{SW} \cdot C_{TOT} \cdot V_{DD,opt}^2 \left[1 + \left.\frac{E_{lkg}}{E_{dyn}}\right|_{MEP}\right] \qquad (4.18)$$

where, considering that $f(V_{DD}) \approx 1$ and $\alpha_{X_{off}} \ll 1$ in (4.13), the energy-optimum ratio E_{lkg}/E_{dyn} at the MEP is given by

$$\left.\frac{E_{lk}}{E_{dyn}}\right|_{MEP, sub-threshold} \approx ILDR \cdot e^{-\frac{V_{DD,opt}}{n\frac{kT}{q}}} \approx 0.2 + \frac{17}{ILDR^{0.75}} \qquad (4.19)$$

In (4.19), an empirical approximate expression has been introduced to facilitate its estimate at design time, and its error is within 12% for low values of *ILDR* (down to 150), as plotted in Fig. 4.23. Observe that E_{lkg}/E_{dyn} at the MEP is independent from the chosen (single) threshold voltage, as expected from the considerations in Sect. 4.3.4. In other words, sub-threshold designs that differ only for the threshold voltage choice have the same leakage percentage contribution, other than the same $V_{DD,opt}$ (see Sect. 4.3.4). Accordingly, the MEP is hence chosen based on the performance target rather than energy. Also, this means that the MEP voltage can be estimated at design time even before choosing the transistor flavor (and hence before actual implementation).

Compared to the overall energy budget, E_{lkg} at the MEP needs to be 40–50% for very fast/active designs (*ILDR* $\leq$ 150), around 15–30% for more typical designs (*ILDR* > 150 but still in sub-threshold). Previous work on joint supply/ threshold voltage and sizing optimization showed that energy optimality is achieved when E_{lkg} is about one third of the overall energy (Markovic et al. 2004; Nose et al. 2000; Patil). Accordingly, these results hold in sub-threshold region only for relatively slow and inactive designs, from Fig. 4.23.

As discussed in Sect. 4.3.4, very fast and active designs have low $V_{DD,opt}$, which often times falls below V_{min}. In these cases, minimum-energy and reliable operation is

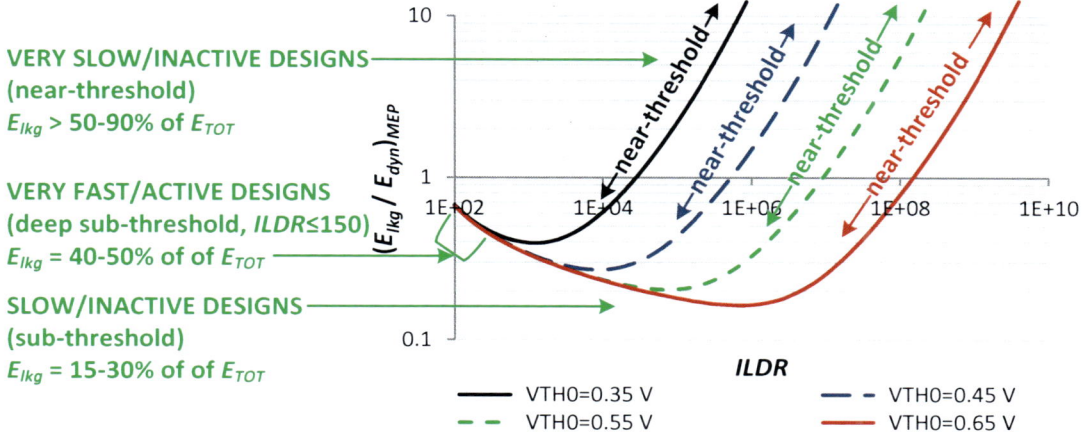

VERY SLOW/INACTIVE DESIGNS (near-threshold)
$E_{lkg} > 50\text{-}90\%$ of E_{TOT}

VERY FAST/ACTIVE DESIGNS (deep sub-threshold, $ILDR \leq 150$)
$E_{lkg} = 40\text{-}50\%$ of of E_{TOT}

SLOW/INACTIVE DESIGNS (sub-threshold)
$E_{lkg} = 15\text{-}30\%$ of of E_{TOT}

VTH0=0.35 V   VTH0=0.45 V
VTH0=0.55 V   VTH0=0.65 V

Fig. 4.23 Leakage-dynamic energy ratio at the MEP vs. *ILDR*, obtained through numerical minimization of (4.13a)

achieved at $V_{DD} = V_{min} > V_{DD,opt}$. If the extra performance compared to the MEP is not utilized since the design is essentially energy constrained, operation at $V_{min} > V_{DD,opt}$ leads to an increase in E_{dyn} by a factor $(V_{min}/V_{DD,opt})^2$ from (4.9), compared to the MEP. At the same time, a smaller increase in E_{lkg} by a factor $V_{min}/V_{DD,opt}$ is observed (since same clock cycle is assumed in (4.10), and DIBL effect is neglected). In other words, designs with $V_{min} > V_{DD,opt}$ typically have lower E_{lkg}/E_{dyn}, compared to operation at the MEP in Fig. 4.23.

For larger *ILDR*, again the MEP moves to the near-threshold region and the energy-optimum ratio E_{lkg}/E_{dyn} at the MEP increases again when increasing V_{DD}. This is because the increase in $V_{DD,opt}$ at near-threshold voltages determines a much smaller reduction in E_{lkg} compared to sub-threshold, as the gate delay decreases much slower than exponentially from (4.6). Analytically, this is accounted for by the increase in f (V_{DD}) in (4.13), which determines a proportional increase in E_{lkg}/E_{dyn}. Accordingly, E_{lkg} becomes again a substantial fraction of the energy budget when the MEP is pushed at near-threshold voltages or higher (i.e., for large *ILDR*), as shown in Fig. 4.23. This explains why E_{lkg}/E_{dyn} heavily depends on V_{TH} at near-threshold voltages, as in Fig. 4.23. For extremely slow and inactive circuits, the MEP moves to the

above-threshold region, and is definitely dominated by E_{lkg}.

From the above considerations, the energy E_{min} at the MEP monotonically increases when increasing *ILDR*. At sub-threshold voltages, this is due to the increase in the energy-optimal voltage $V_{DD,opt}$ in (4.17), which is certainly more rapid than the reduction in $\left(1 + E_{lkg}/E_{dyn}\big|_{MEP}\right)$ in (4.19). Since both dependencies were found to be unaffected by the threshold voltage in sub-threshold, E_{min} for MEP in sub-threshold is independent of V_{TH} as well. At near-threshold voltages, E_{min} keeps increasing since $V_{DD,opt}$ in (4.17) continues to increase with the same trend as sub-threshold (see Fig. 4.22a), and $\left(1 + E_{lkg}/E_{dyn}\big|_{MEP}\right)$ increases as well. For above-threshold MEP, $V_{DD,opt}$ in (4.17) saturates to an almost constant value, and $\left(1 + E_{lkg}/E_{dyn}\big|_{MEP}\right)$ keeps increasing.

From the above considerations, the energy at the MEP is monotonically degraded when *ILDR* is increased, i.e., for leaky or little active designs. This is essentially due to the increase in $V_{DD,opt}$ (i.e., dynamic energy at the MEP), and the increase in the leakage-dynamic energy ratio at voltages above V_{TH}. More quantitatively, Fig. 4.24 shows that E_{min} increases in an approximately linear fashion when increasing *ILDR*. In

Fig. 4.24 Energy at MEP E_{min} normalized to $\alpha_{SW}C_{TOT}$ vs. *ILDR* for various threshold voltages

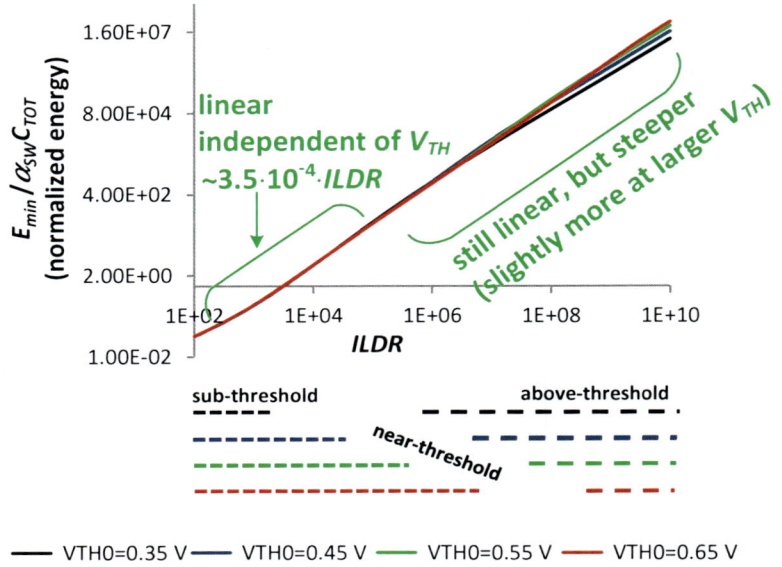

particular, in the sub-threshold region the ratio $E_{min}/\alpha_{SW}C_{TOT}$ is well approximated by $3.5 \cdot 10^{-4}$ $\cdot ILDR$ regardless of the threshold voltage, hence

$$E_{min} \approx 3.5 \cdot 10^{-4} \cdot \alpha_{SW} \cdot C_{TOT} \cdot ILDR \quad (4.20)$$

which is within 20% of exact E_{min} for *ILDR* up to a few tens of thousands. For less typical design with larger *ILDR*, the trend becomes slightly steeper by a factor ranging from to 2.5X to 4X compared to (4.20), for a threshold voltage in the 350–650 mV range. In other words, when the MEP is at near-threshold voltages, E_{min} actually increases when V_{TH} increases, although moderately.

Summarizing these conclusions and those in Sect. 4.3.4, the MEP voltage is unaffected by the (single) threshold voltage when operating in sub- and near-threshold voltages. On the other hand, the energy portion associated with leakage drastically increases when operating at near-threshold voltages, compared to sub-threshold ones. At above-threshold voltages, the MEP voltage becomes a function of V_{TH}, and energy is dominated by leakage. Regardless of the voltage range in which the MEP lies in, the minimum achievable energy monotonically and proportionally increases with *ILDR*.

4.3.6 Sensitivity of Nearly-Minimum Energy to V_{DD} Inaccuracies

From a design standpoint, it is necessary to predict the required accuracy for V_{DD} to achieve nearly-minimum energy per operation, which in turn constraints the design of the voltage regulation circuitry and the power management sub-system. As can be seen from Fig. 4.20a–d, the energy-voltage curve is steeper at the left of the MEP, due to the exponential increase in the leakage energy at low voltages. In other words, an uncertainty $\pm\Delta V_{DD}$ in the supply voltage around the MEP degrades the energy more substantially when it pushes V_{DD} below $V_{DD,opt}$ rather than above (even more so, if performance is considered). Due to the same reason, the energy degradation at the left of the MEP compared to the right becomes more evident for larger ΔV_{DD}.

In nearly-minimum energy designs, the maximum tolerable percentage energy degradation *% energydegradation* compared to the MEP due to the uncertainty in V_{DD} needs to be translated into the specification of the maximum tolerable uncertainty ΔV_{DD}. In sub-threshold region, the resulting tolerable uncertainty ΔV_{DD}

Fig. 4.25 Maximum supply voltage deviation from MEP that maintains the energy degradation within the target (on x-axis) in sub-threshold region (model in (4.21a), (4.21b))

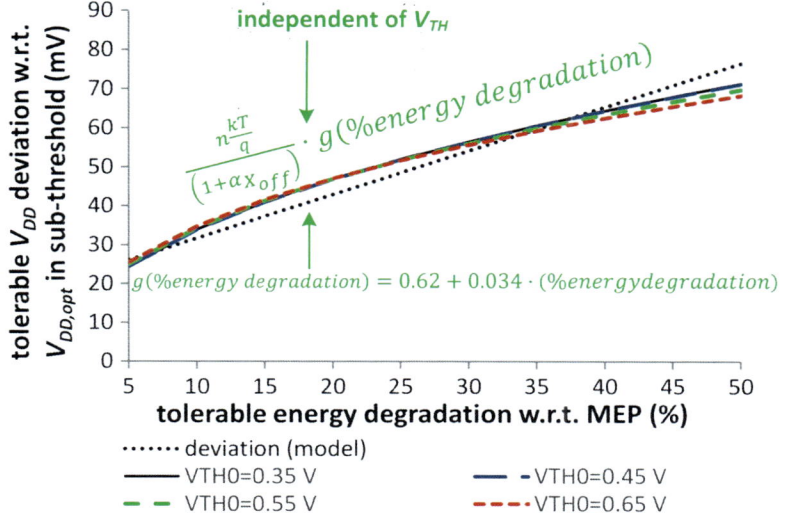

is independent of V_{TH}, since the energy is independent of V_{TH} as well (see Sect. 4.3.4). Since $f(V_{DD}) \approx 1$, Eq. (4.13) can be expressed in a technology-independent manner[10] by defining the normalized voltage $V_{DD,norm} = V_{DD}(1 + \alpha_{X_{off}})/n\frac{kT}{q}$. The maximum deviation $\Delta V_{DD,norm}$ in $V_{DD,norm}$ compared to the value that minimizes the energy can be easily solved numerically. The numerical solution $\Delta V_{DD,norm}$ turns out to be largely independent of $ILDR$, and is hence only a function of % energy degradation. $\Delta V_{DD,norm}$ is well approximated (within 10%) by $0.62 + 0.034 \cdot (\%energydegradation)$, as shown in Fig. 4.25. Accordingly, the maximum tolerable voltage deviation that meets a targeted percentage deviation in sub-threshold region is

$$\Delta V_{DD,subthreshold} \approx \frac{n\frac{kT}{q}}{(1 + \alpha_{X_{off}})} \cdot g(\%energy\ degradation) \qquad (4.21a)$$

$$g(x) = 0.62 + 0.034 \cdot x \qquad (4.21b)$$

Interestingly, from (4.21a), (4.21b), ΔV_{DD} in sub-threshold does not depend on the position of the MEP, and it only depends on technology through subthreshold slope and DIBL coefficient, and on the targeted maximum energy degradation. As an example, Fig. 4.25 plots the maximum tolerable ΔV_{DD} versus the energy degradation in 28 nm, and shows that V_{DD} needs to be set with a precision of about a thermal voltage (25–35 mV) to keep the energy degradation compared to the MEP modest (5–10%). Larger V_{DD} uncertainty (e.g., 1.5–2X the thermal voltage) leads instead to an unacceptably large energy degradation, and should hence be avoided in practical cases.

When the MEP moves to the near-threshold region, a larger voltage deviation can be tolerated for a targeted maximum energy degradation compared to the MEP. This is because $FO4$ and hence E_{lkg} (see Eq. (4.10)) become less sensitive to V_{DD} compared to sub-threshold, as shown in Fig. 4.26. As expected, the tolerable voltage

---

[10] Indeed, eq. (5.13) in sub-threshold region becomes $E_{cycle} \propto V_{DD,norm}^2[1 + ILDR \cdot e^{-V_{DD,norm}}]$, assuming $\alpha_{X_{off}} \approx \lambda_{DIBL}$ (which is generally, since $\lambda_{DIBL} \ll 1$).

Fig. 4.26 Maximum
supply voltage deviation
from MEP vs. *ILDR* for
different V_{TH0} in 28 nm
(target energy degradation
w.r.t. MEP = 10%)

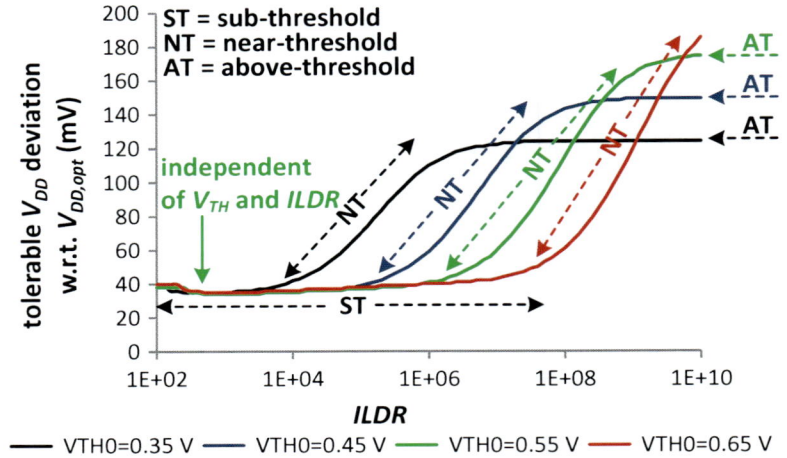

deviation at near-threshold depends on V_{TH0}, as opposed to sub-threshold. This is because the energy in (4.13) depends on V_{TH0} at near threshold (see Sect. 4.3.4), since V_{TH0} defines the voltage range (and hence *ILDR*) in which transistors enter this region. From Fig. 4.26, 4 to 6 thermal voltages can be tolerated with minimal energy penalty at near-threshold voltages.

For larger MEP voltages in the above-threshold region, an even larger voltage deviation around the MEP is tolerable for a given allowed energy degradation. This is because *FO*4 has the minimum sensitivity to V_{DD} across voltages from (4.6). As a consequence of the saturation of the MEP voltage discussed in Sect. 4.3.4, the maximum voltage

deviation saturates as well at above-threshold voltages, as shown in Fig. 4.26. As expected from (4.16), larger V_{TH0} pushes the saturation to higher voltages (and hence *ILDR*). Analytically, the maximum tolerable allowed voltage deviation around the MEP is evaluated by equating (4.13) and the energy at MEP (i.e., voltage in (4.16)) increased by a factor $(1 + \%energy\ degradation/100)$, and solving for V_{DD}. By approximating $\ln\left(e^{\frac{V_{DD}(1+\lambda_{DIBL})-V_{TH0}}{n\cdot(kT/q)}} + 1\right) \approx \frac{V_{DD}(1+\lambda_{DIBL})-V_{TH0}}{n\cdot(kT/q)}$ and $\alpha_{X_{off}} \approx \lambda_{DIBL}$ in (4.13), the maximum voltage deviation at above-threshold voltages results to

$$\Delta V_{DD,above-threshold} = V_{DD,opt}\big|_{ILDR \to \infty} \cdot h\left(\frac{\%energy\ degradation}{100}\right) \qquad (4.22a)$$

$$h(x) = -x + \sqrt{x\cdot(1+x)} \qquad (4.22b)$$

the first of which is plotted in Fig. 4.27 for a 28-nm technology, along with the technology-independent curve in (4.22b). Equation (4.22a), (4.22b) was found to be within 10-20% of the exact solution, for typical threshold voltages. For large V_{TH0} (e.g., 0.65 V), the error increases to 30–40% since ΔV_{DD} in (4.22a) becomes so large that it intrudes the near-threshold region, and the

above calculations hence become inaccurate. From (4.22a), (4.22b) the maximum voltage deviation around MEP above threshold depends on the technology through a proportional dependence on $V_{TH0}/(1 + \lambda_{DIBL})$. In other words, the maximum voltage deviation around the MEP above threshold is a fixed and technology-independent fraction of the MEP voltage, as set by h(% *energy degradation*/100). As opposed to sub-threshold region, the maximum deviation

Fig. 4.27 Maximum supply voltage deviation from MEP that maintains the energy degradation within the target (on x-axis) in above-threshold region (model in (4.22a), (4.22b))

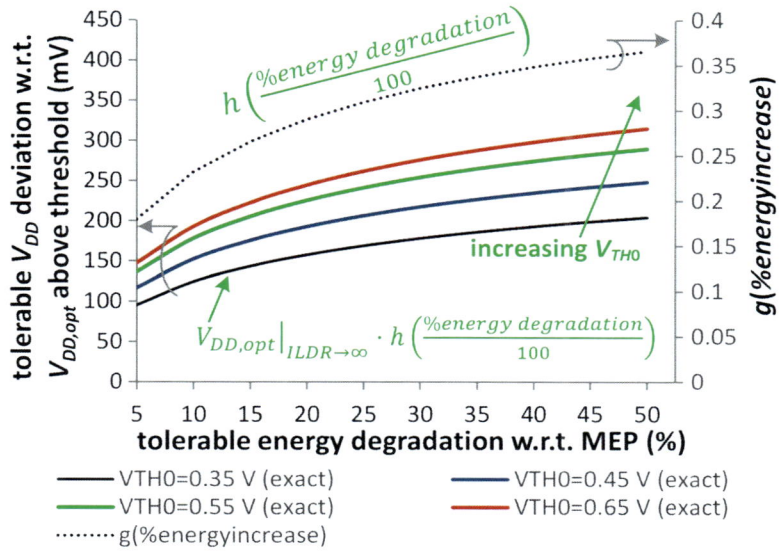

around a MEP lying above threshold depends on V_{TH0}, and larger thresholds further relax the precision requirement on the voltage optimization and delivery. This is because a larger V_{TH0} enlarges the voltage range in which the MEP effectively increases for larger *ILDR* (i.e., from about V_{TH0} up to (4.16)).

In summary, nearly-minimum energy operation requires the voltage to be controlled within approximately one thermal voltage when the MEP is in the sub-threshold region, independently of the position of the MEP. This requirement is substantially relaxed at near threshold, and increases at above threshold until saturation to a value that is proportional to the threshold voltage and sub-linearly related to the tolerable energy degradation. More in detail, deviation increases up to 4 thermal voltages for relatively low V_{TH0}, whereas it increases to more than 6 thermal voltages under large V_{TH0}.

Overall, this suggests that the performance can be increased with modest energy penalty by raising the voltage compared to the MEP, when the latter is at near- or above-threshold voltages. These considerations are summarized in the example in 28 nm in Fig. 4.28, which plots ΔV_{DD} versus *ILDR* for various energy degradation targets and the related analytical models.

4.3.7 Example: ARM Core Operating at Minimum Energy

As a further numerical example, let us apply the above energy and MEP models to the design of the ARM Cortex M0 core in Myers et al. (2015). Table 4.5 reports all technology-, design- and workload-dependent parameters for this design, as obtained from the process design kit and information provided in Myers et al. (2015). The two programs "checksum" and "AES" are considered to consider a wide range of activities, from low (checksum) to high (AES), and hence observe the MEP shift due to different activity factors (the latter has 60% higher activity than the former (Myers et al. 2015)).

The resulting energy curve versus V_{DD} from experimental results in Myers et al. (2015) and the model in (4.13) is plotted in Fig. 4.29. Very good agreement can be observed across the wide voltage range from 0.25 to 1.2 V, with an average error of 1.7%, and an error well below 10% down to 0.3 V. As expected from Sect. 4.3.3 and Fig. 4.21, the MEP for the AES program is pushed to the left of the MEP for the checksum program, due to the higher activity determined by the former one. More quantitatively, the *ILDR* factor in (4.13) and (4.14a), (4.14b) from the

Fig. 4.28 Summary of maximum V_{DD} deviation from MEP that maintains the energy degradation within the target vs. *ILDR* and related models

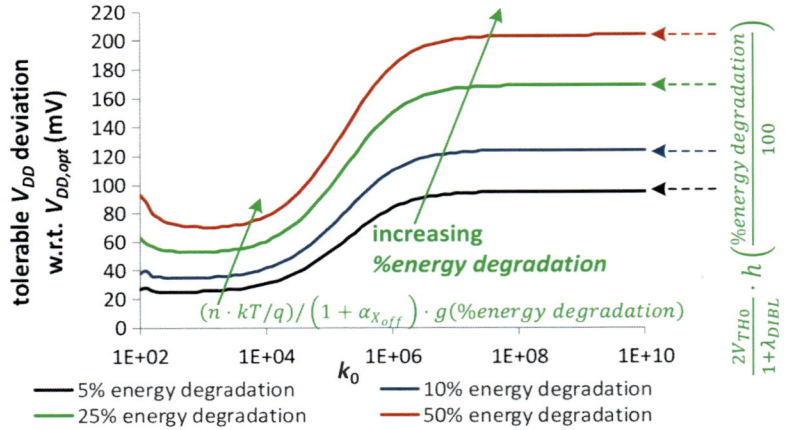

$$(n \cdot kT/q)/\left(1 + \alpha_{X_{off}}\right) \cdot g(\%energy\ degradation)$$

increasing %energy degradation

$$\frac{2V_{THO}}{1+\lambda_{DIBL}} \cdot h\left(\frac{\%energy\ degradation}{100}\right)$$

— 5% energy degradation — 10% energy degradation
— 25% energy degradation — 50% energy degradation

Table 4.5 Technology-, design- and workload-dependent parameters for ARM Cortex M0 Core in Myers et al. (2015)

	Parameter	Value
Technology	*Process*	65 nm
	$V_{DD,nom}$	1.2 V
	V_{TH} @ $V_{DD,nom}$	0.47 V
	λ_{DIBL}	0.095 V/V
	$\alpha_{X_{off}}$	0.087 V/V
	$FO4$ @ $V_{DD,nom}$	60 ps
Architecture/ckt design	LD_{eff}	240[a]
	$\frac{C_{cell}}{C_{in,min}}$	4[b]
	strength	2[b]
Workload	α_{SW} (checksum program)	5%[c]
	α_{SW} (AES program)	8%[c]

[a]Estimated from cycle time at nominal voltage (14.3 ns) and *FO4* at nominal voltage
[b]Typical values for very slow and low-energy designs (changes in a reasonable range do not significantly influence results)
[c]Activity factor in AES obtained via a 60% increase compared to checksum program (Myers et al. 2015)

parameters in Table 4.5 results to 18,000 for the checksum program, and 11,200 for the AES program. The resulting voltage and energy at the MEP are summarized in Table 4.6, for both the experimental results in Myers et al. (2015) and the above models.

From Table 4.6, the estimated MEP voltage of the core in Myers et al. (2015) from Eq. (4.17) is 378 mV for the checksum program, which is close to the measured MEP voltage of 390 mV. The resulting minimum energy estimate of 11.6 pJ from (4.18) is also close to the measured energy of 11.7 pJ (Myers et al. 2015). At the MEP, the leakage energy is estimated to be smaller than the dynamic energy by a factor of 0.21 from eq. (4.19), which is close to the value of 0.22 in Myers et al. (2015). Good agreement of the models is also confirmed for the AES program, from the same table. Finally, the maximum tolerable V_{DD} uncertainty for a 10% energy increase compared to the MEP results to 45 mV from (4.21a), (4.21b), which agrees well with the value of approximately 48 mV in Myers et al. (2015).

4.4 Exploration of MEP Dependence on Logic Depth, V_{TH}, Activity and Ineffectiveness of Leakage Reduction Techniques

In this section, the impact of logic depth, threshold voltage and activity are quantitatively and widely explored by considering the reference circuit in Fig. 4.30, applying the insights gained in Sect. 4.3. The simplicity and regularity of the circuit in 4.30 permits to gain an intuitive

Fig. 4.29 Experimental energy curve vs. V_{DD} in ARM Cortex M0 core (Myers et al. 2015) and energy predicted by the model in (4.13)

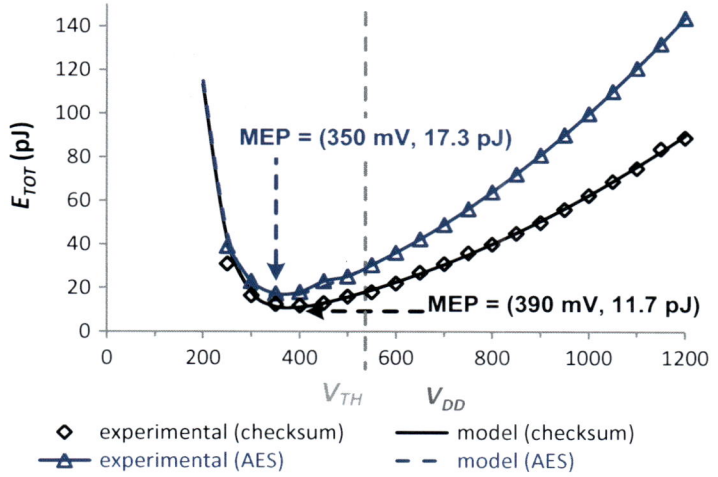

MEP = (350 mV, 17.3 pJ)

MEP = (390 mV, 11.7 pJ)

◇ experimental (checksum)   ── model (checksum)
△ experimental (AES)   ─ ─ model (AES)

Table 4.6 Minimum Energy Point, Leakage/Dynamic Energy Ratio and Maximum Tolerable *VDD* Uncertainty from (Myers et al. 2015) and Above Models

	checksum program		AES program	
	experimental (Myers et al. 2015)	model (equation)	experimental (Myers et al. 2015)	model (equation)
$V_{DD,opt}$	390 mV	378 mV (4.17)	350 mV	360 mV (4.17)
E_{min}	11.7 pJ	11.6 pJ (4.18)	17.3 pJ	16.8 pJ (4.18)
$\left.\frac{E_{lk}}{E_{dyn}}\right|_{MEP}$	0.22	0.21 (4.19)	0.22	0.22 (4.19)
ΔV_{DD} (energy increase = 10%)	45 mV	48 mV (4.21a), (4.21b)	45 mV	48 mV (4.21a), (4.21b)

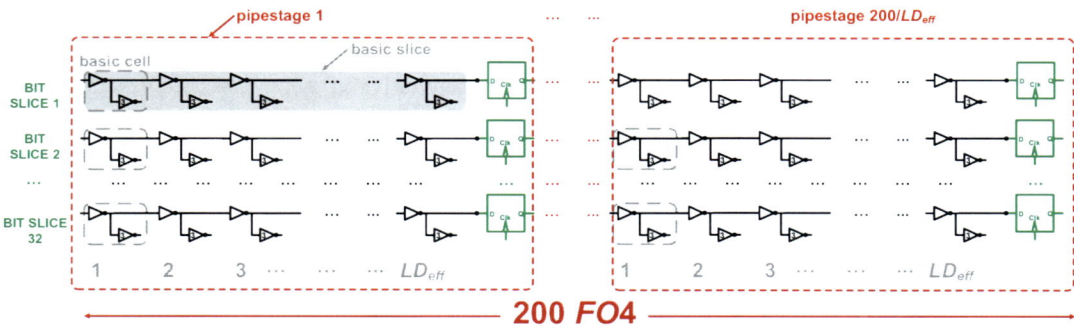

Fig. 4.30 Reference circuit for evaluation of the impact of logic depth, threshold voltage and activity

understanding of the underlying tradeoffs. Such circuit contains 32 slices of inverter gates, each with a fan-out of 4, and with a total logic depth LD_{TOT} (and hence delay by construction) of

$200FO4$, as representative of a relatively complex microprocessor. The slices are interrupted through the insertion of registers, whose number is adjusted to achieve a targeted logic depth

LD_{eff}. Registers are made up of transmission-gate flip-flops, which are customarily encountered in standard cell libraries (Alioto et al. 2015).

4.4.1 Impact of Logic Depth

The heavy impact of E_{lkg} at near- and sub-threshold voltages can be mitigated by adopting microarchitectures with lower logic depth (i.e., deeper pipelining), from (4.10). However, deeper pipelining should be applied judiciously to avoid a significant increase in the clocking overhead, which might offset some of the benefit brought by reduction in E_{lkg}. In the following, we will assume that the additional clocking cost of meeting the timing constraints with lower logic depth is modest (which is typically true in microarchitectures with $LD_{eff} \geq 25$ $FO4/cycle$).

The reference circuit in Fig. 4.30 has an overall energy per cycle equal to

$$E_{cycle} = \alpha_{SW} \cdot C_{TOT} \cdot V_{DD}^2 + \frac{LD_{TOT}}{LD_{eff}} \cdot E_{REG} + V_{DD} \cdot I_{off} \cdot FO4 \cdot LD_{eff} \qquad (4.23)$$

where it was assumed that pipestages are perfectly balanced (i.e., the number of pipestages is $\frac{LD_{TOT}}{LD_{eff}}$). In (4.23), E_{REG} is the energy consumed by a single register, and $\frac{LD_{TOT}}{LD_{eff}}$ represents the number of registers in the above circuit. From (4.23), an energy-optimal logic depth exists at a given V_{DD}, and its expression is readily found to be

$$LD_{eff} = LD_{TOT} \sqrt{\frac{E_{REG}}{V_{DD} \cdot I_{off} \cdot LD_{TOT} \cdot FO4}} = LD_{TOT} \sqrt{\frac{E_{REG}}{E_{lkg}}} \qquad (4.24)$$

where (4.10) was used to express E_{lkg}. From (4.24), the optimal logic depth is determined by the balance between the clocking and the leakage energy, as a larger number of registers leads to an increase in the former and a decrease in the latter. Such tradeoff is not really observed in traditional low-power (above-threshold) designs, as the leakage energy is usually kept a small fraction of the overall budget through several techniques (Narendra and Chandrakasan 2006), and the amount of pipelining is mainly defined by the performance target, or the dynamic energy-performance tradeoff under dynamic voltage scaling. Instead, nearly-minimum energy designs require a careful management of the clocking-leakage energy tradeoff, due to their strong interdependence (Alioto 2012). In addition, this means that energy-centric (micro)architectures need to be tailored around the targeted operation voltage, and traditional architectures conceived for nominal voltage operation tend to be energy inefficient at low voltage. In other words, ultra-low power architectures need to be deeply rethought to truly enable nearly-minimum energy operation, as discussed in Chap. 3.

Quantitatively, eq. (4.24) suggests that the energy-optimal pipeline depth LD_{TOT}/LD_{eff} is given by the square root of the leakage-clocking energy ratio. Considering the large contribution of E_{lkg} at near-threshold voltages, the theoretical energy-optimal pipedepth tends to be quite small. In (4.24), the energy cost of all registers E_{REG} is assumed to be the same, since it refers to the simple reference circuit in Fig. 4.30. In more general architectures, the number of flip-flops per register, and hence the energy cost of a register, increases super-linearly under higher pipedepths (Chinnery and Keutzer 2007). Indeed,

Fig. 4.31 Energy normalized to value at nominal voltage vs. V_{DD} for logic depth of 25FO4, 50FO4 and 100FO4

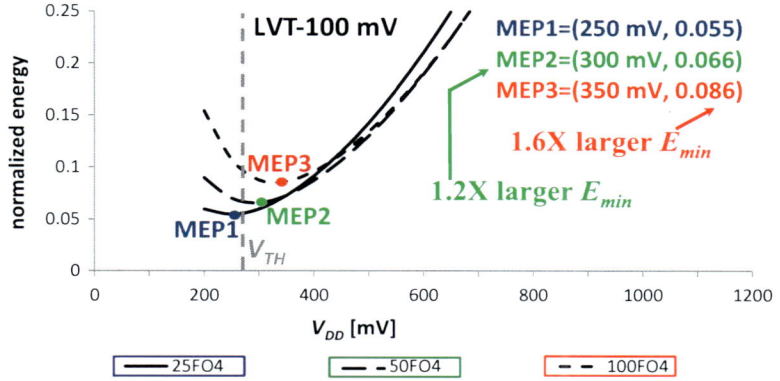

the overall number of flip-flops in a digital module increases according to a power law (LD_{TOT}/LD_{eff})^{LGF}, where $LGF > 1$ is the Latch Growth Factor, which is mainly defined by the specific function implemented (Srinivasan et al. 2002). Hence, the energy-optimal logic depth in general architectures tends to be moderately larger than predicted by (4.24).

On the low side, the energy-optimal logic depth is practically limited by the rapidly increasing clocking energy cost at small logic depths (i.e., deep pipelines). Typically, low logic depths in the order of 20–25 FO4 or smaller have a disproportionately large energy cost in the clock network at ultra-low voltages, and require non-straightforward clock network design approaches. The necessity of "fast" circuit and architectural designs[11] with deep pipelining at ultra-low voltages was first shown in Jeon et al. (2013), where an aggressive 17FO4 logic depth was adopted in a 1024-point complex FFT processor. The adoption of such deep pipeline led to 17.7 nJ/transform at $V_{DD,opt} = 270$ mV in 65-nm CMOS, which was a 3.6X lower energy than previous state of the art. However, this required

some non-trivial clocking technique to avoid timing violations under the unavoidably large variations (see Sect. 4.7), such as 2-phase latch clocking, custom latches with embedded logic, aggressive hold fix buffer insertion, and shallow clock network (3 levels, for the reasons clarified in Sect. 4.10).

The resulting energy curve versus V_{DD} for different logic depths in the reference circuit in Fig. 4.30 is reported in Fig. 4.31 in 28-nm CMOS. As expected, increasing the logic depth to the practical lower bound of 25FO4 to larger logic depths of 50 and 100FO4 leads to a significant 20% and a considerable 60% energy increase at the MEP. This is respectively due to a 2X and 4X leakage energy increase, due to the larger logic depth from (4.10). From (4.14a), (4.14b), such increase in E_{lkg} leads to a 2X (4X) increase in $ILDR$, which from (4.15) translates into an increase in the MEP voltage of approximately 35 mV and 65 mV (Fig. 4.31 discretizes voltages in 50-mV step, and hence results to 50 and 100 mV).

Finally, it should be observed that true minimum-energy operation actually requires a complex optimization that involves logic depth, voltage and transistor sizing. Unfortunately, no thorough methodology and no CAD support is currently available for this purpose, hence such joint optimization is still an open research question. A qualitative treatment of this problem will be presented in Sect. 4.11, to gain an insight into this fundamental design problem.

---

[11] Here, "fast" refers to the clock cycle normalized to $FO4$ (i.e., LD_{eff}), rather than the absolute clock cycle. This choice is motivated by the need for characterizing the design regardless of the specific voltage and hence $FO4$. Indeed, low values of $T_{CK}/FO4$ identify designs that would be fast at nominal and any other voltage, regardless of $FO4$. On the other hand, ultra-low voltage operation makes the absolute T_{CK} large simply because of the increase in $FO4$, not because of the design itself.

Fig. 4.32 Energy normalized to value at nominal voltage vs. V_{DD} for single- and multi-VTH design (50% HVT, 50% LVT cells, logic depth of 25 $FO4$ 10% activity)

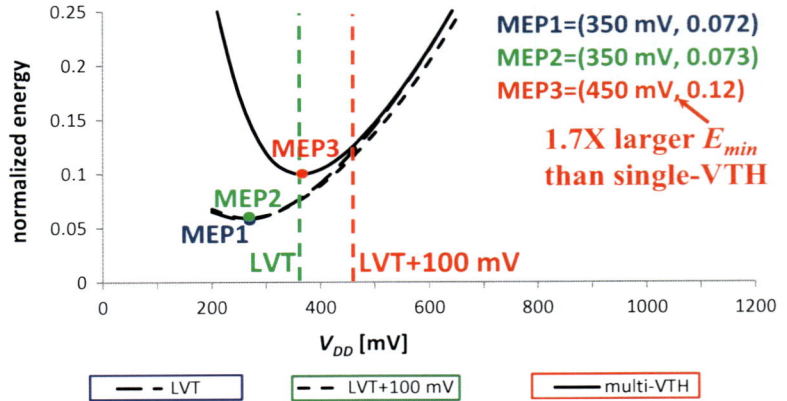

MEP1=(350 mV, 0.072)
MEP2=(350 mV, 0.073)
MEP3=(450 mV, 0.12)
1.7X larger E_{min} than single-VTH

4.4.2 Impact of Threshold Voltage and Activity

The impact of the threshold voltage on the reference circuit in Fig. 4.30 is shown in Fig. 4.32. As expected from Sect. 4.3.4, the MEP voltage and energy are the same for different threshold voltages under a single-V_{TH} design, being in the sub-threshold region. On the other hand, mixing the two threshold voltages leads to a substantially larger MEP energy (by 1.7X in this specific case). This suggests that multi-V_{TH} design is not really advantageous at near- and sub-threshold voltages, and it should hence be avoided. Thorough analysis and justification of this observation will be provided in the next section.

The effect of activity is depicted in Fig. 4.33, which once again confirms that the MEP moves to the left when the dynamic energy is increased, as was observed in Fig. 4.21. More quantitatively, the increase in the activity factor from 3% to 10% (20%) leads to a 3.3X (6.6X) decrease in *ILDR*, which from (4.15) translates into a decrease in the MEP voltage of 51 mV and 81 mV (the latter is not precisely visualized in Fig. 4.33, as voltages are discretized in 50-mV step).

To provide a broader view on the impact of the above parameters onto the MEP position, Fig. 4.34a–c plot the statistical distribution of the MEP voltage for several different activities and logic depths, respectively for a very low, relatively low and relatively high V_{TH}. From this figure, the MEP lies in the sub-threshold

region for most of the designs, and it is pushed into in the near-threshold region only for very low threshold voltages (see Fig. 4.34a). Figure 4.34d–f show the contribution of the leakage energy as a fraction of the overall energy for the same threshold voltages. From this figure, E_{lkg} accounts for 40% of the total energy or more in most of the designs, and tends to be larger under lower threshold voltages. According to Fig. 4.23, this is because the MEP is pushed to near-threshold voltages at low V_{TH}, and the resulting E_{lkg} can be as high as 70% of the total energy in some designs (see Fig. 4.34d).

4.5 Ineffectiveness of Traditional Leakage Reduction Techniques

This section shows that traditional leakage reduction techniques (e.g., stacking, multi-V_{TH}) are far less effective at near-threshold voltages, thus posing a challenge on leakage management at such voltages.

Transistor stacking has been extensively exploited to reduce leakage in above-threshold circuits (Narendra and 2006), as the off-stacking factor[12] is typically much higher than the on-stacking factor. In other words, the series

[12] The off (on) stacking factor is defined as the factor by which the transistor current of an off (on) single transistor is reduced due to the series connection of multiple transistors having the same size.

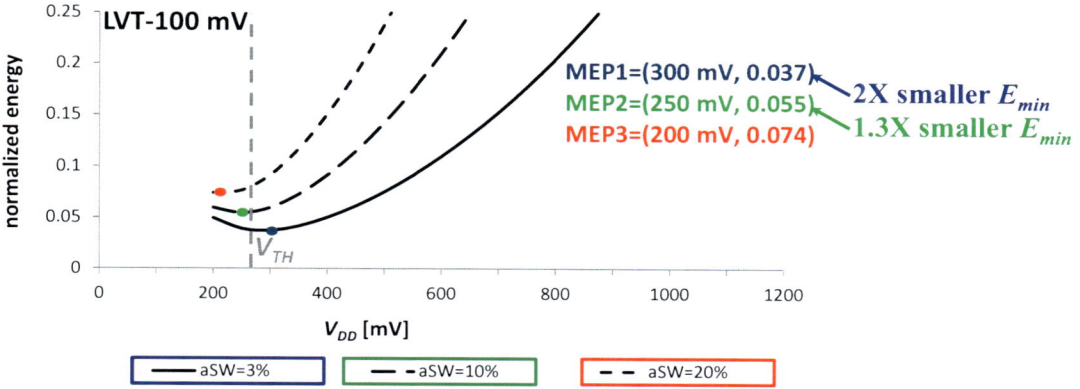

Fig. 4.33 Energy normalized to value at nominal voltage vs. V_{DD} for different activity values (logic depth of $25FO4$)

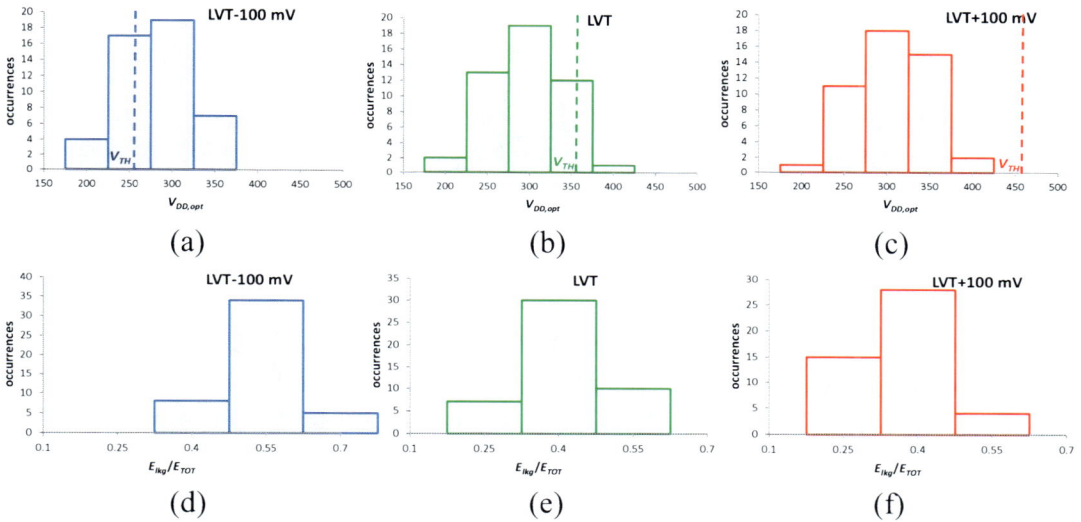

Fig. 4.34 Statistics on the MEP voltage $V_{DD,opt}$ for the reference circuit in Fig. 4.30 across different values of activity and logic depth for three threshold voltages (a)–(c). Ratio between leakage energy and total energy for same threshold voltages (d)–(f)

connection of multiple transistors reduces the leakage current much more heavily than the on-current (i.e., performance). As shown in Fig. 4.35, this is true in the above-threshold region, where the off-stacking factor for 2 to 4 transistors is larger than the on-stacking factor by an order of magnitude. At lower voltages, the on-stacking factor tends to moderately increase, for the reasons discussed in Sect. 4.2.3. At the same time, the off-stacking factor decreases exponentially when reducing V_{DD} (Narendra and Chandrakasan 2006). Indeed, the off-stacking factor for two stacked transistors can be expressed as

$e^{\alpha_{X_{off}} \cdot V_{DD}/(nkT/q)}$ (Narendra and Chandrakasan 2006), where

$$\alpha_{X_{off}} = \lambda_{DIBL} \cdot \frac{1 + \lambda_{DIBL}}{1 + 2\lambda_{DIBL}} \ll 1 \qquad (4.25)$$

and approximately the same dependence is observed for a larger number of stacked transistors, as shown in Fig. 4.35. In other words, the off-stacking factor $X_{stack,off}$ is proportional to $e^{\alpha_{X_{off}} \cdot \frac{V_{DD}}{nkT/q}}$ (Narendra and Chandrakasan 2006), hence it can be expressed as

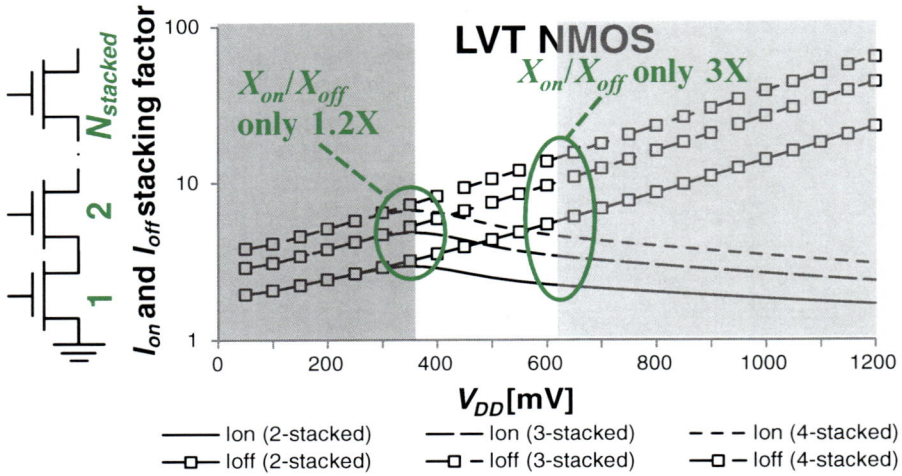

Fig. 4.35 On- and off-stacking factor X_{on} and X_{off} vs. V_{DD} for 2, 3 and 4 stacked transistors in 28 nm

Fig. 4.36 Multi-VTH approach and critical path shift from LVT to HVT paths when scaling down V_{DD}

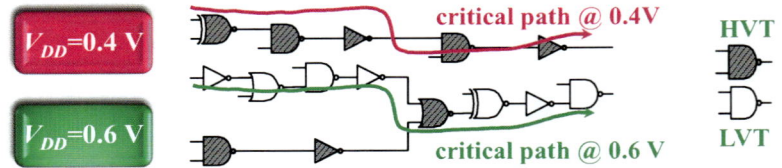

$$X_{stack,off} = X_{stack,off}\big|_{V_{DD,nom}} \cdot e^{\alpha_{X_{off}} \cdot \frac{V_{DD} - V_{DD,nom}}{n \cdot kT/q}} \quad (4.26)$$

which tends to be very accurate across all voltages (within 2% in 28 nm, according to Fig. 4.35).

From Fig. 4.35, the off-stacking factor at near-threshold voltages is no longer much larger than the on-stacking factor, hence no significant leakage reduction is actually allowed for a given performance penalty. In other words, transistor stacking is rather ineffective in counteracting leakage at near-threshold voltages, as opposed to common low-power wisdom (i.e., above threshold).

As another traditional leakage reduction technique, let us consider the adoption of multiple threshold voltages, as depicted in Fig. 4.36 for the simple case of a design with two thresholds (i.e., low and high V_{TH}). In multi-V_{TH} designs, cells in critical paths are LVT to meet the performance requirement, whereas cells in non-critical paths are replaced by the HVT counterparts. At above-threshold voltages, such replacement does

not really degrade performance thanks to its weak dependence on V_{TH}, while it certainly reduces the leakage current thanks to its strong dependence on V_{TH} from (4.8). In other words, the multi-V_{TH} approach offers a favorable tradeoff between performance and leakage in traditional low-power designs operating above threshold. On the contrary, performance becomes very sensitive to V_{TH} at near-threshold voltages as discussed in Sect. 4.2.4, and the HVT cells are slowed down much more substantially than LVT when V_{DD} is dynamically down-scaled (see Figs. 4.5 and 4.6). As a consequence, non-critical HVT paths at a given voltage (e.g., 0.6 V in Fig. 4.36) actually become critical[13] when down-scaling V_{DD} (e.g., 0.4 V in

---

[13] This is unavoidable in real designs, as overall energy-performance optimization aims to equalize the delay of different paths (De Micheli 1994), so that non-critical paths can be down-sized to reduce their energy, while maintaining the same performance target (Narayanan et al. 2010).

Fig. 4.36). In other words, the clock cycle of a multi-V_{TH} design at lower voltages is significantly larger than a single-LVT design, thus leading to a leakage energy increased compared to the latter one, from (4.10). At the same time, the leakage current of a multi-V_{TH} design is significantly larger than a single-HVT design, as the LVT cells in the design have a considerably larger leakage (typically more than an order of magnitude increase when moving from a threshold value to the immediately lower one (International Technology Roadmap for Semiconductors 2013)).

From the above considerations, multi-V_{TH} designs suffers from substantially larger leakage energy compared to single-V_{TH} designs for two concurrent reasons, under dynamic voltage scaling. Hence, multi-V_{TH} approaches actually deteriorate the energy efficiency of VLSI circuits, and should be always avoided in favor of single-V_{TH} designs. The choice of the single V_{TH} has been discussed in Sect. 4.2.4. As an example, this is shown in Fig. 4.32, where the multi-V_{TH} design of the reference circuit in Fig. 4.30 is found to be 1.7X less energy efficient than the single-V_{TH} designs. In terms of energy-performance tradeoff at the MEP, Fig. 4.37 confirms that the multi-V_{TH} design is essentially as slow as the single-HVT

design, in spite of its significantly larger energy consumption.

Similar considerations hold for other traditional leakage reduction techniques, such as power gating (Flynn et al. 2007). At above-threshold voltages, power gating is well-known to provide substantial leakage reductions due to two different mechanisms. First, the sleep transistor size can be much lower than the overall effective transistor width of the power gated circuit, as only a fraction of the cells are active at a given time. Since the relative strength of the sleep and the power gated transistors is maintained at low voltages, this reduction mechanism is essentially maintained at near-threshold voltages. Second, the sleep transistor (see Fig. 4.38a) is able to provide its large on-current during active mode ($\overline{sleep} = 0$), whereas it delivers only its off-current during sleep mode ($\overline{sleep} = 1$). Such reduction is clearly more pronounced for larger I_{on}/I_{off} ratio, which is traditionally obtained by using HVT devices for sleep transistors at above-threshold voltages. At near-threshold voltages, the transistor I_{on}/I_{off} ratio is severely degraded (by 1–2 orders of magnitude) as shown in Fig. 4.38b. Hence, the leakage reduction enabled by power gating at near-threshold voltages is worsened by at least

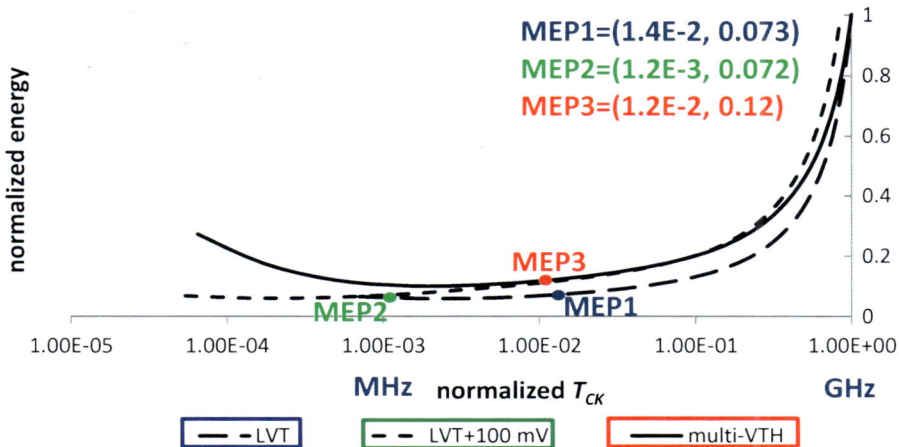

Fig. 4.37 Energy vs. clock frequency (both normalized to value at nominal voltage) for single- and multi-VTH design, and logic depth of $25FO4$

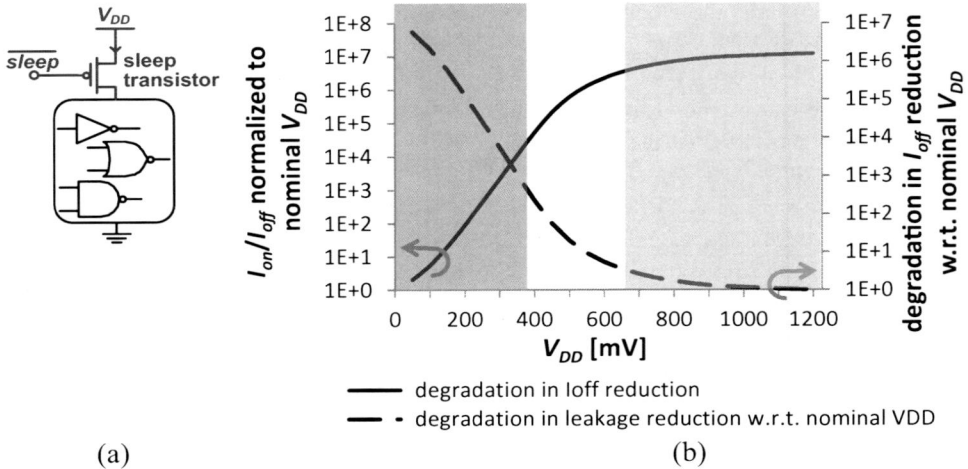

Fig. 4.38 (a) Power gating scheme, (b) I_{on}/I_{off} ratio of sleep transistor in 28 nm

one order of magnitude, compared to above threshold. Such degradation in the effectiveness of power gating at near-threshold voltages can be partially recovered by boosting the gate voltage of the sleep transistor (Myers et al. 2015). Indeed, boosting its gate voltage only during active mode significantly increases I_{on}, while maintaining the same I_{off}. At near-threshold voltages, the sleep transistor I_{on}/I_{off} ratio (and hence the effectiveness of power gating) can be further improved by using thick-oxide (i.e., I/O) NMOS transistors whose gate is powered at the large I/O voltage (e.g., 1.8 V instead of 1 V). In this case, such I_{on}/I_{off} improvement is achieved at the expense of a larger energy and slower transient to turn on the sleep transistor, and hence to switch from sleep to active mode.

4.6 Challenges: Performance

As discussed in Sect. 4.2, operation at near-threshold voltages entails a ~10X penalty in terms of $FO4$ and hence performance, compared to the same architecture operating at nominal voltage. For sub-100 nm technologies, $FO4$ at near-threshold voltages is typically in the order of few hundreds of picoseconds. For reasonable architectures with a logic depth of up to several tens of $FO4$, this translates into a cycle time in

the order of nanoseconds. Hence, throughputs in the order of hundreds of MOPS (Millions of Operations per Second) are easily achievable by near-threshold microprocessor cores. Such level of performance achievable near the threshold is actually acceptable for (or can exceed) the typical requirements of IoT systems, at least in the most frequent operation modes and in most of the practical applications. Higher performance might be needed occasionally in some applications, or customarily for compute-intensive ones, such as computer vision or real-time pattern recognition.

Sustained throughputs that are higher than hundreds of MOPS can always be obtained through appropriate architectures at near-threshold voltages (e.g., multi-core) and specialized hardware, as discussed in Chap. 3. Occasional performance boosts can be achieved through wide dynamic voltage scaling, i.e., by raising V_{DD} from the MEP to the nominal voltage (Chandrakasan et al. 2010). Such temporary voltage up-scaling permits to increase the performance by one (two) order(s) of magnitude, when the MEP is in the near-threshold (sub-threshold) region. This performance increase is achieved at the expense of an increase in the energy per operation, as summarized in Fig. 4.39 for several integrated prototypes (Abouzeid et al. 2012; Gammie et al. 2011; Hsu et al. 2012; Jain et al. 2012; Kaul et al. 2012;

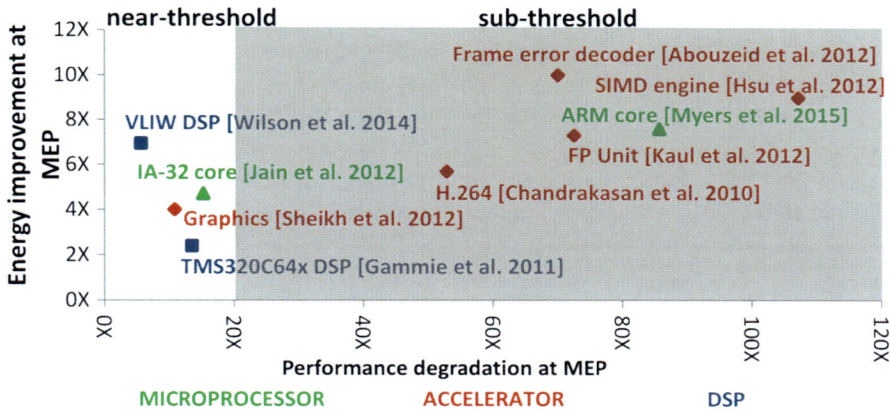

Fig. 4.39 Energy improvement at MEP vs. performance degradation at MEP, as compared to operation at nominal voltage

Fig. 4.40 (**a**) RC wire delay, (**b**) relative scaling of gate and wire delay vs. V_{DD}

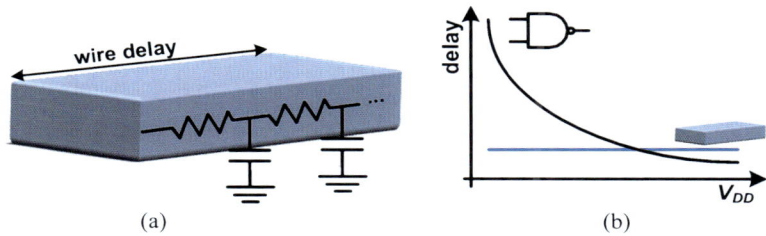

Myers et al. 2015; Sheikh et al. 2012; Wilson et al. 2014; Jacquet et al. 2013). From this figure, this energy increase is more pronounced when the MEP is in the sub-threshold region, due to the larger voltage difference between the MEP and the nominal voltage.

As another property of circuits operating near the MEP, the gate delay dominates over the wire delay, as shown in Fig. 4.40. Indeed, they might be comparable at nominal voltage in realistic VLSI architectures, due to the significant resistive, capacitive, and sometimes inductive parasitics of metal wires. However, operation at the MEP voltage determines a substantial increase in the gate delay, while keeping the wire delay constant. Hence, the wire delay is no longer a challenge in circuits with nearly-minimum energy, which certainly simplifies the design, the circuit modeling and the timing closure.

The reduced wire-to-gate delay ratio around the MEP has also important consequences on the choice of the architecture, and the way the latter is mapped into the physical level. Indeed, VLSI architectures for nearly-minimum energy need to be different from traditional low-power architectures (i.e., for above-threshold operation). More specifically, signals can be propagated through a wider silicon area compared to nominal voltage operation. In detail, for unrepeated wires a ~3X longer distance[14] can be covered by the same wire at near-threshold voltages, as compared to the same circuit operating at nominal voltage, when maintaining the wire delay a fixed fraction of the clock cycle. For similar reasons, repeated wires require 3X fewer repeaters per wire unit length since the optimal distance between

---

[14] This is due to the well-known quadratic dependence of the RC wire delay on its length (Weste and Harris 2011), and assuming a 10X *FO4* degradation at the MEP compared to the nominal voltage.

repeaters is proportional to the square root of *FO*4 (Weste and Harris 2011), thus improving the route-ability. At the same time, the size of each repeater at MEP needs to be increased by 3X compared to nominal voltage, as its performance-optimal size is proportional to *FO*4 (Weste and Harris 2011). Since the number of repeaters is reduced by the same factor by which their area is increased, the energy and area cost of intra-chip global communication in designs around the MEP remains approximately the same. In summary, VLSI architectures for nearly-minimum energy can afford more global communications and larger modules (e.g., shared caches), compared to traditional low-power architectures. Such profound difference in the communication-computation energy/performance tradeoff requires the adoption of innovative architectures, as discussed in Chap. 3.

4.7   Challenges: Variations

In this section, the impact of variations is analyzed in the context of circuits operating around the MEP. In general, process, voltage and temperature variations as well as aging impose an additional timing margin that stretches the clock cycle as shown in Fig. 4.41. This conservative approach preserves correct functionality and performance specifications even in the worst-case die and environmental conditions.

The above cycle time margin resulting from variations translates into an increase in the energy per operation, as faster chips are forced to operate as slowly as the worst-case die and at the same voltage (which is higher than needed). Figure 4.42 shows the voltage increase required by the circuit in Fig. 4.30 to maintain a given clock cycle under a given clock cycle margin, as well as the resulting energy increase, assuming a logic depth of 25FO4, 10% activity, and LVT transistors. From Fig. 4.42, the voltage increase imposed by variations is fairly linear with the cycle time margin, and tends to be larger at higher nominal operating voltages. This is because *FO*4 (i.e., the cycle time from Sect. 4.3.2) is less sensitive to voltage increases at higher operating voltages, and hence requires larger increase to achieve a given percentage

Fig. 4.41 Nominal cycle time and additional margin accounting for variations

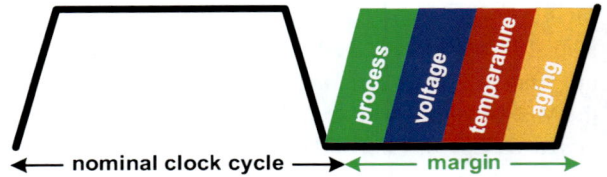

Fig. 4.42 Required V_{DD} to sustain a given performance specification vs. cycle time margin (i.e., factor by which the cycle time needs to be increased due to variations), and resulting energy increase due to variations

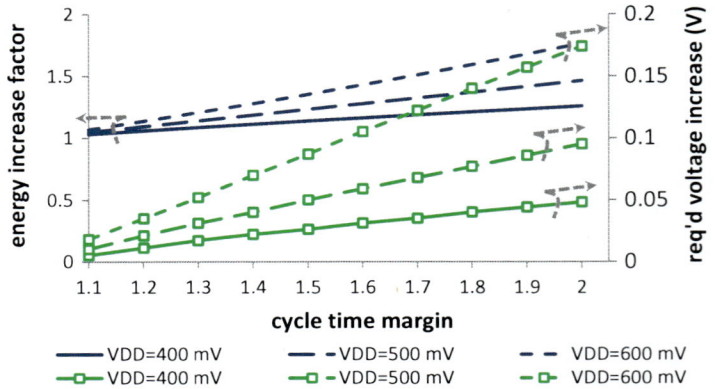

performance improvement. From Fig. 4.42, a typical 1.5X–2X cycle time increase requires a voltage increase by 100–200 mV to sustain the same performance as the nominal corner, which leads to an energy increase by a factor of 1.5X–2X. In other words, variations at near threshold entail a very large energy cost, which can negate the advantages of operating at the MEP. Accordingly, variations need to be accounted for in first place in the design of near-threshold circuits rather than an afterthought, as discussed more in detail in the following. Similar considerations hold for the sensitivity to soft errors, which is somewhat increased at near threshold voltages, compared to nominal voltage. On the other hand, operation at near-threshold voltages suppresses most of aging and reliability issues, such as Bias Temperature Instability, Hot Carrier Injection, Time Dependent Dielectric Breakdown. Indeed, such phenomena are all exponentially dependent on the supply voltage, and its reduction to near-threshold voltage substantially mitigates them.

4.7.1   Process Variations

Random (within-die) process variations are well known to be responsible for a major fraction of the cycle time margin, having a much heavier effect than fully correlated (die-to-die) variations (Orshansky et al. 2008). Indeed, threshold voltage variations at low voltages are dominated by random dopant fluctuations (Alioto et al. 2010; Orshansky et al. 2008), and their effect requires much more sophisticated feedback schemes that are immune to transistor mismatch (e.g., with

timing error detection or prediction (Bowman et al. 2009; Bowman et al. 2011; Das et al. 2009; Khayatzadeh et al. 2016; Zhang et al. 2016)), rather than corner-based adaptive voltage scaling and body biasing techniques (Gregg and Chen 2007; Meijer and Pineda de Gyvez 2012; Martin et al. 2002; Olivieri et al. 2005; Tschanz et al. 2002). Accordingly, our analysis in the following will be focused on random variations.

At low voltages, process variations determine a much larger path delay variations than above-threshold voltages due to two phenomena:

1. the variability of the gate delay defined as the ratio between the standard deviation and mean value increases significantly
2. the probability distribution function (PDF) of such delay is no longer Gaussian for short paths, and has a longer tail on the right side.

Regarding the first phenomenon, the variability of the critical path delay is mainly due to the intrinsically larger variability of the transistor I_{on} current (see Eq. (4.6)). This is mostly due to the larger impact of the threshold voltage, and hence of its variations, at lower voltages (see Sect. 4.2.1). More quantitatively, the delay variability is approximately equal to the variability in I_{on} from (4.6). If the nominal threshold voltage V_{TH} is subject to a variation ΔV_{TH} that is Gaussian distributed with zero mean and standard deviation $\sigma_{V_{TH}}$, from (4.3) to (4.4) the variability of I_{on} for above- and sub-threshold voltages is readily found to be

$$\left.\frac{\sigma_{\tau_{PD}}}{\mu_{\tau_{PD}}}\right|_{above-threshold} \approx \left.\frac{\sigma_{I_{on}}}{\mu_{I_{on}}}\right|_{above-threshold} = \frac{\sigma_{V_{TH}}}{V_{DD} - V_{TH}} \tag{4.27a}$$

$$\left.\frac{\sigma_{\tau_{PD}}}{\mu_{\tau_{PD}}}\right|_{sub-threshold} \approx \left.\frac{\sigma_{I_{on}}}{\mu_{I_{on}}}\right|_{sub-threshold} = \sqrt{e^{\left(\frac{\sigma_{V_{TH}}}{n \cdot v_t}\right)^2} - 1} \tag{4.27b}$$

For example, for the typical values $\sigma_{V_{TH}} = 35$ mV and $V_{TH} = 0.4$ V in 28 nm CMOS, the gate delay variability turns out to be 7% at 0.9 V, and

124% in the sub-threshold region (e.g., 0.3 V). In other words, the delay variability in sub-threshold is an order of magnitude larger

Fig. 4.43 Variability of *FO*4 normalized to value at 1.2 V vs. V_{DD} (28 nm CMOS)

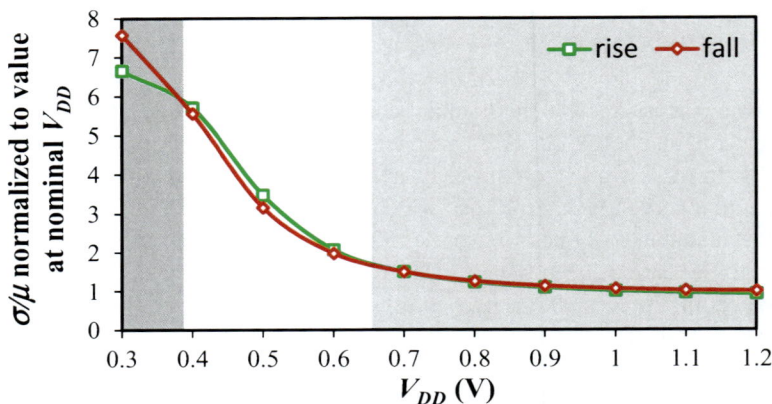

than above threshold. At near-threshold voltages, the variability is somewhat intermediate.

Figure 4.43 plots the variability of the *FO*4 delay normalized to the value at nominal voltage in 28 nm CMOS. From this figure, the gate delay variability increases when V_{DD} is reduced, and becomes 2X–6X larger than at nominal V_{DD} at near threshold, and about an order of magnitude larger in the deep sub-threshold region. A typical delay variability of around 6–8% at nominal voltage in 28 nm translates into a sizable delay variability of various tens of percentage points at near threshold. To achieve a parametric yield of approximately 99%, three standard deviations are needed, hence the margin for a single gate can easily be 100%, which entails an unfeasibly large margin in Fig. 4.41 (i.e., an unacceptably high energy cost from Fig. 4.42, which easily offsets the energy benefit of operating at near-threshold voltages). When the MEP is in sub-threshold region, such margin becomes even higher.

Regarding the second phenomenon that was observed above, the statistical delay distribution of a single gate is no longer Gaussian when operating at near- and sub-threshold voltages (Alioto 2012; Alioto et al. (in press); Gammie et al. 2011). In the sub-threshold region, I_{on} and hence the gate delay are lognormally distributed due to the exponential dependence of I_{on} on the threshold voltage in (4.4), being the latter Gaussian distributed. As shown in Fig. 4.44, the

lognormal distribution has a much longer tail compared to the Gaussian distribution, at same standard deviation. This leads to a considerable increase in the number of standard deviations needed as design margin to meet a given yield target, as shown in Table 4.7. For example, from this table the worst-case gate delay margin across 99.9% of the cases is three standard deviations for Gaussian (i.e., above threshold), and twenty standard deviations for lognormal (i.e., sub-threshold). In the near-threshold region, the distribution is somewhat intermediate between above- and sub-threshold, and hence it is neither perfectly Gaussian nor lognormal. This is shown in Fig. 4.45a–c, which show the quantile-quantile (Q–Q) plot (Walpole et al. 2006) of the statistical *FO*4 delay sample in 28 nm CMOS versus the theoretical quantiles of a Gaussian distribution with same mean and standard deviation. The deviation from a straight line (i.e., perfect Gaussian behavior) of the Q–Q plot becomes noticeable at near threshold (see Fig. 4.45b), and is substantial at sub-threshold voltages (see Fig. 4.45c). Figure 4.45d confirms the *FO*4 lognormal distribution in sub-threshold.

The above considerations of non-Gaussian delay distribution at low voltages hold for single logic gates, and can be extended to short paths, i.e., paths that can be problematic in terms of hold time violations rather than setup time. Accordingly, short paths and hold fix at sub- and near-threshold voltages requires a much wider design

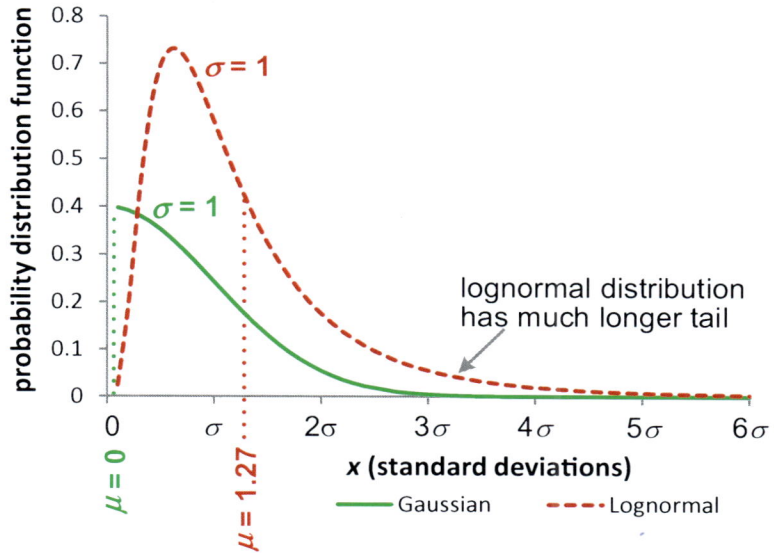

Fig. 4.44 Probability density function of Gaussian and lognormal distribution at same standard deviation ($\sigma = 1$)

Table 4.7 Number of V_{TH} Standard Deviations beyond the Mean to Achieve Given Yield Target in Gaussian and Lognormal

Yield target	Single gate		Logic path (lognormal gate delay)			
			Logic depth ↓			
	Gaussian	Lognormal	4 gates	8 gates	16 gates	32 gates
84%	$\sigma_{V_{TH}}$	$e \cdot \sigma_{V_{TH}}$	$1.07 \cdot \sigma_{V_{TH}}$	$1.05 \cdot \sigma_{V_{TH}}$	$1.03 \cdot \sigma_{V_{TH}}$	$1.02 \cdot \sigma_{V_{TH}}$
97.7%	$2\sigma_{V_{TH}}$	$e^2 \cdot \sigma_{V_{TH}} \approx 7.4 \cdot \sigma_{V_{TH}}$	$2.07 \cdot \sigma_{V_{TH}}$	$2.05 \cdot \sigma_{V_{TH}}$	$2.03 \cdot \sigma_{V_{TH}}$	$2.02 \cdot \sigma_{V_{TH}}$
99.87%	$3\sigma_{V_{TH}}$	$e^3 \cdot \sigma_{V_{TH}} \approx 20.1 \cdot \sigma_{V_{TH}}$	$3.07 \cdot \sigma_{V_{TH}}$	$3.05 \cdot \sigma_{V_{TH}}$	$3.03 \cdot \sigma_{V_{TH}}$	$3.02 \cdot \sigma_{V_{TH}}$
99.997%	$4\sigma_{V_{TH}}$	$e^4 \cdot \sigma_{V_{TH}} \approx 54.6 \cdot \sigma_{V_{TH}}$	$4.07 \cdot \sigma_{V_{TH}}$	$4.05 \cdot \sigma_{V_{TH}}$	$4.03 \cdot \sigma_{V_{TH}}$	$4.02 \cdot \sigma_{V_{TH}}$
99.99997%	$5\sigma_{V_{TH}}$	$e^5 \cdot \sigma_{V_{TH}} \approx 148.4 \cdot \sigma_{V_{TH}}$	$5.07 \cdot \sigma_{V_{TH}}$	$5.05 \cdot \sigma_{V_{TH}}$	$5.03 \cdot \sigma_{V_{TH}}$	$5.02 \cdot \sigma_{V_{TH}}$
99.9999999%	$6\sigma_{V_{TH}}$	$e^6 \cdot \sigma_{V_{TH}} \approx 403.4 \cdot \sigma_{V_{TH}}$	$6.07 \cdot \sigma_{V_{TH}}$	$6.05 \cdot \sigma_{V_{TH}}$	$6.03 \cdot \sigma_{V_{TH}}$	$6.02 \cdot \sigma_{V_{TH}}$

margin, compared to above-threshold. In other words, the timing margin against hold time violations at low voltages tends to be very large compared to nominal voltage, and hence requires a much larger number of hold fix buffers.

On the other hand, long logic paths have a Gaussian delay distribution even in sub-threshold voltages. This is because of the Central Limit theorem, which guarantees that the sum of non-Gaussian random variables rapidly tends to a Gaussian distribution, when increasing the number of variables being summed (Walpole et al. 2006) (e.g., the number of logic gates whose delays are added to derive the critical path delay). This is quantitatively shown in Table 4.7 for 4, 8, 16 and 32 equal cascaded gates, which are individually assumed to have a lognormal delay distribution, as relevant to the sub-threshold region. Indeed, this table shows that margin in terms of standard deviations is essentially the same as an ideal Gaussian distribution even for a relatively short path of 4 cascaded gates, and is closer for a larger number of gates. This means that the clock cycle distribution is Gaussian at any voltage, and hence the margin in terms of standard deviations is the same as nominal voltage. In other words, the timing margin against setup time violations at low voltages scales like $FO4$, as opposed to hold violations.

Fig. 4.45 Q–Q plots of
*FO*4 distribution (y axis)
in 28 nm CMOS at
(**a**) 1.2 V, (**b**) 0.6 V,
(**c**) 0.3 V with normal
distribution on the x axis.
(**d**) at 0.3 V with lognormal
distribution on the x axis
(100,000 Monte Carlo
runs)

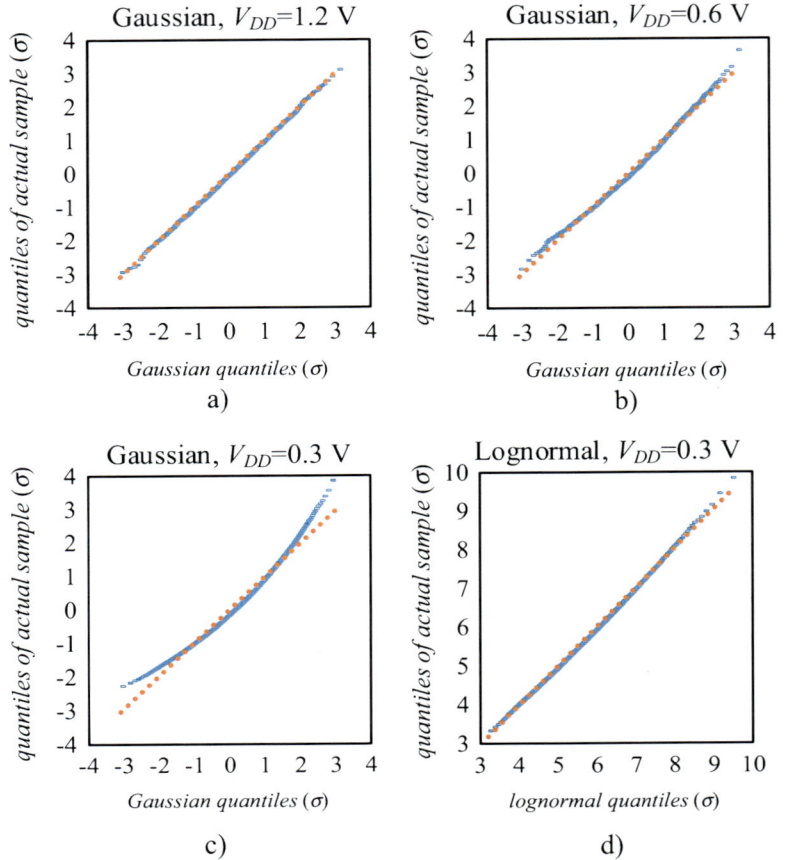

4.7.2 Voltage and Temperature Variations

Voltage variations have a heavy impact on the clock cycle margin, due to the strong sensitivity of I_{on} and hence the gate delay on V_{DD} (see Sect. 4.2.1). Figure 4.46 plots the cycle time margin associated with a typical 5% and 10% voltage drop of the circuit in Fig. 4.30 in 28 nm CMOS. As all gate delays scale approximately like *FO*4 when V_{DD} changes, this example is representative of any logic path. From this figure, a large cycle time margin of 20–50% is imposed by supply variations, if not kept under strict control. As discussed in Sect. 4.10, supply variations in systems designed for nearly-minimum energy operation are dominated by fluctuations in the output voltage of the regulator providing the supply. Accordingly, supplies for minimum-energy operation need to be designed with quite stringent specifications on voltage stability across temperatures, as well as line and load regulation.

Temperature variations in circuits designed for minimum energy have an effect that is quite different from above-threshold circuits. Indeed, larger temperatures lead to a substantial increase in the energy per operation at low voltages, due to the large contribution of the leakage energy (see Sect. 4.3.2). Such effect is more pronounced in architectures with larger leakage energy, e.g., with larger logic depth. Figure 4.47 plots the energy versus V_{DD} of the circuit in Fig. 4.30 at different temperatures (27 °C and 70 °C), and for logic depths widely ranging from 25*FO*4 to 100*FO*4. From this figure, the minimum energy

Fig. 4.46 Plot of the cycle time margin vs. V_{DD} (assuming cycle time scaling to be proportional to $FO4$)

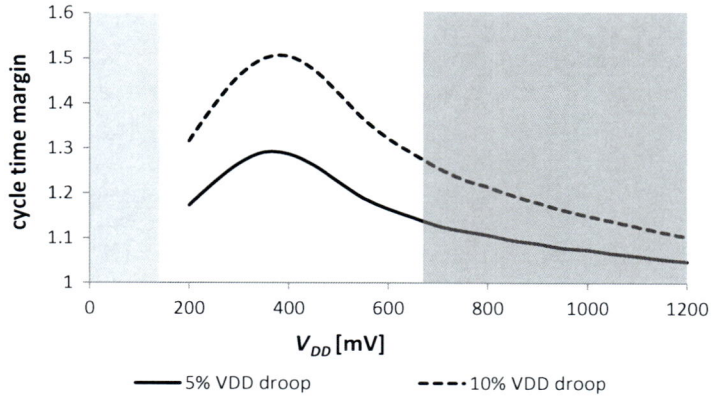

Fig. 4.47 Impact of temperature on minimum energy vs. V_{DD} for different logic depths (10% activity, very low V_{TH})

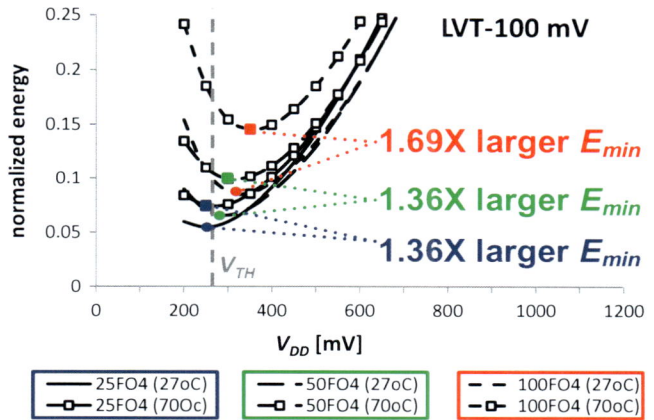

is heavily influenced by the operating temperature, as it is increased by a factor of 1.3X for well-designed architectures with reasonable logic depth, and 1.7X for less energy-efficient and leakier architectures.

As further difference compared to traditional above-threshold low-power designs, the performance of circuits operating around the MEP actually benefits from increased temperature. Indeed, the I_{on} transistor current is much more sensitive to the threshold voltage rather than the carrier mobility, hence it increases at larger temperatures. Figure 4.48 shows the $FO4$ delay versus V_{DD} for various threshold voltages. From this figure, a temperature raise from 27 °C to 70 °C leads to a 1.4X–2X $FO4$ reduction at V_{DD} equal to the threshold voltage V_{TH0} in

(4.7). Such effect is less pronounced at higher threshold voltages, as the corresponding higher supply voltage emphasizes the carrier velocity saturation and the mobility degradation due to high-field operation, which in turn weaken the dependence of I_{on} on V_{TH0}. At above-threshold voltages around 700–800 mV, the effect of temperature is insignificant. At larger voltages, the temperature has the traditional inverse effect on the performance, and is much weaker (e.g., 2% change due to a temperature change from 27 °C to 70 °C) than near threshold. Hence, unless the operating temperature range set by the application is narrow (e.g., indoor applications), active compensation of temperature variations is essential in any integrated system aiming at minimum-energy operation.

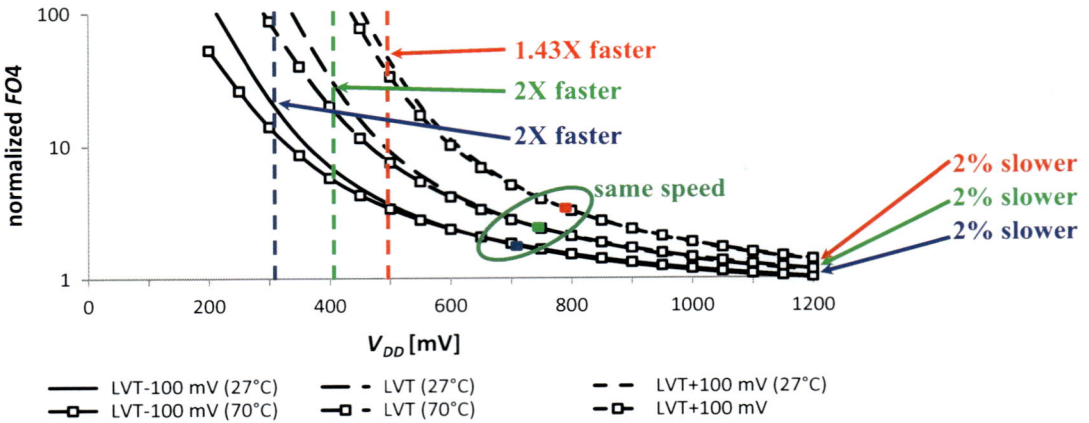

Fig. 4.48 Impact of temperature on performance vs. V_{DD} for different threshold voltages ($FO4$ is normalized to the value at nominal voltage for lowest V_{TH})

Table 4.8 Dependence of the delay variability in logic paths, cells and transistors

		Design parameter	Delay variability dependence
Logic path		logic depth LD	$\propto \frac{1}{\sqrt{LD}}$
Standard cell		$N_{stacked}$	$\propto \frac{1}{\sqrt{N_{stacked}}}$
Transistor		channel width W	$\propto \frac{1}{\sqrt{strength}}$

4.8 The Leakage-Variability Tradeoff

Operation around the MEP introduces a tradeoff that is not encountered in traditional low-power above-threshold designs, namely the variability-leakage tradeoff. This is an unavoidable tradeoff that constrains the design at all levels of abstraction and is tightly linked to the averaging effect of additive variations, as discussed below.

At the gate level, a logic path with logic depth LD has a delay that is the sum of LD delays, as depicted in Table 4.8. The resulting path delay variability is inversely proportional to $\sqrt{LD}$ thanks to the averaging effect of the random variations across cascaded gates (Alioto et al. 2010; Merrett et al. 2010) (i.e., more cascaded gates reduce the overall delay variability thanks

to better averaging across a larger number of cells). Hence, the reduction of the delay variability would require the adoption of microarchitectures with larger logic depths. On the other hand, larger logic depths increase the clock cycle and hence the leakage energy per cycle from (4.10). In other words, the mitigation of delay variations comes at the cost of a higher leakage energy, and vice versa. Such tradeoff is very specific to operation at near- and sub-threshold voltages, due to the much more important contribution of the leakage energy, as opposed to above-threshold designs.

At the cell circuit level, a similar tradeoff is encountered when the number of stacked transistors is considered in a standard cell (Alioto et al. 2010; Merrett et al. 2010) (i.e., the cell fan-in). Indeed, the variability of the I_{on} current delivered by the cell to the load, and hence the cell delay variability, is inversely proportional to $\sqrt{N_{stacked}}$ as in Table 4.8. Thus, the mitigation of variations through more stacked transistors comes at the cost of larger delay (see considerations on stacking in Sect. 4.2.3) and hence larger leakage energy from (4.10).

At the transistor level, from Table 4.8 wider transistors exhibit smaller I_{on} variability thanks to the Pelgrom's law (Pelgrom et al. 1989). Hence cells with larger strength have smaller delay variability, as the latter is inversely proportional to $\sqrt{strength}$. Again, the delay variability mitigation comes at the cost of larger leakage energy, as the leakage current drawn by a cell is proportional to its strength.

Summarizing, the variability-leakage tradeoff is an unescapable challenge in the design of circuits and systems operating around the MEP, as opposed to traditional low-power above-threshold designs. This tradeoff involves all levels of abstraction, and needs to be constantly taken care of during the design process. Such tradeoff can be broken by introducing innovative design techniques that do not purely rely on timing margining, as discussed later in this chapter.

4.9 Near-Threshold Cell Libraries

Designing cell libraries for operation around the MEP certainly helps manage the peculiar

tradeoffs observed at near-threshold voltages in a more efficient manner, at the cost of additional design and characterization effort.

Since performance is not the main objective, near-threshold cell libraries can be designed with short standard cells (e.g., 7 metal tracks), as transistors typically do not need to be wide, as depicted in Fig. 4.49. In near-threshold designs, taller cells (e.g., 10–12 metal 1 tracks) can achieve higher performance but lead to a significant area efficiency degradation, and longer interconnects, thus degrading energy efficiency.

The composition of near-threshold libraries does not need to be as wide as libraries for above-threshold (i.e., higher performance) operation. Indeed, cell versions with very large strength can be suppressed, as they are typically not used due to the more relaxed performance constraints. Similarly, cells with large fan-in need to be eliminated, since they suffer from disproportionately larger delay, as discussed in Sect. 4.2.3. Typically, libraries with around 100 cells are adequate for near-threshold designs. From observations of prior designs, the energy reduction obtained through a custom near-threshold library can be in the order of 20% (Gemmeke et al. 2013; Gammie et al. 2011), compared to a pruned out conventional library for above-threshold voltages (see below).

The circuit design of cells is affected by near-threshold operation in terms of sizing as well.

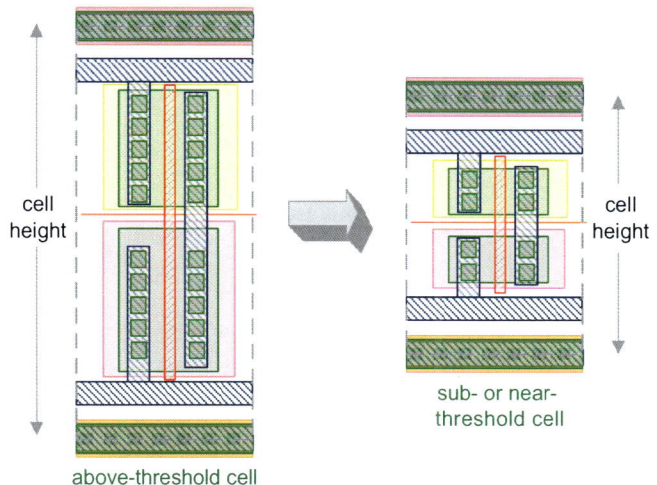

Fig. 4.49 Near-threshold cells are shorter than typical above-threshold cells

Indeed, minimum transistor size needs to be skipped in technologies that are significantly affected by Narrow Channel Effects (see Sect. 4.2.2), to avoid the related increase in the transistor threshold voltage.

Near-threshold libraries might need to be enriched with cells that are normally not available in above-threshold libraries. For example, cells with thick-oxide transistors might be needed for always-on blocks (see Chap. 1) that need to be very low leakage, or connected directly to 3.6-V LiIon batteries (see Chap. 15). Being particularly critical in terms of the minimum voltage V_{min} assuring correct operation, flip-flops usually need to be thoroughly redesigned to achieve adequate functional yield at low voltages. This is usually achieved through circuit techniques that eliminate the potential current contention between transistors (Jain et al. 2012; Kim et al. 2014a). V_{min} is further reduced by replacing conventional dynamic circuits (e.g., periphery in register files) by their static CMOS counterparts. As summarized in Fig. 4.50, V_{min} is determined by several contributions arising at the process and circuit level, and is certainly dominated by variations (Alioto 2012).

As an alternative option, existing cell libraries designed for above-threshold regions can be reused at lower voltages, after proper pruning to eliminate the cells that suffer from robustness issues or particularly pronounced delay increase (Alioto 2010; Wang et al. 2006). In a given library designed for above-threshold voltages, the number of usable cells at lower voltages decreases when reducing V_{DD}, as fewer cells can operate reliably at lower voltages. Typically, as summarized in Fig. 4.51, the suppression of cells with a high fan-in (e.g., 4) leads to approximately 100-mV V_{min} reduction (Gemmeke et al. 2013).

4.10 Clock and Supply Networks for Near-Threshold Operation

The design of clock networks for near-threshold designs is very different from above-threshold networks, due to the very different balance between clock repeater and wire delay, and the clock skew is determined by different dominant mechanisms (Alioto 2014; Lin et al. 2017; Seok et al. 2011; Tolbert et al. 2011). At above-threshold voltages, several levels of clock

Fig. 4.50 Breakdown of minimum supply voltage V_{min} of logic gates ensuring correct operation (Alioto 2012)

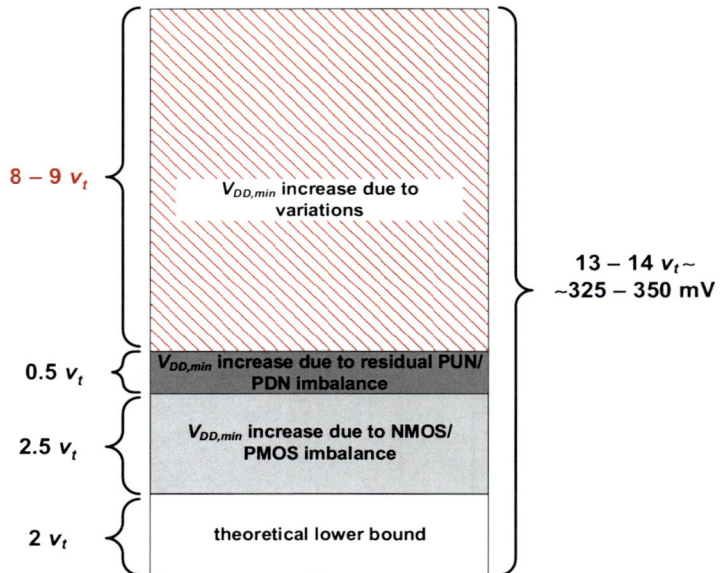

$8 - 9\ v_t$ — $V_{DD,min}$ increase due to variations

$13 - 14\ v_t \sim$ $\sim 325 - 350$ mV

$0.5\ v_t$ — $V_{DD,min}$ increase due to residual PUN/PDN imbalance

$2.5\ v_t$ — $V_{DD,min}$ increase due to NMOS/PMOS imbalance

$2\ v_t$ — theoretical lower bound

Fig. 4.51 Percentage of library cells operating correctly vs. V_{DD} (Gemmeke et al. 2013)

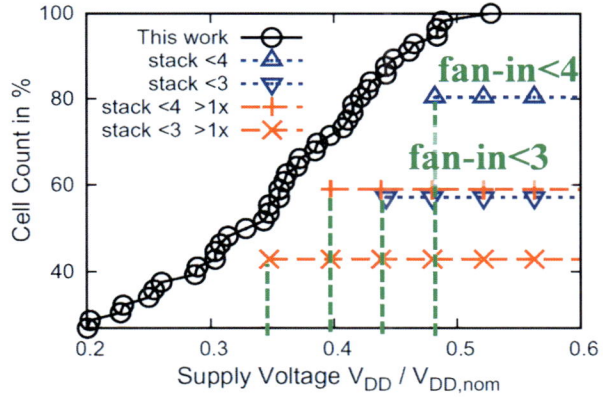

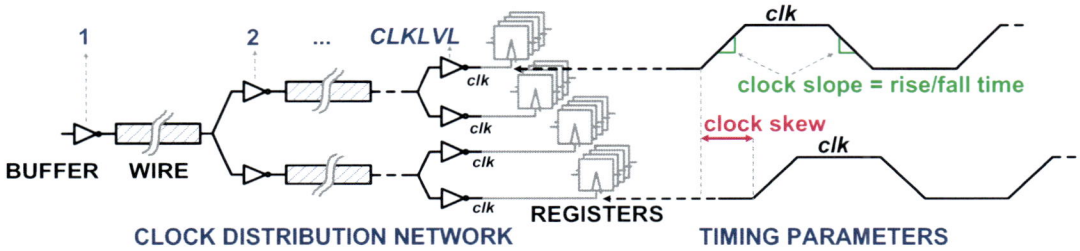

Fig. 4.52 General clock network structure and related timing parameters (Alioto 2014)

repeaters are needed to frequently interrupt wires to limit the related RC time constant and hence clock slope through the wires (Xanthopoulos 2009) (see Fig. 4.52). Indeed, excessive clock slope induces large random delay variations in the clock repeaters at intermediate nodes of the clock network, and degrades flip-flop nominal timing parameters (as well as its variations) when considering the sinks of the same network. In other words, the significant wire RC delay and its impact on clock skew through the clock slope justifies the adoption of deep above-threshold clock networks.

At sub- and near-threshold voltages, the gate delay becomes much larger than the wire delay (see Sects. 4.2.3 and 4.6), hence the clock slope through wires is no longer an issue, and the random skew is dominated by the intrinsic variations in the clock repeaters. According to the Central Limit theorem (Walpole et al. 2006), the random skew standard deviation is proportional to the square root of the number of

cascaded repeaters, i.e., of the depth of the clock network. Accordingly, shallow networks need to be used at sub- and near-threshold voltages, so that the dominant skew contribution due to the number of clock repeaters is reduced.

From the above considerations, the design the clock network at a given voltage leads to a skew degradation at the other end of the voltage range. As an example, Fig. 4.53 plots the skew of a clock network in 28 nm that has been designed at 1.2 V and used at lower voltages. This figure shows that the skew at low voltages becomes several $FO4$ and even exceeds $10FO4$ in sub-threshold. This means that the skew in a clock network used in a wide range of voltages becomes a large fraction of typical cycle time targets of energy-efficient designs (see Sect. 4.4.1). In other words, using a clock network in a wide voltage range leads to significant performance degradation (or energy efficiency, if V_{DD} is increased to recover the lost performance). Similarly, the clock skew easily exceeds

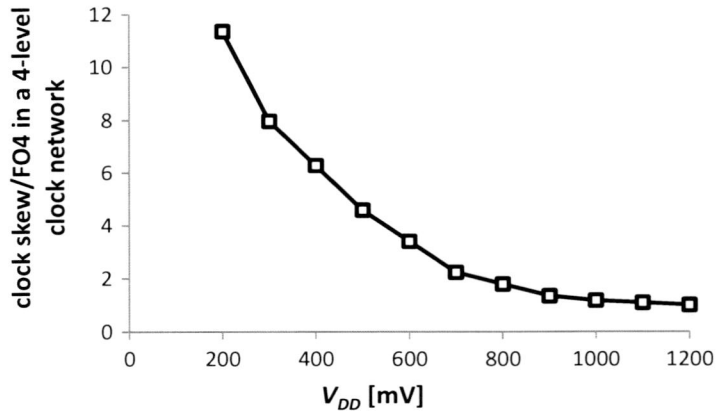

Fig. 4.53 Clock skew of a sample clock network in 28 nm designed at 1.2 V (normalized to $FO4$) (Alioto 2014)

the available hold margin (Alioto et al. 2015), thus leading to timing failures at low voltages. In other words, the clock skew degradation at low voltages typically defines V_{min}. Similar trends are observed when designing the clock network at low voltages and running at above-threshold voltages.

From the above considerations, the design of the clock network of integrated systems operating in a wide voltage range entails a fundamental tradeoff between the performance at above-threshold voltages, and the ability to scale down to low voltages. Various approaches have been proposed to make this tradeoff more favorable, and mitigate the skew-energy penalty imposed by the adoption of deep or shallow clock networks. For example, moderately deep networks with long-channel LVT buffers have been proposed in Myers et al. (2015). Design methodologies have been introduced in Seok et al. (2011), Tolbert et al. (2011), Zhao et al. (2012) to optimally design clock networks, although for a single low voltage. Techniques for adaptive point-to-point interconnects with regenerative drivers have been also proposed (Kim et al. 2014b; Wang et al. 2015), although they cannot be used for clock networks and are not supported by commercial EDA tools. Voltage-adaptive delay insertion across different clock domains was introduced in (Jain et al. 2012; Tokunaga et al. 2014) to mitigate the inter-domain skew (e.g., between processor and

memory), although no adaption to voltage has been performed within each clock domain. Clock network adaptation to a wide range of voltages with each clock domain has been demonstrated in Lin et al. (2017), where the clock network topology is reconfigured to minimize the skew at each specific voltage.

Regarding the supply network, voltage drops are less of a concern at near-threshold voltages and below, as the I_{on} transistor current is at least an order of magnitude lower than at nominal voltage. Accordingly, the current density drawn by the digital circuit is reduced by the same amount, and hence issues related to voltage drops across the supply network are largely mitigated. This partially alleviates the problem of the stronger impact of V_{DD} fluctuations at near-threshold voltages, due to the larger sensitivity of performance (see Sect. 4.7.2). This translates in a relaxed requirement on the supply rail width in the cell library, which can help slightly reduce the cell height. For analogous reasons, the lower clock frequency of near-threshold circuits makes the effect of the wire parasitic inductance negligible. Finally, the peak current absorbed by near-threshold circuits is also reduced by an order of magnitude, compared to above-threshold operation. Hence, the size of decoupling capacitors to keep V_{DD} fluctuations within a targeted band can be reduced by the same amount, thus saving area and improving the utilization factor of the module under design.

4.11 Perspectives and Trends

In summary, near-threshold circuits pose challenge and opportunities that are significantly differ from conventional above-threshold low-power circuits. Counteracting leakage in spite of the inefficacy of conventional leakage reduction techniques (see Sect. 4.5) requires a radically different approach that maximizes the opportunities to reduce leakage when transistors are not being used. This can be accomplished by introducing fine-grain power domains that can be power gated (e.g., with gate boosting to improve its effectiveness, as in Sect. 4.5) or voltage scaled to mitigate the leakage contribution of unused transistors. Power domains are typically coarse and of the size of at least an entire microprocessor, whereas such fine-grain power domains have the size of sub-blocks or execution units (e.g., ALU), or even finer (e.g., individual operators in the ALU). Although such approach certainly enhances the chances to turn off transistors, its direct application leads to significant area/energy/performance overhead. The latter is due to the need for additional power domain control circuitry, as well as isolation/clamping cells for power gating and level shifters (see Chap. 9) at the boundary of each domain.

Fine-grain voltage domains are also a highly promising approach in near-threshold circuits. Indeed, the ability to distribute different voltages with fine granularity maximizes the opportunities to correct variations in paths that turn out to be critical due to random variations, while reducing the energy in all other domains. The effectiveness of fine-grain voltage domains is further enhanced by the strong sensitivity of performance on V_{DD} (see Sect. 4.2), which ensures that voltage boosting is kept small (e.g., 100–200 mV) in all practical cases. For example, selective boosting can be used to reduce the general V_{min} of the circuit, while raising the voltage of the small portion of the circuit that needs to operate at higher voltages (Tokunaga et al. 2014). As another example, (Muramatsu et al. 2011) leverages such small voltage difference across voltage domains by suppressing level shifters altogether, so that the voltages can be freely assigned to very small domains to compensate variations where they arise, while avoiding the otherwise large overhead of level shifters. The Panoptic approach (Putic et al. 2009) introduces both spatial and temporal fine granularity by using multiple sleep transistors that also dynamically connect sub-blocks to three different supply voltages. The sleep transistors serve the purpose of reducing leakage of unused sub-blocks, and assign them the minimum possible voltage for the task at hand when used.

Variations can also be exploited rather than added as design margin, when an adequately large number of replicas of a given block are available on the same chip. For example, Raghunathan et al. (2013) introduces the concept of "cherry picking" among many on-chip cores, which consists in the post-silicon selection of the most energy-efficient cores while keeping others off. This permits to maximize the energy efficiency by leveraging the inevitable random variations, rather than tolerating them, at the cost of area due to the partial utilization of cores. Observe that full utilization would not be allowed anyway in practical cases, due to the "dark silicon" issue (see Chap. 1) that is determined by the chip power constraint.

In general, variations can be mitigated at different times, from design time to testing, chip boot time and run-time, as summarized in Fig. 4.54. At design time, all variation contributions need to be incorporated into the design (e.g., cycle time) margin, as they are not known upfront. The margin is lowered at testing time, as process variations are known and can hence be suppressed, whereas voltage, temperature and aging-induced variations are need to be included (as they will be defined later at in-field operation). At boot time, aging can be compensated as well. The margin is made very small and virtually removed when variations are compensated at run-time, i.e., when all process, temperature, (slow) voltage variations are known. Obviously, the cost of such detection and compensation of variations increases when moving from design to run-time.

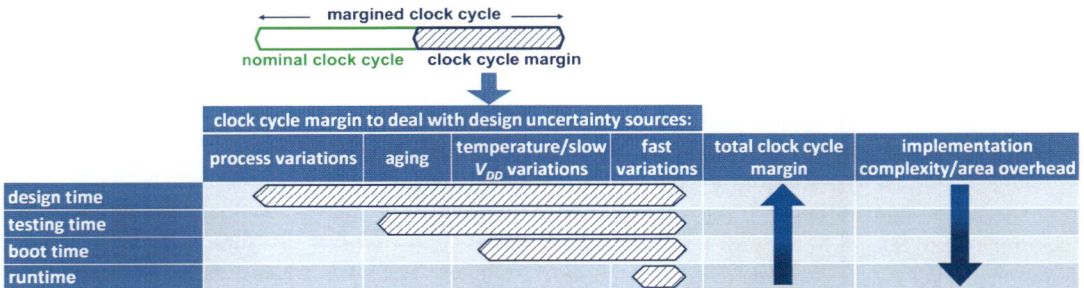

Fig. 4.54 Summary of techniques to counteract variations at different time, and resulting cycle margin and overhead

Due to very large design margin required at near-threshold voltages (see Sect. 4.7), adequate yield and energy efficiency certainly require the adoption of run-time compensation of variations. This is typically performed through timing error detection and correction (EDAC) methods, which have been investigated since early 2000s (Ernst et al. 2003), EDAC methods sense the timing margin at run time by detecting timing failures, so that the system can be tuned to operate at nearly-zero margin (Ernst et al. 2003). This permits to run at the highest possible frequency at given voltage, or at the minimum possible voltage at given frequency. Hence, error detection and correction improves the energy efficiency of circuits operating at any voltage, typically by 1.3–1.45X (see references below).

Error detection can be performed through canary circuits and Tunable Replica Circuits (TRCs) mimicking critical path variations, and hence predicting the occurrence of timing violations with high (but not 100%) level of confidence, at rather low overhead (Bowman et al. 2011). However, tracking the critical path across a wide range of voltages is difficult, and hence such methods are more appropriate for operation on a narrow range. Also, since TRCs try to replicate the critical path, they cannot completely eliminate the design margin. *In-situ* error detection is performed by inserting timing sensors to detect true timing failures, which typically entails significant area overhead. Several *in-situ* error detection methods have been proposed, such as Razor (Ernst et al. 2003), Razor II (Das et al. 2009), EDS (Bowman et al. 2009; Bowman et al. 2011), ERSA (Leem et al.

2010), Bubble Razor (Fojtik et al. 2013). However, their overhead is in the order of various (if not several) tens of percentage points, and hence an order of magnitude larger than TRCs, which has prevented their adoption in commercial chips. Recently, very low-overhead (i.e., percentage points) *in-situ* approaches have been demonstrated, such Razor-Lite (Kwon et al. 2014) and iRazor (Zhang et al. 2016) for processors, and RazorSRAM for on-chip memories (Khayatzadeh et al. 2016). Being very lightweight, these approaches promise a much wider adoption of *in-situ* error detection in mass produced chips.

Another very promising direction to further reduce the energy per computation is offered by its tradeoff with quality. As discussed in Chap. 1, quality can be defined in different ways depending on the application and the sub-system under design. In a processing sub-system, quality is related to accuracy in terms of precision in case of arithmetic tasks, misclassification rate in classification tasks, or effective number of bits in an Analog-to-Digital Converter (ADC). The concept is far more general than approximations (e.g., approximate computing), in that it applies to a broad range of types of tasks and applications, and the tradeoff between quality and energy is dynamic and based on quality sensing (see example in Sect. 1.6.2).

Based on the concepts described in Sect. 1.6.2, energy-quality scalability has been introduced in many different sub-systems and levels of abstraction. For example, the first energy-quality SRAM memory has been

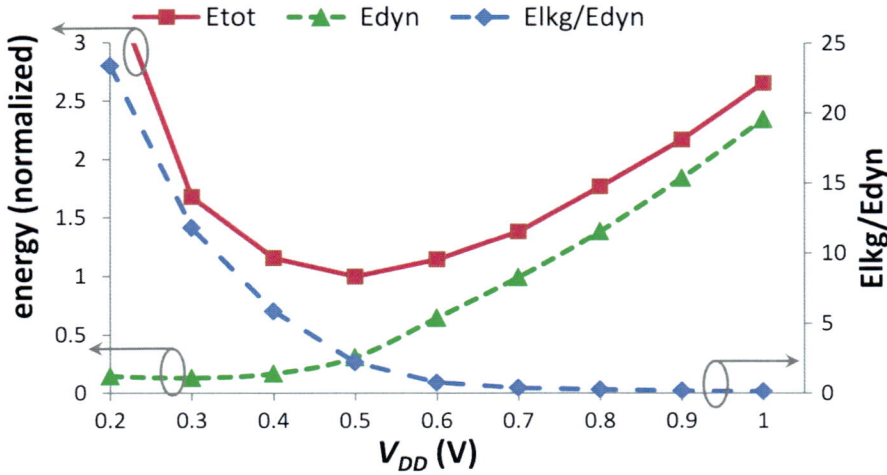

Fig. 4.55 Energy curve vs. V_{DD} in a 32-bit multiplier in 28-nm technology

introduced in Frustaci et al. (2015), where occasional faults (e.g., bitcells with inadequate write- or read-ability) occur in the array. The scalability comes from a bit-level management of the tradeoff between the bit error rate and the energy, by adjusting assist techniques (see Chap. 5) differently for different positions and in a dynamically scalable manner. This is beyond traditional memories where assist is uniformly applied to all bit positions and to fully suppress errors, which entails a substantial error cost. Similarly, selective Error Correction Codes have been introduced in the SRAM, to favor the robustness of the bits carrying the highest information content (e.g., MSBs in video processing applications), while saving on the other bit positions. Overall, this approach leads permits to improve the general quality by spending some energy in selected bit positions, thus enabling much more aggressive scaling on all positions and hence achieving quadratic benefit. Energy reductions of 2X have been demonstrated compared to traditional voltage scaling, at iso-quality (Frustaci et al. 2016). The same general concept has been applied to several other sub-systems, such as ADCs with dynamically scalable resolution (Freyman et al. 2014; Yip et al. 2011). In this case, when the application can tolerate a reduction in the ADC resolution, a more than 2X energy reduction is gained when

the resolution is reduced by one bit, leading to an exponential energy saving.

Finally, the presence of a minimum-energy point (MEP) actually poses a fundamental challenge in terms of energy scalability when the system is operating at the right of the MEP. Indeed, the MEP tends to be a flat minimum as discussed in Sect. 4.3.2, which in turns translates into insignificant energy savings when the voltage is scaled down from values at the right of the MEP towards the MEP itself. As an example, from Fig. 4.55 there is almost no energy saving when scaling from 0.6 V (i.e., at the right of the MEP) down to 0.5 V (i.e., the MEP), due to the flatness of the MEP. In other words, the voltage scalability of the design (i.e., its ability to operate at very low voltages) does not translate in an actual energy scalability (i.e., the ability to reduce energy when scaling down the voltage). To preserve energy scalability, the energy curve in Fig. 4.55 needs to be steep rather than flat, which is achieved only if the operating voltage is far enough on the right side of the MEP. For example, quadratic benefit is observed in this figure, at voltages from 0.8 V to 1 V. Conversely, to achieve good energy scalability at a given low voltage (e.g., 0.5 V), the MEP needs to be pushed to the left of this targeted voltage (e.g., 0.3 V). In other words, innovation is needed to move the MEP where needed, depending on the operating voltage. At above-threshold voltages, the MEP can lie at a fairly high voltage, while not being a problem since the dominance of the dynamic energy still assured a quadratic benefit when downscaling V_{DD}. When a near-threshold voltage is targeted and further voltage scaling needs to be applied, the MEP needs to be dynamically moved to the left to make the energy curve steeper, and again achieve a nearly-quadratic energy benefit. We believe that this is one of the fundamental challenges that needs to be addressed to further improve the energy efficiency of low-voltage integrated systems for IoT.

Acknowledgement The authors acknowledge the kind support by the MOE2014-T2-2-158 grant from the Singaporean Ministry of Education.

References

F. Abouzeid, S. Clerc, B. Pelloux-Prayer, F. Argoud, P. Roche, 28 nm CMOS, energy efficient and variability tolerant, 350 mV-to-1.0 V, 10 MHz/700MHz, 252bits frame error-decoder, in *Proceedings of ESSCIRC 2012* (Bordeaux, France, Sept. 2012), pp. 153–156

M. Alioto, G. Scotti, A. Trifiletti, A novel framework to estimate the path delay variability via the fan-out-of-4 metric. IEEE Trans. Circuits Syst—Part I (in press)

M. Alioto, Understanding DC behavior of subthreshold CMOS logic through closed-form analysis. IEEE Trans. Circuits Syst.—part I **57**(7), 1597–1607 (2010)

M. Alioto, Ultra-low power VLSI circuit design demystified and explained: a tutorial. IEEE Trans. Circuits Syst.—part I (invited) **59**(1), 3–29 (2012)

M. Alioto, Challenges and techniques for ultra-low voltage logic with nearly-minimum energy. in *Short course at VLSI Symposium 2014*, Hawaii 10 June 2014

M. Alioto, G. Palumbo, M. Pennisi, Understanding the effect of process variations on the delay of static and Domino logic. IEEE Trans. VLSI Syst. **18**(5), 697–710 (2010)

M. Alioto, Guest editorial for the special issue on "Ultra-low-voltage VLSI circuits and systems for green computing. IEEE Trans. Circuits Systems—part II **59**(12), 849–852 (2012)

M. Alioto, E. Consoli, G. Palumbo, *Flip-Flop Design in Nanometer CMOS—From High Speed to Low Energy* (Springer, Berlin, 2015)

Z. Bo, D. Blaauw, D. Sylvester, K. Flautner, Theoretical and practical limits of dynamic voltage scaling, in *Proceedings of DAC* (2004), pp. 868–873

K.A. Bowman, J.W. Tschanz, N.S. Kim, J.C. Lee, C.B. Wilkerson, S.-L. Lu, T. Karnik, V. De, Energy-efficient and metastability-immune resilient circuits for dynamic variation tolerance. IEEE J. Solid-State Circuits **44**, 49–63 (2009)

K.A. Bowman, J.W. Tschanz, S.-L. Lu, P. Aseron, M. Khellah, A. Raychowdhury, B. Geuskens, C. Tokunaga, C. Wilkerson, T. Karnik, V. De, A 45 nm resilient microprocessor core for dynamic variation tolerance. IEEE J. Solid-State Circuits **46**(1), 194–208 (2011)

T. Burd, T. Pering, A. Stratakos, R. Brodersen, A dynamic voltage scaled microprocessor system, in *IEEE ISSCC Dig. Tech. Papers* (Feb. 2015), pp. 294–295

A. Chandrakasan, D. Daly, D. Finchelstein, J. Kwong, Y. Ramadass, M. Sinangil, V. Sze, N. Verma, Technologies for ultradynamic voltage scaling. Proc. IEEE **98**(2), 191–214 (2010)

D. Chinnery, K. Keutzer, *Closing the Power Gap between ASIC & Custom* (Springer, Berlin, 2007)

J. Crop, E. Krimer, N. Moezzi-Madani, R. Pawlowski, T. Ruggeri, P. Chiang, M. Erez, Error detection and recovery techniques for variation-aware CMOS computing: a comprehensive review. J. Low Power Electron. Appl. **1**, 334–356 (2011)

S. Das, C. Tokunaga, S. Pant, W.-H. Ma, S. Kalaiselvan, K. Lai, D.M. Bull, D.T. Blaauw, Razor II: in situ error detection and correction for PVT and SER tolerance. IEEE J. Solid-State Circuits **44**(1), 32–48 (2009)

G. De Micheli, *Synthesis and Optimization of Digital Circuits* (McGraw Hill, New York, 1994)

R. Dreslinski, M. Wieckowski, D. Blaauw, D. Sylvester, T. Mudge, Near-threshold computing: reclaiming Moore's law through energy efficient integrated circuits. Proc. IEEE **98**(2), 253–266 (2010)

C. Enz, E. Vittoz, *Charge-Based MOS Transistor Modeling: The EKV Model for Low-Power and RF IC Design* (Wiley, New York, 2006)

D. Ernst, N.S. Kim, S. Das, S. Pant, R. Rao, T. Pham, C. Ziesler, D. Blaauw, T. Austin, K. Flautner, T. Mudge, Razor: a low-power pipeline based on circuit-level timing speculation, in *Proceedings of MICRO-36* (Dec. 2003), pp. 7–18

D. Flynn, R. Aitken, A. Gibbons, K. Shi, *Low Power Methodology Manual* (Springer, New York, 2007)

M. Fojtik, D. Fick, Y. Kim, N. Pinckney, D. Harris, D. Blaauw, D. Sylvester, Bubble razor: eliminating timing margins in an ARM Cortex-M3 processor in 45 nm CMOS using architecturally independent error detection and correction. IEEE J. Solid-State Circuits **48**(1), 66–81 (2013)

L. Freyman, D. Fick, M. Alioto, D. Blaauw, D. Sylvester, A 346 µm2 VCO-based, reference-free, self-timed sensor interface for cubic-millimeter sensor nodes in 28 nm CMOS. IEEE J. Solid-State Circuits **49**(11), 2462–2473 (2014)

F. Frustaci, M. Khayatzadeh, D. Blaauw, D. Sylvester, M. Alioto, SRAM for error-tolerant applications with dynamic energy-quality management in 28 nm CMOS. IEEE J. Solid-State Circuits **50**(3), 1310–1323 (2015)

F. Frustaci, D. Blaauw, D. Sylvester, M. Alioto, Approximate SRAMs with dynamic energy-quality management. IEEE Trans. VLSI Syst. **24**(6), 2128–2141 (2016)

G. Gammie, N. Ickes, M. Sinangil, R. Rithe, J. Gu, A. Wang, H. Mair, S. Datla, R. Bing, S. Honnavara-Prasad, L. Ho, G. Baldwin, D. Buss, A. Chandrakasan, U. Ko, A 28 nm 0.6 V low-power DSP for mobile applications, in *ISSCC Digest of Technical Papers (ISSCC)* (San Francisco, Feb. 2011)

T. Gemmeke, M. Ashouei, B. Liu, M. Meixner, T.G. Noll, H. de Groot, Cell libraries for robust low-voltage operation in nanometer technologies. Solid-State Electron. **84**, 132–141 (2013)

J. Gregg, T.W. Chen, Post silicon power/performance optimization in the presence of process variations using individual well-adaptive body biasing. IEEE Trans. VLSI Syst. **15**(3), 366–376 (2007)

S. Hanson, B. Zhai, D. Blaauw, D. Sylvester, A. Bryant, X. Wang, Energy optimality and variability in sub-threshold design, in *Proceedings of ISLPED 2006* (2006), pp. 363–365

S. Hanson, B. Zhai, K. Bernstein, D. Blaauw, A. Bryant, L. Chang, K.K. Das, W. Haensch, E.J. Nowak, D.M. Sylvester, Ultralow-voltage, minimum-energy CMOS. IBM J. Res. & Dev. **50**(4/5) (2006), pp. 469–490

S. Hanson, B. Zhai, M. Seok, B. Cline, K. Zhou, M. Singhal, M. Minuth, J. Olson, L. Nazhandali, T. Austin, D. Sylvester, D. Blaauw, Exploring variability and performance in a sub-200-mV processor. IEEE J. Solid-State Circuits **43**(4), 881–890 (2008)

D. Harris, R. Ho, G.-Y. Wei, M. Horowitz, The Fanout-of-4 inverter delay metric, unpublished manuscript http://citeseerx.ist.psu.edu/viewdoc/download? doi=10.1.1.68.831&rep=rep1&type=pdf

S. Hsu, A. Agarwal, M. Anders, S. Mathew, H. Kaul, F. Sheikh, R. Krishnamurthy, A 280 mV-to-1.1 V 256b reconfigurable SIMD vector permutation engine with 2-dimensional shuffle in 22 nm CMOS, in *ISSCC Digest of Technical Papers (ISSCC)* (San Francisco, Feb. 2012)

International Technology Roadmap for Semiconductors: 2013 edition. http://www.itrs.net (2013)

D. Jacquet et al., 2.6GHz ultra-wide voltage range energy efficient dual A9 in 28 nm UTBB FD-SOI, in *IEEE Symposium on VLSI Circuits Dig. Tech. Papers* (June 2013)

S. Jain et al., A 280 mV-to-1.2 V wide-operating-range IA-32 processor in 32 nm CMOS, in *IEEE ISSCC Dig. Tech. Papers* (Feb. 2012), pp. 66–67

D. Jeon, M. Seok, C. Chakrabarti, D. Blaauw, D. Sylvester, A super-pipelined energy efficient sub-threshold 240 MS/s FFT core in 65 nm CMOS. IEEE J. Solid-State Circuits **47**(1), 23–34 (2013)

H. Kaul, M.A. Anders, S.K. Mathew, S.K. Hsu, A. Agarwal, F. Sheikh, R.K. Krishnamurthy, S. Borkar, A 1.45 GHz 52-to-162GFLOPS/W

variable-precision floating-point fused multiply-add unit with certainty tracking in 32 nm CMOS, in *IEEE ISSCC Dig. Tech. Papers*, (Feb. 2012), pp. 182–183

M. Khayatzadeh, M. Saligane, J. Wang, M. Alioto, D. Blaauw, D. Sylvester, A reconfigurable dual port memory with error detection and correction in 28 nm FDSOI, in *IEEE ISSCC Dig. Tech. Papers* (Feb. 2016), pp. 310–311

Y. Kim, W. Jung, I. Lee, Q. Dong, M. Henry, D. Sylvester, D. Blaauw, A static contention-free single-phase-clocked 24T Flip-Flop in 45 nm for low-power applications. in *IEEE ISSCC Dig. Tech. Papers* (Feb. 2014)

S. Kim, M. Seok, Reconfigurable interconnect-driving technique for ultra-dynamic-voltage-scaling systems, in *IEEE ACM International Symposium on Low Power Electronics and Design (ISLPED)* (2014)

I. Kwon, S. Kim, D. Fick, M. Kim, Y.-P. Chen, D. Sylvester, Razor-lite: a light-weight register for error detection by observing virtual supply rails. IEEE J. Solid-State Circuits **49**(9), 2054–2066 (2014)

L. Leem, H. Cho, J. Bau, Q.A. Jacobson, S. Mitra, ERSA: error resilient system architecture for probabilistic applications, in *Proceedings of DATE 2010* (Dresden, Germany, Mar. 2010), pp. 1560–1565

L. Lin, S. Jain, M. Alioto, Reconfigurable clock networks for random skew mitigation from sub-threshold to nominal voltage, in *IEEE ISSCC Dig. Tech. Papers* (Feb. 2017)

M. Alioto, G. Scotti, A. Trifiletti, A novel framework to estimate the path delay variability via the Fan-Out-of-4 metric. IEEE Trans. Circuits Syst.—part I

D. Markovic, V. Stojanovic, B. Nikolic, M.A. Horowitz, R.W. Brodersen, Methods for true energy-performance optimization. IEEE J. Solid-State Circuits **39**(8), 1282–1293 (2004)

S.M. Martin, K. Flautner, T. Mudge, D. Blaauw, Combined dynamic voltage scaling and adaptive body biasing for lower power microprocessors under dynamic workloads, in *Proceedings of ICCAD'02* (Nov. 2002), pp. 721–725

M. Meijer, J. Pineda de Gyvez, Body-bias-driven design strategy for area- and performance-efficient CMOS circuits. IEEE Trans. VLSI Syst. **20**(1), 42–51 (2012)

M. Merrett, Y. Wang, M. Alioto, M. Zwolinski, Design metrics for RTL level estimation of delay variability due to intradie (random) variations, in *Proceedings of ISCAS 2010* (Paris (France), May 2010), pp. 2498–2501

A. Muramatsu, T. Yasufuku, M. Nomura, M. Takamiya, H. Shinohara, T. Sakurai, 12% power reduction by within-functional-block fine-grained adaptive dual supply voltage control in logic circuits with 42 voltage domains, in *37th European Solid-State Circuits Conference (ESSCIRC)* (Helsinki (Finland), Sep. 2011), pp. 191–194

J. Myers, A. Savanth, R. Howard, R. Gaddh, P. Prabhat, D. Flynn, An 80nW retention 11.7pJ/cycle active sub-threshold ARM Cortex®-M0+ sub-system in

65 nm CMOS for WSN applications, in *IEEE ISSCC Dig. Tech. Papers* (Feb. 2015), pp. 144–145

S. Narayanan, J. Sartori, R. Kumar, D.L. Jones, Scalable stochastic processors, in *Proceedings of DATE 2010* (Dresden, Germany, Mar. 2010), pp. 335–338

S. Narendra, A. Chandrakasan (eds.), *Leakage in Nanometer CMOS Technologies* (Springer, Berlin, 2006)

K. Nose, T. Sakurai, Optimization of VDD and VTH for low-power and high-speed applications, in *Proceedings of ASPDAC* (Jan. 2000), pp. 469–474

M. Olivieri, G. Scotti, A. Trifiletti, A novel yield optimization technique for digital CMOS circuits design by means of process parameters run-time estimation and body bias active control. IEEE Trans. VLSI Syst. **13** (5), 630–638 (2005)

M.M. Orshansky, S. Nassif, D. Boning, *Design for Manufacturability and Statistical Design* (Springer, Berlin, 2008)

D. Patil, M. Horowitz, Joint supply, threshold voltage and sizing optimization for design of robust digital circuits. http://vlsiweb.stanford.edu/papers/JointVddVthSizing.pdf

M. Pelgrom, A. Duinmaijer, A. Welbers, Matching properties of MOS transistors. IEEE J. Solid-State Circuits **24**(1), 1433–1439 (1989)

M. Putic, L. Di, B. H. Calhoun, J. Lach, Panoptic DVS: a fine-grained dynamic voltage scaling framework for energy scalable CMOS design, in *Proceedings of ICCD 2009* (Lake Tahoe, CA, Oct. 2009), pp. 491–497

B. Raghunathan, Y. Turakhia, S. Garg, D. Marculescu, Cherry-picking: exploiting process variations in dark-silicon homogeneous chip multi-processors, in *Proceedings of DATE 2013* (Grenoble, France, Mar. 2013), pp. 39–44

Y.K. Ramadass, A.P. Chandrakasan, Minimum energy tracking loop with embedded DC–DC converter enabling ultra-low-voltage operation down to 250 mV in 65 nm CMOS. IEEE J. Solid-state Circuits **43**(1), 256–265 (2008)

T. Sakurai, R. Newton, Alpha-power law MOSFET model and its applications to CMOS inverter delay and other formulas. IEEE J. Solid-State Circuits **25**(2), 584–594 (1990)

W. Sansen, *Analog Design Essentials* (Springer, New York, 2006)

M. Seok, D. Blaauw, D. Sylvester, Robust clock network design methodology for ultra-low voltage operations, in *IEEE Transactions on Emerging Selected Topics Circuits Sys*tems, vol. 1(2) (2011)

F. Sheikh, S. Mathew, M. Anders, H. Kaul, S. Hsu, A. Agarwal, R. Krishnamurthy, S. Borkar, A 2.05 GVertices/s 151 mW lighting accelerator for 3D graphics vertex and pixel shading in 32 nm CMOS, in *IEEE ISSCC Dig. Tech. Papers* (Feb. 2012), pp. 178–179

V. Srinivasan, D. Brooks, M. Gschwind, P. Bose, V. Zyuban, P.N. Strenski, P.G. Emma, Optimizing pipelines for power and performance, in *Proceedings*

of *International Symposium on Microarchitectures* (2002), pp. 333–344

I. Sutherland, B. Sproull, D. Harris, *Logical Effort: Designing Fast CMOS Circuits* (Morgan-Kaufmann, Burlington, 1999)

C. Tokunaga, J.F. Ryan, C. Augustine, J.P. Kulkarni, Y.-C. Shih, S.T. Kim, R. Jain, K. Bowman, A. Raychowdhury, M.M. Khellah, J.W. Tschanz, V. De, A graphics execution core in 22 nm CMOS featuring adaptive clocking, selective boosting and state-retentive sleep, in *ISSCC Digest of Technical Papers (ISSCC)* (San Francisco, CA, Feb. 2014)

J.R. Tolbert, X. Zhao, S.K. Lim, S. Mukhopadhyay, Analysis and design of energy and slew aware subthreshold clock systems. IEEE Trans. CAD **30**(9), 1348–1358 (2011)

J.W. Tschanz, J.T. Kao, S.G. Narendra, R. Nair, D.A. Antoniadis, A.P. Chandrakasan, V. De, Adaptive body bias for reducing impacts of die-to-die and within-die parameter variations on microprocessor frequency and leakage. IEEE J. Solid-State Cicuits **37**(11), 1396–1042 (2002)

Y. Tsividis, *Operational Modeling of the MOS Transistor*, 2nd edn. (McGraw-Hill, New York, 1999)

R.E. Walpole, R.H. Myers, S.L. Myers, K. Ye, *Probability & Statistics for Engineers & Scientists* (Prentice Hall, Englewood Cliffs, 2006)

A. Wang, A. Chandrakasan, 180-mV subthreshold FFT processor using a minimum energy design methodology. IEEE J. Solid-State Circuits **40**(1), 310–319 (2005)

A. Wang, B.H. Calhoun, A. Chandrakasan, *Sub-threshold design for ultra low-power systems* (Springer, Berlin, 2006)

J. Wang, N. Pinckney, D. Blaauw, D. Sylvester, Reconfigurable self-timed regenerators for wide-range voltage scaled interconnect, in *Proceedings of ASSCC 2015* (Nov. 2015)

N. Weste, D. Harris, *CMOS VLSI Design*, 4th edn. (Pearson Education, Upper Saddle River, 2011)

R. Wilson et al., A 460 MHz at 397 mV, 2.6 GHz at 1.3 V, 32b VLIW DSP, embedding FMAX tracking, in *IEEE ISSCC Dig. Tech. Papers* (Feb. 2014), pp. 452–453

T. Xanthopoulos, *Clocking in Modern VLSI Systems* (Springer, New York, 2009)

M. Yip, A. Chandrakasan, A resolution-reconfigurable 5-to-10 b 0.4-to-1 V power scalable SAR ADC, in *IEEE ISSCC Dig. Tech. Papers* (2011), pp. 190–191

Y. Zhang, M. Khayatzadeh, K. Yang, M. Saligane, M. Alioto, D. Blaauw, D. Sylvester, iRazor: 3-transistor current-based error detection and correction in an ARM Cortex-R4 processor, in *IEEE ISSCC Dig. Tech. Papers* (Feb. 2016), pp. 160–161

X. Zhao, J.R. Tolbert, S. Mukhopadhyay, S.K. Lim, Variation-aware clock network design methodology for ultralow voltage (ULV) circuits. IEEE Trans. CAD **31**(8), 1222–1234 (2012)

W. Zhao, Y. Ha, M. Alioto, Novel self-body-biasing and statistical design for near-threshold circuits with ultra energy-efficient AES as case study. IEEE Trans. VLSI Syst. **23**(8), 1390–1401 (2015)

Energy Efficient Volatile Memory Circuits for the IoT Era

5

Jaydeep P. Kulkarni, James W. Tschanz, and Vivek K. De

This chapter addresses the challenges involved in designing energy efficient embedded Static Random Access Memory (SRAM) circuits for the IoT era. It discusses memory design for wide voltage range operation using 6 Transistor (6T), 8T and 10T bitcells and novel circuit assist techniques. In addition, it discusses future memory designs using emerging nano-wire FET, Tunnel FET, III-V FET, and monolithic 3-D technologies.

5.1 Introduction and Challenges in Embedded Memory Design for IoT

5.1.1 Introduction

Technological advances and form factor driven cost reduction have resulted in tremendous growth of computing devices. As shown in Fig. 5.1, the number of internet-connected devices deployed worldwide is projected to grow to 50 billion by 2020.

If continued, this trend might result in tens of billions of computing systems consisting of personal computers, desktop machines, smartphones, wearables and many units connected to the internet; collectively known as the Internet of Things (IoT). This dramatic growth in compute devices results in a data explosion (known as 'data deluge') and would require millions of Zettabytes of memory in order to perform this large scale of computing (Semiconductor Industry Association 2015). Therefore, memories play a critical role in future energy efficient computing systems. This chapter addresses the challenges involved in designing energy efficient embedded memory circuits operating across a wide voltage range, and also presents design examples using emerging device technologies.

5.1.2 SRAM Scaling Trends

Aggressive scaling of transistor dimensions with each technology generation has resulted in increased integration density and improved device performance. The SRAM bitcell area has been scaled by ~0.5× over each process generation as shown in Fig. 5.2. This area scaling is achieved by various lithographic and circuit innovations such as thin-cell layouts, high-K metal gate technology, tri-gate geometry, leakage-reduction techniques, and low voltage assist techniques (Wang et al. 2012; Yoshinobu et al. 2003).

J.P. Kulkarni (✉) • J.W. Tschanz • V.K. De
Circuit Research Lab, Intel Corporation,
Hillsboro, OR, USA
e-mail: jaydeep.p.kulkarni@intel.com

© Springer International Publishing AG 2017
M. Alioto (ed.), *Enabling the Internet of Things*, DOI 10.1007/978-3-319-51482-6_5

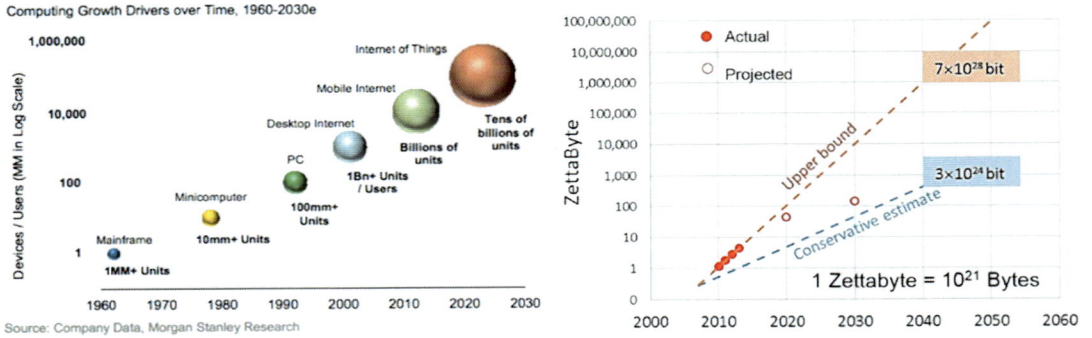

Fig. 5.1 Projected growth of IoT devices and the required memory capacity

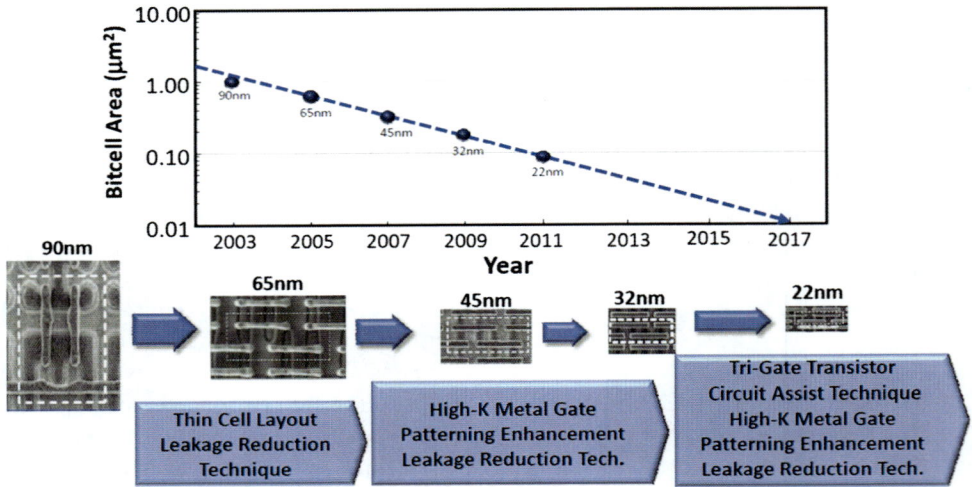

Fig. 5.2 6T Static Random Access Memory (SRAM) bitcell scaling trend

5.1.3   SRAM Design Octagon

Energy-efficient SRAM design in nanoscale technology needs to address several process, design, architecture, and reliability issues as shown in Fig. 5.3. Process technology issues relate to the restricted design rules, diffusion-notch free layout, proximity effects, stress effects, and threshold voltage (V_T) optimization. It impacts the 6T bitcell sizing/area, read/write stability, 6T+ bitcells, and single/multi-port bitcell requirements. These process optimizations and design requirements govern the array efficiency and bit density.

Circuit techniques focus on lowering the minimum successful operating supply voltage (also

known as V_{MIN}) using various read/write assist techniques. Another important aspect is leakage reduction during active as well as idle mode. This is particularly important for IoT designs that in general have very strict power budgets and at the same time can exhibit very low array activity due to duty-cycled or burst-mode styles of operation.

Reliability is also an important aspect of modern SRAM design. The effects of bias temperature instability (BTI), time-dependent dielectric breakdown (TDDB), hot carrier injection (HCI), random telegraph noise (RTN), erratic bits, and radiation-induced soft errors limit the operating voltage range and lifetime of the SRAM array.

Architectural techniques such as redundancy, parity protection, error correcting codes (ECC),

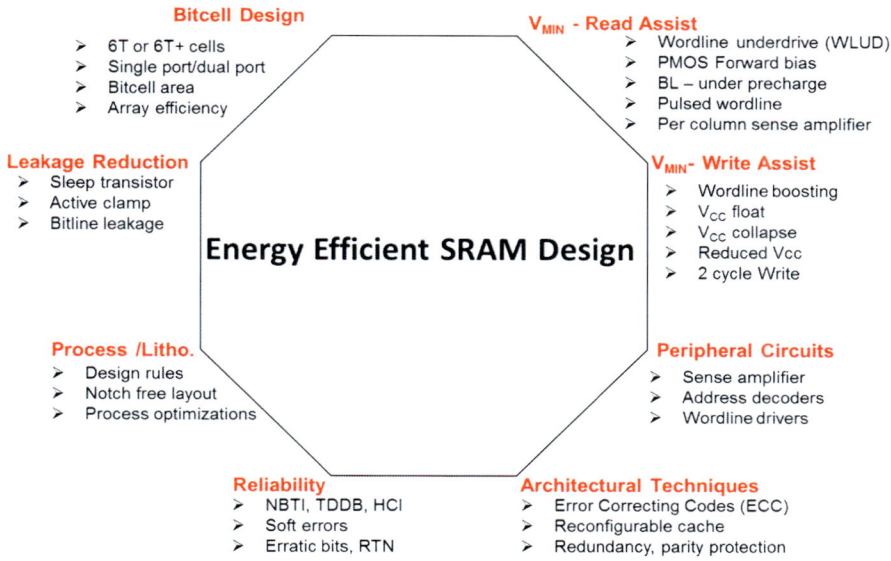

Fig. 5.3 Energy-efficient SRAM design requirements

and reconfigurable caches are used along with the circuit techniques to address these reliability issues. In this chapter, we focus on the circuit techniques for enabling energy-efficient SRAM operation across a wide voltage range specifically needed for IoT applications.

5.1.4 SRAM Parametric Failures and V_{MIN}

Supply voltage scaling has remained the major focus of energy-efficient memory design. However, as the supply voltage is reduced the sensitivity of the circuit parameters to process variations increases. Nanoscale SRAM bitcells having minimum-sized transistors are vulnerable to inter-die as well as intra-die process variations. Intra-die process variations include random dopant fluctuation (RDF), line edge roughness (LER), and other manufacturing variations that may result in a threshold voltage mismatch between adjacent transistors in a memory cell (Bhavnagarwala et al. 2001). Coupled with inter-die and intra-die process variations, supply voltage scaling is limited due to various memory failures (i.e. read failure, retention failure, access time failure, and write failure) also known as minimum operating voltage (V_{MIN}) (Mukhopadhyay et al. 2005; Khellah et al. 2008).

The 'de facto' memory bitcell used in modern SRAM designs is a 6-transistor cell consisting of a cross-coupled inverter pair. SRAM cells are sized to satisfy the conflicting design requirements of read stability and write-ability. For a read stability optimized bitcell as shown Fig. 5.4, pass gate transistors (AXL, AXR) are sized weaker than the pull-up and pull-down transistor.

During a read operation, when a wordline (WL) is asserted, the voltage at the node storing logic 0 (Node V_L in left Fig. 5.4) rises above V_{SS} due to voltage divider action between the pass-gate AXL and pull-down NL transistors. At nominal supply voltage, the V_L node voltage rise during read operation is not significant enough to flip the bitcell contents (shown in blue color). However due to process variations, especially for a bitcell operating at lower supply voltages, the voltage rise at V_L node can be higher than the trip point of the other inverter and can flip the V_R node, resulting in a read failure event (shown in red color). Read failure can also occur during a dummy-read scenario (half-select) in a column-interleaved design.

For the write operation, the design requirement is that the bitcell nodes should flip easily. In a write-optimized bitcell, the pass gate is

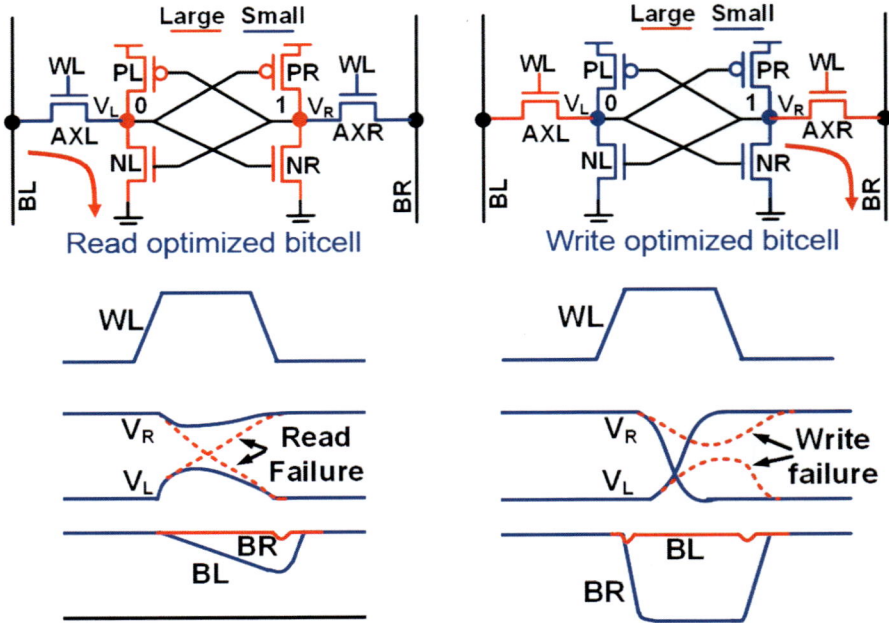

Fig. 5.4 Read and write failure events in a 6T SRAM bitcell

stronger than the pull-up and pull-down devices. At nominal supply voltage, when write data is applied on the bitline (BL) and bitline complement (BR), the pass gate connected to the grounded bitline (BR) pulls down the bitcell node (V_R in right Fig. 5.4) below the trip point of the other inverter (V_L) resulting in the flip of the storage bit (node V_L) and a successful write operation. However due to process variations especially at lower supply voltage, the pass gate AXR can be weaker than the pull-up device PR and may not lower the V_R node voltage below the switching threshold of the other side inverter resulting in a write failure event. Write failure can also happen if the wordline pulse is not long enough for the bitcell to flip the internal nodes.

5.2   6T SRAM Circuit Techniques

To improve the operating voltage range of SRAM arrays in the presence of process variations and to satisfy the conflicting design requirement of read vs. write stability, various circuit assist techniques have been explored. The key idea is to bias different nodes of the 6T

bitcells (bitcell-V_{CC}/V_{SS}, WL, BL) appropriately to favor a read or a write operation.

5.2.1   Read Assist Techniques

As explained in the earlier section, a read failure is caused by increase in the access transistor drive strength compared to the cross-coupled inverter pair due to low voltage operation and/or process variations. The read assist circuits try to reinforce access transistors to be weaker than the cross-coupled inverter pair. The node biasing techniques include raising the bitcell-V_{CC} and/or lowering bitcell-V_{SS} of the cross-coupled inverter pair. These techniques also include wordline under-drive (WLUD) or operating the entire bitcell at a higher supply voltage compared to the peripheral circuits (Mann et al. 2010; Khellah et al. 2008) (Figs. 5.5 and 5.6).

5.2.2   Write Assist Techniques

Contrary to a read operation, a write failure is caused by decrease in the access transistor drive strength compared to the cross-coupled inverter

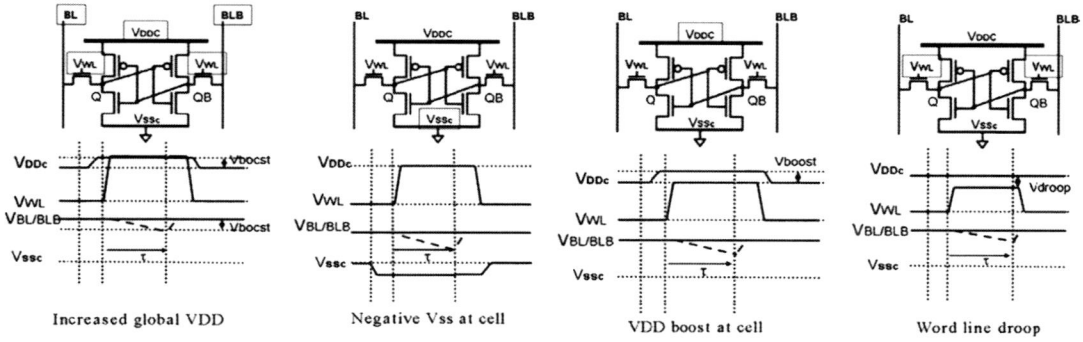

Fig. 5.5 6T SRAM read assist techniques

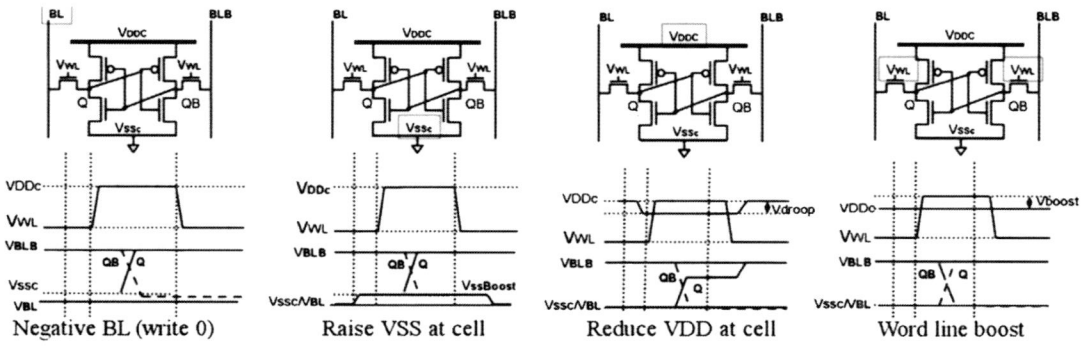

Fig. 5.6 6T SRAM write assist techniques

pair due to low voltage operation and/or process variations. The write assist circuits try to reinforce the access transistor to be stronger than the cross-coupled inverter pair. The node biasing techniques include lowering the bitcell-V_{CC} and/or raising the bitcell-V_{SS} of the cross-coupled inverter pair. Write assist techniques also include wordline boosting or negative bitline approaches (Mann et al. 2010; Khellah et al. 2008).

5.3 8T SRAM Circuit Techniques

5.3.1 Benefits of Decoupled Read/ Write Operation

The fundamental conflicting design constraints of read-stability vs. write-ability in the 6T SRAM bitcell limits low voltage operation and

requires power and area hungry V_{MIN} assist techniques. The operating voltage can be reduced substantially by adding a dedicated read port using two extra transistors. This 8T bitcell (Fig. 5.7) is commonly used in microprocessor cores for performance-critical low-level caches and multi-ported register-file arrays.

The 8T bitcell offers fast read and write operation, dual-port capability, and generally lower V_{MIN} than the 6T SRAM bitcell. By using a decoupled single-ended read port with domino-style hierarchal read bit-line, the 8T cell features a fast read evaluation path without causing access disturbance that limits read-V_{MIN} in the 6T bitcell. Using the 8T cell in a half-select-free architecture eliminates pseudo-reads during partial writes, hence enabling write-V_{MIN} optimization independent of read. It is well known that both with-in-die (WID) and die-to-die (D2D) device parameter variations are getting worse

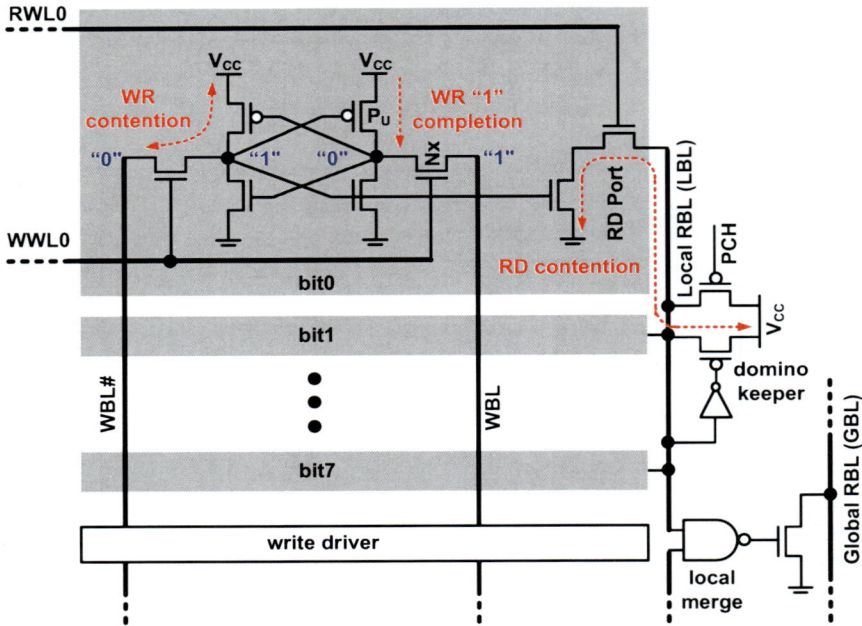

Fig. 5.7 8T bitcell array organization with non-interleaved columns

with feature size scaling. Unfortunately, the typical approach of sizing up the 8T read and write ports to mitigate process variation has limited V_{MIN} returns.

In the read case, using larger NMOS read port helps when reading a "1" by reducing contention with the PMOS domino keeper on the local bit-line (LBL). To compensate for degraded noise margin that comes with a larger (and leakier) read port, the PMOS keeper needs to be relatively upsized for good reading of a "0", resulting in diminishing read-V_{MIN} returns with continued upsizing. Similarly, contention between PMOS pull-up (P_U) and the NMOS pass (N_X) impacting write-V_{MIN} is reduced by sizing up N_X. However, P_U needs to be relatively sized up as well since at a given point a weak P_U will limit the completion of write within a given WL pulse. Only 10% total V_{MIN} reduction is attained through optimal cell upsizing at a cost of ~25% increase in array area (Raychowdhury et al. 2010). Therefore, read and write assis circuit that can achieve V_{MIN} reduction at a minimal area impact are necessary.

5.3.2 Wordline Boosting as V_{MIN} Assist Technique

Wordline boosting is an effective technique for lowering the read and write V_{MIN} of 8T bitcell arrays with minimal area and power overheads. The key idea is to selectively boost critical V_{MIN}-limiting nodes of the 8T bitcell, allowing the majority of the array peripheral circuits and remaining logic block to operate at much lower voltage thereby lowering the overall chip V_{MIN}.

There are various ways to achieve a boosted voltage for the wordline, such as charge-pump based boosting, capacitive coupling, and using separate high-V_{CC} rail. All three schemes rely on one or more combinations of (1) read wordline (RWL) boosting, (2) write wordline (WWL) boosting, and/or bitcell boosting. Boosting RWL and bitcell-V_{CC} enable larger read "ON" current without forcing a larger PMOS keeper. Boosting WWL helps write-V_{MIN} for two reasons—improving contention without upsizing N_X (or lowering its V_{TH}), and improving completion by writing a "1" from the other side. Unlike V_{CC} collapse, WWL boosting does not degrade the dynamic retention margin of unselected cells on the same column.

5.3.2.1 Charge Pump Based Wordline Boosting

In this approach, an embedded charge pump is used to generate a higher voltage to boost RWL and WWL (V_{BOOST} generation) (Raychowdhury et al. 2010). Figure 5.8 shows a 2 KB 8T SRAM bank with a locally-integrated charge pump (CP) and associated wordline level shifter circuits. The CP itself is divided into ten identical units placed in the layout slices created by the level shifters in the global IO of the 2 KB macro. The boosting ratio is adjusted based on the built-in read-ability and write-ability sensor. The ideal boosting ratio (V_{BOOST}/V_{CC}) under no load current (I_{LOAD}) is $2V_{CC}$.

The actual boosting ratio is lower, however, as determined by I_{LOAD} from all active and inactive level shifters, boosting clock (BCLK) frequency (F_{BCLK}), and boosting capacitance (C_{CP}). To minimize the load current requirement on the charge pump design, a two-stage level shifter is used as the wordline driver. Unlike the conventional (DCVS) level shifter, where a "0"-to-V_{BOOST} transition is all supplied by the V_{BOOST} rail, the two-stage level shifter performs this transition in two steps. In the first step, "0"-to-V_{CC} is supplied by M_{P1} at which point M_{P2} kicks in to supply the remaining V_{CC}-to-V_{BOOST}. This significantly reduces the maximum charge pump load current allowing the boosting ratio to increase to ~1.6×, compared to only 1.1× using DCVS level shifter.

Figure 5.9 shows measurement results from a 45 nm test chip achieving V_{MIN} improvement of

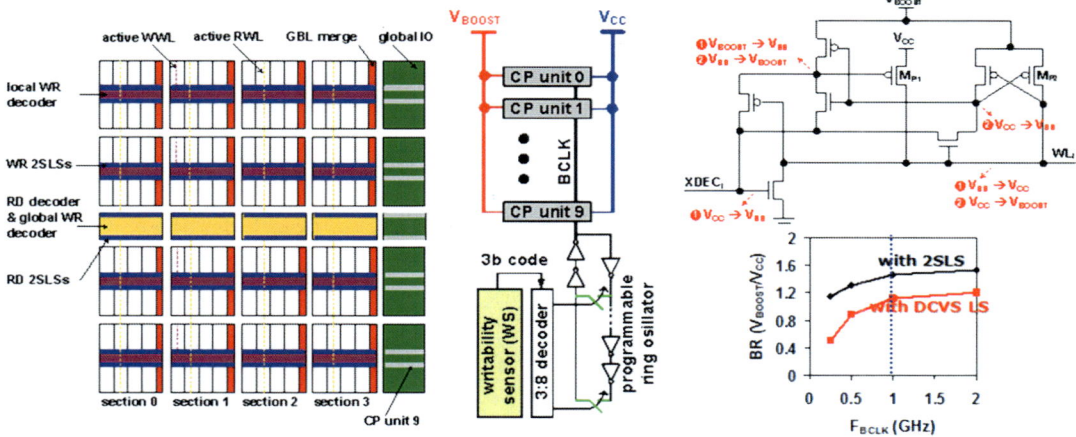

Fig. 5.8 8T SRAM array with distributed charge pump and two-step level shifters

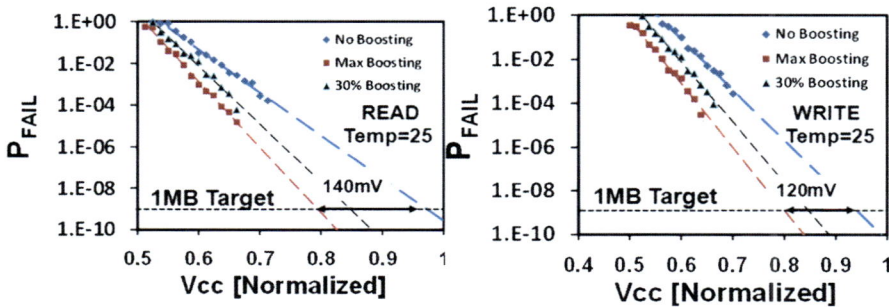

Fig. 5.9 Charge pump based boosting: Measured bit failures vs. V_{CC} extrapolated to 1 MB array

140 mV (120 mV) for read (write) for a 1 MB target array size. The area overhead due to distributed charge pump and two-stage level shifters is significant (around ~25%).

5.3.2.2 Capacitive Coupling with Self-Induced-Collapse

The adoption of charge pump based boosting requires careful design of the charge pump and two-stage level shifters. Furthermore, dynamically turning the charge pump off (or putting it in a drowsy mode) during inactive periods is necessary for net power savings. As an alternative to the charge bump, the capacitive coupling (CC) scheme achieves write-wordline (WWL) boosting using write-bitline (WBL) to WWL coupling. This eliminates the need for a power-hungry charge pump as well as complex level shifters. The basic idea is to take advantage of the large G-S/D capacitance to create a boosted voltage on the WWL without using any charge pump, level shifter or high voltage supply (Kulkarni et al. 2010). There are two locations on the WWL where this capacitance can be found (Fig. 5.10). The first is at the WWL interface to the PMOS/NMOS devices of the final

WWL driver (C1), while the second is at the WWL interface to the cell's NMOS WR devices (C2 and C3).

To enable use of the first capacitance, the input of the WWL driver is asserted normally (high-to-low) to create a $0 \rightarrow V_{CC}$ transition on the WWL. After a short delay, the input is de-asserted (goes from low to high) turning off the top PMOS (but without turning on the bottom NMOS)-effectively floating the WWL. This in turn excites the G-to-D capacitance from the PMOS transistor of the WL driver creating about 3–5% coupling to the floating WWL. To enable use of the second capacitance, both WBL and WBLx are pre-discharged and, depending on data polarity; one of the bit-lines makes a $0 \rightarrow V_{CC}$ transition. This rising transition on the bitline and the internal bitcell node is capacitively coupled to the floated WWL, boosting it by ~20% of V_{CC}. This scheme is scalable to any number of bits per WWL with the scaling of the WWL driver size and the per-bit coupling capacitance.

A beneficial side effect of pre-discharging both the WBL and WBLx prior to the write operation is a self-induced collapse (SIC) of the

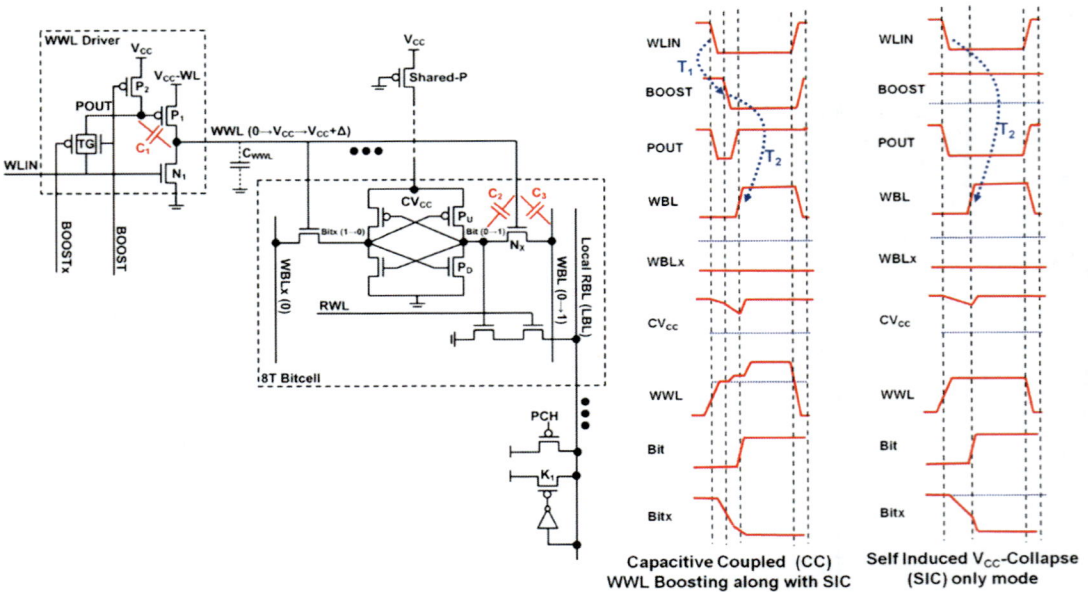

Capacitive Coupled (CC) WWL Boosting along with SIC

Self Induced V_{CC}-Collapse (SIC) only mode

Fig. 5.10 Capacitive coupled WWL boosting with self-induced-collapse (SIC)

virtual bitcell voltage when the WWL is asserted. While SIC is inherent with the CC boost technique, it can also be used alone as a low-overhead write-V_{MIN} reduction technique, and is an effective alternative when WWL boosting violates gate-oxide reliability limits at high voltage. The SIC-only mode can be enabled by keeping the boost signal high (Fig. 5.10), ensuring that the WWL is not floated or boosted during the write operation.

Figure 5.11 shows the measured write failure rate versus supply voltage for CC boost and SIC-only modes. The slope of the write P_{FAIL} curve is governed by the SIC magnitude, WWL boost ratio, and how late the WBL transitions with respect to WWL activation. Extrapolating P_{FAIL} data to 1 MB array size demonstrates 140 mV reduction in write-V_{MIN} for CC boost and 80 mV reduction for SIC-only mode at optimal timing settings at 1.6 GHz operating frequency.

At lower frequencies, V_{MIN} savings increase to 180 mV for CC boost and 130 mV for

SIC-only mode (Fig. 5.12). Both SIC and CC boost incur an increase in array power when run at the nominal voltage as baseline due to additional switching of the WBLs, bitcell V_{CC}, and overhead circuitry. However, both techniques enable V_{MIN} scaling beyond the baseline. Total array power savings when operating at lower V_{MIN} are 12% for SIC and 27% for CC boost (Fig. 5.12).

5.3.2.3 Dual-V_{CC} Design

Dual-V_{CC} based boosting selectively increases the voltage of critical nodes in an 8T-bitcell while incurring no array area overhead. A separate voltage $V_{BOOST} \leq V_{MAX}$, supplied externally or generated locally from a fixed high input voltage rail (V_{IN}) using a step-down voltage regulator (VR), is used to "boost" selected read/write wordlines (R/WWLs) and cell-V_{CC} (during read only) as shown in Fig. 5.13 (Kulkarni et al. 2013).

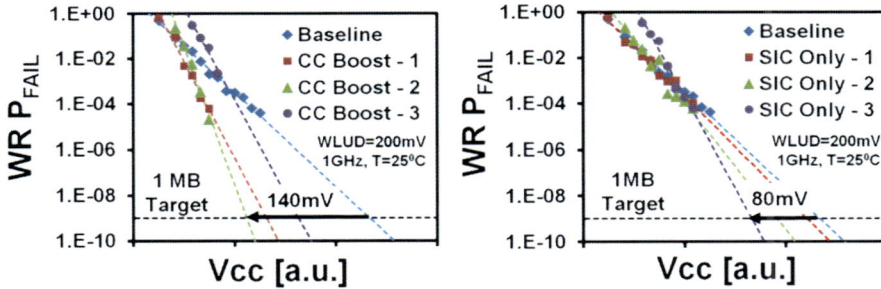

Fig. 5.11 Measured bit failure rate vs. V_{CC} for capacitive coupling and self-induced collapse technique

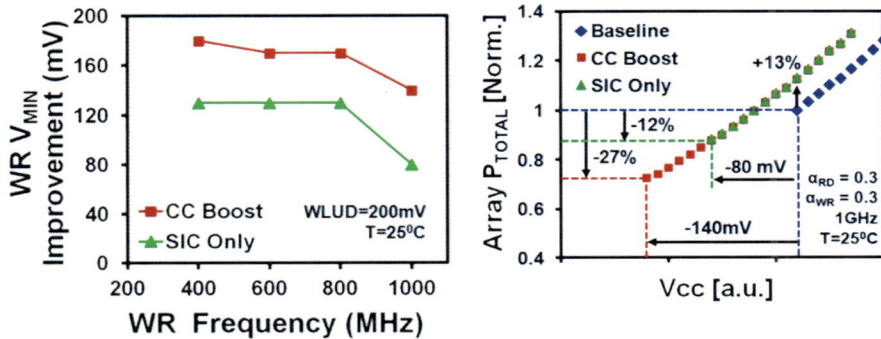

Fig. 5.12 Measured V_{MIN} improvement with frequency and power measurement

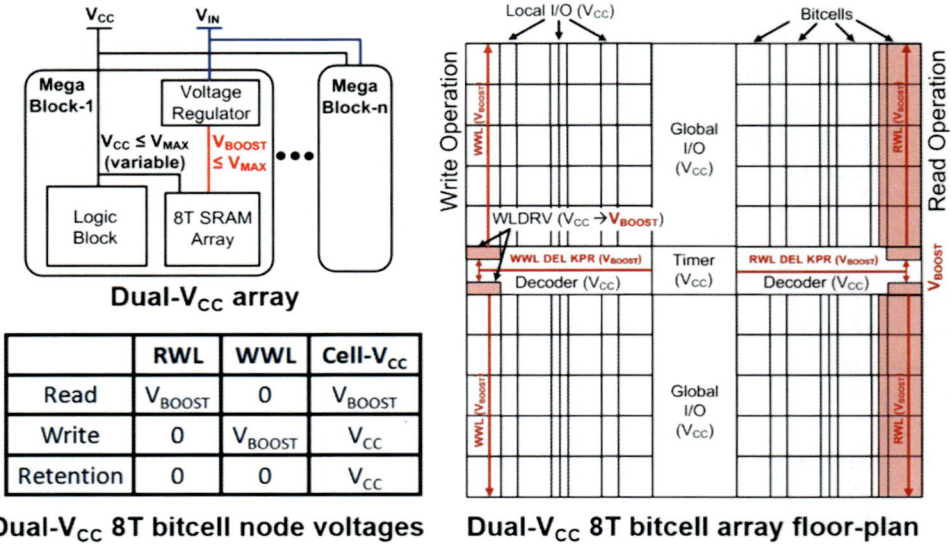

Fig. 5.13 Dual-V_{CC} approach and array floorplan

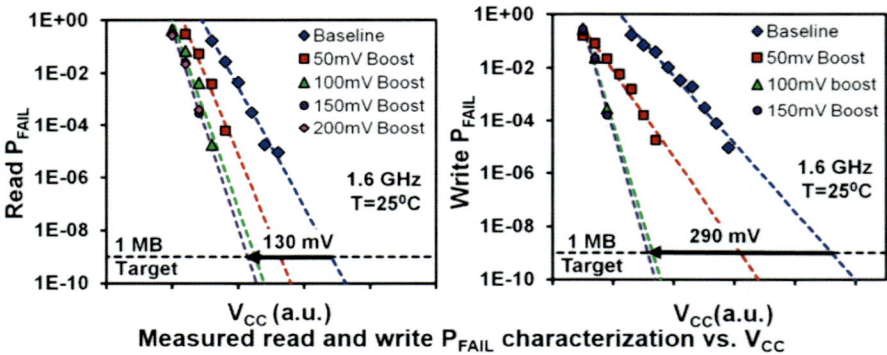

Fig. 5.14 Measured read and write failures vs. V_{CC} for varying boosting levels

All remaining array circuits such as R/WWL pre-decoder, pre-charge logic, local and global bitline (LBL/GBL) sensing, timer, and column- I/O drivers are connected to the variable $V_{CC} \leq V_{MAX}$ that is shared with core logic operating across a wide voltage range. By decoupling the V_{MIN}-limiting 8T bitcell from remaining array and core logic, overall chip V_{MIN} can be reduced, thus improving energy efficiency. During a read operation, selected RWL and associated bitcells are switched to Vboost to enable overdrive of the read-port transistor stack. This alleviates keeper contention and also improves bitline evaluation delay compared to the baseline

single-V_{CC} design. During a write operation, selected bitcells remain at V_{CC} while the WWL is boosted to mitigate contention between the pass NMOS and pull-up PMOS in the bitcell. WWL boosting also aids write completion by passing a strong "1" through the pass NMOS. A dynamic level-shifting NAND WL decoder replaces the static single-V_{CC} NAND implementation while fitting in the same area.

Measurement results from a 22 nm tri-gate CMOS process show 130 mV read-V_{MIN} improvement and 290 mV write-V_{MIN} improvement when extrapolated to a 1 MB target array size at 1.6 GHz (Fig. 5.14). As boosting

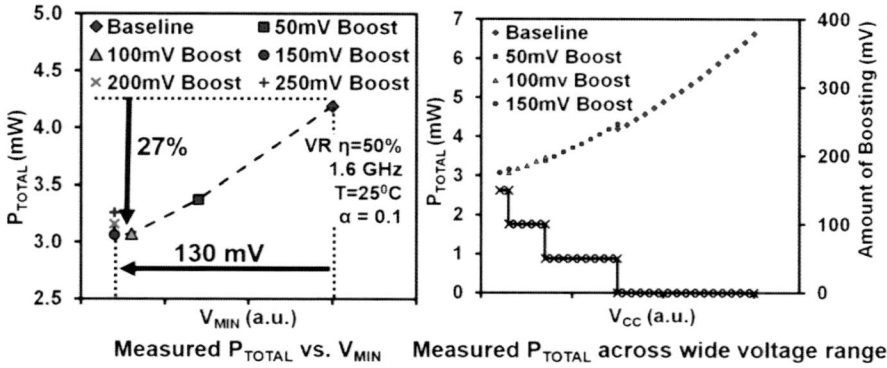

Fig. 5.15 Measured total power vs. V_{MIN} and total power vs. supply voltage

magnitude is increased, the array V_{MIN} is progressively reduced and achieves an optimum of 130 mV improvement for 150 mV of boosting operating at 1.6 GHz, resulting in 27% lower power. Any further boosting does not improve the array V_{MIN} as it is now limited by peripheral circuits, and results in an increase in power dissipation. Operation of the dual-V_{CC} 8T bitcell SRAM across a wide voltage range is achieved by gradually increasing V_{BOOST} value as V_{CC} is scaled down (Fig. 5.15).

5.3.3 Adaptive and Resilient Techniques

In modern microprocessor design, the register file operating voltage (V) and frequency (F) are limited by the delay of the precharge-evaluate read critical path. Furthermore, additional V/F guardbands are applied to account for worst-case dynamic variations such as voltage droops, temperature fluctuations, and aging-induced degradation. However, since most systems usually operate at nominal conditions, the fixed V/F guardbands for infrequent dynamic variations significantly limit the best-achievable performance and energy efficiency. To reduce these guardbands in flip-flop based static CMOS logic units, replica-based approaches such as Tunable Replica Circuits (TRC) have been proposed (Tschanz et al. 2009). In this approach, a set of replica circuits are calibrated to match the critical

path pipeline stage delay, and timing errors due to dynamic variations are detected by double-sampling the TRC outputs. The key requirement is that the TRC must always fail before the critical path fails. These circuit techniques are used in conjunction with architecture features to support instruction replay (to recover from a timing error) or dynamic clocking techniques that prevent failure of the critical paths.

The alternative in-situ approach for timing error detection uses Error Detection Sequentials (EDS) in the critical paths of the pipeline stage. Timing errors are detected by a double-sampling mechanism using a flip-flop and a latch (Bowman et al. 2009; Ernst et al. 2003). These error detection techniques, however, cannot be directly used for two-phase precharge-evaluate domino read critical paths in high-performance RF arrays since the data outputs are valid only during the evaluate phase. For error detection in RF arrays, replica-based techniques such as Tunable Replica Bits (TRB) have been proposed (Raychowdhury et al. 2011). In this approach, a set of replica memory bits are tuned at test time so that in the presence of dynamic variations the TRBs fail before the worst-case memory bit fails.

An alternate approach implements in-situ Timing Margin Detector (TMD) and Timing Error Detector (TED) circuits for domino read paths in 8T-bitcell arrays (Kulkarni et al. 2015). TMD circuit enables voltage and frequency adaptation to low-frequency voltage variations, temperature and aging, as well as to excessive

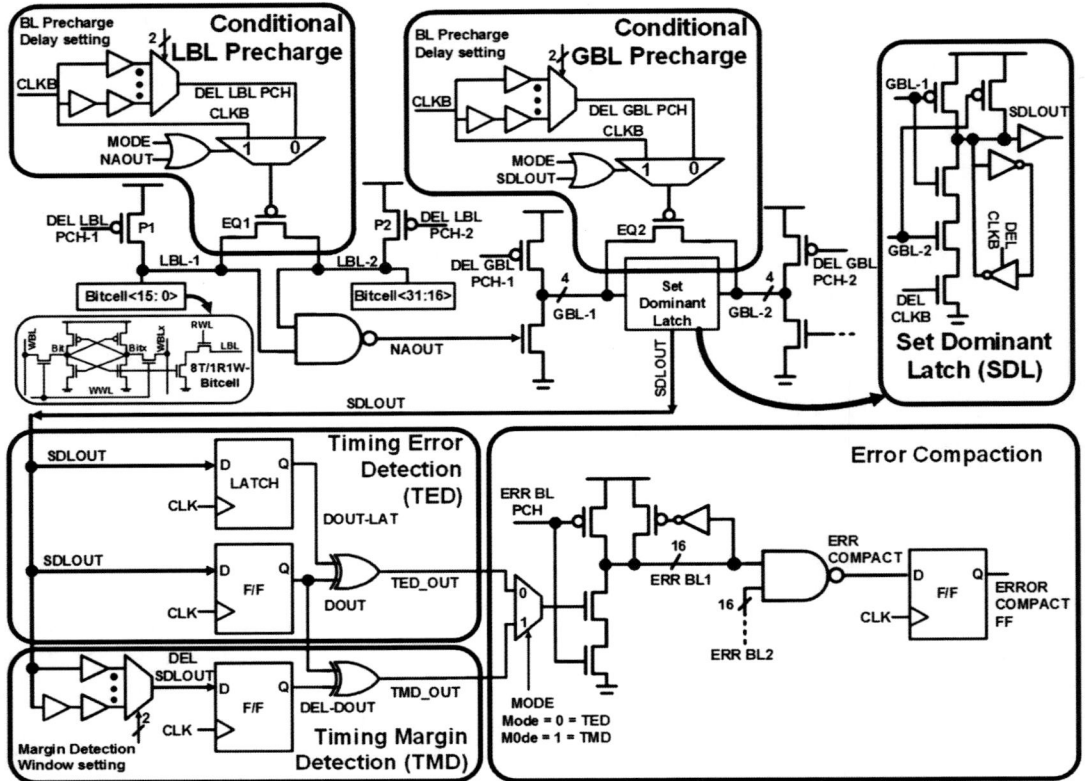

Fig. 5.16 In-situ timing margin and timing error detection along with error compaction circuits for high-performance adaptive and resilient domino register file design

persistent timing errors produced by certain data access patterns. The timing margin is detected by double-sampling the array read output and its delayed version at the same clock edge as shown in Fig. 5.16.

TED circuits enable resiliency to timing errors triggered by local high-frequency voltage droops and nominally random data access in the presence of within-die (WID) delay variations. The timing errors are detected by double-sampling the read output with a flop+latch comparison (Fig. 5.16). The sensing errors in the precharge/evaluate domino read path are converted into timing errors using conditional delayed bitline precharge without affecting the subsequent precharge operation. An error compaction circuit combines the error output from every bit slice to generate a one-bit error-compact signal for the entire register file array. TMD/TED techniques

incur 6–13% area overhead and 0.2–0.3% power overhead for a 4 KB sub-array, but this area overhead can be amortized over a larger array size.

Figure 5.17 shows the read timing error measurements at 440 mV under voltage and temperature fluctuations. The read failure measurements are extrapolated to 10^{-6} to estimate the required frequency guardbands for a 1 Mb target array size. With 10% voltage droop, the operating frequency is lowered by 8%. With temperature variation from 100 to 25 °C, the frequency further reduces by 7%. The aging guardband is estimated to be 3% based on a representative ring oscillator delay degradation.

Therefore a baseline design with process, voltage, temperature and aging guardband would operate at 860 MHz at 440 mV. As frequency is pushed higher using the TMD + TED scheme,

Fig. 5.17 Measured read failures vs. frequency and throughput improvement with adaptive and resilient approach

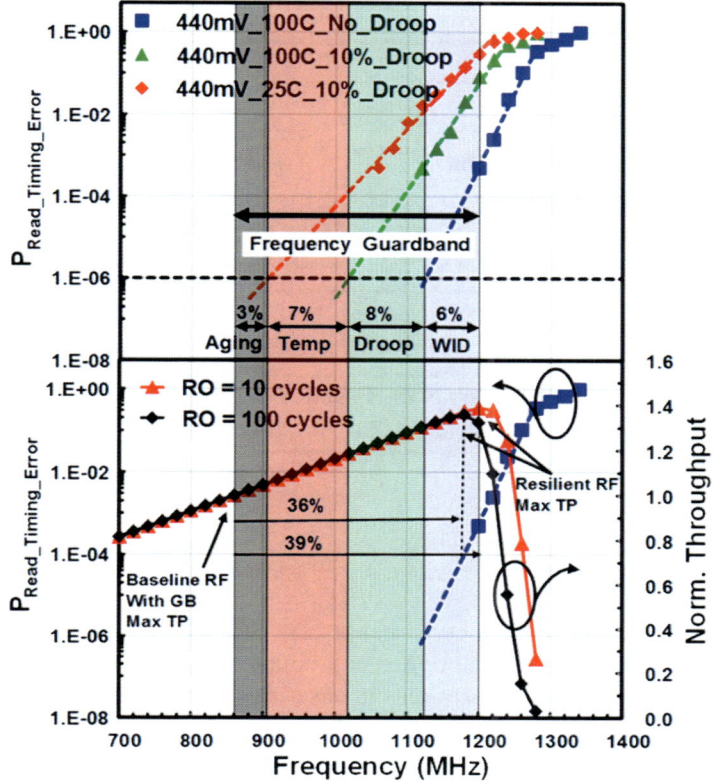

throughput first increases proportionally, and then peaks when the error rate and the corresponding recovery overheads become too large. As the recovery overhead increases, throughput improvement further reduces. The maximum frequency can be improved by 36–39% using the TMD/TED based V/F adaptation and error recovery via replay, depending on the number of recovery overhead cycles. With TED, frequency can be pushed by 6%, thus compensating for WID variation impacts across random bitcell access patterns.

These 22 nm tri-gate CMOS testchip measurement results demonstrate that the throughput gain is eventually limited by the replay overheads. Maximum achievable throughput gain observed is 21%. For a given throughput (0.9 GOPS) the energy efficiency is improved by 65–67% with a peak energy efficiency 409 GOPS/W.

Figure 5.18 shows the unified framework for the adaptive and resilient static CMOS logic

block along with the domino register file operating at the same V_{CC} and frequency. The TRC + EDS address the dynamic variations in the static CMOS logic block. For the domino RF array, TMD addresses the slow variations and generates an output to the V/F adaptation controller. The adaptation controller modulates the clock generator and/or voltage regulator to adapt the voltage and/or frequency applied to the design. On the other hand, TED addresses the fast dynamic variations that occur too quickly to be mitigated by V/F adaptation. The TED circuits give an output to the error response controller which initiates a replay mechanism at increased supply voltage or reduced clock frequency. An error-rate tracker is also included to monitor the error rate. If the error rate exceeds a certain threshold, a signal (ERTe) is sent to the adaptation controller which adapts voltage and frequency to minimize the overheads due to frequent replay mechanisms. In this way,

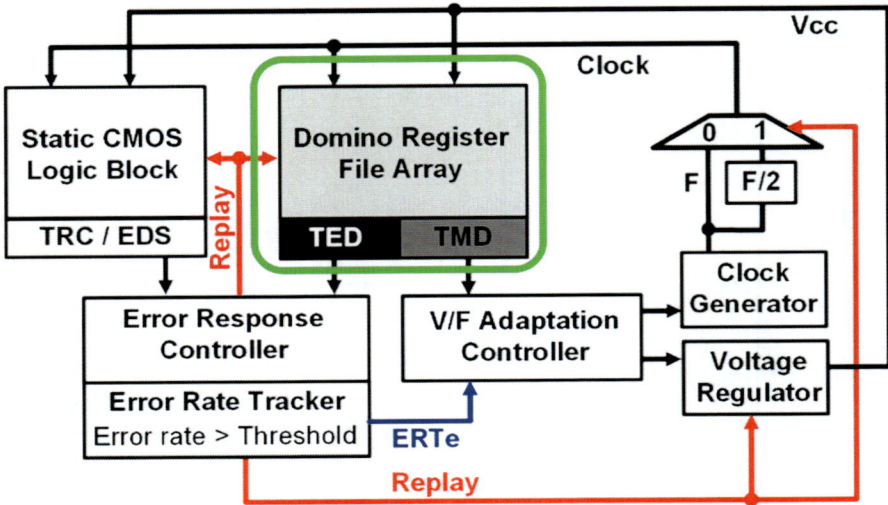

Fig. 5.18 Unified framework for adaptive and resilient logic+register file array operating on same voltage and frequency domain

energy-efficient high-performance 8T SRAM based register files featuring in-situ timing margin and timing error detection circuits can enable an adaptive and resilient framework for the entire block operating on the same voltage and frequency domain.

5.4   10T SRAM for Sub-threshold Operation

Most IoT systems exhibit a burst mode of operation followed by long idle periods. For always-ON blocks performing continuous sensing, minimizing the state retention leakage becomes an important design consideration. For such always-ON blocks, ultra-low voltage memory operation is critical for extending the battery lifetime. Various SRAM bitcells have been explored for sub-threshold region operation.

5.4.1   Sub-threshold SRAMs

For sub-threshold operation, various 6T, 8T bitcells with decoupled read and write operations have been proposed (Zhai et al. 2007; Chang et al. 2007; Verma et al. 2007). Single-ended

10T bitcells are similar to the single-ended 8T bitcell except for the read port configurations as shown in Fig. 5.19. Additional transistors are used to control for the read bitline leakage (Calhoun et al. 2006; Kim et al. 2007). Noguchi et al. have proposed a single-ended transmission gate 10T bitcell (Noguchi et al. 2008) in which the bitcell contents are buffered using an inverter and then transferred to the read bitline whenever the bitcell is accessed. The use of a transmission gate eliminates domino-style read-bitline sensing. Thus read bitline does not require precharge and keeper transistors. Also if the same data is accessed in consecutive read cycles, the charging/discharging of the read-bitline is reduced. A differential 10T bitcell with two separate ports for read-disturb-free operation has also been reported (Noguchi et al. 2008). Chang et al. have proposed a read-disturb-free differential 10T bitcell which is suitable for bit-interleaved architecture (Chang et al. 2008). A similar 10T cell with a column-assist technique is also reported (Okumura et al. 2009). However, series-connected write access transistors degrade the write-ability of the bitcell and require write-assist circuits such as wordline boosting for a successful write operation.

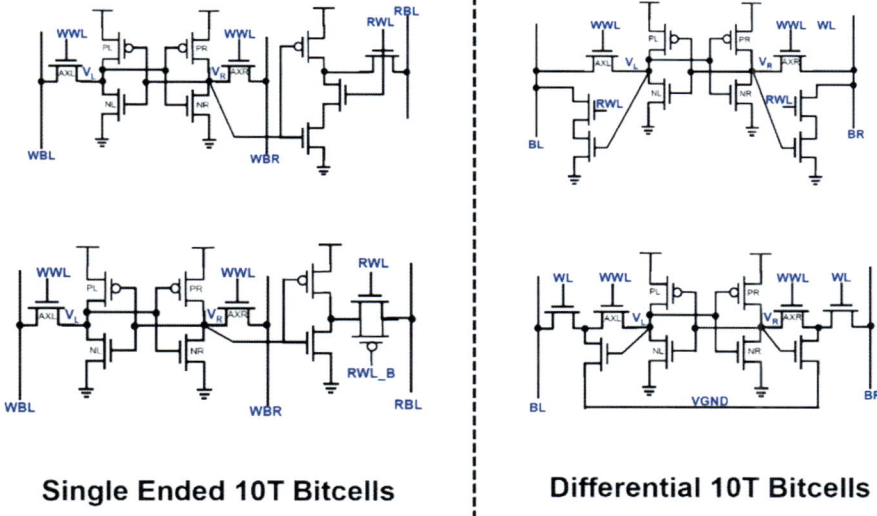

Fig. 5.19 10T sub-threshold SRAM bitcell topologies

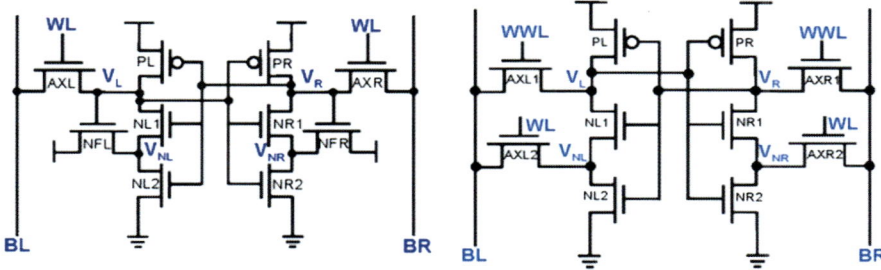

Fig. 5.20 10T Schmitt Trigger-1 (*ST-1 on left*) and Schmitt Trigger-2 (*ST-2 on right*) SRAM bitcell topologies

5.4.2 Schmitt Trigger SRAMs

To improve the stability of the cross-coupled inverter pair for ultra-low voltage state retention in always-On domains of an IoT system, Schmitt Trigger based SRAM topologies have been proposed (Kulkarni et al. 2007; Kulkarni and Roy 2012). A Schmitt trigger is used to modulate the switching threshold of an inverter depending on the direction of the input transition. In the Schmitt trigger SRAM bitcells, the feedback mechanism is used only in the pull-down path.

Figure 5.20 (left side) shows the schematics of Schmitt Trigger-1 (ST-1) bitcell. ST-1 bitcell utilizes differential sensing with 10 transistors,

1 wordline (WL) and 2 bit-lines (BL/BR). Transistors PL-NL1-NL2-NFL form one ST inverter while PR-NR1-NR2-NFR form another ST inverter. Feedback transistors NFL/NFR raise the switching threshold of the inverter during the $0 \rightarrow 1$ input transition enabling the Schmitt Trigger action. The positive feedback from NFL/NFR adaptively changes the switching threshold of the inverter depending on the direction of the input transition. During a read operation, (with $V_L = 0$ and $V_R = 1$, for example) due to voltage divider action between the access transistor and the pull down NMOS, the voltage of V_L node rises. If this voltage is greater than the switching threshold (trip point) of the V_R

inverter, the contents of the cell can get flipped resulting in a read failure event. In order to avoid a read failure, the feedback mechanism should increase the switching threshold of the inverter PR-NR1-NR2. Transistor NFR raises the voltage at node V_{NR} and increases the switching threshold of the inverter storing '1'. Thus Schmitt trigger action is used to preserve the logic '1' state of the memory cell.

Figure 5.20 (right side) shows the schematics of the Schmitt Trigger-2 (ST-2) bitcell utilizing differential sensing with 10 transistors, two wordlines (WL/WWL) and two bit-lines (BL/BR). The WL signal is asserted during read as well as the write operation, while WWL signal is asserted during the write operation. During the retention mode, both WL and WWL are OFF. In the ST-2 bitcell, feedback is provided by separate control signal (WL) unlike the ST-1 bitcell, wherein feedback is provided by the internal nodes. In the ST-1 bitcell, the feedback mechanism is effective as long as the storage node voltages are maintained. Once the storage nodes start transitioning from one state to another state, the feedback mechanism is lost. To improve the feedback mechanism, a separate control signal WL is employed for achieving stronger feedback. This enables robust operation compared to 6T and ST-1 bitcells operating at very low supply voltages.

5.4.3   Static Memory Arrays

Another approach to achieve very low voltage operation is to design the SRAM bitcell to avoid contention during the write operation, and to employ hierarchical read bitline to avoid leakage due to unselected bits in order to achieve high I_{ON}/I_{OFF} ratio. Worst-case process variations make it difficult to satisfy both read and write conditions at sub-threshold voltages. A latch-based write scheme with C^2MOS tri-state inverters can be more effective for sub-threshold operation (Wang and Chandrakasann 2005).

The read operation of the memory in sub-threshold can be challenging due to bitline leakage, where the leakage through the pull-down devices causes the dynamic bitline to drop. Because clock speed is not the key metric in this application, a sense-amplifier-based read-bitline for fast read accesses is not needed. In a conventional dynamic bitline design, the keeper/precharge transistor needs to be upsized to mitigate the leakage due to unselected bitcells. However it reduces effective read current due to lower I_{READ}/I_{OFF} ratio (Fig. 5.21).

Instead, a hierarchical read-bitline which segments the bitline by using a 2-to-1 multiplexers can be used. Segmenting reduces parallel leakage for each level of the hierarchy and the effect of process variations is mitigated.

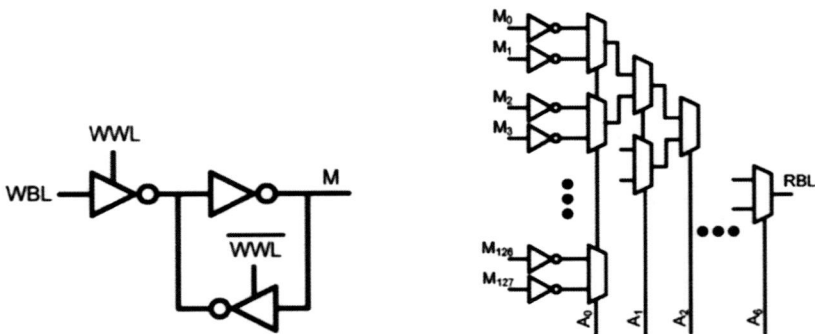

Latch based single ended write port   2:1 Mux based static read port

Fig. 5.21 Static memory design using latch bitcell and hierarchical mux based read path

These multiplexers can be designed to avoid stacked devices and sneak leakage paths by inserting inverters between each level of hierarchy.

5.5   SRAMs Using Emerging Technologies

Many alternative technologies such as nanowire Field Effect Transistors (FET), Tunnel FETs, and III–V semiconductor FETs having superior Ion/Ioff characteristics compared to the silicon MOSFTs have been explored for energy efficient low voltage SRAM designs.

5.5.1   Nanowire FET

Nanowire transistors with gate all around structure exhibit significant reduction in V_T variation with lower channel dopant concentration as well as superior short-channel control. Figure 5.22 shows nano-wire FET TEM cross-section and $0.039~\mu m^2$ 6T SRAM bitcell SEM image (Chen et al. 2009). Nano-wire FET is fabricated using Nano Injection Lithography (NIL) technique which doesn't use photoresist as well as any masks. This patterning technology utilizes gas-phase reaction activated with finely-controlled electron beam to deposit the desired

nanometer scale hard mask for subsequent etching process. Mask-less lithography reduces the number of process steps, while photoresist-free technology is more immune to light/electron interference, thus resulting in less proximity effects and better spacing resolution. A Dynamic V_{DD} Regulator (DVR) is used to improve the bitcell stability. DVR increases bitcell voltage during read operation and lowers the read disturb voltage.

5.5.2   Tunnel FET

The Tunneling Field Effect Transistor (TFET) is a quantum mechanical device in which electron transport is governed by the quantum tunneling across the source-channel junction instead of thermionic barrier modulation as in MOSFETs. Figure 5.23 shows $0.069~\mu m^2$ hybrid 6T SRAM bitcell schematic and SEM fabricated in 65 nm process technology (Nirschl et al. 2005). Pull-down (PD) NMOS devices are formed using TFETs. Pass gate (PG) transistor requires bi directional current conduction capability due to read-0 vs. write-0 scenarios. Hence TFETs are not used for PG but only for PD devices.

An all-TFET SRAM design could be challenging due to the unidirectional current conduction in TFETs, complicating its use as a SRAM access transistor. Dual-port 8T SRAM bitcells

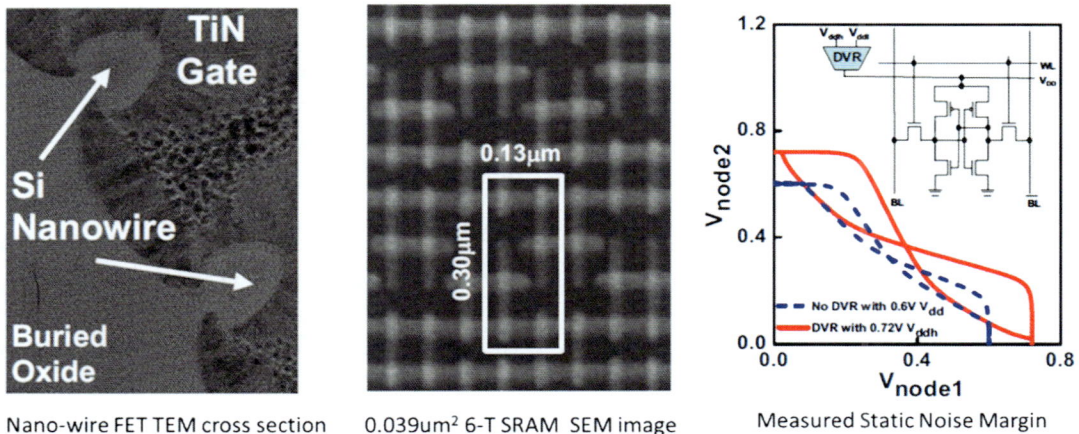

Nano-wire FET TEM cross section $0.039um^2$ 6-T SRAM SEM image   Measured Static Noise Margin

Fig. 5.22 Nanowire FET structure, 6T SRAM bitcell SEM image and read stability analysis

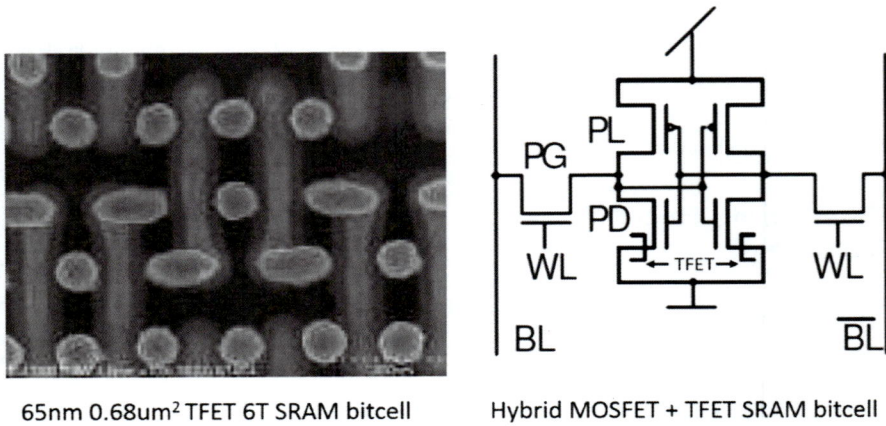

65nm 0.68um² TFET 6T SRAM bitcell   Hybrid MOSFET + TFET SRAM bitcell

Fig. 5.23 TFET based 6T SRAM schematics and SEM image

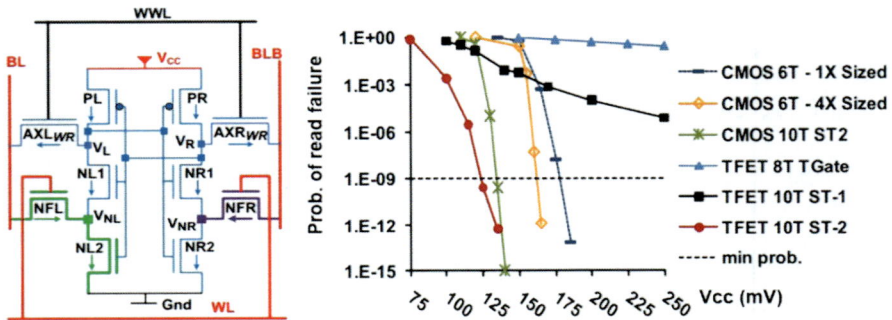

Fig. 5.24 TFET based Schmitt Trigger-2 bitcell and read stability comparison

with two TFET access transistors having opposite current directionality are used for read and write operation. Schmitt Trigger-2 bitcell topology can be realized with TFETs as shown in Figure 5.24 (Saripalli et al. 2011). Statistical simulations show extreme low voltage read operation for TFET based ST-2 bitcell compared to the CMOS based 6T, 10T and TFET dual port 8T bitcells. Therefore, TFET devices with better Ion/Ioff at low supply voltages could be a suitable candidate for ultra-low voltage SRAM designs in always-ON modules of an IoT system.

5.5.3 III–V Quantum Well FET

Compound semiconductor-based quantum well FETs are a promising candidate to realize high-performance yet low-voltage SRAM designs. Figure 5.25 shows Indium Antimonide (InSb) semiconductor having highest electron mobility used to form NFET device. This could be used to form the read port of the hybrid 8T SRAM bitcell (Kulkarni and Roy 2008). The higher Ion current in InSb can be leveraged to improve the read bitline delay up to 60% compared to the Silicon CMOS based 8T SRAM.

5.5.4 Monolithic 3-D Integration

The demand for higher SRAM density is continually growing due to increased design complexity (application processor, modem, and graphics on the same die) and also due to increased die cost in dimensionally-scaled advanced CMOS

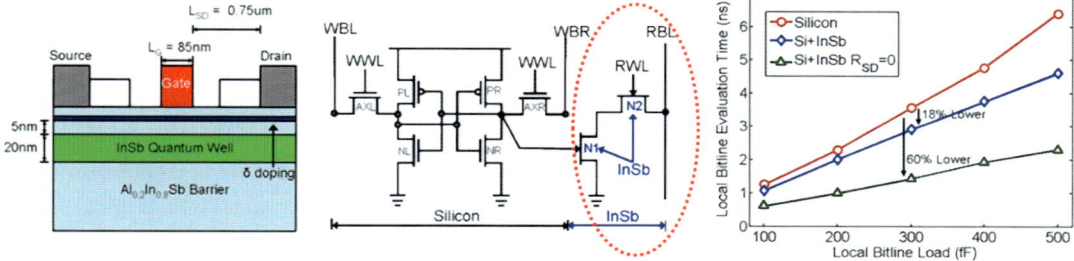

Fig. 5.25 InSb based quantum well FET structure, hybrid Si+InSb 8T SRAM bitcell and bitline delay comparison

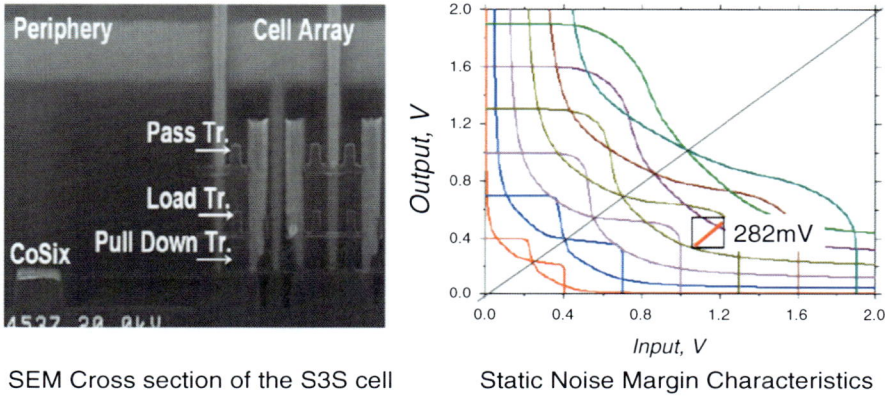

SEM Cross section of the S3S cell   Static Noise Margin Characteristics

Fig. 5.26 Monolithic SRAM bitcell SEM image and read stability measurements

technologies. Towards this goal, many approaches for monolithic 3-D integration of SRAM bitcells have been reported. Figure 5.26 shows the scanning electron microscope (SEM) cross section of Stacked Single crystal Silicon (S^3) SRAM cell fabricated using 65 nm process technology achieving 0.16 μm^2 bitcell area showing almost 3× better bitcell density compared to the conventional 65 nm planar bitcell (Jung et al. 2005). Load PMOS and pass gate NMOS transistors are stacked on the planar pull-down transistors in different layers. The Static Noise Margin (SNM) achieved is 282 mV at $V_{CC} = 1.2$ V operation.

5.6   Summary

In this chapter we have shown that there are a wide variety of circuit and device techniques that can be applied for energy-efficient, low-voltage memory arrays. These techniques improve the dynamic operating range of the memory by enhancing the periphery to apply optimum access voltages (dynamic assist, dual-V_{CC} arrays), or optimizing the bitcell itself for low-voltage operation (10T bitcells, static memory arrays), or by embedding resiliency techniques to dynamically detect and respond to variations. These techniques differ widely in terms of area overhead, complexity, ease of integration, and power and voltage reduction that can be achieved. Therefore it is important to analyze the unique requirements for each memory array in a particular application before selecting the best implementation for that array (Table 5.1).

In this section we give a high-level summary of the key metrics for the techniques which have been detailed in this chapter. In addition to these circuit techniques, emerging technologies (such as nanowires, tunnel FETs, and quantum well FETs) can also provide power/performance

Table 5.1 Qualitative comparison of various 6T, 8T, 10T SRAM V_{MIN} and guardband reduction techniques

Technique	Ref.	Array type	Key goal	Overhead[a]	Example results[a]
Read/write assist	Khellah et al. (2008) and Mann et al. (2010)	6T SRAM	V_{MIN} reduction	Varies	Varies
Wordline boosting: charge pump	Raychowdhury et al. (2010)	8T SRAM	V_{MIN} reduction	~25% area	120–140 mV V_{MIN} reduction
Wordline boosting: capacitive coupling	Kulkarni et al. (2010)	8T SRAM	V_{MIN} reduction	5–11% area	140 mV write V_{MIN} reduction
Dual-Vcc design	Kulkarni et al. (2013)	8T SRAM	V_{MIN} reduction	Extra supply rail	130 mV V_{MIN}, 27% lower power
Subthreshold SRAMs	Verma et al. (2007); Calhoun et al. (2006); and Kim et al. (2007)	10T SRAM bitcells	V_{MIN} reduction	~50% area	Sub-500 mV operation
Schmitt trigger SRAM	Kulkarni et al. (2007) and Kulkarni and Roy (2012)	10T SRAM bitcells	V_{MIN} reduction	~100% area	~100 mV lower V_{MIN}
Static memory	Wang and Chandrakasann (2005)	Latch-based bitcell	V_{MIN} reduction	>2–3× area	Sub-500 mV operation
Resiliency: tunable replica bits	Raychowdhury et al. (2011)	6T, 8T, 10T	V_{MIN}, guardband reduction	~5% area	9% V_{MIN}, 7.5% power reduction
Resiliency: in-situ timing monitors	Kulkarni et al. (2015)	8T	V_{MIN}, guardband reduction	6–13% area	~36% frequency increase

[a]Note: overheads and results for these techniques are strongly dependent on process technology, array design characteristics, optimization target, etc. The numbers shown here are meant to be only representative examples

improvements at the low voltages required for many IoT devices.

5.7 Trends and Perspectives

The IoT domain presents both a challenge and opportunity for advances in low-power, low-voltage memory arrays. Power requirements for many IoT devices are extremely low, requiring memory arrays with low standby power as well as low access (read and write) power. Voltage scaling of traditional SRAM arrays has slowed or even stopped with technology scaling, which points to the need for advanced circuit techniques for energy efficient and reliable on-die memory for the IoT space. At the same time, new and emerging workloads of special importance to IoT-for example, always-on audio and visual recognition workloads, cognition, etc.-require ever-increasing amounts of memory storage and bandwidth. Clearly, advancements in memory design are needed to enable these usages for IoT.

The techniques discussed here also point to the need for cross-layer optimization to reap maximum benefit for these advanced memory techniques while minimizing cost. IoT SoCs can contain hundreds of embedded memory arrays, each with its own unique size, arrangement, access pattern, and power/performance requirements. Detailed architecture and system simulations are needed early in the design phase to determine the optimum method for implementing each memory array. Dynamic assist or multi-voltage techniques require area-efficient and energy-efficient power delivery circuits such as on-die voltage generators, charge pumps, and dynamic power management. Resiliency techniques that detect timing margins or timing errors show the promise to drastically reduce array operating voltages, but require architecture support (such as instruction replay) to obtain the full benefit. Looking further towards the future, dramatic gains in memory energy efficiency can be obtained by over-scaling the voltage and allowing a small number of bits to fail. Traditionally, these failures are either avoided or handled via error-correction techniques such as ECC for large memory arrays. However, there are a class of emerging workloads such as convolutional neural networks that can tolerate infrequent errors in certain computations. These types of workloads, coupled with the necessary resiliency and monitoring techniques to ensure memory operation within the desired bit error range, could provide the next big gains in energy efficient operation.

Acknowledgments Authors would like to thank Muhammad Khellah, Arijit Raychowdhury, Keith Bowman, Carlos Tokunaga, Dinesh Somasekhar, Bibiche Geuskens, Tanay Karnik, and Shekhar Borkar for insightful technical discussions. Authors would also like to thank Greg Taylor, Richard Forand and Matthew Haycock for encouragement and support. This research was, in part, funded by the U.S. Government (DARPA). The views and conclusions contained in this document are those of the authors and should not be interpreted as representing the official policies, either expressed or implied, of the U.S. Government.

References

A. Bhavnagarwala, X. Tang, J. Meindl, The impact of intrinsic device fluctuations on CMOS SRAM cell stability. IEEE J. Solid State Circ. **36**(4), 658–665 (2001)

K. Bowman, J. Tschanz, C. Wilkerson, S.-L Lu, T. Karnik, V. De, S. Borkar, Circuit techniques for dynamic variation tolerance, in *Proceedings of the 46th Annual Design Automation Conference* (2009), pp. 4–7

B.H. Calhoun, A.P. Chandrakasan, A 256 kb Sub-threshold SRAM in 65 nm CMOS, in *Proceedings of the International Solid State Circuits Conference* (2006), pp. 628–629

L. Chang, Y. Nakamura, R.K. Montoye, J. Sawada, A.K. Martin, K. Kinoshita, F.H. Gebara, K.B. Agarwal, D.J. Acharyya, W. Haensch, K. Hosokawa, D. Jamsek, A 5.3 GHz 8T-SRAM with operation down to 0.41 V in 65 nm CMOS, in *Proceedings of the VLSI Circuit Symposium* (2007), pp. 252–253

I. Chang, J.-J. Kim, S. Park, K. Roy, A 32 kb 10T subthreshold SRAM array with bit-interleaving and differential read scheme in 90 nm CMOS, in *Proceedings of the International Solid State Circuits Conference* (2008), pp. 628–629

H.-Y. Chen, C.-C. Chen, F.-K. Hsueh, J.-T. Liu, C.-Y. Shen, C.-C. Hsu, S.-L. Shy, B.-T. Lin, H.-T. Chuang, C.-S. Wu, C. Hu, C.-C. Huang, F.-L. Yang, 16 nm functional 0.039 μm^2 6T-SRAM cell with nano

injection lithography, nanowire channel, and full TiN gate, in *Proceedings of International Electron Device Meeting (IEDM)* (2009), pp. 958–960

D. Ernst, N.S. Kim, S. Das, S. Pant, R. Rao, T. Pham, C. Ziesler, D. Blaauw, T. Austin, K. Flautner, T. Mudge, Razor: a low-power pipeline based on circuit-level timing speculation, *IEEE/ACM MICRO-36* (2003), pp. 7–18

S.-M. Jung, Y. Rah, T. Ha, H. Park, C. Chang, S. Lee, J. Yun, W. Cho, H. Lim, J. Park, J. Jeong, B. Son, J. Jang, B. Choi, H. Cho, K. Kim, Highly cost effective and high performance 65 nm S^3 (Stacked Single-crystal Si) SRAM technology with $25F^2$, 0.16 μm^2 cell and doubly stacked SSTFT cell transistors for ultra high density and high speed applications, *Symposium on VLSI Technology* (2005), pp. 220–221

M.M. Khellah, A. Keshavarzi, D. Somasekhar, T. Karnik, V. De, Read and write circuit assist techniques for improving Vccmin of dense 6T SRAM cell, in *Proceedings of the International Conference on Integrated circuit Design and Technology* (2008), pp. 185–189

T.-H. Kim, J. Liu, J. Keane, C.-H. Kim, A high-density subthreshold sram with data-independent bitline leakage and virtual ground replica scheme, in *Proceedings of the International Solid State Circuits Conference* (2007), pp. 330–331

J.P. Kulkarni, K. Roy, Technology circuit co-design for ultra-fast InSb quantum well transistors. IEEE Trans. Electron Dev. **55**, 2537–2545 (2008)

J.P. Kulkarni, K. Roy, Ultralow-voltage process-variation-tolerant schmitt-trigger-based SRAM design. IEEE Trans. VLSI Syst. **20**(2), 319–332 (2012)

J.P. Kulkarni, K. Kim, K. Roy, A 160 mV robust schmitt trigger based sub-threshold SRAM. IEEE J. Solid State Circ. **42**(10), 2304–2313 (2007)

J. Kulkarni, B. Geuskens, T. Karnik, M. Khellah, J. Tschanz, V. De, Capacitive-coupling wordline boosting with self-induced VCC collapse for write V_{MIN} reduction in 22-nm 8T SRAM, *International Solid State Circuits Conference (ISSCC)* (2010), pp. 234–235

J.P. Kulkarni, M. Khellah, J. Tschanz, B. Geuskens, R. Jain, S. Kim, V. De, Dual-Vcc 8T-bitcell SRAM array in 22 nm Tri-gate CMOS for energy efficient operation across wide dynamic voltage range, *VLSI Circuit Symposium (VLSI Symp)* (2013), pp. C126–C127

J. P. Kulkarni, C. Tokunaga, P. Aseron, T. Nguyen Jr., C. Augustine, J. Tschanz, V. De, A 409 GOPS/W adaptive & resilient domino register file in 22 nm Tri-Gate CMOS featuring in-situ timing margin & error detection for tolerance to within-die variation, voltage droop, temperature & aging, *International Solid State Circuits Conference (ISSCC)* (2015), pp. 82–83

R.W. Mann, J. Wang, S. Nalam, S. Khanna, G. Braceras, H. Pilo, B.H. Calhoun, Impact of circuit assist methods on margin and performance in 6T SRAM. Solid State Electron. **54**(11), 1398–1407 (2010)

S. Mukhopadhyay, H. Mahmoodi, K. Roy, Modeling of failure probability and statistical design of SRAM array for yield enhancement in nanoscaled CMOS. IEEE Trans. Comput. Aided Des. **24**(12), 1859–1880 (2005)

H. Noguchi, S. Okumura, Y. Iguchi, H. Fujiwara, Y. Morita, K. Nii, H. Kawaguchi, M. Yoshimoto, Which is the best dual-port SRAM in 45-nm process technology?—8T, 10T single end, and 10T differential, *IEEE International Conference on Integrated Circuit Design and Technology* (2008), pp. 55–58

Th. Nirschl, St. Henzler, J. Fischer, A. Bargagli-Stoffi, M. Fulde, M. Sterkel, P. Teichmann, U. Schaper, J. Einfeld, C. Linnenbank, J. Sedlmeir, C. Weber, R. Heinrich, M. Ostermayr, A. Olbrich, B. Dobler, E. Ruderer, R. Kakoschke, K. Schrüfer, G. Georgakos, W. Hansch, D. Schmitt-Landsiedel, The 65 nm tunneling field effect transistor (TFET) 0.68 μm^2 6T memory cell and multi-V_{th} device, *35th European Solid-State Device Research Conference* (2005), pp. 173–176

S. Okumura, Y. Iguchi, S. Yoshimoto, H. Fujiwara, H. Noguchi, K. Nii, H. Kawaguchi, M. Yoshimoto, A 0.56-V 128 kb 10T SRAM using Column Line Assist (CLA) scheme, in *Proceedings of the International Symposium on Quality Electronics Design (ISQED)* (2009), pp. 659–663

A. Raychowdhury, B. Geuskens, J. Kulkarni, J. Tschanz, K. Bowman, T. Karnik, S.-L. Lu, V. De, M. Khellah, PVT & Aging Adaptive Word-Line Boosting for 8T SRAM Power reduction, *International Solid State Circuits Conference (ISSCC)* (2010)

A. Raychowdhury, B. Geuskens, K. Bowman, J. Tschanz, S.-L. Lu, T. Karnik, M. Khellah, V. De, Tunable replica bits for dynamic variation tolerance in 8T SRAM arrays, IEEE Journal of Solid State Circuits, **46**(4), (2011)

V. Saripalli, S. Datta, V. Narayanan, J.P. Kulkarni, Variation-tolerant ultra low-power hetero-junction tunnel FET SRAM Design, *7th International Symposium on Nanoscale Architectures (NANOARCH)* (2011), pp. 45–52

Semiconductor Industry Association (SIA) and Semiconductor Research Corporation (SRC) report on "Rebooting the IT revolution" (2015)

J. Tschanz, K. Bowman, S. Walstra, M. Agostinelli, T. Karnik, V. De, Tunable replica circuits and adaptive voltage-frequency techniques for dynamic voltage, temperature, and aging variation tolerance, *Digest of Technical Papers, Symposium on VLSI Circuits* (2009), pp. 112–113

N. Verma, A.P. Chandrakasan, 65 nm 8T Sub-Vt SRAM employing sense-amplifier redundancy, in *Proceedings of International Solid State Circuits Conference* (2007), pp. 328–329

A. Wang, A. Chandrakasann, A 180-mV subthreshold FFT processor using a minimum energy design methodology. IEEE J. Solid State Circ. **40**(1), 310–319 (2005)

Y. Wang, Robust SRAM Design in Nanoscale CMOS Circuit and Technology, *ISSCC Forum on embedded memories* (2012)

N. Yoshinobu, H. Masahi, K. Takayuki, K. Itoh, Review and future prospects of low-voltage RAM circuits. IBM J. Res. Dev. **47**(5/6), 525–552 (2003)

B. Zhai, D. Blaauw, D. Sylvester, S. Hanson, A sub-200 mV 6T SRAM in 0.13 μm CMOS, in *Proceedings of the International Solid State Circuits Conference* (2007), pp. 332–333

On-Chip Non-volatile Memory for Ultra-Low Power Operation

6

Meng-Fan Chang

This chapter addresses trends and challenges in the development of on-chip (embedded) non-volatile memory (NVM) for ultra-low power operation. Various NVM technologies have been introduced, including Flash, OTP/MTP, resistive RAM, and phase-change memory (PCM). In the following, we examine some of the challenges in the design of circuits used for read and write operations. Future trends in ultra-low-power NVM are also discussed.

6.1 Operational Requirements of On-Chip NVM for IoT

In this section, we outline the requirements of on-chip NVM for IoT devices under low-power operation and provide a review of current state-of-the-art on-chip NVMs.

6.1.1 Operation of On-Chip NVM for IoT and Normally-Off Applications

Numerous energy-efficient systems employ intelligent power on-off schemes to reduce system standby power, as shown in Fig. 6.1. This is a particularly important issue when dealing with

battery-powered or energy-harvester-powered IoTs or wearable devices equipped with nanometer chips, which are particularly susceptible to leakage current. Many energy-efficient chips employ on-chip (embedded) nonvolatile memory (eNVM) for the storage of programs and critical data while in power-off mode (Yano et al. 2012; Hatanaka et al. 2007; Hidaka et al. 2011; Zwerg et al. 2011; Wang et al. 2012c, 2014; Chang et al. 2011a, b, 2012a, 2014b, 2015b; Chiu et al. 2010; Yamamoto et al. 2009; Huang et al. 2014; Eshraghian et al. 2010; Matsunaga et al. 2012; Li et al. 2014). At present, embedded Flash (eFlash) is the most mature and reliable solution for the mass production of eNVM. Several other emerging memory technologies, such as Phase Change Memory (PCM) (Wen et al. 2011; Rizzi et al. 2014; Pozidis et al. 2013; Boniardi et al. 2014), Spin-Torque Transfer Magnetic RAM (STT-MRAM) (Kim et al. 2011; Kitagawa et al. 2012; Wang et al. 2012b; Tsunoda et al. 2013), and ReRAM devices (memristor or resistive RAM, ReRAM) (Chang et al. 2011a, b, 2012a, 2013a, b, c, 2014a, b, c, 2015b; Chiu et al. 2010; Yamamoto et al. 2009; Huang et al. 2014; Eshraghian et al. 2010; Banno et al. 2014; Xue et al. 2013; Wataru et al. 2011; Fackenthal et al. 2014; Liu et al. 2013b; Kawahara et al. 2012; Wang et al. 2012a; Koveshnikov et al. 2012; Zhang et al. 2014; Lee et al. 2010; Hsieh et al. 2013; Sakotsubo et al. 2010; Jo et al. 2014; Serb et al. 2014; Yang et al. 2012; Nardi et al. 2011),

M.-F. Chang (✉)
National Tsing Hua University, Hsinchu City, Taiwan
e-mail: mfchang@ee.nthu.edu.tw

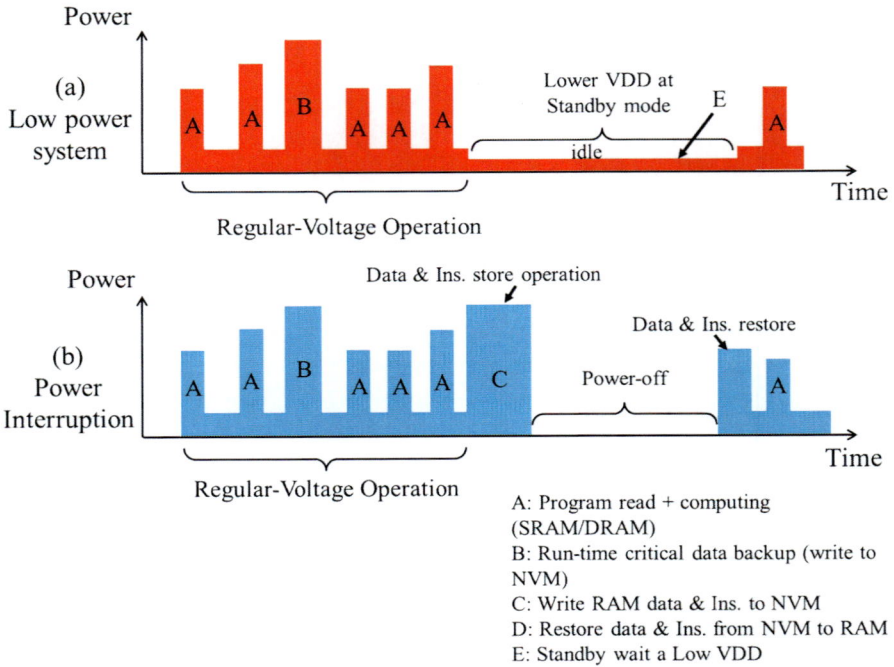

Fig. 6.1 Underlying concepts in the use of (**a**) low-voltage or (**b**) power interrupt to reduce power consumption

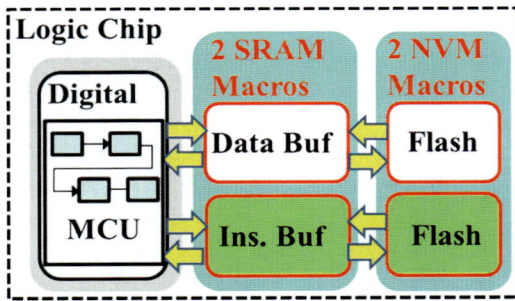

Fig. 6.2 Operation of NVM in high-performance IoT devices

also are good candidates for IoT devices. In Sect. 6.2, we discuss on-chip NVM devices for IoT applications.

As shown in Fig. 6.2, most high performance IoT (HP-IoT) devices require large-capacity eNVM for the storage of long stretches of program-code and large volumes of data. This commonly involves the use of two discrete volatile memory devices (i.e., SRAM or eDRAM macros) to provide access to instructions and data. During power-on procedures, program-code

stored in eNVM is uploaded to on-chip instruction buffer (Frustaci et al. 2014; Chen et al. 2014; Oh et al. 2014; Hamzaoglu et al. 2014; Song et al. 2014), whereas the data is uploaded to an on-chip data buffer, such as SRAM or DRAM (Frustaci et al. 2014; Chen et al. 2014; Oh et al. 2014; Hamzaoglu et al. 2014; Song et al. 2014). In computation mode, the volatile memory macros provide access to program-code and perform data buffering while the eNVM is used for run-time data back-up. During power-down procedures, all critical data is flushed to eNVM.

As shown in Fig. 6.3, most low-cost IoT devices use eNVM for power-off storage as well as power-on program-code access as a means of reducing reduce chip area. Eliminating the SRAM instruction macro means that eNVM must perform frequent-read and infrequent-write actions, thereby necessitating a reduction in the power consumption associated with read operations in order to reduce overall power consumption.

Figure 6.4 illustrates the standby energy consumption of a SRAM macro at the data-retention voltage (DRV) and that of the memory power interrupt approach (eNVM-based two-macro scheme) for a given memory capacity. Standby energy consumption of two-macro scheme includes the energy consumed by data backup operations (SRAM read out + eNVM write in) as well as in the restore operation (eNVM read out + SRAM write in). Compared with an SRAM macro held at a DRV, the memory power interrupt approach (one power-off and one-power on operation) using a eNVM-based two-macro scheme consumes less standby energy when the standby period exceeds the break-even time (BET). BET refers to the time required for a eNVM-based two-macro device to break even in power consumption when balanced

against a low-voltage on-chip SRAM during sleep-mode, where VDD is biased at DRV. BET is shorter in more advanced nanometer technologies due to an increase in SRAM leakage current. This makes the two-macro eNVM-based two-macro scheme more energy efficient than DRV-biased SRAM in cases of extended system standby period.

6.1.2   Requirements of On-Chip NVM for Ultra-Low Power Operation

In this subsection, we discuss the requirements of on-chip NVM to reduce the power consumption of various applications involving IoT devices.

Figure 6.5 presents a simplified illustration of a typical on-chip NVM macro, which includes a cell array, row drivers, input-output (IO) circuits, timing controller, and high-voltage generator.

In a cell array, each NVM cell is connected to three signals: word-line (WL), source-line (SL), and bit-line (BL). The row drivers include row-decoders, level-shifters, and drivers for the word-line and source-lines.

The IO circuit includes column multiplexors as well as read-path and write-path sub-circuits.

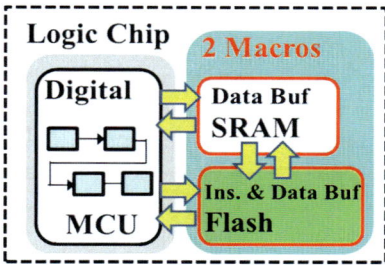

Fig. 6.3 Operation of NVM in low-cost IoT devices

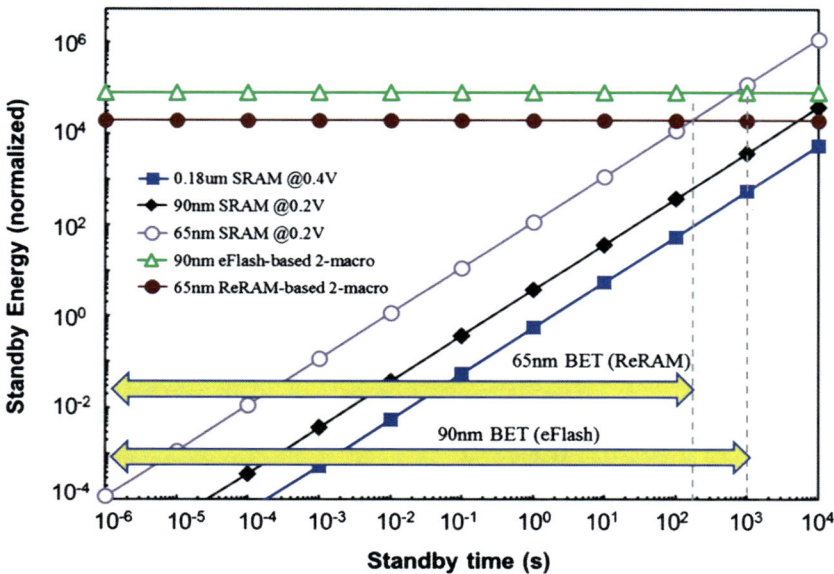

Fig. 6.4 Standby energy consumption of a SRAM macro and two-macro schemes

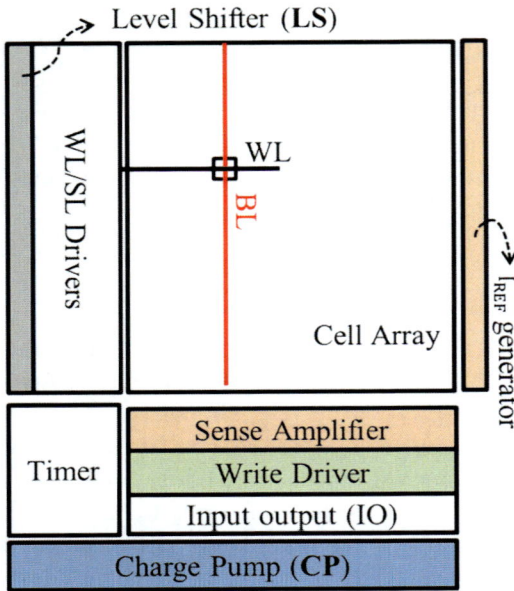

Fig. 6.5 Simplified structure of typical on-chip NVM macros

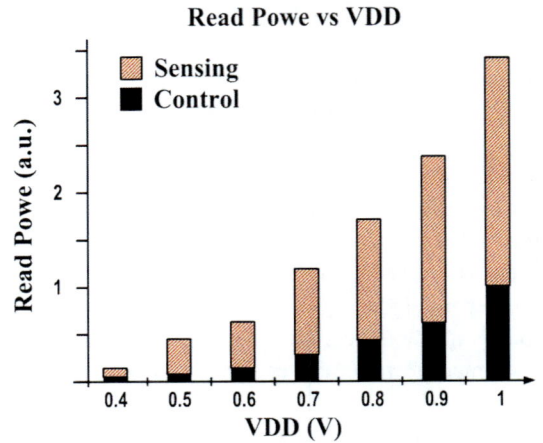

Fig. 6.6 Power breakdown in read operation of an NVM macro

The read-path includes output buffers and sense amplifiers. The write-path includes write-drivers and input buffers.

Most high-voltage generators employ an on-chip charge-pump circuit to generate the required voltage, which is higher than the supply voltage (VDD) for write operations in NVM devices.

As discussed in Sect. 6.1.1, low-cost IoT devices require eNVM to reduce power consumption when performing read operations. Figure 6.6 presents the normalized power breakdown of a typical read operation in an embedded Flash memory. The following approaches can be used to reduce read power: (1) low bit-line bias (precharge) voltage to suppress power consumption by the cell array. (2) Small input-offset sense amplifier to reduce the DC-current consumption time and read delay time.

Many energy-efficient and energy-harvesting-powered devices (Zwerg et al. 2011; Wang et al. 2012c; Tang et al. 2014) must deal with a limited supply voltage and strict power budget. Implementing low VDD is an effective approach to reduce dynamic power consumption through the suppression of voltage swing associated with

control signal toggling and data access for a processor. As shown in Fig. 6.7, NVM with low-voltage read operations further reduces power consumption, particularly in IoT devices with frequent-read-seldom-write operations.

In Sect. 6.3, we discuss various design challenges and outline a number of solutions that have been proposed for low-power read schemes.

Figure 6.8 illustrates the normalized power breakdown in the write operation of a typical embedded Flash memory device. Most NVM devices require high voltages and excessive time for write operations, such that most of the write power is consumed by sub-blocks using DC-current or through continuous toggling throughout the write period. The sub-blocks that consume much of the write power include row-drivers (level-shifters), write-drivers, and the charge-pump circuit. This means that an on-chip NVM with low write power should employ low write-energy NVM devices as well as means of suppressing power consumption of sub-blocks.

Power integrity is another important concern for most chips (Jain et al. 2011; Chang et al. 2015e), particularly those used in low-voltage IoT devices. Supply noise can induce functional failures in supply-sensitive circuits, such as on-chip mixed-signal and memory circuits

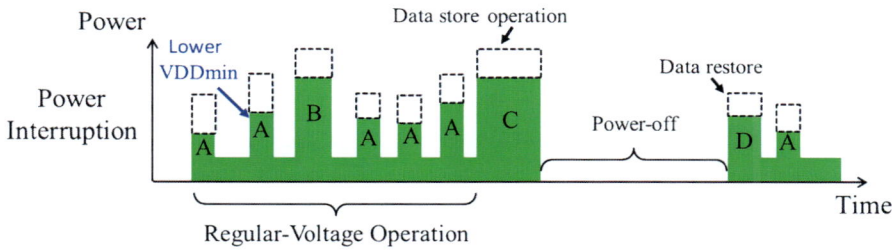

Fig. 6.7 Use of low-voltage to reduce dynamic power consumption in IoT chips

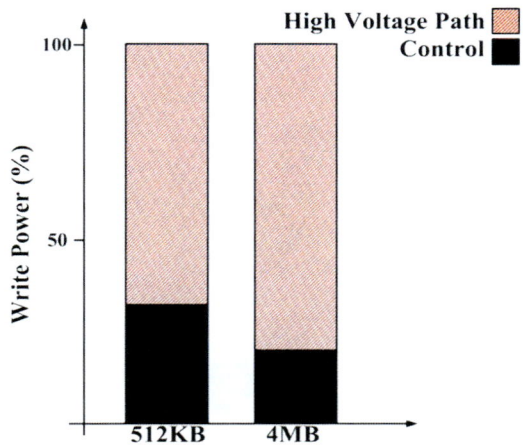

Fig. 6.8 Normalized power breakdown of write operation in a typical embedded Flash memory device

(Chang and Yang 2009; Larsson et al. 2001a; Mezhiba and Friedman 2004). Supply noise can also affect program-verify operations and the threshold voltage distribution of NVM cells. These issues underline the importance of developing low peak-current CPs for 3D-ICs and 3D-memory modules.

In Sect. 6.4, we discuss the design challenges and outline a number of solutions that have been proposed for low-power write schemes.

6.2   On-Chip NVM Devices

In this section, we introduce various mass-production-ready on-chip NVM devices for IoT applications, including embedded Flash and one-time-programmable (OTP) devices. Two emerging memory devices with high signal ratio, resistive memory (ReRAM) and phase-change memory (PCM), are also introduced.

6.2.1   Embedded Flash (eFlash) Memory

On-chip Flash memory cells employ two basic structures: split-gate or stack-gate. Both of these structures feature a floating-gate (FG) beneath a control-gate (CG) for the storage of charge used to alter the threshold voltage (V_{TH-MC}) of an N-channel transistor memory cell. The charge can be stored in the form of electrons or holes (Zhang 2009). Flash memory performs two write operations: program and erase. Program operations involve the injection of charge into the FG to achieve a high V_{TH-MC} in order to yield a tiny amount of cell current ($I_{CELL-PRG}$) in read mode. Erase operations remove the negative charge from the FG in order to reduce V_{TH-MC} and generate substantial cell current (I_{CELL}) in read mode. These cell structures employ either a floating-poly-gate or charge-trapping technology (Kuo et al. 1998; Kamiya et al. 1982; Kianian et al. 1994; Liu et al. 1997; Takahashi et al. 1999; Eitan et al. 1999; Chen et al. 1997; Lee et al. 2006a; Yater et al. 2007).

Figure 6.9 presents a simplified illustration of a typical one-transistor (1T) stack-gate Flash device in which the control-gate and floating-gate are stacked vertically. The 1T stack-gate achieves a compact cell area and is commonly utilized in discrete memory products, such as NOR and 2D NAND Flash memories.

In program operations (Zhang 2009), a portion of the channel hot electron (CHE), generated by the source-to-drain current, is injected into the FG through the bottom oxide in order to negatively charge the FG, thereby increasing the V_{TH-MC}. In erase operations, high voltage is applied to the channel (substrate) to

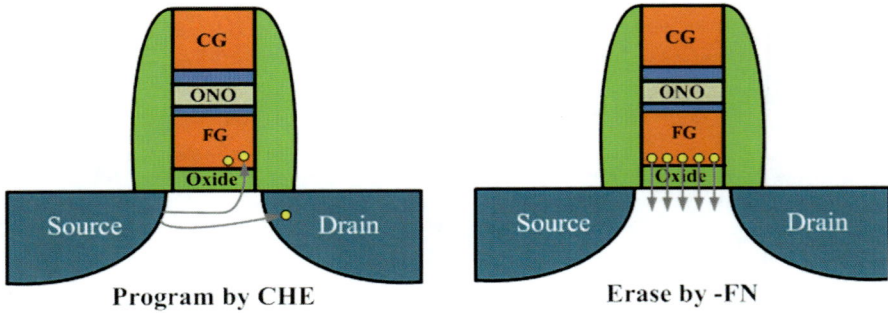

Fig. 6.9 Simplified structure of typical one-transistor (1T) stack-gate Flash device

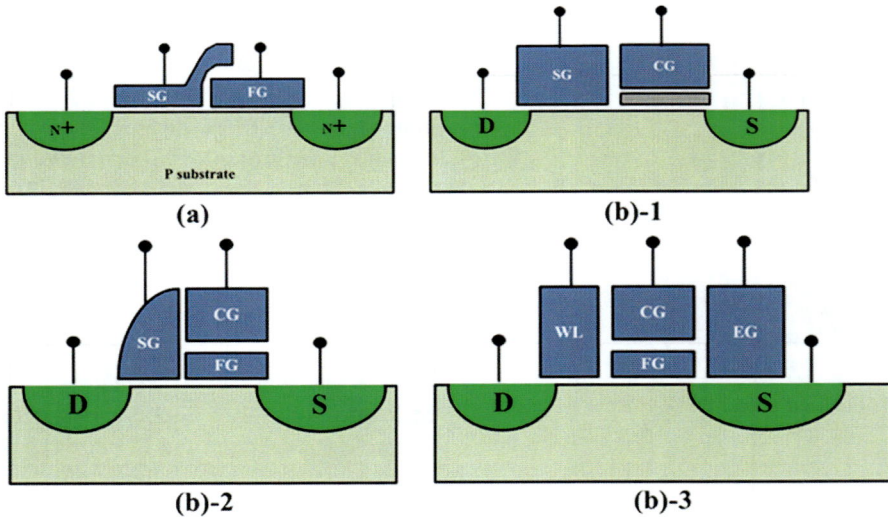

Fig. 6.10 Simplified structure of (**a**) conventional and (**b**) recently developed split-gate Flash cells

initiate the Fowler-Nordheim (FN) electron tunneling mechanism between the FG and the channel. The negative charge is then removed from the FG through the bottom tunneling oxide in order to lower the V_{TH-MC}.

This 1T stack-gate cell features a compact cell size but suffers from issues related to (1) high program current and (2) over-erase. High program current necessitates the consumption of high write power as well as a charge-pump circuit with large area overhead. Over-erase results in ultra-low V_{TH-MC} and considerable bit-line leakage current from unselected erased cells during read operations, thereby degrading the sensing margin. Unfortunately, most on-chip NVM devices have less memory capacity than do

commodity NVMs, thereby precluding the use of complex program-verify schemes to avoid the issue of over-erase.

Since the early 1980s, numerous split-gate cells (Kianian et al. 1994; Yater et al. 2007; Van Houdt et al. 1992; Ahn et al. 1998; Huang et al. 2000; Mih et al. 2000; Tsouhlarakis et al. 2001; Wang et al. 2005; Saha 2007; Cho et al. 2006; Kotov et al. 2013; Kono et al. 2013) have been proposed. Figure 6.10 illustrates the simplified structure of conventional and recently developed split-gate Flash cells. In conventional split-gate cells, a portion of the floating gate is placed beneath the control gate, such that the channel of a memory cell transistor is controlled by the control gate (wordline) as well as the floating gate. This causes the split-gate Flash memory cell to act as two transistors operating in serial, equivalent to 1.5T per cell. This 1.5T serial-transistor structure makes it possible to turn off unselected erased cells using the control gate, even in cases of a negative V_{TH-MC} caused by the floating gate. Unlike stack-gate cells, split-gate cells are immune to over-erase, which eliminates the need to boost the wordline (WL) voltage for fast read operations. In recent split-gate Flash cells, most of the area of the floating-gate is located beneath the control gate. A side-wall select gate (SG) or erase gate (EG) can also be added to facilitate the inclusion of high-voltage and low-voltage areas within a split-gate cell (Kono et al. 2013).

A source-side injection (SSI) program scheme reduces the current required by split-gate cells as well as the power consumed in program operations. Moreover, the small gap between the floating gate and control gate enables rapid erase operations via poly-to-poly FN tunneling operations (Kianian et al. 1994) or hot hole injection via band-to-band tunneling (BTBT) (Eitan et al. 2000). Small program current, fast erase operations, and immunity to over-erase have made split-gate flash cells popular for on-chip (embedded) applications, particularly those used in low-power applications.

6.2.2 Embedded OTP/MTP Memory

For IoT devices that do not require frequent code modification, low-cost logic-process-compatible on-chip one-time-programmable (OTP) provide a viable alternative for on-chip NVM. Generally, OTP cells cannot be used for more than 100 program/erase cycles.

Two types of OTP cells are commonly used for logic processes: floating-gate (FG) and anti-fuse (AF) based OTP.

Logic process compatibility means that FG-based OTP (FG-OTP) cannot use stack-gate or split-gate structures, as in eFlash. Thus, OTP usually uses a two-transistor (2T) cell structure, in which one transistor functions as the FG, while the other transistor functions as the control gate (CG). The program operation of FG-OTP usually uses a channel hot electron injection (CHI) mechanism for the injection of negative charge into the FG, such that the FG-transistor ends up with a high threshold voltage (V_{TH}). The erase operation of FG-OTP uses either UV light or the Fowler-Nordheim (FN) electron tunneling mechanism to reduce the negative charge on the FG, which means that a FG-transistor has low V_{TH}. Due to the non-permanence of programming behavior, most FG-OTP devices are capable of multiple program/erase cycles. Figure 6.11 presents an example of FG-based OTP.

Anti-fuse based OTP (AF-OTP) is another popular solution for cost-sensitive chips that do not require repeated program operations. Most AF-OTP cells initially have high impedance between the gate and channel. During program operations, a strong electrical field is applied across the gate oxide. With sufficient program time and electric field, the gate oxide breaks down, thereby reducing the impedance between the gate and channel. This gate oxide breakdown structure provides highly secure mechanism for the storage of information. Once the information is stored, it is nearly impossible for it to be

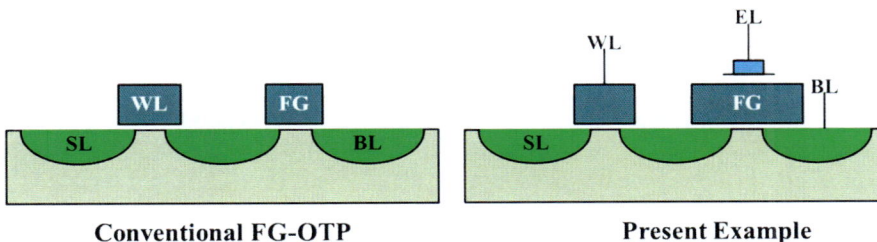

Conventional FG-OTP   Present Example

Fig. 6.11 Typical structures used in FG-OTP cells

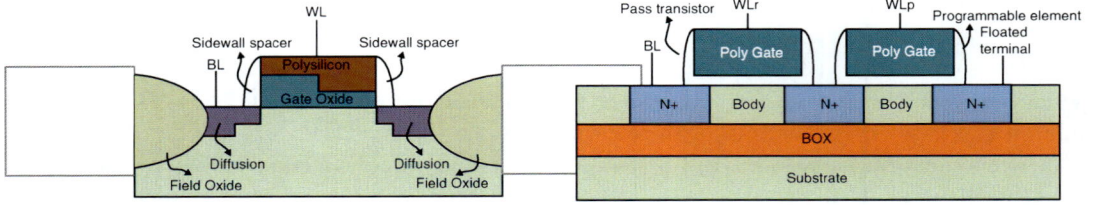

Fig. 6.12 Typical structure used in AF-OTP cells

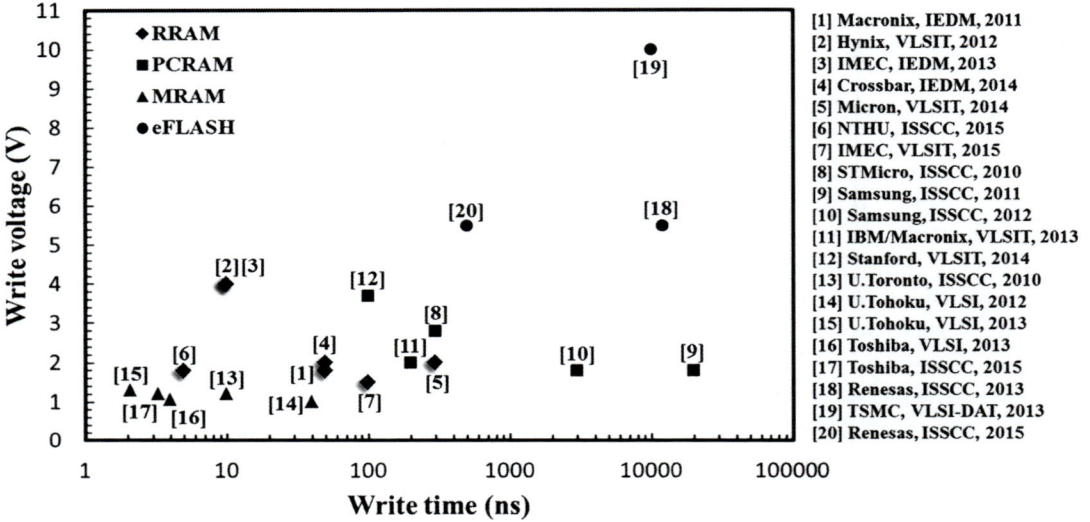

Fig. 6.13 Write time vs. write voltage in recent NVM devices

changed without completely destroying the data. However, the gate oxide cannot be recovered following breakdown; therefore, the AF-OTP can be programmed only once. This limits the testing coverage of any chip using AF-OTP. Figure 6.12 presents several examples of AF-based OTP.

6.2.3 Emerging Resistive-Type ReRAM for Embedded Applications

As Fig. 6.13 shows, a variety of emerging resistive-type memory devices have been proposed to achieve faster write times than those of on-chip flash memory with lower write-voltage and write-energy consumption. Those resistive-type memory devices include phase change memory (PCM) (Wen et al. 2011; Rizzi et al. 2014; Pozidis et al. 2013; Boniardi et al. 2014), resistive RAM (ReRAM) (Chang et al. 2011a, b, 2012a, 2013a, b, c, 2014a, b, c, 2015b; Chiu et al. 2010; Yamamoto et al. 2009; Huang et al. 2014; Eshraghian et al. 2010; Banno et al. 2014; Xue et al. 2013; Wataru et al. 2011; Fackenthal et al. 2014; Liu et al. 2013b; Kawahara et al. 2012; Wang et al. 2012a; Koveshnikov et al. 2012; Zhang et al. 2014; Lee et al. 2010; Hsieh et al. 2013; Sakotsubo et al. 2010; Jo et al. 2014; Serb et al. 2014; Yang et al. 2012; Nardi et al. 2011), and spin-torque transfer magnetic RAM (STT-MRAM) (Kim et al. 2011; Kitagawa et al. 2012; Wang et al. 2012b; Tsunoda et al. 2013). PCM and ReRAM both feature a high resistance-ratio (R-ratio) but suffer from limited endurance. STT-MRAM has high endurance but a small R-ratio (or TMR ratio). The R-ratio

($=R_{HRS}/R_{LRS}$) represents the ratio between the cell resistance values in high resistance state (HRS, R_{HRS}) vs. the values in a low resistance state (LRS, R_{LRS}). In the following subsection, we introduce two high R-ratio resistive-type NVM devices, PCM and ReRAM.

ReRAM and PCM devices are capable of performing two direct overwrite operations: SET and RESET. The SET operation changes the ReRAM device from a high-resistive state (HRS) to a low-resistive state (LRS). The RESET operation changes the ReRAM/PCM device from low resistance (R_L, LRS) to high resistance (R_H, HRS). For most embedded applications, ReRAM/PCM macros employ a one-transistor one-resistor (1T1R) structure.

The materials used in ReRAM devices vary widely with regard to write-speed and write-current performance. The write behavior of most ReRAM devices is either bipolar or unipolar.

Figure 6.14 illustrrates the simplified cell structure of 1T1R ReRAM cells using back-end-of-line (BEOL) processing. Bipolar and unipolar ReRAM devices can be designed with the same cell array structure, in which all 1T1R ReRAM cells in a given row share the same word-line (WL) and source-lines (SLs), while all cells in a given column share the same bit-line (BL).

Figure 6.15 presents the IV-curves and bias conditions of bipolar and unipolar ReRAM devices. The switching of cell resistance states in bipolar ReRAM devices (Chang et al. 2013a, b, 2014a; Banno et al. 2014; Xue et al. 2013; Wataru et al. 2011; Fackenthal et al. 2014; Liu et al. 2013b; Kawahara et al. 2012; Wang et al. 2012a; Koveshnikov et al. 2012; Zhang et al. 2014; Lee et al. 2010; Jo et al. 2014; Serb et al. 2014; Yang et al. 2012; Nardi et al. 2011) is achieved by providing different voltage polarities across the two terminals (top and bottom plates) for SET and RESET operations. The switching of cell resistance state in unipolar ReRAM devices (Chang et al. 2013c, 2014a, c; Hsieh et al. 2013; Sakotsubo et al. 2010) is achieved by providing the same voltage polarities but with different amplitudes across the two terminals of the ReRAM device for SET and RESET operations.

It should be noted that in bipolar-ReRAM cells, the polarity of the bias voltage for read operations is the same as for either a SET or RESET operation. As a result, bipolar ReRAM devices suffer from read disturbance when under high stress voltage (V_R) and/or extended stress time (T_R). As shown in Fig. 6.16, the cell read current (I_{CELL}) of an HFO-based bipolar ReRAM device (Lee et al. 2010) gradually decreases with an increase in T_R at $V_R = 0.5$ V. This means that LRS resistance increases slowly (is disturbed) during the read process. Low bit-line bias voltage (V_{BL}) is required to prevent read disturbance when reading from bipolar ReRAM cells.

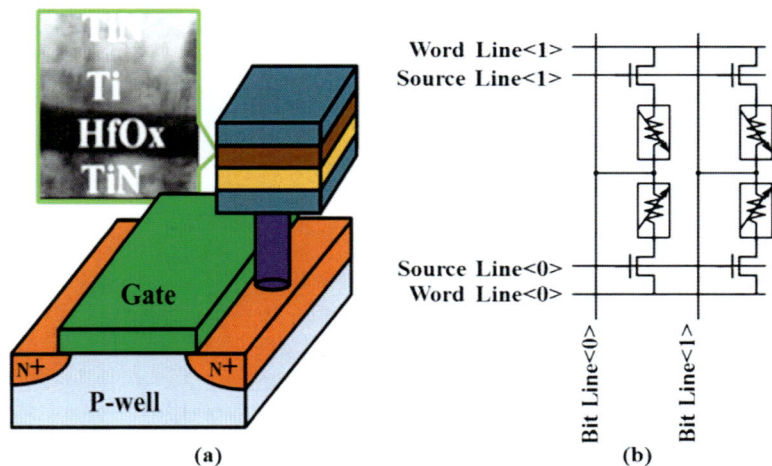

Fig. 6.14 Simplified diagrams of (**a**) cell structure and (**b**) mini-array of 1T1R ReRAM cells

Fig. 6.15 IV-curves of (**a**) bipolar and (**b**) unipolar ReRAM devices

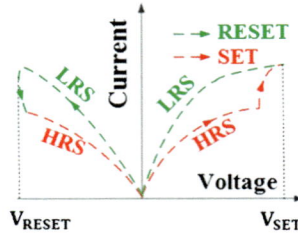

(a)-1

Operation	SET	RESET	READ
WL	V_{WL_SET}	V_{WL_RESET}	V_{WL_READ}
SL	V_{SET}	0V	0V
BL	0V	V_{RESET}	V_{READ}

(a)-2

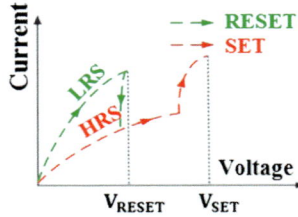

(b)-1

Operation	SET	RESET	READ
WL	V_{WL_SET}	V_{WL_RESET}	V_{WL_READ}
SL	V_{SET}	V_{RESET}	0V
BL	0V	0V	V_{READ}

(d)-2

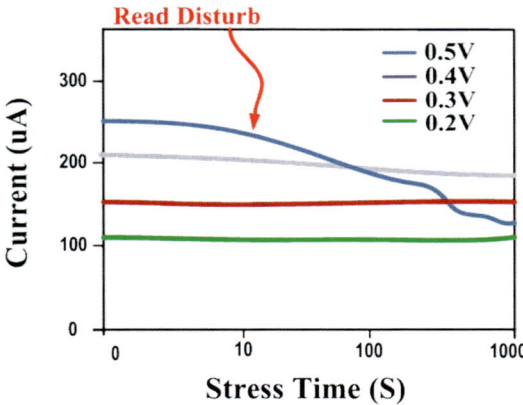

Fig. 6.16 Cell read current vs. stress time of HFO-based ReRAM device under various stress voltages

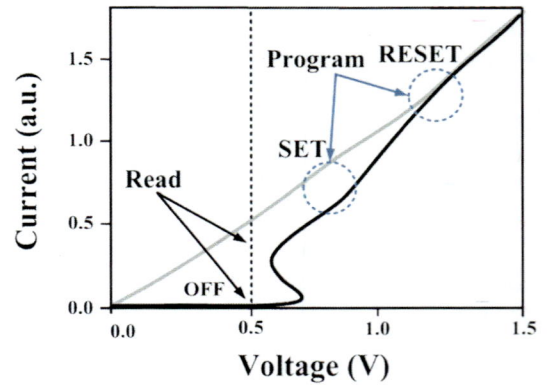

Fig. 6.17 Typical current–voltage (*I–V*) curves of PCM devices

6.2.4 Emerging Resistive-Type ReRAM for Embedded Applications: PCM

Phase Change Memory (PCM) cells are based on a chalcogenide compound produced from Ge, Sb, and Te (GST). These devices exploit the large resistance contrast between crystalline and amorphous phases to enable the storage of data. Despite the fact that the fundamental principles

of PCM were first reported in the 1960s (Ovshinsky 1968), only in the past two decades has it become possible to produce PCM chip-level devices with performance comparable to that of other contemporary memory technologies (Lai et al. 2003). In this subsection, we introduce the basic operations of PCM and outline a number of recent developments.

Figure 6.17 presents typical current–voltage (*I–V*) curves of a PCM device, revealing two important characteristics: (1) threshold switching

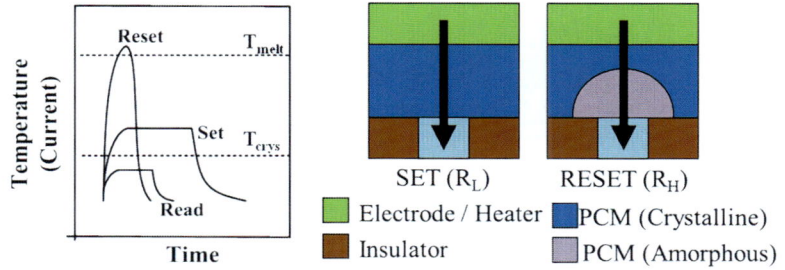

Fig. 6.18 Conventional mushroom-type PCM cell and conceptual illustration of SET and RESET operations in PCM

and (2) phase transformation. Threshold switching occurs when the applied voltage increases beyond the voltage snapback point at V_{th}, whereupon the material enters a conductive state due to impact ionization and carrier recombination (Pirovano et al. 2004). Phase transition to a crystalline or amorphous state depends on programming pulse conditions and can be induced by a further increase in the current amplitude.

The transition from an amorphous state to a crystalline state is commonly referred to as a SET operation, the reverse of which is a RESET operation. Figure 6.18 presents a conceptual illustration of SET, RESET, and Read operations. In a SET operation, the application of an electrical pulse heats the PCM cell to above its crystallization temperature (T_{CRYS}), whereupon the slow-quench time allows the GST material to return to a crystalline state. Conversely, RESET operations involve heating the PCM cell to above its melting temperature (T_{MELT}) and abruptly cutting off the heat source in order to melt-quench the material, thereby leaving it in an amorphous state. For read operations, cell resistance is measured by passing an electrical current small enough to prevent disturbing the existing state of the material.

Researchers have recently focused on improving device characteristics through the use of material engineering and innovative device structures. Those structures include (a) conventional mushroom-type device structures (Shih et al. 2008), (b) dash-type confined cells (Im et al. 2008), (c) thermally-confined bottom electrodes (Wu et al. 2011), and (d) sidewall bottom electrodes (Wu et al. 2015). Several

researchers have reported that critical device performance indicators (e.g., switching speed and thermal stability) are largely a function of Ge-Sb-Te composition and doping. Those studies sought to engineer GeSbTe ternary alloys along isoelectronic tie lines and the Ge/Sb_2Te_3 tie lines in the search for high performance materials (Cheng et al. 2011, 2012).

6.3 Read Circuits for On-Chip NVM

In this section, we outline the general behavior and introduce the key sub-circuits used in the read operations of on-chip NVM. The read operations in this section include current-mode and voltage-mode read schemes. We also explore various challenges in the development of read circuits for low-power on-chip NVM and review a number of advanced techniques used in the development of read circuits.

6.3.1 General Operation and Sub-circuits for Read Operations

On-chip NVM commonly employs one of two read schemes: voltage mode and current mode sensing (Chang et al. 2015a).

6.3.1.1 Current-Mode Sensing

Figure 6.19 illustrates the circuit structure and waveforms of a typical current-mode sensing scheme, in which the bit-line (BL) voltage is maintained at a constant value (BL clamping voltage or V_{PRE}) throughout the sensing period.

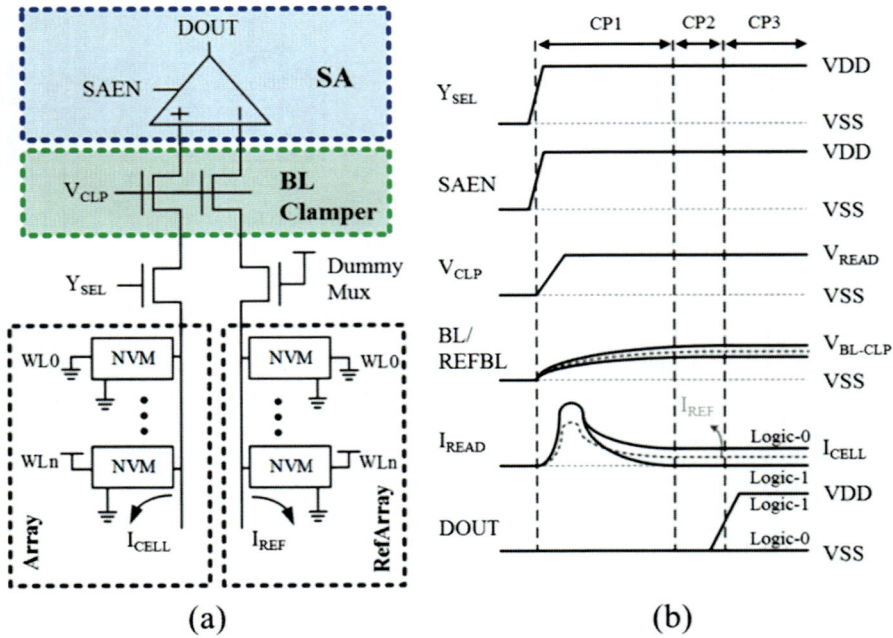

Fig. 6.19 (a) Circuit structure and (b) waveforms of typical current-mode sensing scheme

The operation of current-mode sensing proceeds through three phases: bit-line precharge, current development, and current comparison. In the bit-line precharge phase (CP1), a precharge transistor is turned on to raise the bit-line voltage from 0 V to the precharge voltage (V_{PRE}) or bit-line clamping voltage (V_{BL-CLP}). In the I_{CELL} development phase (CP2), V_{BL-CLP} varies the cell read current (I_{CELL}) in accordance with the resistance of the accessed memory cell. The I_{CELL} of a logic-0 cell (I_{CELL-0}) is larger than that of a logic-1 cell (I_{CELL-1}) under the same V_{PRE} or V_{BL-CLP}. Finally, in the current comparison phase (CP3), a current-mode sense amplifier (CSA) is used to compare the I_{CELL} values using a reference current (I_{REF}), whereupon a digital output is generated at data-out (DOUT).

Figure 6.20 presents the two bit-line clamping methods commonly used for current-mode sensing: static and dynamic. In static clamping, a clamping transistor (CLP) is controlled through the application of constant V_{BL-CLP}. In dynamic bias clamping, the bit-line voltage is amplified and fed back to the gate of the CLP. Dynamic biasing provides a larger CLP gate swing and

consequently faster bit-line bias speeds, at the cost of additional area and power overhead.

6.3.1.2 Voltage-Mode Sensing

Figure 6.21 presents the read path and a conceptual waveforms used in a typical voltage-mode sensing scheme. The operation of voltage-mode sensing proceeds through three phases: bit-line precharge, bit-line voltage development, and voltage comparison. In the bit-line precharge phase (VP1), a precharge transistor is turned on to raise the BL voltage from 0 V to a precharge voltage (V_{PRE}). In the bit-line developing phase (VP2), bit-line voltage (V_{BL}) begins falling in accordance with the cell resistive state of the accessed memory cell. When a logic-1 (program/HRS) cell is read, the cell read current (I_{CELL}) of the logic-1 cell (I_{CELL-1}) is small, such that the bit-line is maintained near V_{PRE}. When a logic-0 (erase/LRS) cell is read, a larger I_{CELL} (I_{CELL-0}) causes the bit-line to drop more rapidly, thereby generating a bit-line voltage swing (V_{BLS}) exceeding that of the logic-1 cell. Finally, in the voltage comparison phase (VP3), a voltage-mode sense amplifier (VSA) is used to

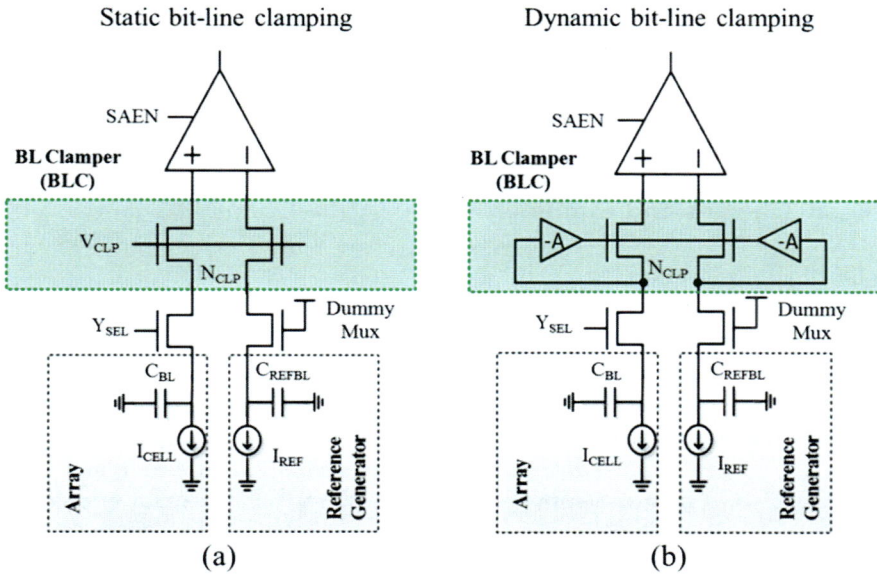

Fig. 6.20 Typical (**a**) static and (**b**) dynamic bit-line clamping schemes

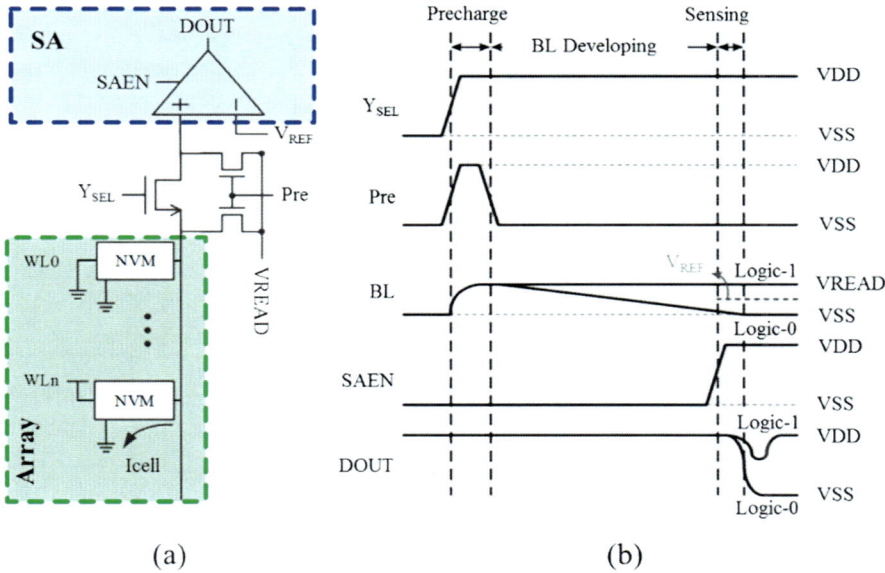

Fig. 6.21 (**a**) Circuit structure and (**b**) waveforms of typical voltage-mode sensing scheme

compare V_{BL} with a reference voltage (V_{REF}), whereupon a digital output is generated at data-out (DOUT).

6.3.2 Challenges in Current-Mode Read Schemes and Examples

In this subsection, we explore some of the challenges associated with the development of current-mode read schemes aimed at achieving small cell current, low-voltage operation, and high-speed read operations. Examples of advance circuit techniques are also reviewed.

6.3.2.1 Challenges in Current-Mode Sensing

BL Clamping Voltage Vs. Read-Disturb
Current mode sensing requitres that the bit-line be clamped at a constant voltage ($V_{BL\text{-}CLP}$). Limitations in clamping voltage are determined by the read disturb behavior of eNVM devices. To ensure robust, high-speed read operations, $V_{BL\text{-}CLP}$ should be as high as possible in order to compensate for bit-line leakage current

($I_{BL\text{-}LEAK}$) from unselected cells as well as the input offset of the CSA. However, process variation can lead to local fluctuations in $V_{BL\text{-}CLP}$.

Generation of Reference Current Vs. Process Variation
As described above, data stored in a flash cell is accessed using a current-sense amplifier (CSA) to compare the cell current (I_{CELL}) of the accessed NVM cell with a reference current (I_{REF}). However, a fixed I_{REF} is unable to match the I_{CELL} fluctuations across various processes, voltages and temperature (PVT) conditions. This has led to the application of replica cell schemes (Kobayashi et al. 1990; Bauer et al. 1995; Micheloni et al. 2003; Le et al. 2004; Ogura et al. 2006; Chung et al. 2007; Chang and Shen 2009) to enable the generation of a I_{REF} capable of tracking variations in I_{CELL} across a range of PVT conditions. Figure 6.22 presents a nominal replica-cell current sensing circuit for eNVM. This replica array comprises inactive as well as active replica eNVM cells. An active replica cell emulates regular eNVM cells and provides a replica cell current, referred to as $I_{CELL\text{-}REPLICA}$. Inactive replica cells preserve neighboring

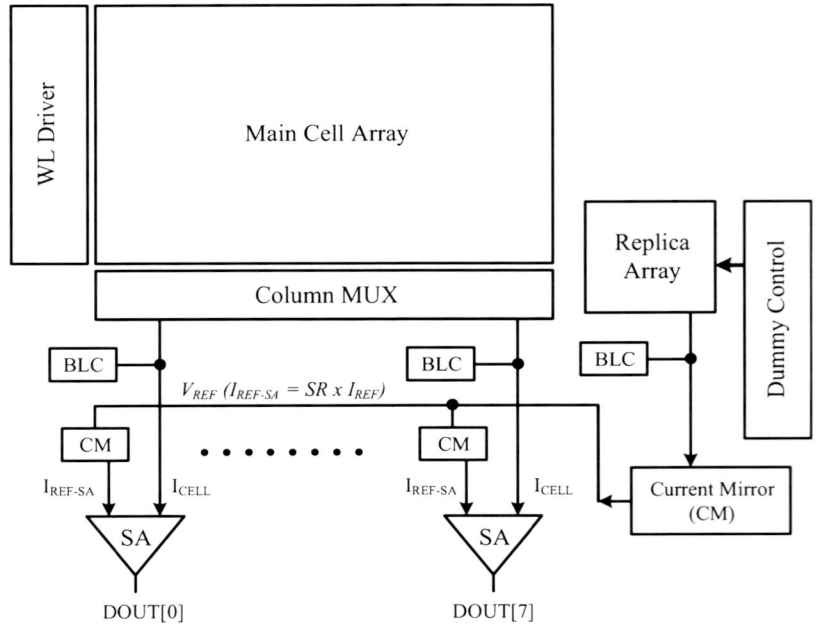

Fig. 6.22 Nominal replica-cell current sensing circuit for eNVM

geometric patterns, as in a regular cell array, in order to minimize mismatch between I_{CELL} and $I_{CELL\text{-}REPLICA}$. Inactive replica cells also provide the same parasitic load for a replica bit-line as for a regular bit-line.

In cases where a hard defect or significant process variation occurs in a replica cell, the resulting reference current (I_{REF}) would have inaccurate values or vary considerably across dies. This would subsequently leads to read failure or long read access times and increases power consumption. Accordingly, the generation of read reference current is a crucial issue in the design of eNVM.

Low-Voltage Operation Vs. Voltage Head Room

Figure 6.23 outlines the concept of voltage budget in a current-mode sensing scheme. Efforts to reduce the supply voltage (VDD) in current-mode sensing are constrained by the voltage headroom of bit-line voltage clamper (BLC) and the diode-connected current-mirror (CM) circuitry. As shown in Fig. 6.22, a 3 µA bit-line current requires 400 mV of headroom for a typical 65 nm current-mirror circuit (Chang et al. 2013c). Low VDD suppresses the effective voltage on the BL (V_{BL}) as well as the voltage across the eNVM device (V_{NVM}). This leads to small cell read current, which greatly reduces the sensing margins and read speeds when VDD is low.

6.3.2.2 Advanced Circuit Techniques for Current-Mode Read Operations

Generation of Reference Current with Narrow Distribution

Most floating-gate (FG)-based replica cells connect selected floating gates to their control gates at the edge of the replica array. This is done to avoid the need for additional erase cycles and thereby maintain the voltage potential of the floating gate in a replica cell following multiple access cycles. The sharp corners and special shape of the floating-gate layer in split-gate flash technology make it difficult to maintain uniformity and low resistance in the metal-to-FG contacts and floating-gate poly. Therefore, the resistance between the control-gate line (dummy word-line) and the floating gate ($R_{DWL\text{-}FG}$) of accessed replica cells varies across rows/dies. Differences in $R_{DWL\text{-}FG}$ affect the turn-on times of the selected replica cell and produce fluctuations in read reference current (I_{REF}) settling time, which can result in ringing at the data-out (DOUT), long-address access times (T_{AC}), and increased power consumption. FG-always-high and pre-stable current sensing (PSCS) schemes have been proposed to overcome $R_{DWL\text{-}FG}$ induced fluctuations in I_{REF} settling time, as shown in Fig. 6.24.

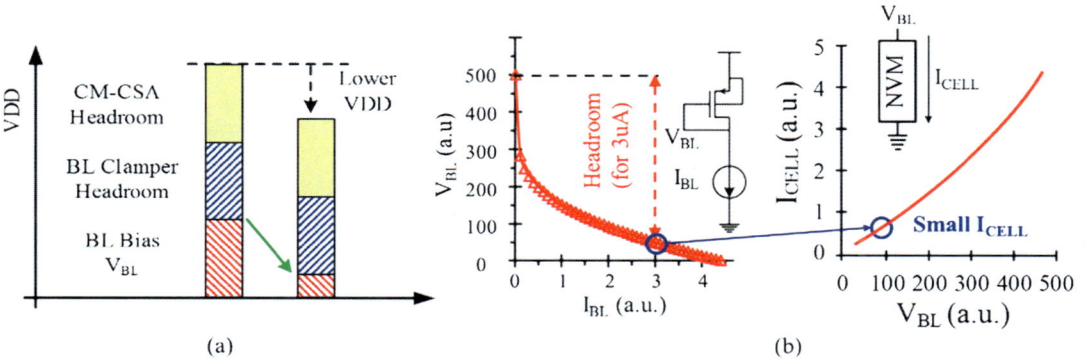

Fig. 6.23 (a) Concept of voltage budget and (b) bit-line voltage vs. bit-line current in current-mode sensing scheme

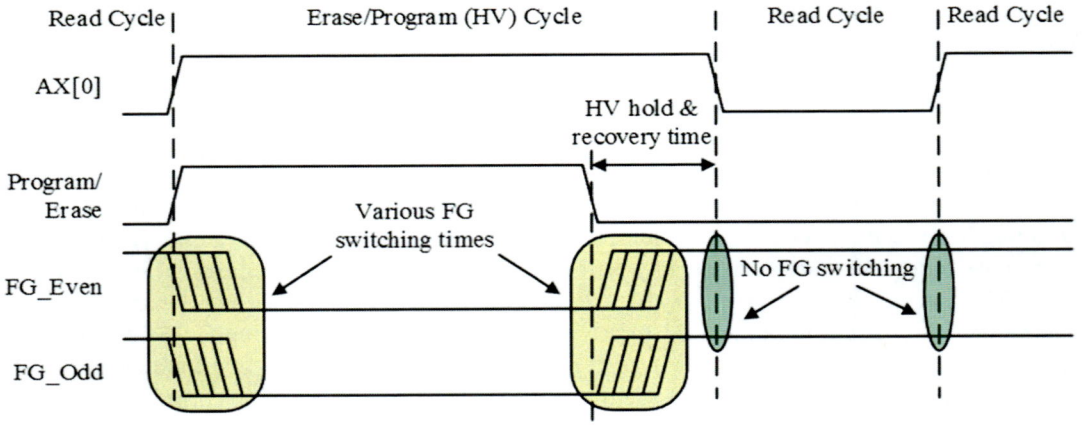

(a)

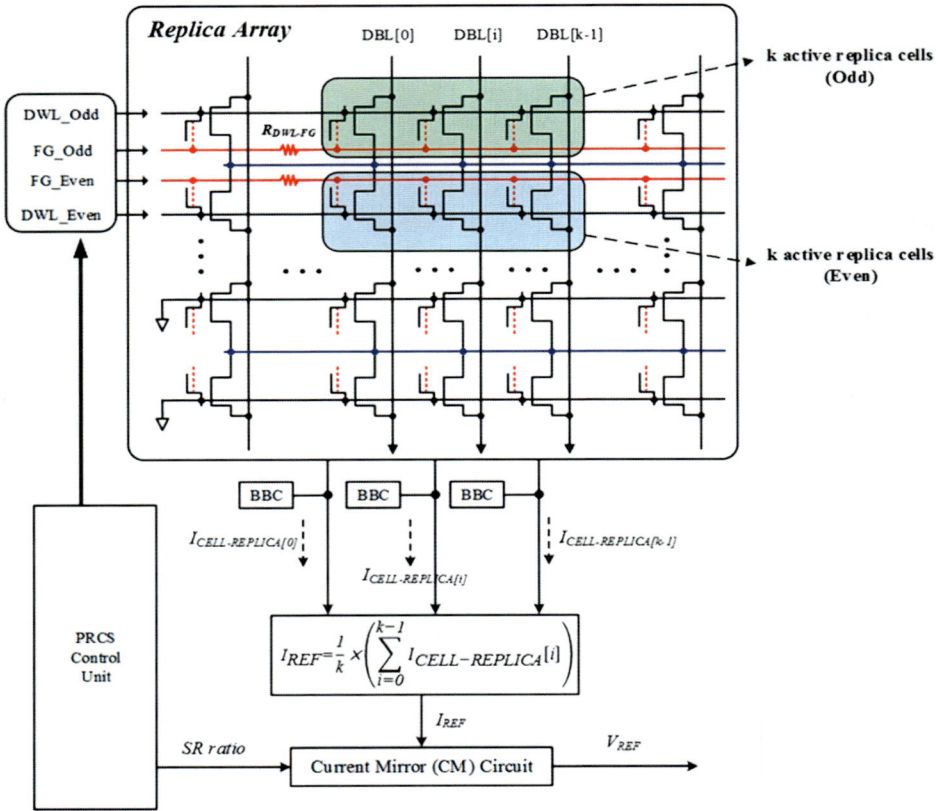

(b)

Fig. 6.24 (**a**) FG-always-high and (**b**) pre-stable current sensing (PSCS) replica cell schemes

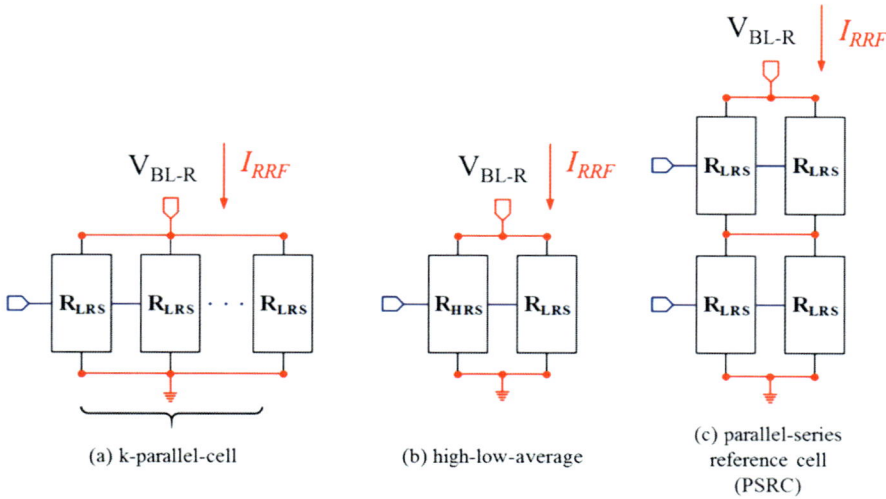

Fig. 6.25 I_{REF}-generation schemes based on replica-cells for resistive-type memory: (**a**) k-parallel-cell, (**b**) high-low-average, and (**c**) parallel-series reference cell (PSRC) schemes

FG-always-high (Chih et al. 2002) is a straightforward circuit in which the floating gate is kept high at all times, making this a non-switching approach. However, this approach exposes the FG-oxide of a replica cell to longer voltage-stress times than that of regular Flash cells, which are not exposed to high FG voltage in most applications. This means that accessed replica cells in the FG-always-high scheme are more susceptible to the effects of aging than are nominal cells. Moreover, the read reference current generation circuit in the FG-always-high scheme consumes power even during non-read cycles (erase and program cycles), due to the fact that one of the replica cells is always on.

In the PSCS scheme (Chang and Shen 2009), the turn-on of I_{REF} generation and the switching of the floating gate in the replica-cell occur in the final period of an erase/program cycle or during the power-up period, rather than at the beginning of every read cycle, as in conventional flash memories. The time required to complete the erase/program operation is generally far longer than the worst-case settling time of I_{REF} in foundry-provided eFlash macros (Datasheet 2001, 2005). Thus, the FG switching time of replica cells and the settling time of I_{REF} can be

hidden in the ending period of an erase/program cycle prior to subsequent read cycles. The replica cells in PSCS undergo turn-on only in read cycles; therefore, they experience less voltage-stress and consume less power than the case in the FG-always-high scheme. Moreover, with multiple replica-cells turning on simultaneously (Chang and Shen 2009), the distribution of I_{REF} can be narrowed.

Several I_{REF}-generation schemes based on replica-cells have been proposed for resistive-type emerging memory, including k-parallel-cell, high-low-average, and parallel-series reference cell (PSRC), as shown in Fig. 6.25. The k-parallel-cell scheme provides the average I_{CELL} of k replica LRS cells. High-low-average schemes provide the average I_{CELL} of LRS and HRS cells, as long as the R-ratio is small. However, in cases of large variation in cell resistance, k-parallel-cell and high-low-average schemes both suffer from wide I_{REF} distribution, due to the fact that I_{REF} is dominated by the ultra-low-resistance of the replica cell (a tail bit). In contrast, PSRC eliminates the dependence of I_{REF} on a single tail-bit reference cell by having two parallel LRS sets connected in serial. This enables the PSRC scheme to narrow the distribution of I_{REF} associated with variations in R_{HRS}/R_{LRS}.

Current-Mode Sensing with Small Input-Offset

Several CSA schemes with small input-offset have been proposed for high-speed or small cell-current eNVMs.

Figure 6.26 presents a small-offset CSA sampling scheme using threshold voltage (V_{TH}) (Javanifard et al. 2008). The V_{TH} of the NMOS latch (M1/M2) is stored in capacitors C1 and C2 to provide offset compensation. This scheme succeeds in achieving small input offset but requires considerable area overhead due to its use of numerous switches.

Figure 6.27 presents two CSA schemes based on current-sampling (IS-CSA). As shown in Fig. 6.27a, unlike the V_{TH}-sampling scheme, IS-CSA (Chang et al. 2013a, b) employs two capacitors (C1/C2) to store the gate-source voltage difference (V_{GS}) of critical transistors (M1/M2) in order to yield the I_{CELL} and I_{REF}. IS-CSA then uses the sampled I_{CELL} and I_{REF} for second-stage amplification operations, which are insensitive to variations in the V_{TH} of M1/M2. As with IS-CSA, time-differential IS-CSA (TD-IS-CSA) (Jefremow et al. 2013) uses a single capacitor for the sampling of I_{REF}, as shown in Fig. 6.27b. In the subsequent timing phase, TD-IS-CSA compares the sampled I_{REF} with I_{CELL} in order to achieve robust sensing with small input offset.

Figure 6.28 shows two digital-calibration-based CSAs (DC-CSA) designed to enable high-speed sensing without the need for run-time offset-cancellation operations. A digital offset cancellation CSA (DOC-CSA) (Kono et al. 2013) was proposed for the application of additional offset-cancel current (Ioc) to the differential inputs (LBLL or LBLR) of the comparator. The amount of Ioc is digitally calibrated according to the mismatch of the comparator in the wafer testing stage. In the calibration-based asymmetric-voltage-biased CSA (AVB-CSA) (Chang et al. 2015c) in Fig. 6.28b, a pair of asymmetric precharge voltages (V_{AP-CP} and V_{AP-RP}) are applied to the differential node (CP and RP nodes) of a latch-type comparator prior to the sensing operation. V_{AP-CP} and V_{AP-RP} are calibrated during wafer testing in accordance with the offset voltage of the read-path detected at nodes CP and RP. Accordingly, AVB-CSA suppresses the offset caused by the comparator as well as the offset produced along the entire read-path (I_{OS-SUM}).

Low-Voltage Current-Mode Sensing

Figure 6.29 illustrates a low-voltage, body-drain driven (BDD) CSA (Chang et al. 2013c). A BDD-based current mirror (CM) circuit requires smaller voltage headroom than do conventional diode-connected current mirrors. The small voltage headroom on read circuits enables BDD-CSA to assign most of the voltage budget to the voltage across NVM devices, thereby producing I_{CELL} sufficient for sensing even at low VDD.

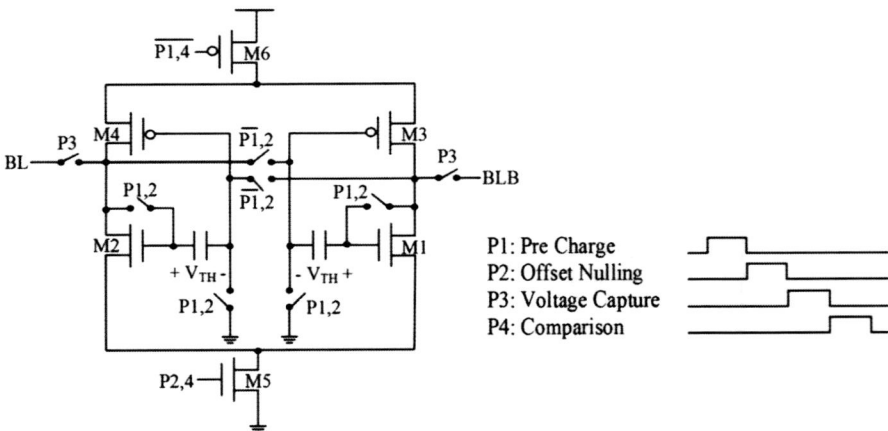

Fig. 6.26 Small-offset CSA using threshold-voltage (V_{TH}) sampling scheme

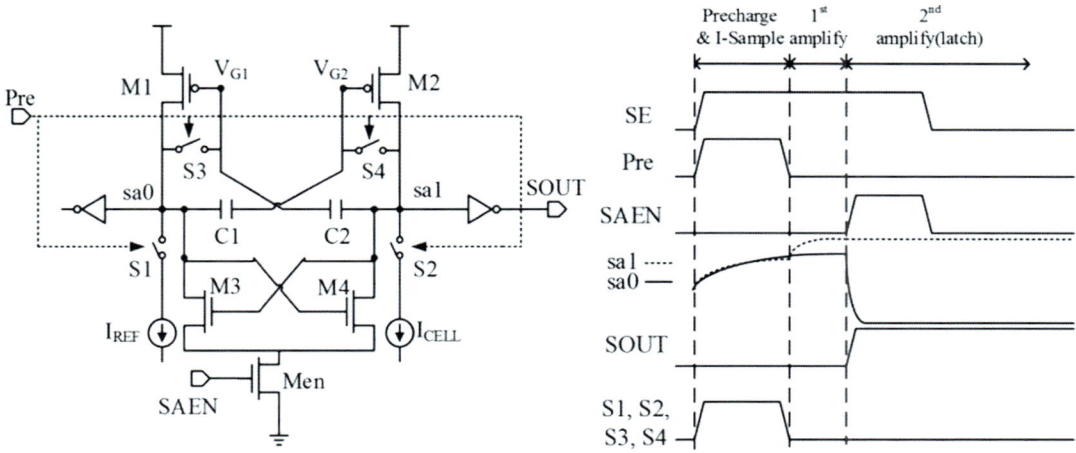

(a) 2-capacitor IS-CSA

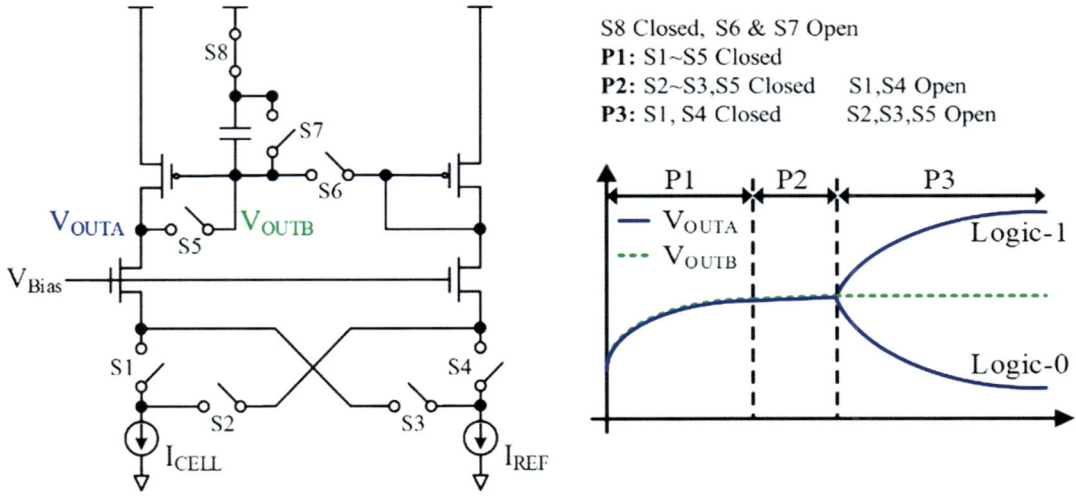

(b) single-capacitor time-differential IS-CSA

Fig. 6.27 Two CSAs based on current-sampling (IS-CSA): (**a**) two-capacitor IS-CSA and (**b**) single-capacitor time-differential IS-CSA

6.3.3   Challenges in Voltage-Mode Read Schemes and Examples

In this subsection, we investigate some of the challenges associated with the development of voltage-mode read schemes for small cell current, low-voltage operation, and high-speed read. Examples of advanced circuit techniques are also reviewed.

6.3.3.1 Challenges in Voltage-Mode Sensing

On-Off Current Ratio Vs. Sensing Margin
SRAM and logic-ROM are able to generate the large bit-line voltage swings required for read-0 operations, while maintaining near-zero bit-line voltage swings for read-1 operations. This can be attributed to their small read-1 cell

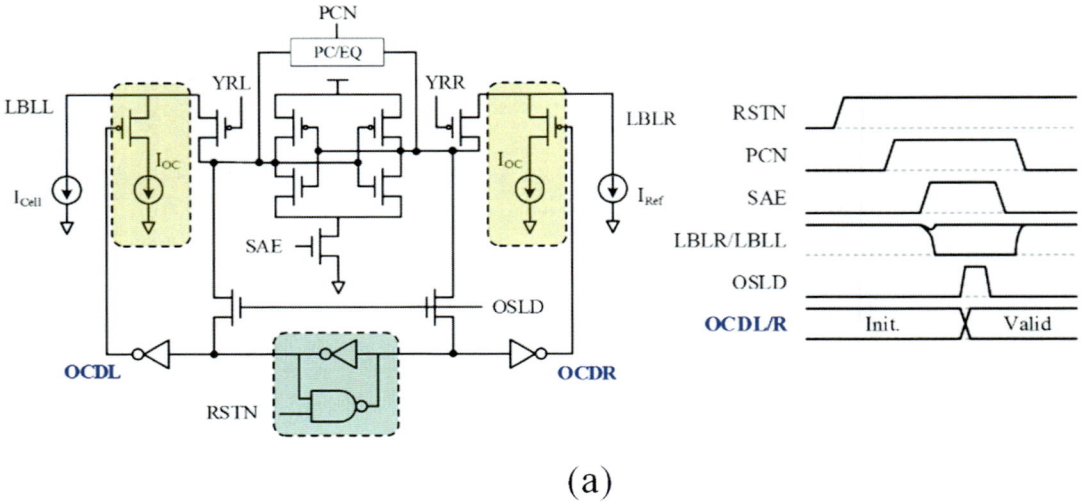

(a)

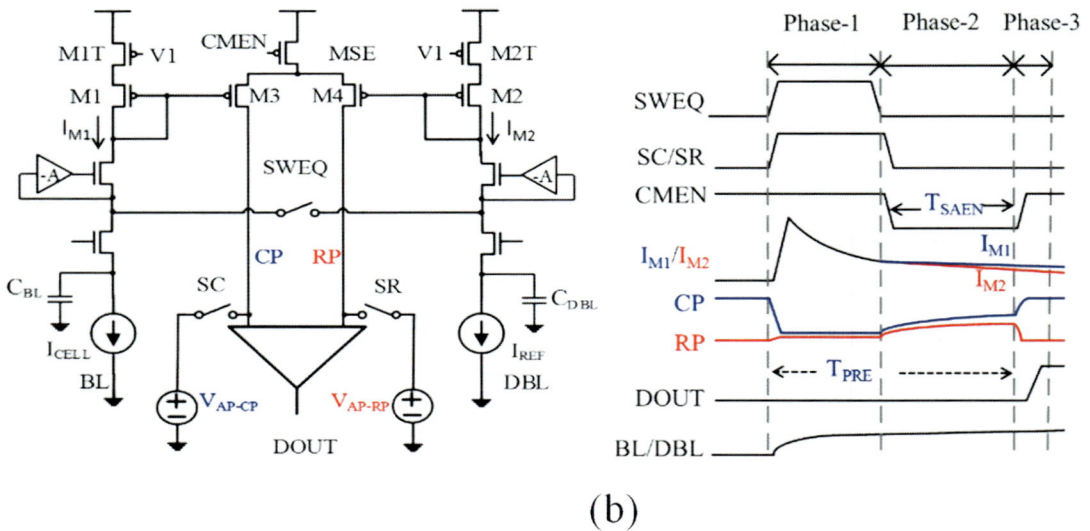

(b)

Fig. 6.28 Two digital-calibration based CSAs (DC-CSA): (**a**) digital offset cancellation CSA (DOC-CSA), and (**b**) asymmetric-voltage-biased CSA (AVB-CSA)

current (I_{CELL-1}) and high on-off current ratio (I-ratio = I_{CELL-0}/I_{CELL-1}) between read-1 and read-0 (I_{CELL-0}) cell currents. Thus, most SRAM and logic-ROM devices use conventional voltage-mode sense amplifiers (VSA) for differential sensing or single-end sensing using a reference voltage (V_{REF}). In typical VSA, the V_{REF} is placed at the mid-point between the bit-line voltages when reading tail logic-0 and logic-1 cells. Thus, the sensing margin (V_{SM}) is half (or less) of the ΔV_{BLS_TAIL}.

Process variation subjects the tail-bits of logic-1 eNVM cells to higher I_{CELL-1}, compared to those of typical cells, which tends to be close to zero and far below I_{CELL-0}. This is because tail eFlash cells have a lower threshold voltage (V_{THP}) for programmed cells, or tail resistive eNVM cells have low HRS resistance. As shown in Fig. 6.30, eNVM suffers from small difference in bit-line voltage between tail bits when reading logic-0 and logic-1 cells. Same as read-0 operation, the bit-line that is used to read a

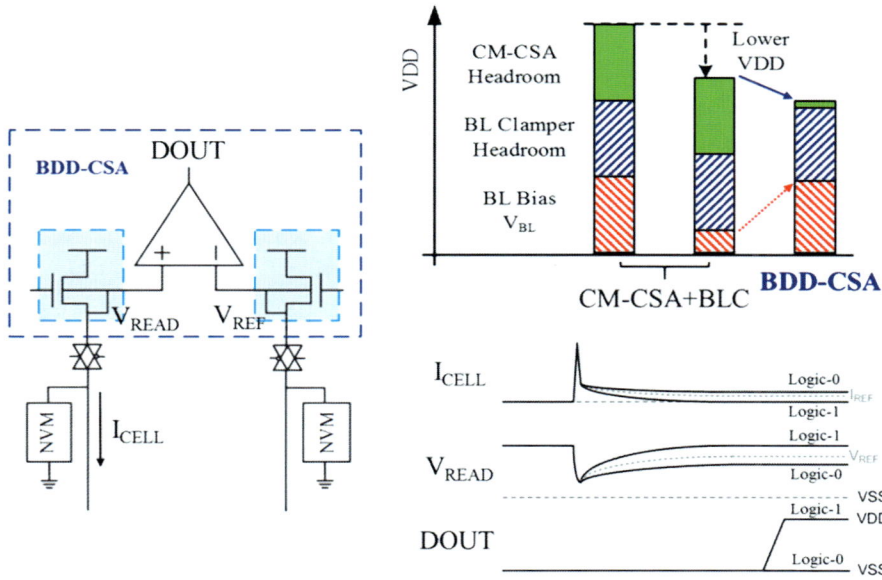

Fig. 6.29 Low voltage, body-drain driven CSA (BDD-CSA)

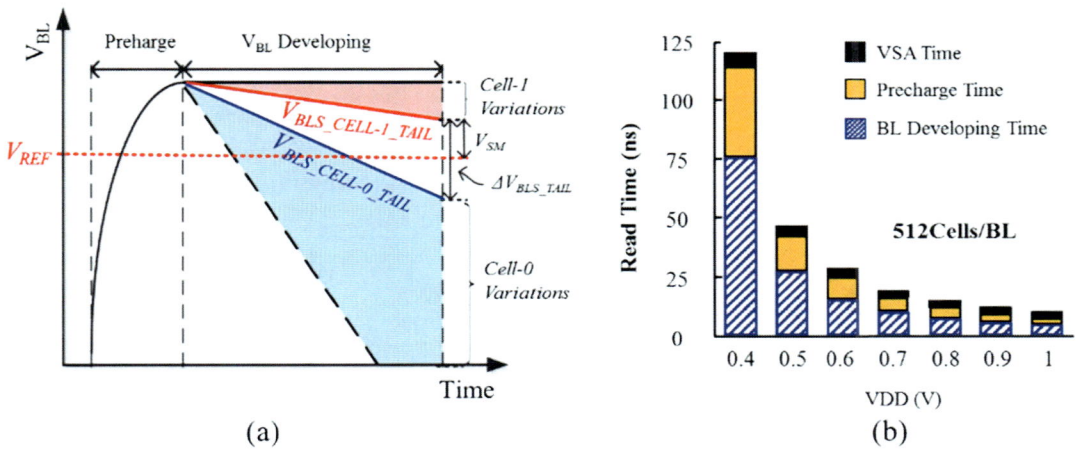

Fig. 6.30 (a) Variations in BL voltage swing during read operations; (b) timing breakdown vs. VDD in a typical VSA

tail logic-1 cell has a greater drop in voltage (below the bit-line precharge voltage, V_{PRE}) with an increase in bit-line developing time (T_{BL}). Read-1 and read-0 operations both undergo dynamic voltage swings on bit-lines proportional to T_{BL}; therefore, eNVM is unable to use conventional fixed-value V_{REF} schemes (Verma and Chandrakasan 2008; Kim et al. 2009; Zhai et al. 2008; Chang et al. 2009a; Morita et al. 2007; Chen et al. 2006) or some

novel VSAs (Takeuchi et al. 2007; Byeon et al. 2005; Chang et al. 2009b; Marotta et al. 2010), which assume that the read-1 bit-line voltage is maintained at close to V_{PRE} or VDD.

Read Access Time Vs. VDD

Figure 6.30b presents a breakdown in the timing of a typical voltage-mode sensing scheme employing a 65 nm eNVM memory macro with a bit-line length of 512 cells. Lowering the VDD reduces the driving strength of the transistors, which can lead to an increase in macro-level read access times (T_{AC}) and a reduction in V_{SM}. It is worth noting that under a low VDD, T_{AC} is dominated by the bit-line precharge time (T_{PRE}) and bit-line developing time (T_{BL}), rather than the response time of a VSA. This effect is more obvious in macros with a long BL length. Thus, efforts to improve the speed and yield of voltage-mode sensing under low VDD should focus on reducing the T_{BL} while maintaining a V_{SM} sufficient to resist the input offset voltage (V_{OS}) of the VSA.

6.3.3.2 Advanced Circuit Techniques for Voltage-Mode Read Operations

Low-Voltage Voltage-Mode Sensing
Figure 6.31 illustrates the concept of low-VDD swing-sample-and-couple (SSC) VSA (Chang et al. 2014c). Unlike conventional VSA, which

utilizes only $0.5 \times \Delta V_{BLS\_TAIL}$ for the sensing margin, SSC-VSA has V_{REF} placed below the bit-line voltage of the tail-LRS cell and uses a switch-capacitor-like circuit to enable the full utilization of the ΔV_{BLS_TAIL} for the sensing margin (V_{SM}). This increment in V_{SM} (nearly double) leads to a reduction in VDDmin and makes it possible to achieve read speeds exceeding those of conventional VSA at low VDD.

6.4 Write Circuits for On-Chip NVM

In this section, we outline the general behavior and key sub-circuits used in the write operations of on-chip NVM. The read operations discussed in this section include current-mode and voltage-mode read schemes. We also explore various challenges in the development of read circuits for low-power on-chip NVM and review a number of advanced read circuit techniques.

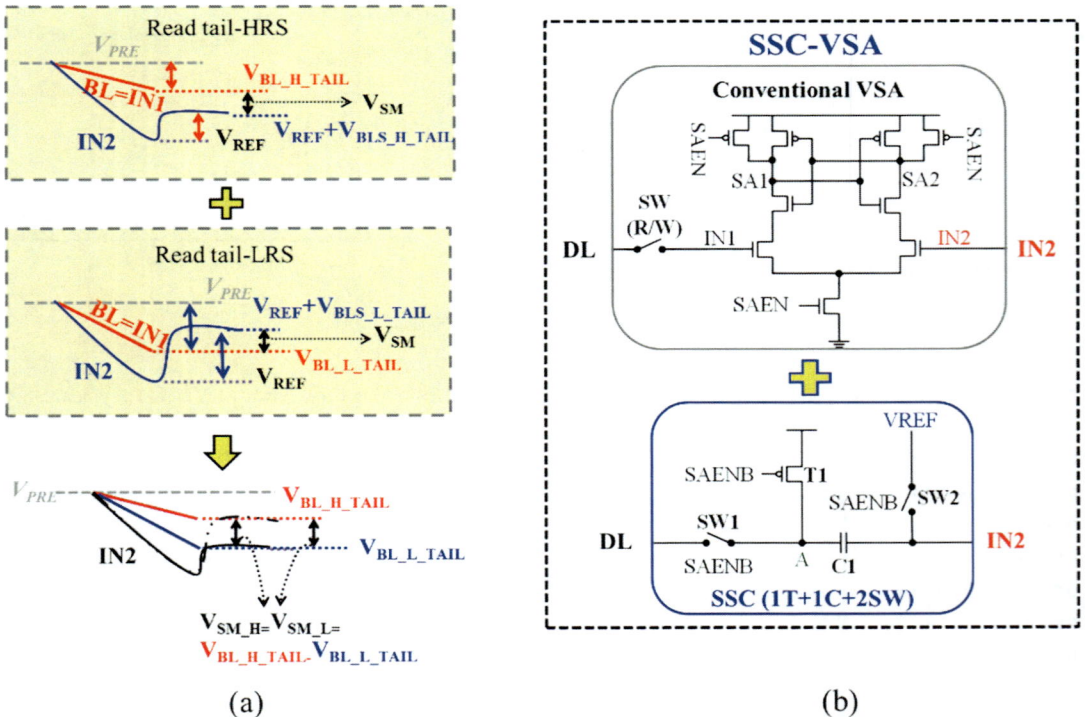

(a)   (b)

Fig. 6.31 (a) Conceptual illustration and (b) circuitry of a low-VDD swing-sample-and-couple (SSC) VSA

6.4.1 General Operations and Sub-Circuits for Write Operations

Figure 6.32 illustrates the structure of the write-path in a typical eNVM macro. The key sub-circuits used for write operations include level-shifters, write-drivers, and charge-pump circuits.

Most eNVMs use multiple voltages higher than the supply-voltage (VDD) for write operations. A charge-pump (CP) circuit is used for the conversion of VDD to several high voltage levels in order to provide the output current required for write operations. Write drivers are employed only to provide the write voltage required for the cell array. The drivers for word-lines (WL) or source-lines (SL), which are used in read as well as write operations, vary in their output voltages across operation modes. As a result, level-shifters are commonly used for WL/SL drivers in NVMs.

It should be noted that the high voltages and currents generated by charge pump circuits are limited and somewhat inefficient. This means that the current load on the high-voltage path should be minimized in order to reduce power consumption by the eNVM macro. The resulting low-power eNVM requires low-DC/leakage current for consumption by the level-shifters and write-drivers. A small-peak-current energy-efficient CP is also important when low-power eNVM macros are embedded on a chip.

6.4.2 Challenges in Level-Shifters

Level shifters are embedded in eNVMs for the conversion of logic signals (0 V or VDD) generated in row decoders to either 0 V or voltages higher than VDD (VDDH). The VDDH could be any write voltage required for NVM devices. Generally, level-shifters (LSs) pose challenges in the following two areas: (1) operations with a low input voltage (VDDL) and high VDDH during write operations; (2) large DC current consumption by conventional low-VDDL level shifters in unselected rows, resulting in significant charge loss in the VDDH path. These difficulties can lead to an increase in the current-load and power consumption of on-chip charge-pump circuits, particularly in NVMs with a large number of rows.

In the following, we examine a number of existing level-shifters:

6.4.2.1 Half-Latch Level Shifter (HL-LS)

Figure 6.33 illustrates the circuit and waveform of common half-latch level shifter (HL-LS) (Wooters et al. 2010; Chang et al. 2011c), which includes two pull-down NMOS (M_{NL} and M_{NR}) and a pair of cross-coupled PMOS (M_{PL} and M_{PR}) transistors as well as an input signal inverter (INV1) and an output buffer (BUF). When SEL = 1, M_{NL} is on, which pulls down the voltage at node VOB (V_{OB}) to 0 V. This turns on M_{PR} and raises the voltage at node VOT (V_{OT}) to VDDH. Thereafter, M_{PR} remains off due to cross-coupled feedback, wherein the output (V_O) of HL-LS is VDDH. When SELb = 1, the turning on of M_{NR} creates a discharge current (I_D) that lowers V_{OT} to below VDDH. If I_D is

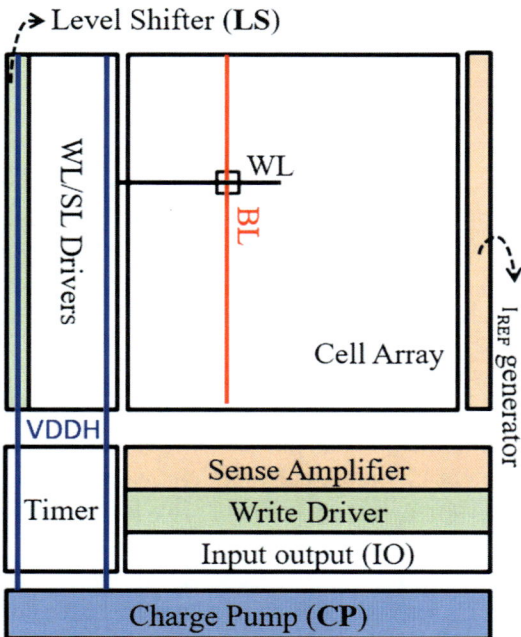

Fig. 6.32 Structure of write-path in a typical eNVM macro

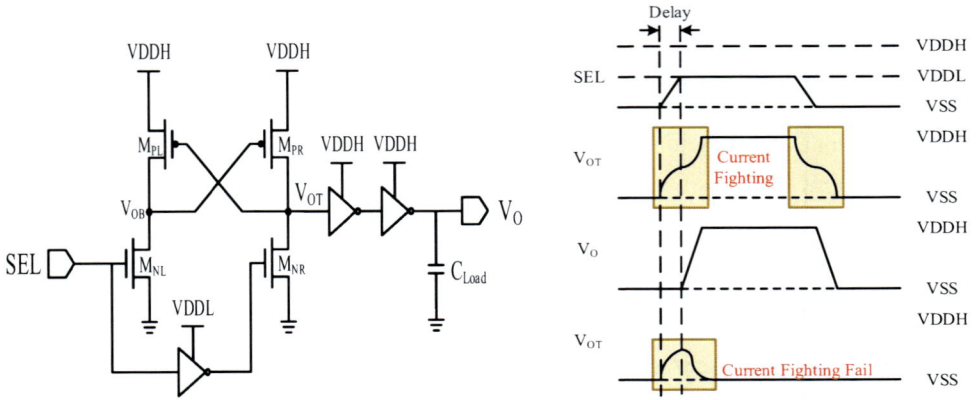

Fig. 6.33 Circuit and waveform of a half-latch level shifter (HL-LS)

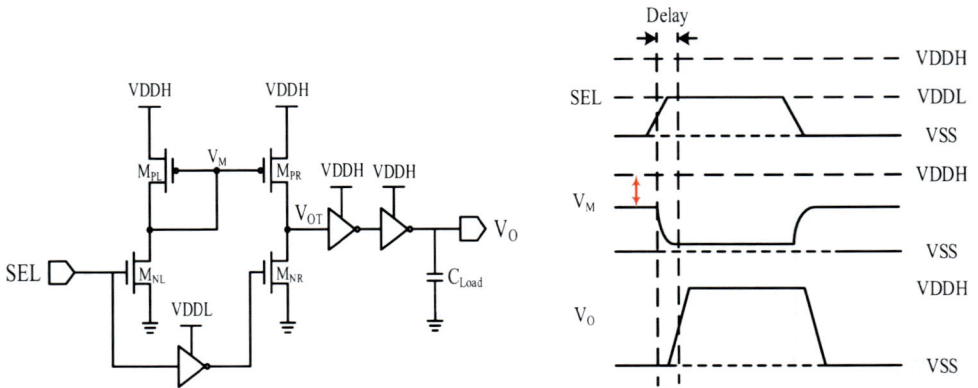

Fig. 6.34 Circuit and waveform of a current-mirror based level shifter (CM-LS)

large, the fight for current between M_{PR} and M_{NR} can cause M_{NR} to pull down V_{OT} to 0 V. When VDD is low (i.e., near or below the threshold voltage of M_{NL}/M_{NR}), I_D is too small to win the fight for current over M_{PL}/M_{PR}. This prevents HL-LS from applying low VDDL and requires that ultra-large transistors be used for M_{NL}/M_{NR} in HL-LS.

6.4.2.2 Current-Mirror Based Level Shifter (CM-LS)

As shown in Fig. 6.34, current-mirror (CM)-based level shifters (Wooters et al. 2010) (CM-LS) were developed to reduce VDDL for low-VDD applications. CM-LS comprises two pull-down NMOS (M_{NL} and M_{NR}) transistors, a pair of PMOS current-mirror (M_{PL} and M_{PR})

transistors, an inverter, and an output buffer (BUF). When SEL = 1, the fact that M_{NL} is on lowers the voltage at node VM (V_M) and then turns on M_{PR}. This, in turn, raises the voltage at node VOT (V_{OT}) to VDDH. When SEL = 0, M_{NL} is off and V_M is slowly pulled down to VDDH-$V_{TH\text{-}MPL}$ by M_{PL}, which significantly reduces the current flowing through M_{PR}. This enables M_{NR} to pull down V_{OT} to below the trig-point of the output buffer without incurring serious current-fighting behavior. This makes it possible for CM-LS to achieve a low VDDL using a smaller M_{NL}/M_{NR} than that required for HL-LS.

However, DC current remains in selected and unselected CM-LSs. When SEL = 1 (selected row), the DC current flows through M_{PL} and M_{NL} (I_{MPL}). When, SEL = 0 (unselected rows),

DC leakage current flows through the weakly-turn-off M_{PR} (I_{MPL}) because its gate voltage is VDDH-$V_{TH\text{-}MPL}$. In the case of an eNVM with k rows, the DC leakage current on the VDDH path consumed by CM-LS is "$I_{MPL} + (k-1) \cdot I_{MPL}$". When k is large, the large current-load on VDDH prevents the on-chip charge-pump from providing current and voltage sufficient for write operations. A Wilson CM LS (Lütkemeier et al. 2010) has been proposed for the suppression of DC current at SEL = 1; however, DC leakage current at SEL = 0 still remains.

6.4.2.3 Pseudo-Diode-Mirrored Level Shifter (PDM-LS)

A pseudo-diode-mirrored level shifter (PDM-LS) was proposed (Chang et al. 2014) to achieve low VDDL while suppressing DC current, as shown in Fig. 6.35. The PDM-LS consists of two pull-down NMOS (M_{NL} and M_{NR}) transistors, a pseudo-diode PMOS transistor (M_{PS}), a current-cut-off PMOS transistor (M_{PM}), and pseudo PMOS current-mirror (M_{PL} and M_{PR}) transistors, an inverter, and a buffer. One source/drain terminal of M_{PS} is connected to the gate of M_{PL} (V_{ML}), while the gate and the other source/drain terminal of M_{PS} are connected to the gate of M_{PR} and the drain of M_{NL} (node VM), respectively. M_{PM} is placed between M_{PL} and M_{NL} and controlled by VOT.

When SEL = 1, M_{NL} is on, which pulls down V_M to 0 V, which turns on M_{PR}, thereby charging the voltage at node VOT (V_{OT}) to VDDH. When $V_M = 0$ V, VM becomes the drain terminal of M_{PS}, causing M_{PS} to form a diode-connected structure, in which V_{ML} is equal to the V_{TH} of M_{PS} ($V_{TH\text{-}MPS}$). "V_{OT} = VDDH" causes M_{PM} to turn off, which subsequently cuts off the DC current flowing through M_{PL}. When SEL = 0, SELb = 1 and M_{NR} turn on, thereby pulling down V_{OT}. M_{PM} is then turned on slightly and VM is charged by M_{PL}. Meanwhile, MPS enters cut-off mode because in this operation, its source terminal is VM. VM then rises to a level approaching VDDH, which is higher than the level of CM-LS, thereby strongly suppressing the DC leakage current flowing through M_{PR}. Thus, by cutting off the DC current at SEL = 1 and suppressing DC leakage current at SEL = 0, the PDM-LS is able to operate consuming far less DC current than that required for conventional CM-LS, while maintaining a low minimum VDDL (VDDLmin).

6.4.3   Challenges in Write-Drivers

In a typical eNVM cell array, the cells selected for write operations are in the same row and share the same WL and SL. As a result, all cells in that row have the same voltage bias on WL

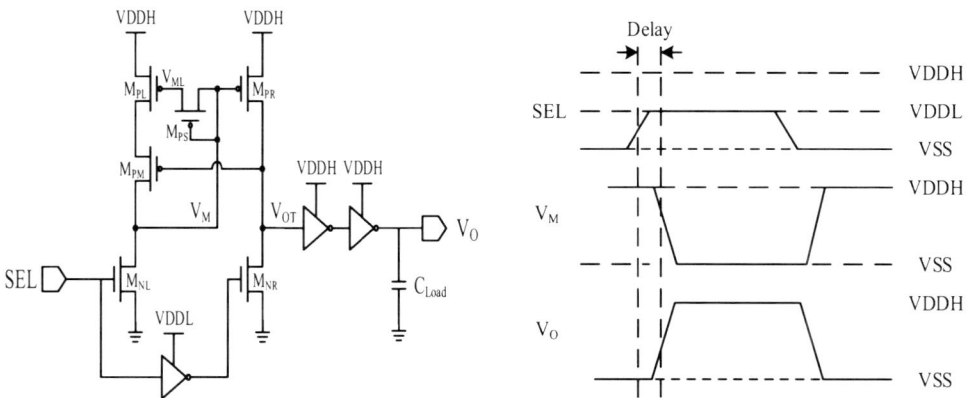

Fig. 6.35 Circuit and waveform of a pseudo-diode-mirrored level shifter (PDM-LS)

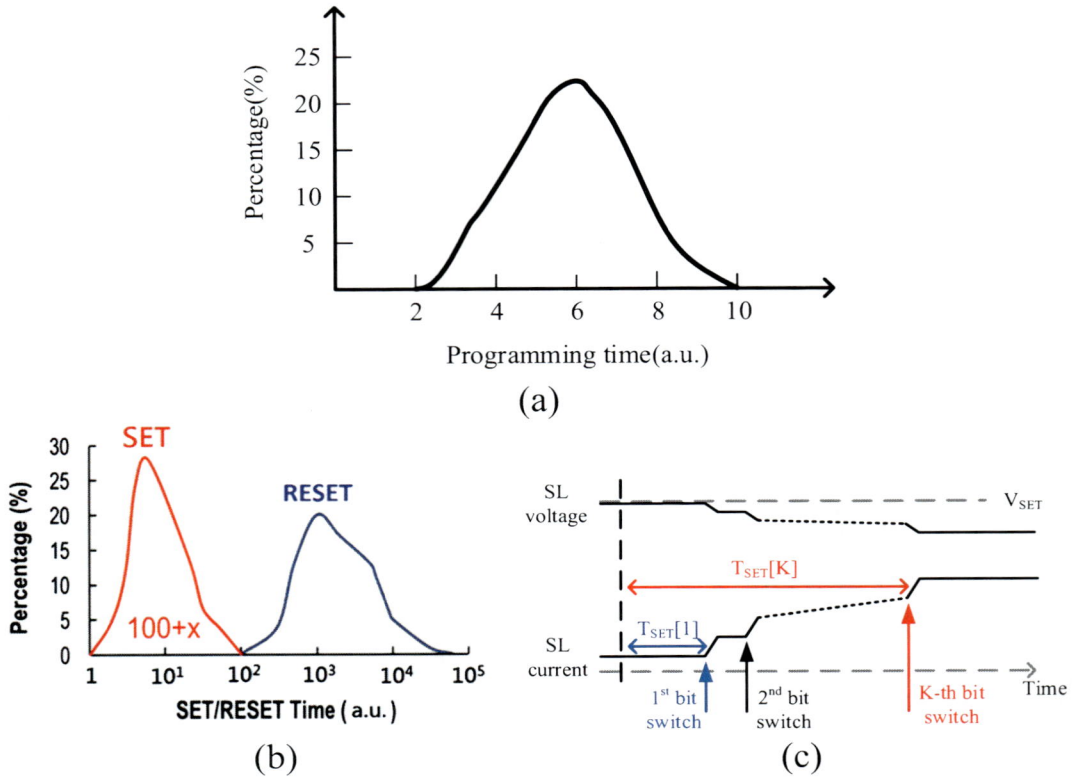

Fig. 6.36 Write times of logic-process (**a**) OTP and (**b**) resistive-type eNVM devices; (**b**) waveforms associated with the voltage and current in a high-voltage path (VDDH) during program operations

and SL. As shown in Fig. 6.36a, process variation can cause considerable variability in the write times required for NVM devices, such as program times (T_{PROG}) or SET (T_{SET})/RESET (T_{RESET}) times. Covering the tail bits (with slower $T_{PROG}/T_{SET}/T_{RESET}$) requires that the voltage bias for word-line and source-line be maintained until the tail bits have completed logic-state switching. For OTP and resistive-type memory, the switching of an eNVM device from HRS to LRS (program or SET operation) causes a considerable flow of DC current ($I_{DC\text{-}CELL}$) between the two terminals of an NVM device. As a result, eNVM devices with shorter T_{PROG}/T_{SET} consume large $I_{DC\text{-}CELL}$ from the high-voltage supply source (VDDH). As shown in Fig. 6.36b, a large $I_{DC\text{-}CELL}$ results in a waste of energy and degrades the output voltage of the on-chip voltage generator as well as the source-line voltage. This can lead to write failure in cells (tail bits) requiring longer T_{PROG}/T_{SET}.

Bit-by-bit write-termination schemes (Halupka et al. 2010; Xue et al. 2013; Chang et al. 2014c) have been proposed for the suppression of $I_{DC\text{-}CELL}$ by monitoring the switching of each accessed eNVM device and then initiating the termination of the Program/SET operation as soon as it has been successfully written. In the following, we review three write-termination schemes: (1) OP-based current-mode, (2) negative-resistance based current-mode write-termination, and (3) voltage-mode write-termination.

6.4.3.1 OP-Based Current-Mode Write-Termination (OP-CWT) Scheme

In (Xue et al. 2013), a current-mode write-termination (OP-CWT) scheme based on operational amplifiers was proposed to perform termination operations for the SET and RESET operations of resistive NVM devices, as shown in Fig. 6.37. This scheme uses operational amplifiers (OP),

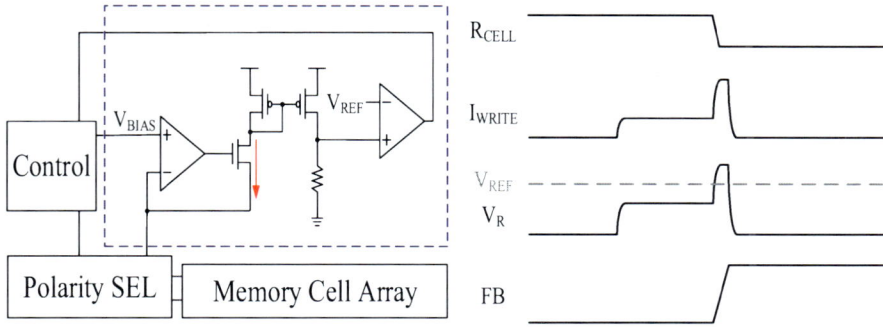

Fig. 6.37   Circuits and waveforms of OP based current-mode write-termination (OP-CWT) scheme

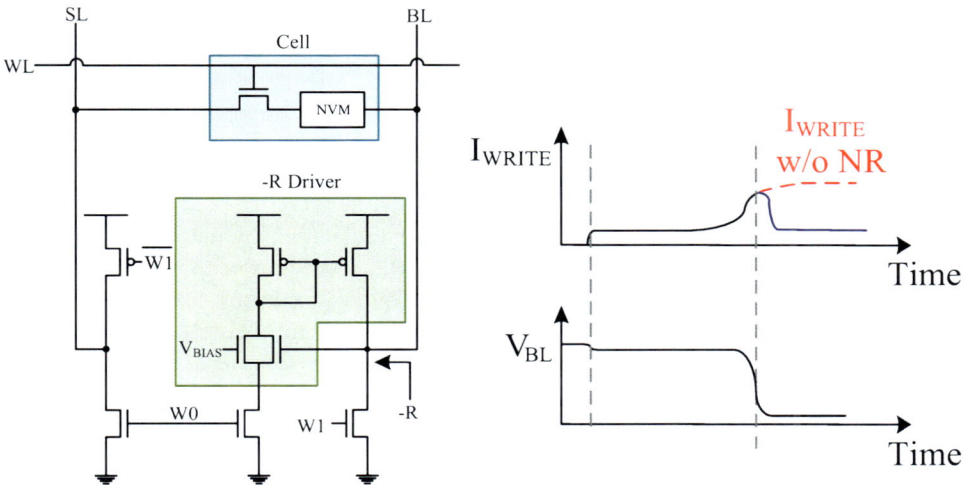

Fig. 6.38   Circuits and waveforms of negative-resistance based current-mode write-termination (NR-CWT) scheme

current-mirror (CM) circuits, and bias circuits to monitor the write current during SET and RESET operations. The high-voltage bias circuit is disabled to terminate the SET operation of an accessed cell. This OP-CWT scheme is effective in suppressing $I_{DC\text{-}SET}$; however, the use of Ops/CMs incurs significant area overhead, slows the response time, and leads to the consumption of considerable quantities of DC current.

6.4.3.2 Negative-Resistance Based Current-Mode Write-Termination (NR-CWT) Scheme

Figure 6.38 illustrates the negative-resistance based current-mode write-termination (NR-CWT) scheme proposed in (Halupka et al. 2010).

NR-CWT employs a current-mirror based negative-resistance ($-R$) circuit to provide feedback for the modulation of write current. When writing '1' (LRS $\rightarrow$ HRS), the WR1 signal is high, the WR0 signal is low, and the current flows through the NVM cell from SL to BL. When writing '0' (switch HRS to LRS), WR1 signal is low, the WR0 signal is high, and the current flows through the NVM cell from BL to SL. The $-R$ driver reflects into the bit-line (BL) a quantity of current proportional to the cell resistance of the NVM cell (R_{NVM}). Once the write '0' is completed and R_{NVM} reaches the LRS resistance value, the increase in current flowing through the NVM cell pulls down the BL voltage. A lower BL voltage reduces the flow of current

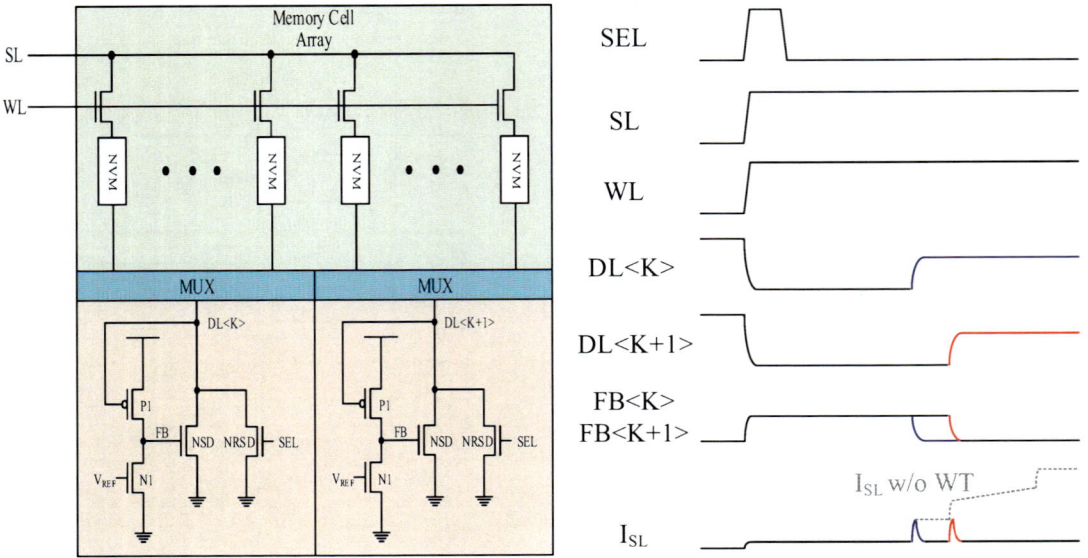

Fig. 6.39 Circuits and waveforms of voltage-mode write-termination scheme

through the current mirror, which lowers the BL voltage considerably, thereby forming a positive feedback loop. The current mirror and positive feedback loop moderate (rather than terminate) the write current, because current continues flowing through the −R driver, even after the write-0 operation is complete.

6.4.3.3 Voltage-Mode Write-Termination (VWT) Scheme

Figure 6.39 illustrates the voltage-mode write-termination (VWT) scheme devised in (Chang et al. 2014c) for the SET operation. This scheme uses only four transistors to monitor bit-line voltage and reuses the $I_{DC-CELL}$ to raise the bit-line voltage in order to terminate the SET operation of an accessed cell. The use of voltage-mode operation and reuse of I_{DC-SET} minimize the area overhead, reduce the power consumption, and improve the response time of write-termination operations.

6.4.4 Challenges in Charge-Pump Circuits

eNVMs use an on-chip charge pump (CP) for the conversion of power supply voltage (VDD)

to several high voltage levels (VDDH1, VDDH2, ...) in order to provide write voltage and output current sufficient for write operations. Unfortunately, the inefficiency of voltage conversion leads to the consumption of a great deal of power. Thus, the charge pump circuit consumes a significant portion of the write power used in an eNVM macro.

Many IoT devices use on-chip NVM macros integrated with on-chip sensors, mixed-signal blocks, and SRAM macros (Chang and Yang 2009; Larsson et al. 2001b; Mezhiba and Friedman 2004), all of which are sensitive to supply noise. In most IoT chips, on-chip sub-blocks share the same power supply source (i.e., battery, energy harvesters, or power generators). Supply noise can induce functional failures in supply-sensitive sub-blocks and even affect write operations and threshold voltage distribution in NVM cells (Chang and Shen 2009; Chang et al. 2015c). Supply noise can also degrade the minimum VDD (VDD_{min}) of the sub-blocks on a chip. Charge-pump circuits require a pumping-capacitor (PC) with large-capacitance for each stage, in order to obtain a sufficiently large output current (I_{OUT}). This greatly increases the switching power required for the clock buffers. The switching of the clock buffer also generates large peak current, which results in the generation of a great deal of supply noise. Thus, the charge-pump circuits in low-power eNVMs face two major challenges: increasing power efficiency, and reducing the consumption of peak current in order to avoid inducing noise at the chip level.

In the following, we review a number of charge-pumps commonly used for eNVM macros.

6.4.4.1 Two-Phase Charge-Pump (2P-CP) Circuit

Figure 6.40 illustrates a conventional two-phase Dickson-type charge-pump (2P-CP) circuit (Dickson 1976; Palumbo and Pappalardo 2009; Witters et al. 1989; Kuriyama et al. 1992; Atsumi et al. 1994). In a 2P-CP, each pumping stage (PS) includes a diode-connected NMOS transistor (NSW) and a pumping-capacitor (PC). Even and odd stages employ opposite clock phases (P1 and P2) for the alternate transfer of boosted charge to subsequent stages. A 2P-CP necessitates a threshold-voltage (V_{TH}) difference between the source and drain terminals of the NSW and is therefore limited with regard to maximum pumping output voltage (V_{OUT}) and power efficiency.

6.4.4.2 Four-Phase Charge-Pump (4P-CP) Circuit

Figure 6.41 illustrates a four-phase Dickson-type CPs (4P-CPs), which was developed to increase the overall power efficiency. A nominal 4P-CP comprises two pumping-stages (PSs) and four clock phases (P1 ~ P4) in a single block. Each pumping-stage includes a diode-connected NMOS transistor (NSW) for charge transfer and a gate switch (GSW) to control the gate of NSW (node GNSW). The 4P-CP eliminates the V_{TH} drop between the source and drain terminals of each NSW by applying a voltage higher than that of 2P-CP at the gate of NSW (V_{GNSW}).

Figure 6.41b illustrates the relationship between P1 ~ P4 in a nominal 4P-CP. The overlapping period between P1-rising and P3-falling is listed as T_{OVER-L}. The overlapping period between P3-rising and P1-falling is listed as T_{OVER-R}. When P1 rises, the voltage at node-A (V_A) increases. During T_{OVER-L} period, V_A is transferred to the V_{GNSW} of the even stage (V_{GNSW-E}). At this point, P4 rises to pump V_{GNSW-E} to a voltage higher than V_A, thereby allowing the transfer of the charge at node-A to node-B without a drop in V_{TH}. During T_{OVER-R} period, P1 is pulled down to discharge V_{GSW-E} in order to turn off the NSW of the even stage (NSW-E). At the same time, the V_{GSW} of the

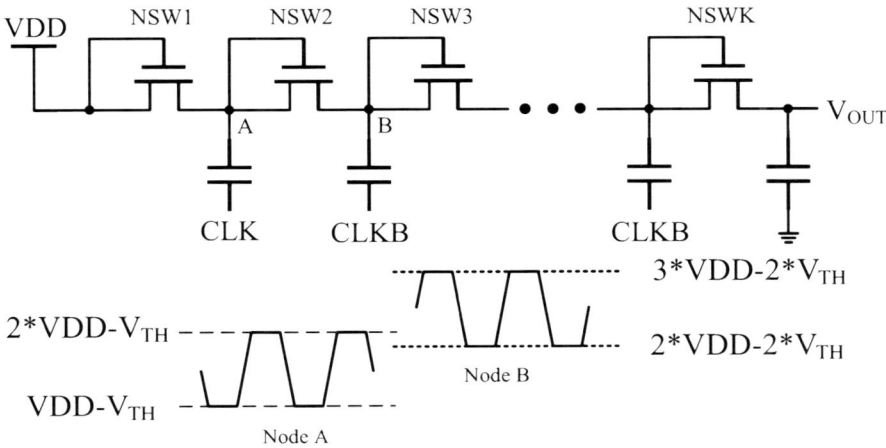

Fig. 6.40 Circuit and waveforms of typical two-phase charge-pumps (2P-CP)

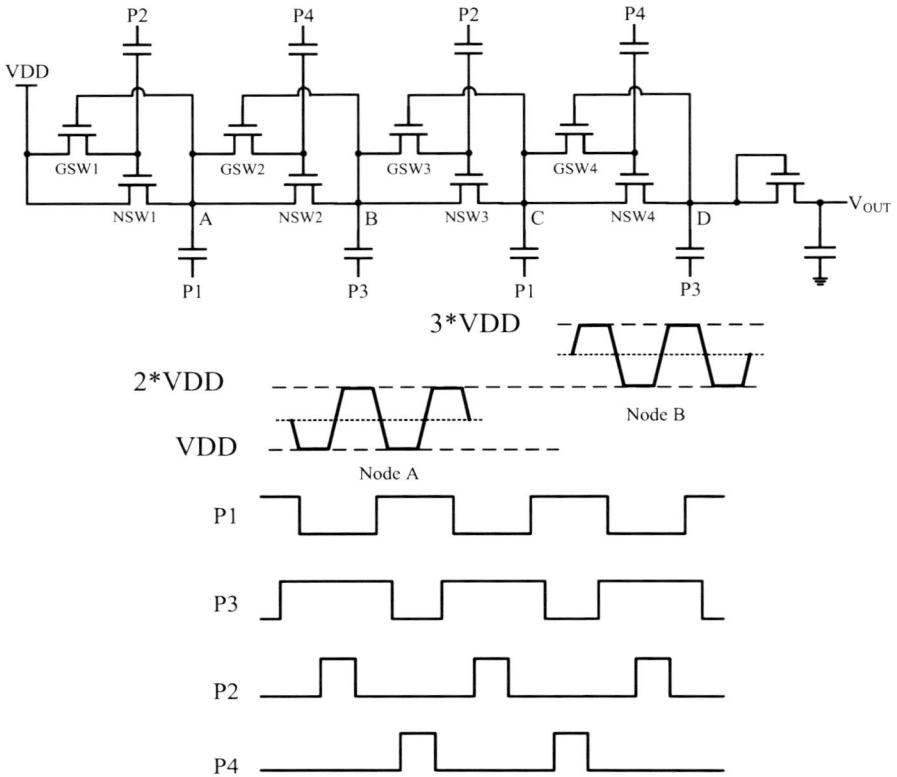

Fig. 6.41 Circuit and waveforms of typical four-phase charge-pump (4P-CP)

odd stage (V_{AGSW-O}) of the following block (i.e., BK2) is raised by node-B of the previous block (i.e., BK1).

In conventional 4P-CP, T_{OVER-L} and T_{OVER-L} are given the same period in order to ensure consistency in V_A-to-V_{GSW} and cross-stage charge transfer times across stages within a single block (i.e., PS1-PS2 or PS3-PS4) or across blocks (i.e., PS2-PS3).

6.4.4.3 Cross-Coupled Charge-Pump (CC-CP) Circuit

As shown in Fig. 6.42, the cross-coupled charge-pump (CC-CP) proposed in (Ker et al. 2006) comprises two branches (branch A and B). Each branch comprises k pairs of NMOS-PMOS (MN-MP) connected in serial. The NMOS-PMOS pairs are cross-couple connected between the two branches. Branches A and B alternate in the pumping of output voltage. During the charge

transfer operation, the gate voltage of an NMOS pass-gate is one VDD higher than its terminal connected to the previous stage. As a result, the NMOS is on, thereby providing efficient charge transfer behavior, while the PMOS is off, thereby cutting off the charge flow-back path.

6.4.4.4 Two-Step Clocking Charge-Pump (TSC-CP) Circuit

Figure 6.43 illustrates the two-step clocking charge-pump (TSC-CP) scheme presented in (Lauterbach et al. 2000). Conventional CPs use single voltage steps for pumping clocks wherein the energy delivered by the clock buffers equals ($Q \times VDD$). The TSC-CP scheme employs two voltage steps for the pumping clock, which means that the energy delivered by the clock buffers in TSC-CP is equal to ($3/4 \times Q \times VDD = 1/2 \times Q \times VDD + Q \times VDDQ + 1/2 \times Q \times VDD$).

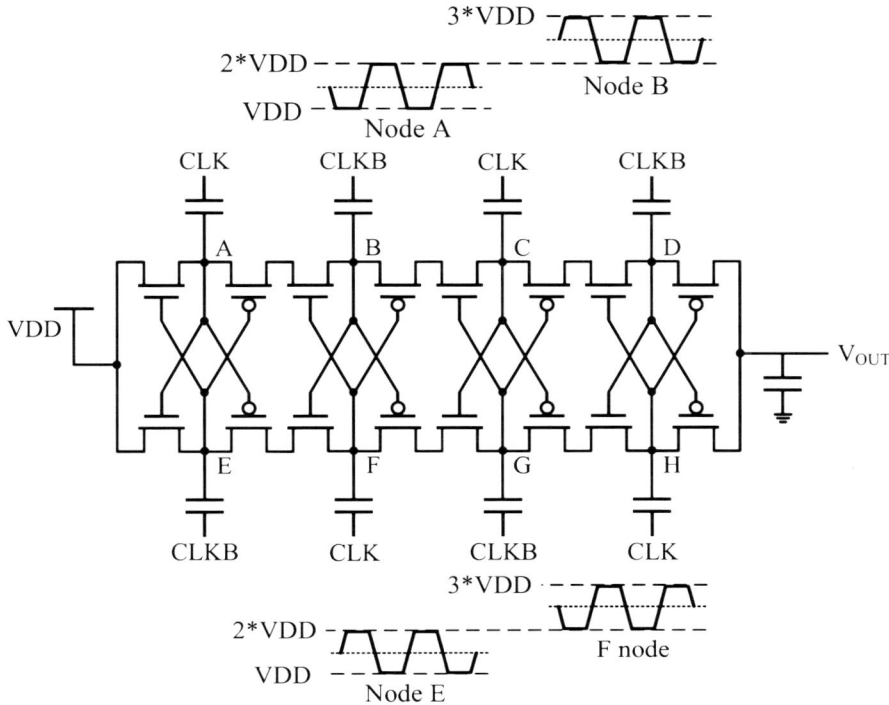

Fig. 6.42 Circuit and waveforms of a typical cross-couple charge-pump (CC-CP)

This results in energy consumption below that of a conventional CP.

The use of equalization between the up-pumping capacitor (i.e., C1) and down-pumping capacitor (i.e., C2) in neighboring stages means that the TSC-CP needs to charge the up-pumping capacitor using VDD/2, instead of VDD, as in conventional CPs. This leads to a further reduction in power consumption. Combining two-step clocking with capacitor equalization makes it possible for the TSC-CP to reduce energy consumption far below that of conventional CP.

6.4.4.5 Low-Peak-Current Split-asymmetrically-Shifted Clocking Charge-Pump (SASC-CP) Circuit

One effective approach to reduce the peak current of a charge-pump involves preventing the simultaneous occurrence of all switching-related activities in multiple pumping blocks. However, the clock phases of all pumping blocks in conventional charge-pumps are shared and symmetrical. Thus, the timing relationship between P1 ~ P4 in each pumping block should be fixed in order to avoid charge-feedback activity within a single pumping block (intra-block) or between pumping blocks (inter-block).

As shown in Fig. 6.44, a split asymmetrically shifted clocking (SASC) scheme was proposed in Chang et al. (2015f) to (1) prevent simultaneous switching activities across pumping blocks and (2) maintain the same timing relationship between the local four-phase clocking of each pumping block.

Splitting the four-phase clocks between pumping blocks enables the discrete control of charge boosting (pumping) processes at inter-block and intra-block levels. The local four-phase clocks of each pumping block switch after those of the previous pumping block (its input stage) due to T_{DELAY}. To prevent a drop in the pumping efficiency of each pumping block, SASC-CP employs different timing for T_{OVER-R} and T_{OVER-L}. It should be noted that in SASC-CP, T_{OVER-R} is not as critical as it

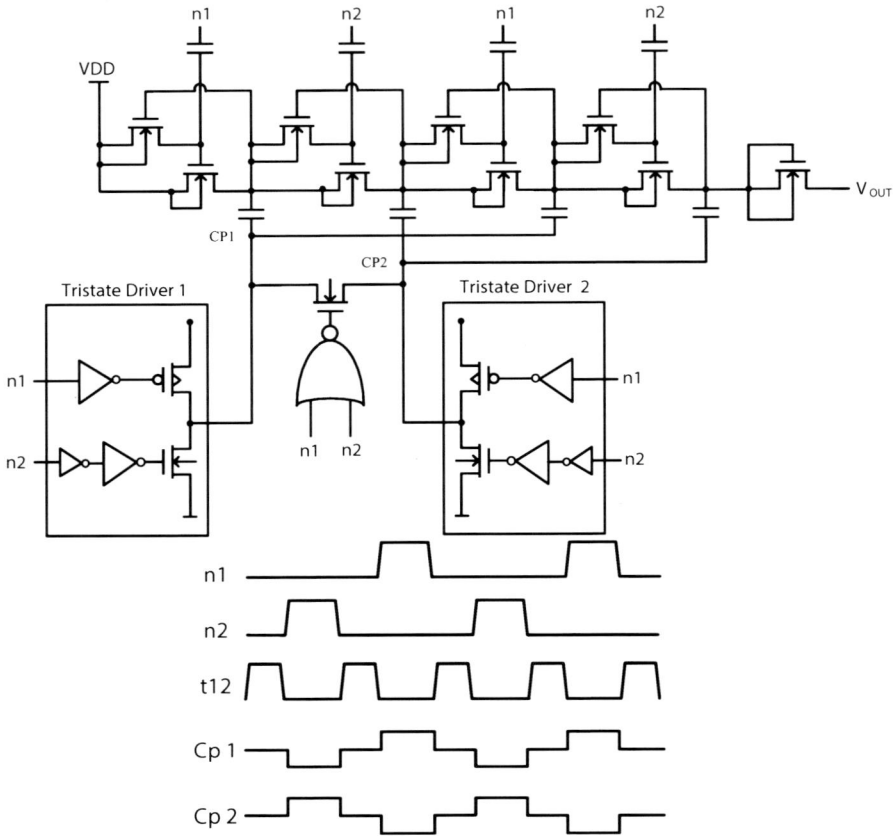

Fig. 6.43 (a) Circuit and (b) waveform of two-step clocking charge-pump (TSC-CP) scheme

would be in conventional charge-pumps. This makes it possible to employ a shorter period for T_{OVER-R} than that used in conventional charge-pumps without affecting intra-block pumping behavior. The timing relationship between T_{OVER-R} and T_{OVER-L} is presented as $T_{OVER-L} = T_{DELAY} + T_{OVER-R}$.

Figure 6.44b presents the clock generation scheme for SASC-CP using one clock generator for the global clocks (GP1 ~ GP4). A local delay buffer (LDB) is placed above each pumping block to provide local four-phase clocks (P1 ~ P4). Accordingly, the RC delays of P1 ~ P4 are far smaller than in conventional charge-pumps. This also makes it possible to use a higher clock frequency with SASC-CP than would be possible with conventional charge-pumps.

6.5 Perspectives and Trends

In this section, we review various state-of-the-art on-chip NVMs and discuss recent trends in low-power on-chip NVM for IoT applications. We also explore extended applications of NVM devices, beyond the issue of memory usage for IoT applications.

6.5.1 State-of-the-Art On-Chip NVMs and Trends in On-Chip NVM for IoT Devices

IoT devices employ a wide range of technology nodes to accommodate a wide range of applications and cost structures. Figure 6.45

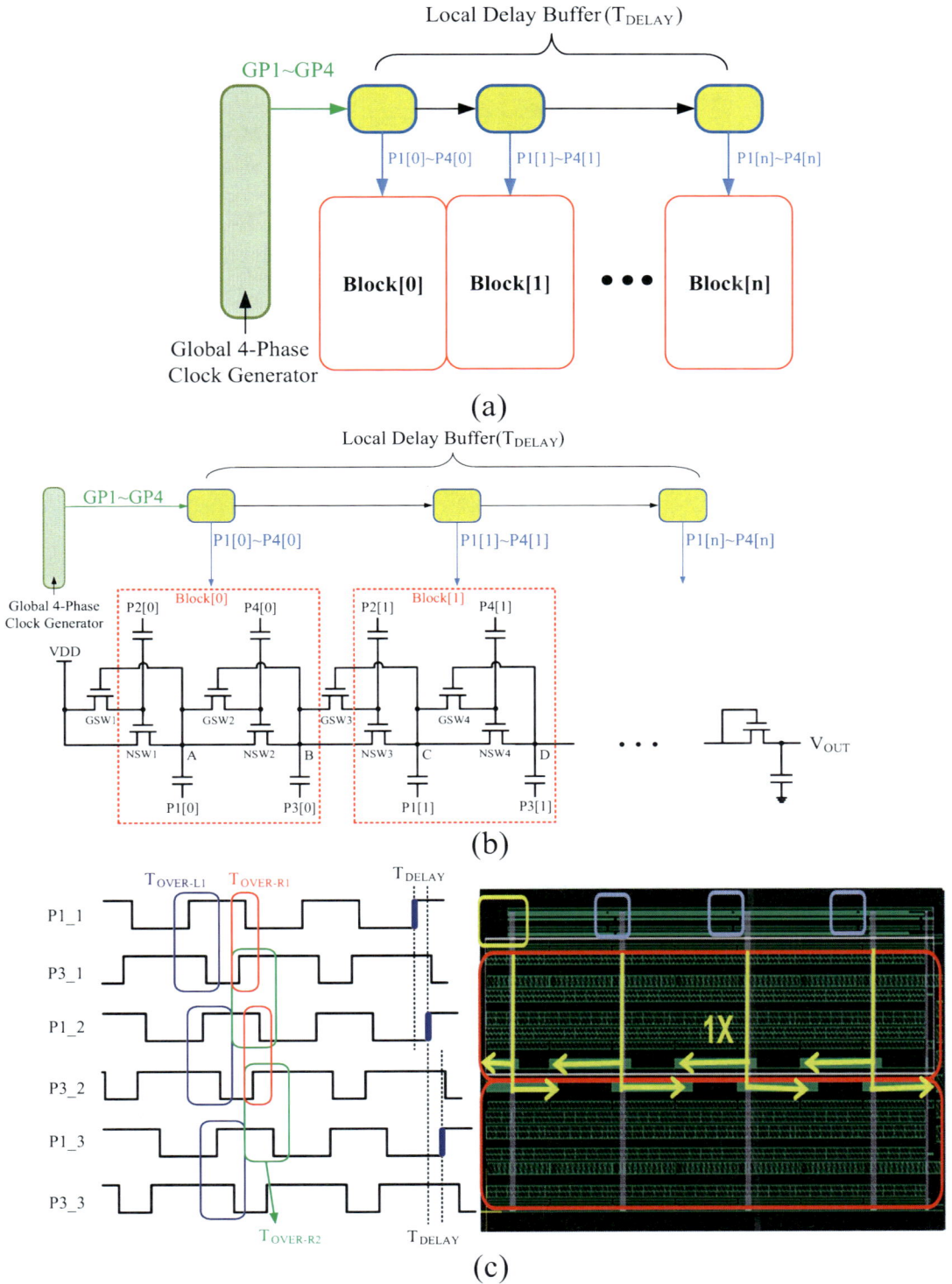

Fig. 6.44 (**a**) Concept, (**b**) circuit, (**c**) waveform and layout used in split asymmetrically shifted (SAS) clocking scheme

Fig. 6.45 Read power vs. technology nodes for eFlash and emerging memory solutions

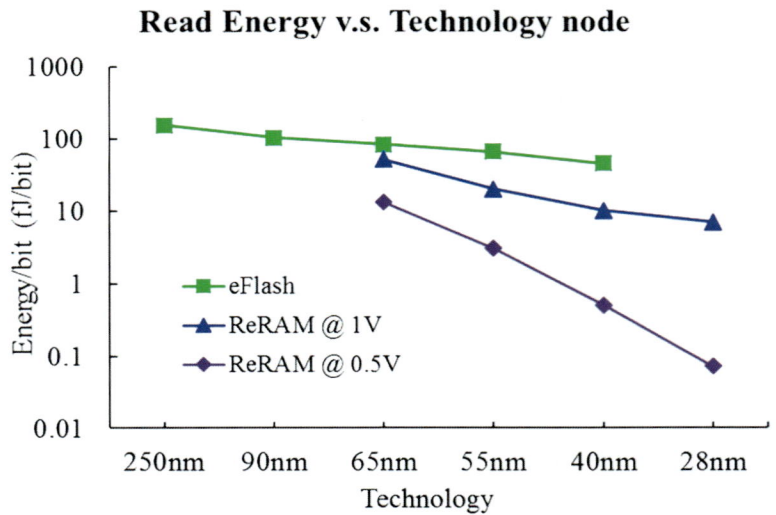

presents the read energy versus technology nodes of recent eFlash macros (Jefremow et al. 2012; Liu et al. 2013a; Kono et al. 2013, 2014; Cho et al. 2013; Yu et al. 2013; Taito et al. 2015; Yamauchi et al. 2015a, b). State-of-the-art on-chip eFlash macros (at 28 nm nodes) consume less read energy than do the eFlash macros at the nodes of mature processes (i.e., 0.25 μm node). This significant reduction in read energy is due primarily to the scaling of devices and supply voltages (VDD).

Due to their low write voltage requirements, ReRAM and PCM macros do not require the high-voltage transistors found in eFlash macros for peripheral circuits. This makes it possible for ReRAM and PCM macros (Chien et al. 2010; Chang et al. 2012b, 2014c; Fackenthal et al. 2014; De Sandre et al. 2010; Chung et al. 2011; Close et al. 2011; Choi et al. 2012) to employ a lower supply voltage (VDD) for read operations and thereby achieve a read energy lower than that of eFlash macros. As a result, ReRAM and PCM macros are more practical than eFlash for IoT devices using low supply voltages to reduce read energy consumption.

Figure 6.46 presents the write energy versus technology nodes of recently reported NVM macros. The write energy consumed by state-of-the-art on-chip eFlash macros (at 28 nm nodes) is slightly smaller than that of eFlash macros at the nodes of more mature processes (i.e., 0.25 μm node). The small scaling ratio in write energy for eFlash across process nodes is due to the fact that similar write (program and erase) mechanisms are used for program and erase operations. In contrast, some emerging resistive-type memories require far less write energy than does eFlash, due to lower write voltages and faster write times.

The state-of-the-art processes used in eFlash memory are a few generations behind the most advanced logic processes, due to the lengthy time required for the development of processes to deal with the multiple gates (floating gate, control gate, select gate, and erase gate) found in eFlash cells. With the migration of advanced logic processes away from planar transistors to FinFET transistors, it is becoming increasingly difficult to integrate conventional floating-gate based eFlash technology with FinFET processes. This has led to the need for on-chip NVM solutions that are logic-process compatible with applicability to nanometer processes. ReRAM and PCM are promising eNVM solutions for advanced nanometer processes due to their inclusion of back-end processes (post-BEOL processes) to decouple NVM device processing from front-end processes (transistors).

Fig. 6.46 Write power
versus technology nodes
for eFlash and emerging
memory solutions

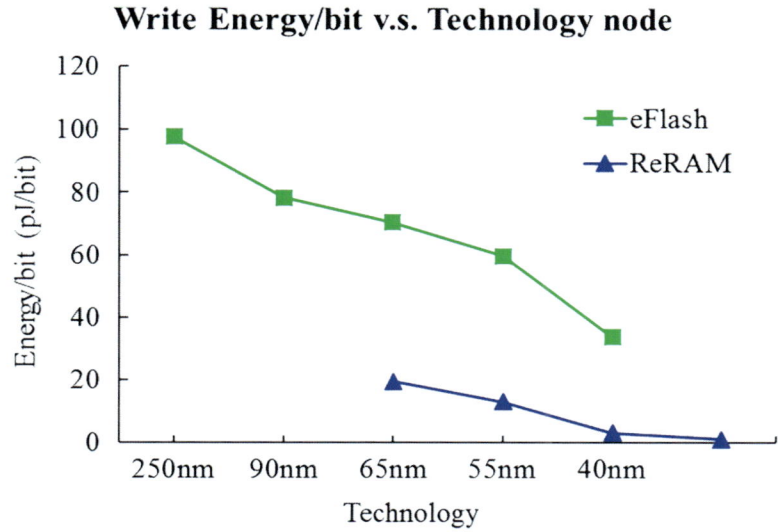

Write Energy/bit v.s. Technology node

6.5.2   Extension of NVM devices to IoT Devices

Recent emerging memory devices have been employed as NVM macros and combined with SRAM or TCAM to produce hybrid-mode memories, such as nonvolatile SRAM (nvSRAM) (Ohsawa et al. 2013; Wang et al. 2006; Sheu et al. 2013; Yamamoto et al. 2009; Chiu et al. 2010; Wei et al. 2014; Lee et al. 2015) and nonvolatile TCAM (nvTCAM) (Li et al. 2013; Matsunaga et al. 2012; Huang et al. 2014; Chang et al. 2015d).

As discussed in Sect. 6.1, many energy-efficient chips employ SRAM for computing and NVM for power-off storage in order to reduce standby current. Unfortunately, this two-macro (SRAM+NVM) scheme consumes a great deal of energy during power interruptions and results in slow store (power-off) and restore (power-on) operations, due to the word-by-word serial transfer of data.

nvSRAMs perform bit-to-bit data transfer between SRAM and NVM devices within a single cell and are capable of block-level parallel data transfer with store/restore operations faster than that found in two-macro schemes. Figure 6.47 presents several examples of recent nvSRAM macros.

Conventional SRAM-based TCAM (Hayashi et al. 2013; Arsovski et al. 2013; Yabuuchi et al. 2014; Do et al. 2014; Jeloka et al. 2015; Nii et al. 2015) uses two 6T-SRAMs as content-storage units and 4T comparison logics to perform data-matching comparisons. Figure 6.48 shows several recent nvTCAM macros. The use of NVM as a content storage unit makes it possible to reduce the search energy, standby power, and area of nvTCAM. It also enables search speeds comparable to those of SRAM-based TCAM. Moreover, the NVM devices in nvTCAM are able to perform storage as well as comparison functions, thereby eliminating the need for the movement of data between volatile and NVM devises, as in nvSRAM. This one-macro nvTACM scheme enables fast low-power power-off and power-on operations for IoT devices.

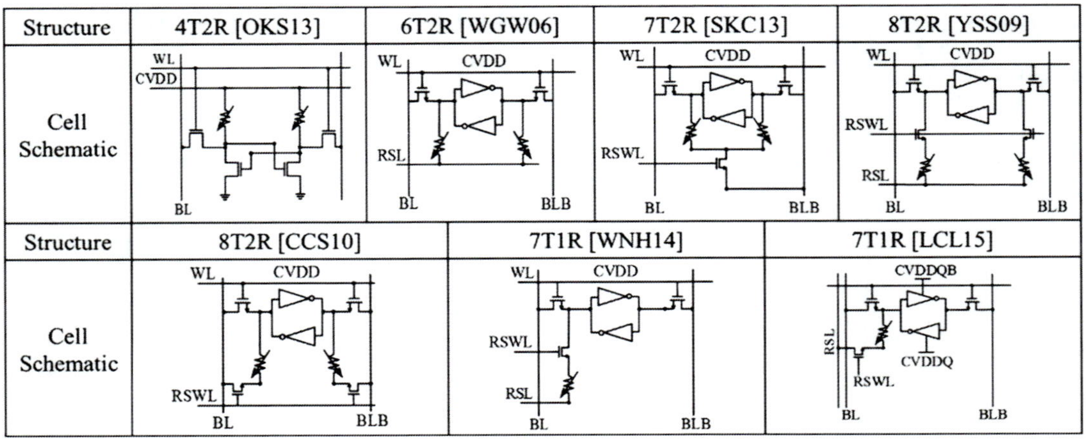
Structure
4T2R [OKS13]
6T2R [WGW06]
7T2R [SKC13]
8T2R [YSS09]
Cell Schematic
WL
CVDD
BL
BLB
RSL
RSWL
Structure
8T2R [CCS10]
7T1R [WNH14]
7T1R [LCL15]
Cell Schematic
CVDDQB
CVDDQ

Fig. 6.47 Several examples of recent nvSRAM cells

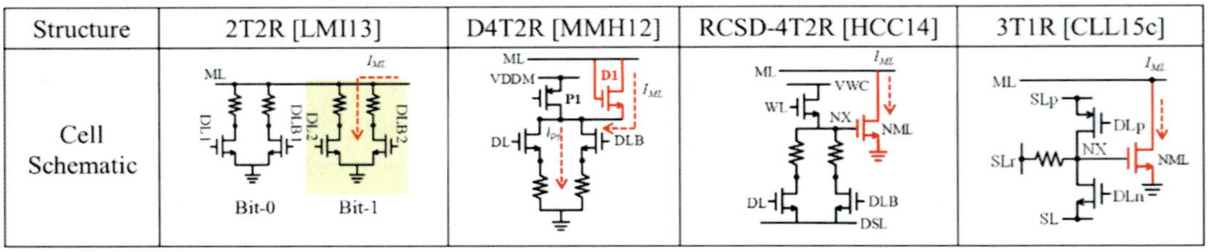
Structure
2T2R [LMI13]
D4T2R [MMH12]
RCSD-4T2R [HCC14]
3T1R [CLL15c]
Cell Schematic
ML
DL1
DLB1
DL2
DLB2
Bit-0
Bit-1
VDDM
D1
P1
DL
DLB
VWC
WL
NX
NML
DSL
SLp
DLp
SLr
DLn
SL

Fig. 6.48 Several recent nvTCAM cells

References

B.J. Ahn, J.H. Sone, J.W. Kim, I.H. Choi, D.M. Kim, A simple and efficient self-limiting erase scheme for high performance split-gate flash memory cells. IEEE Electron Device Lett. **19**, 438–440 (1998)

I. Arsovski, T. Hebig, D. Dobson et al., A 32 nm 0.58-fJ/bit/search 1-GHz ternary content addressable memory compiler using silicon-aware early-predict late-correct sensing with embedded deep-trench capacitor noise mitigation. IEEE J. Solid State Circuits **48**(4), 932–939 (2013)

S. Atsumi, M. Kuriyama, A. Umezawa et al., A 16-Mb Flash EEPROM with a new self-data-refresh scheme for a sector erase operation. IEEE J. Solid State Circuits **29**(4), 461–468 (1994)

N. Banno, M. Tada, T. Sakamoto, et al., A fast and low-voltage Cu complementary-atom- switch 1 Mb array with high-temperature retention, *IEEE Symposium on VLSI Technology (VLSIT) Dig. Tech. Papers* (2014), pp. 1–2

M. Bauer, R. Alexis, G. Atwood, et al., A multilevel-cell 32 Mb flash memory, *IEEE International Solid-State Circuits Conference (ISSCC) Dig. Tech. Papers*, 15–17 Feb. 1995 (1995), pp.132–133

M. Boniardi, A. Redaelli, C. Cupeta, et al., Optimization metrics for Phase Change Memory (PCM) cell architectures, *IEEE International Electron Devices Meeting (IEDM) Dig. Tech. Papers* (2014), pp. 29.1.1–29.1.4

D.S. Byeon, S.S. Lee, Y.H. Lim et al., An 8 Gb multi-level NAND flash memory with 63 nm STI CMOS process technology. ISSCC Dig. Tech. Pap. **1**, 46–47 (2005)

M.F. Chang, S.J. Shen, A process variation tolerant embedded split-gate flash memory using pre-stable current sensing scheme. IEEE J. Solid-State Circuits **44**(3), 987–994 (2009)

M.-F. Chang, S.-M. Yang, Analysis and reduction of supply noise fluctuations induced by embedded ROM. IEEE Trans. VLSI Syst. **17**(6), 758–769 (2009)

I.J. Chang, J.J. Kim, S.P. Park et al., A 32 kb 10T sub-threshold SRAM array with bit-interleaving and differential read scheme in 90 nm CMOS. IEEE J. Solid State Circuits **44**(2), 650–658 (2009a)

S.H. Chang, S.K. Lee, S.J. Park, et al., A 48 nm 32 Gb 8-level NAND flash memory with 5.5 MB/s program throughput, *IEEE International Solid-State Circuits Conference (ISSCC) Dig. Tech. Papers* (2009), pp. 240–241, 241a

M.-F. Chang, P.-F. Chiu, S.-S. Sheu, Circuit design challenges in embedded memory and resistive RAM (RRAM) for mobile SoC and 3D-IC, in *Proceedings of the IEEE Asia and South Pacific Design Automation Conference (ASP-DAC)* (2011), pp. 197–203

M.-F. Chang, P.-F. Chiu, W.-C. Wu, C.-H. Chuang, S.-S. Sheu, Challenges and trends in low-power 3D die-stacked IC designs using RAM, memristor logic, and resistive memory (ReRAM), in *IEEE International Conference on ASIC (ASICON)* (2011), pp. 299–302

I.J. Chang, J.J. Kim, K. Kim et al., Robust level converter for sub-threshold/super-threshold operation: 100 mV to 2.5 V. IEEE TransactionsVLSI Syst. **19**(8), 1429–1437 (2011c)

M.-F. Chang, C.-H. Chuang, M.-P. Chen, et al. Endurance-aware circuit designs of nonvolatile logic and nonvolatile SRAM using resistive memory (memristor) device, in *Proceedings of the IEEE Asia and South Pacific Design Automation Conference* (2012), pp. 329–334

M.F. Chang, C.W. Wu, C.C. Kuo, et al., A 0.5 V 4 Mb logic-process compatible embedded resistive RAM (ReRAM) in 65 nm CMOS using low-voltage current-mode sensing scheme with 45 ns random read time, *IEEE International Solid-State Circuits Conference (ISSCC) Dig. Tech. Papers* (2012), pp. 434–436

M.-F. Chang, S.-S. Sheu, K.-F. Lin, et al., A high-speed 7.2-ns read-write random access 4-Mb embedded Resistive RAM (ReRAM) macro using process-variation- tolerant current-mode read schemes, *IEEE Journal of Solid-State Circuits* (2013), pp. 878–891

M.F. Chang, S.J. Shen, C.C. Liu et al., An offset-tolerant fast-random-read current-sampling-based sense amplifier for small-cell-current nonvolatile memory. IEEE J. Solid State Circuits **48**(3), 864–877 (2013b)

M.F. Chang, C.W. Wu, C.C. Kuo et al., A low-voltage bulk-drain-driven read scheme for Sub-0.5 V 4 Mb 65 nm logic-process compatible embedded resistive RAM (ReRAM) macro. J. Solid-State Circuits **48**(9), 2250–2259 (2013c)

M.-F. Chang, C.-C. Kuo, S.-S. Sheu, et al. Area-efficient embedded Resistive RAM (ReRAM) macros using logic-process Vertical-Parasitic-BJT (VPBJT) switches and read-disturb-free temperature-aware current-mode read scheme, *IEEE J. Solid-State Circuits (JSSC)* (2014), pp. 908–916

M.-F. Chang, A. Lee, C.-C. Kuo, et al. Challenges at circuit designs for resistive-type nonvolatile memory and nonvolatile logics in mobile and cloud applications, in *Proceedings of the IEEE International Conference on Solid-State and Integrated Circuit Technology (ICSICT)* (2014), pp. 1–4

M.-F. Chang, J.J. Wu, T.F. Chien, et al., 19.4 embedded 1 Mb ReRAM in 28 nm CMOS with 0.27-to-1 V read using swing-sample-and-couple sense amplifier and self-boost-write-termination scheme, *IEEE International Solid-State Circuits Conference (ISSCC) Dig. Tech. Papers* (2014), pp. 332–333

M.F. Chang et al., A low-power subthreshold-to-superthreshold level-shifter for sub-0.5V embedded resistive RAM (ReRAM) macro in ultra low-voltage chips, in *2014 IEEE Asia Pacific Conference on Circuits and Systems (APCCAS)*, Ishigaki, 2014, pp. 695–698

M.F. Chang, A. Lee, P.C. Chen et al., Challenges and circuit techniques for energy-efficient on-chip nonvolatile memory using memristive devices. IEEE J. Emerg. Select. Top. Circuits Syst. **5**(2), 183–193 (2015a)

M.-F. Chang, A. Lee, C.-C. Lin, et al. Read circuits for resistive memory (ReRAM) and memristor-based nonvolatile Logics, in *Proceedings of the IEEE Asia and South Pacific Design Automation Conference (ASP-DAC)* (2015), pp. 569–574

M.F. Chang, Y.F. Lin, Y.C. Liu et al., An asymmetric-voltage-biased current-mode sensing scheme for fast-read embedded flash macros. IEEE J. Solid State Circuits **50**(9), 2188–2198 (2015c)

M.F. Chang, C.C. Lin, A. Lee, et al., A 3T1R nonvolatile TCAM using MLC ReRAM with Sub-1 ns search time, *IEEE International on Solid-State Circuits Conference (ISSCC) Dig. Tech. Papers* (2015), pp. 1–3

M.-F. Chang, W.-Y. Lu, S.-J. Shen, M.-P. Chen, C.-S. Lin, S.-S. Sheu, C.-H. Hung, Y.-S. Yang, Y.-J. Kuo, S.-N. Hung, H.-T. Lue, C.-H. Shen, J.-M. Shieh, Supply-variation-resilient nonvolatile 3D IC and 3D memory using low peak-current on-chip charge-pump circuits, in *Proceedings of the IEEE Electron Devices and Solid-State Circuits (EDSSC)* (2015), pp. 118–121

M.F. Chang, W.Y. Lu, S.J. Shen, et al., Supply-variation-resilient nonvolatile 3D IC and 3D memory using low peak-current on-chip charge-pump circuits, in *Proceedings of the IEEE Conference on Electron Devices and Solid-State Circuits* (2015), pp. 118–121

W.-M. Chen, C. Swift, D. Roberts, K. Forbes, J. Higman, B. Maiti, W. Paulson, K.-T. Chang, A novel flash memory device with split gate source side injection and ono charge storage stack (SPIN), *Symp. VLSI Technology Dig. Tech. Papers* (1997), pp. 63–64

J.H. Chen, L.T. Clark, T.H. Chen, An ultra-low-power memory with a subthreshold power supply voltage. IEEE J. Solid State Circuits **41**(10), 2344–2353 (2006)

Y.-H. Chen, W.-M. Chan, W.-C. Wu, et al., A 16 nm 128 Mb SRAM in high-κ metal-gate FinFET technology with write-assist circuitry for low-VMIN applications, in *IEEE International Solid-State Circuits Conference (ISSCC) Dig. Tech. Papers* (2014), pp. 238–239

H.Y. Cheng, T.H. Hsu, S. Raoux, J.Y. Wu, P.Y. Du, M. Breitwisch, Y. Zhu, E.K. Lai, E. Joseph, S. Mittal, R. Cheek, A. Schrott, S.C. Lai, H.L. Lung, C. Lam, A high performance phase change memory with fast switching speed and high temperature retention by engineering the GexSbyTez phase change material, *IEEE International Electron Devices Meeting (IEDM) Dig. Tech. Papers* (2011), pp. 3.4.1–3.4.4

H.Y. Cheng, J.Y. Wu, R. Cheek, S. Raoux, M. BrightSky, D. Garbin, S. Kim, T.H. Hsu, Y. Zhu, E.K. Lai, E. Joseph, A. Schrott, S.C. Lai, A. Ray, H.L. Lung, C. Lam, A thermally robust phase change memory by engineering the Ge/N concentration in $(Ge, N)_x Sb_y Te_z$ phase change material, *IEEE International Electron Devices Meeting (IEDM) Dig. Tech. Papers* (2012), pp. 31.1.1–31.1.4

W.C. Chien, Y.R. Chen, Y.C. Chen, et al., A forming-free WOx resistive memory using a novel self-aligned field enhancement feature with excellent reliability and scalability, *IEEE International Electron Devices Meeting (IEDM) Dig. Tech. Papers* (2010), pp. 19.2.1–19.2.4

Y.D. Chih, C.H. Wang, C.H. Kuo, Reference cell circuit for split gate flash memory. U.S. Patent 6,396,740, 28 May 2002

P.-F. Chiu, M.-F. Chang, S.-S. Sheu, et al., A low store energy, low VDD_{min}, nonvolatile 8T2R SRAM with 3D stacked RRAM devices for low power mobile applications, *Symposium on VLSI Circuits (VLSIC) Dig. Tech. Papers* (2010), pp. 229–230

C.Y.-S. Cho, M.-J. Chen, C.-F. Chen, P. Tuntasood, D.-F. Fan, T.-Y. Liu, A novel self-aligned highly reliable sidewall split-gate flash memory. IEEE Trans. Electron Dev. **53**, 465–473 (2006)

C.Y.-S. Cho, J.C. Wang, L. Huang, et al., A 55-nm, 0.86-Volt operation, 75 MHz high speed, 96 µA/MHz low power, wide voltage supply range 2 M-bit split-gate embedded Flash, in *Proceedings of the IEEE International Symposium on VLSI Design, Automation, and Test (VLSI-DAT)* (2013), pp. 1–4

Y.D. Choi, I.H. Song, M.H. Park, et al., A 20 nm 1.8 V 8 Gb PRAM with 40 MB/s program bandwidth, *IEEE International Solid-State Circuits Conference (ISSCC) Dig. Tech. Papers* (2012), pp. 46–48

C.C. Chung, H.C. Lin, Y.T. Lin, A multilevel read and verifying scheme for Bi-NAND flash memories. IEEE J. Solid State Circuits **42**(5), 1180–1188 (2007)

H. Chung, B.H. Jeong, B.J. Min, et al., A 58 nm 1.8 V 1 Gb PRAM with 6.4 MB/s program BW, *IEEE International Solid-State Circuits Conference (ISSCC) Dig. Tech. Papers* (2011), pp. 500–502

G.F. Close, U. Frey, J. Morrish, et al., A 512 Mb phase-change memory (PCM) on 90 nm CMOS achieving 2b/cell, *Symposium on VLSI Circuits (VLSIC) Dig. Tech. Papers* (2011), pp. 202–203

Datasheet, "sfc 0064_08b9_he" Taiwan Semiconductor Manufacturing Company (TSMC) (2001)

Datasheet, AF64K8AF25, v1.0 1st Silicon Sdn. Bhd. (X-Fab) (2005)

G. De Sandre, L. Bettini, A. Pirola, et al., A 90 nm 4 Mb embedded phase-change memory with 1.2 V 12 ns read access time and 1 MB/s write throughput, *IEEE International Solid-State Circuits Conference (ISSCC) Dig. Tech. Papers* (2010), pp. 268–269

J.F. Dickson, On-chip high-voltage generation in MNOS integrated circuits using an improved voltage multiplier technique. IEEE J. Solid State Circuits **11**(3), 374–378 (1976)

A.T. Do, C. Yin, K. Velayudhan et al., 0.77 fJ/bit/search content addressable memory using small match line swing and automated background checking scheme for variation tolerance. IEEE J. Solid State Circuits **49**(7), 1487–1498 (2014)

B. Eitan, P. Pavan, I. Bloom, E. Aloni, A. Frommer, D. Finzi, Can NROM, a 2-bit, trapping storage BVN cell, give a real challenge to floating gate cells, in *Proc. Int. Conf. Solid State Devices and Materials* (1999), pp. 522–524

B. Eitan, P. Pavan, I. Bloom, E. Aloni, A. Frommer, D. Finzi, NROM: a novel localized trapping, 2-bit nonvolatile memory cell. IEEE Electron Device Lett. **21**(11), 543–545 (2000)

K. Eshraghian, K.-C. Cho, O. Kavehei, et al., Memristor MOS Content Addressable Memory (MCAM): hybrid architecture for future high performance search engines, *IEEE Transactions on Very Large Scale Integration (VLSI) Systems* (2010), pp. 1407–1417

R. Fackenthal, M. Kitagawa, W. Otsuka, et al., 19.7 A 16 Gb ReRAM with 200 MB/s write and 1 GB/s read in 27 nm technology, in *IEEE International Solid-State Circuits Conference (ISSCC) Dig. Tech. Papers* (2014), pp. 338–339

F. Frustaci, M. Khayatzadeh, D. Blaauw, D. Sylvester, M. Alioto, A 32 kb SRAM for error-free and error-tolerant applications with dynamic energy- quality management in 28 nm CMOS, *IEEE International Solid-State Circuits Conference (ISSCC) Dig. Tech. Papers* (2014), pp. 244–245

D. Halupka, S. Huda, W. Song, et al., Negative-resistance read and write schemes for STT-MRAM in 0.13 µm CMOS, *IEEE International Solid-State Circuits Conference (ISSCC) Dig. Tech. Papers* (2010), pp. 256–257

F. Hamzaoglu, U. Arslan, N. Bisnik, et al. A 1 Gb 2 GHz embedded DRAM in 22 nm Tri-Gate CMOS technology, *IEEE International Solid-State Circuits Conference (ISSCC) Dig. Tech. Papers* (2014), pp. 230–231

M. Hatanaka, H. Hidaka, Value creation in SOC/MCU applications by embedded nonvolatile memory evolutions, in *Proceedings of the IEEE Asia Solid-State Circuits Conf. (A-SSCC)* (2007), pp. 38–42

I. Hayashi, T. Amano, N. Watanabe et al., A 250-MHz 18-Mb full ternary CAM with low-voltage matchline sensing scheme in 65-nm CMOS. IEEE J. Solid State Circuits **48**(11), 2671–2680 (2013)

H. Hidaka, Evolution of embedded flash memory technology for MCU, in *IEEE International Conference on IC Design & Technology (ICICDT)* (2011), pp. 1–4

M.-C. Hsieh, Y.-C. Liao, Y.-W. Chin, et al. Ultra high density 3D via RRAM in pure 28 nm CMOS process, *IEEE International Electron Devices Meeting (IEDM) Dig. Tech. Papers* (2013), pp. 10.3.1–10.3.4

K.-C. Huang et al., The impacts pf control gate voltage on the cycling endurance of split gate flash memory. IEEE Electron Device Lett. **21**, 359–361 (2000)

L.Y. Huang, M.F. Chang, C.H. Chuang et al., ReRAM-based 4T2R nonvolatile TCAM with 7x NVM-stress reduction, and 4x improvement in speed-wordlength-capacity for normally-off instant-on filter-based search engines used in big-data processing, *Symposium on VLSI Circuits Dig. Tech. Papers* (2014), pp. 1–2

D.H. Im, J.I. Lee, S.L. Cho, H.G. An, D.H. Kim, I.S. Kim, H. Park, D.H. Ahn, H. Horii, S.O. Park, U-in Chung, J.T. Moon, A unified 7.5 nm dash-type confined cell for high performance PRAM device, *IEEE International Electron Devices Meeting (IEDM) Dig. Tech. Papers* (2008), pp. 1–4

P. Jain, D. Jiao, X. Wang, C.H. Kim, Measurement, analysis and improvement of supply noise in 3D ICs, *Symposium on VLSI Circuits Dig. Tech. Papers* (2011), pp. 46–47

J. Javanifard, T. Tanadi, H. Giduturi, et al., A 45 nm self-aligned-contact process 1 Gb NOR flash with 5 MB/s program speed, *IEEE International Solid-State Circuits Conference (ISSCC) Dig. Tech. Papers* (2008), pp. 424–624

M. Jefremow, T. Kern, U. Backhausen, et al., Bitline-capacitance-cancelation sensing scheme with 11 ns read latency and maximum read throughput of 2.9 GB/s in 65 nm embedded flash for automotive, *IEEE International Solid-State Circuits Conference (ISSCC) Dig. Tech. Papers* (2012), pp. 428–430

M. Jefremow, T. Kern, W. Allers, et al., Time-differential sense amplifier for sub-80 mV bitline voltage embedded STT-MRAM in 40 nm CMOS, *IEEE International Solid-State Circuits Conference (ISSCC) Dig. Tech. Papers* (2013), pp. 216–217

S. Jeloka, N. Akesh, D. Sylvester, et al., A configurable TCAM/BCAM/SRAM using 28 nm push-rule 6T bit cell, *Symposium on VLSI Circuits* (2015), pp. C272–C273

S.H. Jo, T. Kumar, S. Narayanan, W.D. Lu, H. Nazarian, 3D-stackable crossbar resistive memory based on Field Assisted Superlinear Threshold (FAST) selector, *IEEE International Electron Devices Meeting (IEDM) Dig. Tech. Papers* (2014), pp. 6.7.1–6.7.4

M. Kamiya, Y. Kojima, Y. Kato, K. Tanaka, Y. Hayashi, EPROM CellWith High Gate Injection Efficiency, *IEEE International Electron Devices Meeting (IEDM) Dig. Tech. Papers* (1982), pp. 741–744

A. Kawahara, R. Azuma, Y. Ikeda, et al. An 8 Mb multi-layered cross-point ReRAM macro with 443 MB/s write throughput, *IEEE International Solid-State Circuits Conference (ISSCC) Dig. Tech. Papers* (2012), pp. 432–434

M.D. Ker, S.L. Chen, C.S. Tsai et al., Design of charge pump circuit with consideration of gate-oxide reliability in low-voltage CMOS processes. IEEE J. Solid State Circuits **41**(5), 1100–1107 (2006)

S. Kianian, A. Levi, D. Lee, Y. W. Hu, A novel 3 volts-only, small sector erase, high density flash E PROM, *Symp. VLSI Technology Dig. Tech. Papers* (1994), pp. 71–72

T.H. Kim, J. Liu, C.H. Kim, A voltage scalable 0.26 V, 64 kb 8T SRAM with V_{min} lowering techniques and deep sleep mode. IEEE J. Solid State Circuits **44**(6), 1785–1795 (2009)

W.J. Kim, J.H. Jeong, Y. Kim, et al., Extended scalability of perpendicular STT-MRAM towards sub-20 nm MTJ node, in *IEEE International Electron Devices Meeting (IEDM) Dig. Tech. Papers* (2011), pp. 24.1.1–24.1.4

E. Kitagawa, S. Fujita, K. Nomura, et al., Impact of ultra-low power and fast write operation of advanced perpendicular MTJ on power reduction for high-performance mobile CPU, *IEEE International Electron Devices Meeting (IEDM) Dig. Tech. Papers* (2012), pp. 29.4.1–29.4.4

K. Kobayashi, T. Nakayama, Y. Miyawaki et al., A high-speed parallel sensing architecture for multi-megabit flash E²PROMs. IEEE J. Solid State Circuits **25**(1), 79–83 (1990)

T. Kono, T. Ito, T. Tsuruda, T. Nishiyama, T. Nagasawa, T. Ogawa, Y. Kawashima, H. Hidaka, T. Yamauchi, 40-nm embedded split-gate MONOS (SG-MONOS) flash macros for automotive with 160-MHz random access for code and endurance over 10 M cycles for data at the junction temperature of 170 °C. IEEE J. Solid State Circuits **49**, 154–166 (2013)

T. Kono, T. Ito, T. Tsuruda et al., 40-nm embedded Split-Gate MONOS (SG-MONOS) flash macros for automotive with 160-MHz random access for code and endurance over 10 M cycles for data at the junction temperature of 170 C. IEEE J. Solid State Circuits **49**(1), 154–166 (2014)

A. Kotov, Three generations of Embedded SuperFlash split gate cell: scaling progress and challenges, *Leti Innovation Days–Memory Workshop* (2013)

S. Koveshnikov, K. Matthews, K. Min, et al. Real-time study of switching kinetics in integrated 1T/ HfOx1R RRAM: intrinsic tunability of set/reset voltage and trade-off with switching time, *IEEE International Electron Devices Meeting (IEDM) Dig. Tech. Papers* (2012), pp. 20.4.1–20.4.3

C. Kuo, D. Chrudimsky, T. Jew, C. Gallun, J. Choy, B. Wang, S. Pessoney, A 32-Bit RISC microcontroller with 448 K bytes of embedded flash memory, *Int. NonVolatile Memory Technol. Conference* (1998), pp. 28–33

M. Kuriyama, S. Atsumi, A. Umezawa, et al., A 5 V-only 0.6 µm flash EEPROM with row decoder scheme in triple-well structure, *IEEE International Solid-State Circuits Conference (ISSCC) Dig. Tech. Papers* (1992), pp. 152–153

S. Lai, Current status of the phase change memory and its future, *IEEE International Electron Devices Meeting (IEDM) Dig. Tech. Papers* (2003), pp. 10.1.1–10.1.4

P. Larsson, Measurements and analysis of PLL jitter caused by digital switching noise, *IEEE Journal of Solid-State Circuits* (2001), pp. 1113–1119

P. Larsson, Measurements and analysis of PLL jitter caused by digital switching noise, *IEEE J. Solid-State Circuits* (2001), pp. 1113–1119

C. Lauterbach, W. Weber, D. Romer et al., Charge sharing concept and new clocking scheme for power efficiency and electromagnetic emission improvement of boosted charge pumps. IEEE J. Solid State Circuits 35 (5), 719–723 (2000)

B.Q. Le, M. Achter, C.G. Chng et al., Virtual-ground sensing techniques for a 49-ns/200-MHz access time 1.8-V 256-Mb 2-bit-per-cell flash memory. IEEE J. Solid State Circuits 39(11), 2014–2023 (2004)

J.Y. Lee, S.E. Kim, S.J. Song et al., A regulated charge pump with small ripple voltage and fast start-up. IEEE J. Solid State Circuits 41(2), 425–432 (2006a)

H.-Y. Lee, Y.-S. Chen, P.-S. Chen, et al., Comprehensively study of read disturb immunity and optimal read scheme for high speed HfOx based RRAM with a Ti layer, in *Proceedings of the IEEE Symposium on VLSI Technology, Systems and Applications (VLSI-TSA)* (2010), pp. 132–133

A. Lee, M.F. Chang, C.C. Lin, et al., RRAM-based 7T1R nonvolatile SRAM with 2x reduction in store energy and 94x reduction in restore energy for frequent-off instant-on applications, *Symposium on VLSI Circuits (VLSI Circuits)* (2015), pp. C76–C77

J. Li, R. Montoye, M. Ishii, et al., 1 Mb 0.41 μm^2 2T-2R cell nonvolatile TCAM with two-bit encoding and clocked self-referenced sensing, *Symposium on VLSI Technology (VLSIT)* (2013), pp. C104–C105

J. Li, R.K. Montoye, M. Ishii, L. Chang, 1 Mb 0.41 μm^2 2T-2R cell nonvolatile TCAM with two-bit encoding and clocked self-referenced sensing. IEEE J. Solid State Circuits 49, 896–907 (2014)

W. Liu, K.T. Chang, C. Cavins, B. Luderman, C. Swift, K.M. Chang, B. Morton, G. Espinor, S. Ledford, A 2-Transistor Source-Select (2TS) flash EEPROM for 1.8 V-Only applications, *Non-Volatile Semiconductor Memory Worshop* (1997), pp.4.1.1–4.1.3

Y.C. Liu, M.F. Chang, Y.F. Lin, et al., An embedded flash macro with sub-4 ns random-read-access using asymmetric-voltage-biased current-mode sensing scheme, in *Proceedings of the IEEE Asian Solid-State Circuits Conference (A-SSCC)* (2013), pp. 241–244

T.Y. Liu, T.H. Yan, R. Scheuerlein, et al., A 130.7 mm^2 2-layer 32 Gb ReRAM memory device in 24 nm technology, *IEEE International Solid-State Circuits Conference (ISSCC) Dig. Tech. Papers* (2013), pp. 210–211

S. Lütkemeier, U. Ruckert et al., A subthreshold to above-threshold level shifter comprising a wilson current mirror. IEEE Trans. Circuits Syst. II: Exp. Brief. 57 (9), 721–724 (2010)

G.G. Marotta, A. Macerola, A. d'Alessandro, et al., A 3 bit/cell 32 Gb NAND flash memory at 34 nm with 6 MB/s program throughput and with dynamic 2 b/cell blocks configuration mode for a program throughput increase up to 13 MB/s, *IEEE International Solid-State Circuits Conference (ISSCC) Dig. Tech. Papers* (2010), pp. 444–445

S. Matsunaga, S. Miura, H. Honjou, et al., A 3.14 μm^2 4T-2MTJ-cell fully parallel TCAM based on nonvolatile logic-in-memory architecture, *Symposium on VLSI Circuits (VLSIC)* (2012), pp. 44–45

A.V. Mezhiba, E.G. Friedman, Scaling trends of on-chip power distribution noise. IEEE Trans. VLSI Syst 12 (4), 386–394 (2004)

R. Micheloni, L. Crippa, M. Sangalli et al., The flash memory read path: building blocks and critical aspects. IEEE Proc. 91(4), 537–553 (2003)

R. Mih et al., 0.18 m modular triple self-aligned embedded split-gate flash memory, in *Symp. VLSI Technology Dig. Tech. Papers*, 2000, pp. 120–121

Y. Morita, H. Fujiwara, H. Noguchi, et al., An area-conscious low-voltage-oriented 8T-SRAM design under DVS environment, *Symposium on VLSI Circuits Dig. Tech. Papers* (2007), pp. 256–257

F. Nardi, S. Balatti, S. Larentis, D. Ielmini, Complementary switching in metal oxides: Toward diode-less crossbar RRAMs, *IEEE International Electron Devices Meeting (IEDM) Dig. Tech. Papers* (2011), pp. 31.1.1–31.1.4

K. Nii, K. Yamaguchi, M. Yabuuchi, et al., Silicon measurements of characteristics for passgate/pull-down/pull-up MOSs and search MOS in a 28 nm HKMG TCAM bitcell, in *Proceedings of the International Conference on Microelectronic Test Structures (ICMTS)* (2015), pp. 200–203

T. Ogura, M. Hosoda, T. Ogawa et al., A 1.8-V 256-Mb multilevel cell nor flash memory with BGO function. IEEE J. Solid State Circuits 41(11), 2589–2600 (2006)

T.-Y. Oh, H. Chung, Y.-C. Cho, et al., A 3.2 Gb/s/pin 8 Gb 1.0 V LPDDR4 SDRAM with integrated ECC engine for sub-1 V DRAM core operation, in *IEEE International Solid-State Circuits Conference (ISSCC) Dig. Tech. Papers* (2014), pp. 430–431

T. Ohsawa, H. Koike, S. Miura et al., A 1 Mb nonvolatile embedded memory using 4T2MTJ cell with 32 b fine-grained power gating scheme. IEEE J. Solid State Circuits 48(6), 1511–1520 (2013)

S.R. Ovshinsky, Reversible electrical switching phenomena in disordered structure. Phys. Rev. Lett. 21(20), 1450–1455 (1968)

G. Palumbo, D. Pappalardo, Charge pump circuits: an overview on design strategies and topologies. IEEE Circuits Syst. Mag. 10(1), 31–45 (2009)

A. Pirovano, A.L. Lacaita, A. Benvenuti, F. Pellizzer, R. Bez, Electronic switching in phase-change memories. IEEE Trans. Electron Dev. 51(3), 452–459 (2004)

H. Pozidis, N. Papandreou, A. Sebastian, et al., Reliable MLC data storage and retention in phase-change memory after endurance cycling, in *Proceedings of the IEEE International Memory Workshop (IMW)* (2013), pp. 100–103

M. Rizzi, N. Ciocchini, S. Caravati, et al., Statistics of set transition in phase change memory (PCM) arrays, *IEEE International Electron Devices Meeting (IEDM) Dig. Tech. Papers* (2014)

S.K. Saha, Design considerations for sub-90-nm split-gate flash-memory cells. IEEE Trans. Electron Dev. **54**, 465–473 (2007)

Y. Sakotsubo, M. Terai, S. Kotsuji, et al., A new approach for improving operating margin of unipolar ReRAM using local minimum of reset voltage, *IEEE Symposium on VLSI Technology (VLSIT) Dig. Tech. Papers* (2010), pp. 87–88

A. Serb, R. Berdan, A. Khiat, C. Papavassiliou, T. Prodromakis, Live demonstration: a versatile, low-cost platform for testing large ReRAM cross-bar arrays, in *Proceedings of the International Symposium on Circuits and Systems (ISCAS)* (2014), pp. 441

S.S. Sheu, C.C. Kuo, M.F. Chang, et al., A ReRAM integrated 7T2R non-volatile SRAM for normally-off computing application, in *Proceedings of the IEEE Asian Solid-State Circuits Conference (A-SSCC)* (2013), pp. 245–248

Y.H. Shih, J.Y. Wu, B. Rajendran, M.H. Lee, R. Cheek, M. Lamorey, M. Breitwisch, Y. Zhu, E.K. Lai, C.F. Chen, E. Stinzianni, A. Schrott, E. Joseph, R. Dasaka, S. Raoux, H.L. Lung, C. Lam, Mechanisms of retention loss in Ge2Sb2Te5-based phase-change memory, *IEEE Electron Devices Meeting (IEDM) Dig. Tech. Papers* (2008), pp. 1–4

T. Song, W. Rim, J. Jung, et al., A 14 nm FinFET 128 Mb 6T SRAM with VMIN enhancement techniques for low-power applications, *IEEE International Solid-State Circuits Conference (ISSCC) Dig. Tech. Papers* (2014), pp. 232–233

Y. Taito, M. Nakano, H. Okimoto, et al., 7.3 A 28 nm embedded SG-MONOS flash macro for automotive achieving 200 MHz read operation and 2.0 MB/S write throughput at Ti, of 170 °C, *IEEE International Solid-State Circuits Conference (ISSCC) Dig. Tech. Papers* (2015), pp. 1–3

K. Takahashi, H. Doi, N. Tamura, K. Mimuro, T. Hashizume, Y. Moriyama, Y. Okuda, A 0.9 V operation 2-transistor flash memory for embedded logic LSIs, *Symp. VLSI Technology Dig. Tech. Ppaers* (1999), pp. 21–22

K. Takeuchi, Y. Kameda, S. Fujimura et al., A 56-nm CMOS 99-mm2 8-Gb Multi-Level NAND Flash Memory With 10-MB/s Program Throughput. IEEE J. Solid State Circuits **42**(1), 219–232 (2007)

K.-T. Tang, S.-W. Chiu, C.-H. Shih, et al., A 0.5 V 1.27 mW nose-on-a-chip for rapid diagnosis of ventilator-associated pneumonia, *IEEE International Solid-State Circuits Conference (ISSCC)* (2014), pp. 1–2, pp. 420–421

J. Tsouhlarakis, G. Vanhorebeek, G. Vehoeven et al., A flash memory technology with quasi-virtual ground array for low-cost embedded applications. IEEE J. Solid State Circuits **36**(6), 969–978 (2001)

K. Tsunoda, M. Aoki, H. Noshiro, et al. Highly manufacturable multi-level perpendicular MTJ with a single top-pinned layer and multiple barrier/free layers, in *IEEE International Electron Devices Meeting (IEDM) Dig. Tech. Papers* (2013), pp. 3.3.1–3.3.4

J. Van Houdt, P. Heremans, L. Deferns, G. Groeseneken, H.E. Maes, Analysis of the enhanced hot-electron injection in split-gate transistors useful for EEPROM applications. IEEE Trans. Electron Dev. **39**, 1150–1156 (1992)

N. Verma, A.P. Chandrakasan, A 256 kb 65 nm 8T subthreshold SRAM employing sense-amplifier redundancy. IEEE J. Solid State Circuits **43**(1), 141–149 (2008)

Y.-H. Wang, M.-C. Wu, C.-J. Lin et al., An analytical programming model for the draincoupling source-side injection split gate flash EEPROM. IEEE Trans. Electron Dev. **52**, 385–391 (2005)

W. Wang, A. Gibby, Z. Wang, et al., Nonvolatile SRAM Cell, *IEEE International Electron Devices Meeting (IEDM) Dig. Tech. Papers* (2006), pp. 1–4

X.P. Wang, Z. Fang, X. Li, et al., Highly compact 1T-1R architecture (4F2 footprint) involving fully CMOS compatible vertical GAA nano-pillar transistors and oxide-based RRAM cells exhibiting excellent NVM properties and ultra-low power operation, *IEEE International Electron Devices Meeting (IEDM) Dig. Tech. Papers* (2012), pp. 20.6.1–20.6.4

Y.-H. Wang, S.-H. Huang, D.-Y. Wang, et al., Impact of stray field on the switching properties of perpendicular MTJ for scaled MRAM, *IEEE International Electron Devices Meeting (IEDM) Dig. Tech. Papers* (2012), pp. 29.2.1–29.2.4

Y. Wang, Y. Liu, S. Li, D. Zhang, B. Zhao, M.-F. Chiang, Y. Yan, B. Sai, H. Yang, A 3µs wake-up time nonvolatile processor based on ferroelectric flip-flops, in *Proceedings of the European Solid-State Circuits Conference (ESSCIRC)* (2012), pp. 149–152

Y. Wang, Y. Liu, S. Li, X. Sheng, D. Zhang, M.-F. Chiang, B. Sai, X.-S. Hu, H. Yang, PaCC: A parallel compare and compress codec for area reduction in nonvolatile processors, *IEEE Transactions on Very Large Scale Integration (VLSI) Systems* (2014), pp. 1491–1505

O. Wataru, K. Miyata, M. Kitagawa, et al., A 4 Mb conductive-bridge resistive memory with 2.3 GB/s read-throughput and 216 MB/s program-throughput, *IEEE International Solid-State Circuits Conference (ISSCC) Dig. Tech. Papers* (2011), pp. 210–211

W. Wei, K. Namba, J. Han et al., Design of a nonvolatile 7T1R SRAM cell for instant-on operation. IEEE Trans. Nanotechnol. **13**(5), 905–916 (2014)

C.-Y. Wen, J. Li, S. Kim, M. Breitwisch, C. Lam, J. Paramesh, L.T. Pileggi, A non-volatile look-up table design using PCM (phase-change memory) cells, *Symposium on VLSI Circuits (VLSIC) Dig. Tech. Papers* (2011), pp. 302–303

J.S. Witters, G. Groeseneken, H.E. Maes, Analysis and modeling of on-chip high-voltage generator circuit for use in EEPROM circuits. IEEE J. Solid State Circuits **24**, 1372–1380 (1989)

S.N. Wooters, B.H. Calhoun, T.N. Blalock et al., An energy-efficient subthreshold level converter in 130-nm CMOS. IEEE Trans. Circuits Syst. II: Exp. Brief. **57**(4), 290–294 (2010)

J.Y. Wu, M. Breitwisch, S. Kim, T.H. Hsu, R. Cheek, P.Y. Du, J. Li, E.K. Lai, Y. Zhu, T.Y. Wang, H.Y. Cheng, A. Schrott, E.A. Joseph, R. Dasaka, S. Raoux, M.H. Lee, H.L. Lung, C. Lam, A low power phase change memory using thermally confined TaN/TiN bottom electrode, *IEEE International Electron Devices Meeting (IEDM) Dig. Tech. Papers* (2011), pp. 3.2.1–3.2.4

J.Y. Wu, W.S. Khwa, M.H. Lee, H.P. Li, S.C. Lai, T.H. Su, M.L. Wei, T.Y. Wang, M. BrightSky, T.S. Chen, W.C. Chien, S. Kim, R. Cheek, H.Y. Cheng, E.K. Lai, Y. Zhu, H.L. Lung, C. Lam, Greater than 2-bits/cell MLC storage for ultra high density phase change memory using a novel sensing scheme, *Symposium on VLSI Technology (VLSI Technology)* (2015), pp. T94–T95

X.Y. Xue, W.X. Jian, J.G. Yang et al., A 0.13 µm 8 Mb Logic-Based Cu_xSi_yO ReRAM With Self-Adaptive Operation for Yield Enhancement and Power Reduction. IEEE J. Solid State Circuits **48**(5), 1315–1322 (2013)

M. Yabuuchi, Y. Tsukamoto, M. Morimoto, et al., 13.3 20 nm high-density single-port and dual-port SRAMs with wordline-voltage-adjustment system for read/write assists, *IEEE International Solid-State Circuits Conference (ISSCC) Digest of Technical Papers* (2014), pp. 234–235

S. Yamamoto, Y. Shuto, S. Sugahara, Nonvolatile SRAM (NV-SRAM) using functional MOSFET merged with resistive switching devices, in *Proceedings of the IEEE Custom Integrated Circuits Conference (CICC)* (2009), pp. 531–534

T. Yamauchi, Prospect of embedded non-volatile memory in the smart society, in *Proceeding of the IEEE International Symposium on VLSI Technology, Systems and Application (VLSI-TSA)* (2015), pp. 1–2

T. Yamauchi, H. Kondo, K. Nii, Automotive low power technology for IoT society, *Symposium on VLSI Technology (VLSIT), Dig. Tech. Papers* (2015), pp. T80–T81

J.J. Yang et al., Engineering nonlinearity into memristors for passive crossbar applications. Appl. Phys. Lett. **100**, 113501 (2012)

Y. Yano, Take the expressway to go greener, in *IEEE International Solid-State Circuits Conference (ISSCC) Dig. Tech. Papers* (2012), pp. 24–30

J.A. Yater, S.T. Kang, R. Steimle, C.M. Hong, B. Winstead, M. Herrick, G. Chindalore, Optimization of 90 nm split gate nanocrystal non-volatile memory, in *Proceedings of the Non-Volatile Semiconductor Memory Workshop* (2007), pp. 77–78

H.C. Yu, K.F. Lin, K.C. Lin, et al., A 180 MHz direct access read 4.6 Mb embedded flash in 90 nm technology operating under wide range power supply from 2.1 V to 3.6 V, in *Proceedings of the IEEE International Symposium on VLSI Design, Automation, and Test (VLSI-DAT)* (2013), pp. 1–4

B. Zhai, S. Hanson, D. Blaauw et al., A variation-tolerant sub-200 mV 6-T subthreshold SRAM. IEEE J. Solid State Circuits **43**(10), 2338–2348 (2008)

K. Zhang, *Embedded memories for nano-scale VLSIs* (Springer, New York, 2009)

L. Zhang, B. Govoreanu, B. Redolfi, et al., High-drive current (>1 MA/cm^2) and highly nonlinear ($>10^3$) TiN/Amorphous-Silicon/TiN scalable bidirectional selector with excellent reliability and its variability impact on the 1S1R array performance, *IEEE International Electron Devices Meeting (IEDM) Dig. Tech. Papers* (2014), pp. 6.8.1–6.8.4

M. Zwerg, A. Baumann, R. Kuhn, M. Arnold, R. Nerlich, An 82 µA/MHz microcontroller with embedded FeRAM for energy-harvesting applications, *IEEE International Solid-State Circuits Conference (ISSCC) Dig. Tech. Papers* (2011), pp. 334–336

On-Chip Non-volatile STT-MRAM for Zero-Standby Power

7

Xuanyao Fong and Kaushik Roy

In this Chapter, we present spin-transfer torque magnetic random access memory (STT-MRAM) suitable for IoT applications. Its ability to operate at low supply voltages, non-volatility, good endurance, and small bit-cell footprint are especially attractive for IoT applications in which low energy consumption is crucial. We will present the fundamentals of STT-MRAM. The design of the STT-MRAM storage device, memory bit-cell and memory array architecture are also discussed to highlight the benefits STT-MRAM brings to IoT applications, as well as the design issues that need to be considered. We then present a device/circuit/ architecture co-design approach for STT-MRAM. Finally, we will discuss the trends in STT-MRAM and give some perspectives on the future of STT-MRAM design.

random access for IoT applications (Yamauchi et al. 2015). However, STT-MRAM is especially attractive as compared to ReRAM and eFlash due to its relative ease of integration into the back-end-of-line (BEOL) in the CMOS fabrication process, ability to operate at <1.2 V supply voltages, <100 ns read and write delays, good endurance ($>10^{14}$ cycles) and bit-cell footprint as small as 40 F^2 (F is the smallest feature size of the CMOS technology) (ITRS Roadmap 2014). In this chapter, we discuss the modeling, design, and optimization of STT-MRAMs and present some of the potential benefits in relation to IoT applications. As we will see later in this chapter, a highly desirable trait of STT-MRAM for IoT applications is that they only need to be powered when they are being accessed, which results in zero-standby power.

7.1 Introduction

As discussed in Chap. 6, several non-volatile memory technologies such as embedded flash (eFlash) and resistive RAM (ReRAM) are available for implementing memories with truly

7.2 The Magnetic Tunnel Junction and STT-MRAM

The storage device in STT-MRAM is the magnetic tunnel junction or MTJ. The structure of a typical MTJ with in-plane magnetic anisotropy is shown in Fig. 7.1. The MTJ may be visualized as a stack consisting of two nano-magnets sandwiching a tunneling oxide barrier (usually AlO_x or more commonly MgO). One nano-magnet is a soft ferromagnetic layer used to store the information (also called the "*free*"

X. Fong (✉)
National University of Singapore, Singapore 119077, Singapore
e-mail: elefongx@nus.edu.sg

K. Roy
Purdue University, West Lafayette, IN 47907, USA

© Springer International Publishing AG 2017
M. Alioto (ed.), *Enabling the Internet of Things*, DOI 10.1007/978-3-319-51482-6_7

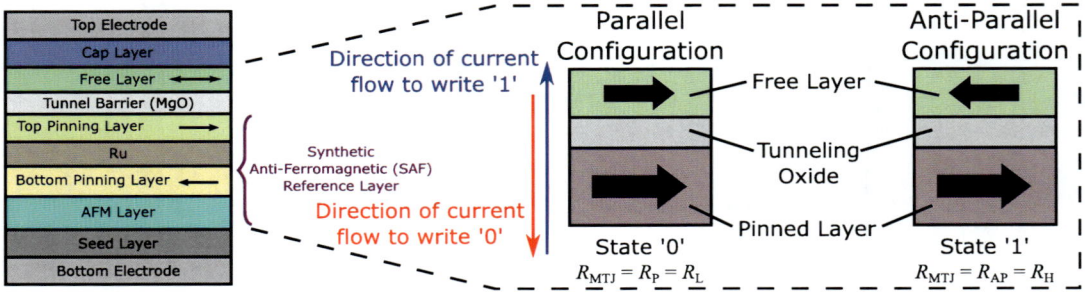

Fig. 7.1 The storage device in the STT-MRAM is the magnetic tunnel junction (MTJ). The typical stack structure of an MTJ with in-plane anisotropy is shown. It is easier to understand the operation of an MTJ by only considering the simple tri-layer stack illustrating the parallel and anti-parallel MTJ configurations. The directions of programming current flow through the bit-cell are shown using colored arrows whereas the black arrows indicate possible magnetization directions of magnetic layers in the MTJ

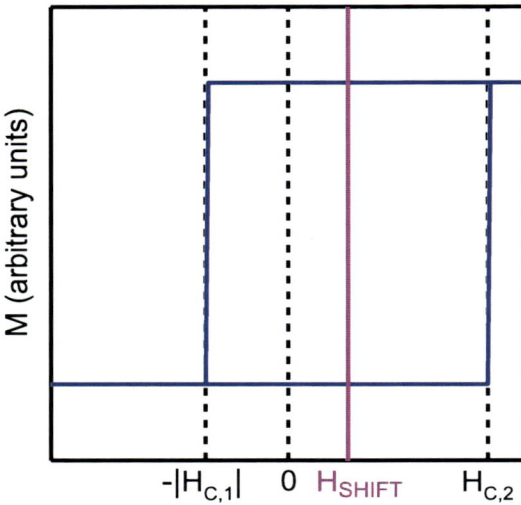

Fig. 7.2 The *M-H* loop of an MTJ without SAF based pinned layer exhibits a shifted hysteresis loop as illustrated here, which corresponds to a degraded retention time

$H_{\text{SHIFT}} = H_{\text{SHIFT}} - |H_{\text{C},1}|)$. However, the thermal stability and retention time of the MTJ is determined by $min\,(|H_{\text{C},1}|, |H_{\text{C},2}|)$. Thus, non-zero H_{SHIFT} reduces the thermal stability and retention time of the MTJ.

An example of a SAF layer is shown in Fig. 7.1. The anti-ferromagnetic (AFM) layer pins the magnetization of the bottom pinning layer via exchange bias effect (Nogués et al. 2005). The top pinning layer is anti-ferromagnetically coupled to the bottom pinning layer via interlayer exchange coupling with a non-magnetic spacer such as *Ru* (Parkin et al. 1990). Exchange coupling increases the magnetic field needed to switch the magnetizations of the coupled pinning layers and effectively pins the magnetizations of both ferromagnetic layers. This allows the magnetization of the top pinning layer to be used as a reference.

The MTJ is designed to be switchable between two stable states. When the magnetization directions of both the free and the pinned layers point in the same direction, the MTJ configuration is called the "*parallel*" state (P). When instead the magnetization directions of the free layer and the pinned layer point in opposite directions, the MTJ configuration is called the "*anti-parallel*" state (AP). An important metric for the MTJ is its resistance-area (*RA*) product (Huai 2008). The *RA* product of the MTJ varies exponentially with the thickness of the tunneling oxide barrier (t_{MgO}) since the mechanism for electron transport is tunneling. The MTJ

layer) whereas the other nano-magnet is a hard ferromagnetic layer for use as a reference layer (also called the "*fixed*" or "*pinned*" layer).

In many MTJ stacks, a synthetic anti-ferromagnetic (SAF) layer is used for the pinned layer to reduce the stray magnetic field. The stray magnetic field may shift the *M-H* loop of the MTJ as illustrated in Fig. 7.2. The *M-H* loop (*M*: magnetization, *H*: applied magnetic field) exhibits a horizontal shift, H_{SHIFT}. $H_{\text{C},1}$ and $H_{\text{C},2}$ are symmetric about H_{SHIFT} (i.e., $H_{\text{C},2} -$

resistance, R_{MTJ}, depends linearly on the cross-sectional area of the MTJ (A_{MTJ}) similar to an Ohmic conductor when t_{MgO} is constant. R_{MTJ} also depends on the relative magnetic polarization of the free layer with respect to the pinned layer. The dependence of R_{MTJ} on magnetic polarization arises due to the difference in density of states around the Fermi energy, E_F, in the ferromagnetic layers (Datta et al. 2012). When the MTJ is in the P configuration, the density of states of like-spins around E_F is very high in the ferromagnetic layers. Conversely, the density of states of like-spins around E_F in the ferromagnetic layers is very low when the MTJ is in AP configuration. Thus, R_{MTJ} is low in the P configuration ($R_{MTJ} = R_P = R_L$) and high in the AP configuration ($R_{MTJ} = R_{AP} = R_H$). This difference in R_{MTJ}, termed the "*tunneling magnetoresistance ratio*" (or *TMR*), is quantified as

$$TMR = \frac{R_{AP} - R_P}{R_P} \times 100\% \qquad (7.1)$$

and is an important metric for the performance of MTJs as memory elements. A larger *TMR* also means that MTJ states can be distinguished more easily. Binary data may then be represented and stored as the resistance state of the MTJ.

The magnetic layers in an MTJ, which may be considered as nano-magnets, are engineered with anisotropy energies to satisfy thermal stability and data retention requirements. The required energy barrier height (E_B, in Joules) and retention time (τ, in seconds) must satisfy:

$$\Delta = \frac{E_B}{k_B T} > ln\left(\frac{m\tau}{\tau_0 \, ln \, 2}\right) \qquad (7.2)$$

where k_B is the Boltzmann constant, T is the temperature in Kelvin, m is the number of bits in the memory and τ_0 is the characteristic time in seconds ($\tau_0 \approx 1$ ns). $\Delta = 40.66$ corresponds to a retention time of about 10 years for $m = 1$. In real STT-MRAM arrays, $\Delta > 70$ may be required (Naeimi et al. 2013). The minimum magnetic field for switching a nano-magnet, H_k, and the energy barrier are related by (Sun 2000):

$$H_k = \frac{2E_B}{\mu_0 M_{sat} V_{FL}} \qquad (7.3)$$

where μ_0 is the permeability of vacuum, and M_{sat} and V_{FL} are the saturation magnetization and volume of the nano-magnet, respectively. E_B of the pinned layer in the MTJ is engineered to be much larger than that of the free layer in the MTJ so that the pinned layer magnetization direction is fixed. As such, only the magnetization direction of the free layer can change during operation. The most common form of anisotropy engineered into the free layer of an MTJ is the *uniaxial anisotropy*. This causes the magnetization of the magnetic layers to have a preferential alignment axis—the magnetization will align along this axis when no external stimulus is present. The *uniaxial anisotropy energy density*, K_{u2}, and E_B of the nano-magnet are related by

$$K_{u2} V_{FL} = E_B \qquad (7.4)$$

Hence, $K_{u2} V_{FL}$ must be kept constant to maintain the same thermal stability when the volume of the nano-magnet is reduced. We will discuss this in more detail in the later sections. In the presence of a stray magnetic field that shifts the *M-H* loop of the MTJ by H_{SHIFT} (as in Fig. 7.2), the effective barrier height becomes

$$E_B = 0.5\mu_0 M_{sat} V_{FL}(H_k - H_{SHIFT}) \qquad (7.5)$$

Hence, the free layer of the MTJ needs to be engineered with a larger K_{u2} to compensate for the barrier height degradation due to H_{SHIFT}.

7.2.1 Spin-Transfer Torque

Nano-scale MTJs may be switched using the spin-transfer torque (STT) phenomenon shown in Fig. 7.3a, which was theoretically predicted by Slonczewski and Berger independently in 1996 (Berger 1996; Slonczewski 1996). Since then many experiments have observed STT switching (Huai et al. 2004; Katine et al. 2000; Myers 1999). STT arises due to the spin property of electrons. When the majority of electron spins in a nano-magnet are aligned in a particular direction, the magnetization of the nano-magnet also points in that direction. This is illustrated by the different density of states for different electron

(a)

(b)

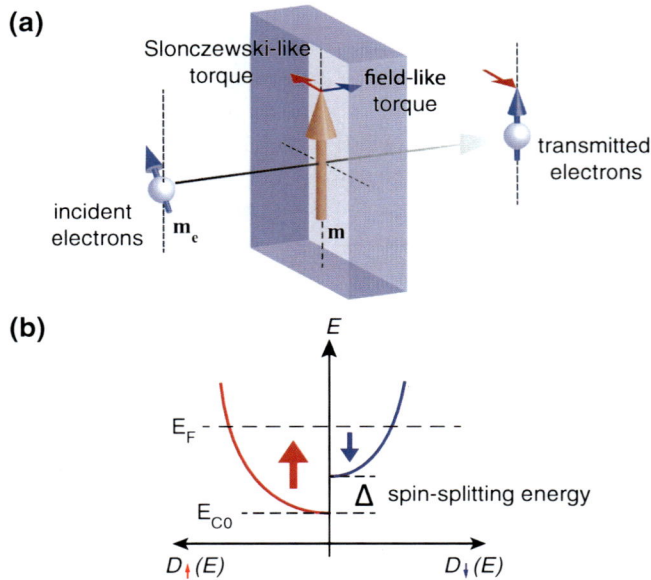

Fig. 7.3 (**a**) When an electron carries spin polarization that is non-collinear with the magnetization of the nano-magnet it is incident on, it exchanges spin angular momentum with the nano-magnet. Consequently, spin-transfer torque (decomposed into Slonczewski-like and field-like components) is exerted on the magnetization of the nano-magnet to align it into the spin polarization direction of the electron. This occurs because magnetism is a result of unequal spin density of states in the nano-magnet as illustrated in (**b**)—the flow of electrons with other spin directions perturb the total spin populations in the nano-magnet, which manifests as a torque on the magnetization of the nano-magnet

spins in Fig. 7.3b. When an electron is incident on a nano-magnet, it experiences an exchange field—the same field that aligns the spin directions of all electrons in the ferromagnet—if its spin polarization direction is non-collinear with the magnetization direction of the nano-magnet. The exchange field exerts a torque on the spin polarization of the electron, aligning it with the magnetization direction of the nano-magnet. Hence, when current flows through the MTJ, the ferromagnetic layers also act as spin filters that polarize the spin direction of electrons constituting the current flow. Whereas the exchange field exerts torque on the spin polarization of the electron, due to conservation of spin angular momentum, an equal and opposite torque (spin-transfer torque) is exerted on the magnetization of the nano-magnet to align it with the spin polarization direction of the electron. Hence, the electrons in a spin-polarized current (i.e., a current in which the majority of electrons are spin polarized in a particular direction) transfer their spin momentum to the nano-magnet and in the process exerts a torque on the magnetization of the nano-magnet that tries to align the magnetization direction of the nano-magnet with the spin polarization direction of the spin-polarized current. The magnetization direction of the nano-magnet may be switched if the torque exerted is large enough to overcome all other energies in the nano-magnet. Furthermore, the rate of spin momentum transfer and the torque exerted are proportional to the rate of electron flow or the current, and determine the switching time. The current or current density needed to achieve a specific switching time is the critical current, I_C, or critical current density, J_C. It is often easier to analyze the MTJ in terms of the *intrinsic switching current density*, J_{C0}, given as (Huai 2008; Sun 2000)

$$J_{C0} = \frac{2e\alpha M_{sat} t_{FL} H_{eff}}{\hbar \eta} \qquad (7.6)$$

Here, e is the elementary charge, α and t_{FL} are the Gilbert damping constant and thickness of the free layer, respectively. $\hbar$ is the reduced Planck constant, η is the spin polarization efficiency factor, and H_{eff} is the effective magnetic field spin-transfer torque must overcome when switching the free layer magnetization.

In an MTJ, it is easier for spin-transfer torque to switch the free layer than to switch the pinned layer because the E_B of the pinned layer is much higher than that of the free layer. Let us consider what happens when electrons are flowing from the pinned layer to the free layer in an MTJ. The pinned layer spin-polarizes the incoming electrons which then flow into the free layer. These electrons are spin-polarized in the direction of the pinned layer magnetization and transfer their spin momentum to the free layer. The spin-transfer torque exerted on the free layer tries to align the free layer magnetization with that of the pinned layer (i.e., MTJ is switched into the P configuration). Consider when electrons flow from the free layer to the pinned layer instead. Electrons entering the free layer from the non-magnetic metal interconnect are not spin-polarized and can have any spin direction. Electrons spin-polarized in direction of the pinned layer magnetization are able to tunnel across the oxide easily. Those with the opposite spin polarization may not tunnel across the oxide easily and accumulate in the free layer. These electrons transfer their spin angular momentum to the free layer and exert a torque that aligns the direction of free layer magnetization opposite to that of the pinned layer (i.e., MTJ is being switched into the AP configuration). Consequently, the spin directions of the electrons become aligned with the magnetization direction of the pinned layer and they may then easily tunnel across the oxide. From this discussion, we can see that the process of parallelizing the MTJ configuration is more efficient than that for anti-parallelizing the MTJ configuration (i.e., I_C and J_C are asymmetric and depends on switching direction (Datta et al. 2012; Ikeda et al. 2010). It has been reported that J_C for anti-parallelizing the MTJ can be 10%–200% larger than for parallelizing the MTJ (Ikeda et al. 2010; Kishi et al. 2008).

As just mentioned, spin-transfer torque, τ_{STT}, is only exerted when the spin polarization direction, $\boldsymbol{p}$, of the electrons incident on the nano-magnet is non-collinear with the magnetization direction of the nano-magnet, $\boldsymbol{m}$. It can be shown that (Salahuddin et al. 2008)

$$\tau_{STT} = \beta[\varepsilon(\mathbf{m} \times \mathbf{p} \times \mathbf{m}) + \varepsilon'(\mathbf{p} \times \mathbf{m})]$$
$$(7.7)$$

where β is a scalar factor proportional to the current flowing through the MTJ. ε and ε' are scalar factors proportional to the strength of Slonczewski-like and field-like torques, respectively. In an MTJ, the spin polarization direction of the electrons is pointing in the magnetization direction of the pinned layer or that of the free layer. The magnetization directions of the two magnetic layers are collinear to maximize the TMR and the distinguishability between the P and AP configurations of the MTJ. Thus, τ_{STT} should be negligible. However, thermal effects perturb the magnetization of the nano-magnets in the MTJ, causing them to be non-collinear and τ_{STT} can be large enough to overcome all other energies in the free layer of the MTJ. Hence, spin-transfer torque switching of the MTJ is a stochastic process since thermal effects are random in nature. The thermal effect may be modeled as a fluctuating magnetic field written as (Brown 1963)

$$\boldsymbol{H}_{EFF} = \xi \sqrt{\frac{\alpha k_B T}{\gamma \mu_0 M_{sat} V_{FL} \delta t}} \qquad (7.8)$$

$\boldsymbol{\xi}$ is a 3-vector whose components are independent Gaussian random variables with zero mean and unit variance. γ is the gyromagnetic ratio

Fig. 7.4 This graph illustrates an example of the pulse width, τ, dependence of J_C needed to program an MTJ with 0.5 success probability. The spin torque generated at low current densities is still able to switch the MTJ state due to heating of the MTJ, which increases the thermal effects

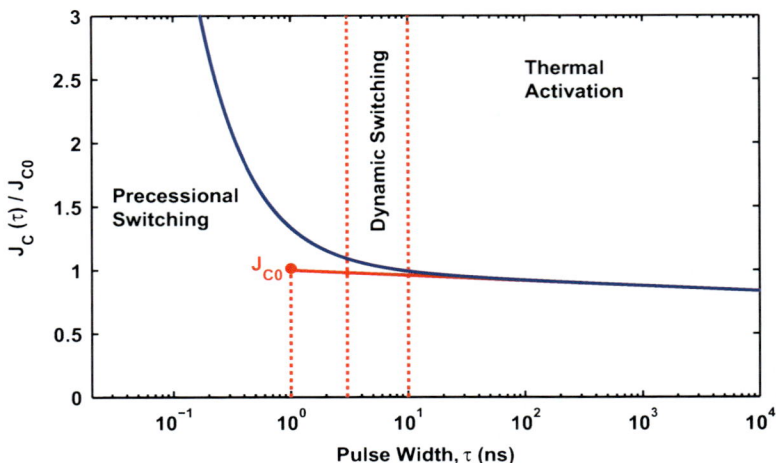

(17.6 MHz/Oe) and δt is the constant time step used in numerical simulation of the MTJ dynamics.

Also, it was found that J_C depends on the pulse width, τ (Huai 2008). An example of this is illustrated in Fig. 7.4. Three switching regimes can be observed: the *precessional*, *thermal* and *dynamic* regimes. In the precessional regime, τ_{STT} completely overcomes all other energies in the free layer to switch the MTJ. Thermal effects only affect the magnetization direction of the free layer in the MTJ just prior to onset of J_C. Thus, the dependence of programming failure on J_C is determined by the distribution of free layer magnetization direction just prior to applying the switching current pulse. In the thermal regime, τ_{STT} alone is unable to overcome all other energies in the free layer. However, thermal effects, which are also increased due to Joule heating by the current flowing through the MTJ, assists τ_{STT} in switching the MTJ configuration. The dependence of programming failure on J_C in the thermal regime is determined by the random time needed for thermal effects to sufficiently reduce the effective barrier height such that τ_{STT} can switch the MTJ configuration. In the dynamic regime, the dependence of programming failure on J_C is determined by both the distribution of free layer magnetization prior to applying J_C and the random time needed for

thermal effects to sufficiently reduce the effective barrier height.

7.2.2 Integrating MTJ with CMOS Technology

When designing STT-MRAM arrays, the fabrication steps need to be understood in order to understand the impact of design choices on the characteristics of the MTJ, the impact of the bit-cell topology on the bit-cell footprint, layout of the memory array in terms of area overhead, and performance and energy overheads due to increased parasitics. Figure 7.5 illustrates one method of integrating MTJs into the back end of the CMOS fabrication process (back-end-of-line, BEOL) developed by Qualcomm (Kang et al. 2014). The MTJs are placed in between metal layers that form the interconnects of the integrated circuit (IC). After the chemical mechanical polishing (CMP) step (step 1) to expose the metal layer on which the MTJs are to be placed, the layers constituting the MTJ stack are deposited (steps 2 and 3). The MTJ stack consists of many thin layers including those needed to form the top and bottom electrodes (TE and BE, respectively), the seed layer for growing high quality magnetic thin films, the magnetic layers needed to form the

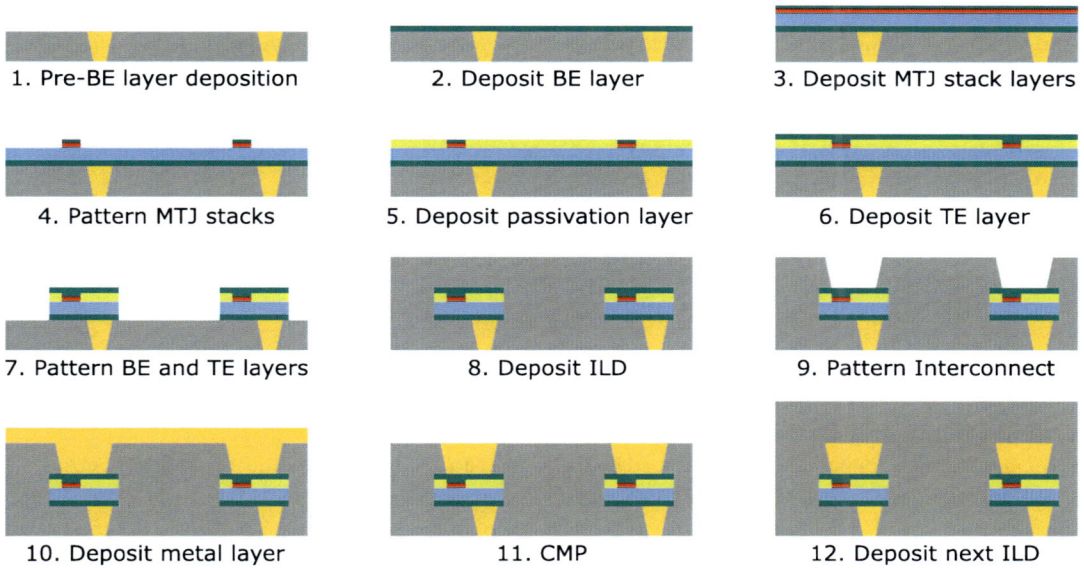

Fig. 7.5 An example flow for fabricating magnetic tunnel junctions in the back-end-of-line (BEOL) of the CMOS fabrication process. BE: bottom electrode, TE: top electrode, ILD: interlayer dielectric, CMP: chemical mechanical polishing

pinned and free layers of the MTJs, and also the tunnel barrier. In Fig. 7.5, the seed layer and layers constituting the pinned layer of the MTJ are deposited on the bottom electrode (BE) before the tunneling oxide layer. The first lithographic step is then applied to pattern the MTJ pillars using the BE as the etch stop layer (step 4). Since the thin film layers constituting the MTJ stack are very sensitive to particle contaminants, a dielectric passivation layer is immediately deposited after MTJ patterning (step 5) to protect them. Particle contamination can be detrimental to MTJ characteristics. A CMP step is performed to expose the capping layer of the MTJs after the dielectric passivation layer is deposited. Thereafter, the TE layer is deposited to contact the capping layer of the MTJs (step 6). The TE and BE of individual MTJ pillars are then defined by lithography and etch (step 7). The next interlayer dielectric (ILD) layer is then deposited (step 8). Lithography and etching are then performed to define the next layer of interconnect (step 9). This layer also serves as an electrical connection to the MTJs. Note that the material for the TE layer have been chosen so that it may be used as an etch stop. The

next metal layer is then deposited (step 10) and a CMP step (step 11) is performed to complete the definition of the metal interconnect layer. The next ILD layer is then deposited (step 12) and the rest of the BEOL fabrication process continues.

7.2.3   Impact of STT-MRAM Integration on IC Design

The STT-MRAM array may be fabricated together with other CMOS circuits in an system-on-chip (SoC) and hence the placement of the MTJ layers needs to account for the interconnect wiring in the rest of the silicon wafer. For example, increasing the separation between the lower metal layers to accommodate the MTJ may increase parasitics in the interconnects in other parts of the IC, which can negatively affect the overall performance of the IC. The MTJs may be formed between the higher interconnect metal layers (such as between the last $1\times$ metal layer and the first $2\times$ interconnect metal layer) where the separation between layers across the IC is large enough accommodate the MTJs. As a result, the minimum pitch between MTJs and hence the integration density of STT-MRAM, may be limited by the minimum pitch between metal interconnects. Furthermore, the via parasitics between the MTJ and the transistors below may be significantly increased if the MTJ is placed too high in the interconnect layers. Hence, process for integrating MTJs may become an important factor not only in determining the characteristics of the MTJ, but also in the overall performance of the IC.

The order of stack deposition may also impact the footprint of the STT-MRAM bit-cell, as we shall see in the next section. The MTJ pinned layer stack may be deposited first followed by the MgO and then the free layer stack. Consider if the pinned layer of the MTJ is connected to the transistor layer below as shown in Fig. 7.6a. This may be achieved by placing the MTJ on top of a stack of vias. If the free layer is to be connected to the transistor below instead as Fig. 7.6b illustrates, the metal layer on top of the TE needs to be extended to one side and then connected to the transistor below through a stack of vias. The additional area needed to accommodate this extension may increase the

footprint of the STT-MRAM bit-cell that requires such an MTJ connection. An alternative scheme is to swap the order of pinned layer and free layer deposition but the impact on the MTJ characteristics depends on the fabrication process.

7.3   Design of the STT-MRAM Bit-Cell

Figure 7.7 shows the topology of the basic one-transistor one-MTJ (1T-1M) STT-MRAM bit-cell. One electrode of the MTJ is connected to the *bit line* (BL) while the other electrode is connected to the access transistor (ATx). ATx is also connected to the *source line* (SL) as shown, and the *word line* (WL) is connected to the gate of ATx. WL is used to turn ATx ON and OFF. The 1T-1M STT-MRAM bit-cells can have two configurations (Kishi et al. 2008; Lin et al. 2009): the "standard" connection (SC) as shown in Fig. 7.7a, and the "reversed" connection (RC) shown in Fig. 7.7b. The bit-cell is accessed for read and for write operations by charging WL to V_{DD} to turn ATx ON. Read operation may then be performed by sensing the resistance between BL and SL. Write operations are performed by setting the voltages on BL and SL to the values shown in Fig. 7.7. Let us consider the SC configuration for example. A bit-cell having MTJ in P configuration stores '0' whereas a bit-cell having MTJ in AP configuration stores '1'. The bit-cell is programmed with '0' by applying V_{DD} and *GND* to BL and SL, respectively. The electrons constituting the write current flow from the pinned layer into the free layer of the MTJ to parallelize the MTJ configuration. A '1' is programmed into the bit-cell by applying V_{DD} and *GND* to SL and BL, respectively. The electrons flow from the free layer to the pinned layer in this configuration, and exert torque that anti-parallelizes the MTJ configuration. Note that the size of ATx, the value of V_{DD}, and the write current pulse width are all designed such that the current flowing through MTJ during write operations is larger than I_C. Compared to the SC bit-cell configuration, the connections of

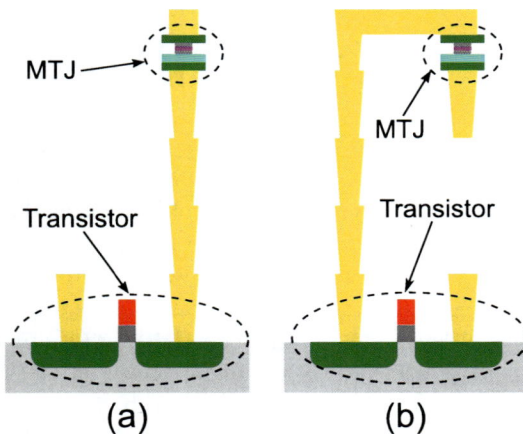

Fig. 7.6 The bottom pinned MTJ may have its pinned layer directly connected to the transistor below as shown in (**a**). If the free layer is to be connected instead and the order of layer deposition cannot be altered, the metal layer above TE has to be extended, as shown in (**b**), and connected to the transistor below through a stack of vias

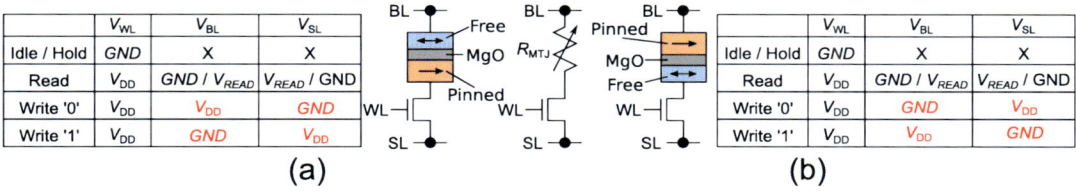

(a) standard

	V_{WL}	V_{BL}	V_{SL}
Idle / Hold	GND	X	X
Read	V_{DD}	GND / V_{READ}	V_{READ} / GND
Write '0'	V_{DD}	V_{DD}	GND
Write '1'	V_{DD}	GND	V_{DD}

(b) reversed

	V_{WL}	V_{BL}	V_{SL}
Idle / Hold	GND	X	X
Read	V_{DD}	GND / V_{READ}	V_{READ} / GND
Write '0'	V_{DD}	GND	V_{DD}
Write '1'	V_{DD}	V_{DD}	GND

Fig. 7.7 The topology of the 1T-1M STT-MRAM bit-cell is shown. The (**a**) "standard" and (**b**) "reversed" connection differs in the way the MTJ is connected to the access transistor. The voltages on the control lines of the bit-cell corresponding to various operations of each bit-cell configuration are shown in the tables

the MTJ are swapped in the RC bit-cell configuration. Hence, the voltages on BL and SL for write operations of the RC bit-cell configuration are also swapped as compared to the SC bit-cell configuration.

During read operations, a sense amplifier is used to sense the MTJ state in the STT-MRAM bit-cell through BL or through SL. Also, a constant voltage or constant current scheme may be used to sense R_{MTJ} (Dorrance et al. 2011; Fong et al. 2012). In the constant voltage scheme, a fixed voltage, V_{RD}, is applied across the bit line and the source line of the STT-MRAM bit-cell and the resulting current flowing through the MTJ, I_{MTJ}, is compared to a reference current, I_{REF}. I_{MTJ} can be either higher or lower than I_{REF}, depending on the resistance state of the MTJ. The advantage of the constant voltage scheme is that I_{MTJ} during read operations may be amplified in the sense amplifier to improve sensing speed. The disadvantage is that the result of the sensing needs to be converted into an output voltage. In the constant current scheme, a fixed current, I_{RD} is passed through the MTJ and the voltage developed across the bit line and the source line, V_{BC}, is compared with a reference voltage, V_{REF}. The constant current scheme has the advantage that the result of the sensing is the voltage domain does not need to be converted. Furthermore, since the MTJ is programmed by passing current through it as we will see later, I_{RD} may be limited to prevent accidental programming of the MTJ during read operations or read-disturb failures. However, the $|V_{BC}\text{-}V_{REF}|$ signal generated by I_{RD} may be too small to be easily sensed.

7.3.1 Design Issues of the 1T-1M STT-MRAM Bit-Cell

Let us now discuss the major design issues of 1T-1M STT-MRAM bit-cells. The source degeneration of ATx during write operations, shared read and write current paths, and single-ended sensing scheme are the three major issues in 1T-1M STT-MRAM bit-cell. As we shall see, these design issues have conflicting design requirements, which constrain the design space of 1T-1M STT-MRAM bit-cell. If not carefully taken care of, these design issues may lead to excessive energy consumption, and degraded STT-MRAM performance and reliability.

7.3.1.1 Source Degeneration of ATx

In order to program the 1T-1M STT-MRAM bit-cell, bi-directional write current flow is required. Figure 7.8 shows the voltages on the circuit nodes of the bit-cell during write operations. When current flows from BL to SL as illustrated in Fig. 7.8a, the overdrive voltage of ATx (V_{GS}) is at V_{DD}. When the direction of current flow is reversed as shown in Fig. 7.8b, the voltage on the source terminal of ATx is at $GND < V_S < V_{DD}$. As such, $V_{GS} < V_{DD}$ and the drive strength of ATx is weakened. Furthermore, the asymmetry in ATx drive strength may be exacerbated by the asymmetry in I_C of the MTJ. The size of ATx may need to be enlarged to ensure successful write operations when passing current from SL to BL. This increases the write current supplied when write current is flowing from BL to SL, which increases energy consumption and may degrade the reliability of the MTJ tunneling oxide barrier. The reliability of

Fig. 7.8 (a) The overdrive voltage, V_{GS}, of the access transistor is at V_{DD} when write current flows from BL to SL. (b) $V_{GS} < V_{DD}$ when write current flows from SL to BL, and the drive strength of the access transistor is reduced

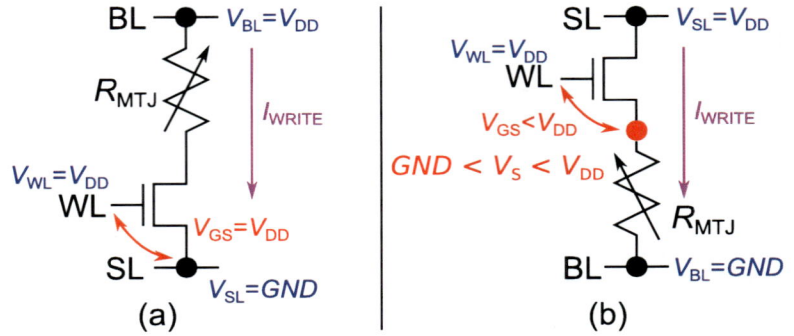

7.3.1.2 Shared Read and Write Current Paths

During write operations, current needs to be passed through the MTJ to exert STT so as to switch the MTJ configuration. During read operations, current flows through the MTJ regardless of whether the sensing scheme used is the constant voltage scheme or constant current scheme. Hence, the current flowing through the MTJ needs to be limited to avoid *disturb failures*. A disturb failure occurs when the MTJ is accidentally programmed during a read operation. Since the circuit nodes in the bit-cell need to be charged or discharged during sensing operations, limiting the amount of current flow through the bit-cell limits the rate at which these circuit nodes are charged or discharged. As a consequence, the sensing speed is reduced. Furthermore, limiting the amount of current flowing through the bit-cell during sensing may also reduce the signal margin available for the sense amplifier to determine the MTJ state. As a result, the sense amplifier may not be able to reliably distinguish between R_P and R_{AP}.

Note that it may be more difficult to limit the current flowing through the MTJ during read operation if the ATx is enlarged to mitigate the source degeneration problem discussed previously. The conflicting requirements for mitigating source degeneration and disturb

failure may significantly constrain the design space. For example, E_B may be increased to reduce disturb failures. Doing so increases I_C, resulting in increased write currents. Hence, design choices must be made carefully to achieve an optimum 1T-1M STT-MRAM design.

7.3.1.3 Single-Ended Sensing Scheme

The sensing scheme for 1T-1M STT-MRAM bit-cells described in the earlier sections uses *single-ended sensing*—the voltage across or current through the bit-cell is compared with a common reference during sensing. Under process variations, the resistance of MTJs will deviate from the nominal value as depicted by the scatter plot in Fig. 7.9. Some bit-cells having MTJ in P configuration may have read current smaller than the reference current, or read voltage larger than the reference voltage. Also, some bit-cells having MTJ in AP configuration may have read current larger than the reference current, or read voltage smaller than the reference voltage. The sense amplifier will sense the bit-cells as having MTJ in AP configuration in the former case whereas the bit-cells are sensed as having MTJ in P configuration in the latter case. This is also called *decision failure*.

7.3.2 Scalability of STT-MRAM

When designing STT-MRAM, it is crucial for STT-MRAM designers to understand how the size of the access transistor and the MTJ impact

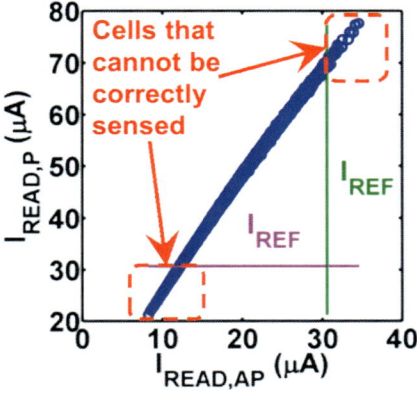

Fig. 7.9 Under process variations, the read currents through the 1T-1M STT-MRAM bit-cells may be distributed as this example plot shows. Each data point on the plot is the read current through a bit-cell when its MTJ is in the P configuration and the AP configuration. The bit-cells falling to the right of and below I_{REF} will not be correctly sensed by the sense amplifier

the characteristics of the STT-MRAM bit-cell. Furthermore, it is desirable to scale down the size of the STT-MRAM bit-cell to achieve high bit density and reduce the cost per bit. The size of the MTJ may need to be varied to achieve this while also ensuring that its resistance is not too large for the access transistor to provide adequate write current.

MTJs may be engineered with two flavors of magnetic anisotropy for satisfying thermal stability and retention time requirements—perpendicular and in-plane magnetic anisotropy (PMA and IMA, respectively). The magnetization of the magnetic layers in the MTJ with IMA points within the plane of the wafer on which the MTJ is deposited. On the other hand, the magnetization of the magnetic layers in the MTJ with PMA points perpendicular to the plane of the wafer. In both flavors of MTJs, the minimum magnetic field to switch the MTJ configuration, H_{k}, and the energy barrier, E_{B}, are related by (Sun 2000):

$$H_{\mathrm{k}} = \frac{2E_{\mathrm{B}}}{\mu_0 M_{\mathrm{sat}} V_{\mathrm{FL}}} \qquad (7.9)$$

where μ_0 is the permeability of vacuum, and M_{sat} and V_{FL} are the saturation magnetization and volume of the nano-magnet, respectively.

For MTJs with IMA, the shape of the nano-magnets are engineered to achieve the required energy barrier height (Devolder 2011). In these nano-magnets, the width is defined by the limits of the process technology. The length and thickness of the nano-magnet are varied to engineer its energy barrier. However, the aspect ratio ($AR =$ length/width) is usually in the range of 2–3 to keep the MTJ footprint small and for manufacturability (Apalkov et al. 2010). The thickness of the nano-magnet is then varied to meet thermal stability requirements. On the other hand, the nano-magnets in MTJs with PMA have other anisotropies, such as interfacial perpendicular anisotropy (Ikeda et al. 2010) or magneto-crystalline anisotropy (Apalkov et al. 2010), that overcome the shape anisotropy. The nano-magnets in MTJs with PMA may be modeled as having only uniaxial anisotropy and

$$H_k = \frac{2K_{u2}}{\mu_0 M_{\mathrm{sat}}} \qquad (7.10)$$

where K_{u2} is the effective uniaxial anisotropy constant.

Scaling analysis of the footprint of IMA and PMA based MTJs conclude that IMA based MTJ has excellent scalability down to 10–20 nm widths for thermal stability factor, $\Delta = 60$ (Apalkov et al. 2010; Devolder 2011). Below this critical width, the geometry of the free layer causes the anisotropy energies in the nano-magnet to favor a PMA configuration. Also, the critical width is increased if the required thermal stability factor is larger. This has motivated a shift in research focus toward PMA based MTJs.

Another crucial characteristic of the PMA based MTJ is that its I_{C0} is constant under size scaling (Apalkov et al. 2010). Thus, the voltage required to switch the MTJ, V_{C0}, scales as

$$V_{\mathrm{C0}} = \frac{4 \times RA \times I_{\mathrm{C0}}}{\pi w^2} \qquad (7.11)$$

where a cylindrical MTJ with diameter w is assumed. The supply voltage required is expected to scale down together with the CMOS technology node (Roadmap 2014), which is important for IoT and other applications requiring low energy consumption. Hence, the *RA* product of the MTJ needs to be scaled down (by reducing the thickness of the tunneling oxide barrier) so that the access transistor is able to supply sufficient write current to the MTJ. However, it was shown that the *TMR* of the MTJ may be substantially degraded when its *RA* product is reduced (Yuasa et al. 2004).

Another important consideration when scaling down the *RA* product of the MTJ is the available voltage signal for the sense amplifier to distinguish between a bit-cell storing an MTJ with P configuration from one storing an MTJ with AP configuration. Let us now derive a minimum magneto-resistance (MR) ratio of the bit-cell needed to meet sensing requirements when the resistance of the access transistor is neglected. The inclusion of the resistance of the access transistor will increase the MR ratio we calculate. We will assume a constant current sensing scheme so as to limit read-disturb failures. Under this scheme, the current being passed through the bit-cell during sensing is limited to a fraction of I_{C0}. The voltage developed across the bit-cell is compared to a reference voltage by a sense amplifier to determine the configuration of the MTJ. Ordinarily, the difference in voltage that the sense amplifier sees needs to be at least 50 mV. This means the difference in voltage due to the difference in MTJ resistance is $\delta V \geq 0.1$. Hence, for a cylindrical MTJ with diameter w, the MR ratio has to satisfy:

$$\text{MR ratio} \geq \frac{0.1(0.25\pi w^2)}{\kappa I_{C0} RA} \quad (7.12)$$

κI_{C0} is the current passed through the bit-cell to sense the MTJ configuration. For a cylindrical MTJ with 50 nm diameter and $\kappa I_{C0} = 5$ μA, the MR ratio needs to be larger than 300% for $RA \leq 13$ Ω-μm². Fortunately, the required MR ratio decreases rapidly when the MTJ diameter is scaled down. For $w \leq 30$ nm and $RA \geq 6$ Ω-μm², the required MR ratio is less than 250%. Note that in the preceding analysis, we have assumed that I_{C0} does not scale down with w.

An alternative scheme that relaxes the requirement to scale the *RA* product of the MTJ is to scale down I_{C0} as the footprint of the MTJ is scaled down. The options for doing so, from analyzing Eq. (7.6), are to: (1) decrease the damping factor, α, (2) decrease M_{sat}, or (3) to improve the polarization efficiency factor, η. Reducing M_{sat} may require a change of material for the free layer of the MTJ, whereas increasing η also requires careful optimization of the interface between the free layer and tunneling oxide barrier in the MTJ. It was found recently that α and the interfacial magnetic anisotropy energy density may be optimized in a PMA based MTJ having an FeB free layer sandwiched between two MgO based tunneling oxide barriers (Tsunegi et al. 2014). Reducing α is promising because both the randomness of STT switching and I_{C0} are reduced. The magnitude of the random thermal field depends on $\sqrt{\alpha}$ (see Eq. (7.8)) and hence, reducing α also reduces the stochasticity of STT based switching in MTJs. Thus, recent research works are focused on reducing α. However, the STT-MRAM designer must ensure that the required MR ratio to meet sensing requirements is achievable when I_{C0} is reduced.

7.3.3 Alternative STT-MRAM Bit-Cell Topologies

As we saw in the preceding sections, one of the most challenging aspect of STT-MRAM design is the conflicting design requirements for read operations and for write operations—improving write operation in the 1T-1M STT-MRAM bit-cell requires degrading the read operation and vice versa. However, alternative bit-cell topologies may open new pathways for optimizing the design of STT-MRAM. In the following sections, we will present several alternative STT-MRAM bit-cell topologies that are able to mitigate the conflicting design requirements in STT-MRAM.

7.3.3.1 The 2T-1M STT-MRAM Bit-Cell

The motivation for using the 2T-1M STT-MRAM bit-cell topology may be understood by first analyzing the 1T-1M STT-MRAM bit-cell. The major failure mechanisms in the

1T-1M STT-MRAM bit-cell are decision, disturb, and write failures (Fong et al. 2012). Decision and disturb failures were briefly described in Sect. 7.3.1. *Write failure* occurs when the MTJ configuration is not successfully programmed during the write operation. The optimization methodology proposed in (Fong et al. 2012) sizes the width of the access transistor (ATx) so as to optimize the total failure probability as shown in Fig. 7.10. In this example, the STT-MRAM bit-cell is designed using a commercial 45 nm CMOS technology with MTJs that have 40×100 nm elliptical cross-section area. The read operation may be designed to select the dominant read failure mechanism. For example, disturb failure is the dominant read failure mechanism when the read operation uses a constant voltage sensing scheme that has a large read voltage. In the example shown in Fig. 7.10, the decision failure probability, P_{DECISION}, is minimized when ATx width is 908 nm. However, when the dominant mechanism for read failure is decision failure, the ATx width needs to be increased to reduce the write failure probability (P_{WRITE}) and optimize the total failure

probability. As a result, the minimum P_{DECISION} cannot be achieved.

Alternatively, the design constraint can be relaxed by noting that multi-finger transistors are typically used to implement very wide transistors. Multi-finger transistors are multiple transistors connected in such a way that their gate, source, and drain terminals are shared. When multi-finger transistors are used in the STT-MRAM bit-cell design, the effective access transistor width may be varied using two word lines instead of one as illustrated in Fig. 7.11, which results in the 2T-1M STT-MRAM bit-cell topology (Li et al. 2010). During write operations, both word lines are turned ON and OFF in unison to supply maximum current through the MTJ. During read operation, Word Line 2 keeps M2 OFF whereas Word Line 1 switches M1 ON and OFF. The width of M1 may then be optimized for decision failures (908 nm in our example), while the width of M2 is made as large as required to fulfill write failure, array area, and array capacity requirements. Hence, the P_{WRITE} of the 2T-1MTJ bit-cell may be optimized without worsening P_{DECISION}.

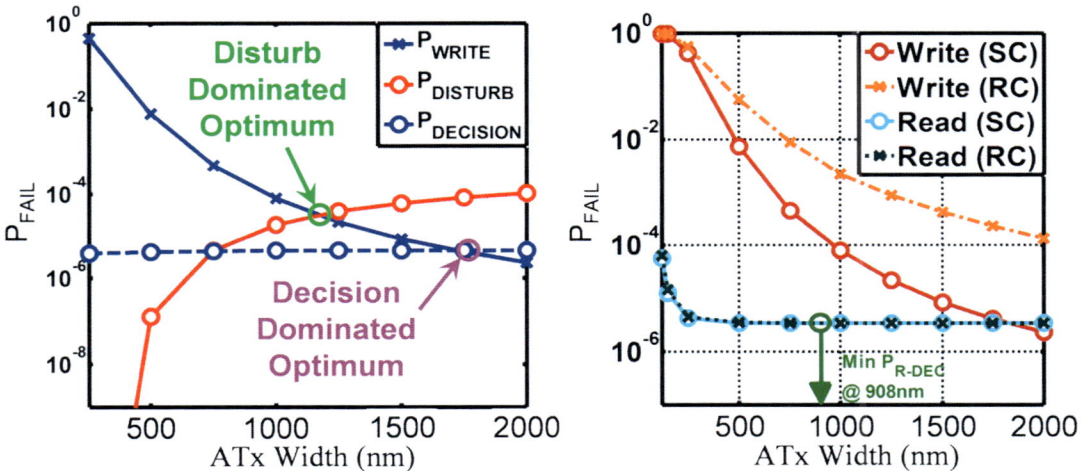

Fig. 7.10 The read-decision, read-disturb, and write failure rates of an example 1T-1M STT-MRAM bit-cell design are shown on the left. The optimum width for the access transistor (ATx) depends on whether the dominant failure mechanism for read operations is decision failure or disturb failure. When the ATx width is small, write failure probability is significantly larger than that for read operations. Based on the optimization methodology proposed in (Fong et al. 2012), the optimum ATx width is 1829 nm, as shown on the right, when decision failure is the dominant read failure mechanism

Fig. 7.11 The schematic of the (**a**) 1T-1M and (**b**) 2T-1M STT-MRAM bit-cells. If the 1T-1M STT-MRAM bit-cell requires an ATx with large width, the ATx may be formed by connecting two ATx's with smaller width in parallel. The 2T-1M STT-MRAM topology is implemented by connecting the gates of the ATx's to different word lines, enabling independent control to each ATx

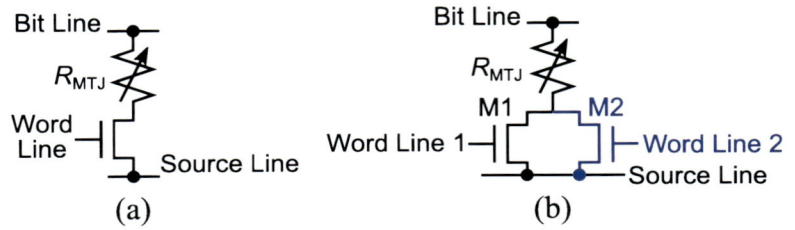

7.3.3.2 Nonvolatile SRAM Using MTJs

The write current requirement of the STT-MRAM bit-cells presented thus far may be relaxed by using a longer write current pulse width. Consequently, the write access delays of these bit-cells may be very long (possibly ≥ 1 μs), and are unsuitable for IoT applications that require faster write access delays. Several alternative STT-MRAM bit-cell structures (see Fig. 7.12) have been proposed in the literature to target IoT applications that require faster access delays (Abe et al. 2004; Ohsawa et al. 2012; Yamamoto and Sugahara 2009). The proposed bit-cell structures are similar to the conventional 6T SRAM bit-cell structure shown in Fig. 7.12. MTJ0 and MTJ1, which store complementary data (i.e., one stores '0' and the other stores '1'), are used to skew the cross-coupled inverters in the SRAM cell to implement a non-volatile SRAM (NV-SRAM) bit-cell. When the NV-SRAM cell is powered off, the MTJs are first programmed with the state of the cell. When the NV-SRAM is powered back on later, the MTJs skew the cross-coupled inverters so that the state the NV-SRAM returns to the state before it was powered off. The advantage of NV-SRAM is that since the SRAM bit-cell topology is preserved, only the array peripheral circuitry for write operations need to be modified to meet write requirements. Furthermore, the differential nature of the bit-cell structure enables fast self-referenced differential read operations for

fast read access delays. Another advantage of NV-SRAM is that only the bit-cells that are being accessed need to be powered ON—the rest may be powered OFF to save on standby power, which is highly desirable for IoT applications.

Although NV-SRAMs are able to achieve short access delays, their restore operations depend on the ability of the MTJs to skew the cross-coupled inverters. Hence, the proposed NV-SRAM bit-cells may be very sensitive to mismatches in the characteristics of the MTJ and the CMOS transistors. Furthermore, the sizes of the transistors in the bit-cells shown in Fig. 7.12 may need to be enlarged to meet write current requirements. The proposed NV-SRAM bit-cells illustrated in Fig. 7.12 also need extra transistors as compared to the 1T-1M STT-MRAM bit-cell. Moreover, both MTJs in the NV-SRAM bit-cells need to be programmed which results in high write energy consumption.

7.4 STT-MRAM Array Architecture and Layout

In the preceding sections, the design of STT-MRAM bit-cells and the MTJ have been discussed. We have seen that at the bit-cell level, it is extremely challenging to design STT-MRAM to meet the performance achievable by 6T SRAMs. The write and read access

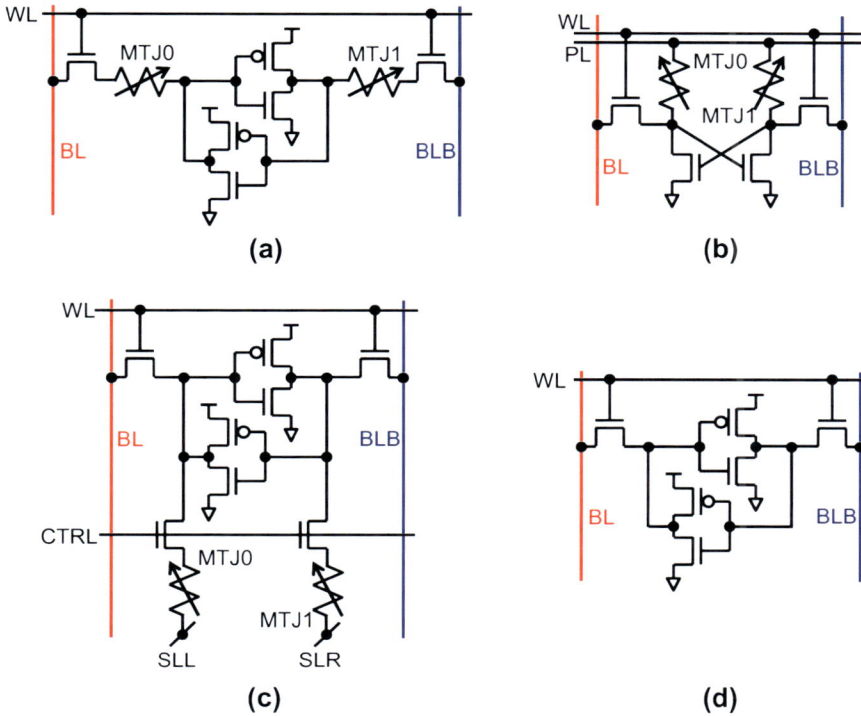

Fig. 7.12 The structures of NV-SRAMs proposed (**a**) in (Abe et al. 2004), (**b**) in (Ohsawa et al. 2012), and (**c**) in (Yamamoto and Sugahara 2009) are shown. The structure of (**d**) a volatile 6T SRAM is also shown for comparison

delays may be limited by large I_C needed and the single-ended nature of the sensing scheme, respectively. Although the STT-MRAM bit-cell does not outperform 6T SRAM, STT-MRAM arrays have several advantages over their 6T SRAM counterparts (Park et al. 2012).

The logical organization of memories stores data words sequentially as shown in Fig. 7.13a. This may result in a memory footprint that is not feasible for physical implementation. Alternatively, several data words are stored in every row as shown in Fig. 7.13b. *Bit-interleaving* (Park et al. 2011) is used to reduce the wiring in the peripheral circuitry that selects the array columns during data access. Note that bit-interleaving is commonly applied to 6T SRAM arrays to mitigate soft errors due to particle strikes. When a data word is being accessed, the word line corresponding to the address of the data word is turned ON. The address of the data

word also selects the columns to be accessed in the array.

The schematic and corresponding layout of an array of 1T-1M STT-MRAM bit-cells are shown in Fig. 7.14a, b, respectively. The bit-cells are arranged into rows and columns where a word line turns on the ATx's of bit-cells connected to the same row. The bit-cells connected along the column stores the data bit of the data word stored in the corresponding row. Due to the nonvolatile nature of STT-MRAM, only the accessed bit-cells need to be powered ON. Power is not supplied to the rest of the bit-cells, which saves leakage power consumption in the array. Note that the STT-MRAM array based on NV-SRAM bit-cells may also be organized in a similar manner so that they may be powered OFF when idle.

Now, let us compare the access operation of a 6T SRAM array to that of the STT-MRAM array. Figure 7.15a illustrates the access operation to a

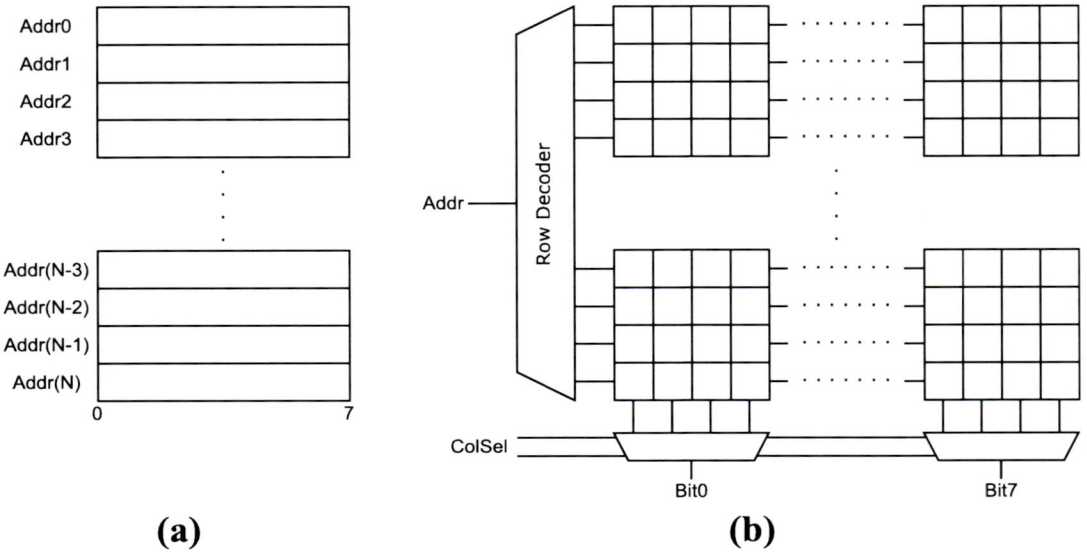

(a)   **(b)**

Fig. 7.13 An example showing the (**a**) logical and (**b**) physical organization of a memory consisting of N 8-bit wide data words (Bit0 to Bit7). Each row stores 4 data words. Bit-interleaving (i.e., the bit-cells storing the same bit position of each data word are located together physically as a group) is used to reduce the wiring to the column selection multiplexers shown in (**b**). These multiplexers are two-way multiplexers that allow Bit0 to Bit7 to behave as input and output

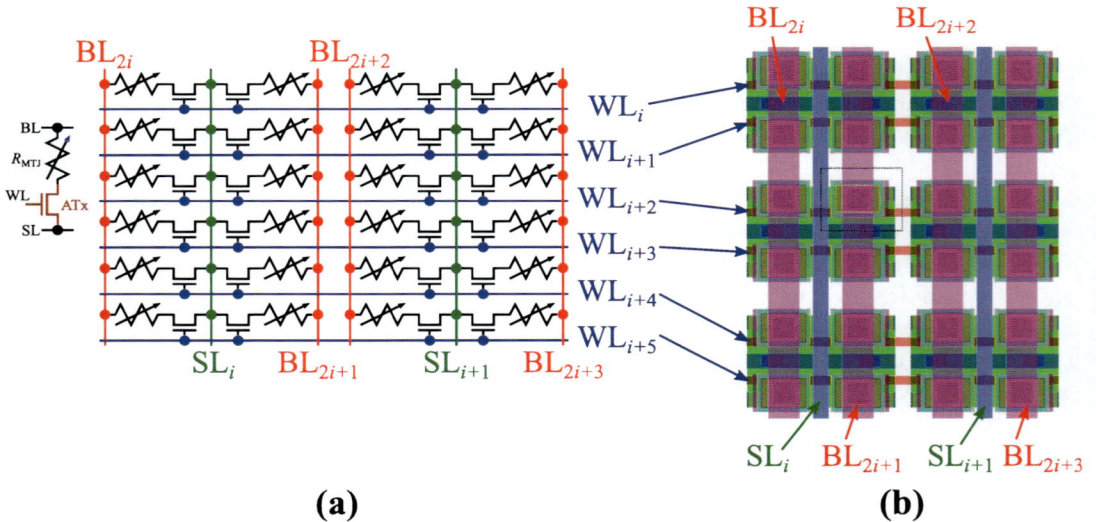

(a)   **(b)**

Fig. 7.14 This figure illustrates the (**a**) schematic and (**b**) array layout of an example array of 1T-1M STT-MRAM bit-cells. An individual 1T-1M STT-MRAM bit-cell is shown on the left. Only 6 rows × 4 columns are shown. Note that every SL is shared between two columns to reduce the total array area. In this layout, the pitch between MTJs limits the lowest achievable layout area

a

b

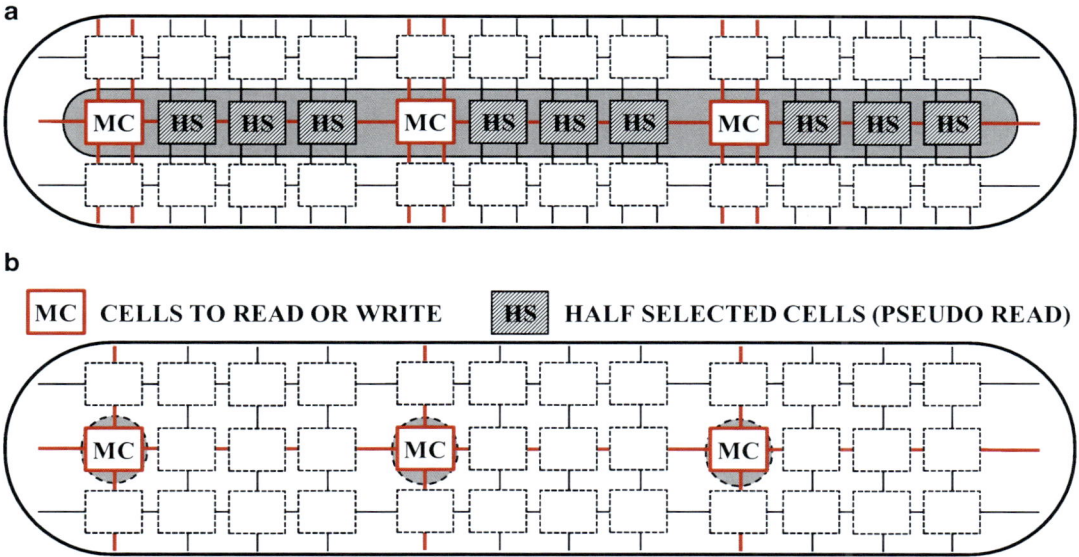

Fig. 7.15 (**a**) 6T SRAM arrays have the half select issue because power must be continuously supplied to all bit-cells to retain data. Thus, unaccessed columns in the selected row need to be biased in the pseudo-read condition to prevent disturb failures. (**b**) Non-volatile STT-MRAM cells do not have the half select issue because power is only supplied to bit-cells that are being accessed

row in a 6T SRAM array. When a row is accessed, only the columns corresponding to the word being selected are accessed. The bit lines connected to the other columns must be precharged to put bit-cells connected to them in the *pseudo-read condition*. Doing so ensures that the data stored in those bit-cells are not accidentally overwritten (i.e., the *half-select issue*). This is required because the SRAM bit-cells need to be continuously powered to retain data. Since STT-MRAM is nonvolatile, power does not need to be supplied to the bit-cells that are not accessed as Fig. 7.15b shows. Hence, the half-select issue present in the 6T SRAM array is absent in the STT-MRAM array. Furthermore, only the peripheral circuitry consumes standby power, leading to significant energy savings that is desirable for IoT application.

Further comparisons show that although the 6T SRAM bit-cell outperforms the 1T-1M STT-MRAM bit-cell, the 1T-1M STT-MRAM based cache may outperform the 6T SRAM based cache (Park et al. 2012). This is illustrated by the cache comparisons reproduced from (Park et al. 2012) in Fig. 7.16. First, the 1T-1M

STT-MRAM based cache may have as much as $3\times$ higher capacity than its 6T SRAM counterpart at the same bit-cell array footprint as shown in Fig. 7.16a. As the cache capacity increases, the read and write latencies of the 6T SRAM based cache increases faster than the STT-MRAM based cache as shown in Fig. 7.16b, c. This is because the access latencies are dominated by the delays needed to charge and discharge parasitic capacitances in the cache. The 1T-1M STT-MRAM has a smaller bit-cell footprint than the 6T SRAM bit-cell. Hence, the parasitic line capacitances increase much more quickly in 6T SRAM cache than the STT-MRAM cache when its capacity is increased. Consequently, the read and write energies of 6T SRAM cache increases faster than their 1T-1M STT-MRAM counterpart as plotted in Fig. 7.16d, e. The increase in read and write energies of 6T SRAM cache with increasing cache capacity is exacerbated by the half-select issue discussed earlier. The biggest advantage of the 1T-1M STT-MRAM based cache over its 6T SRAM counterpart is the lower leakage power consumption graphed in Fig. 7.16f.

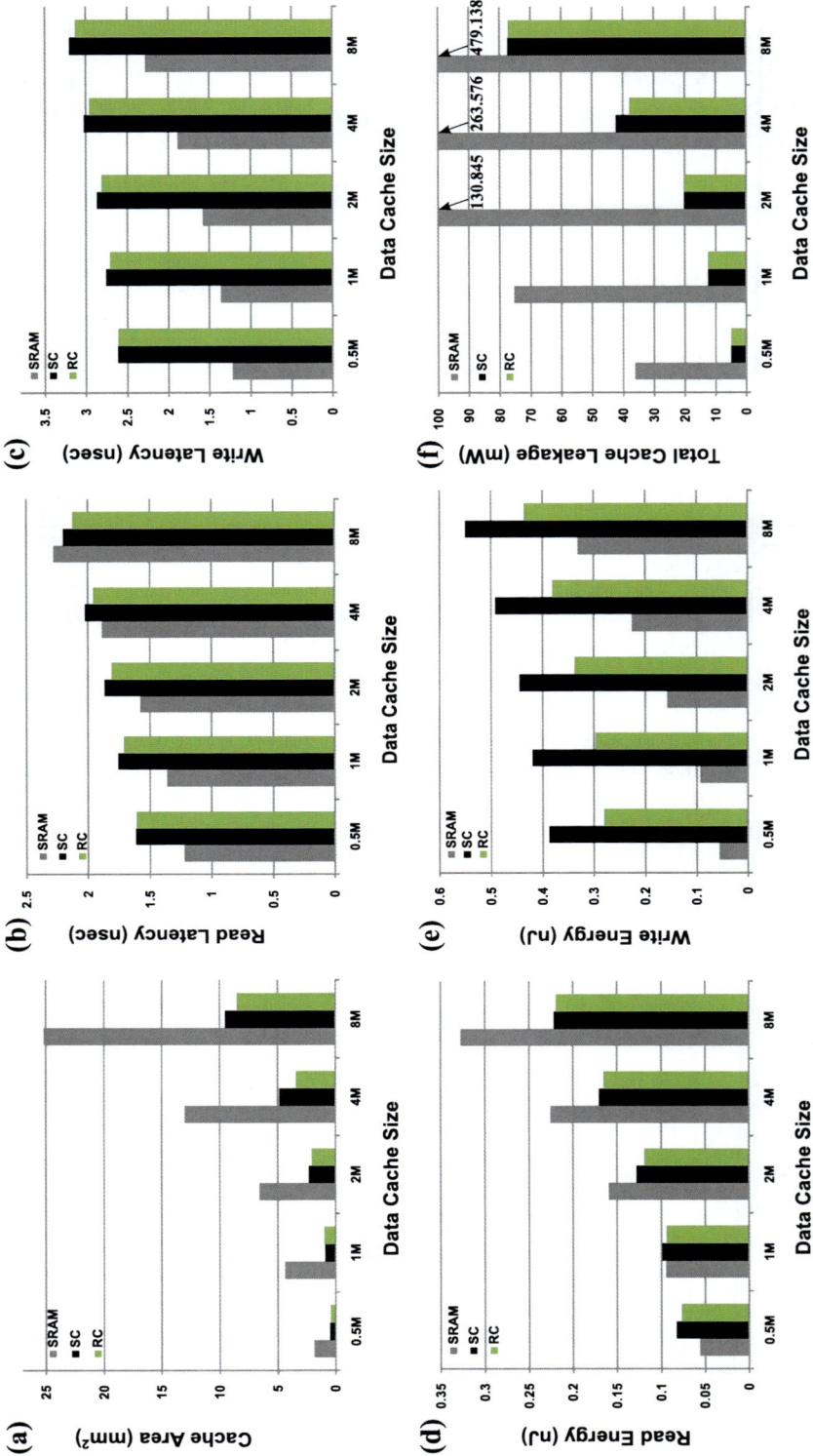

Fig. 7.16 (a) Array area of SRAM and STT-MRAM based data caches (4-way, 64B cache line, B = Byte, M = Mega Byte), (b) read latency and (c) write latency, (d) read energy per operation and (e) write energy per operation, and (f) total leakage power

Fig. 7.17 Total energy consumption versus cache utilization for SRAM and for STT-MRAM based caches for various levels of cache utilization. Note that due to the difference in cache capacity, the cache utilization of the STT-MRAM based caches (SC and RC) is only a quarter of that in the 6T SRAM based cache is the data stored is at most 0.5 MB

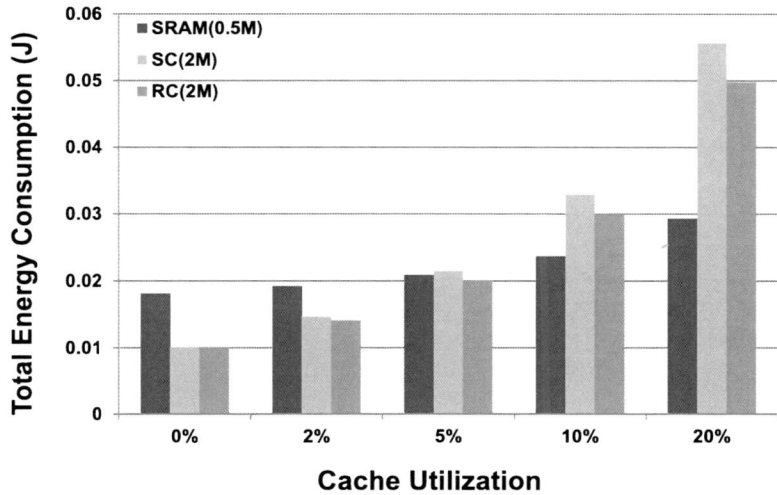

Real world applications affect how the cache is accessed (number of read and write operations, number of cycles during which the cache is idle, etc.) as well as the cache utilization (amount of data used by the program that is stored in the cache as a fraction of the total cache capacity). The comparison between 1T-1M STT-MRAM based cache and 6T SRAM based cache in (Park et al. 2012) analyzes the performance of the caches using SPEC2000 benchmarks that emulate real world applications. The total energy consumption versus cache utilization for caches based on 6T SRAM, standard connection (SC) and reversed connection (RC) 1T-1M STT-MRAM bit-cells is plotted in Fig. 7.17. Consider when the data that needs to be stored in cache is 0.1 MB. This corresponds to 20% cache utilization in the 6T SRAM based cache whereas the cache utilization is only 5% in the STT-MRAM based caches. The results show that the total energy consumption of the STT-MRAM based caches is 25%–35% lower than that of the 6T SRAM based cache.

At this juncture, we would like to highlight several design techniques proposed in the literature to further reduce the energy consumption of STT-MRAM arrays. The common source line technique was proposed to reduce the bit-cell footprint of the 1T-1M STT-MRAM bit-cell, which also reduces the parasitic line capacitances

that need to be charged and discharged during memory accesses (Zhao et al. 2012). Techniques have also been proposed in the literature to improve the energy efficiency of write operations in STT-MRAM, which may also improve the reliability of the tunneling oxide barrier in the MTJ as explained in Sect. 7.3.1. Examples of such techniques include the stretched write cycle (Augustine et al. 2012), the balanced write scheme (Lee et al. 2012), and write optimization techniques discussed in (Kim et al. 2012). The STT-MRAM array architecture may also be modified to implement early write termination (Zhou et al. 2009), partial line update (Park et al. 2012), and write biasing techniques (Ahn et al. 2013; Jung et al. 2013; Mao et al. 2014; Rasquinha et al. 2010; Wang et al. 2013b). The objective of these array architecture modifications is to ensure that write operations into the STT-MRAM array are performed only when they are required, and in an energy efficient manner. Some proposed array architecture design techniques even exploit the asymmetry in write characteristics of STT-MRAM bit-cells to improve the write energy efficiency of the STT-MRAM array (Kwon et al. 2013; Sun et al. 2012).

Although significant improvements to STT-MRAM may be achieved using the design techniques presented earlier, they may not be the most optimum. In the next section, we present a device/circuit/array architecture co-design technique that optimizes the design of the STT-MRAM array for energy efficiency in the presence of process variations. This technique leverages insights from earlier work that retention time requirements may be relaxed in certain applications (Jog et al. 2012; Li et al. 2013a, b; Sun et al. 2011, 2014; Smullen et al. 2011) and that error-correcting codes (ECC) may be used to mitigate retention time errors as well as write failures and disturb failures that are caused by thermal effects (Xu et al. 2009). This is particularly useful for caches used in some IoT applications.

7.5 ECC and Device/Circuit/ Architecture Co-design for Low-Power and Higher Reliability

It is important to note that many IoT applications do not need 10 years of data retention time, which is the requirement of most other memory applications. For example, the IoT system may periodically wake up from sleep mode, sample new data, process the new data with that gathered when it was previously awake, store results and needed data in memory, and discard old data before returning to sleep mode. In such cases, the memory only needs to retain data long enough so that it is still available when the system next wakes up. To save on leakage energy, a nonvolatile memory technology is desirable for implementing caches for these applications. Although dynamic RAM (DRAM) may be used, it may need to be frequently refreshed because its retention time is tens to hundreds of milliseconds, whereas the retention time for IoT applications maybe several days to several weeks. Hence, DRAM based caches may lead to high energy consumption in systems for IoT applications. Another requirement for the memory system is that the any overhead in energy or latency due to transitioning between sleep and active modes must be sufficiently low compared to the leakage energy savings.

STT-MRAM may be designed to fulfill the design requirements of memories for IoT applications. The design strategy proposed in (Pajouhi et al. 2015) exploits the fact that the retention time requirement in STT-MRAM is tunable (either by changing the shape anisotropy or interfacial anisotropy of MTJs with in-plane and perpendicular magnetic anisotropy, respectively). In the proposed approach, STT-MRAM is designed with a retention time target that is significantly relaxed from the conventional target of 10 years so as to save write energy. Note that it is difficult to achieve 10 year retention time target for large STT-MRAM arrays (Kwon et al. 2015; Naeimi et al. 2013). Moreover, the disturb failure probability of STT-MRAM is increased when the retention time target is reduced. The design methodology proposed in (Pajouhi et al. 2015) mitigates the increased disturb failure rate using error correction codes (ECC). ECC may also be used to mitigate failures caused by process variations (Kwon et al. 2015; Xu et al. 2009; Yang et al. 2015).

To exemplify the aforementioned STT-MRAM design approach, let us consider an example of STT-MRAM cache design with 10 years retention time requirement as done in (Pajouhi et al. 2015). The graph of 10 year retention failure probability versus the thermal stability factor for different memory word lengths shown in Fig. 7.18a imply the tradeoff between retention time and I_{C0}, derived as:

$$I_{C0} = \frac{e\alpha}{\hbar\eta}\left(4\Delta k_B T + H_{Demag}\mu_0 M_{sat}V\right) \quad (7.13)$$

Here, α, M_{sat}, and V are the Gilbert damping factor, saturation magnetization, and volume, respectively, of the free layer of the MTJ. η is the spin polarization efficiency describing the spin-transfer torque generated per unit current. $\hbar$, e and μ_0 are the reduced Planck's constant, elementary charge and permeability of free space, respectively. H_{Demag} is the demagnetizing field that spin-transfer torque needs to overcome in the free layer of the MTJ. In the MTJ with perpendicular magnetic anisotropy (PMA), $H_{Demag}\mu_0 M_{sat}V \ll E_B$ and hence the

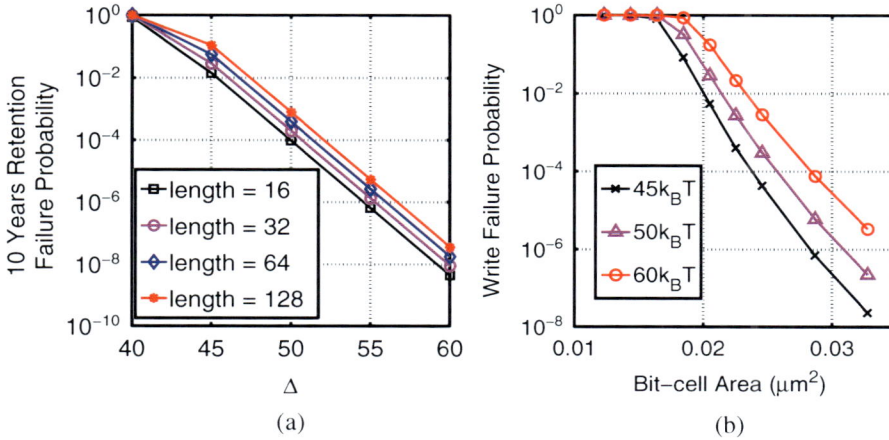

Fig. 7.18 (**a**) The retention failure probability versus thermal stability factor for different word lengths. (**b**) The write failure probability versus bit-cell layout area for different thermal stability factors

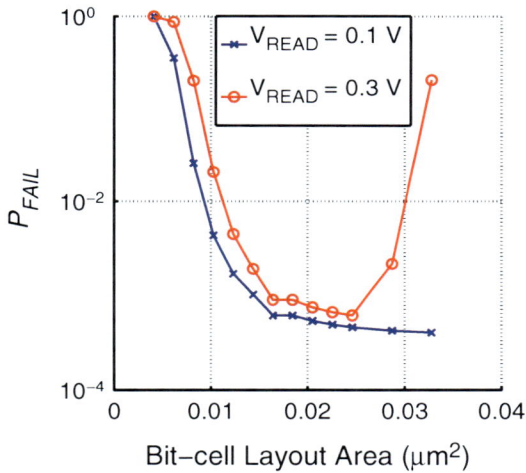

Fig. 7.19 The probability of read failure for constant voltage sensing scheme with $V_{READ} = 0.1$ and 0.3 V

demagnetizing field term in Eq. (7.13) may be neglected. Due to reduced I_{C0} by reducing Δ, the write failure probability of an STT-MRAM bit-cell decreases under the same bit-cell bias conditions as graphed in Fig. 7.18b. Note that the following analysis assumes that the MTJ resistance is independent of Δ, which may not be generally true.

The biasing conditions of the bit-cell also determine the read failure probability of the STT-MRAM array. Figure 7.19 graphs the read failure probability versus bit-cell layout area for STT-MRAM bit-cells designed with MTJs having $\Delta = 45$. The constant voltage scheme is used, and the read failure probabilities are graphed for bit-cell read voltage of $V_{READ} = 0.1$ V and $V_{READ} = 0.3$ V. The word line voltage is $V_{DD} = 1.0$ V. In the bit-cell layouts used, the bit-cell area is dominated by the access transistor. Thus, the resistance of the access transistor is large compared to that of the MTJ when the bit-cell area is small. This results in significant read failures because the bit-cell resistance when the MTJ is in P configuration is not substantially different than when the MTJ is in AP configuration (also called *decision failures*). When the bit-cell area is large, the resistance of the MTJ is dominant compared to that of the access transistor. If V_{READ} is large, the state of the MTJ may be accidentally switched during read operations (also called *disturb failures*). In Fig. 7.19, $V_{READ} = 0.3$ V results in significant number of disturb failures when the bit-cell area is larger than ~0.025 μm^2. This is not the case when $V_{READ} = 0.1$ V, and the read failure probability approaches the minimum due to decision failure. To simplify the following analysis, we will consider the case when $V_{READ} = 0.1$ V. Other parameters of the memory array that we analyze are listed in Table 7.1.

When the STT-MRAM is designed without ECC and MTJs of different Δ may be chosen, the optimum Δ depends on the bit-cell layout area (Pajouhi et al. 2015). Figure 7.20a graphs the total failure probability of the STT-MRAM bit-cell versus its layout area for different Δ. When the bit-cell area is small, write failure is dominant because the access transistor is unable to supply enough write current to the MTJ. The total failure probabilities approach the minimum determined by retention failure as the layout area of the bit-cell is increased. For bit-cell area ranging from ~0.0175 to ~0.025 μm^2, bit-cell having MTJs with $\Delta = 60$ has higher write failure probability than those having MTJs with $\Delta = 50$. Hence, write failure is still dominant and a small Δ is preferred. However, MTJ with $\Delta = 45$ is not suitable because the total failure

probability of the bit-cells using them is limited by retention failure. If the bit-cell area can be larger than ~0.025 μm^2, retention failure is the dominant failure mechanism and MTJ with higher Δ is preferred. This result implies that at fixed STT-MRAM array area budget, the desired cache capacity and the total array failure probability needs to be jointly considered to select the Δ for the MTJ.

The graph of total failure probability versus bit-cell layout area of STT-MRAM arrays implemented with different ECC schemes and Δ of the MTJ is shown in Fig. 7.20b. A trend similar to Fig. 7.20a may be observed. The ECC schemes encode data words as code words before writing into the array so as to correct for bit errors due to read, write or retention failures. The length of these code words are longer than that of data words as shown in Table 7.1. Due to stronger error correction capability of the Bose-Chaudhuri-Hocquenghem (BCH) code utilized, the minimum achievable total failure probability for STT-MRAM implemented with BCH code may be lower than that implemented with Hamming code. Note that the minimum achievable total failure probability in the STT-MRAM arrays implemented with ECC schemes is much lower than the STT-MRAM implemented without ECC. As the bit-cell layout area is increased, the point at which the total failure probability of

Table 7.1 Parameters of the memory array used in our example

CMOS Technology	32 nm
MTJ cross-section	Equivalent to 64 × 64 nm
Total Capacity	64 Mega-byte
Length of stored data word	64 bits
Hamming Code	71 bit code words, 1 bit error correction
BCH Code	78 bit code words, 2 bit error correction

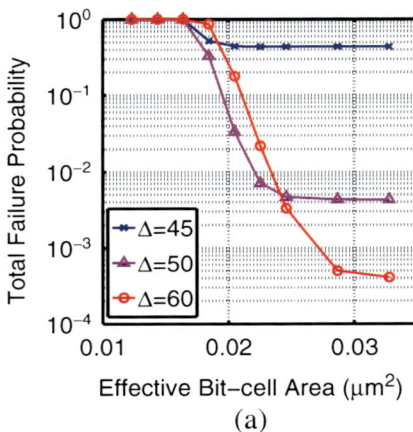

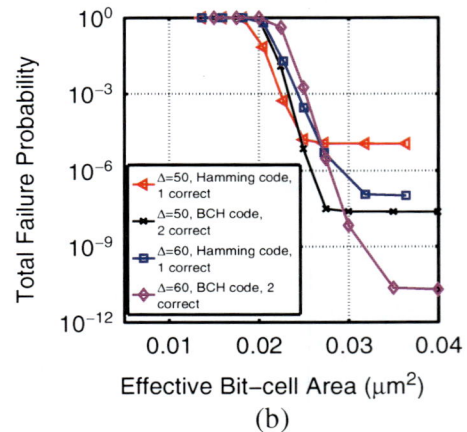

(a)   (b)

Fig. 7.20 Total failure probability of STT-MRAM arrays implemented (**a**) without ECC, and (**b**) with ECC based on Hamming code and BCH code with single- and double-error correction capability, respectively

Table 7.2 Possible design choices based on device/circuit/array co-design methodology presented

1	Minimize failure probability, no bit-cell area constraint	Use MTJ with $\Delta = 60$, use BCH code ECC
2	Bit-cell layout area < 0.0225 μm^2, minimize failure probability	Use MTJ with $\Delta = 50$, use Hamming code ECC
3	0.025 $\mu m^2 <$ bit-cell layout area < 0.0285 μm^2, minimize failure probability	Use MTJ with $\Delta = 50$, use BCH code ECC

the arrays start decreasing depends on the Δ of the MTJ used, as well as the ability to correct for bit errors. For the arrays implemented without ECC, the one using MTJs with $\Delta = 45$ are more severely affected by retention failures than the array using MTJs with $\Delta = 50$. Consequently, it is not as apparent that the total failure probability of the array using MTJs with $\Delta = 45$ decreases earlier than the array using MTJs with $\Delta = 50$.

Figure 7.20b also shows there are differences between the STT-MRAM arrays implemented with Hamming code and those implemented with BCH code. One difference is the point at which the total failure probability starts decreasing. As mentioned earlier, write failure is dominant when the bit-cell layout area is small. Since fewer bits needs to be written into the array implemented with Hamming code for ECC as compared to the array implemented with BCH code for ECC, the probability of write failure is also smaller. Thus, the total failure probability of the array implemented with Hamming code as the ECC scheme decreases earlier than in the array implemented with BCH code as the ECC scheme at the same Δ.

Another difference is the point at which retention failure starts becoming dominant compared to write failures. As Fig. 7.20b shows, in the STT-MRAM arrays using MTJs with $\Delta = 60$, retention failure becomes dominant when the bit-cell layout area is larger than 0.03 μm^2. Retention failure becomes dominant in the STT-MRAM arrays using MTJs with $\Delta = 50$ when the bit-cell layout area is larger than ~0.025 μm^2. Note that for the STT-MRAM implemented without ECC, retention failure becomes dominant when the bit-cell area is larger than ~0.0225 and ~0.0275 μm^2 for the MTJs with $\Delta = 50$ and $\Delta = 60$, respectively. Hence, by considering the array layout area

budget, total failure probability, and desired capacity, the optimum STT-MRAM array design can be selected. Table 7.2 lists some possible considerations and the optimum STT-MRAM array design corresponding to them. A similar analysis based on the flow we just described may be applied to design STT-MRAM arrays that fulfill the requirements of IoT systems.

7.6   Other Uses of MTJ for Sensing, Random Number Generation

We have discussed the design and optimization of STT-MRAM for IoT applications and presented several unique characteristics of STT-MRAM earlier. Several works in the literature exploit these unique characteristics to embed additional functionality within the STT-MRAM array while incurring minimal area overhead and near zero performance penalty (Ahn et al. 2013; Fukushima et al. 2014, 2015; Fong et al. 2015; Lee et al. 2013; Zhang et al. 2014a, b, 2015). These additional functionality are particularly interesting for ultralow power IoT systems. In the following sections, we will briefly discuss three design techniques that embed new functionality in STT-MRAM. We discuss how STT-MRAM may be used as truly random number generator (TRNG) and as physically unclonable function (PUF). TRNGs and PUFs may be used as on-chip security hardware, which is needed in IoT applications. Finally, we present a methodology for embedding read-only memory (ROM) in an STT-MRAM array. The embedded ROM stores data which may be different from that stored in RAM, and may be used to accelerate functions that use ROM, such as many digital signal processing (DSP) functions.

7.6.1 STT-MRAM as True Random Number Generators

The STT-MRAM write process is stochastic as we have discussed in Sect. 7.2.1 and may be exploited for random number generation. Since thermal disturbance is used as the entropy source, the STT-MRAM bit-cell may be used as a truly random number generator (TRNG) (Ahn et al. 2013; Fukushima et al. 2014). On-chip security hardware, which is crucial for IoT applications, may use TRNGs for generating high quality encryption keys. Hence, STT-MRAM that can also function as high quality TRNG with little overhead may be a very cost effective solution to enable on-chip security hardware suitable for IoT applications.

The operation concept of the STT-MRAM based TRNG illustrated in Fig. 7.21a is as follows. When a random number is requested, data stored in a row of STT-MRAM bit-cells is read out and stored in a buffer. Next, the random number is generated by performing a special write operation to the same row of bit-cells. During this step, a predetermined data (either all '0' or all '1') is first written to every bit-cell in the row with 100% probability of success.

Thereafter, the complementary data is written to every bit-cell in the row (all '1' if all '0' were written previously and vice versa). When complementary data is being written, the current supplied to each bit-cell is such that the state of the bit-cell has 50% probability of switching. After the write operation is completed, the data stored in the bit-cells is read out and supplied to the circuitry requesting the random number. Finally, the data stored in the buffer is written back into the bit-cells to restore them to the previous state.

Note that three write operations are required in the TRNG scheme just described. The energy consumption of this TRNG scheme may be very high due to the high write energy of STT-MRAM bit-cells, which was discussed in Sect. 7.4. Figure 7.21b shows that the overall write energy may be reduced by reducing the number of write operations (Ahn et al. 2013). Once data has been read out of the bit-cells and stored in a buffer, we may immediately generate the random number by overwriting the bit-cells storing '0' with '1', and those storing '1' with '0'. The current supplied to each bit-cell during this write operation is such that the probability of write failure in each bit-cell is 50%. After the random number is generated, we compare it to

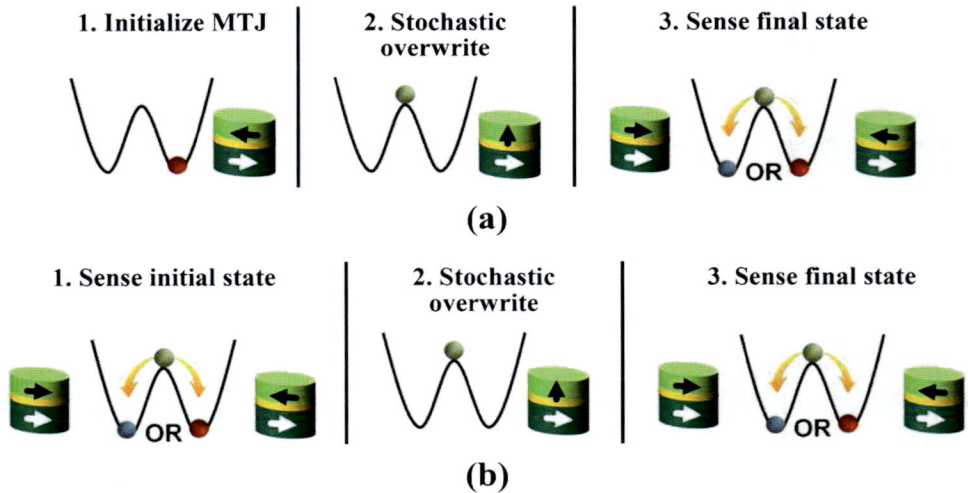

(a)

(b)

Fig. 7.21 These steps show how STT-MRAM may be used as a truly random number generator by exploiting the stochastic nature of its write operation. (a) The MTJ may be initialized to some predetermined state prior to random number generation, which incurs large energy dissipation. (b) The initialization step in (a) is redundant and may be eliminated to reduce energy consumption

the data stored in the buffer to identify the bit-cells whose data that has been modified by the random number generation process. The data in these bit-cells are then restored from the buffer.

Although the STT-MRAM based TRNG may generate very high quality random numbers (Ahn et al. 2013), it may be very challenging to choose the write current needed to implement a process variation tolerant random number generation process. Due to process variations, the amount of write current supplied to each bit-cell during the random number generation process may result in switching probabilities that are not exactly 50%. This issue may be exacerbated if the TRNG scheme with fewer write operations is used. Depending on the degradation of the switching probability in the random number generation process, strong postprocessing techniques may be required to ensure the quality of random numbers generated by STT-MRAM based TRNGs is sufficiently high (Dichtl 2008; Lacharme 2008; Von Neumann 1951).

7.6.2   Physically Unclonable Functions Using STT-MRAM

Another interesting aspect of STT-MRAM is that the process variations in the bit-cells may be exploited to implement a memory-based physically unclonable function (PUF) (Zhang et al. 2014a, b, 2015). PUFs may be used as on-chip security hardware that generates chip-unique keys for IoT applications. The input and output of the PUF are called the *challenge* and *response*, respectively. Since it is desired that the PUF generate chip-unique keys, the set of challenge-response pairs (CRPs) of a PUF must be unique (the *uniqueness* criterion). During operation, the set of CRPs for each PUF is fixed (the *reliability* criterion). To ensure that the PUF is indeed unclonable, the response of a PUF to any challenge exploits variations in the fabrication process so that it is random and not known at design time (the *randomness* criterion).

A memory-based PUF (MemPUF) may be implemented using 1T-1M STT-MRAM bit-cells by exploiting the single-ended nature of the sensing scheme (Zhang et al. 2014a). Every bit-cell in the STT-MRAM array may be grouped into pairs where one is the *data* cell and the other is the *reference* cell as shown in Fig. 7.22a. The challenge is given as the address to the STT-MRAM array and the response will be generated. In the flow chart shown in Fig. 7.22b, the data stored in the STT-MRAM is first copied to a buffer when the array operates as a PUF. When a challenge is given and a response is required, the data and reference

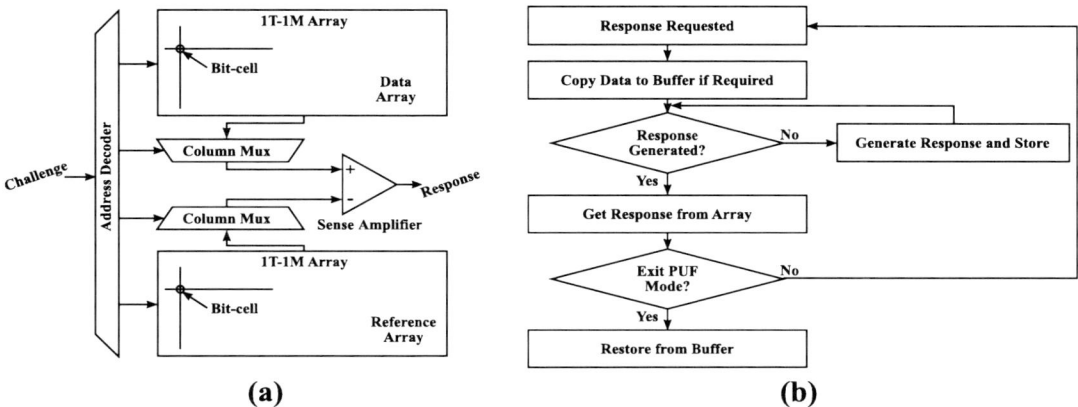

(a)   (b)

Fig. 7.22 (a) The schematic showing how an STT-MRAM array consisting of a data array and a reference array, which are identical 1T-1M bit-cell arrays, may be used as a physically unclonable function (PUF). (b) The flow chart shows how the PUF functionality may be used without overwriting the data stored in the STT-MRAM array

cells selected by the challenge are written with '0'. The response is generated by comparing the resistance of the data cell with its companion reference cell. Since the data cell and reference cell are storing the same values, the result of the resistance comparison depends on the process variations in the bit-cells. The corresponding bit in the response is '0' if the resistance of the data cell is smaller and '1' otherwise. When the STT-MRAM resumes operation as RAM, the data stored in the buffer is restored into the array.

The abovementioned scheme may not be reliable due to thermal perturbations. Since the data and reference cells store the same value during response generation, thermal perturbations in the magnetic layers of the MTJs in the bit-cells, which are random by nature, may affect the result of the resistance comparison. As a result, the response to the challenge may be unreliable (i.e., response to the challenge is not deterministic). The proposed STT-MRAM based MemPUF in (Zhang et al. 2014a) overcomes this design issue by using two operational phases: the *enrolment* and the *regeneration* phase. During the enrolment phase, every bit-cell in the array is written with '0'. Then, the resistance of every data cell is compared with its companion reference cell. The cell with the larger resistance is overwritten with a '1' and the enrolment phase ends. During the regeneration phase, the resistance of the data cell selected by the challenge is compared to that of its companion reference cell. The result of the comparison gives the corresponding bit of the response to the challenge. Hence, the operation of the proposed STT-MRAM based MemPUF must begin with an enrolment phase. The reliability of the proposed STT-MRAM based MemPUF is enhanced during the enrolment phase by ensuring that the data cell and reference cell stores complementary data. Furthermore, the CRPs generated may depend on whether '1' or '0' was written during the first step of the enrolment phase. If the *TMR* variation is sufficiently large and random, CRPs generated by writing '1' during the first step in the enrolment phase may be uncorrelated with those generated by writing '0' during the first step in the enrolment phase. This

may be exploited to expand the set of CRPs to improve its resilience against attacks.

Although promising as on-chip security hardware for IoT applications, there are several disadvantages of the MemPUF scheme just described. First, the reliability of the MemPUF may be degraded by disturb failures. The currents flowing through the bit-cells during response generation must be sufficiently small that the states of the data cell and the reference cell are not accidentally switched. Next, to utilize the MemPUF as a RAM as well, a small buffer is needed. Prior to enrolment, the data stored in the selected data and reference cells are first copied into the buffer. The data in the buffer is restored to the data and reference cells once the MemPUF response is no longer needed, incurring additional write energy consumption.

7.6.3 Embedding Read-Only Memory in STT-MRAM

As illustrated earlier in Fig. 7.14, every column in the STT-MRAM array consists of a bit line to which bit-cells in the column are connected to. The physical connection of the STT-MRAM bit-cells in the array may be exploited such that every bit-cell stores a RAM bit in their constituent MTJ as well as a read-only memory (ROM) data (Fong et al. 2015; Lee et al. 2013). The embedding of ROM in a column of the array is enabled by an additional bit line as illustrated by Fig. 7.23. The layout of an example STT-MRAM bit-cell array without and with embedded ROM is shown in Fig. 7.24a, b, respectively. Note that when the size of the access transistor is large, the additional bit line may be placed without changing the footprint of the bit-cell array. Each bit-cell in the column is connected to either BL0 or BL1. Bit-cells connected to BL0 store a ROM value of '0' whereas those connected to BL1 store a ROM value of '1'.

In the ROM-embedded STT-MRAM array shown in Fig. 7.23, a pair of pass transistors connects BL0 and BL1 to the RAM mode peripheral circuitry. During RAM mode operations, the pair of pass transistors are turned ON. Hence,

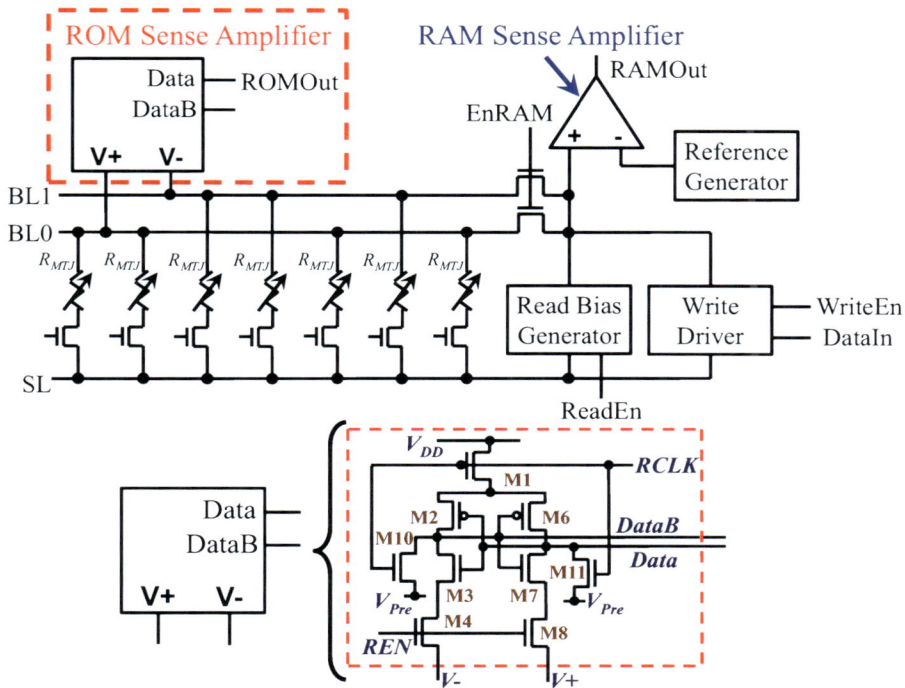

Fig. 7.23 This schematic shows a column of an STT-MRAM array with embedded read-only memory (ROM). The physical connection of each bit-cell to BL0 or BL1 is used to store ROM data in each bit-cell. RAM data is stored using the MTJ in each bit-cell and hence can be different from the ROM data

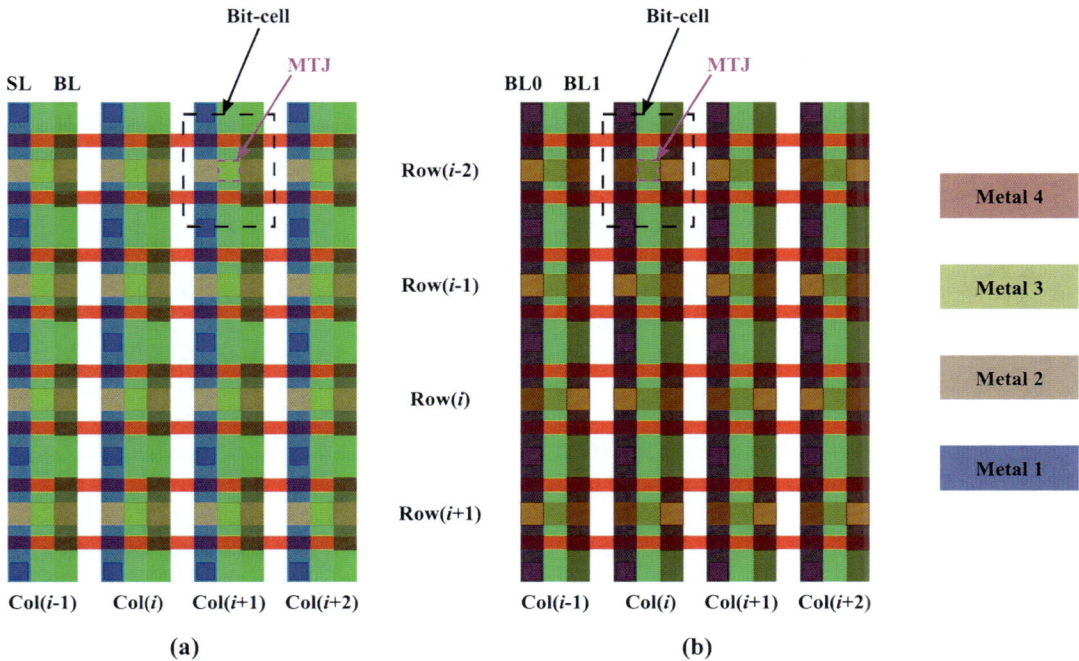

Fig. 7.24 The layout of an STT-MRAM bit-cell array (**a**) without embedded ROM and (**b**) with embedded ROM. ROM data is stored as the via connection from the MTJ to BL0 or BL1

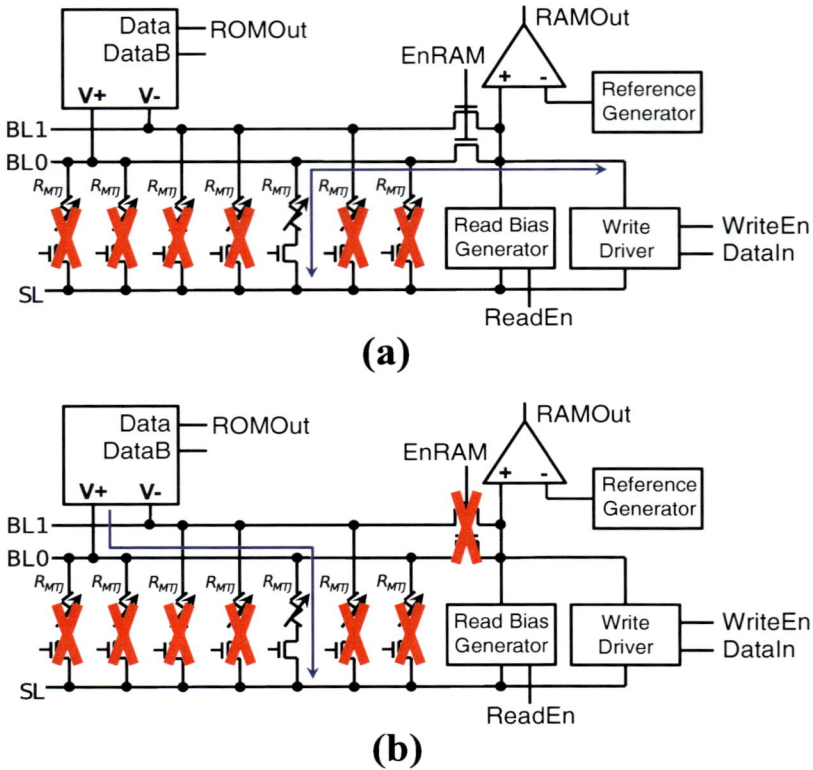

Fig. 7.25 Current path through the ROM-embedded STT-MRAM array during ROM operation when the bit-cell selected stores (**a**) ROM data '0' and (**b**) ROM data '1'

BL0 and BL1 are electrically shorted during RAM mode operations. Consequently, the peripheral circuitry is able to access any selected bit-cell regardless of whether it is connected to BL0 or BL1. During ROM mode operations, the pair of pass transistors is turned OFF. The SL of the column is then grounded. The access transistor of the selected bit-cell is turned ON next, followed by the ROM mode sensing latch. Figure 7.25a, b show the current path in the ROM-embedded STT-MRAM array during ROM operation when a bit-cell storing ROM data '0' and ROM data '1' is selected, respectively. Note that of the two bit lines, the one connected to the selected bit-cell has a much smaller resistance to SL compared to the other bit line. Hence, during ROM mode sensing operation, the output of the sensing latch that senses the bit line connected to the selected bit-cell is easily discharged to *GND* whereas the other is

charged to V_{DD} by the cross-coupled inverter action in the latch. Note that because RAM mode operations require BL0 and BL1 to be electrically connected whereas ROM mode operations require BL0 and BL1 to be electrically disconnected, RAM mode and ROM mode operations to the same column of bit-cells cannot occur simultaneously.

The ROM embedded in the STT-MRAM may be used to store instructions and data closer to the processor core, and applications utilizing the ROM may be accelerated even though RAM mode and ROM mode operations cannot occur simultaneously. Note that without the embedded ROM, instructions and data are stored off-chip and need to be fetched on-chip before the processor core may operate on them (Fong et al. 2015; Lee et al. 2013). The cost of accessing off-chip memory may greatly outweigh the cost due to not being able to perform RAM mode and ROM mode operations simultaneously. This is the case in the benchmark programs analyzed in (Fong et al. 2015; Lee et al. 2013) and it was observed that applications exploiting the ROM embedded in STT-MRAM may be accelerated by as much as 30%.

7.7 Perspectives and Trends

As we saw in this chapter, STT-MRAM may be suitable for a wide array of IoT applications. Comparisons of STT-MRAM with other eNVM technologies in Chap. 6 show that the low voltage and high speed of STT-MRAM is suitable for IoT applications. Thus, STT-MRAM may be useful for implementing memory systems for IoT applications with unstable power supplies. The non-volatility and relatively fast access speed of STT-MRAM may also enable IoT systems that can quickly transition between sleep and active states, which significantly improves energy efficiency. Another attractiveness of STT-MRAM is that its unique characteristics may also be exploited to embed functionality, which is useful for IoT applications, in the memory array with few overhead as discussed in Sect. 7.6. However, such cost savings must be weighed against the increase in STT-MRAM array design complexity as well as the implications on array test methodology and failure rate.

The immediate STT-MRAM design challenges that need to be addressed are the high write energy consumption, and the severely limited design space imposed by the two-terminal nature of the MTJ (Augustine et al. 2010, 2011; Mojumder and Roy 2012). Several works investigated alternative spintronic device structures that are based on the MTJ concept (Braganca et al. 2009; Fong and Roy 2013; Fong et al. 2013, 2014; Huda and Sheikholeslami 2013; Kim et al. 2013; Mojumder et al. 2011a, b, 2013; Shiota et al. 2012; Wang et al. 2012, 2013a; Wang and Chien 2013). Of these device proposals, the two-terminal device based on voltage-controlled magnetic anisotropy (Shiota et al. 2012; Wang et al. 2012, 2013a; Wang and

Chien 2013) and the multi-terminal device based on spin Hall effect or spin-orbit torque (Kim et al. 2013; Wang et al. 2013a) have garnered much research interest recently.

The two-terminal device exploiting voltage-controlled magnetic anisotropy has a stack composition that is very similar to that in the conventional MTJ structure we presented earlier (Shiota et al. 2012; Wang et al. 2012, 2013a; Wang and Chien 2013). Hence, the device integration issues are not significantly different from that for the conventional MTJ. It was found that the perpendicular magnetic anisotropy of the CoFeB free layer in the MTJ may be modulated by the voltage-induced electric-field at the CoFeB/MgO interface. When the applied voltage reduces the perpendicular magnetic anisotropy, the hysteresis loop of the CoFeB free layer narrows. Due to the stray field from the pinned layer, the hysteresis loop is not centered about zero applied magnetic field. Hence, the applied voltage may narrow the hysteresis loop to the point that the stray field from the pinned layer aligns the free layer magnetization parallel to that of the pinned layer. The magnetization of the free layer may be aligned anti-parallel to that of the pinned layer using a magnetic field (Wang et al. 2012).

Alternatively, current-induced spin-transfer torque, just like in the conventional MTJ, may be used to switch the state of the MTJ (Wang and Chien 2013). When this is the case, the polarity of the voltage applied across the MTJ must be designed such that the current flowing through the MTJ tries to anti-parallelize the MTJ configuration. The magnitude of the applied voltage is then used to determine whether the stable MTJ state is anti-parallel (for small applied voltage) or parallel (for large applied voltage) (Wang and Chien 2013). A major disadvantage of this scheme is that although the current required to program the MTJ is reduced, there is a significant increase in MTJ resistance. As a result, the voltage needed to program the MTJ may be significantly larger than that available for IoT applications. Furthermore, this may limit the write energy reduction. Hence, the major focus of research on voltage-controlled magnetic

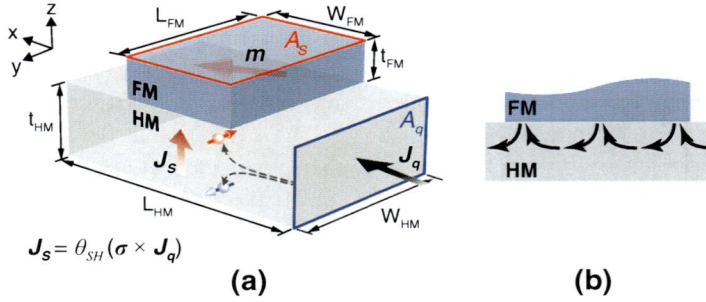

$$J_s = \theta_{SH}(\sigma \times J_q)$$

(a)   **(b)**

Fig. 7.26 (**a**) Current-induced torque is exerted on a magnetic layer adjacent to a heavy metal as shown here. (**b**) One possible mechanism is the spin Hall effect. After exchanging spin angular momentum with the adjacent magnetic layer, the spin polarization direction of an electron may be realigned by the spin Hall effect in the heavy metal and gets reinjected into the magnetic layer. Hence, an electron may transfer multiple units of spin angular momentum to the magnetic layer

anisotropy is on devising a method and/or device structure that allows the MTJ state to be switched without the need for current flow or magnetic fields.

Unlike the data storage devices using voltage-controlled magnetic anisotropy, those utilizing spin-orbit torque or spin Hall effect are still current-driven by nature (Kim et al. 2013; Wang et al. 2013a). In these devices, the MTJ grown on top of a heavy metal (or HM, such as Pt (Miron et al. 2011), β-Ta (Liu et al. 2012), β-W (Pai et al. 2012), CuBi (Niimi et al. 2012), or CuIr (Niimi et al. 2011)) with its free layer in direct contact with HM. Consider the structure shown in Fig. 7.26a. When charge current flows in HM, a spin current (i.e., a current in which all constituent electrons are identically spin-polarized) that flows perpendicular to the direction of charge current flow is generated via the spin Hall effect. In this spin current, electrons carrying opposite spin polarizations flow in opposite directions. Furthermore, the spin polarization direction of the electron is perpendicular to both charge and spin current flow directions. When the HM is in contact with the free layer of an MTJ, the spin current can get injected into the free layer of the MTJ from HM, which exerts spin-transfer torque on the magnetization of the free layer. After transferring spin angular momentum to the free layer of the MTJ, the spin polarization of the electron may get realigned and injected into the free layer again.

Hence, an electron may transfer multiple units of spin angular momentum to the free layer of the MTJ as illustrated by Fig. 7.26b. Note that since the free layer of the MTJ is metallic, some charge current is shunted through it. Due to structural asymmetry and strong spin-orbit interaction, electrons at the interface between HM and the free layer may also experience a Rashba field (Gambardella and Miron 2011). The spin Hall effect and Rashba field contributes to the spin-orbit torque that is exerted on the magnetization of the free layer of the MTJ.

The spin Hall effect is promising for reducing the write energy in STT-MRAM because the spin current generated, I_s, can be controlled by engineering the geometry of the structure in Fig. 7.26a. I_s is given as:

$$I_s = \theta_{SH}\frac{A_s}{A_q}I_q\sigma \qquad (7.14)$$

where A_q and A_s are as shown in Fig. 7.26a, σ is the electron spin polarization direction, and θ_{SH} is the spin Hall angle of the HM. It was also experimentally demonstrated that spin-orbit torque may be used to switch the magnetization of a free layer with PMA (Yu et al. 2014). Furthermore, the current for programming the MTJ is not passed through the tunneling oxide of the MTJ and hence the reliability of the tunnel oxide barrier is improved. This is crucial in maintaining high *TMR*. The resistance of the

HM layer may also be reduced to lower the voltage required to drive the write current, which is desirable for reducing write energy. Because the device structure that enables spin-orbit torque is compatible with that having voltage-controlled magnetic anisotropy, a device structure that utilizes voltage-controlled magnetic anisotropy for assisting spin-orbit torque switching may be the key to achieving STT-MRAM for ultralow power IoT applications.

References

K. Abe, S. Fujita, T. H. Lee, Architecture of three-dimensional circuit using nanoscale memory devices, Eur. Micro Nano Syst., Noisy le Grand, France, 2004. TIMA, Grenoble, France (2004), pp. 225–229

J. Ahn, S. Yoo, K. Choi, Write intensity prediction for energy-efficient non-volatile caches, *Int. Symp. Low Power Electron. Des.* (2013), pp. 223–228

D. Apalkov, S. Watts, A. Driskill-Smith, E. Chen, Z. Diao, V. Nikitin, Comparison of scaling of in-plane and perpendicular spin transfer switching technologies by micromagnetic simulation. IEEE Trans. Magn. **46**(6), 2240–2243 (2010)

C. Augustine, A. Raychowdhury, D. Somasekhar, J. Tschanz, K. Roy, V. K. De, Numerical analysis of typical STT-MTJ stacks for 1T-1R memory arrays, *Int. Electron Devices Meet.* (2010), pp. 22.7.1–22.7.4

C. Augustine, A. Raychowdhury, D. Somasekhar, J. Tschanz, V. De, K. Roy, Design space exploration of typical stt mtj stacks inmemory arrays in the presence of variability and disturbances. IEEE Trans. Electron Devices **58**(12), 4333–4343 (2011)

C. Augustine, N.N. Mojumder, X. Fong, S.H. Choday, S.P. Park, K. Roy, Spin-transfer torque mrams for low power memories: perspective and prospective. IEEE Sens. J. **12**(4), 756–766 (2012)

L. Berger, Emission of spin waves by a magnetic multilayer traversed by a current. Phys. Rev. B **54**(13), 9353–9358 (1996)

P.M. Braganca, J.A. Katine, N.C. Emley, D. Mauri, J.R. Childress, P.M. Rice, E. Delenia, D.C. Ralph, R.A. Buhrman, A three-terminal approach to developing spin-torque written magnetic random access memory cells. IEEE Trans. Nanotechnol. **8**(2), 190–195 (2009)

W.F. Brown, Thermal fluctuations of a single-domain particle. Phys. Rev. **130**(5), 1677–1686 (1963)

D. Datta, B. Behin-Aein, S. Datta, S. Salahuddin, Voltage asymmetry of spin-transfer torques. IEEE Trans. Nanotechnol. **11**(2), 261–272 (2012)

T. Devolder, Scalability of magnetic random access memories based on an in-plane magnetized free layer. Appl. Phys. Exp. **4**(9), 093001 (2011)

M. Dichtl, Bad and good ways of post-processing biased physical random numbers, in *Fast Software Encryption* (Springer, Berlin, 2008), pp. 137–152

R. Dorrance, F. Ren, Y. Toriyama, A. A. Hafez, C. K. Yang, D. Markovic, Scalability and design-space analysis of a 1T-1MTJ memory cell, in *IEEE/ACM Int. Symp. Nanoscale Archit.* (2011), pp. 32–36

X. Fong, K. Roy, Complementary polarizers STT-MRAM (CPSTT) for on-chip caches. IEEE Electron Device Lett. **34**(2), 232–234 (2013)

X. Fong, S.H. Choday, K. Roy, Bit-cell level optimization for non-volatile memories using magnetic tunnel junctions and spin-transfer torque switching. IEEE Trans. Nanotechnol. **11**(1), 172–181 (2012)

X. Fong, K. Roy, Low-power robust complementary polarizer STTMRAM (CPSTT) for on-chip caches, in *5th IEEE Int. Mem. Work.* (2013), pp. 88–91

X. Fong, R. Venkatesan, A. Raghunathan, K. Roy, Non-volatile complementary polarizer spin-transfer torque on-chip caches: a device/circuit/systems perspective. IEEE Trans. Magn. **50**(10), 1–11 (2014)

X. Fong, R. Venkatesan, D. Lee, A. Raghunathan, K. Roy, Embedding read-only memory in spin-transfer torque mram based on-chip caches. IEEE Trans. Very Large Scale Integr. Syst. **24**(3), 992–1002 (2016)

A. Fukushima, T. Seki, K. Yakushiji, H. Kubota, H. Imamura, S. Yuasa, K. Ando, Spin dice: a scalable truly random number generator based on spintronics. Appl. Phys. Exp. **7**(8), 083001 (2014)

A. Fukushima, K. Yakushiji, H. Kubota, S. Yuasa, Spin dice (physical random number generator using spin torque switching) and its thermal response, in *IEEE Magn. Conf.* (2015), pp. 1–1

P. Gambardella, I.M. Miron, Current-induced spin-orbit torques. Philos. Trans. A. Math. Phys. Eng. Sci. **369**, 3175–3197 (2011)

Y. Huai, Spin-transfer torque MRAM (STT-MRAM): challenges and prospects. AAPPS Bull. **18**(6), 33–40 (2008)

Y. Huai, F. Albert, P. Nguyen, M. Pakala, T. Valet, Observation of spin-transfer switching in deep submicron-sized and low-resistance magnetic tunnel junctions. Appl. Phys. Lett. **84**(16), 3118–3120 (2004)

S. Huda, A. Sheikholeslami, A novel STT-MRAM cell with disturbance-free read operation. IEEE Trans. Circ. Syst. I, Reg. Papers **60**(6), 1534–1547 (2013)

S. Ikeda, K. Miura, H. Yamamoto, K. Mizunuma, H.D. Gan, M. Endo, S. Kanai, J. Hayakawa, F. Matsukura, H. Ohno, A perpendicular-anisotropy CoFeB-MgO magnetic tunnel junction. Nat. Mater. **9**(9), 721–724 (2010)

A. Jog, A. K Mishra, C. Xu, Y. Xie, V. Narayanan, R. Iyer, C. R. Das, Cache revive: architecting volatile STT-RAM caches for enhanced performance in CMPs, in *Proceedings of the 49th Annu. Des. Autom. Conf.—DAC'12* (2012), p. 243

J. Jung, Y. Nakata, M. Yoshimoto, H. Kawaguchi, Energy-efficient spin-transfer torque RAM cache exploiting additional all-zero-data flags, in *Int. Symp. Qual. Electron. Des.* (2013), pp. 216–222

S. H. Kang, X. Li, S. Gu, K. Lee, X. Zhu, STT MRAM magnetic tunnel junction architecture and integration (2014)

J. Katine, F. Albert, R. Buhrman, E. Myers, D. Ralph, Current-driven magnetization reversal and spin-wave excitations in Co/Cu/Co pillars. Phys. Rev. Lett. **84** (14), 3149–3152 (2000)

Y. Kim, S. K. Gupta, S. P. Park, G. Panagopoulos, K. Roy, Write-optimized reliable design of STT MRAM, in *Proceedings of the 2012 ACM/IEEE Int. Symp. Low power Electron. Des.—ISLPED'12* (2012), p. 3

Y. Kim, S.H. Choday, K. Roy, DSH-MRAM: differential spin hall MRAM for on-chip memories. IEEE Electron Device Lett. **34**(10), 1259–1261 (2013)

T. Kishi, H. Yoda, T. Kai, T. Nagase, E. Kitagawa, M. Yoshikawa, K. Nishiyama, T. Daibou, M. Nagamine, M. Amano, S. Takahashi, M. Nakayama, N. Shimomura, H. Aikawa, S. Ikegawa, S. Yuasa, K. Yakushiji, H. Kubota, A. Fukushima, M. Oogane, T. Miyazaki, K. Ando, Lower-current and fast switching of a perpendicular TMR for high speed and high density spin-transfer-torque MRAM, in *2008 I.E. Int. Electron Devices Meet.* (2008), pp. 1–4

K. Kwon, S.H. Choday, Y. Kim, K. Roy, AWARE (Asymmetric Write Architecture with REdundant blocks): a high write speed STTMRAM cache architecture. IEEE Trans. Very Large Scale Integr. Syst. **1**, 1–1 (2013)

K. Kwon, X. Fong, P. Wijesinghe, P. Panda, K. Roy, High-density & robust STT-MRAM array through device/circuit/architecture interactions. IEEE Trans. Nanotechnol., **14**(6), 1024–1034 (2015)

P. Lacharme, Post-processing functions for a biased physical random number generator, in *Fast Software Encryption* (Springer, Berlin, 2008), pp. 334–342

D. Lee, S. K. Gupta, K. Roy, High-performance lowenergy STT MRAM based on balanced write scheme, in *Proceedings of the 2012 ACM/IEEE Int. Symp. Low power Electron. Des.—ISLPED'12* (2012), p. 9

D. Lee, X. Fong, K. Roy, R-MRAM: A ROM-embedded STT MRAM cache. IEEE Electron Device Lett. **34** (10), 1256–1258 (2013)

J. Li, P. Ndai, A. Goel, S. Salahuddin, K. Roy, Design paradigm for robust spin-torque transfer magnetic RAM (STT MRAM) from circuit/architecture perspective. IEEE Trans. Very Large Scale Integr. Syst. **18**(12), 1710–1723 (2010)

Q. Li, J. Li, L. Shi, C. J. Xue, Y. Chen, Y. He, Compiler-assisted refresh minimization for volatile STT-RAM cache, in *2013 18th Asia South Pacific Des. Autom. Conf.* (2013), pp. 273–278

J. Li, L. Shi, Q. Li, C.J. Xue, Y. Chen, Y. Xu, W. Wang, Low-energy volatile STT-RAM cache design using

cache-coherence-enabled adaptive refresh. ACM Trans. Des. Autom. Electron. Syst. **19**(1), 1–23 (2013b)

C. J. Lin, S. H. Kang, Y. J. Wang, K. Lee, X. Zhu, W. C. Chen, X. Li, W. N. Hsu, Y. C. Kao, M. T. Liu, M. Nowak, N. Yu, 45 nm low power CMOS logic compatible embedded STT MRAM utilizing a reverse-connection 1T/1MTJ cell, in *2009 I.E. Int. Electron Devices Meet.* (2009), pp. 1–4

L. Liu, C.-F. Pai, Y. Li, H.W. Tseng, D.C. Ralph, R.A. Buhrman, Spin-torque switching with the giant spin hall effect of tantalum. Science **336**(6081), 555–558 (2012)

M. Mao, G. Sun, Y. Li, A. K. Jones, Y. Chen, Prefetching techniques for STT-RAM based last-level cache in CMP systems, in *2014 19th Asia South Pacific Des. Autom. Conf.* (2014), pp. 67–72

I.M. Miron, K. Garello, G. Gaudin, P.-J. Zermatten, M.V. Costache, S. Auffret, S. Bandiera, B. Rodmacq, A. Schuhl, P. Gambardella, Perpendicular switching of a single ferromagnetic layer induced by in-plane current injection. Nature **476**(7359), 189–193 (2011)

N.N. Mojumder, K. Roy, Proposal for switching current reduction using reference layer with tilted magnetic anisotropy in magnetic tunnel junctions for spin-transfer torque (STT) MRAM. IEEE Trans. Electron Devices **59**(11), 3054–3060 (2012)

N.N. Mojumder, S.K. Gupta, S.H. Choday, D.E. Nikonov, K. Roy, A three-terminal dual-pillar STT-MRAM for high-performance robust memory applications. IEEE Trans. Electr. Devices **58**(5), 1508–1516 (2011a)

N. N. Mojumder, S. K. Gupta, K. Roy, Dual pillar spin transfer torque MRAM with tilted magnetic anisotropy for fast and error-free switching and near-disturb-free read operations, in *69th Device Res. Conf.* (2011), pp. 67–68

N.N. Mojumder, X. Fong, C. Augustine, S.K. Gupta, S.H. Choday, K. Roy, Dual pillar spin-transfer torque mrams for low power applications. ACM J. Emerg. Technol. Comput. Syst. **9**(2), 1–17 (2013)

E.B. Myers, Current-induced switching of domains in magnetic multilayer devices. Science **285**(5429), 867–870 (1999)

H. Naeimi, C. Augustine, A. Raychowdhury, S. Lu, J. Tschanz, STTRAM scaling and retention failure. Intel Technol. J. **17**(1), 54–75 (2013)

Y. Niimi, M. Morota, D.H. Wei, C. Deranlot, M. Basletic, A. Hamzic, A. Fert, Y. Otani, Extrinsic spin Hall effect induced by iridium impurities in copper. Phys. Rev. Lett. **106**(12), 126601 (2011)

Y. Niimi, Y. Kawanishi, D.H. Wei, C. Deranlot, H.X. Yang, M. Chshiev, T. Valet, A. Fert, Y. Otani, Giant spin Hall effect induced by skew scattering from bismuth impurities inside thin film CuBi alloys. Phys. Rev. Lett. **109**, 156602 (2012)

J. Nogués, J. Sort, V. Langlais, V. Skumryev, S. Suriñach, J.S. Muñoz, M.D. Baró, Exchange bias in nanostructures. Phys. Rep. **422**(3), 65–117 (2005)

T. Ohsawa, H. Koike, S. Miura, H. Honjo, K. Tokutome, S. Ikeda, T. Hanyu, H. Ohno, T. Endoh, 1Mb 4T-2MTJ nonvolatile STT-RAM for embedded memories using 32b fine-grained power gating technique with 1.0ns/200ps wake-up/power-off times, in *2012 Symp. VLSI Circuits* (2012), pp. 46–47

C.F. Pai, L. Liu, Y. Li, H.W. Tseng, D.C. Ralph, R.A. Buhrman, Spin transfer torque devices utilizing the giant spin Hall effect of tungsten. Appl. Phys. Lett. **101**(12), 1–5 (2012)

Z. Pajouhi, X. Fong, K. Roy, Device/Circuit/Architecture Co-Design of Reliable STT-MRAM, in *Proc. 2015 Des. Autom. Test Eur. Conf. Exhib.* (2015), pp. 1437–1442

S. P. Park, S. Y. Kim, D. Lee, J.-J. Kim, W. P. Griffin, K. Roy, Column-selection-enabled 8T SRAM array with 1R/1W multi-port operation for DVFS-enabled processors, in *IEEE/ACM Int. Symp. Low Power Electron. Des.* (2011), pp. 303–308

S. P. Park, S. Gupta, N. Mojumder, A. Raghunathan, K. Roy, Future cache design using STT MRAMs for improved energy efficiency, in *Proceedings of the 49th Annu. Des. Autom. Conf.—DAC'12* (2012), p. 492

S.S.P. Parkin, N. More, K.P. Roche, Oscillations in exchange coupling and magnetoresistance in metallic superlattice structures: Co/Ru, Co/Cr, and Fe/Cr. Phys. Rev. Lett. **64**(19), 2304–2307 (1990)

M. Rasquinha, D. Choudhary, S. Chatterjee, S. Mukhopadhyay, S. Yalamanchili, An energy efficient cache design using spin torque transfer (STT) RAM, in *Proceedings of the 16th ACM/IEEE Int. Symp. Low power Electron. Des.—ISLPED'10* (2010), p. 389

ITRS Roadmap (2014)

S. Salahuddin, D. Datta, S. Datta, Key role of non equilibrium spin density in determining spin torque, in *2008 Device Res. Conf.* (2008), pp. 161–162

Y. Shiota, T. Nozaki, F. Bonell, S. Murakami, T. Shinjo, Y. Suzuki, Induction of coherent magnetization switching in a few atomic layers of FeCo using voltage pulses. Nat. Mater. **11**(1), 39–43 (2012)

J.C. Slonczewski, Current-driven excitation of magnetic multilayers. J. Magn. Magn. Mater. **159**, L1–L7 (1996)

C. W. Smullen, V. Mohan, A. Nigam, S. Gurumurthi, M. R. Stan, Relaxing non-volatility for fast and energy-efficient STT-RAM caches, in *2011 I.E. 17th Int. Symp. High Perform. Comput. Archit.* (2011), pp. 50–61

J. Sun, Spin-current interaction with a monodomain magnetic body: a model study. Phys. Rev. B **62**(1), 570–578 (2000)

Z. Sun, X. Bi, H. Li, W.-F. Wong, Z.-L. Ong, X. Zhu, W. Wu, Multi retention level STT-RAM cache designs with a dynamic refresh scheme, in *Proceedings of the 44th Annu. IEEE/ACM Int. Symp. Microarchitecture—MICRO'11* (2011), p. 329

G. Sun, Y. Zhang, Y. Wang, Y. Chen, Improving energy efficiency of write-asymmetric memories by log style write, in *Proceeding of the 2012 ACM/IEEE Int. Symp. Low power Electron. Des., ISLPED'12* (2012), pp. 173–178

Z. Sun, X. Bi, H. Li, W.-F. Wong, X. Zhu, STT-RAM cache hierarchy with multiretention MTJ designs. IEEE Trans. Very Large Scale Integr. Syst. **22**(6), 1281–1293 (2014)

S. Tsunegi, H. Kubota, S. Tamaru, K. Yakushiji, M. Konoto, A. Fukushima, T. Taniguchi, H. Arai, H. Imamura, S. Yuasa, Damping parameter and interfacial perpendicular magnetic anisotropy of FeB nanopillar sandwiched between MgO barrier and cap layers in magnetic tunnel junctions. Appl. Phys. Exp. **7**(3), 033004 (2014)

J. Von Neumann, Various techniques used in connection with random digit. Natl. Bur. Stand. Appl. Math. Ser. **12**, 36–38 (1951)

W.G. Wang, C.L. Chien, Voltage-induced switching in magnetic tunnel junctions with perpendicular magnetic anisotropy. J. Phys. D. Appl. Phys. **46**(7), 74004 (2013)

W.-G. Wang, M. Li, S. Hageman, C.L. Chien, Electric-field assisted switching in magnetic tunnel junctions. Nat. Mater. **11**(1), 64–68 (2012)

K.L. Wang, J.G. Alzate, P. Khalili Amiri, Low-power non-volatile spintronic memory: STT-RAM and beyond. J. Phys. D. Appl. Phys. **46**(7), 074003 (2013a)

J. Wang, X. Dong, Y. Xie, OAP: an obstruction-aware cache management policy for STT-RAM last-level caches, in *Des. Autom. Test Eur. Conf. Exhib.* (2013), pp. 847–852

W. Xu, Y. Chen, X. Wang, T. Zhang, Improving STT MRAM storage density through smaller-than-worst-case transistor sizing, in *Des. Autom. Conf.* (2009), pp. 87–90

S. Yamamoto, S. Sugahara, Nonvolatile static random access memory using magnetic tunnel junctions with current-induced magnetization switching architecture. Jpn. J. Appl. Phys. **48**(4), 043001 (2009)

T. Yamauchi, Prospect of embedded non-volatile memory in the smart society, in *2015 Int. Symp. VLSI Technol. Syst. Appl.* (2015), pp. 1–2

J. Yang, B. Geller, M. Li, T. Zhang, An information theory perspective for the binary STT-MRAM cell operation channel, in *IEEE Trans. Very Large Scale Integr. Syst.* (2015), pp. 1–1

G. Yu, P. Upadhyaya, Y. Fan, J.G. Alzate, W. Jiang, K.L. Wong, S. Takei, S.A. Bender, L.-T. Chang, Y. Jiang, M. Lang, J. Tang, Y. Wang, Y. Tserkovnyak, P.K. Amiri, K.L. Wang, Switching of perpendicular magnetization by spin-orbit torques in the absence of external magnetic fields. Nat. Nanotechnol. **9**, 548–554 (2014)

S. Yuasa, T. Nagahama, A. Fukushima, Y. Suzuki, K. Ando, Giant room-temperature magnetoresistance in single-crystal Fe/MgO/Fe magnetic tunnel junctions. Nat. Mater. **3**(12), 868–871 (2004)

L. Zhang, X. Fong, C.-H. Chang, Z. H. Kong, K. Roy, Feasibility study of emerging non-volatile memory based physical unclonable functions, in *2014 I.E. 6th Int. Mem. Work.* (2014), pp. 1–4

L. Zhang, X. Fong, C.-H. Chang, Z. H. Kong, K. Roy, Highly reliable memory-based physical unclonable function using spin-transfer torque MRAM, in *2014 I.E. Int. Symp. Circuits Syst.* (2014), pp. 2169–2172

L. Zhang, X. Fong, C.-H. Chang, Z.H. Kong, K. Roy, Optimizating emerging nonvolatile memories for dual-mode applications: data storage and key generator. IEEE Trans. Comput. Des. Integr. Circuits Syst. **34**(7), 1176–1187 (2015)

B. Zhao, J. Yang, Y. Zhang, Y. Chen, H. Li, Architecting a common source-line array for bipolar non-volatile memory devices, in *2012 Des. Autom. Test Eur. Conf. Exhib.* (2012), pp. 1451–1454

P. Zhou, B. Zhao, J. Yang, Y. Zhang, Energy reduction for STT-RAM using early write termination, in *IEEE/ACM Int. Conf. Comput. Des.—Dig. Tech. Pap.* (2009), pp. 264–268

Security Down to the Hardware Level

8

Anastacia Alvarez and Massimo Alioto

This chapter introduces the concept of Physically Unclonable Functions (PUFs), their prospects for hardware security in IoT devices, and their interaction with traditional cryptography. Section 8.1 summarizes the background on PUFs, whereas Sect. 8.2 covers the metrics that are commonly used to evaluate PUF performance. Such metrics are used to comparatively review the state of the art on PUFs in Sect. 8.3. Section 8.4 covers vulnerabilities to malicious attacks attempting to clone or mimic a PUF. In the last section, we introduce the novel concept of PUF-enhanced cryptography as a promising direction aiming to merge PUFs and cryptography in a cohesive framework for IoT hardware-level security.

8.1 Physically Unclonable Functions for IoT

The spatial pervasiveness and the prospectively very large number of deployed nodes monitoring the environment, people, shared resources and goods, makes security a fundamental challenge in IoT. Serious security issues are indeed arising in terms of data authenticity, integrity and confidentiality. Indeed, it is typically necessary to assure that the data and the sender are legitimate, the data has been sent uncorrupted, and

oftentimes data needs to be unreadable from an unintended receiver. Accordingly, IoT requires security to be assured down to the hardware level, as the authenticity and the integrity need to be assured also in terms of the hardware implementation of each IoT node (i.e., each node needs to be confirmed to be authentic and intact, while signaling if it has been counterfeited or tampered with).

In the recent past, Physically Unclonable Functions (PUFs) have emerged as potentially highly secure and lightweight solution to ensure data and hardware security, assuring trustworthiness down to the chip level (Mathew et al. 2014; Maes et al. 2012; Rosenblatt et al. 2013; Su et al. 2007; Maes 2012). A PUF is a function that maps an input (digital) challenge to an output (digital) response in a repeatable but unpredictable manner, leveraging on chip-specific random process variations. PUFs are sometimes referred to as "silicon biometrics", i.e. something equivalent to a "chip fingerprint" that is unique for each die. As such, it eliminates the need to store any key, as the latter is naturally generated and embedded into the chip during its manufacturing. This avoids the need for key programming (e.g., via fuses or e-Flash), and makes IoT nodes less prone to the many existing attacks that uncover the content of memories (Nedospasov et al. 2013), as discussed below.

PUFs are used for chip identification and authentication (Rosenblatt et al. 2013; Su et al.

A. Alvarez (✉) • M. Alioto
National University of Singapore, Singapore, Singapore
e-mail: anastacia@u.nus.edu

2007; Maes 2012; Alvarez et al. 2015; Gassend et al. 2002), secure key storage and lightweight encryption (Mathew et al. 2014; Xu et al. 2014), hardware-entangled cryptography (Sadeghi and Naccache 2010) and identification of malicious hardware (Maes 2013). Chip identification and authentication are typically performed by preliminarily storing all challenge-response pairs (CRPs) of the chip PUF in a secure database, during a first enrollment phase. These (or a subset thereof) are used to verify the response of the chip to a given challenge during in-field operation, making sure not to reuse CRPs to reduce susceptibility to cloning, and counteract replay attacks. Figure 8.1 shows an illustration of the enrollment process and chip authentication.

To keep data secure during transmission, it is typically encrypted using a key that is stored externally, or in an on-chip non-volatile memory

(NVM). Unfortunately, storing the key off chip or in an on-chip NVM facilitates the recovery of the key by other parties. Indeed, several studies have shown that NVM are prone to attacks and easy to read out (Samyde et al. 2002; Kömmerling et al. 1999). PUFs replace the conventional key storage, and hence offer superior robustness against invasive attacks, as they do not really store information since they recreate the keys only while the chip is being powered on.

8.2   PUF Properties and Metrics

Ideally, an array of PUF bitcells generate chip-specific keys that are:

- unpredictable, leveraging on on-chip random process variations

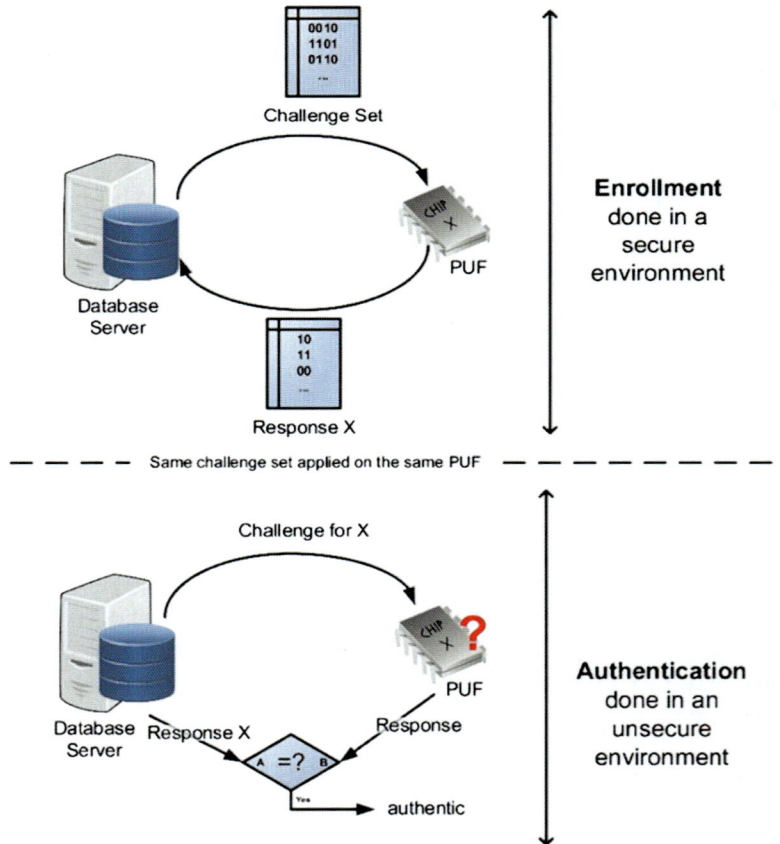

Fig. 8.1 Illustration of typical chip enrollment and subsequent in-field authentication using challenge-response pairs (CRPs) from PUFs

- repeatable, by amplifying random variations, while rejecting global variations and noise (Maes et al. 2012)
- not directly accessible or measurable externally, once the enrollment phase is completed.

There are two main types of PUFs: weak PUFs and strong PUFs. Weak PUFs have limited number of challenge-response pairs, making them equivalent to random key generators that are typically used for encryption and decryption. Weak PUFs essentially provide chip ID, whereas strong PUFs offer a very large number of challenge-response pairs (CRPs), each for one-time use. Given the long lifespan required by IoT applications, PUFs with very large number of CRPs (and therefore large area) are very expensive and typically infeasible. As a numerical example, Table 8.1 shows an example of the cost for a PUF with 256-bit key in 65 nm (Alvarez et al. 2015; Alvarez et al. 2016), whose cost invariably exceeds the overall IoT node cost.

Given the fundamental PUF properties, such as stability, repeatability, uniqueness and randomness (Maes 2013), and knowing the statistical nature of process variations, several metrics have been introduced to quantify the quality of PUF bitcells. In the following, such metrics are summarized in Table 8.2, where typical values based on current literature are also reported.

In detail, any PUF output should ideally remain the same under fluctuating environmental conditions (e.g., voltage, temperature), and at any process corner. Actual PUFs are not able to provide perfectly stable outputs, due to non-perfect rejection of noise, global and environmental variations. Stability is measured by counting all bits that become unstable across repeated PUF evaluations and environmental conditions, within the specified range of voltage and temperature of operation.

Repeatability (or reproducibility) and uniqueness are measured from the Hamming Distance (HD) across several measurements of PUF keys. Such measurements are compared to a reference "golden" key (Maes 2013) that is taken as the first measurement under nominal conditions. Repeatability is the average intra-PUF Hamming Distance (HD) between the golden key and several key evaluations with the same challenge in the same chip, under different environmental conditions. By definition, highly reproducible PUFs should have low intra-PUF Hamming distance (ideally zero). Uniqueness, on the other hand, is taken as the average inter-PUF HD between the golden key and key evaluations from different chips under the same PUF input (Mathew et al. 2014). The inter-PUF HD should be close to the ideal value equal to half the length of the PUF key (e.g., the ideal inter-PUF HD of a 256-bit key is 128). Alternatively, the fractional Hamming Distance (FHD) can be used to

Table 8.1 Example of SRAM PUF silicon cost (assumed to be 5 cents/mm^2, with area/bit representative of very dense SRAM PUFs)

(Encrypted) data transmitted every	PUF capacity (MB)	PUF area (mm^2)	Silicon cost (US$)
1 h	5	24	1.2
10 min	32	147	7.4
1 min	320	1478	74

Table 8.2 PUF metrics and typical values

Metric	Measured by	Typical value	Ideal value
Stability	Unstable bits	1–60%	0
Repeatability	Intra-PUF FHD	0.8–15%	0
Uniqueness	Inter-PUF FHD	30–60%	50%
Identifiability	Inter/intra HD	5–80	∞
Randomness	0/1 bias	40–60%	50%

quantify reproducibility and uniqueness (Maes et al. 2012), where the Hamming distance is simply expressed as a percentage of the key length, or the number of bits N in a PUF key (ideal inter-PUF FHD is 50%). Identifiability quantifies the distinguishability of a PUF instance to other instances, and is loosely taken as the ratio of the inter-PUF and intra-PUF HD (on the assumption that it is both repeatable and unique), where a larger value is desired (Maes 2013; Mathew et al. 2014; Yang et al. 2015).

Figure 8.2 shows an example of probability distribution function of reproducibility (intra-PUF FHD) and uniqueness (inter-PUF FHD). A perfectly identifiable PUF ideally has no intersection between the inter-PUF and intra-PUF curves, which means that a single PUF response is enough to determine whether the chip is authentic or not. In practical cases, the two curves in Fig. 8.2 have an intersection, and an optimal decision threshold needs to be chosen to determine whether a given PUF is identifiable. As shown in Fig. 8.2, such decision threshold is set by the point where Type I and Type II errors are minimized. Type I error is the false positive, where an invalid key is accepted as a valid one. Type II error, on the other hand, is the false negative, where a valid key is discarded as an invalid one.

Regarding chip authentication, false rejection rate (FRR) and false acceptance rate (FAR) can be used as relevant metrics to assess its quality and security level (Stanzione et al. 2009; Lim et al. 2005). Referring to Fig. 8.2, FRR corresponds to the probability of having an output with FHD under the false negative area, whereas FAR corresponds to the area under the false positive area. Accordingly, the PUF yield Y can be defined as the probability that no authentication error occurs during the lifetime of a given PUF chip, i.e. $Y = 1 - FAR - FRR$. The bit error rate (BER) or the percentage of unstable bits can also be used as a metric of the quality of chip authentication, when the whole array is considered, rather than dividing the array into keys of length N.

Another important property of PUFs is the randomness of its responses, as needed to ensure

Fig. 8.2 Sample Inter- and Intra-PUF FHD showing decision threshold and Type I (false positive) and Type II (false negative) errors

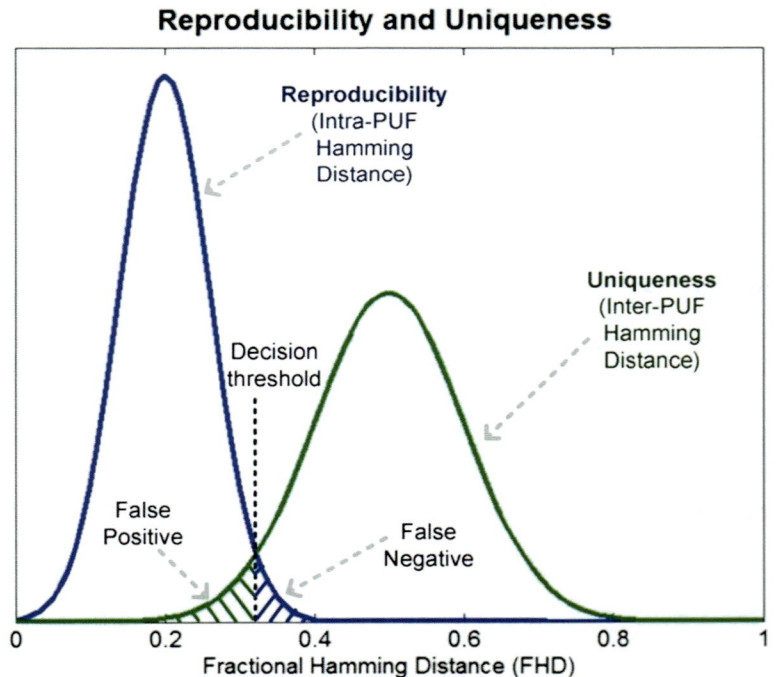

their unpredictability. Randomness is routinely quantified through the statistical characterization in terms of 0/1 bias (defined as the probability of having a 1 in a PUF output bit (Yu et al. 2012)), the entropy (Mathew et al. 2014), and more thoroughly through the NIST randomness tests (Rukhin et al. 2010) (see below). To quantify the randomness of PUF responses across different positions of PUF bitcell within the die, the autocorrelation function (ACF) is routinely used to detect repeating or correlated patterns among different responses (Mathew et al. 2014; Yang et al. 2015). The correlation between PUF output bits is generally due to layout-dependent variations (Alvarez et al. 2015, 2016; Li et al. 2015). Visually, randomness can be represented in the form of the speckle diagram shown in Fig. 8.3, where each pixel represents a PUF bitcell and the PUF output 0's (1's) are represented with black (white) pixels. In Fig. 8.3, the distribution looks somewhat random (i.e., there are no clear patterns) and the 0/1 bias is also close to ideal value of 0.5.

The NIST statistical test suite (Rukhin et al. 2010) is a set of tests to quantify the randomness of a stream of bits. Version 2.1.2 contains 15 tests, each one exercising one property to test randomness. The simplest test is the frequency test, which computes the 0/1 ratio of the whole bitstream. For each of the tests, certain parameters need to be preliminarily set (e.g., length of bitstream n, block size M). Table 8.3 shows the complete list of the tests and parameters to be set.

IoT devices are tightly energy constrained, since they are either battery operated or energy harvested, hence the energy consumption of the PUF is another important metric. To abstract the energy from the PUF organization and size, the most commonly adopted metric is the energy per bit, obtained by dividing the average energy per access by the number of bits within the key. The energy per bit of existing PUFs typically ranges from tens of fJ/bit to tens of pJ/bit (Alvarez et al. 2015, 2016).

Due to the stringent cost requirement in IoT nodes (including silicon area), another important PUF metric is the effective area per bit, as obtained by considering the actual number of available PUF bits obtained after removing unstable bits, and including the area cost of the circuitry performing post-processing on the raw PUF output (see later). Robustness to ageing and chip lifetime are assessed through accelerated ageing tests (Puntin et al. 2008; Stanzione et al. 2009; Selimis et al. 2011). Modeling complexity, in terms of the number of brute force trials needed to model the PUF, can likewise be used to characterize PUFs (Stanzione et al. 2009).

8.3 PUF Topologies and State of the Art

The concept of PUFs have been introduced in the early 2000s, and they have been initially referred to as ICID (Lofstrom et al. 2000), Physical One-Way Functions (POWF) (Pappu et al.

□ 1 ■ 0 INV_PUF: Pr(1) = 0.5072

Fig. 8.3 Sample speckle diagram from (Alvarez et al. 2015)

Table 8.3 NIST statistical test suite

NIST test	Description	Minimum stream length n	Other parameters
Frequency Test	Takes ratio of number of 1's and 0's	100	–
Frequency test within a block	Ratio of 1's and 0's with M-bit block	100	$M \geq 20$ $M > 0.01n$
Runs test	Relative oscillation of bit stream	100	–
Longest Run of Ones	Length of longest consecutive 1's with a block	128	M (set based on present n)
Binary Matrix Run	Rank of disjoint sub-matrix	$38 \cdot M \cdot Q$	M, Q
DFT	Detect periodic features	10^3	–
Non-overlapping Template	Detect occurrence of patterns in an m-bit window	10^6	$m = [2,10]$
Overlapping Template	Detects occurrence of patterns, with overlaps included	10^6	$m = [2,10]$
Universal Statistical Test	Number of bits between matching patterns	387,840	$L = [6,16]$ $Q = 10 * 2^L$
Linear Complexity Test	Length of equivalent LFSR	10^6	M
Serial Test	Detect frequency of overlapping patterns	–	$m < \log_2 n - 2$
Approximate Entropy	Detect frequency of overlapping patterns	–	$m < \log_2 n - 5$
Cumulative Sums	Random walk	100	–
Random Excursions Test	Random walk cycle	10^6	–
Random Excursions Variant Test	Deviations from a random walk	10^6	–

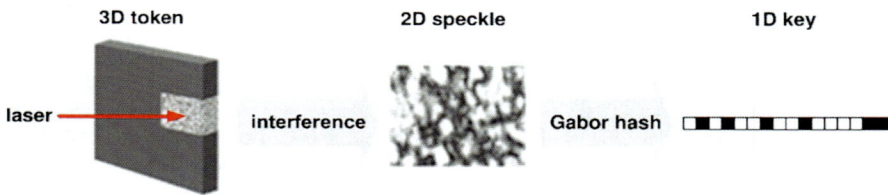

Fig. 8.4 Illustration of the Physical One-Way Function from a non-homogenous material (Pappu et al. 2002)

2002), or Physical Random Functions (PRF) (Gassend et al. 2002), among others. ICID uses an array of MOSFET to generate the random values from random process mismatch, via FET drain current. The physical one-way function was proposed as a solution to the need for a one-way function (easy to evaluate but difficult to invert) for cryptographic applications. The approach uses a laser to scatter light through an inhomogeneous structure (at some precise angle, which serves as the challenge), as shown in Fig. 8.4. The resulting optical speckle diagram is hashed to obtain the key. Most of the literature has then reverted to silicon-based solutions,

leveraging the low-cost and high-volume capability of CMOS chips.

Most of the existing silicon PUFs can be classified as either delay-based or memory-based PUFs (Gassend et al. 2002; Guajardo et al. 2007; Suh et al. 2007; Kumar et al. 2008). In delay-based PUFs, bits are generated by comparing the delay of two nominally identical paths. The sign of the random delay difference between the two delays determines the output bit. One of the earliest implementation of such a concept is the ring oscillator (RO) PUF (Gassend et al. 2002; Suh et al. 2007), whose digital output is determined by the relative frequency of each selected pair of nominally identical ring oscillators. Figure 8.5a shows the general diagram of a ring oscillator PUF, where the challenge selects two of the available ring oscillators, and the corresponding response depends on whether the frequency of the first selected oscillator is greater than the second or not. Knowing that these inverter chain ring oscillators tend to be very sensitive to environmental conditions, several techniques have been introduced to improve the high native instability rate, and poor statistical quality of this pair-wise comparison. Some of these techniques include the adoption of k-sum or 1-out-of-k masking techniques (Suh et al. 2007; Lee et al. 2004).

Another delay-based PUF is the arbiter PUF (Suh et al. 2007; Lim et al. 2005; Lee et al. 2004), as shown in Fig. 8.5b. It compares the delay of two delay lines, and suffers from the same limitations as the RO PUF (Gassend et al. 2002; Lee et al. 2004). A improved delay-based version was recently proposed (Yang et al. 2015), based

on the oscillation collapse in an even-stage ring of delay-adjustable stages. The delay is set by an applied input (PUF challenge) via inverter replica multiplexing. The native instability of PUF outputs was substantially reduced at the cost of much higher energy and the need for CTAT biasing.

All above delay-based PUFs are also intrinsically vulnerable to PUF modeling attacks, which can capture and clone the content of the entire PUF with very low effort. Indeed, the PUF output is dictated by the overall PUF delay, which is in turn defined by the sum of the delays of cascaded stages. Since each stage delay is fixed (although unpredictable), identifying all stage delays from the analysis of the PUF outputs entails only a linear complexity, making the PUF easy to clone (Rührmair et al. 2010).

In memory-based PUFs, a bistable structure of two cross-coupled inverters is used to generate the output bits. They leverage on the natural tendency of cross-coupled inverters to resolve

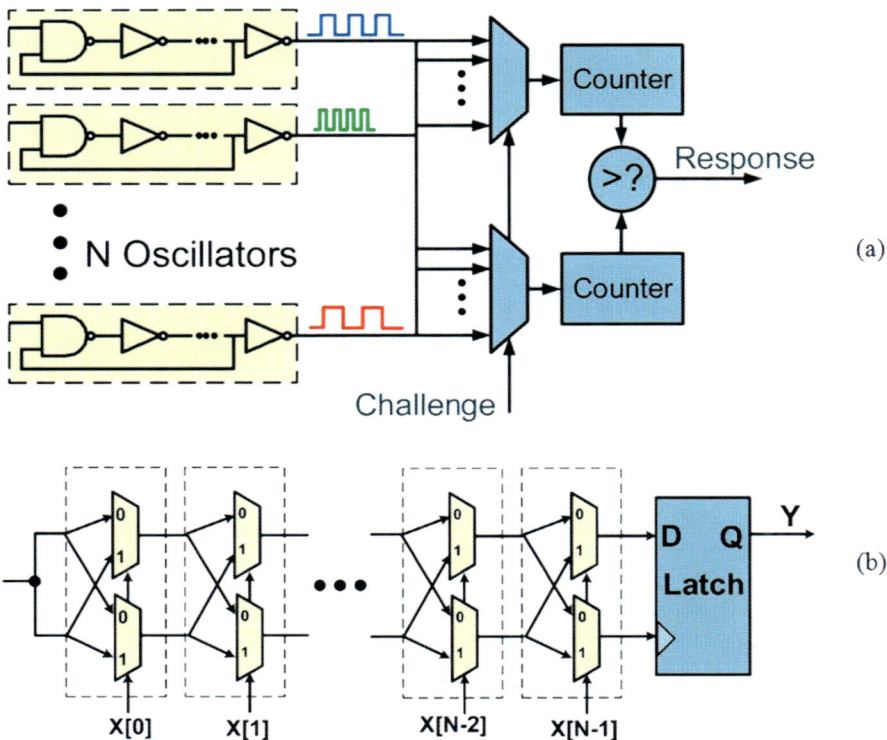

Fig. 8.5 Delay-based PUFs: (**a**) ring oscillator (RO) PUF (Suh et al. 2007), (**b**) arbiter PUF (Lee et al. 2004)

to a preferred state at the power-up, as determined by their asymmetry due to random variations (Suh et al. 2007). For example, SRAM PUFs leverage this property in SRAM bitcells (Guajardo et al. 2007; Holcomb et al. 2009). Other similar PUFs are the Latch PUF (Su et al. 2007), DFF PUF (Maes et al. 2008), butterfly PUF (Kumar et al. 2008), and the buskeeper PUF (Simons et al. 2012), which is similar to the SRAM PUF albeit without the write ability, as access transistors are removed since PUF bitcells are read-only. The butterfly PUF (Kumar et al. 2008) follows the same concept of leveraging on the unstable state of cross-coupled inverters. It was proposed for implementation in an FPGA and uses the available cross-coupled latches instead of inverters, as shown in Fig. 8.6. The operation starts by asserting the *excite* signal, thereby forcing the PUF to be in the unstable state. This signal is then released and after a few clock cycles, *out* signal settles to its natural stable state determined by the random variations in the related logic gates.

The recent literature on memory-based PUFs and their experimental characterization has shown that PUFs typically have poor stability (Schrijen et al. 2012), and are highly vulnerable to semi-invasive attacks such as electrical and optical probing (Nedospasov et al. 2013). The same vulnerability to semi-invasive attacks is found in other PUFs that rely on the same principle, such as senseamp (Bhargava et al. 2010; Bhargava and Mai 2014). For such PUFs, reasonable levels of stability are typically achieved through substantial temporal redundancy at the expense of energy consumption (Helinski et al. 2009). Other proposed PUFs are based on (1) the glitch generated in digital paths, although they suffer from high instability rates (Suzuki et al. 2010), (2) the difference in leakage current generated by nominally identical transistors, although at the cost of large energy due to the necessary circuitry for current/voltage references and opamp (Ganta et al. 2011), (3) DRAM errors under different wordline voltage, although such PUFs are highly vulnerable to non-invasive attacks (Rosenblatt et al. 2013), (4) open or short connection in vias (Choi et al. 2014), (5) oxide breakdown in OTP ROMs (Liu et al. 2010), (6) capacitance mismatch (Tuyls et al. 2006; Roy et al. 2009; Wan et al. 2015), (7) the variations in the supply network resistance, although this requires the generation of very large currents (Helinski et al. 2009).

A hybrid PUF was proposed in (Satpathy et al. 2014; Mathew et al. 2014, 2016) combining delay and metastability as sources of randomness. The basic bitcell is shown in Fig. 8.7, where bistability is forced through the pre-charge transistors controlled by *clk0* and *clk1*. The randomness in delay is introduced through the clock skew between *clk0* and *clk1*. In order to reduce unstable bits, significant temporal majority voting is employed. Soft dark bit masking was also used in (Satpathy et al. 2014) by modulating the load in the *bit* and *bit'*, and masking bits that become unstable with the change in the load. Indeed, load modulation simply injects controlled perturbation in the stability of the PUF bitcell, which in turn permits to identify the truly stable bitcells that do not change output even in the presence of such perturbation.

In order to achieve adequate native stability in spite of voltage and temperature fluctuations, authors in (Li et al. 2015) proposed to use a proportional-to-absolute-temperature (PTAT) as a bitcell. Figure 8.8 shows the bitcell and the

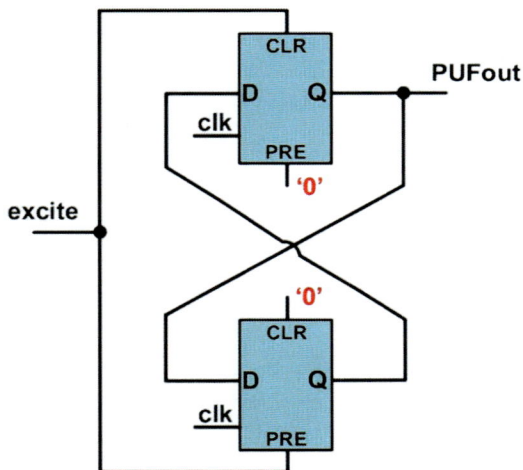

Fig. 8.6 Butterfly PUF (Kumar et al. 2008)

Fig. 8.7 Metastability-based PUF (Mathew et al. 2014)

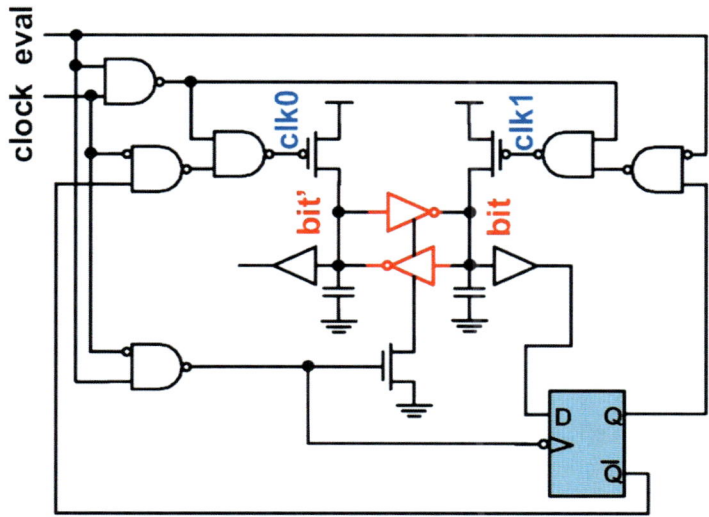

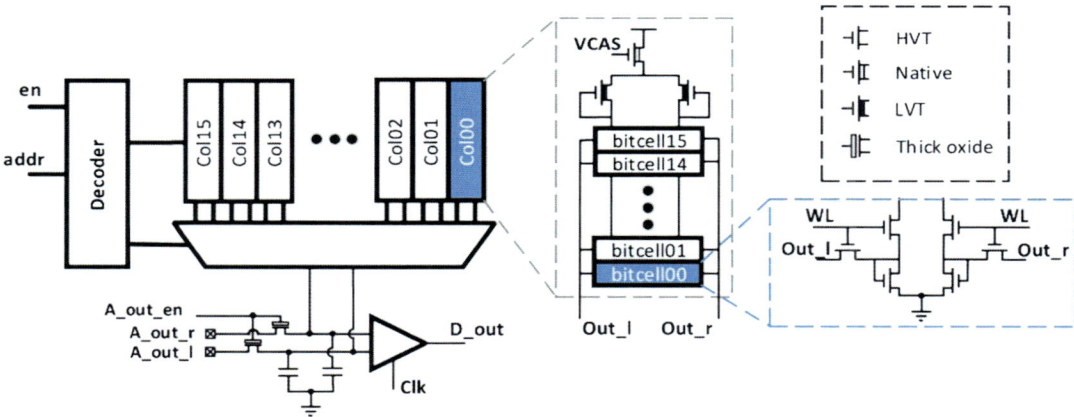

Fig. 8.8 PTAT-based PUF (Li et al. 2015)

architecture and principle of operation of the PTAT-based PUF. As seen in the figure, the PUF bitcell output is determined by the sign of the difference between *Out_l* and *Out_r*, both of which are independent of voltage and temperature. Aside from the high resiliency against voltage and temperature variation, another noteworthy feature of this PUF is its good area efficiency (only 727 F^2/bit, being F the minimum feature size of the adopted process), as enabled by the shared header per column.

A class of static mono-stable PUFs (Alvarez et al. 2015, 2016) for extremely low energy operation and low native instability rate was recently proposed, which relies on the amplification of random transistor mismatch through two complementary current mirrors. Figure 8.9 shows two implementations of the same general concept. Figure 8.9a shows the INV_PUF bitcell implementation of this concept, which comprises the cascode current mirrors M1–M4 and M5–M8. The two 1:1 current mirrors see the same current flowing through their respective input transistors (M3 and M5), and tend to mirror it to their output transistors (M4 and M6, respectively). Without mismatch, M4 and M6 would

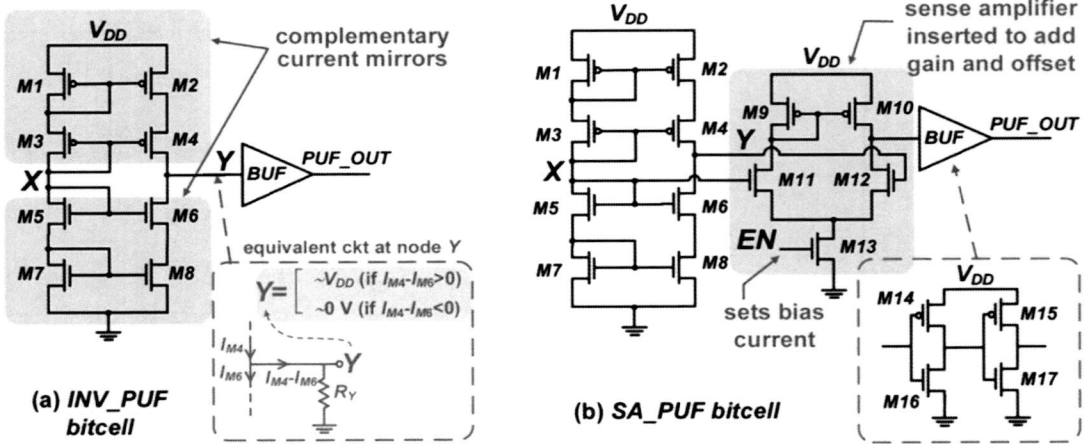

Fig. 8.9 Static mono-stable PUFs (Alvarez et al. 2016) (**a**) INV_PUF and (**b**) SA_PUF

conduct the same saturation current ($I_{M4,SAT}$ and $I_{M6,SAT}$), and node Y would assume the same voltage as node X (e.g., $V_{DD}/2$), due to the symmetry of the topology in Fig. 8.9a. However, random mismatch between M1–M2 and M7–M8 makes these currents unpredictably different. The large output small-signal impedance R_Y at node Y (Fig. 8.9a) translates the difference in such currents into a large voltage deviation. Accordingly, the voltage at node Y becomes essentially V_{DD} if $I_{M4,SAT}-I_{M6,SAT}>0$, or ground if $I_{M4,SAT}-I_{M6,SAT}<0$. Thus a digital output that is dominantly defined by the random mismatch between the two current mirrors is generated. In Fig. 8.9b, the alternative SA_PUF topology adds a sense amplifier (transistors M9–M13) after M1–M8 to further increase the voltage gain (and thus reduce the number of unstable bits) and introduce additional random mismatch through the sense amp offset.

Table 8.4 compares various PUFs in terms of the above mentioned metrics, with the best performing PUF for each metric being highlighted in bold. As can be seen from this table, most PUFs are typically affected by relatively poor randomness/statistical quality (see, e.g., bias) (Maes et al. 2012; Maes 2013; Yu et al. 2012) and up to 30% instability rate (Maes et al. 2012; Alvarez et al. 2015, 2016; Bhargava and Mai 2014).

A more complete list of fabricated PUFs can be found in the new public PUF database (Alioto and Alvarez 2016). Extracted trends in terms of native instability rate, area, and energy are shown in Fig. 8.10. From Fig. 8.10a, the metastability-based PUFs have the worst native instability rate, while the monostable PUFs exhibit the best native instability rate. The high native instability rate in metastability-based PUFs is reduced through post-processing and other stability enhancement techniques that increase testing time (i.e., cost). For the rest of the PUFs, the native instability rate has slightly increased over the years.

From Fig. 8.10b, the area per bit is highest for delay-based PUFs, due to the large number of stages required to (1) limit the oscillation frequency to acceptable values that can be distinguished by the subsequent circuitry, (2) to mitigate the instability rate of individual ring oscillators via k-sum or 1-out-of-k masking (Suh et al. 2007; Lee et al. 2004). In general, the area efficiency of PUF bitcells has improved over time, especially due to the adoption of more digital approaches that offer better density than analog ones. Analog PUF bitcells have an opposite trend, as their area tends to increase over time, when area is normalized to the square of the minimum feature size of the technology. This is mostly because of their analog nature, which

Table 8.4 Comparison of different PUFS

PUF	SRAM_PUF*	RO_PUF*	ICID (Lofstrom et al. 2000)	Arbiter (Lim et al. 2005)	Latch (Su et al. 2007)	Mathew et al. (2014)	PTAT (Li et al. 2015)	Yang et al. (2015)	INV_PUF (Alvarez et al. 2016)
Technology	65 nm	65 nm	0.35 μm	0.18 μm	0.13 μm	22 nm	65 nm	40 nm	65 nm
Stability (% native unstable bits at nominal condition)	16.66	18.16	1.3	9.8	3.04	30	7.1	12.5	2.34
V-T variation	0.6–1 V	0.4–0.5 V	1.1–5 V, −25 to 250 °C	1.8 V ±2%, 27–70° C	0.9–1.2 V	0.7–0.9 V		0.7–0.9 V	0.6–1 V, 25–85 °C
% error with VT variation	55.73	53.9	5	4.82	5.46875	30		12.5	5.72
Energy (pJ/bit)	1.1	0.4748	8333.3333	0.17125	0.93	0.19	1.1	17.75	0.015
Area (F^2/bit)	306	39,000	1708	708,403	4369	9628	727	2062	6000
Randomness (bias = probablity of 1)	0.6141	0.5023				0.4805	0.4928		0.5016
Uniqueness (mean inter-PUF FHD)	0.3321	0.4738	0.4911	0.3800	0.5055	0.5100	0.5001	0.5007	0.5014
Repeatability (mean intra-PUF FHD)	0.0602	0.0458	0.0134			0.0268	0.0057	0.0101	0.0034
Identifiability (inter-PUF/intra-PUF FHD)	6	10	37			19	88	50	149
NIST test						PASS	PASS	PASS	PASS
Entropy	0.9903	0.9947				0.9997	0.9998		0.9967
Autocorrelation function @ 95% confidence	0.0156	0.0884				0.01	0.0188	0.0283	0.0363

*Data taken from (Alvarez et al. 2016)

Fig. 8.10 Trend of (**a**) native instability rate, (**b**) area per bit (normalized to F^2, F being the minimum feature size of the CMOS process), (**c**) energy per bit for different PUFs implemented in custom PUF chips (Alioto and Alvarez 2016) (normalized to the energy E_{inv} consumed by a minimum-sized inverter in a single transition)

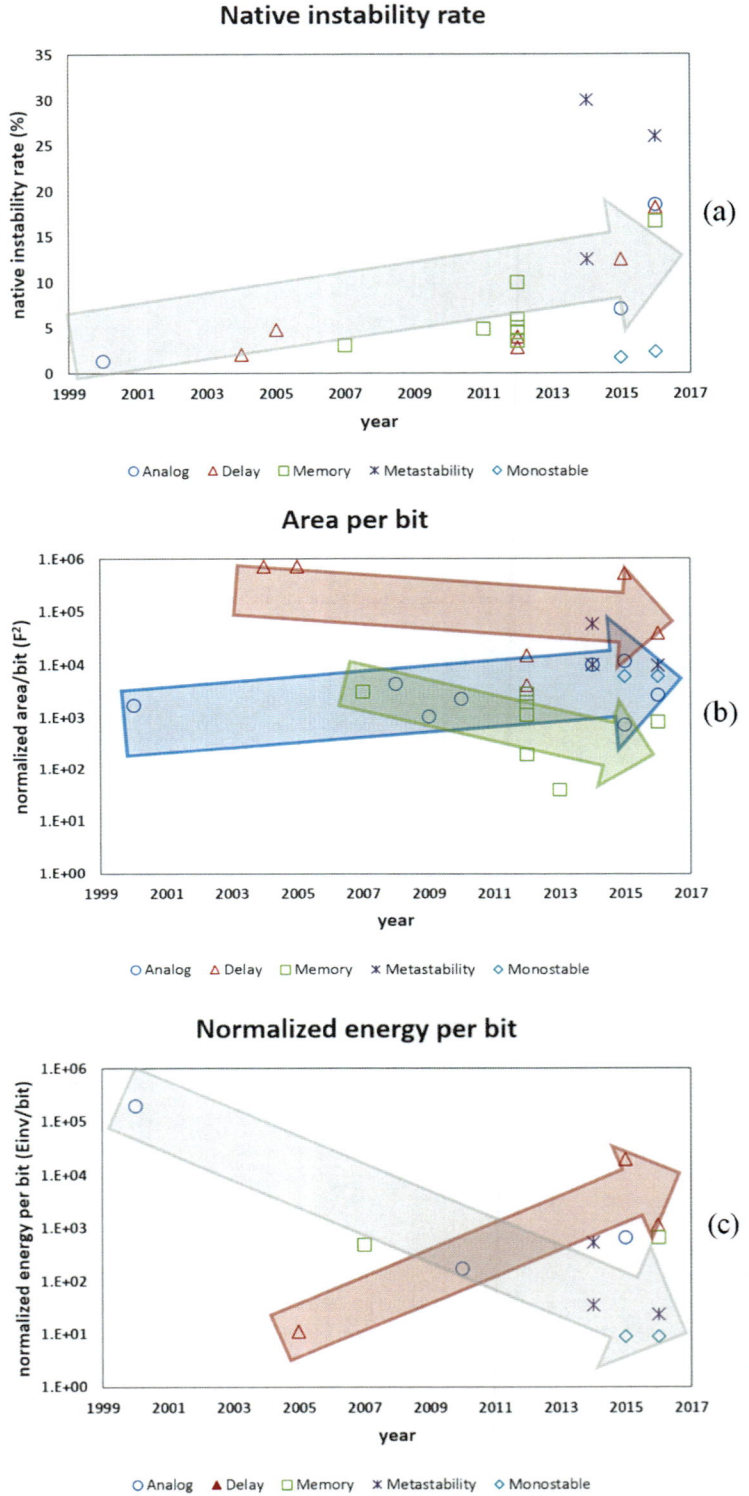

does not really enable shrinking with finer technologies.

From Fig. 8.10c, the energy per bit is improving, thanks to the adoption of more energy-aware PUFs. The circuit improvements in terms of energy definitely dominate the benefits of mere technology scaling. This is shown by Fig. 8.10c, which plots the energy normalized to the energy consumed by a minimum-sized inverter in the same technology, and hence represents a technology-independent metric. Interestingly, from Fig. 8.10c delay-based PUFs are an exception, as they tend to have larger energy per bit over the years. This is due to the need for a larger number of ring oscillators or oscillations to maintain acceptable stability, in spite of the progressively worse native stability in 8.10a.

Some prior work enables the capability to assure a well-defined stability safety margin at the output word level (Yu et al. 2012), as a form of robustness assurance against individual bit instability. Other prior work focuses on improving the stability of PUF bitcells without quantitative stability assurance at run-time. For example, introducing burn-in hardening in (Mathew et al. 2014) improves stability at the expense of significantly longer testing time, which conflicts with the very low cost requirement of IoT nodes (see Chap. 1). Another way to improve the statistical quality and suppress a limited number of unstable bits is through digital post-processing, at the expense of substantially larger silicon area and energy. The post-processing block can be a mixture of the following techniques:

• Error Correcting Code (ECC), which introduces a large area/energy overhead especially for high levels of targeted security, as its complexity grows exponentially in applications requiring wider PUF outputs; post-processing also leaks information and makes the PUF more vulnerable to physical attacks (Bhargava and Mai 2014). Various ECCs were used (Yu et al. 2012), such as 2D Hamming (Gassend et al. 2002), BCH (Suh et al. 2007; Mathew et al. 2016), two-stage ECC (Bösch et al. 2008), soft-decision ECC (Maes et al. 2009a, b), Index-Based Syndrome

(Yu et al. 2011), Code-Offset Syndrome (Gassend et al. 2002; Yu and Devadas 2010; Suh et al. 2007; Dodis et al. 2008; Paral et al. 2011; Eiroa et al. 2012), pattern matching techniques (Paral et al. 2011), and fuzzy extractors (Selimis et al. 2011; Dodis et al. 2008)

• temporal majority voting across repeated PUF readings, which typically slow down and increase the energy per access by more than an order of magnitude (Mathew et al. 2014, 2016; Li et al. 2015)

• on-the-fly PUF bitcell masking (Satpathy et al. 2014), and PUF redundancy (Yang et al. 2015; Suh et al. 2007), which skips the bitcells that are found to be unstable at testing time by storing the bit error map in an additional volatile memory array (Mathew et al. 2014; Bhargava and Mai 2014; Karpinskyy et al. 2016); this approach may introduce significant area/energy overhead, and considerably widens the opportunities to perform successful invasive attacks (e.g., interfering with PUF operation by writing on the additional memory).

Figure 8.11 shows an example where ECC is used to improve the reliability of the PUF (Gassend et al. 2002). In this implementation, redundant information is generated for each challenge-response pair, to allow the correction of the PUF output. The ECC overhead is ~14 kgates, which is about an order of magnitude bigger than the PUF array itself. Similarly, in (Rahman et al. 2014), ECC encoder was shown to have an area of ~3–12 kgates, with the ECC decoder requiring an even larger area of ~20–75 kgates. Detection of instability was proposed in (Karpinskyy et al. 2016) during the PUF response generation, and in (Satpathy et al. 2014) at boot time, as depicted in Fig. 8.12.

8.4   PUF Vulnerability Analysis

Existing PUF solutions suffer from various limitations that have limited their adoption in

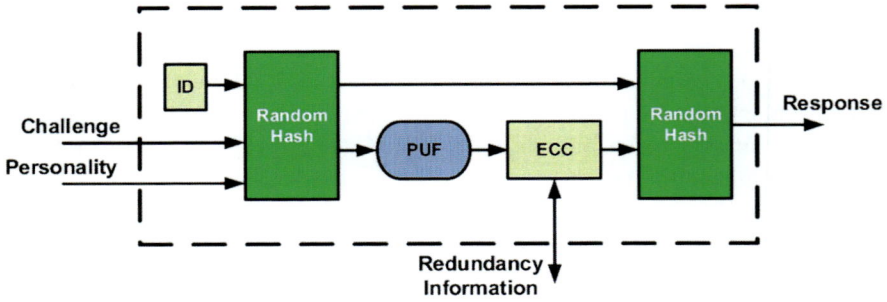

Fig. 8.11 Block diagram of an improved PUF that utilizes ECC to improve the PUF reliability (Gassend et al. 2002)

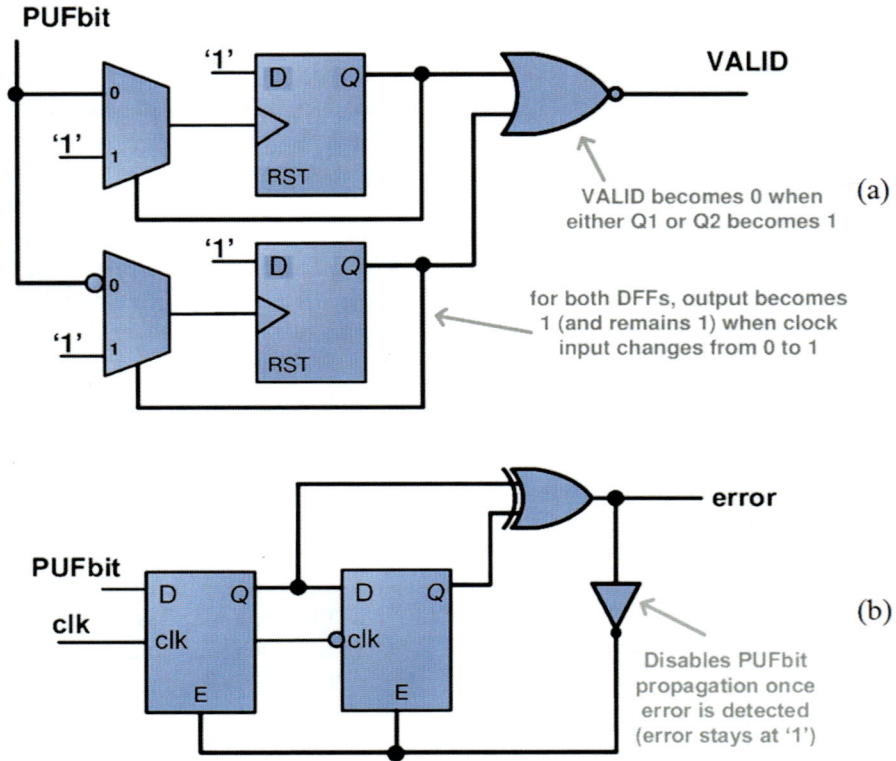

Fig. 8.12 Possible circuits for runtime error detection: (**a**) glitch detector from (Karpinskyy et al. 2016); (**b**) dark bit masking from (Satpathy et al. 2014)

real products. Indeed, to date there are only a few PUFs available in the market, such as (ICTK, Co. Ltd. 2014; Intrinsic-ID 2016; Invia PUF 2016; QuantumTrace 2013; Verayo Inc. 2013). For example, delay- and metastability-based PUFs are very sensitive to voltage/temperature variations, aging and noise (Maes et al. 2012), and are very hard to verify at design time in terms of output statistics and randomness. Hence, they typically require multiple silicon runs to reliably assess a given design. Glitch based PUFs are not yet mature, and are well known to be rather unstable and complex. Leakage-based PUFs are sensitive to environmental variations and need extra circuitry for voltage and current biasing, which are prohibitively costly in terms of area,

Fig. 8.13 Attacks to PUFs
versus level of invasiveness
and level of abstraction

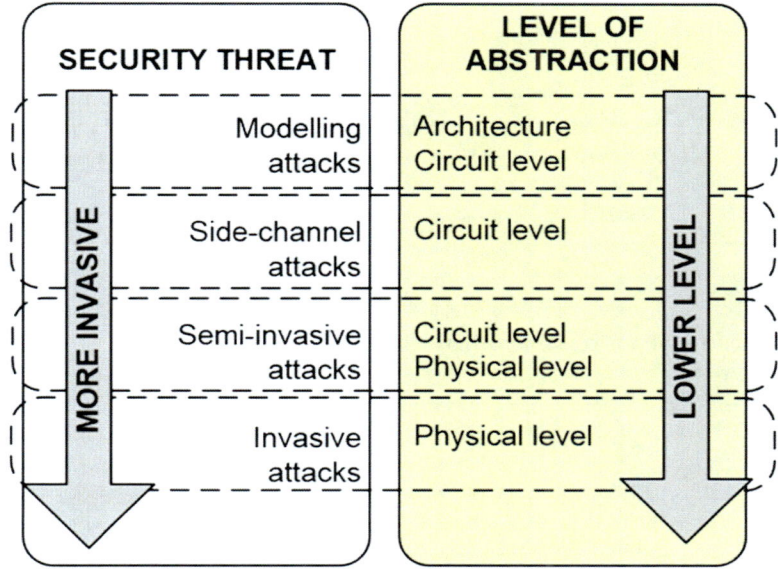

energy and design effort. Memory-based PUFs
are strongly technology dependent and hence not
technology-portable, and suffer from bit flipping
and data remanence (Eiroa et al. 2012). DRAM
error maps are well known to have obvious correlation between different responses, thus drastically weakening the unpredictability of
responses (Rosenblatt et al. 2013). Both delay-
and supply network-based PUFs are very vulnerable to modeling attacks (Rührmair et al. 2013).

The trustworthiness of a PUF is defined by its
resistance to attacks that aim to impersonate,
replicate or recover portions of the PUF bits.
These attacks can be active (injecting fault into
the design) or passive (simply observing). They
can also be classified as invasive (meaning
requiring depackaging the chip to see the design
or probe internal signals) or non-invasive. The
most representative attacks to PUFs are
summarized in Fig. 8.13, from the least to the
most invasive. Modeling attacks are passive
non-invasive, and involve only the observation
of transmitted information and successive trials
to impersonate the device by leveraging the small
search PUF key space or poor randomness, or by
recording and reusing previous CRPs (Sadeghi
and Naccache 2010) (e.g., man-in-the-middle

attacks). Delay-based PUFs are prone to these
types of attacks. In (Rührmair et al. 2013), for
example, they were able to show more than 95%
prediction rate using machine learning to model
the arbiter and ring oscillator PUFs.

For identification purposes, a "strong PUF"
with large number of CRPs is clearly needed to
limit the effectiveness of man-in-the-middle
cryptanalytic attacks, but unfortunately all practically viable strong PUFs are very vulnerable to
modeling attacks (Sadeghi and Naccache 2010;
Helinski et al. 2009), and hence unsuitable for
moderate-to-high levels of security.

Side-channel attacks aim to identify the PUF
key employed by the chip through its correlation
with the measured power consumption (e.g.,
DPA (Kocher et al. 1999), correlation attacks
(Brier et al. 2004) and Leakage Power Analysis
(Alioto et al. 2014, 2010a; Giorgetti et al. 2007)
or electromagnetic emissions (Mangard et al.
2007)). These attacks are performed on the execution of the cryptographic algorithm that uses
the key that the attacker is trying to retrieve.
Differential Power Analysis (DPA) was proposed in 1998 (Kocher et al. 1999). Figure 8.14
shows a sample trace of the whole DES operation, where the 16 repetitive pulses correspond to

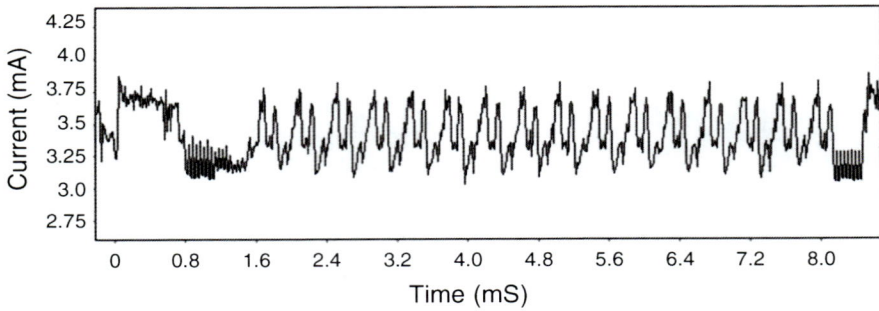

Fig. 8.14 SPA traces showing 16 rounds of DES

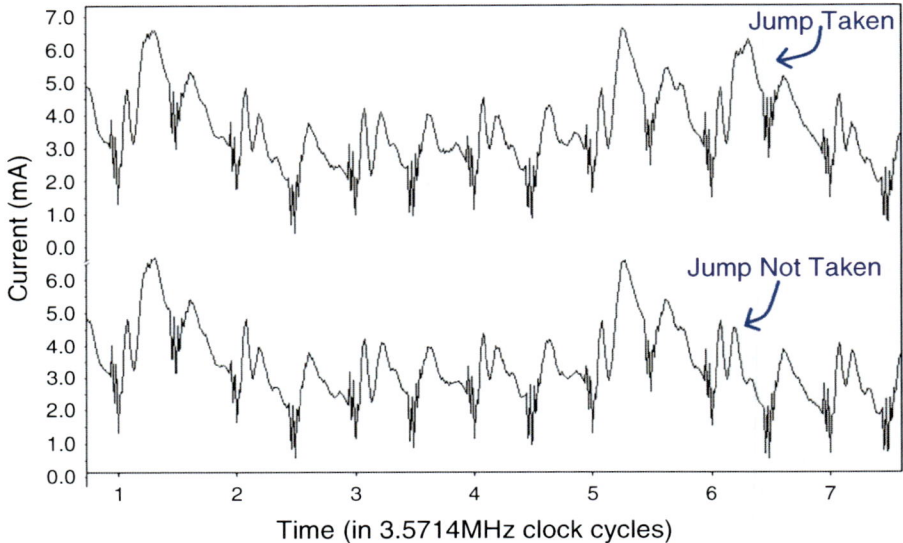

Fig. 8.15 SPA single round trace showing difference in power consumption

the 16 rounds of DES. Measurements are performed during encryption (or decryption) under different plaintexts, in order to see the difference in power consumptions in the different stages of the algorithm. A closer view of the same pulse in Fig. 8.14 could also reveal more about the data, as shown in Fig. 8.15. From the figure, the difference in power consumption in cycle 6 is due to a jump instruction (where jump is taken in the top plot and not taken in the bottom plot). The idea of DPA is to retrieve the key of the cipher using divide and conquer approach by guessing portions of the key, thereby breaking the exponential complexity of deciphering the key and reducing the number of required trials. For each trial, the cipher's power consumption estimated under two key bit guesses is compared, or correlated to the actual power. The difference of two traces (see, e.g., Fig. 8.15) has a spike in cycle 6 when such key bit is used in the algorithm execution, and is almost zero elsewhere. Similar procedure is applied to specific operations in the algorithm in order to help identify whether the initially assumed key is correct or not. A power model for DPA attacks on symmetric-key cryptographic algorithms implemented using static logic was proposed in (Alioto et al. 2010b). The model predicts the effectiveness of DPA attacks and the conditions for which the circuit becomes vulnerable to these attacks.

As DPA attacks target the cryptographic core rather than the PUF itself, one way to prevent them is to mask the power consumption of different operations. This can be done through a change in the algorithm, masking data (Canright et al. 2008), or resorting to codeword encoding (Merli et al. 2013), among others. The use of different logic styles, such as the sense amplifier based logic (SABL) (Tiri et al. 2002) and dual-rail pre-charge (DRP) circuits, such as wave differential dynamic logic (WDDL) (Tiri et al. 2004), masked dual-rail pre-charge logic (MDPL) (Popp et al. 2005) and dual-rail random switching logic (DRSL) (Chen et al. 2006), was shown to be effective in masking operations by maintaining almost the same power consumption regardless of the operation and processed data (Schaumont et al. 2007; Monteiro et al. 2011).

Semi-invasive attacks (e.g., fault attacks) aim to interfere with the circuit operation by introducing glitches and injecting faults that expose data that would otherwise be securely processed internally. Fault injection and timing attacks can be avoided by consistency checking, at the expense of additional runtime and/or area. Laser scanning was used in (Holcomb et al. 2009) and (Nedospasov et al. 2013) to read out the state values of memory elements.

Invasive attacks aim to physically observe (e.g., reverse engineering) or modify the chip physical implementation (e.g., probing, fibbing), and can be counteracted through secure coprocessors (Smith and Weingart 1999). This solution, however, is very expensive both in terms of area and energy, as the coprocessor has to be powered at all times. The work in (Wan et al. 2015) proposes to counteract invasive attack by laying out metal wires on top of transmission lines to switch capacitor circuitry. By doing this, the capacitance of the sampling capacitor changes during invasive attacks, making the PUF output invalid.

The targeted level of trustworthiness defines an adequate set of attacks that need to be counteracted in a PUF design, although to date only individual and fragmented techniques have been proposed. As research challenge in the area of PUFs, a comprehensive set of techniques would be needed to meet a given level of trustworthiness, with each being allocated to the appropriate level of abstraction to meet a security level target at low cost.

8.5 Novel Concept of PUF-Enhanced Cryptography, Trends and Perspectives

Despite the recent and broad interest in PUFs, they have made a very limited impact on real applications to date due to several challenges that seriously hinder PUF trustworthiness, as above mentioned and summarized below:

- PUF responses can be very unstable (i.e., not repeatable), thus requiring a large cost in terms of testing time (post-silicon masking, PUF hardening), area or energy overhead (to suppress unstable bits at design time, and to include expensive post-processing blocks). Additional post-processing circuits also make PUF more vulnerable to physical attacks (e.g., side-channel, probing), as they become part of the PUF itself, and hence introduce additional backdoors to the PUF
- there is no available methodology to systematically assure a level of security (e.g., bit randomness, stability) at design/verification time. This prohibits the provability of PUF trustworthiness at design time, requires repeated silicon runs to converge to a targeted and provable security level, and drastically prolongs the design cycle and time to market
- existing solutions only target a specific type of security threat and level of abstraction (mostly circuit level). Hence, they cannot address the security challenges posed by different types of attacks, and do not introduce across-level solutions (e.g., from PUF layout to architecture and software)
- the PUF design is essentially a manual design process, which prohibits design automation and technology portability, and entails very low design productivity
- specifically for IoT nodes, constant node-to-node communications and data transmission

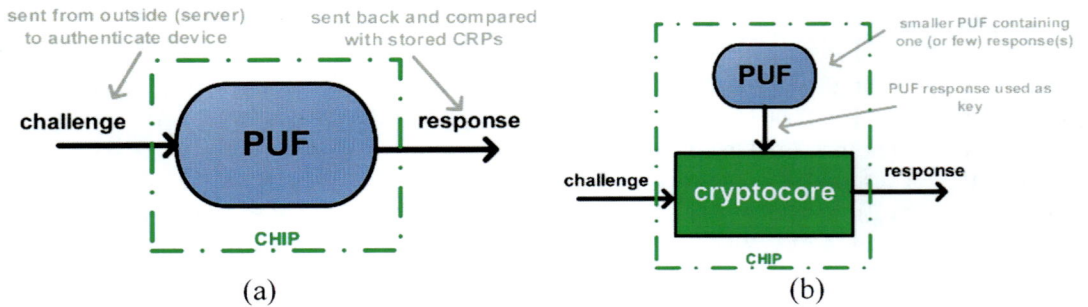

Fig. 8.16 (**a**) PUF as key generator, (**b**) crypto-core-based strong PUF

requires regular authentication. This translates into and unacceptably large number of CRPs and therefore PUF capacity (see Table 8.1). Public-key cryptography to establish trust cannot mitigate such problem, due to its prohibitive area and energy cost in IoT nodes.

So far, we have treated PUFs as secure random key generators for chip identification through conventional challenge-response pairs, as illustrated in Fig. 8.16a. A more promising approach implements a strong PUF through a crypto-core (e.g., AES) using the PUF key as encryption key, and then treating input/output values as CRPs (Bhargava and Mai 2014), as illustrated in Fig. 8.16b. In (Bhargava and Mai 2014), the introduction of AES increased the area by $4.6\times$ compared to the PUF array, but increased the number of available CRPs exponentially. By using an AES design that is designed specifically for IoT applications (Zhao et al. 2015), the power consumption of AES is well below 1 μW and the area cost is reduced by $3\times$, thus becoming very affordable for the same exponential increase in CRPs.

For IoT node-to-node communications, the concept of combining a PUF with a crypto-core can also be used to reduce the circuit complexity and energy required for continuous authentication, thereby reducing the required PUF capacity at a given level of security. Conventional node-to-node communication is illustrated in Fig. 8.17, where CRPs are used to authenticate both nodes each time data is transferred between them.

Instead, a more efficient security scheme is introduced in Fig. 8.18. In this "PUF-enabled node-to-node communication" scheme, secure PUF key exchange is enabled at the authentication phase through cryptography. After one-time authentication, both nodes can communicate with each other securely through encryption and decryption using the exchanged keys, and without server assistance (therefore not needing a large CRP database). This makes communication over complex networks scalable, as the database is involved only at the first communication between nodes. As can be seen in the figure, node-to-node communication is simplified through the joint use of PUF and cryptography, which permit to securely exchange keys over an insecure channel, and avoiding the very energy- and area-hungry public-key cryptography. Such interesting and synergistic use of PUFs and cryptography is here introduced and named "PUF-enhanced cryptography".

Another interesting ramification of PUF-enhanced cryptography is the ability to substantially strengthen the security of a crypto-core against cryptanalytic attacks, by appropriately embedding a PUF into it. As illustrated in Fig. 8.19, PUF-enhanced cryptography goes beyond the traditional scheme of securely storing a single crypto-key, and permits to extend the crypto-key compared to the size imposed by the crypto-algorithm, thus making it stronger against cryptanalytic attacks. Traditionally, key extension is not possible since its length is dictated by the encryption standard. However, in PUF-enhanced cryptography, a PUF with capacity

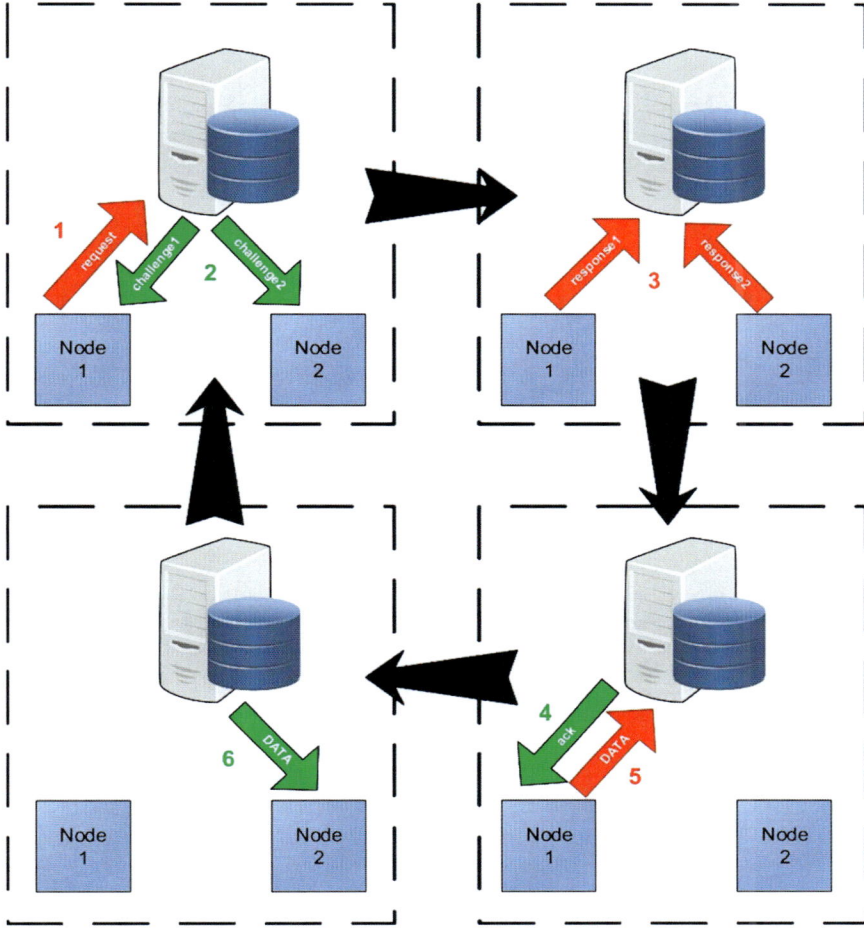

Fig. 8.17 Conventional node-to-node data transfer through server, which needs to constantly assist the two nodes during their communications

larger than the key is used to generate repeatable but unpredictable new keys that are combined with the conventional user key to generate the fixed length enhanced key used by the on-chip crypto-core. To this aim, the key enhancer in Fig. 8.19 is introduced to dynamically concatenate the user and PUF keys, and then compress them into the pre-defined length. Although the key enhancer in Fig. 8.19 is shown to be outside the crypto-core (i.e., without interfering with conventional operation), it can also extend to the inside of the latter, and operate across several blocks of plaintext. The encryption sequence is initialized by the user key, and then managed by a key enhancer. The key enhancer can likewise be a simple finite state machine, which generates time-varying challenges to a PUF, or a lightweight cipher itself (Shiozaki et al. 2015). As a result, as opposed to the traditional scheme that uses a single private key, the PUF-enhanced cryptography scheme in Fig. 8.19 actually uses a larger set of keys, whose number is basically limited by the desired PUF capacity.

From an attacker point of view, guessing the private crypto-key of a typical cryptography system requires an effort that is (exponentially) defined by the size of the single key size. Instead, in the PUF-enhanced cryptography scheme in Fig. 8.19, the search space for the crypto-key is enlarged by the capacity of the PUF, thus easily making the key search unfeasible even under

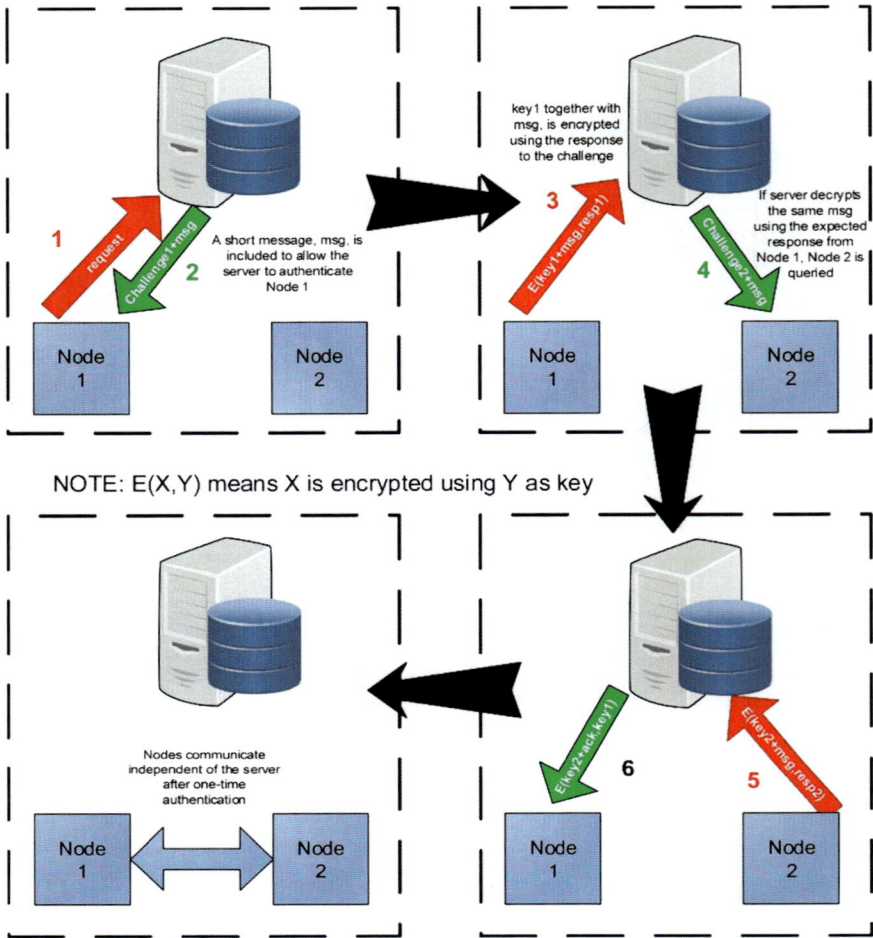

Fig. 8.18 PUF-enabled key exchange and node-to-node communication

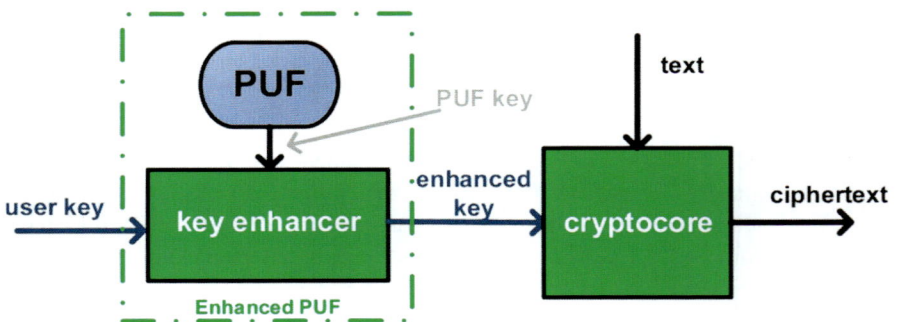

Fig. 8.19 The new concept of PUF-enhanced cryptography (key is continuously enriched with PUF key).

very powerful equipment and computing resources. In practical cases, the PUF-enhanced cryptography permits to drastically strengthen the security of an existing algorithm with (1) limited area cost, thanks to the exponential increase of the size of the key search space, under PUF capacity extension, and (2) no throughput penalty, since the generation of the PUF output is generally much faster than encryption. When using PUFs like in (Alvarez et al. 2016), the latter property is enabled by the intrinsically high speed of the PUF architecture, since PUF bits are always available at the output and only need to be routed to the circuitry that consumes them.

The above mentioned dynamic change of the key over time is a tool to improve the strength of PUF-enhanced cryptography against cryptoanalytic attacks. In the case of IoT devices relying on energy harvesting, changing keys becomes a necessity as dictated by the availability of supply. For example, in (Aysu and Schaumont 2016) key generation is divided into several phases and precomputation is done whenever supply available, and intermediate results are stored, for use in the next phase.

In summary, PUF-enhanced cryptography permits to drastically enhance the security of a crypto-core by leveraging its synergy with a PUF, to generate time-varying crypto-keys instead of having a fixed one. In addition, the adoption of such PUF to enhance the crypto-algorithm also permits to easily scale up the level of security on demand. Indeed, the level of security defines the number of PUF words that are needed, and hence it only affects the periodicity of the key enhancer for a given PUF capacity. Also, the PUF unambiguously authenticates the die that the crypto-core runs on. In addition, the addition of a PUF to a crypto-core generally entails a very small energy overhead, as the energy per bit of a PUF is typically two to three orders of magnitude smaller than a crypto-core. Very similar considerations hold for the area efficiency. These features are particularly interesting in the context of the Internet of Things, as they make crypto-algorithms and crypto-cores affordable in terms of area and energy, thus enabling continuous and ubiquitous security. When a much higher level of security is occasionally needed, the PUF enhancement permits to further scale it up at a very low area/energy cost.

Acknowledgement The authors acknowledge the kind support by the MOE2016-T2-1-150 grant from the Singaporean Ministry of Education.

References

M. Alioto, L. Giancane, G. Scotti, A. Trifiletti, Leakage power analysis attacks: a novel class of attacks to nanometer cryptographic circuits. IEEE Trans. Circuits Syst. I **57**(2), 355–367 (2010a)

M. Alioto, M. Poli, S. Rocchi, Differential power analysis attacks to precharged buses: a general analysis for symmetric-key cryptographic algorithms. IEEE Trans. Dependable Secur. Comput. **7**(3), 226–239 (2010b)

M. Alioto, S. Bongiovanni, M. Djukanovic, G. Scotti, A. Trifiletti, Effectiveness of leakage power analysis attacks on DPA-resistant logic styles under process variations. IEEE Trans. Circuits Syst. I Regul. Pap. **61**(2), 429–442 (2014)

A. Alvarez, W. Zhao, M. Alioto, 15 fJ/bit static physically unclonable functions for secure chip identification with <2% native bit instability and 140x inter/intra puf hamming distance separation in 65 nm. IEEE Int. Solid-State Circuits Conf. **5**, 256–258 (2015)

A.B. Alvarez, W. Zhao, M. Alioto, Static physically unclonable functions for secure chip identification with 1.9–5.8% native bit instability at 0.6–1 V and 15fJ/bit in 65 nm. IEEE J. Solid State Circuits **60**(5), 1–4 (2016)

A. Aysu, P. Schaumont, Precomputation methods for hash-based signatures on energy-harvesting platforms. IEEE Trans. Comput. **65**(9), 2925–2931 (2016)

M. Bhargava, K. Mai, An efficient reliable PUF-based cryptographic key generator in 65 nm CMOS. Design Autom. Test Europe Conf. Exhibition **1**, 1–6 (2014)

M. Bhargava, C. Cakir, K. Mai, Attack resistant sense amplifier based PUFs (SA-PUF) with deterministic and controllable reliability of PUF responses, in *IEEE International Symposium on Hardware-Oriented Security and Trust (HOST)* (2010) pp. 106–111

C. Bösch, J. Guajardo, A.R. Sadeghi, J. Shokrollahi, P. Tuyls, Efficient helper data key extractor on FPGAs. Lect. Notes Comput. Sci. (including Subser. Lect. Notes Artif. Intell. Lect. Notes Bioinformatics) **5154 LNCS**, 181–197 (2008)

E. Brier, C. Clavier, F. Olivier, Correlation power analysis with a leakage model, in *Cryptographic Hardware and Embedded Systems* (2004), pp. 16–29

D. Canright, L. Batina, A very compact 'perfectly masked' S-box for AES, in *Lecture Notes in Computer Science* (2008), pp. 446–459

Z. Chen, Y. Zhou, Dual-rail random switching logic: a countermeasure to reduce side channel leakage, in *Cryptographic Hardware and Embedded Systems (CHES)* (2006), pp. 242–254

B.D. Choi, T.W. Kim, D.K. Kim, Zero bit error rate ID generation circuit using via formation probability in 0.18 μm CMOS process. IET J. Mag. **50**(12), 876–877 (2014)

Y. Dodis, R. Ostrovsky, L. Reyzin, A. Smith, Fuzzy extractors: how to generate strong keys from biometrics and other noisy data. SIAM J. Comput. **38**(1), 97–139 (2008)

S. Eiroa, J. Castro, M. Martínez-Rodríguez, E. Tena, P. Brox, I. Baturone, Reducing bit flipping problems in SRAM physical unclonable functions for chip identification, in *IEEE International Conference on Electronics, Circuits, and Systems (ICECS)* (2012), pp. 392–395

D. Ganta, V. Vivekraja, K. Priya, L. Nazhandali, A highly stable leakage-based silicon physical unclonable functions, in *International Conference on VLSI Design* (2011), pp. 135–140

B. Gassend, D. Clarke, M. van Dijk, S. Devadas, Silicon physical random functions, in *ACM Conference on Computer and Communications Security (CCS)* (2002), p. 148

J. Giorgetti, G. Scotti, A. Simonetti, A. Trifiletti, Analysis of data dependence of leakage current in CMOS cryptographic hardware, in *Great Lakes Symposium on VLSI (GLSVLSI)* (2007), pp. 78–83

J. Guajardo, S.S. Kumar, G. Schrijen, P. Tuyls, FPGA intrinsic PUFs and their use for IP protection, in *Lecture Notes in Computer Science*, ed. by P. Paillier, I. Verbauwhede (Springer, Heidelberg, 2007), pp. 63–80

R. Helinski, D. Acharyya, J. Plusquellic, A physical unclonable function defined using power distribution system equivalent resistance variations, in *ACM/IEEE Design Automation Conference* (2009), pp. 676–681

D.E. Holcomb, W.P. Burleson, K. Fu, Power-up SRAM state as an identifying fingerprint and source of true random numbers. IEEE Trans. Comput. **58**(9), 1198–1210 (2009)

ICTK, Co. Ltd. (2014), http://www.ictk.com/servicenproduct/puf

Intrinsic-ID, SRAM PUF: the secure silicon fingerprint, in *White Paper* (2016)

Invia PUF IP (2016), http://invia.fr/infrastructure/physical-unclonable-function-PUF.aspx

B. Karpinskyy, Y. Lee, Y. Choi, Y. Kim, M. Noh, S. Lee, Physically unclonable function for secure key generation with a key error rate of 2E-38 in 45 nm smart-card chips, in *IEEE International Solid-State Circuits Conference (ISSC)* (2016), pp. 158–160

P. Kocher, J. Ja, B. Jun, Differential power analysis. Lect. Notes Comput. Sci. **1666**, 388–397 (1999)

O. Kömmerling, M.G. Kuhn, Design principles for tamper-resistant smartcard processors, in *USENIX Workshop on Smartcard Technology* (1999), pp. 9–20

S.S. Kumar, J. Guajardo, R. Maes, G. Schrijen, P. Tuyls, The butterfly PUF protecting IP on every FPGA, in *IEEE International Workshop on Hardware-Oriented*

Security and Trust (HOST) (2008), no. 71369, pp. 67–70

J.W. Lee, B. Gassend, G.E. Suh, M. van Dijk, S. Devadas, A technique to build a secret key in integrated circuits for identification and authentication applications, in *Symposium on VLSI Circuits* (2004), pp. 176–179

J. Li, M. Seok, A 3.07 μm^2/bitcell physically unclonable function with 3.5% and 1% bit-instability across 0 to 80 °C and 0.6 to 1.2 V in a 65 nm CMOS, in *IEEE Symposium on VLSI Circuits, Digest of Technical Papers* (2015), pp. 250–251

D. Lim, J.W. Lee, B. Gassend, G.E. Suh, M. Van Dijk, S. Devadas, Extracting secret keys from integrated circuits. IEEE Trans. Very Large Scale Integr. Syst. **13**(10), 1200–1205 (2005)

N. Liu, S. Hanson, D. Sylvester, D. Blaauw, OxID: on-chip one-time random ID generation using oxide breakdown, in *Symposium on VLSI Circuits* (2010), pp. 231–232

K. Lofstrom, W.R. Daasch, D. Taylor, IC identification circuit using device mismatch. IEEE Int. Solid-State Circuits Conf. **46**(8), 1999–2000 (2000)

M. Alioto, A. Alvarez, Physically Unclonable Function database (2016), http://www.green-ic.org/pufdb

R. Maes, *Physically Unclonable Functions: Construction, Properties and Applications* (Springer, London, 2013)

R. Maes, Physically unclonable functions: constructions, properties and applications. Katholieke Universiteit Leuven (2012)

R. Maes, P. Tuyls, I. Verbauwhede, Intrinsic PUFs from flip-flops on reconfigurable devices, in *Workshop on Information and System Security* (2008), no. 71369, pp. 1–17

R. Maes, P. Tuyls, I. Verbauwhede, A soft decision helper data algorithm for SRAM PUFs, in *IEEE International Symposium on Information Theory* (2009), pp. 2101–2105

R. Maes, P. Tuyls, I. Verbauwhede, "Low-overhead implementation of a soft decision helper data algorithm for SRAM PUFs, in *Cryptographic Hardware and Embedded Systems (CHES)* (2009), pp. 1–15

R. Maes, V. Rozic, I. Verbauwhede, P. Koeberl, E. van der Sluis V. can der Leest, Experimental evaluation of physically unclonable functions in 65 nm CMOS, in *European Solid State Circuit Conference (ESSCIRC)* (2012), pp. 486–489

S. Mangard, E. Oswald, T. Popp, *Power Analysis Attacks: Revealing the Secrets of Smart Cards* (Springer, New York, 2007)

S.K. Mathew, S.K. Satpathy, M.A. Anders, H. Kaul, S.K. Hsu, A. Agarwal, G.K. Chen, R.J. Parker, R.K. Krishnamurthy, V. De, A 0.19pJ/b PVT-variation-tolerant hybrid physically unclonable function circuit for 100% stable secure key generation in 22 nm CMOS. Digest Tech. Pap. - IEEE Int. Solid-State Circuits Conf. **2**(c), 278–280 (2014)

S. Mathew, S. Satpathy, V. Suresh, M. Anders, H. Kaul, A. Agarwal, S. Hsu, G. Chen, R. Krishnamurthy, V. De, A 4fJ/bit delay-hardened Physically unclonable function circuit with selective bit destabilization in 14 nm trti-gate CMOS, in *Symposium on VLSI Circuits* (2016), pp. 248–249

D. Merli, F. Stumpf, G. Sigl, Protecting PUF error correction by codeword masking (2013), pp. 1–16

C. Monteiro, Y. Takahashi, T. Sekine, "Resistance against power analysis attacks on adiabatic dynamic and adiabatic differential logics for smart cards, in *International Symposium on Intelligent Signal Processing and Communication Systems (ISPACS)* (2011), pp. 1–5

D. Nedospasov, J.P. Seifert, C. Helfmeier, C. Boit, Invasive PUF analysis, in *Workshop on Fault Diagnosis and Tolerance in Cryptography (FDTC)* (2013), pp. 30–38

R. Pappu, B. Recht, J. Taylor, N. Gershenfeld, Physical one-way functions. Science **297**, 2026–2030 (2002)

Z.S. Paral, S. Devadas, Reliable and efficient PUF-based key generation using pattern matching, in *IEEE International Symposium on Hardware-Oriented Security and Trust (HOST)* (2011), no. 978, pp. 128–133

T. Popp, S. Mangard, Masked dual-rail pre-charge logic: DPA-resistance without routing constraints, in *Cryptographic Hardware and Embedded Systems (CHES)* (2005), pp. 172–186

D. Puntin, S. Stanzione, G. Iannaccone, CMOS unclonable system for secure authentication based on device variability, in *European Solid State Circuit Conference (ESSCIRC)* (2008), pp. 130–133

QuantumTrace, LLC PUF IP Product (2013), http://www.quantumtrace.com/Products/IP/PUF%20IP/

M.T. Rahman, D. Forte, J. Fahrny, M. Tehranipoor, ARO-PUF: an aging-resistant ring oscillator PUF design, in *Design, Automation & Test in Europe Conference & Exhibition (DATE)* (2014), pp. 1–6

S. Rosenblatt, D. Fainstein, A. Cestero, J. Safran, N. Robson, T. Kirihata, S.S. Iyer, Field tolerant dynamic intrinsic chip ID using 32 nm high-K/metal gate SOI embedded DRAM. IEEE J. Solid State Circuits **48**(4), 940–947 (2013)

D. Roy, J.H. Klootwijk, N.A.M. Verhaegh, H.H.A.J. Roosen, R.A.M. Wolters, Comb capacitor structures for on-chip physical uncloneable function. IEEE Trans. Semicond. Manuf. **22**(1), 96–102 (2009)

U. Rührmair, F. Sehnke, J. Sölter, G. Dror, S. Devadas, J. ürgen Schmidhuber, Modeling attacks on physical unclonable functions, in *Proceedings of ACM Conference on Computer and Communications Security* (2010), pp. 237–249

U. Rührmair, J. Sölter, F. Sehnke, X. Xu, A. Mahmoud, V. Stoyanova, G. Dror, J. Schmidhuber, W. Burleson, S. Devadas, PUF modeling attacks on simulated and silicon data. IEEE Trans. Inf. Forensics Secur. **8**(11), 1876–1891 (2013)

A. Rukhin, J. Soto, J. Nechvatal, M. Smid, E. Barker, S. Leigh, M. Levenson, M. Vangel, D. Banks, A. Heckert, J. Dray, S. Vo, A statistical test suite for random and pseudorandom number generators for cryptographic applications. Natl. Inst. Stand. Technol. **800–22**(Rev 1a), 131 (2010)

A.-R. Sadeghi, D. Naccache (eds.), *Towards Hardware-Intrinsic Security: Foundations and Practice* (Springer, Berlin, 2010)

D. Samyde, S. Skorobogatov, R. Anderson, J.-J. Quisquater, On a new way to read data from memory, in *International IEEE Security in Storage Workshop* (2002), pp. 65–69

S. Satpathy, S. Mathew, J. Li, P. Koeberl, M. Anders, H. Kaul, G. Chen, A. Agarwal, S. Hsu, R. Krishnamurthy, 13fJ/bit probing-resilient 250 K PUF array with soft dark-bit masking for 1.94% bit-error in 22 nm tri-gate CMOS," in *European Solid State Circuit Conference (ESSCIRC)* (2014), pp. 239–242

P. Schaumont, K. Tiri, Masking and dual-rail logic don't add up, in *Cryptographic Hardware and Embedded Systems (CHES)* (2007), pp. 95–106

G.-J. Schrijen, V. Van Der Leest, Comparative analysis of SRAM memories used as PUF primitives, in *Design, Automation & Test in Europe Conference & Exhibition (DATE)* (2012), pp. 1319–1324

G. Selimis, M. Konijnenburg, M. Ashouei, J. Huisken, H. De Groot, V. Van Der Leest, G.J. Schrijen, M. Van Hulst, P. Tuyls, "Evaluation of 90nm 6T-SRAM as physical unclonable function for secure key generation in wireless sensor nodes, *Proceedings of IEEE International Symposium on Circuits Systems* (2011), pp. 567–570

M. Shiozaki, T. Kubota, T. Nakai, A. Takeuchi, T. Nishimura, T. Fujino, Tamper-resistant authentication system with side-channel attack resistant AES and PUF using MDR-ROM. in *IEEE International Symposium on Circuits and Systems (ISCAS)* (2015), pp. 1462–1465

P. Simons, E. Van Der Sluis, V. Van Der Leest, Buskeeper PUFs, a promising alternative to D Flip-Flop PUFs, in *IEEE International Symposium on Hardware-Oriented Security and Trust (HOST)* (2012), pp. 7–12

S.W. Smith, S. Weingart, Building a high-performance, programmable secure coprocessor. Comput. Networks **31**(8), 831–860 (1999)

S. Stanzione, G. Iannaccone, Silicon physical unclonable function resistant to a 10^25-trial brute force attack in 90 nm CMOS, in *Symposium on VLSI Circuits* (2009), pp. 116–117

Y. Su, J. Holleman B. Otis, A 1.6pJ/bit 96% stable chip-ID generating circuit using process variations, in *Digest of Technical Papers - IEEE International Solid-State Circuits Conference (ISSCC)* (2007), pp. 406–408

G.E. Suh, S. Devadas, Physical unclonable functions for device authentication and secret key generation, in *ACM/IEEE Design Automation Conference* (2007), pp. 9–14

G.E. Suh, C.W. O'Donnell, S. Devadas, Aegis: a single-chip secure processor. IEEE Des. Test Comput. **24**(6), 570–580 (2007b)

D. Suzuki, K. Shimizu, The glitch PUF: a new delay-PUF architecture exploiting glitch shapes, in *Workshop on Cryptographic Hardware and Embedded Systems (CHES)* (2010), pp. 366–382

K. Tiri, M. Akmal, I. Verbauwhede, A dynamic and differential CMOS logic with signal independent power consumption to withstand differential power analysis on smart cards, in *European Solid-State Circuits Conference (ESSCIRC)* (2002), pp. 403–406

K. Tiri, I. Verbauwhede, A logic level design methodology for a secure DPA resistant ASIC or FPGA implementation, in *Design, Automation & Test in Europe Conference & Exhibition (DATE)* (2004), pp. 246–251

P. Tuyls, G.-J. Schrijen, B. Škorić, J. van Geloven, N. Verhaegh, R. Wolters, Read-proof hardware from protective coatings, in *Cryptographic Hardware and Embedded Systems (CHES)* (2006), pp. 369–383

Verayo Inc. (2013), http://www.verayo.com/tech.php

M. Wan, Z. He, S. Han, K. Dai, X. Zou, An invasive-attack-resistant PUF based on switched-capacitor circuit. IEEE Trans. Circuits Syst. I **62**(8), 2024–2034 (2015)

T. Xu, J.B. Wendt, M. Potkonjak, Matched digital PUFs for low power security in implantable medical devices, *2014 I.E. International Conference on Healthcare Informatics* (2014), pp. 33–38

K. Yang, Q. Dong, D. Blaauw, D. Sylvester, A physically unclonable function with BER $< 10^{-8}$ for robust chip authentication using oscillator collapse in 40 nm CMOS, in *IEEE International Solid-State Circuits Conference (ISSCC)* (2015), pp. 254–256

M.M. Yu, S. Devadas, Secure and robust error correction for physical unclonable functions. IEEE Des. Test Comput. **27**(1), 48–65 (2010)

M.M. Yu, D.M. Raihi, R. Sowell, S. Devadas, Lightweight and secure PUF key storage using limits of machine learning, in *Workshop on Cryptographic Hardware and Embedded Systems* (2011), pp. 358–373

M.M. Yu, R. Sowell, A. Singh, D.M. Raihi, S. Devadas, Performance metrics and empirical results of a PUF cryptographic key generation ASIC, in *IEEE International Symposium on Hardware-Oriented Security and Trust (HOST)* (2012), pp. 108–115

W. Zhao, Y. Ha, M. Alioto, Novel self-body-biasing and statistical design for near-threshold circuits with ultra energy-efficient AES as case study. IEEE Trans. VLSI Systems **23**(8), 1390–1401 (2015)

Design Methodologies for IoT Systems on a Chip

9

David Flynn, James Myers, and Seng Toh

This chapter addresses the approaches and methodologies appropriate to energy-constrained SoC design, implementation and verification using standard multi-voltage Electronic Design Automation tools, rather than resorting to full-custom circuit approaches. The Physical-IP libraries, memories and power-management components required to address both active-mode energy and deep-sleep state retention power are introduced, followed by a case study addressing the specific challenges of optimizing a micro-processor subsystem for Near- and Sub-Threshold Voltage operation. As well as system level power management the implementation and verification of clock distribution and system timing closure are covered in detail.

9.1 Example Activity Profiles for IoT Sensor Nodes

For IOT "edge-nodes" such as Wireless Sensor Nodes the typical activity and power profile is shown in Fig. 9.1.

The height of the bar indicates the relative current consumed or power dissipated and the width is indicative of the duration. The system is typically optimized for minimum residual cur-

rent in between periodic sensing, data processing and data transmission: this is annotated STANDBY in the figure, and in many systems this is the predominant impact on battery life.

1. The sensing activity is normally periodic and triggered by some form of real-time sample request at a controlled data collection rate. The height and width are conceptually marked as the activities labeled SENSE in the figure.
2. Some form of data processing step, such as filtering, or anomaly or limit detection, is often initiated after a certain number of samples have been buffered. The duration and current profile may be data-dependent, and is shown annotated as PROCESS.
3. A Wireless Sensor Node will typically package or compress data to minimize the energy required to transmit the data, and in many systems the transmission time-slots are pre-scheduled at specific times dependent on the wireless access protocol and scheduler (maybe in a sensor hub or base-station) at a rate that is independent of the data-sampling rate. This is shown with arbitrary power and duration as TRANSMIT in the figure.

Regardless of the specific waveform amplitude and activity profiles the key elements required to be minimized in design and implementation are:

D. Flynn (✉) • J. Myers • S. Toh
ARM Ltd, Cambridge, UK
e-mail: David.Flynn@arm.com

© Springer International Publishing AG 2017
M. Alioto (ed.), *Enabling the Internet of Things*, DOI 10.1007/978-3-319-51482-6_9

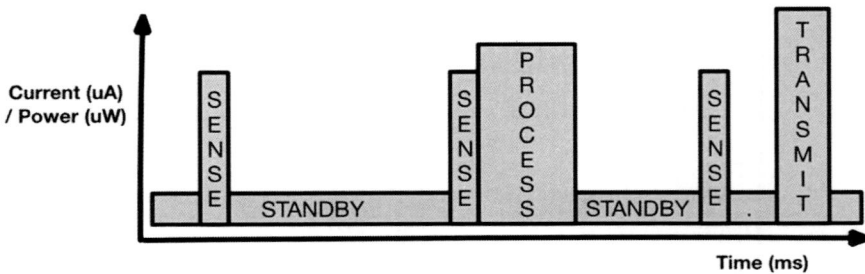

Fig. 9.1 Example active and standby profile for wireless sensor node

- Leakage and state retention energy—the integration of power over time for the Standby component when the main circuitry is inactive.
- Dynamic energy consumption when clocks are enabled to specific components required for active computation or communication.
- Peak active current—which is often the limiting factor in small on-chip power regulation schemes.
- Both active and leakage power consumption for the "always-on" circuitry such as the timer or Real-Time Clock (RTC) that provides the wake-up event scheduling.

- The PMOS and NMOS power gates are optimized for I_{ON}/I_{OFF} ratio, but the series on-resistance typically impacts circuit performance despite the off-current savings so in industrial usage only one is typically used. Footer NMOS power gating is shown in (b).

To mitigate peak current inrush when turning on the power gate there are various ways to build resistive turn-on networks but one approach is to support a threshold-voltage drop using a PMOS transistor in the case of footer-switched VVSS rail, as shown in (c). A logic-0 drive on both **PWR** and **nDROWSE** controls enables this mode of operation.

9.2   Static Power Reduction

9.2.1   Power Gating

The primary technique for static power reduction is power gating (Mutoh et al. 1996), which is well supported in Multi-Voltage EDA tools. Figure 9.2 illustrates the theory and practice:

- Early academic research focused on Multi-Threshold CMOS power gating, MTCMOS (a), where high-threshold "header" and "footer" switches are added to create gated "virtual" VDD and VSS rails, labeled VVDD and VVSS. The logic is powered when **PWR** control to footer is logic-1 and **nPWR** to header switch is logic-0.

9.2.2   Power Gating and Well-Bias

In the case of full MTCMOS power gating, as shown in Fig. 9.3a with the P- and N-wells are explicitly annotated, the VVDD and VVSS virtual switched rails collapse towards a mid-rail voltage with symmetric reverse bias to the switch P- and N-channel logic transistors. With the addition of "drowsy" threshold-voltage transistors it is possible to provide a mode which holds the logic sub-threshold with symmetric well bias (**DROWSE**=1, **nDROWSE**=0) and can support quick wake from sleep with 3× lower wake energy (Mistry et al. 2014).

This is a special case of well-bias for standby which is simple to implement without requiring multi-dimensional standard cell characterization

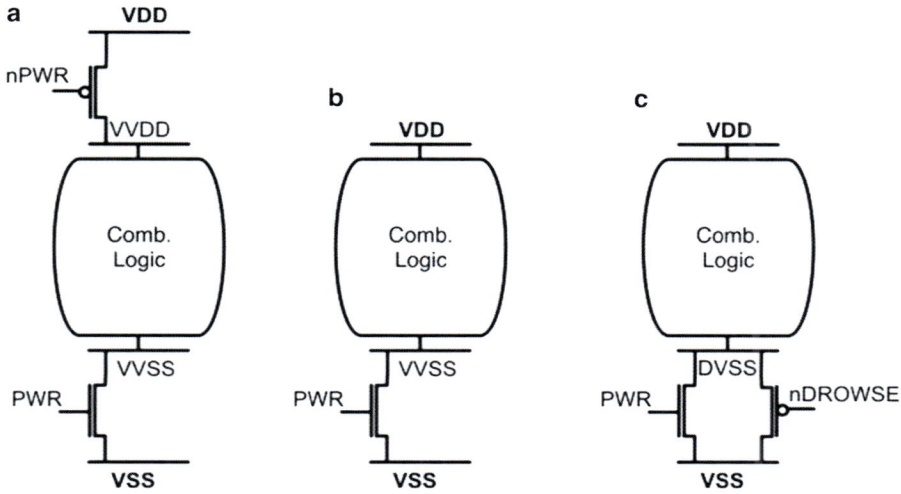

Fig. 9.2 Power gating (**a**) MTCMOS, (**b**) footer, (**c**) "drowsy" scaling

Fig. 9.3 (**a**) Power gating and well-bias, (**b**) drowsy rail power gating

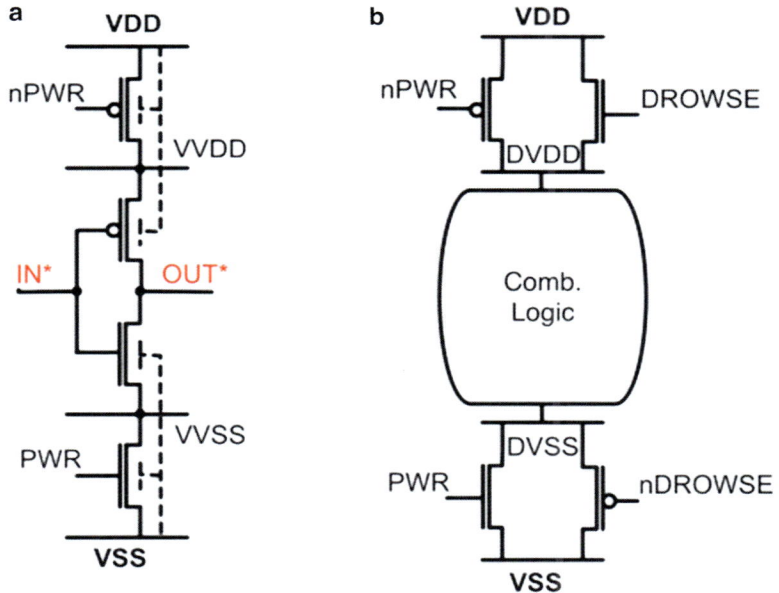

as needed for forward-body-bias active modes. While traditional reverse-body-bias techniques are effective in older bulk process nodes it is a challenge in smaller IoT designs to generate boosted P-well and N-well voltages without expending more active power than the leakage that is saved.

9.2.3 Boosted-Gate Switches for Low-Voltage Power Gating

As discussed in Sect. 5.5, in near- and sub-threshold designs the voltage headroom to drive power gate controls with logic-level voltage swings results in compromised I_{ON} switch

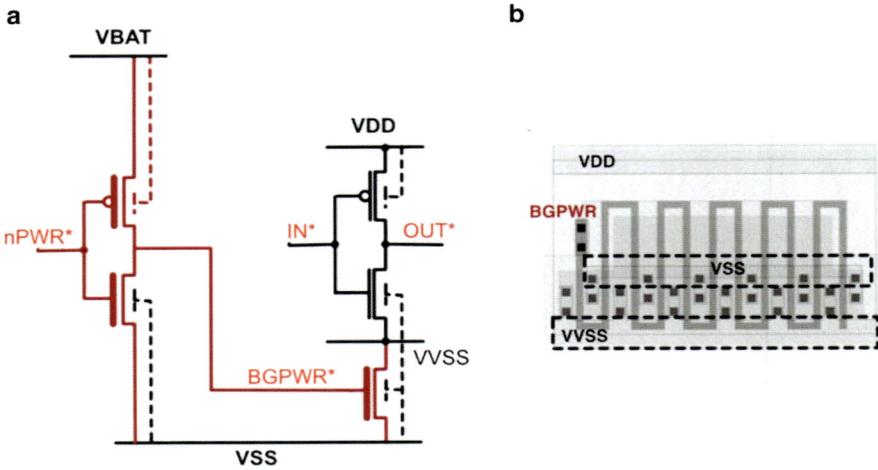

Fig. 9.4 (**a**) BG-CMOS footer power gating, (**b**) footer power gating cell

behavior. An effective technique to ensure highly effective power gating for reduced voltage logic rails is to use high threshold power gates with boosted gate voltage (Stan 1998). An optimal implementation can be achieved by building the control buffering and the footer power gates with Thick-Gate-Oxide (TGO) devices which can be operated from the unregulated battery (or super-capacitor in the case of battery-less systems) voltage rail, shown as **VBAT** in Fig. 9.4a. This requires care in the implementation flow to handle the extra higher-voltage rail distribution, but this is all low-current from a control signal perspective and results in highly effective distributed sub-circuit power gating.

Figure 9.4b also shows an example of the standard cell abstraction used for power gates that can be cleanly deployed in EDA implementation flows. The switch shares the **VDD** supply row architecture but connects the standard-cell ground rail as a switched virtual **VVSS** track. The global **VSS** supply is via-ed down to the power gate from the thick-metal ground mesh, and the switch is laid out as multiple fingers of switch where device length and width are tuned for best I_{ON}/I_{OFF} ratio.

9.2.4 Clamping and Isolation

Power gating provides effective leakage current reduction when sub-circuits are powered down,

but signals at the boundaries collapse to non-logic levels. In order to prevent crow-bar currents flowing in logic down-stream of a power gated block or region, specialized isolation or clamp cells are provided which are powered from the global **VDD** and **VSS** rails and provide the equivalent of AND- or OR-gate signal clamping. IEEE1801 power intent supports explicit association of high or low isolation signals to interface nets (IEEE Standard for Design and Verification of Low Power Integrated Circuits). Example cells are shown in Fig. 9.5.

9.2.5 State Retention with Power Gating

Power gating provides leakage power savings by switching off a sub-circuit, but any state is lost and the sub-circuit needs to be reset or reinitialized after the power is turned back on. For circuits such as fixed function accelerators with only transient state this is usually acceptable. But in many cases the loss of all current state is too costly such that either the architectural state must be saved away before power gating and restored after power is turned on (which may have considerable latency and energy cost), or the state is preserved in-place and maintained for a much smaller leakage cost.

Fig. 9.5 (a) Clamp-low cell, (b) clamp-high cell

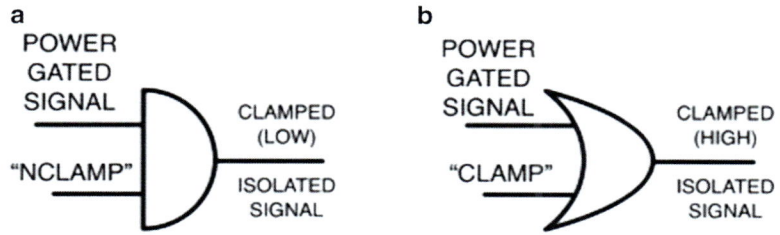

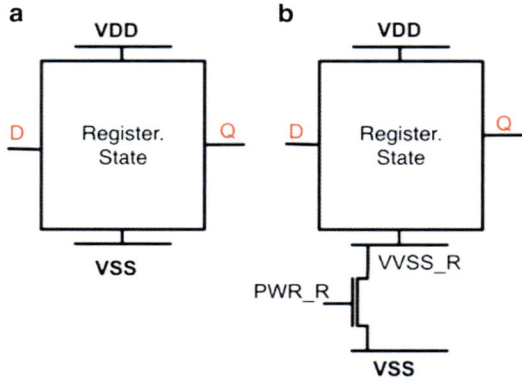

Figure 9.6 illustrates the basic approach of either providing "always-on" power to register state (not supplied from the gated virtual standard cell rails) in (a), or adding a second gated retention rail "VVSS_R" which can be independently controlled by the PWR_R control signal to allow state to be retained for certain periods and turned off fully for deep sleep modes.

Fig. 9.6 (a) Always-on register, (b) independently gated state retention

which can be either fully on (PWR_R=1), drowsy voltage scaled (PWR_R=0, nDROWSE_R=0) or fully off (PWR_R=0, nDROWSE_R=1). Careful validation of the voltage scaled retention reliability must be evaluated for the technology used, but this can provide valuable retention current savings for designs with a large number of registers.

Figure 9.7b illustrates a further optimization where only the slave latch portion of the master-slave flip-flop register is retained, while the input stage, master latch and output driver are all powered from the switched standard-cell VVSS rail, but a separate drowsy-voltage-scaled retention virtual ground rail is provided to the slave latch only (Flynn et al. 2012). While retention registers and power gating have been shown to save 95% of the standby leakage power, drowsy retention can achieve a further 50% reduction.

In both cases there is a need to add minimal clamping circuitry around the register or slave latch to protect this for floating inputs, especially clocks and resets, and this can be added to the cell or supported across a group of registers controlled by a "retention isolation" signal.

9.2.6 State Retention with Power Gating Optimizations

To minimize the state retention currents incurred in standby mode states when retaining state is required there are optimizations that can be appropriate depending on the voltage headroom and register stability at scaled voltages (Kumagai et al. 1998). Figure 9.7 shows the concept applied to standard master-slave registers (a). A drowsy state virtual ground rail, labeled **DVSS** is shown

9.2.7 Physical Considerations for Independently Gated State Retention

While power intent formats such as CPF and UPF (IEEE 1801) support an arbitrary number of switched supplies, there is an assumption of a single default supply pair for each power domain. Power aware EDA tools expect this to be routed as the standard cell main rail. While always-on buffers (for buffering power gate, clamp or retention controls) are a well understood exception

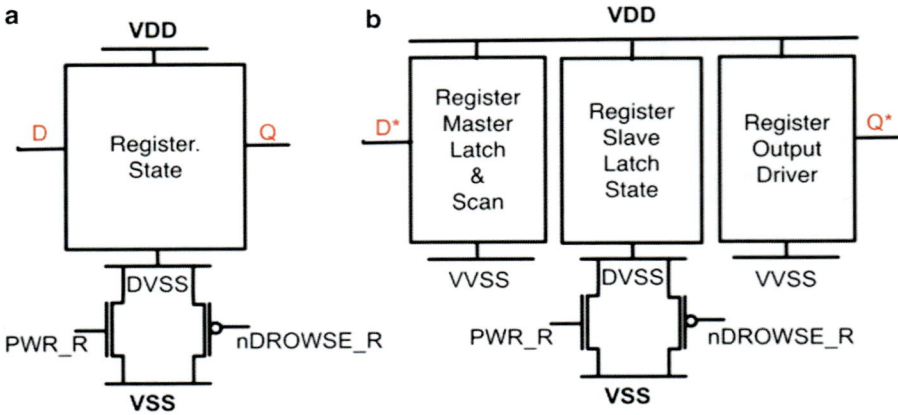

Fig. 9.7 (**a**) Drowsy voltage retention, (**b**) slave latch drowsy retention

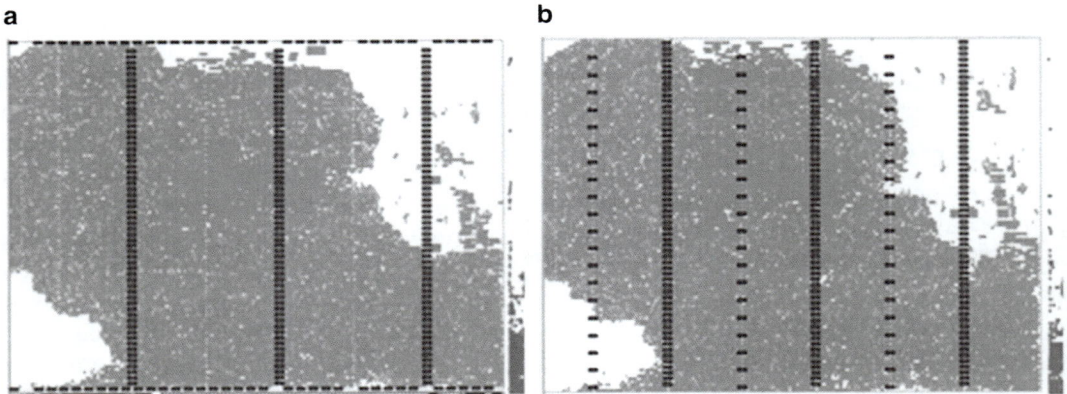

Fig. 9.8 ARM Cortex-M0+ power domain cell placement highlighting power gates (**a**) retention power gates implemented as *top/bottom* rows, (**b**) retention power gates implemented as columns with dedicated standard cell

where power supply effectively bypasses the local power gates, this is more complex for independently gated state retention as two or more switches must be placed within a single floorplan region.

The scheme illustrated in Fig. 9.6b can be implemented in two ways. The main logic power gate is usually distributed throughout the floorplan as many small power gate cells, such that the default VVSS for stateless logic can be driven onto the standard cell rails most effectively. If the retention power gate is to be implemented with the same library power gate cell, then it must be confined to dedicated rows to

prevent VVSS and VVSS_R from shorting together. For smaller power domains this is best implemented as top and bottom rows shown in Fig. 9.8a, but larger domains may also require pairs of dedicated rows through the center to reduce power grid voltage droop.

The other option is to use an alternate power gate cell, which does not drive onto the standard cell main rail but a retention rail instead. This has the advantage of allowing distributed placement as with the logic power gate, shown as additional half-density columns in Fig. 9.8b.

In power domains with a high concentration of retention registers it may be beneficial to route this retention rail across the entire design, although this consumes an entire routing track and may not be feasible with the highest density six or seven track standard cell libraries where pin access is a challenge. In other cases it may be best to connect power gate VVSS_R directly up to a power grid and use on-demand routing to connect down to individual standard cells as required.

Note that the drowsy schemes illustrated in Fig. 9.7 intentionally short the outputs of the PMOS & NMOS footers, so the floorplan does not increase in complexity beyond the switched retention case.

9.2.8 Sequencing of Power Gating Controls

Power gating typically has to be controlled by a state machine powered in a relatively always-on voltage domain. The control signals that are required to drive the power gating and isolation clamping described previously need to have explicit sequencing and these are the ports that get connected to the IEEE 1801 inferred power controls (Keating et al. 2007). Figure 9.9 illustrates the standard sequencing into power-gated mode and then waking up back to active mode. On a request to sleep, first the clock is stopped, then the reset asserted (not strictly necessary but maintains symmetry), the isolation clamping of outputs is asserted and then the power-gating network turned off.

In the figure a power-gating acknowledge signal is shown which is valuable to ensure that the timing required to power down and back up the switched network is handled correctly by design (e.g. to avoid the condition when a wake up occurs just after the power gating is turned off and power is un-driven momentarily). Although comparators or Schmitt circuits may be used to assert this PWR_ACK when the virtual rail has been charged to a target voltage, in practice it is usually the output of a delay line, synchronous counter, or power gate daisy-chain (Shi et al. 2006).

Larger designs may suffer from high in-rush currents and ground bounce when powering up too quickly, which can cause corruption of retained state or timing errors in active blocks. Analysis of this in-rush current is therefore an important sign-off step for low power designs and is well supported by EDA tools. Where in-rush currents are found to be unacceptably high (a typical target is no greater than active peak currents, although tighter constraints may be required for some classes of design), the most common mitigation approach is to stagger power gate enables, such that a fraction of power gates are used to initialize the virtual rail more slowly. Drowsy power gates (as mentioned in Sect. 9.2.2) are another good option.

On a request to wake-up, power is requested, and only when valid is the reset de-asserted, the isolation clamping turned off and the clock finally re-enabled. In the case of state retention power gating the retain control signal timing is often similar to the isolation NCLAMP control waveform.

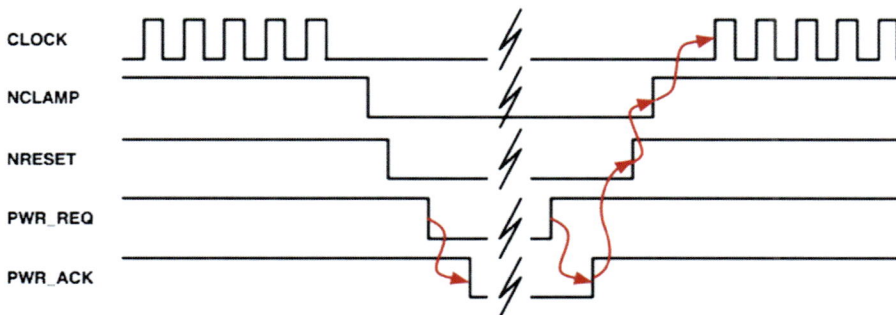

Fig. 9.9 Example clock, clamp, reset and power gate control sequencing

9.3 Active Power Reduction

The management of active power is mainly focused on addressing terms in the familiar CMOS dynamic power proportionality to CV^2F equation. The capacitance term, C, is minimized by striving to keep the circuit small and simple, balancing drive strength and keeping signal routing capacitance to a minimum. The frequency term, F, is addressed by optimizing the circuit implementation for peak required performance and then factoring in both architectural and inferred clock gating to suppress dynamic power dissipation whenever possible. The voltage term, V, is the most valuable control knob given the square-law contribution; in IoT applications the primary focus is to work with low voltage technology, with under-driven super-threshold libraries and memories, or more specialized near-threshold or sub-threshold robust physical IP.

9.3.1 Clock Gating

EDA tools are able to provide transparent clock gating where common sub-expressions in the enable terms of synchronous logic are coded in a clean synthesizable style. Figure 9.10 illustrates the basic scheme for determining groups of registers that share an enable term where the state is defined as sampling or re-circulating values. The multi-bit registers are shielded from the high-toggle-rate clock, marked in red by the inference of a latch and AND gate structure that suppresses clock pulses in cycles when the EN term is inactive (Fig. 9.10b). Figure 9.10c shows the cell abstraction for an Integrated Clock Gate (ICG) that provides the timing and clock

balancing attributes to EDA tools to support clean static timing analysis and clock tree balancing with such gated clocks.

Such ICG elements can also be instantiated in designs to support high-level architectural clock gating where the designer can determine where clock segments can be individually gated explicitly at system level.

9.3.2 Voltage Scaling

Today's production microcontrollers rarely support dynamic voltage and frequency scaling (DVFS) due to the complexities of lightweight OS/SW scheduling, interfacing to off-chip voltage regulators, managing transition periods, identifying optimal voltage-frequency pairs, limited super-threshold voltage headroom, and more. But it is clear that this will be a key area of improvement for future IoT edge-node applications, enabled by integrated voltage regulators (see Chap. 10) and the increased versatility offered by near- and sub-threshold designs.

The physical IP to support this includes cell-libraries that are optimized for the constrained voltage headroom. For near- and sub-threshold operation this usually implies constrained cell architectures, which avoid small length/width devices and minimize transistor stack depths, and register latch and memory bit-cells in particular need to be designed with increased robustness.

The only additional cells required are in the form of level-shifters that manage the voltage drive from low voltages up to higher voltage domains or input/output interface drivers. There are three types of interface with specialized level

Fig. 9.10 Clock gating inference and abstraction

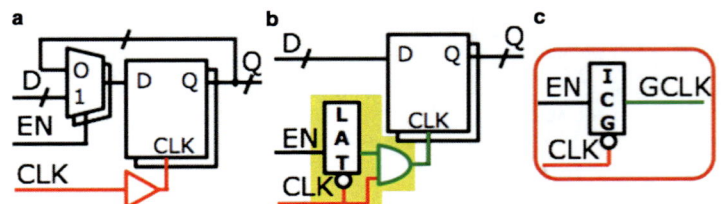

shifters for each: guaranteed low voltage input to high voltage output; guaranteed high voltage input to low voltage output (which may be simply re-characterized buffer cells to enable STA tools to estimate timing correctly); and interfaces where input/output voltages scale independently and may be high/low, low/high, or the same. EDA tools can infer the appropriate level shifters from IEEE 1801 power intent definition and cell-library attributes.

9.3.3   Wake-Up and Power Management Circuits

A special case of active circuits that need to be always-on relative to the processing subsystems are the power management state machines and wake-up sources such as Real-Time-Clock (RTC) alarms. RTC circuits are typically clocked as low as 32 KHz, which dramatically reduces dynamic power compared to other logic running at MHz. Always-on leakage is still a concern and should be reduced by aggressive gate-count reduction and implementation with the lowest leakage devices available. Simple libraries of TGO gates and registers can be beneficial here as these demonstrate up to $100\times$ reduction in leakage compared to regular threshold thin-oxide devices (Taki et al. 2011). The most compelling benefit of TGO libraries however is that they can run directly from unregulated battery voltages, thereby allowing all voltage regulators to be shutdown in deep sleep modes and saving regulator losses which can be significant under very light loading.

9.4   Automated Minimum Energy Design

Conventional EDA tools for automated synthesis, place, and route are usually optimized to produce designs with maximum performance or minimum power. Minimum energy design in general requires achieving both minimum

power without sacrificing performance, especially at the minimum energy point where leakage energy is strongly dependent on performance. This sub-chapter describes how conventional EDA flows may be adapted to achieve a minimum energy design. The impact of key decisions such as standard cell choice and clock design methodology on minimizing energy and cost are also evaluated. Results in this section are derived from a 65 nm R&D sub-threshold ARM® Cortex®-M0+ WSN processing sub-system with prototype 300 mV physical IP.

9.4.1   Implementation Flow

The majority of the implementation methodology is unchanged from a standard EDA flow and no custom tools are required at any stage. Power aware verification of the design is performed using a gate level simulator together with UPF power intent. The flow modifications identified in Fig. 9.11 will be covered in detail below, with the exception of design-for-test and placement steps that are not unusual for a highly power gated design.

9.4.2   Energy Reporting

Energy is the most important metric in this design and needs to be reported in all optimization steps. The tools however only report power. The calculation is simple so custom reporting can be implemented. Leakage energy/cycle is leakage power integrated over the minimal clock period. The subtlety here is that increased leakage power is acceptable, if the corresponding speedup is greater than or equal in magnitude. Dynamic energy/cycle is simply dynamic power divided by clock frequency. The libraries in this example were characterized at five voltages, which allow the majority of the voltage-energy curve to be interpolated.

Fig. 9.11 EDA flow with key modifications for minimum energy design

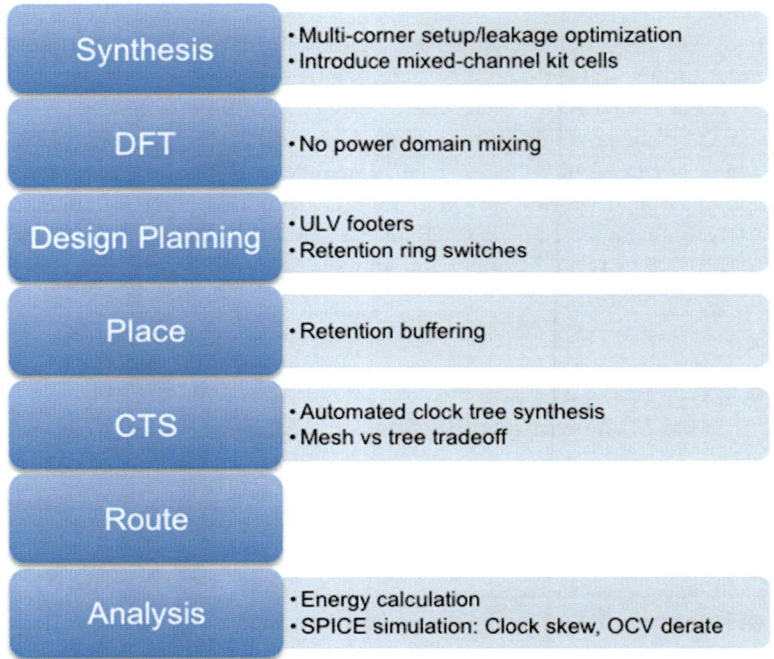

Synthesis
- Multi-corner setup/leakage optimization
- Introduce mixed-channel kit cells

DFT
- No power domain mixing

Design Planning
- ULV footers
- Retention ring switches

Place
- Retention buffering

CTS
- Automated clock tree synthesis
- Mesh vs tree tradeoff

Route

Analysis
- Energy calculation
- SPICE simulation: Clock skew, OCV derate

9.4.3 VT Selection Leakage/Performance Tradeoff

Regular VT (RVT) and low VT (LVT) gates exhibit an 8× difference in performance when operating at sub-threshold voltages while leakage power scales by almost 20× (Fig. 9.12). The amplified performance difference deviates from observed behavior at nominal voltages where performance typically improves by 50% with a 10× leakage power increase by switching to a lower VT choice. Traditional leakage recovery flows that trade-off timing slack on each timing path for leakage reduction are not effective in sub-threshold design as the number of cell swaps that can be taken on each path are limited. Our front-end and back-end flows utilize only RVT gates in order to minimize system leakage. We also utilize an RVT mixed-channel kit (MCK) which has higher performance gates, achieved by optimizing the gate lengths of the transistors. The MCK library cells achieve an average 12% performance improvement at 3×

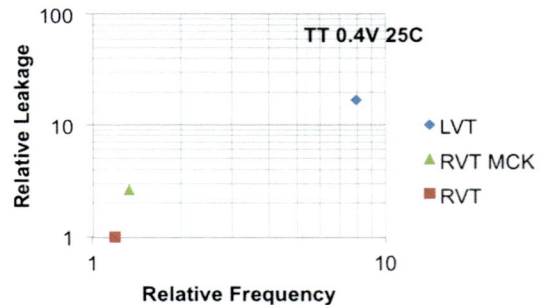

Fig. 9.12 Leakage vs. frequency comparison of various standard cell choices

higher leakage. Using these MCK cells sparingly on the design helps improve performance, resulting in lower leakage energy.

9.4.4 Cell Choices During Optimization Flow

This design ends up with 9.27% of MCK cells (by area) when no constraints are placed on

Table 9.1 Comparison of unconstrained vs. 5% MCK usage constraint

	Unconstrained	5% MCK constraint
Normalized area	1	1.025
Normalized dynamic energy	1	1.060
Normalized leakage energy	1	1.029
Normalized performance	1	1.013
MCK cell area	9.27%	5.95%

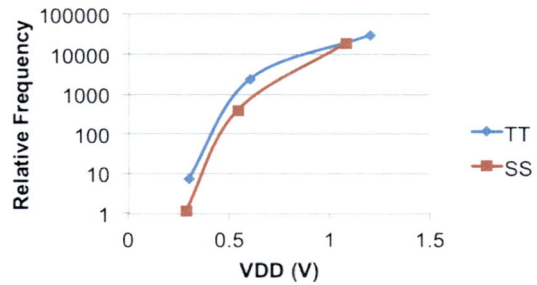

Fig. 9.13 Leakage vs. frequency comparison of various standard cell choices

percentage of MCK cell usage. MCK cell usage constraints applied during synthesis results in a design that has higher area and energy (Table 9.1). This is likely caused by the leakage optimization algorithms in the synthesis tool that try as hard as possible to first implement the design using high VT cells before enabling a minimal amount of low VT cells. This high effort algorithm results in area increase and worse dynamic power. We also experimented with ECO leakage recovery in timing signoff tools but only observed a handful of cell swaps, resulting in miniscule leakage recovery. Based on these experiments, we decided to adopt a simple flow and enable unconstrained use of MCK cells beginning from synthesis.

9.4.5 Optimization Corner Selection for Minimum Energy and Runtime Impact

Battery powered WSN are required to scale voltage and frequency actively in order to meet throughput or latency requirements as well as to minimize energy during low periods of activity. These circuits are therefore required to be functional as VDD is scaled from nominal voltages down to sub-threshold voltages. Figure 9.13 plots the relative frequency of a design at TT and SS global process corners as supply voltage is scaled. Frequency degrades by 4000× across the span of operating voltages while performance degrades by 5× between TT and SS corners. DVFS across such wide operating conditions is

expected to require multi-corner optimization to ensure good performance across corners.

A study was conducted into multi-corner setup optimization vs. single corner setup optimization to determine whether multi-corner optimization is an absolute necessity. Hold timing was still optimized across all corners in these experiments. Figure 9.14 presents the resulting leakage and dynamic energy observed (normalized to single corner setup at SS 1.08 V). Good correlation is observed between the choice of setup corner and optimized leakage energy especially at voltages below 0.6 V. For example, SS 0.54 V minimizes leakage energy around 0.54 V while TT 0.3 V minimizes leakage at 0.3 V. The better performance at the respective voltages results in lower leakage energy, as leakage power is integrated over a shorter period of time. Multi-corner setup optimization appears to produce results that are close to minimum leakage energy across voltages because multi-corner optimization strives to optimize performance across all voltages. Figure 9.14 also plots the Synthesis runtime for the various single corner and multi-corner optimizations. Multi-corner optimization incurs a 4× runtime over single corner, which is quite reasonable considering the various corners that are being considered simultaneously.

9.5 Clock Distribution

Clock distribution is extremely challenging in sub-threshold voltage design due to increased

Fig. 9.15 Comparison of single corner vs. multi-corner optimization

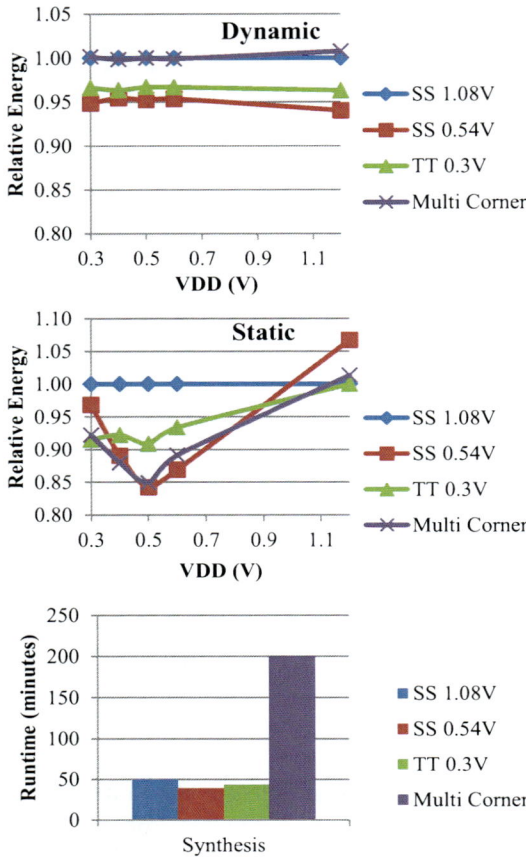

Fig. 9.14 Comparison of power and runtime effects of single corner vs. multi-corner optimization

on-chip-variation (OCV), larger clock latency due to slow buffers, and the requirement for minimizing energy.

9.5.1 OCV Characterization

OCV impacts clock distribution by introducing variation in arrival times to registers and memories. Margins are used to design for this variation, resulting in performance degradation or data-path upsizing to meet performance targets.

Figure 9.15 presents variation in delay through a chain of inverters across different dies. This analysis was performed using SPICE simulation of an extracted netlist of inverters and

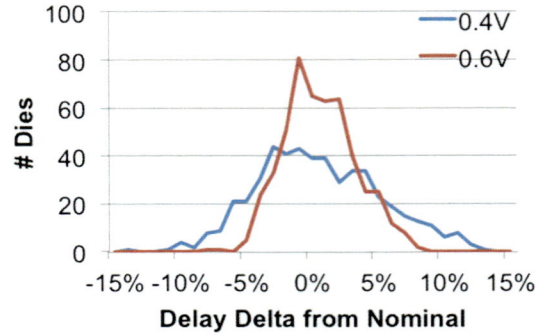

Monte Carlo analysis. The spread in delay increases as voltage is scaled down to sub-threshold voltages because transistor performance is exponentially dependent on threshold voltage. Foundries typically specify OCV derates that are relevant to Vnom +/− 10%. Larger OCV derates are required at sub-threshold voltages in order to margin for the worse variation. An OCV derate is derived from statistical data by multiplying the worst observed sigma by 3.

Clocks are typically distributed using synthesized tree structures or more structured networks like H-trees (Jain 2012) or meshes (SolvNet 2014). Synthesized clock trees are designed and optimized automatically by the tools but tend to exhibit worse performance compared to more structured networks. Clock meshes are quite attractive in sub-threshold design because they can be used to distribute the clock signal across the entire chip without OCV impact. Figure 9.16 illustrates the clock mesh structure that was investigated. Clock meshes essentially distribute a clock source across an entire portion of the design. This is usually achieved using a pre-tree. The outputs of the leaf nodes of the pre-tree are shorted together on the clock mesh to reduce the skew of the clock signal distributed by the pre-tree, creating an almost-ideal clock signal that spans the entire design area. The key to achieving best results from a clock mesh is to reduce the number of gates between the mesh and the clock sinks. We used all clock gates to accomplish this final step, forcing dummy clock gates on clock sinks that

Fig. 9.16 Clock mesh structure

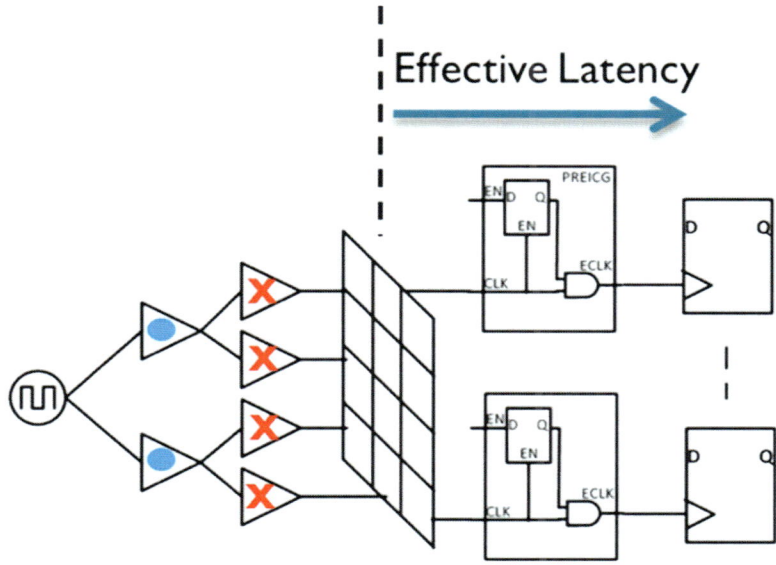

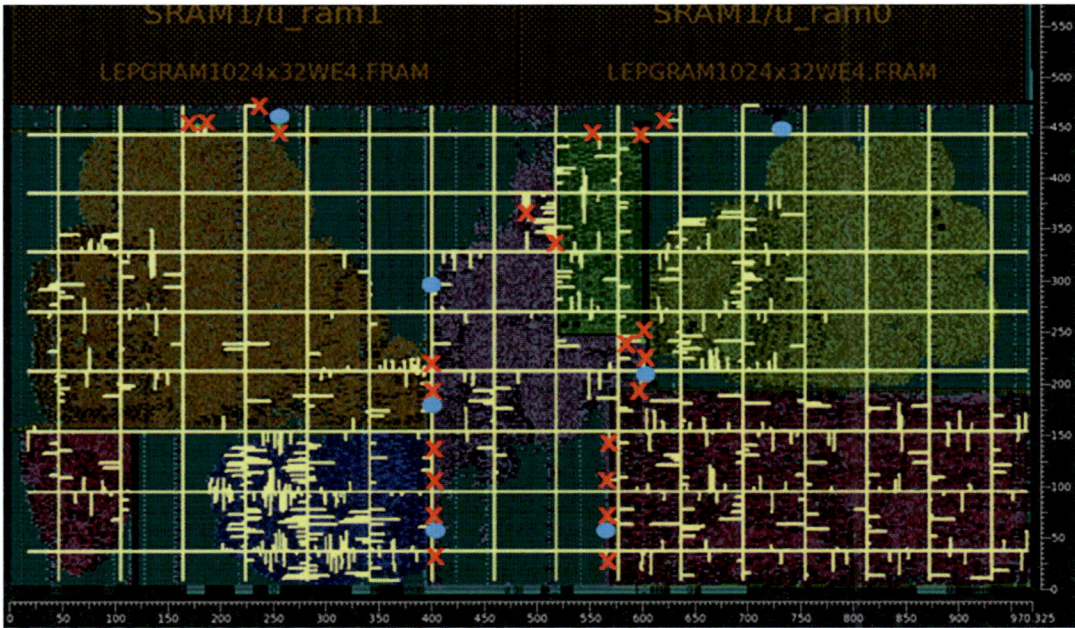

Fig. 9.17 Sub-system standard cell area floorplan, annotated with mesh structure

are not gated. The effective latency of this clock structure is therefore only one gate. Figure 9.17 is a layout view of the clock mesh with respective drivers indicated by respective symbols corresponding to Fig. 9.16. All pre-tree drivers were placed in the middle section of the floorplan because this area corresponds to a power domain that is on in all active modes but can be switched off during retention and power down modes. This ensures that the clock mesh is always alive regardless of the power state combinations of the system. Unfortunately, this also implies that the clock mesh is always consuming dynamic and leakage power.

9.5.2 Clock Tree Synthesis Methodology

Clock trees are the easiest way to distribute clocks with minimal dynamic power and area. Clock trees however suffer from more variability due to the lack of regularity in the structure. This section demonstrates best practices for clock tree synthesis, especially targeting sub-threshold minimum energy clock trees.

Transistor variability is inversely proportional to gate area. Low drive-strength clock cells are therefore expected to exhibit larger variability. Figure 9.18 plots results obtained from OCV characterization of the clock tree. Clock arrival sigma of up to 12% was observed in a clock tree that was not optimized (equating to 36% OCV derate!). Eliminating X1 drive inverters, buffers,

and ICG from the clock cell list reduced sigma to 9%. The area penalty incurred from not using X1 cells is worth the reduction in sigma because the clock tree accounts for less than 1% of total area. Sigma of clock arrival time is further reduced by introducing tighter max-transition constraints on the clock tree.

Automated clock tree synthesis algorithms are typically designed to improve clock-related metrics (latency and skew) at a particular corner. Reducing skew is usually accomplished by adding additional gates to balance delays between paths. We analyzed clock tree metrics observed when using SS 1.08 V CTS corner vs. SS 0.54 V CTS corner (Table 9.2) to determine which strategy produces the best results. Building the clock tree at SS 0.54 V provides minimum clock skew at sub-threshold voltages, compared to the SS 1.08 V CTS corner. This tighter skew was achieved by padding with more clock cells, resulting in 2× larger area. This clock tree however does not scale well with voltage and exhibits a clock skew of 8.9% clock period when measured at the SS 1.08 V corner. The clock tree constructed at the SS 1.08 V corner results in a clock tree that exhibits consistent skew across operating voltages and minimizes area. We have also limited clock tree fanout to 32 which minimizes interconnect delay and helps ensure consistent scaling of all clock endpoints across all operating conditions.

LVT clock cells present an interesting option especially for sub-threshold design due to the 8× better performance (only ~50% better performance is typically observed at nominal voltages). These faster cells will result in up to 8× lower clock latency, which reduces the impact of OCV by the same magnitude. An LVT clock tree will exhibit up to 10× higher leakage power than an RVT clock tree but the improvement in performance can potentially offset the leakage power

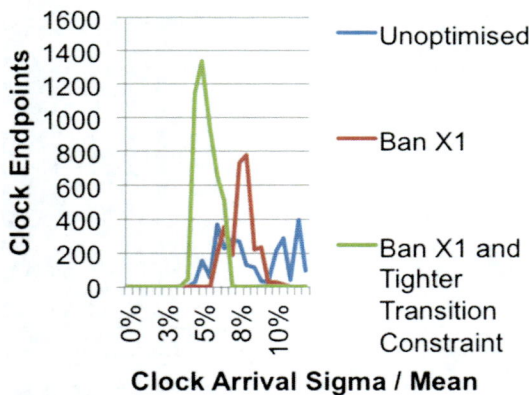

Fig. 9.18 Clock end-point variability across different optimizations (TT 0.4 V)

Table 9.2 Comparison of clock tree metrics synthesized at different CTS corners

| Target CTS corner | Clock skew at respective corner as % of clock period | | | Area (μm^2) | Depth |
	SS 1.08 V	SS 0.54 V	TT 0.4 V		
SS 1.08 V	1.5	2.5	2.1	1587	5–14
SS 0.54 V	8.9	0.9	1.2	3414	13–18

increase, resulting in a net reduction in leakage energy.

Figure 9.19 presents the dynamic energy and leakage energy of the clock distribution network implemented using different strategies. The LVT tree has lower dynamic energy compared to the RVT tree because the transitions are much sharper, resulting in lower short-circuit current. Leakage energy of the clock network however is almost $10\times$ higher than the RVT tree. Table 9.3 presents some additional metrics measured from our WSN sub-system implemented using different clock strategies. The clock latency of the LVT tree is 4% of the clock period while the RVT tree latency is 18% of the clock period. Note that the LVT tree only achieves a design that is at-most 4% faster than the RVT tree, even though clock latency is much lower. The WSN sub-system we have designed is too small to realize the benefits of an LVT clock tree with lower latency. We expect larger designs, where clock latency is a larger fraction of clock period, to exhibit higher performance improvements from an LVT tree due to reduced impact of OCV. Another thing to consider with LVT tree is the cost of the additional VT implant. This could potentially tip the scale in favor of an RVT tree especially in IOT systems where cost is also important.

Our analysis of an RVT clock mesh implemented on the sub-system indicates that the clock mesh consumes significantly more dynamic energy than clock trees. This is because a large portion of the clock structure is always running and can never be gated. Leakage energy of the mesh is slightly higher than an RVT tree due to all the clock gates driving the final stage. Effective clock latency of the mesh is comparable to an LVT tree. The OCV derate of the mesh

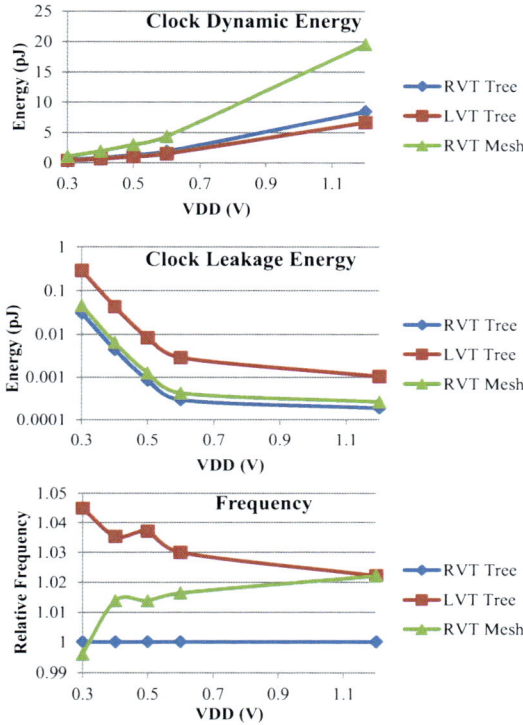

Fig. 9.19 Comparison of clock energy and frequency across clock tree implementations

Table 9.3 Comparison of WSN sub-system energy and performance across clock implementations

TT 0.4 V 25C	RVT tree	LVT tree	RVT mesh
Frequency (kHz)	835	865	847
Dynamic energy (pJ)	8.075	8.025	9.448
Leakage energy (pJ)	0.284	0.311	0.289
Total energy (pJ)	8.359	8.336	9.737
Clock latency (ns)	216	43	163
Effective latency (ns)	216	43	48.1
Clock skew (ns)	77.6	14.7	17.7
Total clock cell area (μm^2)	5940	4999 (0.8% total cell area)	9640
Number of stages	5–13	7–15	1
OCV derate % (TT 0.4 V)	19.5	22.5	66.0
Clock leakage energy (pJ)	0.00436	0.0425	0.00640
Clock dynamic energy (pJ)	0.757	0.628	1.89

(applied only to final ICGs) is much larger than the trees due to the shallow effective clock depth and poor transition on the clock mesh. The RVT clock mesh could be a lower cost alternative to a LVT tree especially for larger designs.

9.6 Perspectives and Trends

The current trend in the industry is towards enabling near-threshold SoCs to be easily and safely designed, implemented, verified and optimized. This is not an easy task however, nor can it be done in isolation. IP providers will have to offer new logical and/or physical IP, EDA tools will need enhancement to support energy optimization and handling of large variation, and silicon foundries will have to provide qualified models. The challenge is in coordinating all of these elements, but there is a real desire and demand for progress, such that the authors are confident that key barriers will be overcome in the next few years.

Looking beyond near threshold, it is clear that there are many innovative and exciting approaches for optimized IoT designs (sub-threshold, adaptive systems, asynchronous, drowsy power gating, non-volatile logic, etc...). Similarly to near-threshold there is often an IP & EDA barrier to the adoption of these cutting edge techniques. Unlike near-threshold however, there can also be an analysis barrier—the system-level cost/benefit tradeoffs of such techniques can be complicated to predict and model. Without progress in system-level exploration and design methodology, otherwise very beneficial technology will continue to be overlooked and fail to gain critical mass or wide adoption.

References

D. Flynn, High performance state retention with power gating applied to CPU subsystems—design approaches and silicon evaluation, HotChips24 2012 Archives HC24.30.p10 (Poster) (2012)

IEEE Standard for Design and Verification of Low Power Integrated Circuits, https://standards.ieee.org/findstds/standard/1801-2013.html

M. Keating, D. Flynn, A. Gibbons, R. Aitken, K. Shi, *Low Power Methodology Manual—For System-on-Chip Design* (Springer, New York, 2007). ISBN 978-0-387-71818-7

S. Jain, A 280 mV-to-1.2 V wide-operating-range IA-32 processor in 32 nm CMOS, in *ISSCC*, 2012

K. Kumagai, H. Iwaki, H. Yoshida, H. Suzuki, T. Yamada, S. Kurosawa, A novel powering-down scheme for low VT CMOS circuits, in *VLSI Circuits, 1998. Digest of Technical Papers. 1998 Symposium on,* Jun 1998 (1998), pp. 44–45

J.N. Mistry, J. Myers, B.M. Al-Hashimi, D. Flynn, J. Biggs, G.M. Merrett, Active mode subclock power gating. IEEE Trans. VLSI Syst. **22**(9), 1898–1908 (2014)

S. Mutoh et al. A 1v multi-threshold voltage CMOS DSP with an efficient power management technique for mobile phone applications, ISSCC1996 (1996), pp. 168–169

K. Shi, D. Howard, Challenges in sleep transistor design and implementation in low-power designs, in *Design Automation Conference, 2006 43rd ACM/IEEE* (2006), pp. 113–116

M. Stan, Low-threshold CMOS circuits with low standby current, in *Proceedings of the International Symposium on Low-Power Electronics and Design* (IEEE/ACM, Monterey, CA, 1998), pp. 97–99

SolvNet: Power Compiler I-2013.12 Update Training (ID: 1506611) (Synopsys Inc., 2014)

D. Taki, T. Shiozawa, K. Ito, Y. Shiba, K. Horisaki, H. Kajihara, T. Yamagishi, M. Sekiya, A. Yamaga, T. Fujita, H. Hara, M. Kuwahara, T. Fujisawa, Y. Unekawa, A 7uW deep-sleep, ultra low-power WLAN baseband LSI for mobile applications, in *Cool Chips XIV, 2011 IEEE*, 20–22 April (2011), pp. 1–3

Power Management Circuit Design for IoT Nodes

10

D. Brian Ma and Yan Lu

This chapter addresses the fundamental structures and operation principles of power management circuits that are commonly used in IoT applications. Following a brief discussion on system design considerations, the chapter reviews the essential power circuit topologies with discussions on key control and operation principles, performance features and drawbacks. With the focus on IoT applications, state-of-the-art works and promising design trends are addressed.

10.1 System Design Overview

Internet of things (IoT) has gained significant research interests in recent years. It is widely believed that the ever-expending network infrastructure of IoT enables unprecedented sensing and control capability over a vast number of interested objects, with applications including, but not limited to, home environmental control, remote healthcare, industrial process monitoring, smart buildings and cities, intelligent transportation,

and so on. By taking advantage of available information and powerful computing capacity of the internet, IoT brings significant social benefits to human life of civilization. Generally speaking, any electronic device with sensing and/or communication ability can work as an IoT node. Apart from well-studied devices such as mobile phones and computers, it is highly anticipated that countless IoT nodes are low-power smart sensors and actuators. For reliability and cost concerns, these devices are preferred to operate autonomously. Hence, the energy used to power these nodes is most likely scavenged from ambient environment, instead of traditional batteries. The main reason is that many IoT nodes have stringent requirement on system volume, whereas batteries can be too bulky for them. In addition, a battery can only store limited amount of energy and thus requires frequent recharge. Due to aging effect, after certain number of recharge, a battery has to be replaced. With countless nodes in the network, maintenance cost can be formidably high. On the other hand, harvesting renewable energy eliminates battery replacement cost and is environment friendly. However, in contrast to a battery-powered system, a self-powered device deals with a much unstable energy source that is sensitive to ambient environment. Effective measures must be taken at device, circuit and system levels to ensure acceptable reliability, energy efficiency and system lifetime.

D. Brian Ma (✉)
University of Texas at Dallas, Richardson,
TX 75080, USA
e-mail: d.ma@utdallas.edu

Y. Lu
University of Macau, Macao, China

© Springer International Publishing AG 2017
M. Alioto (ed.), *Enabling the Internet of Things*, DOI 10.1007/978-3-319-51482-6_10

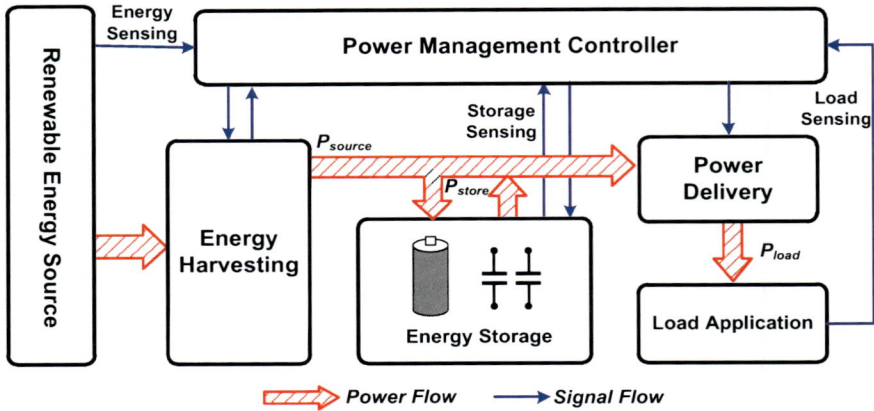

Fig. 10.1 Power management system block diagram for self-powered device

Figure 10.1 illustrates the generic block diagram of a power management system for self-powered applications. It is made up with five major components: energy source(s), energy harvesting module, energy storage module, power delivery module and power management controller. In order to extract the energy from a renewable energy source and convert the extracted energy into electronic format, energy harvesting module becomes an essential part of power management system. To maximize the harvesting efficiency, some best energy harvesting systems undergo a software-hardware co-design process. The software part contributes well-planned harvesting scheme that guides the transducer to extract the highest amount of power from the energy source. Because of the high diversity of energy forms and harvesting transducers, harvesting schemes can be significantly different. For example, maximum power point tracking (MPPT) techniques (Esram and Chapman 2007) are highly popular for photovoltaic solar energy harvesting, whereas resonant frequency and impedance matching schemes are common for piezoelectric energy harvesting (Ottman et al. 2003; Sankman and Ma 2015). As a result, the hardware design of energy harvesting system represents a vast variety of circuit implementations defined by their software schemes. Hence, a separate chapter (Chap. 11) is dedicated to this topic particularly. Another important part of the power management system

is the energy storage module. Because operation conditions of both energy sources and load applications vary continuously in such applications, an energy storage module usually serves as an "energy buffer". When the harvested power is greater than the load power, the excessive power is stored for future use. Due to varying intensity and availability of the energy, a renewable energy source can either fluctuate significantly or offer very low voltage/power level. They cannot be directly used to power an electronic device. Power conditioning module is thus needed to improve the power deliver quality and provide a relatively stable power supply to load applications at a desired voltage level. If the quality of the power is still not acceptable, a power regulation stage would be added. The two stages together accomplish the power delivery to the load application. Here, the load application refers to all the electronic modules that demand electrical power. In particular to IoT applications, as addressed in Sect. 2.1 of Chap. 2, it includes processors/microcontroller units, communication unit as well as sensors/actuators. In a sophisticated system, dictated by intelligent power management schemes such as dynamic voltage/frequency scaling (DVFS), the power delivery can be expended into a sophisticated power management platform with multiple power domains, leveraged by multiple supply voltage levels. In this situation, power management is strategically planned by a global system power optimizer (Sect. 2.7 in Chap. 2), and is often application-specific.

In this chapter, we discuss the power circuits that are essential for power delivery and power management. The discussion starts with a general design consideration on the system level, followed by a comprehensive review of fundamental power circuit topologies and operations. Tailored for the applications like IoT nodes, state-of-the-art works are introduced. The chapter ends with a brief discussion on future trends.

10.2   Fundamentals of Power Converters

Functional modules in IoT power management circuits are built largely based on DC-DC power converters. Such a converter converts an unregulated DC power source to a much regulated one with relatively constant voltage at the output. The output voltage level can be either higher or lower than the input, with either the same or opposite polarity. The stability of the output voltage is a strong indicator of performance. Because the input voltage can fluctuate with ambient environment or battery charge/discharge activities (which are detailed in Chap. 15), the ability of the converter to maintain a constant output voltage is highly expected. This is measured by the line regulation, which is defined as the ratio of the output voltage change ΔV_{OUT} to the input voltage change ΔV_{IN}. In addition, at the output, if the load current I_{OUT} varies, the converter should keep V_{OUT} constant. This behavior is measured by the load regulation, which is defined as the ratio of the output voltage change ΔV_{OUT} to the load current change ΔI_{OUT}. In order to accomplish superior line and load regulation, a DC-DC converter requires a controller, which adaptively controls the power flow according to instant input voltage and load current conditions. The ultimate goal is to retain stable output voltage at the desired level.

Depending on the operation of power transistors, DC-DC converters can be categorized as nonlinear power converters, in which power transistors operate as power switches, and linear (power) regulators, in which power transistors work as voltage- or current-controlled current sources. Furthermore, based on the type of energy storage elements, nonlinear power converters include both inductor based and switched-capacitor based power converters. In general, voltage regulation in a nonlinear power converter involves at least two processes: charging and discharging the energy storage element(s). The purpose of charging process is to extract the energy from input power source and then temporarily store the energy in the energy storage elements. The discharging process is to release the stored energy to load application in a desired power format specified by the output voltage and current levels. If the energy storage element is an inductor or a transformer, the converter is called as a switch mode power converter or a switching converter in short. If the element is a capacitor, the converter is then named as switched-capacitor (SC) power converter.

A linear regulator regulates its output voltage by controlling the ON-resistance of its power transistor, and thus does not generate any output ripples, which can be caused by switching action in nonlinear power converters. Hence, a linear regulator does not require large output capacitor for ripple filtering and its loop bandwidth is in general higher than its nonlinear counterparts. Consequently, it achieves faster transient response. Moreover, a linear regulator is often employed as a post-regulation stage to suppress the ripples generated by nonlinear power converters, and provides a clean supply to noise sensitive loads. The major drawback of linear regulators is that its power transistor bears non-zero voltage and current simultaneously, and thus introduces unavoidable power loss. And this loss is linearly proportional to the dropout voltage from V_{IN} to V_{OUT}.

For electromagnetic wave or vibration energy harvesting systems, AC input energy needs to be converted to DC format, which necessitates the process of rectification. Rectification can be simply achieved by using a diode that only allows the current to flow in one direction and blocks it from going reversely. Low turn-on voltage drop

Schottky diode has been conventionally used in discrete-type AC-DC converters for rectification. However, it is not commonly available in standard CMOS processes, and requires additional mask and thus cost during manufacture. Low cost CMOS rectifier with power transistors replacing the passive diodes gives promising performances in the above mentioned applications, especially at low-input voltage conditions.

10.2.1 Switching Converters

Figure 10.2 illustrates the power stage circuits of four most common switching converters—the buck, the boost, the flyback and the non-inverting buck-boost switching converters. Each consists of an inductor L, a series of power switches, and a capacitor C_L at V_{OUT}. The inductor L serves as the energy storage element. The capacitor C_L filters out switching ripple noise and keeps V_{OUT} constant. The switches, implemented by power transistors or diodes, are periodically turned on and off to charge and discharge the inductor L accordingly. Depending on the input and output voltage levels, a switching converter can be constructed with different topologies. A buck converter in Fig. 10.2a provides a regulated output voltage lower than its

input voltage. If the output voltage should be higher than the input voltage, then a boost converter in Fig. 10.2b can be used. For both converters in Fig. 10.2c, d, the magnitude of the output voltage can be either higher or lower than the input's. However, the polarity of the output voltage is opposite to its input in Fig. 10.2c, but keeps the same in Fig. 10.2d.

In order to provide a constant V_{OUT}, the ON times of the switches should be meticulously controlled. Figure 10.3 illustrates such control actions using a buck converter. Compared to the buck converter in Fig. 10.2a, the main switch S_1 is implemented with a transistor M_P, whereas the secondary switch S_2 is replaced by a diode. M_P is switched ON and OFF periodically, with a switching period T. Assume both the transistor and the diode are ideal, and have zero DC resistance when turned-on. When M_p is ON, it shorts the input V_{IN} to the circuit node V_X, causing a positive inductor voltage $v_L = V_{IN} - V_{OUT}$, if V_{IN} is greater than V_{OUT}. The inductor L is energized and the current i_L rises. Here, the ON-time of M_p is measured as DT, where D is called as the duty ratio. During DT, the energy from the power source is stored in the inductor L. When DT expires, M_P is shut off and the diode is forced ON to short V_X to ground. Hence, $v_L = V_{IN} - V_{OUT}$. The energy stored in the inductor L is released to the load and i_L drops accordingly.

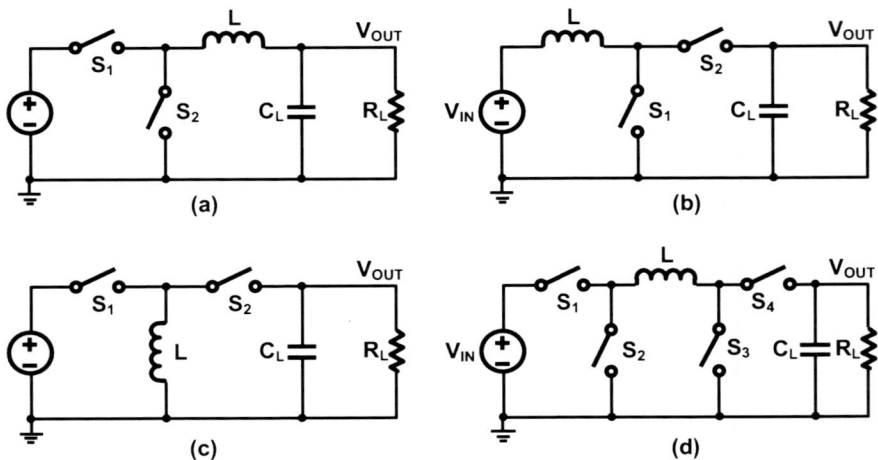

Fig. 10.2 Power stage circuit topologies of switching converters for (**a**) buck, (**b**) boost, (**c**) flyback, and (**d**) non-inverting buck-boost voltage conversion

Fig. 10.3 Switching operation in a buck switching converter

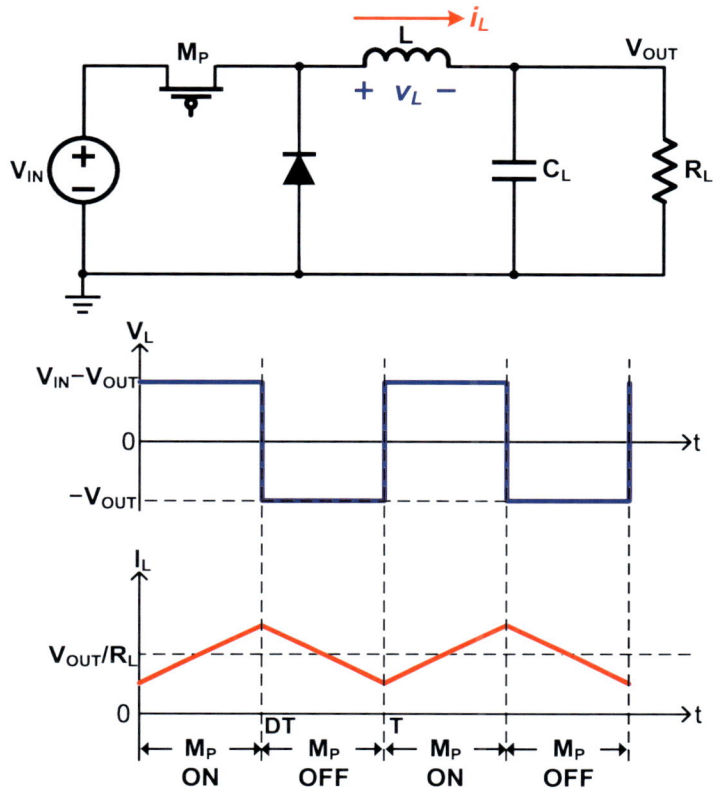

Clearly, if the energy charged into the inductor is equal to the energy demand by the load R_L, the converter enters the steady state. A constant output $V_{OUT} = DV_{IN}$ is achieved.

It is apparent that the voltage regulation of a switching converter is accomplished by controlling the duty ratio D, so that the average of i_L is equal to the load current $\frac{V_{OUT}}{R_L}$. The control method can be a feedback control, or a feedforward control, or a combination of both. With a constant switching frequency, Fig. 10.4 shows a buck converter with two most commonly exploited pulse-width modulation (PWM) feedback controllers. As the converter contains two reactive components L and C_L, either of parameters i_L and V_{OUT} can serve as a control variable to regulate the converter. In Fig. 10.4a, V_{OUT} is used as the feedback signal to determine the duty ratio of M_P. The control is thus considered as voltage-mode control.

In Fig. 10.4b, i_L is employed as the feedback signal. Hence, it is a current-mode control. It should be noted that not all switching converters operate with constant switching period T. For example, in a voltage-mode hysteretic controlled converter of Fig. 10.5, each switching period is not initiated by a clock periodically. Instead, the voltage changes at V_{OUT} determine the status of M_P as well as the switching period T.

With all the converters in Figs. 10.4 and 10.5 managed by feedback controls, a converter can also be regulated with a feedforward control by using the input voltage V_{IN} as the control variable (Smedley and Cuk 1995), as illustrated in Fig. 10.6a. One problem for a feedforward controlled converter, however, is that the output V_{OUT} is not accurately regulated due to the absence of its own feedback signal. The problem can be overcome by employing an additional feedback control to improve the regulation

Fig. 10.4 Switching buck
converter with (**a**) voltage-
mode PWM control, and
(**b**) current-mode PWM
control

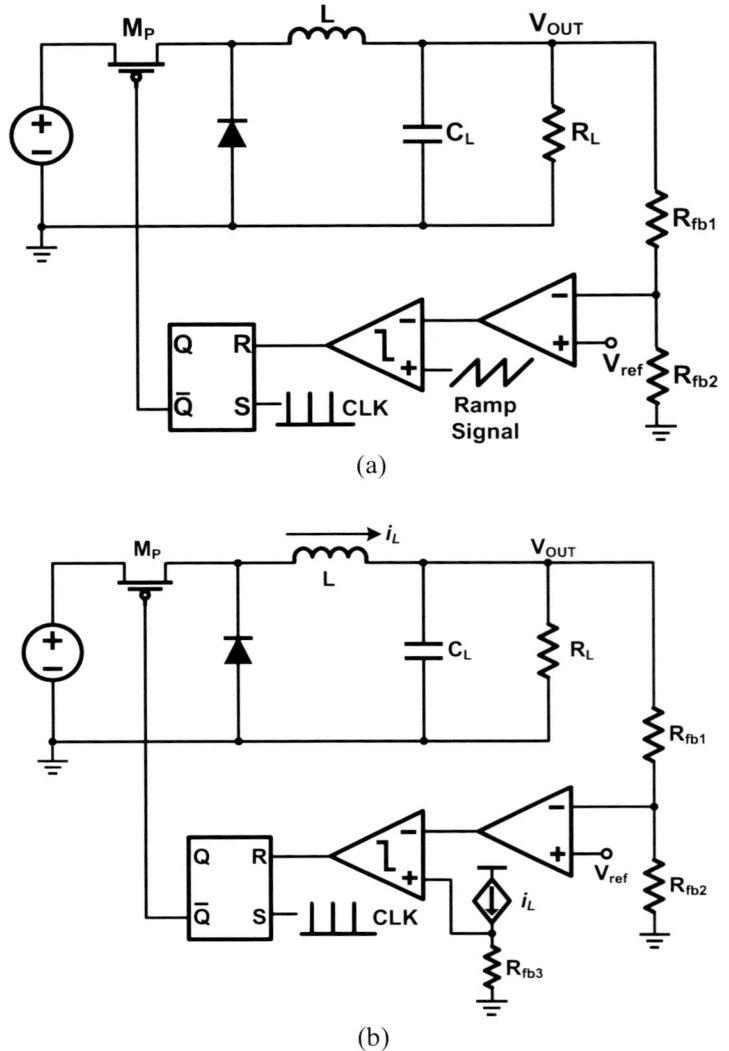

(a)

(b)

Fig. 10.4 Switching buck converter with (**a**) voltage-mode PWM control, and (**b**) current-mode PWM control

accuracy. Figure 10.6b depicts an example of such Ma et al. (2004).

10.2.2 Switched-Capacitor Power Converters

For switching converters introduced in Sect. 10.2.1, the use of inductive devices usually leads to large electromagnetic interference (EMI), which is not desirable for noise sensitive applications. Switched-capacitor (SC) power

converters can serve as an alternative in this situation. Figure 10.7 shows a common SC power converter—voltage doubler, which achieves a voltage conversion gain (CG) of 2. Instead of using an inductor L in a switching converter, a capacitor C_P is used as energy storage element. The converter is controlled by a pair of complementary clock signals, ϕ_1 and ϕ_2. When $\phi_1 = 1$, the switches S_1 and S_2 are ON, the pumping capacitor C_P is charged to V_{IN}. Energy from the input source is temporarily stored in C_P. When $\phi_2 = 1$, S_1 and S_2 are OFF

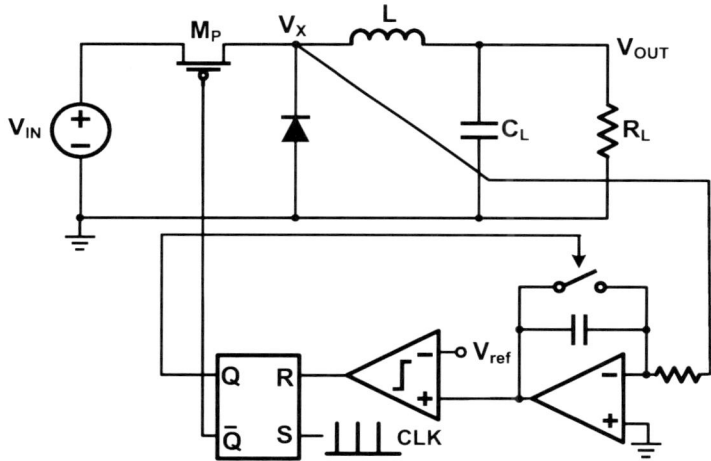

Fig. 10.5 A switching converter with voltage-mode hysteretic control

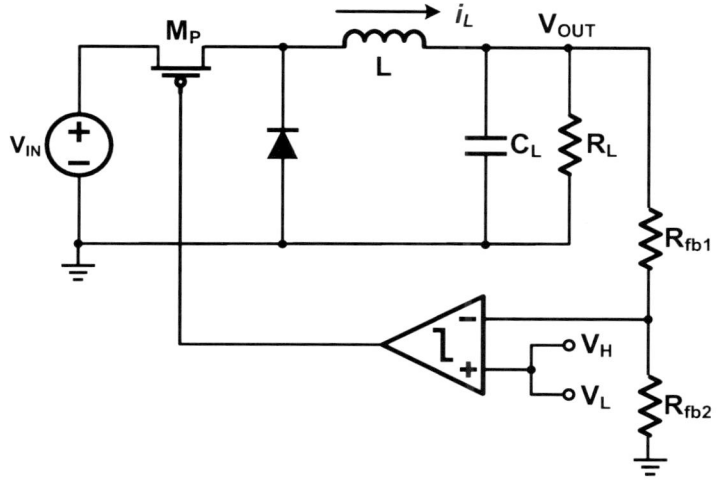

Fig. 10.6 (**a**) A switching control with one cycle control (Smedley and Cuk 1995), and (**b**) a dual-loop one-cycle controller (Ma et al. 2004)

(a)

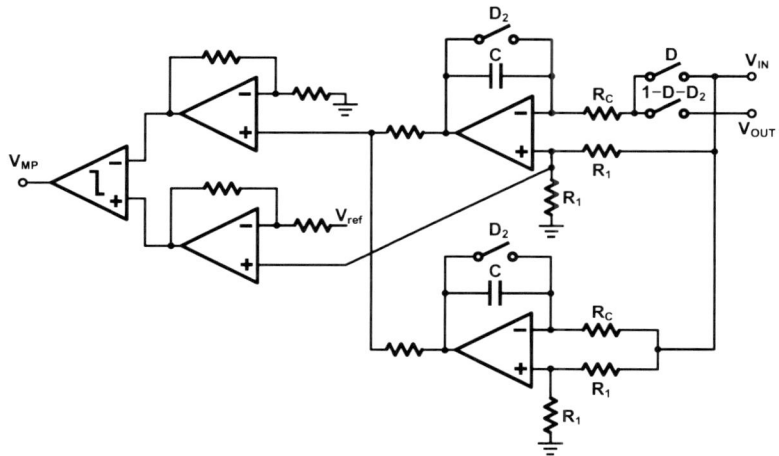

(b)

and S_3 and S_4 is ON. The energy stored in C_P is discharged to the load R_L. At this moment, V_{OUT} is equal to the sum of the voltage across C_P ($=V_{IN}$) and the input voltage V_{IN}, making V_{OUT} = $2V_{IN}$. The circuit is thus called voltage doubler. The control clock signals, ϕ_1 and ϕ_2, can be generated using the circuit in Fig. 10.7b. Note

that it is important to ensure that there are non-overlapping time periods between the two clock signals to avoid large shoot-through currents.

Compared to a switching converter, a SC power converter possesses a fundamental difference in circuit control. The output voltage of a switching converter, as discussed earlier, is regulated by adjusting the duty ratio D of the main power switch. By varying D, the amount of energy per switching period is adjusted. The output voltage is dynamically controlled accordingly. However, the duty ratios of the power switches in Fig. 10.7a are all fixed at 50%. Hence, control methods used in switching converters may not be applicable for SC power converters. In fact, many of SC power converters still remain open-loop today. However, without the presence of feedback control, the output V_{OUT} suffers significant variations and can become fluctuated during load transient periods. A common remedy is to add a linear regulator, as illustrated in Fig. 10.8. The feedback loop of the linear regulator mitigates the load transient challenge by isolating the SC converter from the load. However, because the SC converter has no mechanism to adaptively adjust its load current, if a large load current increase is imposed by R_L, the linear regulator will draw a significant current from V_{OUT}. As a result, V_{OUT} drops significantly. This forces the load supply voltage $V_{OUT,REG}$

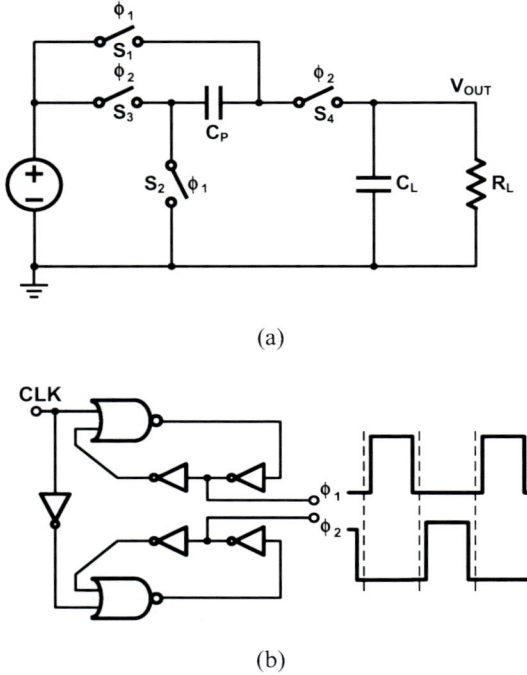

(a)

(b)

Fig. 10.7 (a) SC voltage doubler, with (b) control clock circuit

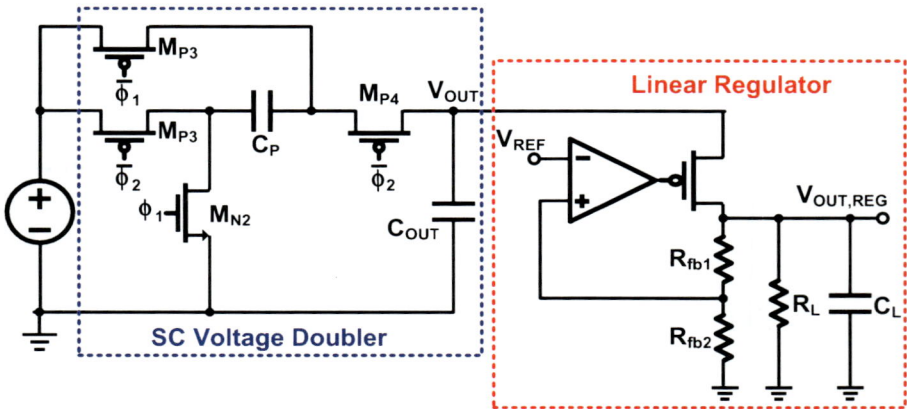

Fig. 10.8 SC voltage doubler followed by a linear regulator for post-regulation

Fig. 10.9 A SC power converter with voltage-mode hysteretic control

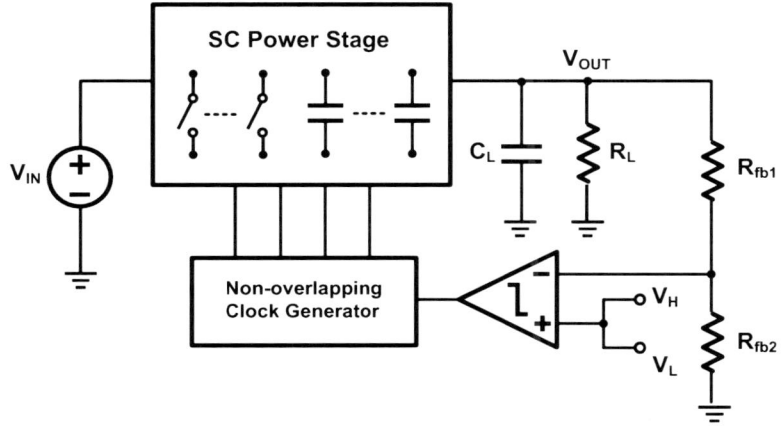

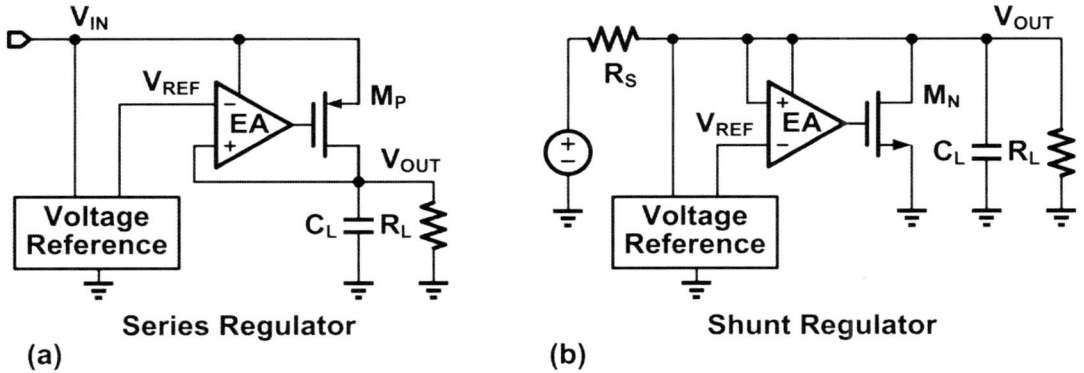

Fig. 10.10 (a) Series regulator and (b) shunt regulator with voltage reference

drops as well. Hence, the problem is not fundamentally solved.

Since the duty ratio of the SC power converters cannot be changed, the switching frequency, which is equal to $1/T$, is used occasionally for closed-loop control. Figure 10.9 demonstrates a circuit implementation of such. Because the equivalent small signal output resistance of a SC power converter is reversely proportional to the switching frequency, the output voltage V_{OUT} can be regulated by adaptively adjusting the switching frequency through the hysteretic feedback control. For example, when load current increases, V_{OUT} drops due to C_L dischaging, which triggers a higher switching frequency. The equivalent resistance of the

converter decreases, allowing more current to be delivered until it meets the new load demand. However, as switching frequency increases, the output resistance of a SC power converter becomes a weak fucntion of the switching frequency. Therefore, the hysteretic control is less effective at heavy load.

10.2.3   Linear Regulators

Both DC-DC and AC-DC converters in power management units generate high levels of switching noise. On the other hand, analog linear regulators, as shown in Fig. 10.10, can filter out the supply noise and provide a clean supply voltage to drive noise sensitive circuits in wireless communication front-end systems and critical paths in VLSI chips, or simply provide a voltage step-down function in low-cost applications. One major difference between a switch mode power converter and a linear regulator is the nature of energy conversion. The power converters use inductors or capacitors to store the energy in one phase and release the energy in the other phase, while the linear regulators just simply consume or dump the extra energy. Thus, the efficiency of a linear series regulator simply approximately equals to V_{OUT}/V_{IN}, assuming the quiescent current consumed by the error amplifier is negligible. The dropout voltage is defined as $V_{IN,MIN}-V_{OUT}$, where $V_{IN,MIN}$ is the minimum input voltage that can maintain the desired output voltage at full load. Therefore, low quiescent current low-dropout (LDO) regulators are favorable in an IoT node due to their low-cost, ripple free, fast transient response and good power supply ripple rejection characteristics.

Since the energy source in energy harvesting scenarios has large variations, when the input current is too high, one more important function for the shunt regulator is to bypass the extra energy to ground to prevent device from breaking down. When an energy storage component is available in the system shown in Fig. 10.1, the shunt regulator is then not needed.

It is also worth noting that digital linear regulator becomes popular in recent years due to its low-voltage operation feature. The operation principle is straightforward and can be illustrated in Fig. 10.11. The digital LDO regulator employs one clocked comparator, one bi-directional shift-register array, and one power transistor array. The comparator compares V_{REF} to V_{OUT} in every clock cycle to decide whether the shift register output bits $D[1:n]$ shift to left (add a "1") or to right (add a "0"). $D[1:n]$ controls the number of turned-on power transistors and consequently the total output current. The clocked comparator can operate at low supply voltage (0.5 V for example) and consumes no DC power, while the other digital cells can operate at low voltage as well. This feature gains the digital LDO regulator potential of being implemented in low-power low-voltage IoT nodes.

The transient response time of the digital LDO regulator is proportional to its clock frequency and the size of each power transistor. Thus, coarse-fine tuning and adaptive clock techniques can be used to improve its transient response without increasing the standby power (Huang et al. 2016). Similar to digitally controlled switching converters, digital LDO regulators also suffer from limit cycle oscillation due to finite resolution of the output current. This problem can be mitigated by adding a one-bit strength feedforward path bypassing the shift register or by introducing a dead zone to the comparator (Huang et al. 2016).

10.2.4 AC-DC Converter

As IoT nodes can also be powered by wireless power transfer (WPT), AC-DC converter (rectifier) is a vital block in both near-field and far-field WPT systems. Note that, rectifiers operating at different frequency ranges, input amplitudes, and output power levels, need specific considerations and dedicated architectures. Thus, researchers have made tremendous efforts to develop wide-range high-efficiency and area-efficient rectifiers in the past decade (Cheng et al. 2016; Guo and Lee 2009; Lu et al. 2011, 2013; Lu and Ki 2014; Lam et al. 2006; Lee and Ghovanloo 2012).

Diode, the simplest semiconductor device, can be easily used to convert AC power into

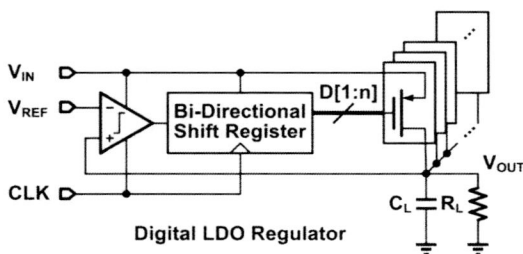

Fig. 10.11 A regular digitally controlled LDO regulator (Okuma et al. 2010)

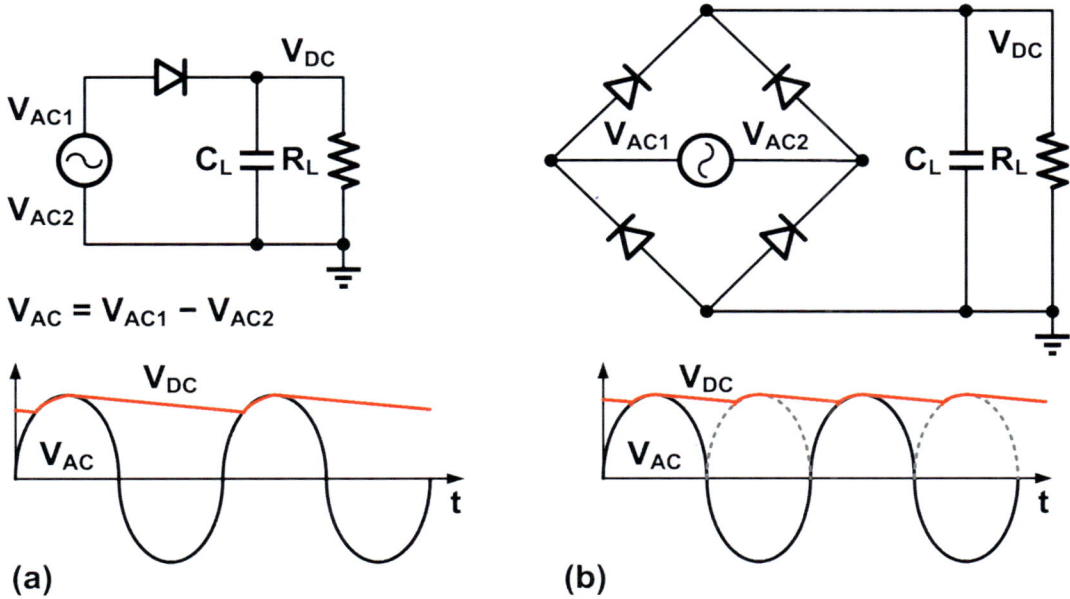

Fig. 10.12 Schematics of (**a**) half-wave rectifier and (**b**) full-wave rectifier

DC power as shown in Fig. 10.12. One terminal of the AC source is tied to ground in the half-wave rectifier as shown in Fig. 10.12a, while both AC terminals are floating in the full-wave rectifier as shown in Fig. 10.12b. The voltage difference between V_{AC1} and V_{AC2}, V_{AC}, is a sinusoidal wave (could be a distorted wave in practice). The half-wave rectifier delivers current from the AC source to the DC output V_{DC} once in every cycle when V_{AC1} is higher than V_{DC}, while the full-wave rectifier delivers current to the output at both positive and negative peak points of V_{AC}. That means the time interval for charge transfer of a half-wave rectifier is only half of that of a full-wave rectifier. Thus, in delivering the same load current with the same output capacitor, the output ripple of the half-wave rectifier is basically two times higher than that of the full-wave rectifier. More importantly, when the input energy is limited like the energy harvesting cases, the equivalent input impedance of the rectifier is different between the half-wave and the full-wave cases since they draw different current from the source. This feature could be a tuning factor for the impedance matching between the AC energy source and the load.

In standard CMOS processes, the diodes can be simply replaced by diode-connected MOS transistors. This is a cost-effective way to implement the passive rectifier. However, the maximum achievable transconductance g_m of the MOS transistors is smaller than that of a real PN junction diode or a Schottky diode. For this reason, a large V_{GS} for MOS transistors is required especially when delivering a large current, meaning that it is less efficient than diodes. If low-threshold voltage transistors are used to reduce the V_{GS}, reverse (sub-threshold) leakage current would increase accordingly, and the stand-by time would be reduced.

For domestic electrical appliances that can use an AC supply voltage of 220 V_{RMS} or 110 V_{RMS}, a passive reconfigurable rectifier (universal rectifier) with 1X (rectifier) mode and 2X (voltage doubler) mode is commonly used as shown in Fig. 10.13. To cater for coupling coefficient k variations with distance and/or orientation in WPT applications, a reconfigurable rectifier may be able to increase V_{DC} without increasing the transmitted power. The similar idea was employed in the loosely coupled inductive power link in Lu et al. (2013) and Lee and

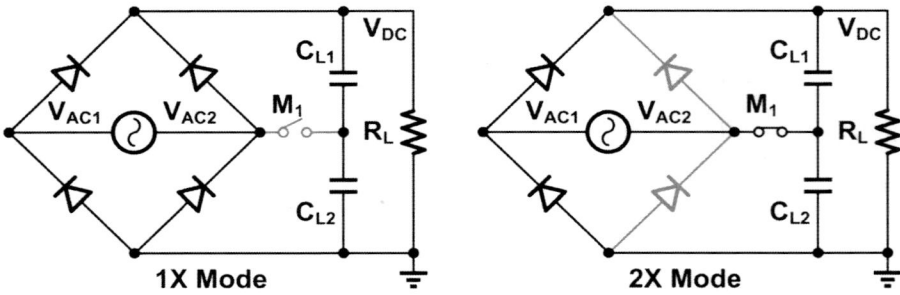

Fig. 10.13 Reconfigurable rectifier (universal rectifier) with 1X (rectifier) mode and 2X (voltage doubler) mode

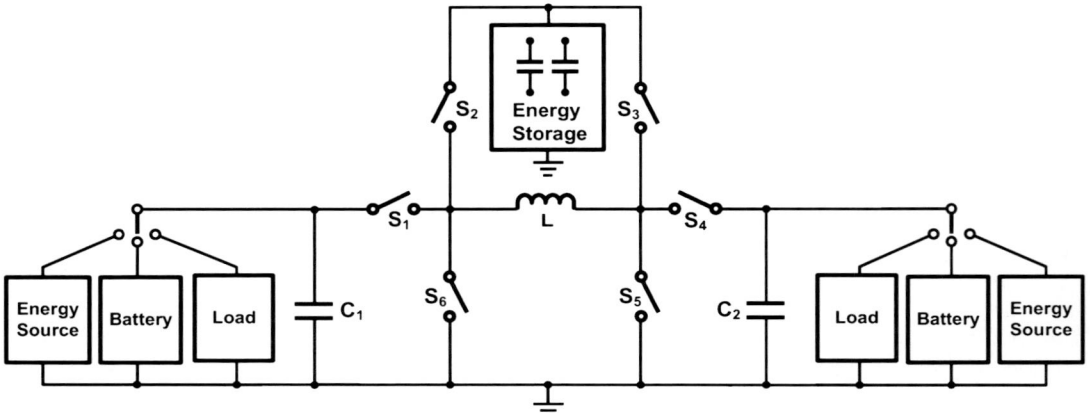

Fig. 10.14 A bi-directional power management unit for self-powered system

Ghovanloo (2012) to extend the coupling range. Note that, the input impedance of the 1X mode ideally is four times larger than that of the 2X mode when driving the same resistive load. Because the 2X mode ideally provides twice output voltage and has twice input-to-output current ratio, making four times input current difference comparing to the 1X mode. This impedance transfer characteristic should be taken into consideration when designing the matching network.

10.3 Control Schemes and Design Trends

10.3.1 Self-Adaptive Architecture

As many IoT nodes are self-powered and rely on renewable energy sources, power management circuits face more challenges than in traditional battery-powered systems. Because power generated through energy harvesting mechanisms is typically low, any power loss matters in these low power devices. Energy conversion efficiency thus rises as the number one design priority. In order to improve the efficiency, traditional techniques used for individual power converters do not suffice. The optimization has to be done from system level and employ much more environmental adaptive operation schemes and circuit architectures. Self-adaptive power circuit architectures thus have been reported (Chen et al. 2010; Sankman et al. 2011).

Figure 10.14 illustrates an example of such self-adaptive power management, which has two distinct differences from a traditional power converter. First, an energy storage unit is included to maximize the utilization of energy and provide a viable way to recycle the energy when excessive. Second, power flow is bi-directional with flexibility of flipping input and output power ports. An advantage of this circuit is that only one inductor is required, leading to a low profile implementation in comparison with traditional circuits. In the meantime, the single-stage architecture benefits the power efficiency. Figure 10.15 illustrates the adaptive reconfiguration Sankman et al. (2011) of the circuit in Fig. 10.14. In this description, the left power port is connected to a power source, whereas the right one is to the load. This situation will change and the power flow becomes reversed, when the source and load swap the ports due to energy availability and load condition. For each direction of power flow, it consists of four operation modes, as shown in Fig. 10.15. In the *Power Mode* (Fig. 10.15a), when the harvested power is greater than the load power, switches S_2 and S_3 are off to disconnect the energy storage element C_S, while the switch pairs S_1/S_5 and S_4/S_6 are complementarily turned on and off to deliver energy to the load. When the harvested power drops too low, C_S is discharged to power the load application in *Power Hungry Mode*. In this mode, S_1 and S_3 are off while switch pairs S_2/S_5 and S_4/S_6 are turned on and off as depicted in Fig. 10.15b. In Fig. 10.15c, if the harvested input power is in excess of load power, the power circuit self-configures into a maximum power point (MPP) tracker to extract power into C_S. In this *Storage Mode*, S_4 is off to isolate the load and switch pairs S_1/S_5 and S_3/S_6 are complementarily turned on and off to deliver harvested energy to C_S. Finally, the dynamic voltage scaling (DVS) configuration is shown in Fig. 10.15d, in which S_1 is off to isolate the input power source, and switch pairs S_2/S_5 and S_4/S_6 are turned on and off to recycle energy from C_O to C_S. Thus, V_{OUT} is reduced to save energy. Once the new V_{OUT} is reached, this *Energy Recycling Mode* can be terminated.

Similarly, self-adaptive architectures can be adopted in SC power converters (Ma and Bondade 2013). Figure 10.16 illustrates a reconfigurable SC converter, which accomplishes both step-up and step down voltage conversions and is compatible with bi-directional power flow operation as well.

10.3.2 Interleaving Multiple-Phase Operation

There are three main reasons to consider multiple-phase operation in a power converter: transient response, ripple reduction and reliability. When the load current changes drastically, it takes time for a converter to respond due to circuit delay and limited loop-gain bandwidth, causing voltage drooping at the output. If loop bandwidth is very limited, the drooping can be severe and the converter takes a long time to recover. A straightforward way to mitigate this issue is to employ parallel architecture. For example, in Fig. 10.17, if there are four identical converters are connected in parallel, the current handing capacity is increased by four times, with equivalently four times increase on loop bandwidth. In theory, an N parallel-cell converter is N times faster than a single cell converter (Kassakian et al. 1990; Perreault and Kassakian 1997).

Interleaving multiple-phase operation also yields great benefit on ripple cancellation. As shown in Fig. 10.17, when each cell converter operates at the same switching frequency but is displaced in phase in each switching period T, due to harmonic cancellation, both voltage and current ripples at the input and output are reduced in magnitude and ripple frequency is pushed to higher level. One critical issue in implementing the interleaving multiple-phase operation is to ensure the phase displacement of each cell is equal, which is also called phase synchronization (Song et al. 2014). A potential phase error can lead to imbalanced current sharing among the cells (Fig. 10.18), causing hot spots and jeopardizing reliability.

Another benefit of interleaving multiple-phase operation is the reliability improvement. If all the cell converters are controlled independently, then the converter as a whole is not centrally controlled. The distributed architecture and operation itself leads to high tolerance to cell failures. Furthermore, similar concept can be applied to each-phase cell converter. Because of large sizes of power devices in the cell, they can be constituted with a series of paralleled small power devices (also called sub-cells), such as

Fig. 10.15 (**a**) Power
Mode, (**b**) Power Hungry
Mode, (**c**) Storage Mode,
and (**d**) Energy Recycling
Mode

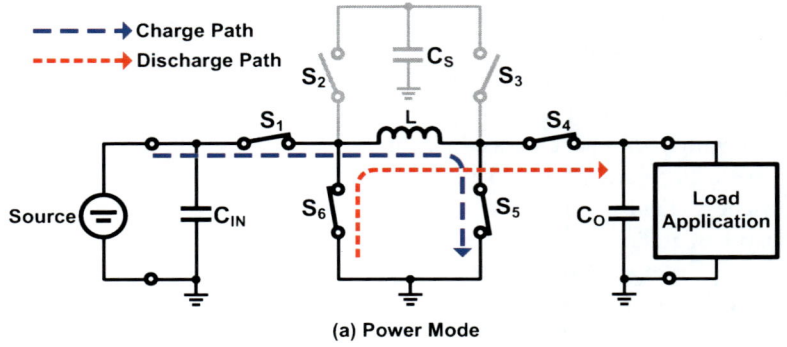

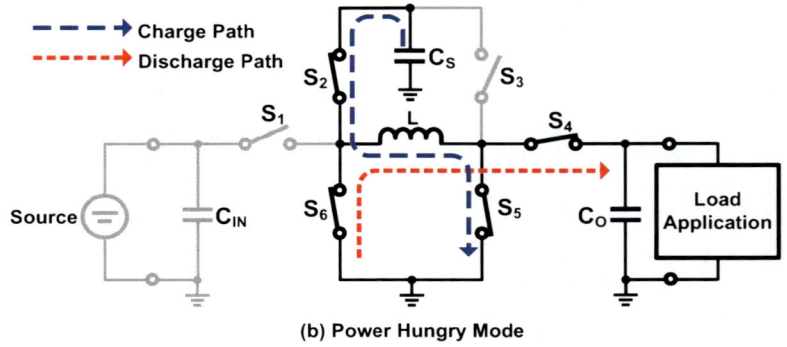

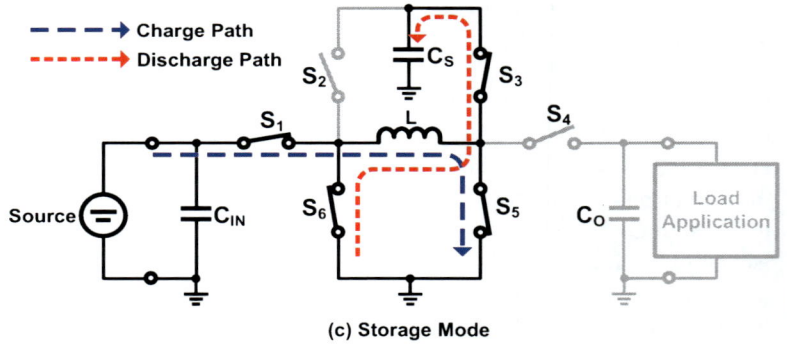

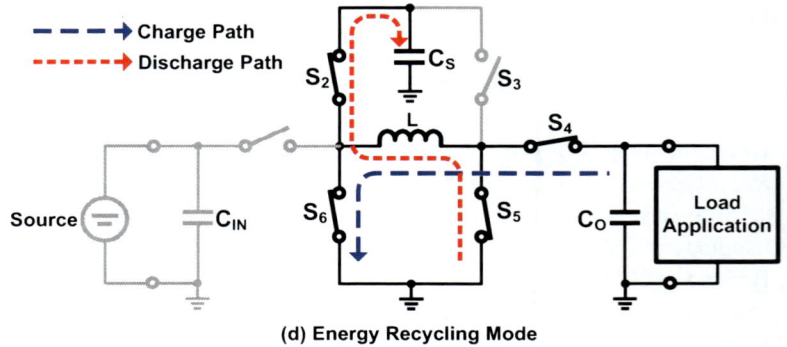

(a) Power Mode

(b) Power Hungry Mode

(c) Storage Mode

(d) Energy Recycling Mode

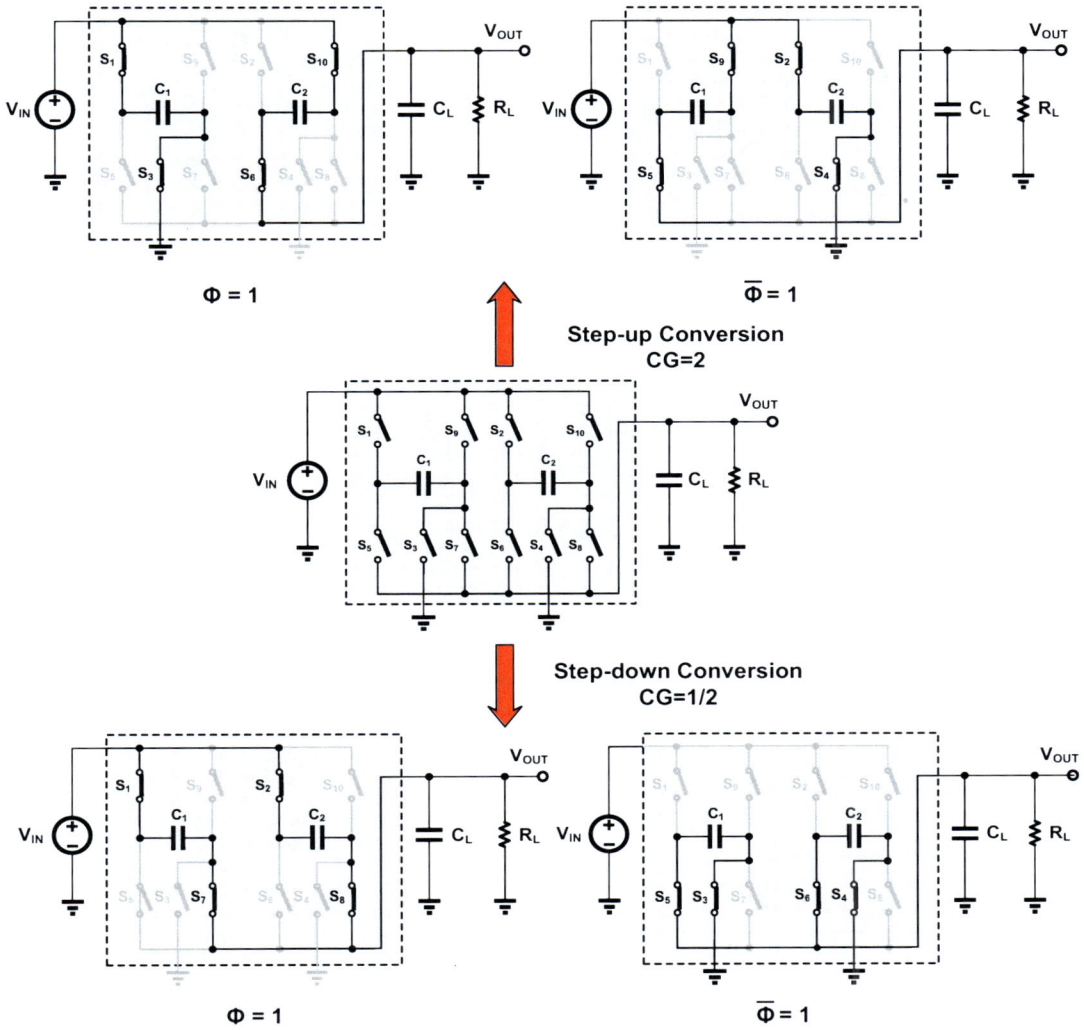

Fig. 10.16 Reconfiguring a SC power converter with CG = 2 or 1/2

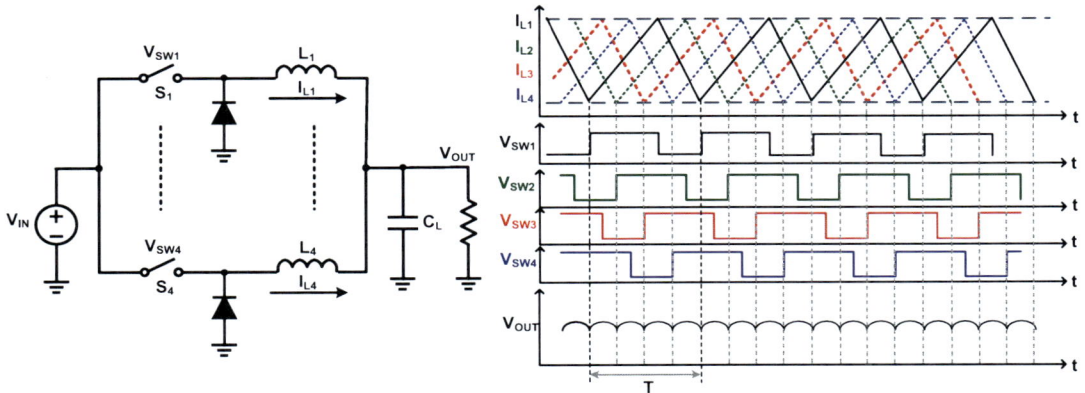

Fig. 10.17 Illustration of a multiple-phase switching converter

Fig. 10.18 Imbalanced
current sharing in a
multiple-phase switching
converter

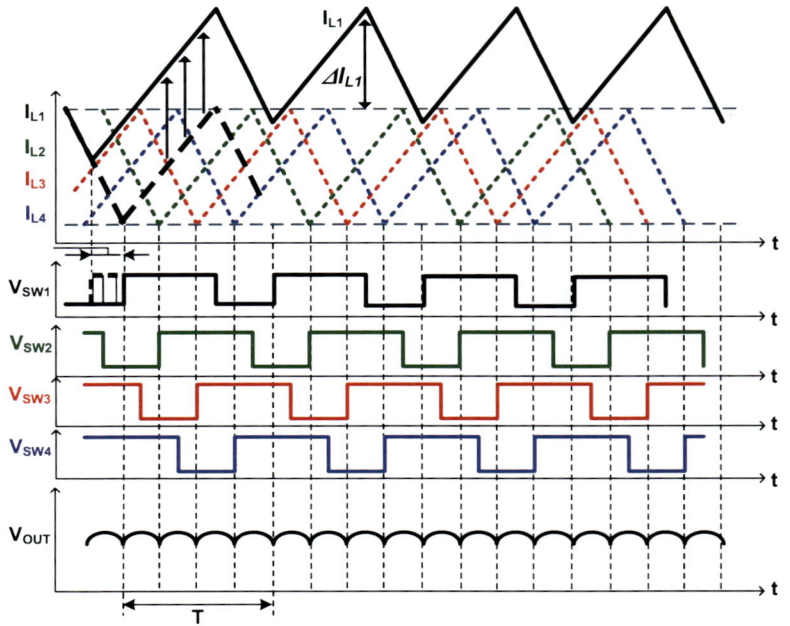

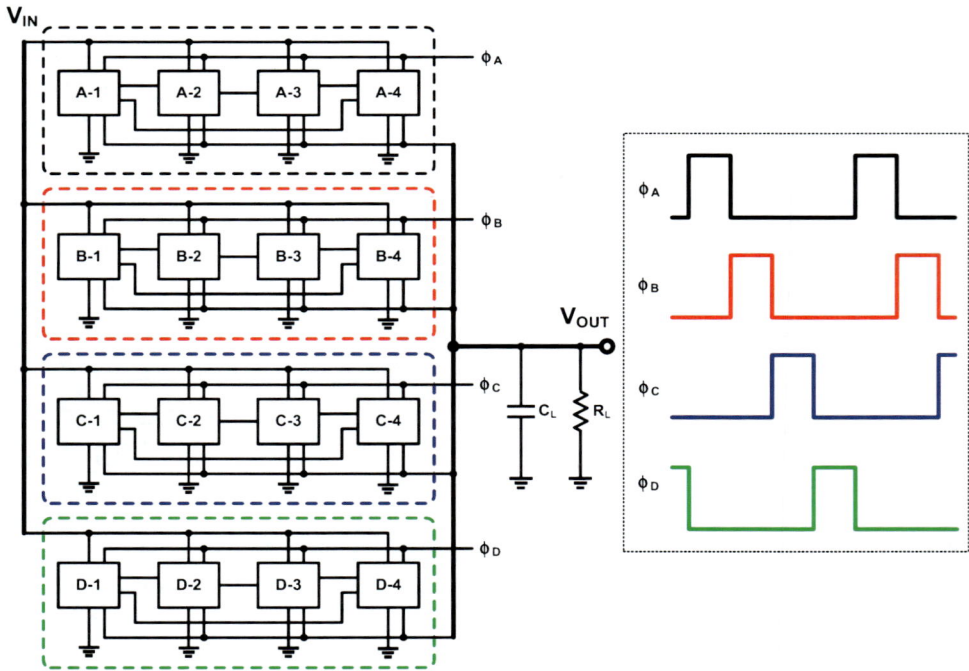

Fig. 10.19 Reconfigurable multiple-phase SC power converter (Ma and Bondade 2013)

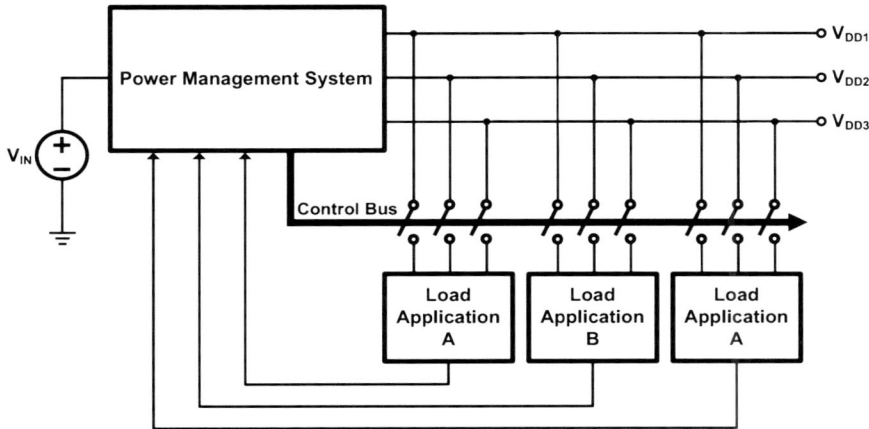

Fig. 10.20 DVFS based power management system with multiple power supplies

A-1 to D-4 in a multiple-phase SC power converter in Fig. 10.19 (Ma and Bondade 2013). Due to the difficulty of implementation, phase displacement cannot be too small in interleaving operation. Hence, the sub-cells (such as A-1 to A-4) retain the same phase (ϕ_A) in each cell (cell A). When one sub-cell fails to function, the other three carry over the total load current designated to the cell, without affecting the operation of others. In addition, the sub-cell arrangement also facilitates dynamic power component sizing, benefiting the efficiency of the converter.

10.3.3 DVFS Capability

In order to mitigate the power crisis in electronic devices, it has been widely believed that power management should be performed not only on device and circuit level, but also on system level. Accordingly, numerous system power management techniques have been reported. To illustrate the effectiveness of the techniques as such, one example is given in Table 15.5 of Chap. 15 by Cymbet Corporation with up to 10.11 power saving ratio. Among these power management techniques, a most popular and effective one is dynamic voltage/frequency scaling (DVFS) (Burd et al. 2000; Cho and Chang 2007; Luo et al. 2007). In this technique, power delivered to each load application is varied adaptively according to the actual load demand.

It directs optimal voltages and frequencies of operation that allow each load application to complete the task right before the deadline, in order to reduce the energy consumption to minimum.

As power supply voltage is a critical control variable in DVFS techniques, traditional battery-powered fixed power supply does not suffice. Instead, it usually relies on power converters to intelligently generate the desired voltages. One common way to do so is to generate multiple discrete supply voltages. As illustrated in Fig. 10.20, based on the power budget and processing deadlines, the DVFS-based power management system instructs the load applications to switch to the supplies that meet the optimization of power and performance.

However, if multiple power converters are employed for such a purpose, system cost, volume and design complexity increase drastically. The use of multiple and bulky power inductors and capacitors require extra PCB area, IC pins, bonding pads. To mitigate this issue, single-inductor multiple-output (SIMO) power converters were developed (Ki and Ma 2001; Ma and Bondade 2010; Ma et al. 2003a, b). As illustrated in Fig. 10.21, a SIMO converter only requires one single inductor and could save the number of power switches by up to 50% and is controlled by a single controller. The cost-effectiveness makes it a popular choice in recent years (Chen et al. 2012; Zhang and Ma 2014, 2015).

Fig. 10.21 A SIMO
power converter for DVFS
power management (Ki and
Ma 2001)

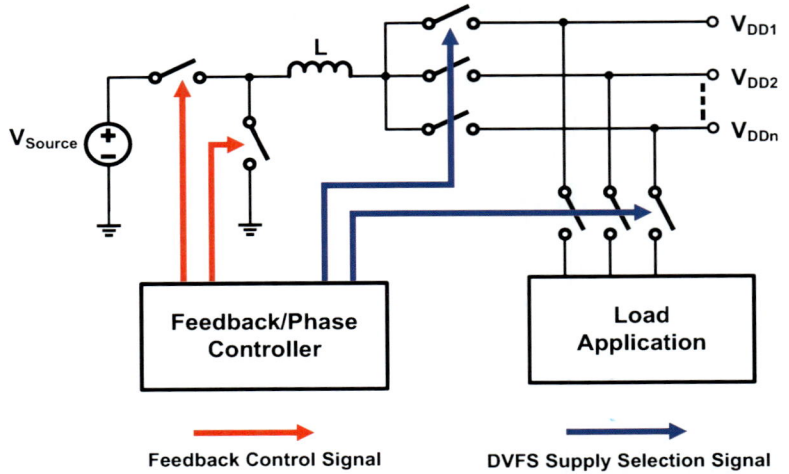

Feedback Control Signal   DVFS Supply Selection Signal

Fig. 10.22 DVFS based
power management system
using a variable-output
buck converter

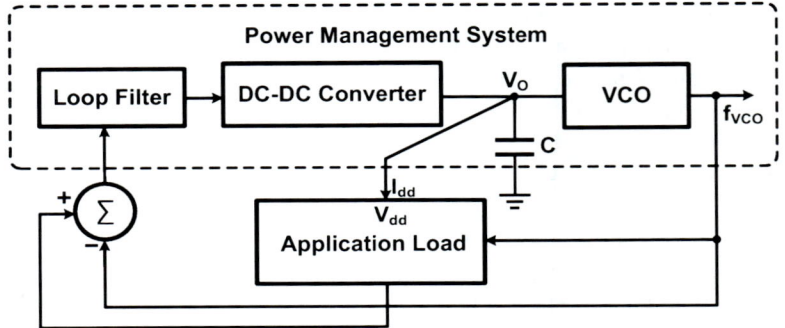

Apart from multiple supply implementations, DVFS techniques can also be accomplished using a single power converter with variable output. For the sake of hardware cost and form factor, this is clearly more economical, because it only requires one single power converter. However, because of the output of the power converter has to continuously vary in a relatively wide voltage range and reasonable transient speed, the design could face some serious challenges. With a wide output range, SC power converters and linear regulators likely suffer from low efficiency. Even for a switching converter, as depicted in Fig. 10.22, its transient response time has to be largely shortened to reduce the latency and computing errors in the application load, which can be a processor running at a much higher operation frequency.

Unlike traditional fixed-output converters that are optimized and compensated at a fixed voltage level, variable output causes stability issues during large dynamic transient periods. Robust control methods (Smedley and Cuk 1995; Song et al. 2014; Utkin 1993; Wang and Ma 2012) are expected for such applications.

10.3.4 CMOS Rectifier

CMOS rectifier that does not need off-chip diodes and extra mask is a low-cost solution for the IoT, without degrading performances. Two important parameters in evaluating a rectifier are the voltage conversion ratio M and the power conversion efficiency (PCE). M is defined as

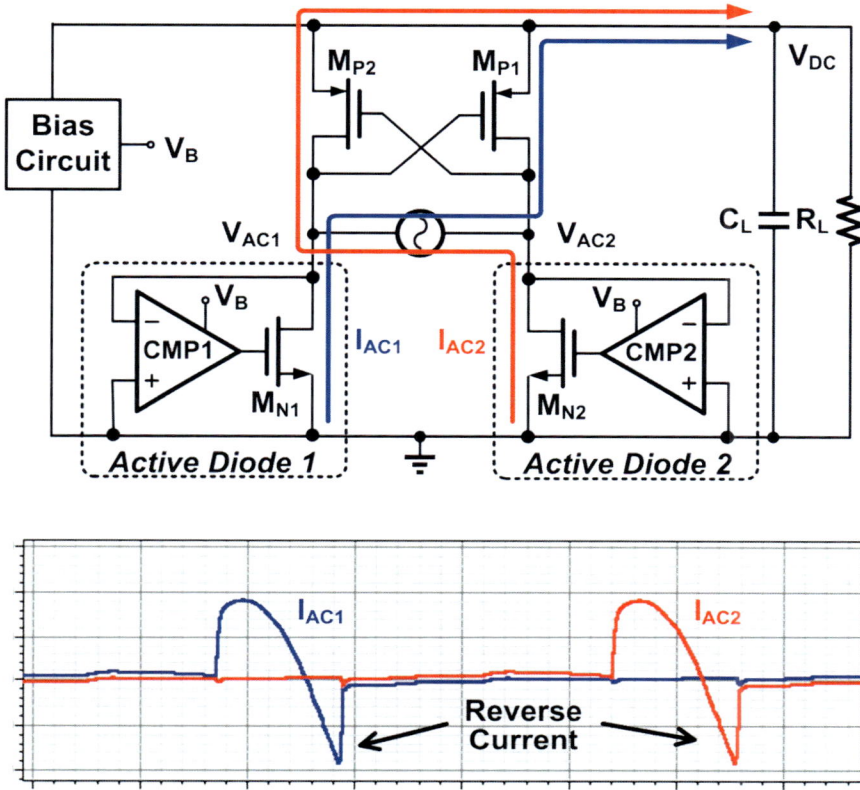

Fig. 10.23 An active full-wave rectifier with simulated AC current waveforms showing the reverse current problem in active rectifier

$$M = \frac{V_{DC}}{|V_{AC}|}, \qquad (10.1)$$

where $|V_{AC}|$ is the amplitude of the input AC signal to the rectifier, and V_{DC} is the averaged rectified output DC voltage. The PCE of an AC-DC converter is defined as

$$\text{PCE} = \frac{P_{OUT}}{P_{IN}} = \frac{V_{DC}^2/R_L}{\frac{1}{N \cdot T} \int_{t_0}^{t_0+N \cdot T} V_{AC}(t) \cdot I_{AC}(t)\mathrm{d}t},$$

$$(10.2)$$

where T is the period of the input sinusoidal wave, N is the number of cycles that are integrated for P_{IN} calculation, and $V_{AC}(t)$ and $I_{AC}(t)$ are the instant voltage and current of the AC source.

For the passive rectifiers, as discussed in Sect. 10.2.3, the diode voltage drop V_D of 0.7 V for typical diodes and 0.3 V for Schottky diodes, limits the voltage conversion ratio (M) and the power conversion efficiency (PCE). An active rectifier that only use CMOS transistors is shown in Fig. 10.23. The high-side diodes in passive rectifier are replaced by two cross-coupled PMOS switches, and the low-side diodes are replaced by two comparator-controlled NMOS switches (active diodes). In this configuration, the voltage drops are reduced from $2V_D$ to $2V_{DS}$ (V_{DS} is the turn-on voltage of the power switches). The operation principle is described as follows: when $V_{AC2} - V_{AC1} > |V_{tP}|$ (threshold voltage of $M_{P1,2}$), M_{P1} is turned on and $V_{AC2} = V_{DC}$; and then V_{AC1} swings below the ground voltage, the comparator CMP1 turns on the

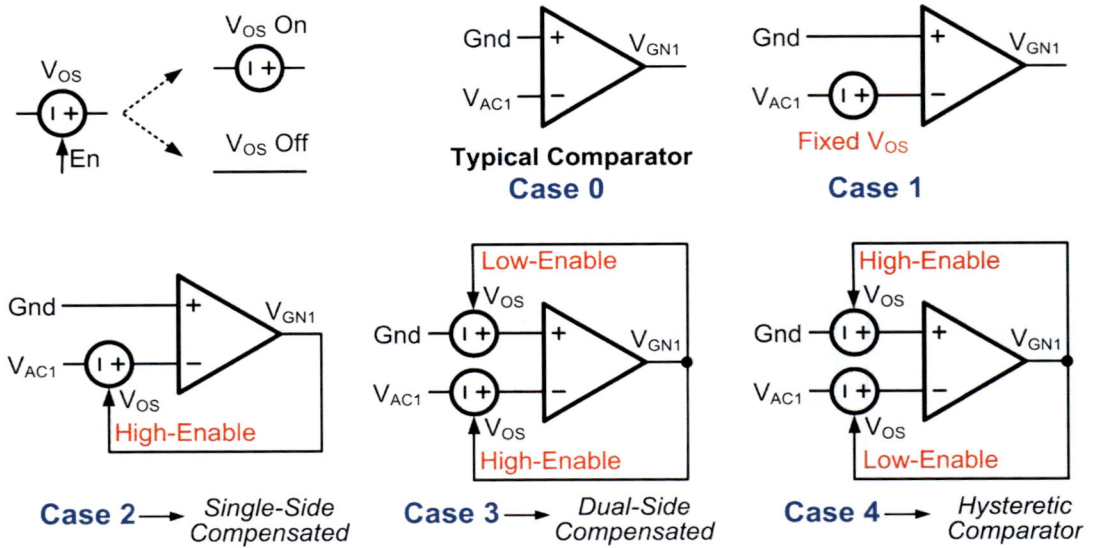

Fig. 10.24 Comparator schemes for reverse current control, and the hysteretic comparator

switch M_{N1}, I_{AC1} charges up V_{DC} by V_{AC}. After V_{AC1} swings above zero, M_{N1} is then turned off by CMP1. During the next half of the AC input cycle, the other half of the rectification circuit will conduct in a similar fashion as described above. Thus, by replacing the four diodes of a full-wave rectifier with power transistors, M is significantly increased especially when the input amplitude is low. However, when operating at a high frequency such as 13.56 MHz, the comparator delay and gate-drive buffer delay would affect the efficiency of the rectifier. As demonstrated by the simulated I_{AC} waveforms, the reverse current will occur if the large power switches are not turned off immediately when V_{AC1} or V_{AC2} is higher than the ground voltage. Thus, in this architecture, the main losses are conduction loss, switching loss and comparator static power, if the reverse current is well-controlled.

Many schemes were proposed to compensate the delay of active diodes (Cheng et al. 2016; Guo and Lee 2009; Lu and Ki 2014; Lam et al. 2006; Lu et al. 2013; Lee and Ghovanloo 2012). The rising edge propagation delay of the comparator and buffer, t_{pLH}, will shorten the current conduction time Δt, limiting the highest operation frequency of the active rectifier. On the other

hand, the falling edge delay t_{pHL} forces the power NMOS transistors $M_{N1,2}$ to turn off late, and the charge of the output capacitor will flow back to ground through $M_{N1,2}$, resulting in reverse leakage current. However, $M_{N1,2}$ have to be large to handle large output current, increasing the response time of the active diodes. This problem is more pronounced when the operation frequency is required to be high and the input amplitude $|V_{AC}|$ is low, because delay time of comparators and buffers are inversely proportional to the supply voltage.

Comparators with unbalanced bias currents or asymmetric differential input are used to set an artificial input offset voltage to compensate for the delay and to turn the power switches on and off properly. Prior reverse current control schemes fall into one of the cases sketched in Fig. 10.24. The symbol of artificial input offset with an enable pin is defined as: when the enable bit is high, a non-zero offset voltage is introduced to the comparator; and when the enable bit is low, the offset voltage is zero (a wire). In Case 0, the comparators have no artificial offset (reverse current occurs due to delays). In Case 1, the comparators have fixed artificial offset such that the power switches are turned off earlier, but turned on later. In Case 2, the comparators have dynamic artificial offset such that the power switches are only turned off earlier. In Case 3, the comparators have dynamic artificial offset at both edges such that the power switches are turned on and off earlier. Case 4 is a hysteretic comparator that will not be used in this scenario, it is listed here only as a complementary example of Case 3.

To compare the performance of the above comparator schemes on the low-voltage active rectifiers especially, the metrics Crest Factor (CF) used in evaluating electrical appliances could be also employed here. CF is defined as the ratio between the peak current delivered to the load and the corresponding RMS current

$$CF = \frac{I_P}{I_{RMS}}. \tag{10.3}$$

A higher CF means a higher peak current for the same load condition, and larger power transistors are thus needed to reduce the V_{DS} for achieving high voltage conversion ratios. Routing metal needs to be wider for a higher CF as well. The CF for each case is considered as below.

The scenarios of current conduction of Case 0 to Case 3 are sketched in Fig. 10.25, and discussed as follows. Case 1 is implemented in Guo and Lee (2009), with constant offset introduced to the comparators using unbalanced bias currents. The power NMOS switches are turned off earlier by t_d to eliminate the reverse current; however, they are turned on later by $2t_d$ than the ideal case, and the conduction time Δt_1 is $2t_d$ shorter. For the same load current, the peak current I_{p1} has to be higher, limiting its operation at a higher frequency. The delays get worse when $|V_{AC}|$ is low.

In Lam et al. (2006), self-biased active diodes were employed, and to reduce or eliminate t_d, a reverse current control (RCC) scheme (Case 2) is introduced. However, both the bias current and the operation of the RCC transistor are highly affected by $|V_{AC}|$ and process variations, making the rectifier hard to be optimized over a wide input range. Moreover, as the reverse current control is realized by a time-varying offset, the

artificial offset would disappear right after the power NMOS transistor is turned off. If the RCC transistor turns on prematurely (for example, due to process variations) and turns off the active diode while $|V_{AC}|$ is still higher than V_{DC}, the comparator will go high again in the same cycle. Simulation waveforms of the described scenario are shown in Fig. 10.26. The efficiency would deteriorate with this multiple-pulsing problem, especially in light load condition when switching loss dominates. Thus, SR latches were used to realize a one-shot per cycle logic to handle this multiple-pulsing problem (Lu et al. 2011).

Case 3 was realized in Lee and Ghovanloo (2012), where it uses self-biased active diodes similar to Lam et al. (2006), with an offset-control function for the comparators to compensate for both turn-on and turn-off delays such that Δt_3 could be maximized (lowest CF). It suffers from the same and even worse multiple-pulsing problem as (Lam et al. 2006), as the dynamic offset (±offset voltages) transition of the offset-control circuit is unstable: when the comparator outputs a 1(0), the dynamic offset flips the output to 0(1), and this positive feedback makes the comparator undergoes self-oscillation. This phenomenon is what a hysteretic comparator is designed to avoid. To make the scheme work, Lee and Ghovanloo (2012) adds a delay cell t_{dp} in the offset-control path in addition to its calibration bits. If the delay time t_{dp} is large enough (comparable to Δt), the comparator could be stable, but then the power switch would be turned on for at least the duration of t_{dp} that limits the minimum conduction time ($\Delta t_{min} > t_{dp}$). This property makes its operation more like a constant-on time control at light load condition. As a result, its light load efficiency is degraded.

Case 3 was also realized in Cheng et al. (2016) that a hard one-shot logic allowing the power transistor only being turned-on once per cycle is used to avoid the multiple-pulsing problem. Furthermore, a near-optimum switched-offset compensation was implemented for both on- and off-delays with additional voltage sensing feedback loops. Consequently, the reverse current in Cheng et al. (2016) is eliminated elegantly.

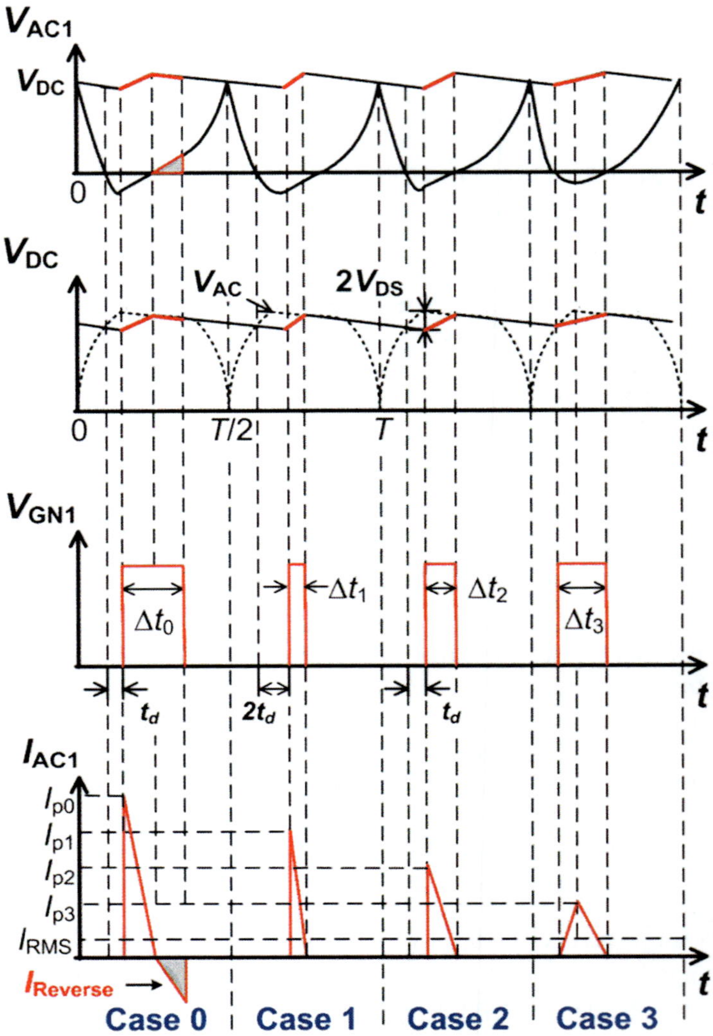

Fig. 10.25 Conceptual
waveforms of the input/
output voltages, power
NMOS gate voltage and
conducted currents for
active rectifiers with
different comparators

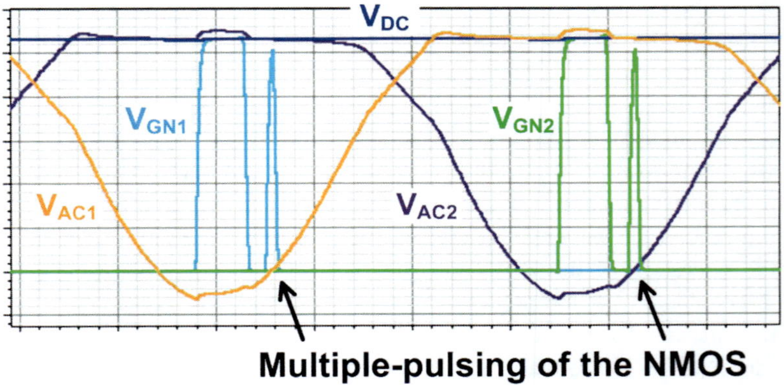

Fig. 10.26 Simulated
waveforms of multiple-
pulsing problem associated
with dynamic offset
schemes

Fig. 10.27 Schematic of the multi-stage step-up rectifier

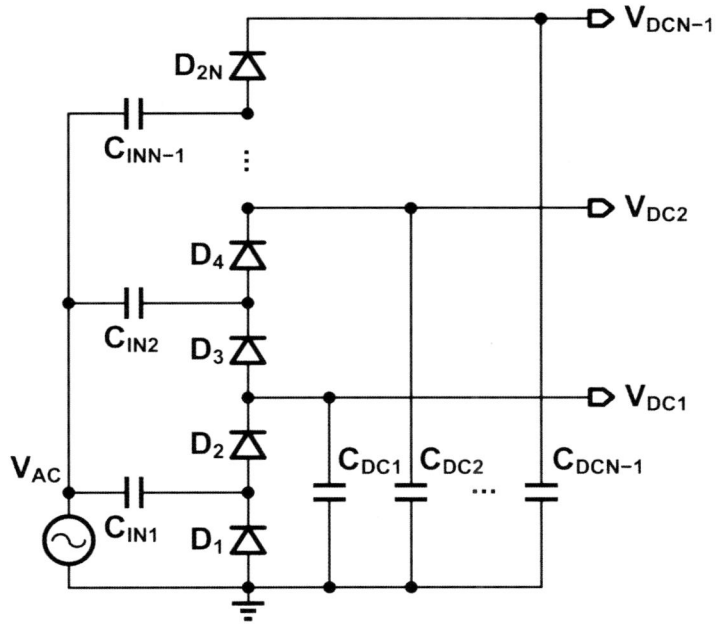

To summarize, Case 2 has longer Δt than Case 1, and in the ideal situation, Case 3 has the largest Δt. However, a comparator with both compensated turn-on and turn-off delays is logically unstable: the hysteresis goes the opposite direction as a normal hysteretic comparator goes, and the robustness of the rectifier is degraded, special logics are needed for a stable operation.

There are some other variations of the CMOS active rectifiers been proposed (Radecki et al. 2011; Xu et al. 2014) for near-field WPT. For example, to operate at a relatively high WPT frequency (150 MHz), the synchronous rectifier (Radecki et al. 2011) uses the delay locked-loop (DLL) to synchronize with the AC input waveform and to generate discrete timing slots for controlling the power switches. In Chung et al. (2012), instead of using the voltage or phase offset to compensate the comparator delay, a positive feedback loop was used to shorten the delay.

RF energy harvesting needs the rectifier to operate in the ultra-high frequency (UHF) range, thus it is commonly known as RF-DC conversion. When the AC-DC conversion considers the problems in voltage domain, the RF-DC conversion prefers to look into circuit in a power

perspective. Since the RF input power level usually is low, multiple step-up stages are needed to boost the output voltage. A Dickson type multi-stage rectifier is shown in Fig. 10.27. Each stage is a voltage doubler, thus the ideal no-load output voltage of an N-stage converter is $2N$ times of the AC input amplitude. An interesting and useful characteristic of the multi-stage rectifier is that the number of stage N does not affect the maximum achievable PCE (Yi et al. 2007). Because each stage draws current from the input source in the charging phase of each cycle, and the stages are stacked in the discharging phase for higher output voltage. Each stage operates in a similar way as the single-stage rectifier, thus the maximum PCE is basically the same.

As mentioned above, diode drop is the main constrain that limits the low-voltage operation of the rectifiers, this problem becomes even severe in the RF-DC senarios. Thus, the cross-connected (CC) CMOS rectifier structure (Facen et al. 2006), as shown in Fig. 10.28a, is a commonly used topology for its low-voltage and auto-switching characteristics. However, a high leakage current will occur at high input conditions because the PMOS and NMOS will

Cross-Connected

V_{TH} Compensated Diode

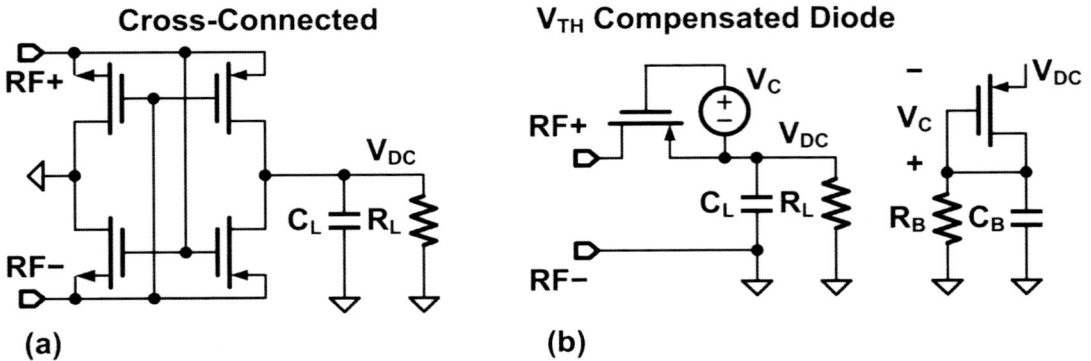

Fig. 10.28 Schematics of (**a**) cross-connected rectifier and (**b**) rectifier with V_{TH} compensation

be turned-on simultaneously during the transition period, which is a similar problem to the shoot-through current in a CMOS inverter. To achieve high heavy load efficiencies requires larger transistor sizes for the rectifier, which will cause more reverse leakage current. Larger transistor size will also increase the parasitic loss during the step up conversion. It has been demonstrated that the voltage conversion ratio of the CC topology could be larger than 80% in Nakamoto et al. (2007), when the input amplitude of the rectifier was higher than 150 mV. Evidently, the CC rectifier provides higher peak PCE than the Dickson-type rectifiers. However, its high-PCE range is narrow due to the above mentioned leakage current problem. A high-PCE range extension technique that employed one high-power path and one low-power path with auto-selection can be used to achieve a wider input power range (Lu et al. 2017). As shown in Fig. 10.28b, another solution (Stoopman et al. 2014) that has much less leakage current is to internally generate a V_{TH} compensation voltage to bias the diode-connected MOSFET. But large values of R_B and C_B are required because the power budget for bias current is tiny, then considerably large area would be occupied in multistage topology.

10.3.5 Supply Noise Rejection

There are many sensory circuits exist in the IoT nodes, and most of them are sensing single-ended

signals that make the power supply rejection ratio of the input buffer a very important specification. Alternatively, reducing the supply noise from power source itself is one equivalent way to reduce the effect of supply noise.

For an LDO regulator, the largest capacitors are the load capacitor C_L and the parasitic gate capacitor C_G of the power transistor. Hence, there are at least two low-frequency (LF) poles on the left-half-plane (LHP): the pole at the output node p_O, and the pole at the gate of the power MOS p_G, as sketched in Fig. 10.29 with either p_O or p_G being the dominant pole. The pole p_O would shift to a lower frequency when the load resistance increases and vice versa. For stability consideration, p_O moves to lower frequency as the load current decreases, which makes the p_O dominant case more stable. On the other hand, the zero-load condition is seldom discussed in many output-capacitor-less designs for stability reason. Instead, a minimum load current is needed to satisfy stability requirements.

By using most of the available capacitance at the output node, p_O is designed to be the dominant pole. Then, a larger load capacitor could better filter out power supply noise and glitches from the pre-stage DC-DC or AC-DC converters, and serves as a buffer for load-transient current variation, resulting in a cleaner V_{OUT}. Also, as shown in Fig. 10.29c, because the output voltage is well regulated by the control loop at low frequency, and the noise is bypassed to ground by C_L at high frequency, the p_O dominant case could provide a good power supply noise rejection

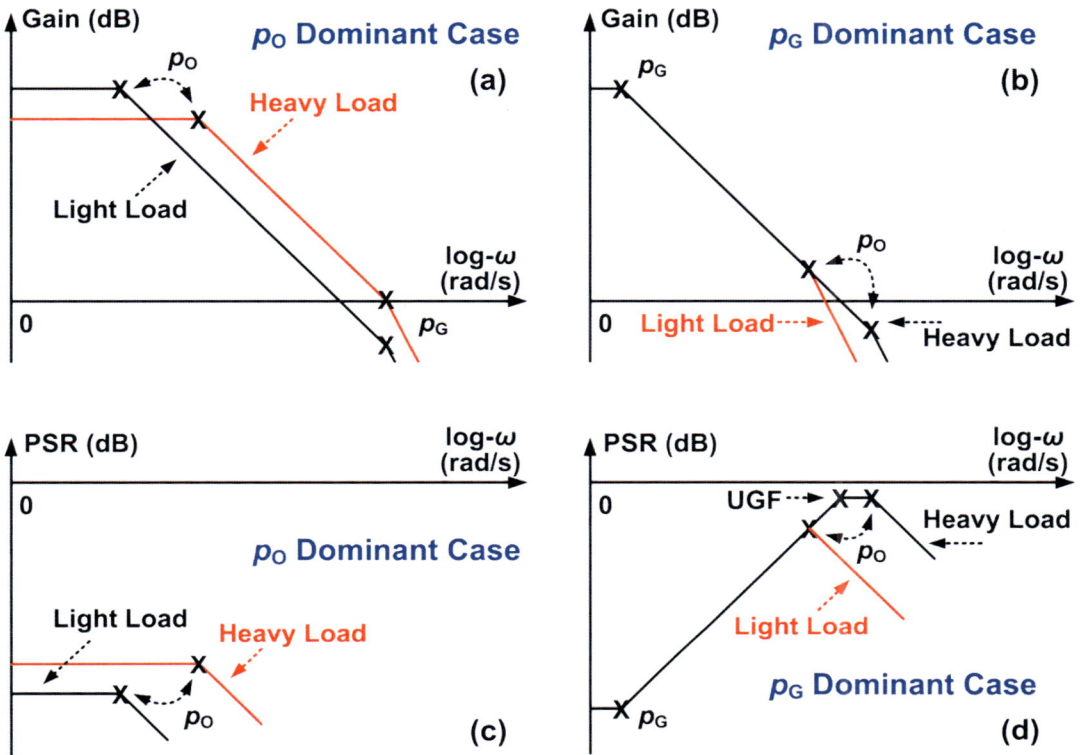

Fig. 10.29 Conceptual frequency response of generic LDO regulators with two low frequency poles: (**a**) open loop-gain with p_O being its dominant pole, (**b**) open loop-gain with p_G being its dominant pole, (**c**) PSR with p_O being its dominant pole, and (**d**) PSR with p_G being its dominant pole

(PSR) across the full-spectrum (Lu et al. 2015). On the other hand, as shown in Fig. 10.29d, the worst case PSR for the p_G dominant case would occur at medium frequency (Gupta et al. 2004). Thus, increasing both the load capacitance and the loop bandwidth (that is, the unity-gain frequency, UGF) would improve the PSR of an LDO regulator.

Meanwhile, PSR can also be enhanced by circuit techniques, like in Park et al. (2014), to replicate the effects from the C_G and to cancel out the input noise. And also, larger dropout voltage usually could give better PSR performance but will sacrifice the efficiency (Zhan and Ki 2014). In a step-down DC-DC converter in cascade with LDO topology, using NMOS as the power transistor driven by the DC-DC input voltage (Lu et al. 2016) could be an easy but effective solution to reject power

supply noise with little area and power overhead.

10.3.6 Passive Components

As mentioned above, passive components including inductors and capacitors are used in power converters for energy storage and transfer purposes. However, passive components are chip or PCB area consuming, and have non-idealities that limit the performances of the converters. Circuit models of both practical inductor and capacitor are given in Fig. 10.30. For the inductor model, it includes a parasitic capacitor C_P which decides the self-resonant frequency (SRF) of the employed inductor. When the inductor operates at frequencies higher than the SRF, the parasitic capacitor path provides lower impedance compared to the inductor path, it starts to act like a

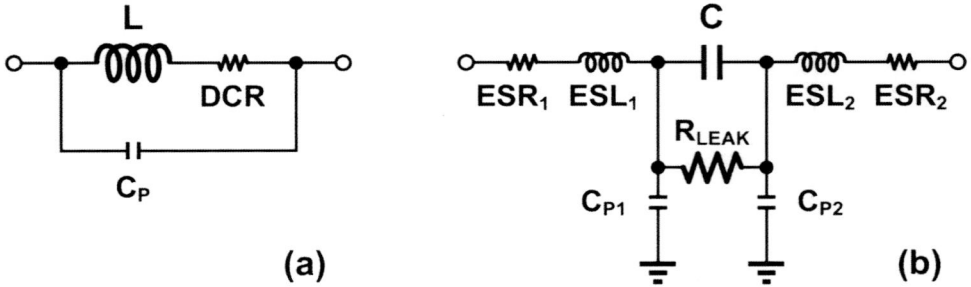

Fig. 10.30 Circuit models of (**a**) a practical inductor and (**b**) a practical capacitor with paracitics

Table 10.1 Examples of state-of-the-art power inductors

Pattern	Model number	Dimensions (mm)	L (μH)	DCR (Ω)	SRF (MHz)	I_{RMS} (mA)	Price (US$)
	PFL1005 (Coilcraft PFL1005 Datasheet 2015)	1.14 × 0.64 × 0.7	1	0.97	460	310	<0.2
	PFL2010 (Coilcraft PFL2010 Datasheet 2015)	2.20 × 1.45 × 1.0	4.7	0.62	66	420	<0.2
	LPS3008 (Coilcraft LPS3008 Datasheet 2015)	2.95 × 2.95 × 0.8	180	9.0	10	120	<0.28
	MSS1038 (Coilcraft MSS1038 Datasheet 2016)	10.2 × 10.0 × 3.8	1000	2.8	3	330	<0.51

capacitor. Meanwhile, the DCR of the inductor will generate I^2R conduction loss. Under the same technology, the value of DCR basically increases with the inductance, because more windings are needed for larger inductance. To reduce DCR, thick wires should be used. But then, the component size will be increased. Thus, there are tradeoffs between the device size, the current rating, and the operating frequency. Since the inductor has poor scalability due to its magnetic property, it has always been one of the main limiting factors for the miniaturized devices such as IoT nodes. A comparison of state-of-the-art power inductor candidates for IoT nodes on their size, electrical parameters, and cost is given in Table 10.1.

Historically, it has been a challenge to integrate power inductors into silicon technology. Major concerns lie in compatibility with CMOS technology, increased fabrication cost and chip area and performance degradation on inductance density and quality factor Q (Boggetto et al. 2002; Gardner et al. 2009; Sullivan et al. 2009). As the alternative, efforts of using package bond-wire inductors (Huang and Mok 2013; Song et al. 2016; Wu et al. 2014) through System in Package (SiP) have been reported to avoid the cost and performance penalty of on-chip inductors, while keeping low system profile. This is also consistent with the global system miniaturization trends and technologies addressed in Chap. 16.

For the capacitor model, parasitic equivalent series resistors (ESRs) and inductors (ESLs) are unavoidable. Both ESR and ESL will generate unwanted additional voltage ripples when the capacitor is being charged or discharged. Also, there will be leakage current conducts through the two terminals of a capacitor. This leakage phenomenon can be modeled as a large resistor

Table 10.2 Examples of state-of-the-art capacitors

Pattern	Type	Dimensions (mm)	C (μF)	Voltage rating (V)	Leakage (μA)	Price (US$)
	Ceramic Capacitor	$1 \times 0.5 \times 0.5$	1	25	n/a	<0.04
	Capacitor Array	$3.2 \times 1.6 \times 1.35$	4×1	10	n/a	<0.15
	Super Capacitor (Maxwell HC series Ultracapacitors Datasheet 2016)	$12 \times 8 \times 8$	10^6	2.7	6	<1

R_{LEAK} across the capacitor. Table 10.2 summaries some state-of-the-art capacitors that can be used in IoT nodes. The capacitor array which contains two or four separate capacitors provides a compact PCB solution, while the super capacitor provides super capacity but suffers from considerable leakage current.

For on-chip implementation, parasitic capacitors (C_{P1} and C_{P2}) coupled to the silicon substrate would be larger compared to the off-chip case. Because the on-chip distances between the capacitor plates and the substrate are short. Nevertheless, the capacitance density has been significantly increased either in advanced nanometer CMOS processes (El-Damak et al. 2013; Jain et al. 2014) or in the post-CMOS deep trench technology (Johari and Ayazi 2009). Thus, the power density of SC power converters has been largely increased in recent years.

10.4 Perspectives and Trends

As semiconductor technology enters the nano regime in recent years, power crisis continues to grow as the number one challenge for system design. The impacts of power management and power circuits have been appreciated far beyond the traditional domain of power electronics. Power circuits, as a hardware platform that enables efficient power management, should be clearly understood by all the design engineers from topology to performance matrix.

For the IoT applications in particular, when selecting the topology of power converters, if the application has rigorous constraints on device volume and switching noise, switched-capacitor power converters are more preferred for fully integration when compared to switching converters, because the required capacitor value of a SC power converter decreases linearly with the output power. If the requirement on device volume is loose, switching converters are more popular for high efficiency and flexibly wider output range. Linear regulator is another indispensable block in power management modules for its clean, area-efficient, and supply ripple rejection characteristics. Also, a linear regulator with fast adaptive output would be favorable for low-duty-ratio IoT nodes. Lastly, wireless powering with near-field or far-field operation is the major intentional (man-made) energy source for passive or active wireless nodes for IoT applications. CMOS rectifier is the most cost-effective and efficient way to convert the received AC power into DC outputs. A wireless power receiver that can acquire multiple frequency bands is in demand since there are various ambient RF energy sources and a couple of wireless power standards co-exist.

In conclusion, power management has become an essential measure to accomplish the optimal performance in modern electronic system designs. Cross-layer power management at device, circuit and system level represents an inevitable trend and leads to a broad spectrum

of research opportunities. At device level, wide
bandgap and 3D power devices that enable much
faster and more efficient operation are in high
demand. Heterogeneous integration of these
devices with mainstream silicon technologies as
well as their interface circuits determines the
timeframe of commercialization. With progressive development on fundamental power circuits,
power circuits in the future will be highly
diversified and system-oriented. Adaptation, efficiency and reliability are the key elements in
evaluating the performances of these circuits.
At the system and application level, the function
of a power circuit should no longer be limited to
voltage regulation. It will bridge global system
power management strategies with local device
operation to accomplish near-optimal energy
efficiency, maximized lifetime and highly elevated reliability.

References

J.M. Boggetto, Y. Lembeye, J.P. Ferrieux, Y. Avenas,
   Micro fabricated power inductors on silicon. IEEE
   Power Electron. Spec. Conf. **3**, 1225–1229 (2002)
T.D. Burd, T.A. Pering, A.J. Stratakos, R.W. Brodersen,
   A dynamically voltage scaled microprocessor system.
   IEEE J. Solid-State Circuits **35**(11), 1571–1580
   (2000)
H. Chen, B. Wei, D. Ma, Energy storage and management
   system with carbon nanotube supercapacitor and
   multi-directional power delivery capability for autonomous wireless sensor nodes. IEEE Trans. Power
   Electron. **25**(12), 897–2909 (2010)
H. Chen, Y. Zhang, D. Ma, A SIMO parallel-string driver
   IC for dimmable LED backlighting with local bus
   voltage optimization and single time-shared regulation loop. IEEE Trans. Power Electron. **27**(1),
   452–462 (2012)
L. Cheng, W.H. Ki, Y. Lu, T.S. Yim, Adaptive on/off
   delay-compensated active rectifiers for wireless power
   transfer systems. IEEE J. Solid-State Circuits **51**(3),
   712–723 (2016)
Y. Cho, N. Chang, Energy-aware clock-frequency assign-
   ment in microprocessors and memory devices for
   dynamic voltage scaling. IEEE Trans. Comput.-
   Aided Des. Integr. Circuits Syst. **26**(6), 1030–1040
   (2007)
H. Chung, A. Radecki, N. Miura, H. Ishikuro, T. Kuroda,
   A 0.025–0.45 W 60%-efficiency inductive-coupling
   power transceiver with 5-bit dual-frequency
   feedforward control for non-contact memory cards.

IEEE J. Solid-State Circuits **47**(10), 2496–2504
   (2012)
Coilcraft LPS3008 Datasheet, revised Sep 2015, http://
   www.coilcraft.com/pdfs/LPS3008.pdf
Coilcraft MSS1038 Datasheet, revised Mar 2016, http://
   www.coilcraft.com/pdfs/mss1038.pdf
Coilcraft PFL1005 Datasheet, revised Sep 2015, http://
   www.coilcraft.com/pdfs/pfl1005.pdf
Coilcraft PFL2010 Datasheet, revised Sep 2015, http://
   www.coilcraft.com/pdfs/pfl2010.pdf
D. El-Damak, S. Bandyopadhyay, A. Chandrakasan, A
   93% efficiency reconfigurable switched-capacitor
   DC-DC converter using on-chip ferroelectric
   capacitors, in *IEEE International Solid-State Circuits
   Conference (ISSCC)*, Feb 2013, pp. 374–375
T. Esram, P.L. Chapman, Comparison of photovoltaic
   array maximum power point tracking techniques.
   IEEE Trans. Energy Convers. **22**(2), 439–449 (2007)
A. Facen, A. Boni, Power supply generation in CMOS
   passive UHF RFID tags, in *Ph.D. Research in Micro-
   electronics and Electronics* (2006), pp. 33–36
D.S. Gardner, G. Schrom, F. Paillet et al., Review of
   on-chip inductor structures with magnetic films.
   IEEE Trans. Mag. **45**(10), 4760–4766 (2009)
S. Guo, H. Lee, An efficiency-enhanced CMOS rectifier
   with unbalanced-biased comparators for
   transcutaneous-powered high-current implants. IEEE
   J. Solid-State Circuits **44**(6), 1796–1804 (2009)
V. Gupta, G.A. Rincon-Mora, P. Raha, Analysis and
   design of monolithic, high PSR, linear regulators for
   SoC applications, in *Proceedings of IEEE International SOC Conference*, Sep 2004, pp. 311–315
C. Huang, P.K.T. Mok, A 100 MHz 82.4% efficiency
   package-bondwire based four-phase fully-integrated
   buck converter with flying capacitor for area reduction. IEEE J. Solid-State Circuits **48**(12), 2977–2988
   (2013)
R. Jain, B.M. Geuskens, S.T. Kim et al., A 0.45–1 V fully-
   integrated distributed switched capacitor DC-DC converter with high density MIM capacitor in 22 nm
   tri-gate CMOS. IEEE J. Solid-State Circuits **49**(4),
   917–927 (2014)
H. Johari, F. Ayazi, High-density embedded deep trench
   capacitors in silicon with enhanced breakdown voltage. IEEE Trans. Compon. Packag. Technol. **32**(4),
   808–815 (2009)
J.G. Kassakian, High frequency switching and distributed
   conversion in power electronic systems, in *6th Conference on Power Electronics and Motion Control*,
   Budapest, Hungary, 1990, pp. 990–994
W.-H. Ki, D. Ma, Single inductor multiple output
   switching converters. IEEE Power Electron. Spec.
   Conf. **1**, 226–231 (2001). Vancouver
Y.-H. Lam, W.-H. Ki, C.-Y. Tsui, Integrated low-loss
   CMOS active rectifier for wirelessly powered devices.
   IEEE Trans. Circuits Syst. II: Express Briefs **53**(12),
   1378–1382 (2006)
H.-M. Lee, M. Ghovanloo, An adaptive reconfigurable
   active voltage doubler/rectifier for extended-range

inductive power transmission. IEEE Trans. Circuits Syst. II: Express Briefs **59**(8), 481–485 (2012)

Y. Lu, W.-H. Ki, A 13.56 MHz CMOS active rectifier with switched-offset and compensated biasing for biomedical wireless power transfer systems. IEEE Trans. Biomed. Circuits Syst. **8**, 334–344 (2014)

Y. Lu, W.-H. Ki, J. Yi, A 13.56 MHz CMOS rectifier with switched-offset for reversion current control, in *IEEE Symposium on VLSI Circuits (VLSIC)*, Jun 2011, pp. 246–247

Y. Lu, X. Li, W.-H. Ki, C.-Y. Tsui, C.P. Yue, A 13.56 MHz fully integrated 1X/2X active rectifier with compensated bias current for inductively powered devices, in *IEEE International Solid-State Circuits Conference (ISSCC)*, Feb 2013, pp. 66–67

Y. Lu, Y. Wang, Q. Pan, W.-H. Ki, C.P. Yue, A fully-integrated low-dropout regulator with full-spectrum power supply rejection. IEEE Trans. Circuits Syst. I: Reg. Pap. **62**(3), 707–716 (2015)

Y. Lu, W. Ki, C. Patrick Yue, An NMOS-LDO regulated switched-capacitor DC–DC converter with fast-response adaptive-phase digital control. IEEE Trans. Power Electron. **31**(2), 1294–1303 (2016)

J. Luo, N.K. Jha, L.S. Peh, Simultaneous dynamic voltage scaling of processors and communication links in real-time distributed embedded systems. IEEE Trans. Very Large Scale Integr. (VLSI) Syst. **15**(4), 427–437 (2007)

M. Huang, Y. Lu, S.-W. Sin, S.-P. U, R. Martins, A fully-integrated digital LDO with coarse-fine-tuning and burst-mode operation. IEEE Trans. Circuits Syst. II: Exp. Briefs **63**(7), 683–687 (2016)

M. Huang, Y. Lu, S.-W. Sin, S.-P. U, R. Martins, W.-H. Ki, Limit cycle oscillation reduction for digital low dropout regulators. IEEE Trans. Circuits Syst. II: Exp. Briefs **63**(9), 903–907 (2016)

D. Ma, R. Bondade, Enabling efficient power management on silicon. IEEE Circuits Syst. Mag. **10**(1), 14–29 (2010)

D. Ma, R. Bondade, *Reconfigurable Switched-Capacitor Power Converters*, 1st edn. (Springer, New York, 2013). ISBN 978-1-4614-4186-1 and 978-1-4614-4187-8 (e-Book)

D. Ma, W.-H. Ki, C.-Y. Tsui, P.K.T. Mok, Single-inductor dual-output CMOS switching converters in discontinuous-conduction mode with time-multiplexing control. IEEE J. Solid-State Circuits **38**(1), 89–100 (2003a)

D. Ma, W.-H. Ki, C.-Y. Tsui, A pseudo-CCM/DCM SIMO switching converter with freewheel switching. IEEE J. Solid-State Circuits **38**(6), 1007–1014 (2003b)

D. Ma, W.-H. Ki, C.-Y. Tsui, An integrated one-cycle control buck converter with adaptive output and dual-loop output error correction. IEEE J. Solid-State Circuits **39**(1), 140–149 (2004)

H. Nakamoto, D. Yamazaki, T. Yamamoto et al., A passive UHF RF identification CMOS tag IC using ferroelectric RAM in 0.35-μm technology. IEEE J. Solid-State Circuits **42**(1), 101–110 (2007)

Y. Okuma, K. Ishida, Y. Ryu, X. Zhang, P.-H. Chen, K. Watanabe, M. Takamiya, T. Sakurai, 0.5-V input digital LDO with 98.7% current efficiency and 2.7-μA quiescent current in 65 nm CMOS, in *IEEE Custom Integrated Circuits Conference (CICC)*, Sep 2010, pp. 1–4

G.K. Ottman, H.F. Hofmann, G.A. Lesieutre, Optimized piezoelectric energy harvesting circuit using step-down converter in discontinuous conduction mode. IEEE Trans. Power Electron. **18**(2), 696–703 (2003)

C.-J. Park, M. Onabajo, J. Silva-Martinez, External capacitor-less low drop-out regulator with 25 dB superior power supply rejection in the 0.4–4 MHz range. IEEE J. Solid-State Circuits **49**(2), 486–501 (2014)

D.J. Perreault, J.G. Kassakian, Distributed interleaving of paralleled power converters. IEEE Trans. Circuits Syst. I **44**(8), 728–734 (1997)

A. Radecki, H. Chung, Y. Yoshida, N. Miura, T. Shidei, H. Ishikuro, T. Kuroda, 6 W/25mm^2 inductive power transfer for non-contact wafer-level testing, in *IEEE International Solid-State Circuits Conference (ISSCC)*, Feb 2011, pp. 230–232

J. Sankman, D. Ma, A 12-μW to 1.1-mW AIM piezoelectric energy harvester for time-varying vibrations with 450 nA IQ. IEEE Trans. Power Electron. **30**(2), 632–643 (2015)

J. Sankman, H. Chen, D. Ma, Supercapacitor-based reconfigurable energy management unit for autonomous wireless sensor nodes, in *IEEE International Symposium on Circuits and Systems*, pp. 2541–2544, Rio de Janeiro, Brazil, May 2011

K. Smedley, S. Cuk, One-cycle control of switching converters. IEEE Trans. Power Electron. **10**(6), 625–633 (1995)

M. Song, J. Sankman, D. Ma, A 6-A, 40-MHz four-phase ZDS hysteretic DC-DC converter with 118 mV droop and 230 ns response time for a 5A/5 ns load transient, in *International Solid-State Circuits Conference*, pp. 80–81, San Francisco, CA, Feb 2014

M. Song, J. Sankman, J. Lee, D. Ma, A 200-MHz 4-phase fully integrated voltage regulator with local ground sensing dual loop ZDS hysteretic control using 6.5nH package bondwire inductors on 65 nm Bulk CMOS, in *IEEE/ACM Design Automation Conference-Asia & South Pacific*, Jan 2016, pp. 9–10

M. Stoopman, S. Keyrouz, H.J. Visser, K. Philips, W.A. Serdijn, Co-design of a CMOS rectifier and small loop antenna for highly sensitive RF energy harvesters. IEEE J. Solid-State Circuits **49**(3), 622–634 (2014)

C.R. Sullivan, Integrating magnetics for on-chip power: challenges and opportunties, in *IEEE Custom Integrated Circuits Conference*, 2009, pp. 291–298

Maxwell HC series Ultracapacitors Datasheet, www.maxwell.com/images/documents/hcseries_ds_1013793-9.pdf. Accessed May 2016

V.I. Utkin, Sliding mode control design principles and applications to electric drives. IEEE Trans. Ind. Electron. **40**(1), 23–36 (1993)

Y. Wang, D. Ma, A 450-mV single fuel cell power management unit with switch-mode quasi-V^2 hysteretic control and automatic start-up on standard 0.35 μm CMOS process. IEEE J. Solid-State Circuits **47**(9), 2216–2226 (2012)

H. Wu, D.S. Gardner, C. Lv, et al., Integration of magnetic materials into package RF and power indcutors on organic substrates for system in package (SiP) applications, in *IEEE Electronic Components and Technology Conference*, May 2014, pp. 1290–1295

H. Xu, M. Lorenz, U. Bihr, J. Anders, M. Ortmanns, Wide-band efficiency-enhanced CMOS rectifier, in *IEEE International Symposium on Circuits and Systems (ISCAS)*, Jun 2014, pp. 614–617

Y. Lu, H. Dai, M. Huang et al., A wide input range dual-path CMOS rectifier for RF energy harvesting, in *IEEE Transactions on Circuits and Systems II: Express Briefs*, **64**, (2017)

J. Yi, W.-H. Ki, C.-Y. Tsui, Analysis and design strategy of UHF micro-power CMOS rectifiers for micro-sensor and RFID applications. IEEE Trans. Circuits Syst. I: Reg. Papers **54**(1), 153–166 (2007)

C. Zhan, W.-H. Ki, Analysis and design of output-capacitor-free low-dropout regulators with low quiescent current and high power supply rejection. IEEE Trans. Circuits Syst. I: Reg. Pap. **61**(2), 625–636 (2014)

Y. Zhang, D. Ma, A fast-response hybrid SIMO power converter with adaptive current compensation and minimized cross-regulation. IEEE J. Solid-State Circuits **49**(5), 1242–1255 (2014)

Y. Zhang, D. Ma, A single-stage solar-powered LED driver using power channel time multiplexing technique. IEEE Trans. Power Electron. **30**(7), 3772–3780 (2015)

Energy Harvesting

11

Ying-Khai Teh and Philip K.T. Mok

This chapter complements other chapters on battery technologies and on-chip DC-DC conversion, and mainly addresses the challenges of designing low power IoT nodes that are powered by energy harvesting sources, which is the key enabling technology to extend battery life and minimize manual battery maintenance, using *in situ* power extraction from the surrounding. Energy harvesting options, circuit concepts, considerations and trade-offs regarding circuit topology, passive component and CMOS process are surveyed. In particular, recent circuit solutions involving non-conventional power management schemes specifically catering for energy-harvesting-assisted IoT systems will be discussed and compared.

11.1 Available Energy Sources for IoT

11.1.1 Classification of Energy Source Characteristics and Classification

Advancement in the thin-film fabrication has successfully produced new generation of energy harvesters at reasonable cost, comes with compact size and able to generate microwatts range of power via surrounding events. An overview of energy harvesters commercially available are given in this section, which include thermoelectric generator (TEG), microbial fuel cell (MFC), photovoltaic cell (PVC) and piezoelectric harvester (PEH). Key classification matrix to characterize the energy sources are defined and outlined, i.e. power profile (current-voltage) and availability (continuous-time, discrete-time), and physical profile. Energy sources which are highly volatile in nature (photovoltaic cell (PVC) in rough, cloudy environment or piezoelectric harvester (PEH) installed in poor vibration frequency band) are considered as discrete-time (DT) energy source; which have high output voltage once ready, but suffers intermittency in term of power availability.

Energy sources such as body-powered TEG and MFC are considered as continuous-time (CT) energy source (Bandyopadhyay and Chandrakasan 2012; Danesh and Long 2011).

Y.-K. Teh
Hong Kong University of Science and Technology, Clear Water Bay, Hong Kong

San Diego State University, San Diego, CA, USA

P.K.T. Mok (✉)
Hong Kong University of Science and Technology, Clear Water Bay, Hong Kong
e-mail: eemok@ust.hk

© Springer International Publishing AG 2017
M. Alioto (ed.), *Enabling the Internet of Things*, DOI 10.1007/978-3-319-51482-6_11

Although the output voltage is low (<300 mV), the available energy is stable and omnipresence.

Table 11.1 classifies these four energy harvesters into according to voltage-current-time matrix. Table 11.2 summarizes key device characteristics of four different energy harvesters under low output power condition (less than 100 μW) i.e. TEG only has 2 °C of temperature difference between the Peltier junctions, MFC with 50 ml of biofuel, PVC is illuminated at 200 lux (indoor lighting condition) and PEH is vibrated at 0.1 g. Figure 11.1 shows the power density of different energy sources, assuming all are at maximum power point. TEG which has the best power density performance has attracted the most research attentions in recent decade.

11.1.2 Continuous-Time Energy Sources (TEG and MFC)

TEG is an interesting energy harvester for its continuous operation against environmental condition change, due to omnipresence of thermal energy gradient (which manifests as temperature difference). State-of-the-art TEGs are fabricated using thin-film deposition technology of Bi_2Te_3 material. Such improvements are desirable for further system miniaturization. The new generation TEG (Micropelt) has smaller device area with increased power density (100 μW/mm^2), compared to conventional bulk material based TEGs (Marlow Industries) (13 μW/mm^2) at temperature difference of 10 °C. Internal resistance (R_{TEG}) of new generation TEG is 400 Ω, which is 80 times larger whereas the absolute output power for conventional TEG is 8 times higher. The output power and power density comparisons of the two TEGs are shown in Fig. 11.2. Simplified circuit model of TEG (Micropelt), and its current-voltage characteristics are shown in Fig. 11.3.

Besides thermoelectric solution, another emerging power source is microbial fuel cells (MFC). MFC harvests electricity from organic dispose such as waste water, which is essentially by-product of catalytic activities of microorganisms over various organic substrates. The nature where MFC derives its energy enables the niche application of waste water treatment and water quality monitor systems.

Table 11.1 Classifications of energy sources

Type	TEG (Micropelt)	MFC Wang et al. (2013)	PVC (Enocean)	PEH (Microgen)
Voltage	Low	Low	High	High
Current	High	High	Low	Low
Time	CT	CT	DT	DT

CT, continuous time; *DT*, discrete time

Table 11.2 Device characteristics of energy sources

Type	TEG (Micropelt)	MFC Wang et al. (2013)	PVC (Enocean)	PEH (Microgen)
V_{OC} (max)	0.3 V	0.6 V	4 V	8 V (rectified)
I_{SC} (max)	750 μA (@ΔT = 2 °C)	2.7 mA	7 μA (@200 lux)	14 μA (@0.1 g)
V_{MPP}[a]	0.15 V (@ΔT = 2 °C)	0.35 V	3 V (@200 lux)	4 V (@0.1 g)
P_{MPP}[b]	56 μW (@ΔT = 2 °C)	440 μW	14 μW (@200 lux)	56 μW (@0.1 g)
Dimension (L × W × H mm^3)	4 × 3 × 1	37 × 37 × 37	35 × 13 × 1	15 × 15 × 6
Power density (μW/mm^3)	4.67	8.68 × 10^{-3}	0.03	0.04

[a]V_{MPP} = Operating voltage of energy source at maximum power point
[b]P_{MPP} = Output power of energy source at maximum power point

Current state-of-the-art MFCs are fabricated using emerging material such as graphene (Wang et al. 2013). It is interesting to note that thin-film TEG operating at low temperature difference (4–5 °C) and graphene-based MFC have similar electrical characteristics, as shown in Fig. 11.4. Nonetheless MFC's power density is approximately three orders lower than TEG's. Electrically both energy sources can be modeled as an ideal voltage source (V_{GEN}) in series with internal resistance (R_S) of a few hundred ohms.

11.1.3 Discrete-Time Energy Sources (PVC and PEH)

Photovoltaic cell (PVC) is capable of converting light directly into electric energy. Most PVCs consist of silicon material. Due to the high cost of the vacuum-based fabrication process of such solar cells, research and development has been directed toward the invention and development of thin film inorganic and emerging inorganic/organic thin film PV solar cells in an attempt to fabricate solar cells out of less-expensive materials and processes. Light is often present for a prolonged period, during which the light energy can be accumulated in storage mechanisms and then used when needed. Simplified circuit model of PVC in (Enocean) and typical output behavior is shown in Fig. 11.5.

Piezoelectric harvester (PEH) with size smaller than a US quarter coin and internal parasitic capacitance of a few nano Farad can be manufactured via commercial CMOS process (Microgen). The generated power is in alternating current (AC) mode and thus requires to be rectified before use. Simplified circuit model of PEH with rectified output in (Microgen) and its typical output behavior is shown in Fig. 11.6. However, as miniaturization continues, the mechanical vibration bandwidth for smaller devices is also getting smaller (a few Hertz). Therefore PEH is sensitive

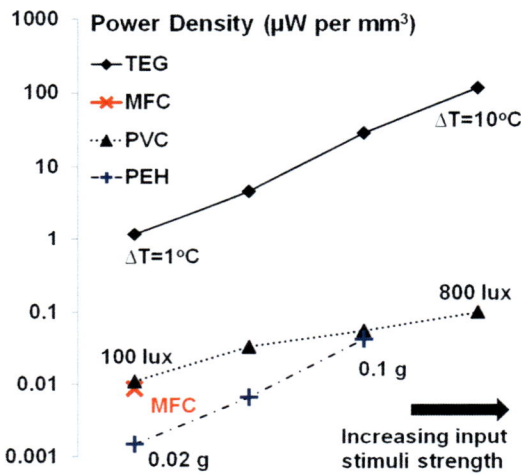

Fig. 11.1 Power density of TEG (Micropelt), MFC (Wang et al. 2013), PVC (Enocean) and PEH (Microgen)

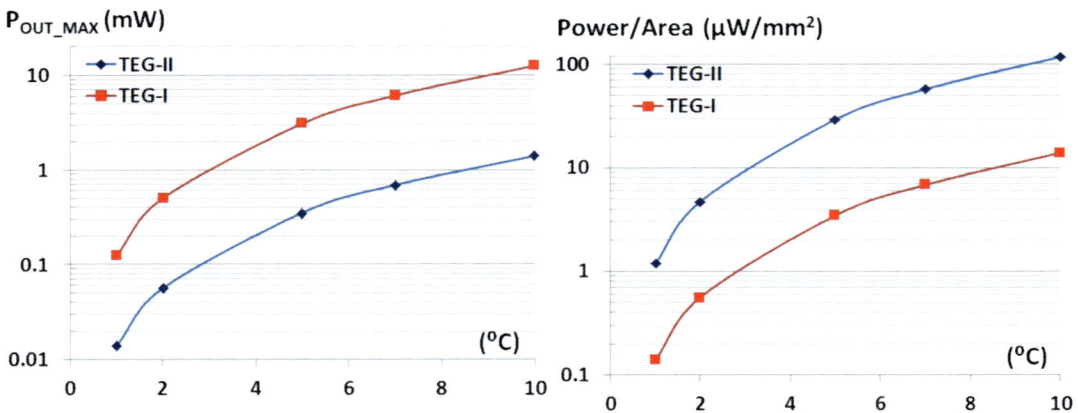

Fig. 11.2 Power and Power density of conventional TEG (TEG-I (Marlow Industries)) and new generation TEG (TEG-II (Micropelt))

Fig. 11.3 Simplified equivalent circuit model and output characteristics of TEG (Micropelt)

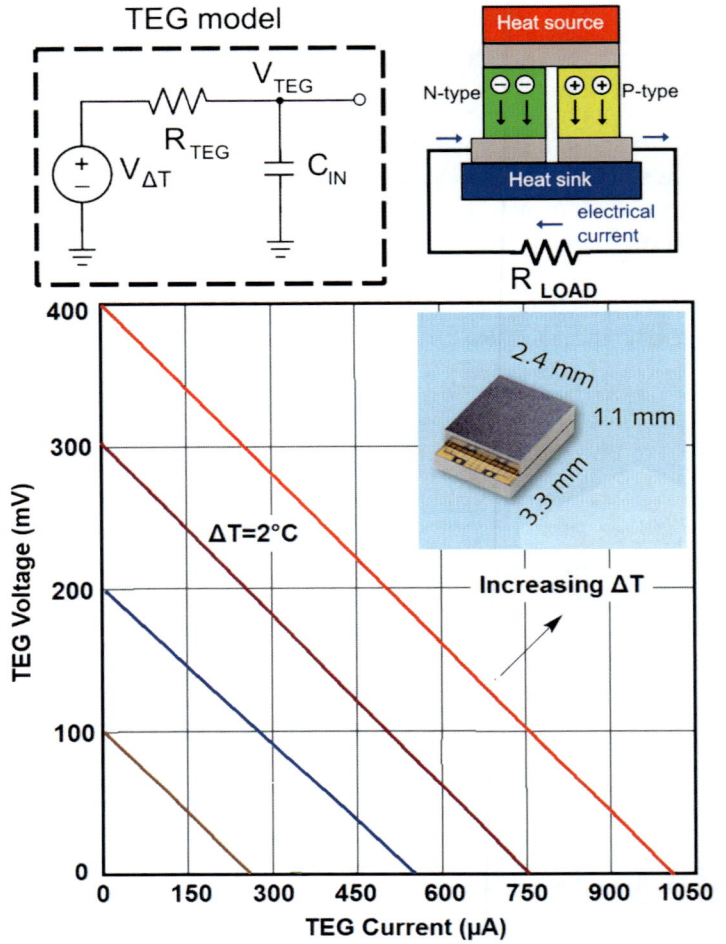

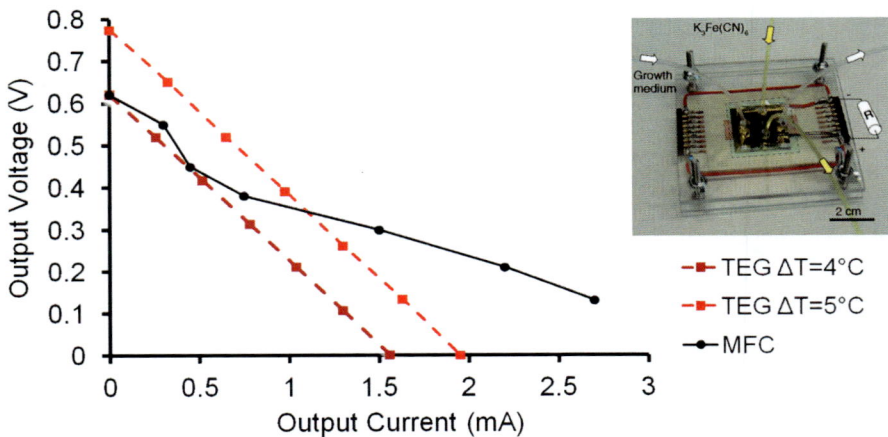

Fig. 11.4 I–V characteristics of TEG (Micropelt) at $\Delta T = 4\,°C$, $\Delta T = 5\,°C$ and MFC (Wang et al. 2013)

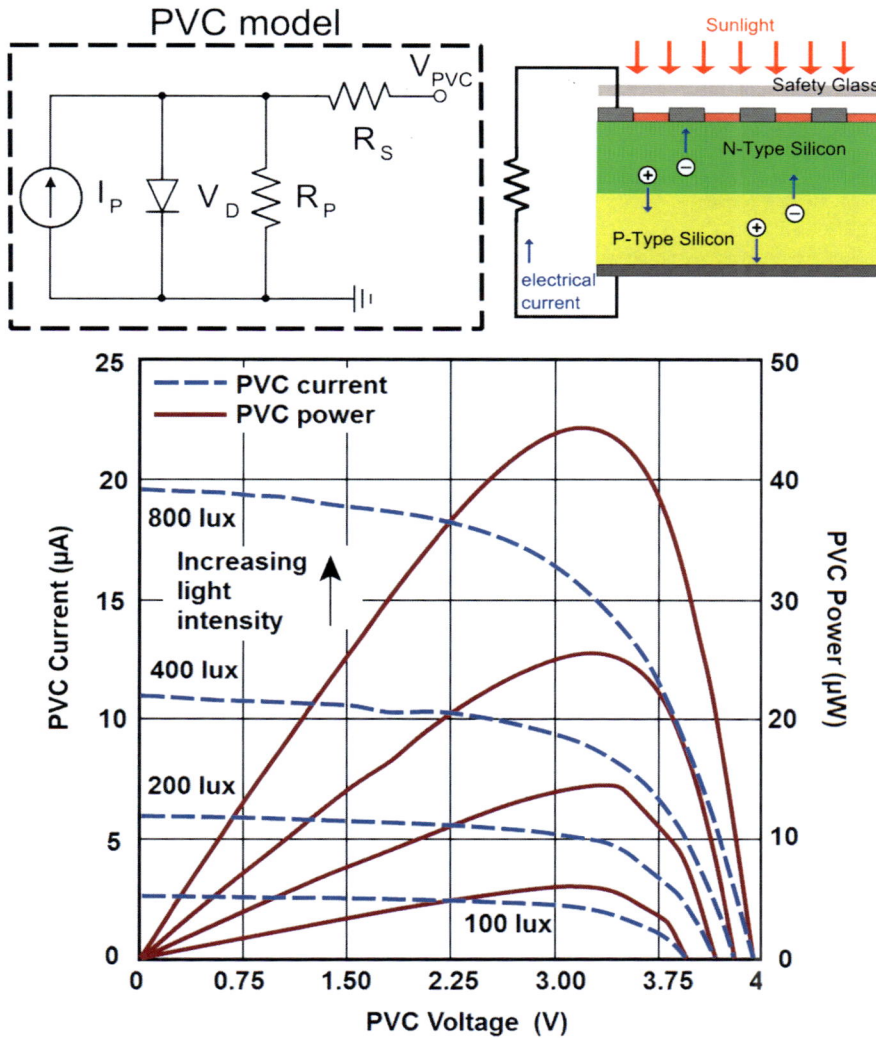

Fig. 11.5 Simplified equivalent circuit model of PVC and its output characteristics (Enocean)

towards environment stimuli and the output power profile is highly volatile.

11.2 Comparison of Battery-Powered and Energy-Harvested Systems

In this section, the difference of battery-powered and energy-harvested systems primarily lies on the known and unknown voltage-current profile, due to different source resistance is highlighted.

The fundamental difference between battery power source and energy harvesting source is that energy harvesters have significantly higher source resistance (much higher than 1 Ω) compared to battery (less than 1 Ω). However, in battery powered application, charge stored in non-rechargeable battery is limited. Before the battery stored charge depletes, power sourcing capability of a battery is much higher than energy harvesting source in general (Table 11.3).

For energy harvesting source, supply of electrical charge is in theory unlimited and most of

Fig. 11.6 Simplified
equivalent circuit model of
PEH and its output
characteristics (Microgen)

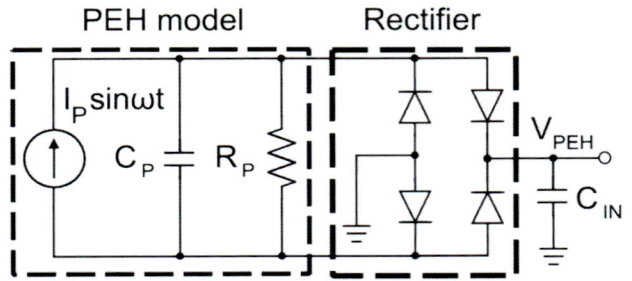

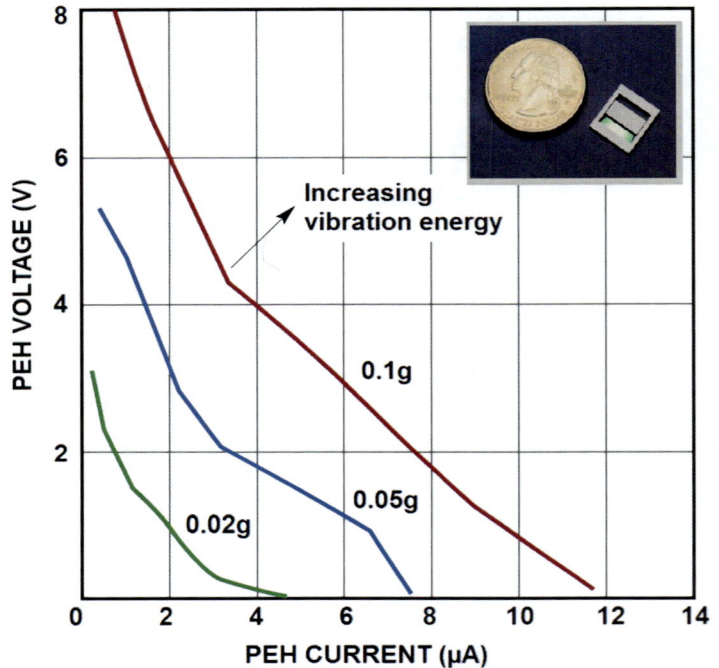

Table 11.3 Energy harvesting source vs. battery

Energy harvesting source	Battery
High internal resistance	Low internal resistance
Unlimited charge in theory	Finite charge
Limited instantaneous power (Duty cycled operation)	High instantaneous power (Duty cycled operation is optional)
DC-DC converter functions:	DC-DC converter functions:
– Output voltage regulation	– Output voltage regulation
– Maximum power point tracking (MPPT)	– No MPPT necessary

the time the energy harvester is charging up the storage capacitor. Before storage capacitor voltage reaches the designed threshold, the regulator is turned off. Having a shunt regulator in this scenario basically acts as charge release valve. The objective is to sense the storage capacitor voltage, once the set threshold is reached, additional charge harvested will be dissipated by shunt regulator. If a series regulator is used instead, the system is prone to overvoltage failure as excess charge will continue to accumulate and build up storage capacitor voltage. Overvoltage protection should be prioritized over power conversion efficiency in this scenario. Hence the automatic protection against overvoltage function of a shunt regulator is indeed attractive for energy harvesting application. High voltage headroom generated by PVC and PEH output can be absorbed by a properly designed shunt regulator such as (Sarkar and Chakrabartty 2013), allowing storage capacitor to be directly charged by these sources.

One key challenge is the design of interfacing power management circuit that efficiently combines many different sources of energy into one. Combining the power from variable sources conventionally requires a sophisticated control system as demonstrated in (Bandyopadhyay and Chandrakasan 2012). The control itself requires additional overhead power, i.e. first sense the availability of power from each source, then compute and optimize the multi-source time multiplexing.

Another challenge is the limit imposed by the low supply voltages in standard CMOS technologies. These constrain the output voltage to be within the defined supply voltage. At higher voltages outside of the nominal range, breakdown mechanisms and hot carrier effects will be present. Hence the power management circuit must prevent the overvoltage problem. On the other end of the spectrum, some energy sources have very low input voltage (<0.2 V), which are lower than transistor threshold voltage. At sub-threshold region, transistors are not fully turned on and conventional voltage converter topologies are no longer valid or suffer poor conversion efficiency.

Typical power consumption pattern of an IoT sensor node is low average power (no active power consumption except watch dog timer) but occasional occurrence of high concentrated burst of power to accomplish a task such as activating external sensors or sending radio packets. If these bursts occur with a low duty cycle such that the total energy needed for a burst can be accumulated between bursts then the output can be maintained entirely by the energy harvesters. An ideal energy harvesting scheme is required to maintain low leakage power of harvested energy during idle period yet able to supply high pulsed power on demand.

11.3 Energy Harvesting and Power Conditioning Circuits

Multi-level supply voltage (V_{DD}) is common for IoT systems implemented using low power modern CMOS process. External I/O interface circuits typically works at 3 V (5 V for legacy devices) and internal digital core circuits typically works at 1 V. Modules such as sensors, analog-to-digital-converters (ADC) and amplifiers always prefer higher voltage headroom to achieve better resolution, gain and linearity (Mak and Martins 2010). For battery-powered applications, such multiple output voltage can be realized using Single-Inductor-Multiple-Output (SIMO) DC-DC converter (Jing and Mok 2013; Ma et al. 2003a, b) or multiple series regulator i.e. low dropout regulator (LDO) (Ho and Mok 2012; Zhan and Ki 2012, 2014). The conventional approaches become unrealistic when the input power approaches microwatts range as the overhead power is too high to jeopardize the entire system operation. Design requirements of energy harvesting system to power a wireless sensor IoT node and the associated challenges hitherto discussed are summarized in Table 11.4.

In previous sections, electrical characteristics of various energy harvesting source and storage capacitors were reviewed. Based on these

Table 11.4 Summaries of design requirements and challenges of IOT wireless sensor node

Input voltage	0–8 V
Input current	0–1 mA
Input power	0–1 mW
Dual output voltage	1 V (Digital), 3 V (Analog, RF and sensors)
Design features required	– Smallest input voltage and power possible
	– Smallest overhead and leakage power possible
	– Highest voltage conversion ratio and efficiency possible
	– Self-start-up during cold-start state
	– Over-voltage protection
	– Maximum power point tracking

properties, this chapter surveys integral system schemes to accommodate multiple types of energy sources simultaneously and minimize storage leakage loss. Each circuit block that can be utilized to implement such systems i.e. linear regulator, switching and switched-capacitor DC-DC converters are stepped through briefly in sequential manner.

Different power converter control schemes will have different dynamic power loss. As rule of thumb, ZCS control is desired to maximize power efficiency. Pulse frequency modulation (PFM) has superior power efficiency than pulse-width modulation (PWM) schemes at light-load. Comparator is one of the most power-consuming circuit blocks in power converter. Designers should use low bias current or clocked comparator to save quiescent power. For energy harvesting source which is steady against environmental change, maximum power point tracking (MPPT) algorithm can be executed in a low duty cycle fashion to save computation power. For energy harvesting source with known, fixed source resistance, timing-based impedance matching MPPT technique can be employed with a comparator-less fully digital approach. This MPPT technique also avoids

interrupting energy harvesting flow since circuit does not need to pause and sample the energy harvester source open circuit voltage from time to time.

Table 11.5 summarizes the characteristics of different DC-DC converter circuits according to conversion possible, peak power efficiency and output voltage control that can be used to construct a multi-input-multi-output energy harvesting power management systems for IoT node.

11.3.1 Linear Regulator in Energy Harvesting Application

As discussed in Chap. 10, there are two possible implementations for linear regulators i.e. series and shunt regulators. In principle these circuits can be designed based on the functional blocks shown in Figs. 11.7 and 11.8 respectively by assuming the internal resistance of energy harvesting source to be series resistance R_S needed in any typical linear regulator realization. Series voltage regulator is more power efficient but lacks the ability to prevent over-voltage. Therefore if shunt regulator is used, constant

Table 11.5 Summary of different DC-DC converters

Type	Conversion direction (Up/Down)	Peak power efficiency	Output voltage
Linear regulator	Down	Low	Adjustable by reference voltage
Switching converter	Up/Down	High	Adjustable by clock duty cycle
Switched- capacitor	Up/Down	Medium	Adjustable by clock duty cycle

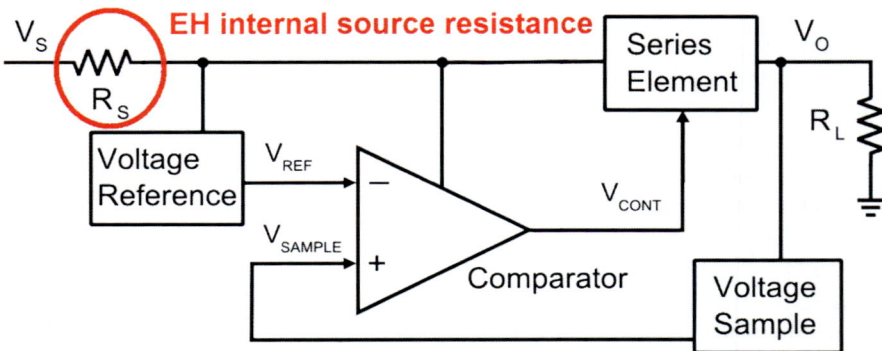

Fig. 11.7 Basic working principle of series voltage regulator

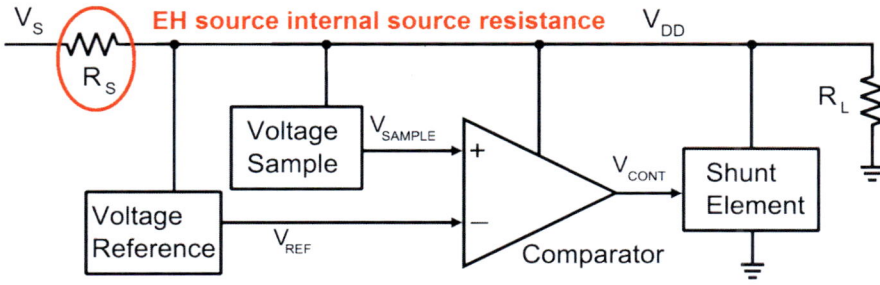

Fig. 11.8 Basic working principle of shunt voltage regulator

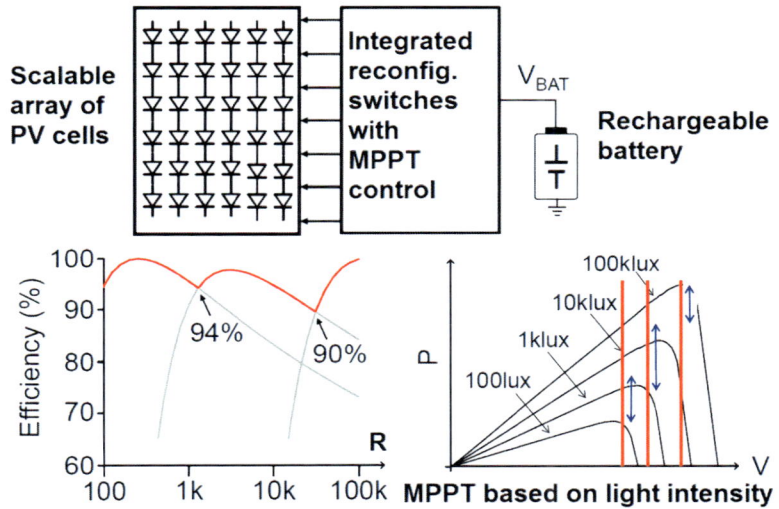

Fig. 11.9 Reconfigurable switch for photovoltaic energy harvesting

power will be drawn from battery despite the load power being low. This is disaster especially for low duty long working hour IoT application because battery charge depletes quickly. This scenario is exactly how the conventional wisdom i.e. series regulator is more efficient is being derived. Shunt voltage regulator is less power efficient but it can protect the circuit from over-voltage, which is useful when interfacing with high voltage low input energy source. However, linear regulator can only step down input voltage to generate a lower output voltage. The best conversion power efficiency possible is V_O/V_S, which implies that this type of regulator is considered efficient only when the ratio of conversion is close to unity.

Voltage Reference element forms the foundation of linear regulator since V_{DD} must be either equal or a multiple of reference voltage V_{REF}.

For good regulation, V_{REF} must be stable across V_{DD} variations. Voltage Sample element monitors V_{DD} and translates it into a level equal to the V_{REF} for a desired regulated voltage. Variations in V_{DD} cause feedback which changes V_{SAMPLE} to some value greater than or less than the V_{REF}. ΔV_{SAMPLE} causes comparator to drive series or shunt element which would respond appropriately to correct the output voltage change.

Besides the textbook classic series and shunt voltage regulator, there is also a unique case of linear (resistive) based power converter circuit worth mentioning, which is a reconfigurable switch to control an array of PVCs (Lee et al. 2016). By using low impedance switches and basic digital logic control, as shown in Fig. 11.9, high power conversion efficiency (>90%) is achieved to control PVC array since there is essentially zero switching loss. This circuit works under the assumption that the total, stacked output voltage of PVC array is always sufficient to direct charge a rechargeable battery.

11.3.2 Switching Converter and Switched-Capacitor (SC) Power Converter in Energy Harvesting Application

As discussed in Chap. 10, switching converter involves three different electrical components i.e. switches S_N, inductor L_N, diode D_N, with the objective of converting a given supply voltage V_{IN}, to output voltage V_{OUT}. There are three basic configurations possible, as illustrated in Fig. 10.2 of previous chapter. Buck converter shown in Fig. 11.2a (Chan and Mok 2011), is suitable for high voltage energy sources (PVC and PEH) where $V_{OUT} < V_{IN}$. Figure 11.2b is a Boost converter shown in Fig. 11.2b (Man et al. 2008), is suitable for low voltage energy sources (TEG and MFC) where $V_{OUT} > V_{IN}$. Buck-boost converter Chew et al. (2013), shown in Fig. 11.2c is suitable for hybrid energy harvester (combination of both high and low voltage sources) where V_{OUT} can be larger than or less than V_{IN}. All output-input voltage relationship of these converters can be adjusted by changing the switching clock duty cycle. The advantage of switching converter is the high conversion power efficiency, which can be up to 90%, achievable across different conversion ratio. The disadvantage is the requirement of off-chip inductor, which is bulky and expensive. Fully integrated on-chip switching converter is a popular on-going research topic (Huang and Mok 2013a, b).

An alternative to using inductor is by shifting to the other reactive circuit element, i.e. capacitor. Switched-capacitor (SC) power converter is also known as charge pump. The main advantage is that fully integrated on-chip implementation is possible and bi-directionality of conversion. For applications where bulky off-chip components must be avoided, charge pump is a good compromise between linear

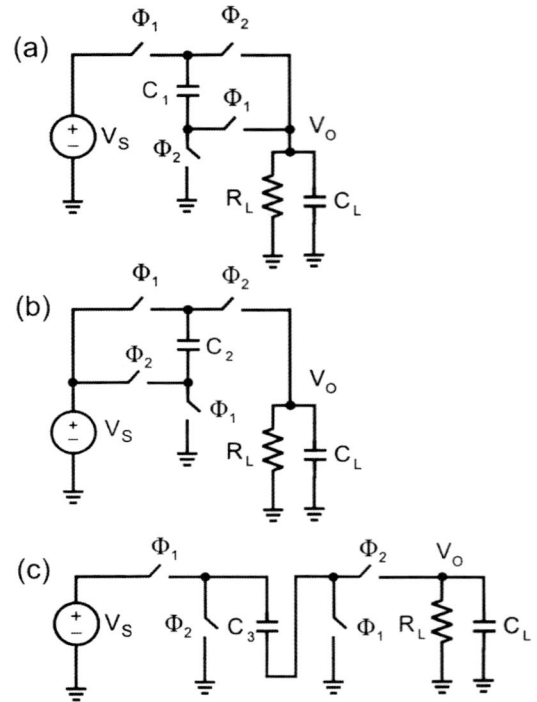

Fig. 11.10 Basic working principle of switched-capacitor DC-DC converter for (**a**) step-down, (**b**) step-up, (**c**) invert

regulator and switching. Similar to the switching counterpart, as shown in Fig. 11.10, there are three basic switched-capacitor configurations, which allow designers to step down (Jiang et al. 2014), step up (Su and Ki 2007) and invert (Wang and Wu 1997) a given voltage. Typically 50% duty 2-phase ($\Phi 1$, $\Phi 2$) clock is used. The invert charge pump is useful in generating negative voltage, which is instrumental in sub-threshold level power converter to suppress start-up system leakage (Kim et al. 2014).

11.3.3 Maximum Power Point Tracking in Energy Harvesting

Due to the volatile nature of energy harvesting source, it is important that the interfacing power management unit maximizes the charge extraction from the source, to minimize system down time before the loading IoT node is ready to

Fig. 11.11 Conventional
maximum power point
tracking algorithm

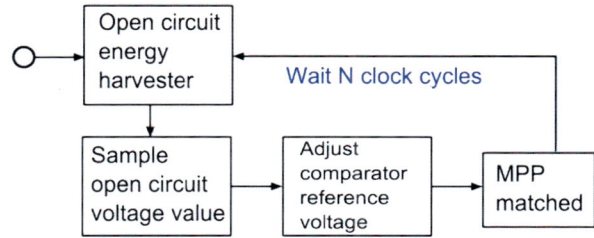

perform the next duty-cycled workload. The maximum power point is reached when the internal resistance of energy harvesting source is matched to the effective load connected. PVC often has non-linear peak output power point but TEG is linear, where the maximum power point is always half of the open-circuit voltage. A typical conventional maximum power point tracking (MPPT) algorithm for thermoelectric harvesting, as shown in Fig. 11.11 for a switching converter (Im et al. 2012) requires sampling of TEG open circuit voltage (V_{OC}), and then modifies the DC-DC converter duty cycle to change the input voltage to match the 0.5 V_{OC}. This approach needs to periodically switch off converter so that V_{OC} can be sampled and updated according to environmental change. The alternative, perturb and observe (P&O) MPPT method is not commonly used in microwatts systems due to the high power consumption of voltage/current sensor.

Since internal resistance R_S is a known constant value with TEG, we can achieve continuous-time tracking of MPP by setting the input impedance to match R_S, without the need to sample V_{OC}. The associated algorithm is shown in Fig. 11.12. This idea effectively leverages on the advantage of a clock-power stage integrated boost converter reported in (Teh and Mok 2014a).

The input impedance of this boost converter can be derived from timing of the self-generating pulse. This assumption is valid as the temperature variation of R_S of thin-film TEG (Micropelt) is less than 1% of its nominal value, even when ΔT is up to 60 °C. However, since ΔT in typical energy harvesting application is small, such variance is acceptable and R_S can be regarded

constant. To ensure absolute variability is less than 5%, simulations based on the assumption R_S varies by 20% (an excessive approximation), the tracked MPP voltage is only off by 5%. For very high performance system, designers need to incorporate additional temperature sensor and look-up table containing the temperature variation information R_S to compensate for all process, voltage and temperature (PVT) variation accurately. Since typical sampling pulse is of a few hundreds μs to ms range, digital clock can be run at MHz (100s of ns) clock range. The power consumption of such on-chip MPPT circuit is estimated to be within the ballpark of 100 μW when active. If MPPT algorithm is allowed to be run at low duty cycle (once for every few second, since temperature change is typically slow), average power can be reduced to microwatts range.

Another continuous-time MPPT method, as shown in Fig. 11.13, is to observe the output power via sampling the voltage accumulated over an on-chip capacitor. Within fixed clock period, the charge pump clock varies according to input voltage and hence the voltage-domain sensing is shifted to time-domain sensing, i.e. the higher power harvester is translated to shorter rising time. This approach effectively circumvents the power hungry voltage/current sensor problem (Liu et al. 2016).

11.3.4 Unique "Investment" Concept in Piezoelectric Harvesting

Kinetic energy in motion is attractive because vibrations are abundant in the environment. This is why piezoelectric transducers are popular today, and because they generate more power

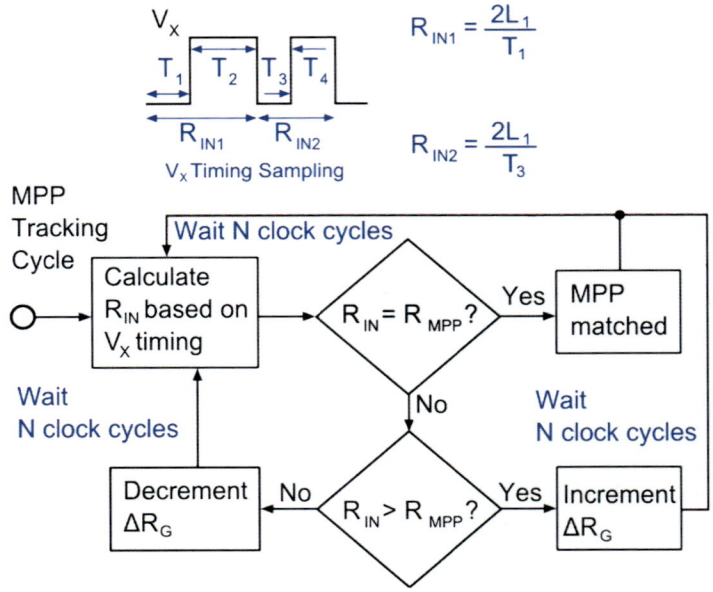

Fig. 11.12 Continuous-time maximum power point tracking algorithm for switching topology based on (Teh and Mok 2014a)

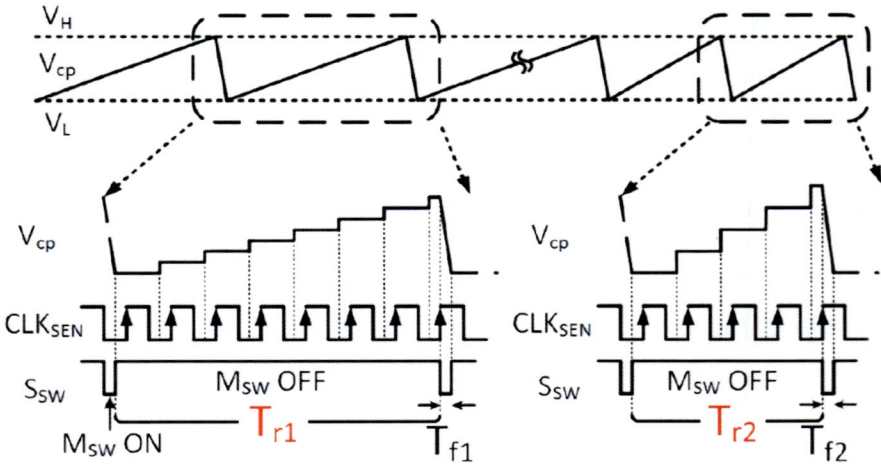

Fig. 11.13 Time-domain maximum power point tracking algorithm for switched-capacitor power converter based on (Liu et al. 2016)

from motion under similar space constraints than their electrostatic and electromagnetic counterparts. In piezoelectric harvesting, there is a unique harvesting technique that is not present in other energy harvesting modes which is the concept of investment. The idea is to strengthen the electrostatic force against which vibrations work. This way, the circuit eventually draws dividend from the investment, i.e. extracting more power from the transducer after investing fixed amount of battery energy into PEH (Kwon and Rincon-Mora 2014). The operation waveforms of such piezoelectric harvesting mechanism are shown in Fig. 11.14.

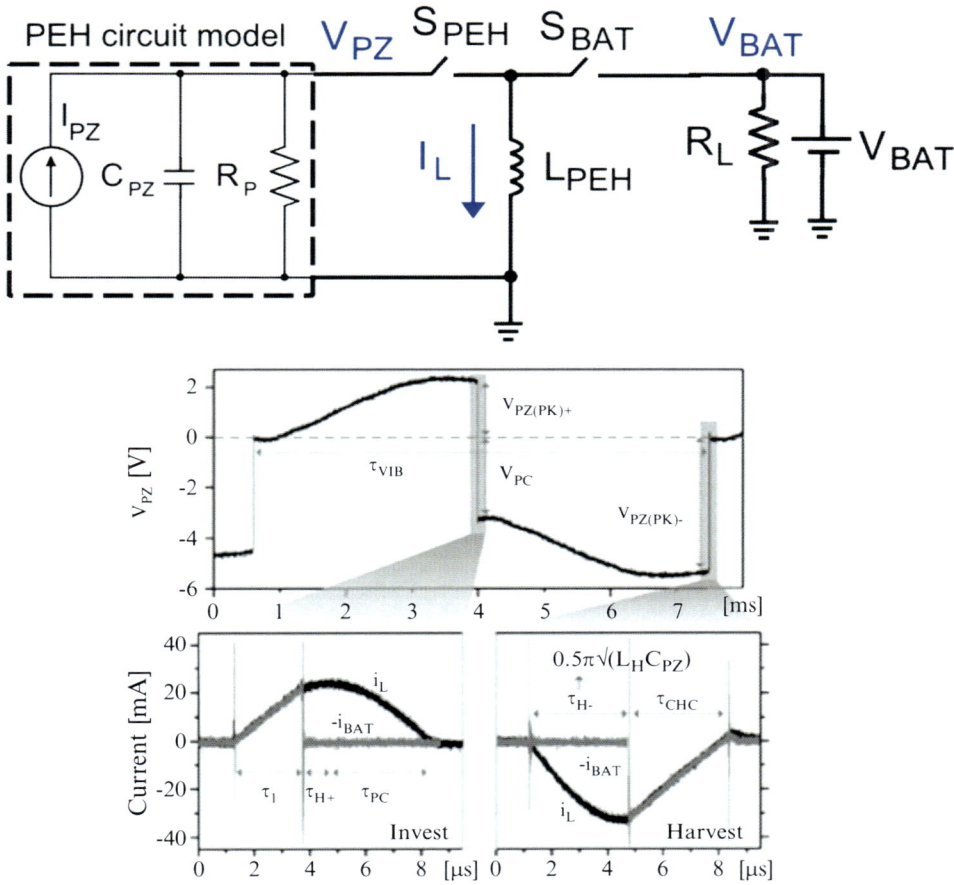

Fig. 11.14 Energy investing concept in piezoelectric harvesting (Kwon and Rincon-Mora 2014)

11.4 Adaptation to Source and Load

11.4.1 One-Shot Input Voltage in Piezoelectric Harvesting

Typically, piezoelectric harvesters generate the most power when they vibrate at their resonant frequency. Unfortunately, motion is not always consistent or periodic. In many applications, in fact, vibrate in response to one-shot shocks, or repeated impact. Many related work such as (Kwon and Rincon-Mora 2014) discussed earlier can still function with one-shot or shocks but are not optimized for such operation mode. One interesting work (Yang et al. 2015a) is to use switched capacitor topology to absorb the high

excitation voltage during shock, and then, as shown in Fig. 11.15. The idea is to stack multiple capacitors in series during high voltage transient, and then redistribute the charge accumulated in parallel by reconfiguring the stacked capacitors, subsequently lowering the accumulated voltage.

11.4.2 Shunt Regulator as Protection Circuit

Under extreme low input stimulus as shown in Table 11.2, the output power from miniaturized PVC and PEH is of microwatts range, albeit output voltage at open circuit condition of these sources can be as high as 8 V. Considering the

Fig. 11.15 Series-parallel switched-capacitor to absorb high voltage over-shoot (Yang et al. 2015a)

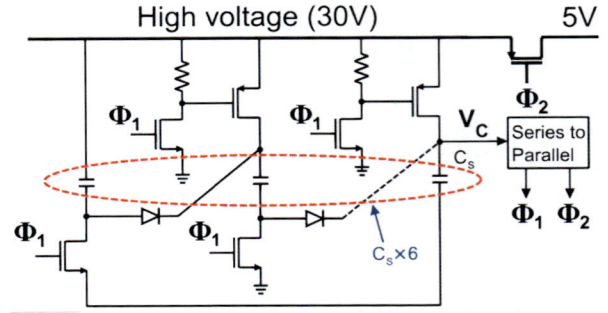

burst mode nature of incoming power, isolation diodes and interfacing power IC are especially vulnerable to breakdown. In standard CMOS 0.13-μm process, the nominal voltage is only 1.2 V for thin oxide transistor and 3.3 V for thick oxide transistor. Obviously input protection circuit needs to be included to avoid overvoltage breakdown caused by such high harvester output voltage. Conventionally series regulator such as low dropout regulator (LDO) is preferred over shunt regulator for higher power efficiency but this topology does not provide any input protection. When input voltage is as high as 8 V, the differential voltage across the dropout transistor is 5 V, if the output voltage is set to 3 V; this condition exceeds the drop out transistor break-down limit. Shunt regulator is a straight forward solution to protect the load from the sudden voltage surge of the energy harvesting source and this cause reduction in power efficiency due to the constant quiescent power drawn by a shunt regulator.

To build a low quiescent shunt regulator, Schmitt-Trigger as shown in Fig. 11.16 can be deemed as the ideal candidate to optimize power consumption in this context (Teh and Mok 2014b). Conventional "problem" i.e. more gain for positive cycle and less gain for negative cycle is paradoxically again becomes favorable for shunt regulator application. Negative cycle of input voltage implies insufficient power from energy harvesting source; subsequently lowering gain of Schmitt-Trigger and cause more current to be diverted to the load instead of being shunted to ground via P_{SHUNT}. During positive cycle of input voltage (more harvested power), Schmitt-Trigger

has higher gain which helps shunting extra current to ground. In the default implementation, R_{COMP} represents the equivalent resistance of transistor drain-body junction leakage component. To further adjust the targeted output voltage after fabrication, the value of resistor R_{COMP} can be adjusted by having an additional bond pad to access the V_{SHUNT} node and connect an additional off-chip resistor. This off-chip resistor is only able to compensate for the process variation whenever R_{COMP} is higher than the simulated value. Note that this circuit is susceptible to CMOS process variation and can only offer coarse voltage regulation with 100 mV margin of error.

11.4.3 Multiple Input Multiple Output Voltage Requirements

Existing energy harvesting systems is commonly powered by single energy harvesting source with photovoltaic cell being the most popular and matured technology. Nevertheless a single source energy harvesting system is often intermittent and unpredictable in term of power generated. Instead of increasing storage capacitor or battery size, recent research direction is to pursue energy harvesting schemes using multiple energy sources. The key challenge is the design of interfacing power management circuit that efficiently combines many energy sources into common storage device. Combining the power from multiple time-varying sources typically requires a sophisticated control system.

A capacitor is analogous to storage reservoir for electrical charge carriers, as a water bucket to

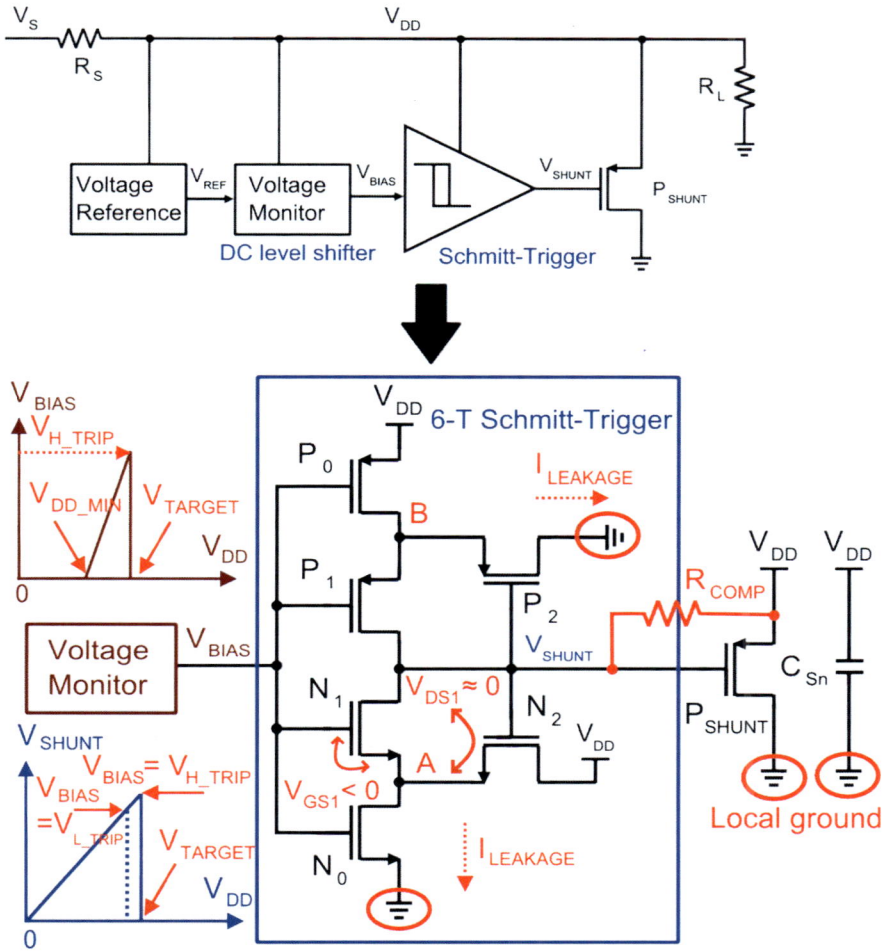

Fig. 11.16 Digital logic based implementation of shunt voltage regulator (Teh and Mok 2014b)

water flow. The capacitor leakage current can be compared to the water seeping through the water bucket. A capacitor will never become fully charged if the leakage current is greater than or equal to the supply current. Designer needs to take extra caution especially when these capacitors are to be charged by microwatts power source as the charging current are comparable to leakage current. Recently, molecularly thin film capacitor of 28 nm thickness (claimed to be better than graphene) and capacitance density of 27.5 $\mu F/cm^2$, which is a factor of three orders better than commercial products in market, was fabricated using oxide nanosheets (Wang et al. 2014). Leakage current versus applied voltage characteristics of such emerging capacitor type is shown in Fig. 11.17, of which leakage current is exponentially dependent on DC bias.

Connecting capacitors in series results in the voltage being split between the capacitors and in turn this is influenced by the leakage current difference between the individual capacitors in a series, as shown in Fig. 11.18. The leakage current component can be modeled as a parallel resistor to the capacitor. The leakage current differences become apparent when the circuits are activated in the form of overvoltage on the component with the lowest leakage current. Since considerable fluctuations are found between individual capacitors (even from the very same production run) in terms of their leakage currents and capacitance, it is possible that large voltage differences may occur and one capacitor in the stacks will sustain higher voltage despite all having the same nominal capacitance. It is therefore important that both capacitance and leakage current differences are balanced in

the system since leakage current can vary up to two order of difference.

The simplest form of balancing circuit is a fixed value resistor parallel to each capacitor. However, this method merely provides balanced voltage but not voltage regulation. To obtain both balancing and regulated voltage, a low quiescent current shunt regulator (with minimal leakage current in its internal voltage reference and parasitic junction diodes) is a better candidate. A proposed scheme essentially emulates a water bucket fountain by envisioning capacitor as individual water bucket; as illustrated in Fig. 11.18 (Teh and Mok 2014b). Incoming charge from high voltage low current (HVLI) energy harvesting sources i.e. PVC and PEH will charge up stack capacitors directly. Once capacitor voltage reaches the set value, excess charge will be pushed to next capacitor below, and eventually to ground if all capacitors are fully charged to preset ΔV value i.e. 1 V. This operation is similar to push pull shunt regulation discussed in (Alon and Horowitz 2008). Diodes can be used to isolate incoming power of PVC and PEH.

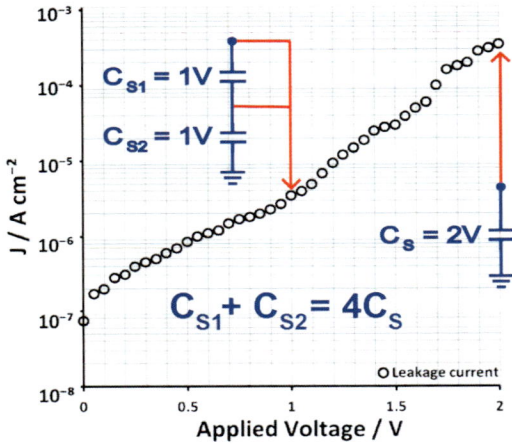

Fig. 11.17 Leakage current versus voltage profiles of the ultrathin capacitor (Wang et al. 2014)

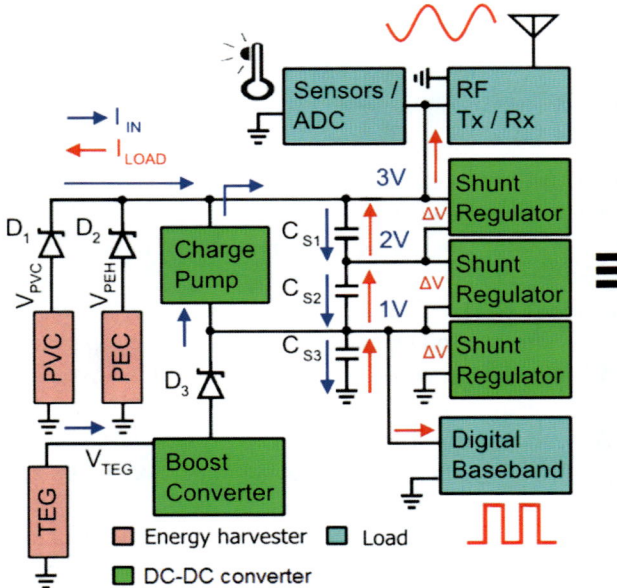

Fig. 11.18 Stacked capacitor energy harvesting scheme (Teh and Mok 2014b)

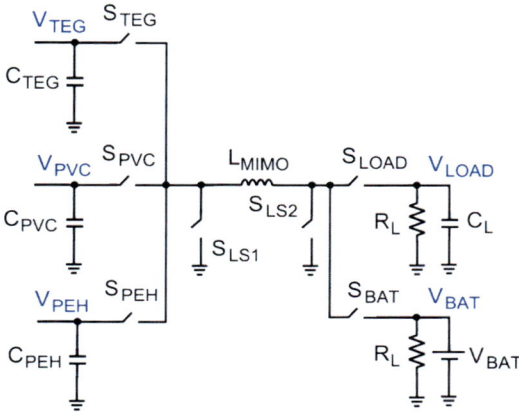

Fig. 11.19 Single-inductor multiple-input-multiple-output energy harvesting scheme (Bandyopadhyay and Chandrakasan 2012)

By combining the idea of divided series capacitor voltage to provide multiple voltage level, by connecting multiple capacitors in series, multiple voltage levels can be generated. Top capacitor (3 V) can power high voltage circuits (I/O, sensors, ADC, analog and RF) and bottom capacitor (1 V) can power low voltage low power digital circuits. In this proposed scheme, high incoming power can be driven directly by the primary power source when it is available. However when ambient power is insufficient, capacitor stacks will act as "flying-battery" configuration (Alon and Horowitz 2008) to supply high power, short interval required by IoT sensor node since each capacitor-shunt regulator pair forms a standalone "battery". All capacitors in the stack therefore share a common DC current when IoT node draws current. Additional capacitor can be installed in parallel fashion if greater energy is demanded.

Besides the stacked capacitor approach, another popular research direction such as (Bandyopadhyay and Chandrakasan 2012) is to pursue energy harvesting schemes using multiple energy sources and simultaneously generate multiple output voltage, using single inductor as the core power conversion element, shown in Fig. 11.19. However, for such single inductor multi-input-multi-output topology, only one source can be polled at a time due to the implicit

time-multiplexing nature and the control is deemed complicated.

11.5 Voltage Limits and Cold Start

The caveat of employing the new generation, seemingly advantageous TEG is the increased R_{TEG}, which has lower output current. Such device characteristics put further design constraint, on top of the low voltage start-up operation, which is already a challenging task to solve. The main problem for a boost converter circuit to self-start-up is the limited voltage level generated by the TEG source, which is often below CMOS transistor's threshold voltage (typically 300–500 mV). For body-wearable application with temperature difference of 1–2 °C, minimum output voltage can be as low as 25 mV. Therefore circuit designers face a catch-22 situation, where the dilemma is to bootstrap the boost converter in generating an output voltage of 1 V or more to power up other peripheral circuits such as sensors, analog-to-digital converter, digital baseband and RF transceiver. Since voltage up conversion always requires oscillation clock (which is generated on-chip or off-chip), as shown in Figs. 11.20 and 11.22. Therefore the solution to start-up problem fundamentally boils down to designing a sub-threshold oscillation clock generator at the lowest voltage possible. Many recent innovations in energy harvesting therefore focus on optimization of cold-start conditions in no-battery or dead-battery situations.

The minimum operation voltage of a binary switching signal transfer limit is known as the Meindl limit (Meindl and Davis 2000) where the theoretical minimum operation supply voltage (V_{DD}) is 36 mV (where 1's and 0's in signals are discernible) for ring oscillator implemented in standard CMOS technology. Another subtle start-up problem arises for systems employing TEG with high R_{TEG}, significant voltage drop across R_{TEG} and low current reduces transistor's gm. This jeopardizes the feedback gain required by oscillators, which need relatively high current

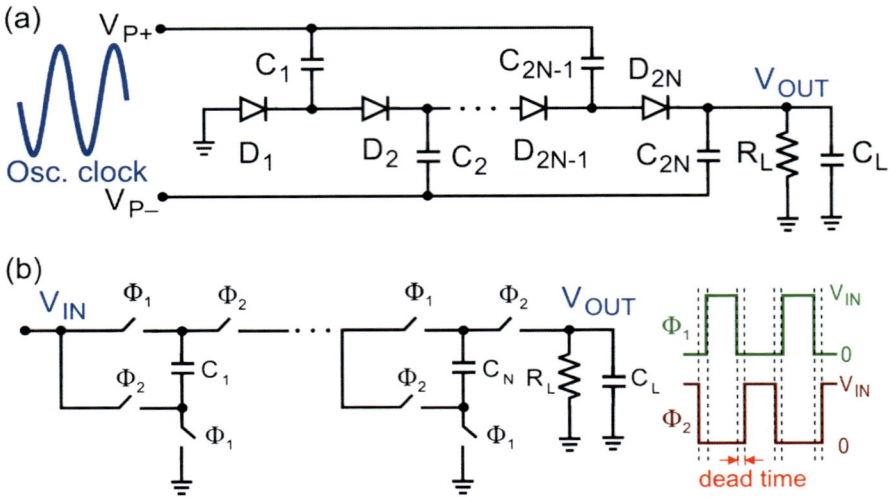

Fig. 11.20 (**a**) RF-pulse driven charge pump (**b**) On-chip clock driven charge pump

to function despite the low V_{DD} requirement (Teh and Mok 2014a).

To accomplish voltage boost conversion, linear regulator which is purely resistive is out of consideration since it can only perform voltage down conversion. Switching converter has the lowest start-up voltage (21 mV) and has the best power efficiency at the cost of bulky external inductor. To achieve the lowest start-up voltage, low threshold or zero threshold transistors should be used. However, power efficiency at higher output voltage will be lowered due to the transistor off-state leakage loss, which is inversely proportional to threshold voltage. From empirical data reported in recent works, CMOS 130 nm process seems to be the optimal choice of technology. Adopting more expensive CMOS process with smaller transistor size, renders diminishing return-over-cost for the sake of achieving lower start-up voltage or better power efficiency due to higher transistor leakage loss. In systems with negative supply voltage (Teh and Mok 2014b), transistor leakage loss can be reduced by applying negative gate voltage. However, larger magnitude of negative bias does not suppress leakage current further due to gate-induced-drain-leakage (GIDL) effect (Chatterjee et al. 2003). For example, the optimal

reverse bias for CMOS 130 nm process is approximately −0.3 V, as shown in Fig. 11.21.

The control circuit of the microwatts power single inductor switching converter is mainly digital-based, using an on-chip oscillator and counter controlled pulse width to achieve zero current switching (ZCS) as shown in Fig. 11.22. Oscillator generates a clock signal, with a particular frequency and duty cycle, which drives the low-side switch, SW_{LS}. The same clock signal is then delayed and used to generate pulses that drive the high-side switch SW_{HS}. The pulse width is controlled by digital counter value. Adaptive dead-time control is employed to minimized synchronization mismatch loss. The minimum start-up voltage of early low voltage TEG-powered switching converter is around 600 mV. However, the peak power efficiency is high, reaching 75% and lowest input voltage is 20 mV once activated (Carlson et al. 2010). Transformer-based switching converters using either high turns-ratio (Im et al. 2012) or unity turn-ratio pulse transformer (Teh and Mok 2014a) further reduce the start-up voltage down to 21 mV, due to minimal overhead control circuit required and the embedment of oscillator into the power stage. With an extra diode and output capacitor, (Teh and Mok 2014a) can be

Fig. 11.21 Off-state leakage current versus reverse bias voltage for CMOS 130 nm transistors

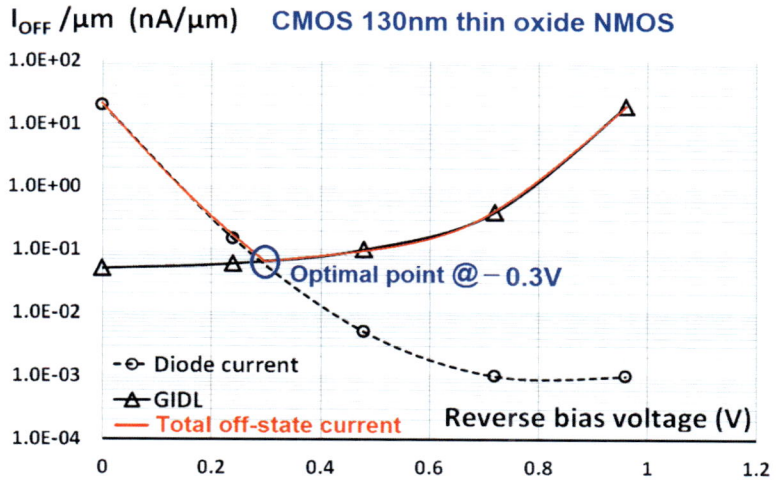

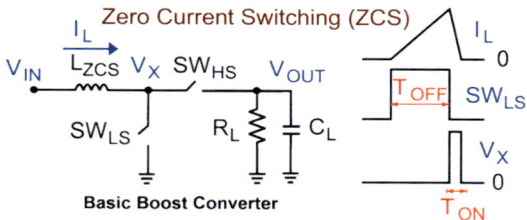

Fig. 11.22 Zero Current Switching (ZCS) control

reconfigured to give bipolar outputs up to ±3 V (Teh and Mok 2014b).

Compared to the conventionally one-stage structure of switching converter, SC power converter usually comprises multiple passive diode-based or active gate driven stages due to the topology-constrained, finite voltage gain per stage, as shown in Fig. 11.20. Similar to switching converter, the minimum start-up of switched-capacitor power converter is bound by the minimal supply voltage required by its on-chip oscillation clock generation. By having more switches and sub-threshold gate drive voltage, SC power converter's power efficiency at low voltage is lower than switching converter, especially for the fully integrated versions due to poor on-chip passives. Design techniques proposed to improve power efficiency of SC converter include adaptive dead-time control which alleviates shoot-through loss, negative voltage gate drive and substrate bias control to reduce

conductance loss (Kim et al. 2014). To eliminate the overhead of separate oscillator circuit, a self-oscillating SC converter is proposed (Jung et al. 2014). Hybrid converter refers to systems which employ both switching and SC converters in the system. LC-tank oscillator is used to generate sinusoidal clock (Bender Machado et al. 2014; Weng et al. 2013), which then drives a Dickson charge pump similar to those found in RFID tag RF-DC rectifier (Yi et al. 2009). (Fuketa et al. 2014) shows a fully integrated version using similar concept using on-chip transformers. Systems in (Chen et al. 2012b; Shrivastava et al. 2014) first self-starts using SC converter, which then drive another switching converter to provide the final output voltage. In (Bandyopadhyay et al. 2014), primary power converter is switching converter-based but SC converter is used to generate separate voltage to reduce high-side switch leakage. In (Teh et al. 2014), SC converter is used to supplement a SI converter, i.e. to transfer extra energy not consumed by the load to another secondary storage capacitor as backup.

Different self-start-up mechanisms reported in recent boost converter works are illustrated in Fig. 11.23. Performance of the representative papers of each start-up mechanism such as minimum start-up voltage, peak power efficiency and component requirements are summarized in Table 11.6. An interesting trend observed in research efforts to reduce start-up power is that

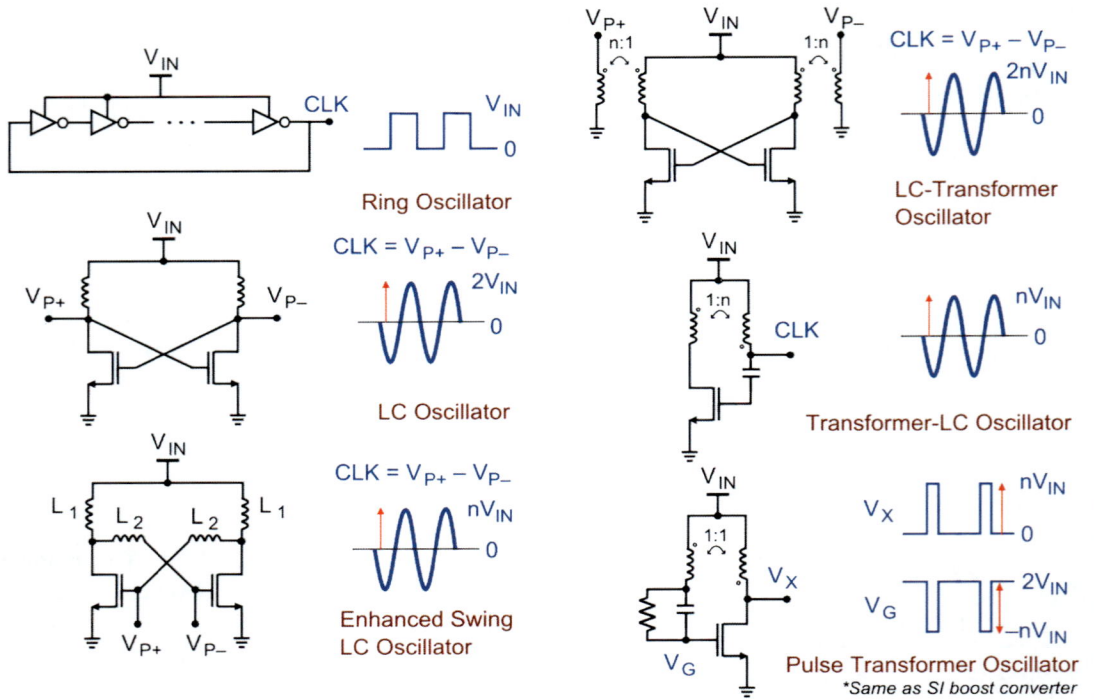

Fig. 11.23 Different self-start-up boost converter circuit topologies

Table 11.6 Performance summary of different topologies

| Start-up mechanism & converter type | Performance | | | |
	Min. start-up voltage	Peak power efficiency	External components	CMOS process
Ring oscillator SC + Switching converter Shrivastava et al. (2014)/Goeppert et al. (2015)	220 mV/ 70 mV	83%/58%	10 μH inductor	130 nm
Ring oscillator SC converter Chen et al. (2012a)	120 mV	39%	None	65 nm
Ring oscillator SC converter Kim et al. (2014)	150 mV	73%	6 × 10 nF capacitors	130 nm
LC oscillator+Switching converter Weng et al. (2013)	50 mV	73%	2 × 2 μH, 27 μH, 100 μH inductors	65 nm
Enhanced swing LC oscillator+SC converter Bender Machado et al. (2014)	10 mV	NA	2 × 9.5 μH, 2 × 950 μH inductors	130 nm
LC-transfromer oscillator+SC converter Fuketa et al. (2014)	85 mV	1.5%	None	65 nm
Transformer-LC oscillator+Switching converter Im et al. (2012)	40 mV	61%	1:60 transformer	130 nm
Pulse transformer oscillator+Switching converter Teh and Mok (2014a)	21 mV	74%	1:1 transformer	130 nm

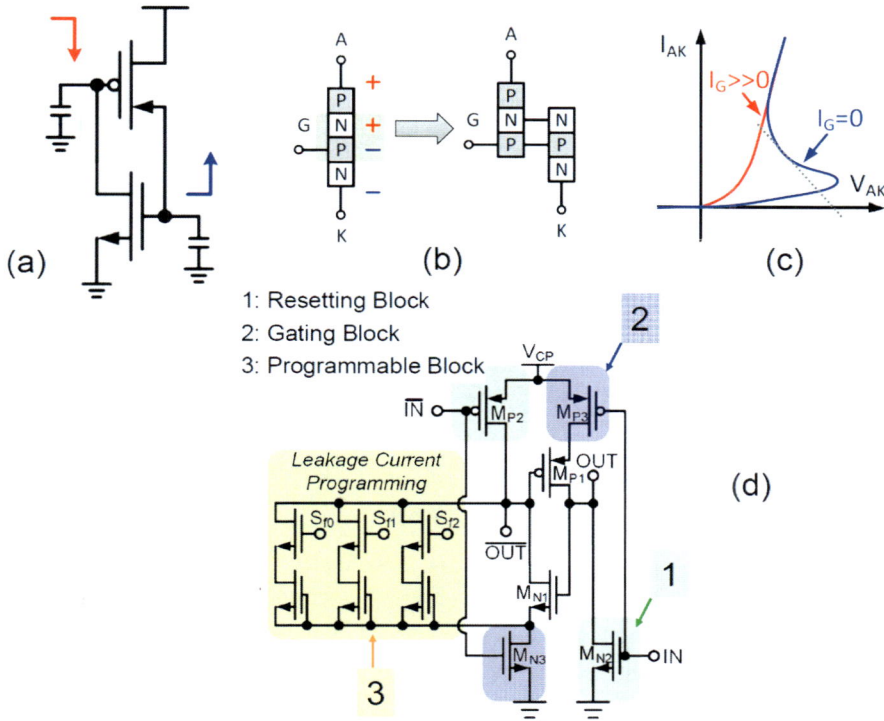

Fig. 11.24 (a) Leakage-based thyristor (b) Parasitic BJT to form equivalent thyristor (c) thyristor I-V curve (d) thyristor-based relaxation oscillator

non-conventional parasitic element in the CMOS process, such as a leakage based thyristor, using parasitic BJT junctions in the CMOS process, as shown in Fig. 11.24 is being used to design on-chip oscillator. The power consumption falls within the ballpark of sub-nanowatts (Liu et al. 2016). Ring oscillator based on a leakage-optimized Schmitt-Trigger cell shown in Fig. 11.25 is also proven the capability to start oscillating at only 70 mV supply voltage (Goeppert et al. 2015), without additional trimming as shown in (Chen et al. 2012b).

11.6 Passives in Energy Harvesting

Transformer-based systems often yield the best (lowest) self-start-up voltage performance, the inter-winding capacitance and series resistance

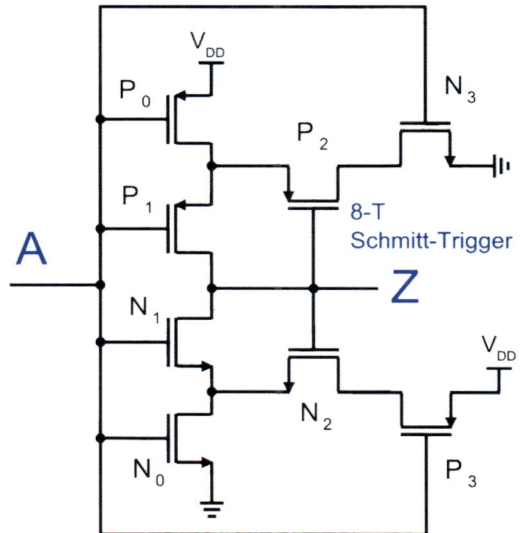

Fig. 11.25 Leakage-optimized Schmitt-Trigger cell (Goeppert et al. 2015)

Table 11.7 Electrical characteristics of high turns-ratio vs. unity turn-ratio transformer

Self-start-up boost converter topology	LC transformer oscillation	Pulse transformer oscillation
Transformer type	Non-standard, vendor specific Coilcraft LPR6235	Standard (Ethernet application) Murata 78601/9C
Turns-ratio	1:100	1:1
Coil resistance	316 Ω	1.3 Ω
Inter-winding capacitance	638 pF	121 pF
Inductance	7.5 μH:75 mH	10 mH:10 mH
Coupling co-efficient (1.0 = perfect)	0.95	0.9999
PCB board space	6 × 6 mm	9 × 9 mm

of transformer coils are increased as transformer size is shrunk, as shown in Table 11.7. High turns-ratio transformer also has more flux leakage than pulse transformer besides the increased parasitic capacitance and resistance. On-chip transformer has lower quality factor (below 20) compared to off-chip inductors which is 80 or higher (Moazenzadeh et al. 2015). Another important external component is Schottky diode, where lower turn-on voltage and smaller junction capacitance is always desired. Besides, due to high voltage conversion ratio (high SI boost converter duty cycle), output capacitor is mostly in self-discharge state during the switching cycle, making capacitor leakage loss an important consideration. Capacitor with higher rated voltage should be used to reduce capacitor leakage loss which is exponentially dependent on biased voltage. Further miniaturization of the system size is yet another research direction. For instance, advanced three-dimension die stacking packaging technology (Yu et al. 2011) can further reduce system board area. Novel wire bonding techniques (Raimann et al. 2012) and new transformer material (Camarda et al. 2015; Raimann et al. 2012) to realize miniaturized transformer on PCB would also be interesting research topics to explore. On-chip capacitors are often associated with bottom plate parasitic that degrades the switched-capacitor power efficiency. However, the availability of deep trench capacitor has managed to push the power efficiency up to 90% levels (Jung et al. 2014).

11.7 Perspectives and Trends on Energy Harvested IoT Nodes

This chapter has presented a general overview of low power energy harvesting and the associated power management circuit techniques reported in the recent decade. Designers need to trade off the choice of circuit topology, external component and CMOS process, based on the application requirement i.e. minimal start-up voltage, power efficiency and system size. The time availability and power profile (determined by source resistance) of energy harvesting source used in the application also limits the choice of circuit topologies. Further innovation and major breakthrough will require first, the advancement of CMOS process with better leakage reduction and second, miniaturized integrated magnetic and capacitive components with higher density and quality factor.

The development of next generation energy harvesters and energy storage using flexible materials is an emerging research topic (Koo et al. 2012; Zhong et al. 2014). Despite enjoying the convenience of being wearable and bendable, these emerging devices also pose new design challenges for circuit designers, due to the interchangeable roles of energy storage and generators in operation, and also the fluctuating electrical characteristics. In recent years, 130 nm and 65 nm technology are commonly chosen for prototype implementation as both are considered as optimum choice of CMOS technology for ultra-low voltage circuit design. Advanced CMOS technologies with shorter channel length have diminishing returns in advantages from the perspective of low voltage analog circuits (Galup-Montoro et al. 2012). However, it is still necessary to explore digital-based reprogrammable power management integrated circuits such as digital LDO (Yang et al. 2015b) which are capable of adapting to such unique device characteristics during in-circuit operation, especially under sub-1 V operations. Further study in designing an advanced reconfigurable charge pump topologies (Jung et al. 2014) which have multiple output, self-driven clock and balances bi-directionality of power flow would also be instrumental.

The control of multi-input-multi-output energy harvesting system typically requires additional overhead power, i.e. first sense the availability of power from each source, then compute and optimize the multi-source time multiplexing on the power converter. Similar to scarcity problem in wireless communication channel allocation, it is foreseen that game theoretic strategies and models can be employed to analyse harvested power supply and demand budget, and to optimize the converter utilization time allocation problem in a multi-source power converter. Hence a software-hardware co-design power management approach is foreseen to be the future trend of energy harvested IoT nodes.

References

E. Alon, M. Horowitz, Integrated regulation for energy-efficient digital circuits. IEEE J. Solid State Circuits **43**(8), 1795–1807 (2008)

S. Bandyopadhyay, A.P. Chandrakasan, Platform architecture for solar, thermal and vibration energy combining with MPPT and single inductor. IEEE J. Solid State Circuits **47**(9), 2199–2215 (2012)

S. Bandyopadhyay, P.P. Mercier, A.C. Lysaght, K.M. Stankovic, A.P. Chandrakasan, A 1.1 nW energy harvesting system with 544pW quiescent power for next-generation implants, in *Proceedings of the IEEE International. Solid State-Circuits Conference (ISSCC)* (2014), pp. 396–397

M. Bender Machado, M. Sawan, M. Cherem Schneider, C. Galup-Montoro, 10 mV—1V step-up converter for energy harvesting applications, in *Proceedings of the Symposium on Integrated Circuits and Systems Design (SBCCI)* (2014), pp. 1–5.

A. Camarda, A. Romani, E. Macrelli, M. Tartagni, A 32 mV/69 mV input voltage booster based on a piezoelectric transformer for energy harvesting applications. Sensor. Actuat. A Phys. **232**, 341–352 (2015)

E.J. Carlson, K. Strunz, B.P. Otis, A 20mV input boost converter with efficient digital control for thermoelectric energy harvesting. IEEE J. Solid State Circuits **45**(4), 741–750 (2010)

M.P. Chan, P.K.T. Mok, Design and implementation of fully integrated digitally controlled current-mode buck converter. IEEE Trans. Circuits Syst. I **58**(8), 1980–1991 (2011)

B. Chatterjee, M. Sachdev, S. Hsu, R. Krishnamurthy, S. Borkar, Effectiveness and scaling trends of leakage control techniques for sub-130 nm CMOS technologies, in *Proceedings of the International. Symposium on Low Power Electronics and Design (ISLPED)* (2003), pp. 122–127

P.-H. Chen, K. Ishida, X. Zhang, Y. Okuma, Y. Ryu, M. Takamiya, T. Sakurai, A 120-mV input, fully integrated dual-mode charge pump in 65-nm CMOS for thermoelectric energy harvester, in *Asia and South Pacific Design Automation Conference (ASP-DAC)* (2012), pp. 469–470.

P.-H. Chen, X. Zhang, K. Ishida, Y. Okuma, Y. Ryu, M. Takamiya, T. Sakurai, An 80 mV startup dual-mode boost converter by charge-pumped pulse generator and threshold voltage tuned oscillator with hot carrier injection. IEEE J. Solid State Circuits **47**(11), 2554–2562 (2012b)

K.W.R. Chew, Z. Sun, H. Tang, L. Siek, A 400 nW single-inductor dual-input-tri-output DC-DC buck-boost converter with maximum power point tracking for indoor photovoltaic energy harvesting, in *IEEE International Solid-State Circuits Conference Digest of Technical Papers* (2013), pp. 68–69.

M. Danesh, J.R. Long, An autonomous wireless sensor node incorporating a solar cell antenna for energy harvesting. IEEE Trans. Microw. Theory Tech. **59**(12), 3546–3555 (2011)

Enocean, Solar Cells ECS300 and ECS310 [Online]. http://www.enocean.com/en/enocean_modules/ecs-300-data-sheet-pdf

H. Fuketa, Y. Momiyama, A. Okamoto, T. Sakata, M. Takamiya, T. Sakurai, An 85-mV input, 50-μs startup fully integrated voltage multiplier with passive clock boost using on-chip transformers for energy harvesting, in *Proceedings of the European Solid State Circuits Conference (ESSCIRC)* (2014), pp.263–266

C. Galup-Montoro, M.C. Schneider, M.B. Machado, Ultra-low voltage operation of CMOS analog circuits:

amplifiers, oscillators, and rectifiers. IEEE Trans. Circuits Syst. II Exp. Briefs **59**(12), 932–936 (2012)

J. Goeppert, Y. Manoli, Inductive boost converter for thermoelectric energy harvesting with fully integrated start-up at 70 mV, in *Proceedings of the European Solid State Circuits Conference (ESSCIRC)* (2015), pp. 233–236.

E.N.Y. Ho, P.K.T. Mok, Wide-loading-range fully integrated LDR with a power-supply ripple injection filter. IEEE Trans. Circuits Syst. II **59**(6), 356–360 (2012)

C. Huang, P.K.T. Mok, A 100 MHz 82.4% efficiency package-bondwire based four-phase fully-integrated buck converter with flying capacitor for area reduction. IEEE J. Solid State Circuits **48**(12), 2977–2988 (2013a)

C. Huang, P.K.T. Mok, An 84.7% efficiency 100-MHz package bondwire-based fully integrated buck converter with precise dcm operation and enhanced light-load efficiency. IEEE J. Solid State Circuits **48**(11), 2595–2607 (2013b)

J.P. Im, S.W. Wang, S.T. Ryu, G.H. Cho, A 40 mV transformer-reuse self-start-up boost converter with mppt control for thermoelectric energy harvesting. IEEE J. Solid State Circuits **47**(12), 3055–3067 (2012)

J. Jiang, Y. Lu, W.H. Ki, Analysis of two-phase on-chip step-down switched capacitor power converters, in *IEEE Asia Pacific Conference on Circuits and Systems (APCCAS)*, Ishigaki Island, Okinawa, Japan (2014), pp. 575–578

X. Jing, P.K.T. Mok, Power loss and switching noise reduction techniques for single-inductor multiple-output regulator. IEEE Trans. Circuits Syst. I **60**(10), 2788–2798 (2013)

W. Jung, S. Oh, S. Bang, Y. Lee, D. Sylvester, D. Blaauw, A 3nW fully integrated energy harvester based on self-oscillating switched-capacitor DC-DC converter, in *Proceedingds of the IEEE International Solid State-Circuits Conference (ISSCC)* (2014), pp. 398–399

J. Kim, P.K.T. Mok, C. Kim, A 0.15V-input energy-harvesting charge pump with switching body biasing and adaptive dead-time for efficiency improvement, in *Proceedings of the IEEE Internatioanl Solid State-Circuits Conference (ISSCC)* (2014), pp. 394–395

M. Koo, K. Park, S. Lee, M. Suh, D.Y. Jeon, J.W. Choi, K. Kang, K.J. Lee, Bendable inorganic thin-film battery for fully flexible electronic systems. Nano Lett. **12**(9), 4810–4816 (2012)

D. Kwon, G.A. Rincon-Mora, A single-inductor 0.35-μm CMOS energy-investing piezoelectric harvester. IEEE J. Solid State Circuits **49**(10), 2277–2291 (2014)

I. Lee, W. Lim, A. Teran, J. Philips, D. Sylvester, D. Blaauw, A > 78% efficient light harvester over 100-to-100klux with reconfigurable PV-cell network and MPPT circuit, in *IEEE International Solid-State Circuits Conference Digest of Technical Papers* (2016), pp. 370–372

X. Liu, E. Sanchez-Sinencio, A single-cycle MPPT charge-pump energy harvester using a thyristor-based vco without storage capacitor, in *IEEE International Solid-State Circuits Conference Digest of Technical Papers* (2016), pp. 364–365

D. Ma, W.H. Ki, C.Y. Tsui, P.K.T. Mok, Single-inductor multiple-output switching converters with time-multiplexing control in discontinuous conduction mode. IEEE J. Solid State Circuits **38**(1), 89–100 (2003a)

D. Ma, W.H. Ki, C.Y. Tsui, A pseudo-CCM/DCM SIMO switching converter with freewheel switching. IEEE J. Solid State Circuits **38**(6), 1007–1014 (2003b)

P.I. Mak, R.P. Martins, High-/mixed-voltage RF and analog CMOS circuits come of age. IEEE Circuits Syst. Mag. **10**(4), 27–39 (2010)

T.Y. Man, P.K.T. Mok, M. Chan, A 0.9-V input discontinuous-conduction-mode boost converter with CMOS-control rectifier. IEEE J. Solid State Circuits **43**(9), 2036–2046 (2008)

Marlow Industries, TG12-2.5 Datasheet [Online]. http://www.marlow.com/media/marlow/product/downloads/tg12-2-5-011/TG12-2.5.pdf

J.D. Meindl, A.J. Davis, The fundamental limit on binary switching energy for terascale integration (TSI). IEEE J. Solid State Circuits **35**(10), 1515–1516 (2000)

Microgen, MEMS-based vibrational energy harvesting [Online]. http://www.microgensystems.co

Micropelt, MPG-D751 Datasheet [Online]. http://www.micropelt.com/down/datasheet_mpg_d651_d751.pdf

A. Moazenzadeh, F.S. Sandoval, N. Spengler, V. Badilita, U. Wallrabe, 3-D Microtransformers for DC-DC on-chip power conversion. IEEE Trans. Power Electron. **30**(9), p5088 (2015)

M. Raimann, A. Peter, D. Mager, U. Wallrabe, J.G. Korvink, Microtransformer-based isolated signal and power transmission. IEEE Trans. Power Electron. **27**(9), 3996–4004 (2012)

P. Sarkar, S. Chakrabartty, Compressive self-powering of Piezo-floating-gate mechanical impact detectors. IEEE Trans. Circuits Syst. Regul. Pap. **60**(9), 2311–2320 (2013)

A. Shrivastava, D. Wentzloff, B.H. Calhoun, A 10mV-input boost converter with inductor peak current control and zero detection for thermoelectric energy harvesting, in *Proceedings of the IEEE Custom Integrated Circuits Conference (CICC)* (2014), pp. 1–4

F. Su, W.H. Ki, Design strategy for step-up charge pumps with variable integer conversion ratios. IEEE Trans. Circuits Syst. II **54**(5), 417–421 (2007)

Y.K. Teh, P.K.T. Mok, Design of transformer-based boost converter for high internal resistance energy harvesting sources with 21 mV self-start-up voltage and 74% power efficiency. IEEE J. Solid State Circuits **49**(11), 2694–2704 (2014a)

Y.K. Teh, P.K.T. Mok, A stacked capacitor multi-microwatts source energy harvesting scheme with

86 mV minimum input voltage and ±3 V bipolar output voltage. IEEE J. Emerging Sel. Top. Circuits Syst. **4**(3), 313–323 (2014b)

Y.K. Teh, P.K.T. Mok, A bipolar output voltage pulse transformer boost converter with charge pump assisted shunt regulator for thermoelectric energy harvesting, in *Proceedings of the IEEE Midwest Symposium on Circuits and Systems (MWSCAS)* (2014), pp. 37–40

C.C. Wang, J.C. Wu, Efficiency improvement in charge pump circuits. IEEE J. Solid State Circuits **32**(6), 852–860 (1997)

H. Wang, G. Wang, Y. Ling, F. Qian, Y. Song, X. Lu, S. Chen, Y. Tong, Y. Li, High power density microbial fuel cell with flexible 3D graphene-nickle foam as anode. Nanoscale **5**(21), 10283–10290 (2013)

C. Wang, M. Osada, Y. Ebina, B.-W. Li, K. Akatsuka, K. Fukuda, W. Sugimoto, R. Ma, T. Sasaki, All-nanosheet ultrathin capacitors assembled layer-by-layer via solution-based processes. ACS Nano **8**(3), 2658–2666 (2014). doi:10.1021/nn406367p

P.S. Weng, H.Y. Tang, P.C. Ku, L.H. Lu, 50 mV-input batteryless boost converter for thermal energy harvesting. IEEE J. Solid State Circuits **48**(4), 333–341 (2013)

J. Yang, M. Lee, M.J. Park, S.Y. Jung, J. Kim, A 2.5-V, 160-μJ-output piezoelectric energy harvester and power management IC for Batteryless Wireless Switch (BWS) applications, in *2015 Symposium on VLSI Circuits Digest of Technical Papers* (2015), pp. 282–283

F. Yang, P.K.T. Mok, A 0.6-1V input capacitor-less asynchronous digital LDO with fast transient response achieving 9.5b over 500mA loading range in 65-nm CMOS, in *European Solid State Circuits Conference (ESSCIRC)*, Graz, Austria (2015), pp. 180–183

J. Yi, W.H. Ki, P.K.T. Mok, C.Y. Tsui, Dual-power-path RF-DC multi-output power management unit for RFID tags, in *Proceedings of the IEEE Symposium on VLSI Circuits (VLSIC)* (2009), pp. 200–201

A. Yu, J.H. Lau, S.W. Ho, A. Kumar, W.Y. Hnin, W.S. Lee, M.C. Jong, V.N. Sekhar, V. Kripesh, D. Pinjala, S. Chen, C.-F. Chan, C.-C. Chao, C.-H. Chiu, C.-M. Huang, C. Chen, Fabrication of high aspect ratio TSV and assembly with fine-pitch low-cost solder microbump for Si interposer technology with high-density interconnects. IEEE Trans. Compon. Packag. Manuf. Technol. **1**(9), 1336–1344 (2011)

C. Zhan, W.H. Ki, An output-capacitor-free adaptively biased low-dropout regulator with subthreshold undershoot-reduction for SoC. IEEE Trans. Circuits Syst. I **59**(5), 1119–1131 (2012)

C. Zhan, W.H. Ki, Analysis and design of output-capacitor-free low-dropout regulators with low quiescent current and high power supply rejection. IEEE Trans. Circuits Syst. I **61**(2), 625–636 (2014)

J. Zhong, Y. Zhang, Q. Zhong, Q. Hu, B. Hu, Z.L. Wang, J. Zhou, Fiber-based generator for wearable electronics and mobile medication. ACS Nano **8**(6), 6273–6280 (2014). doi:10.1021/nn501732z

Ultra-Low Power Analog Interfaces for IoT

12

Jerald Yoo

This chapter addresses the challenges and design strategies in Analog Front-End (AFE) interface circuit design with an umbrella of IoT. A stringent energy constraint in IoT means the circuit specification must take into account the energy-efficient operation. Also, at the same constraint, the dynamic and static offset/noise compensation should be done effectively.

12.1 Introduction

Internet-of-Things (IoT) has variety of applications, including but not limited to, bio-medical sensors, MEMS, environmental sensors and sensor networks. The interface circuits bridges between the physical world and the electrical signal, and it is a crucial component in IoT system. Common constraints on these applications are limited form factor and the consequent limited power/energy sources. Therefore, it is very important we understand the constraints and draw the specification of the interface circuit that meets the IoT needs.

12.1.1 Unique Environment

Let us cover in details several important challenges that Interface circuit faces under IoT environment. These challenges lead to the design requirements and trade-offs.

Limited Power/Energy Source: As an example, a state-of-the-art MEMS piezoelectric power generator, thermoelectric generators (TEG) or photovoltaic cells (PV) is capable of generating ~100s of μW in a form factor smaller than 1 cm^3; let us not forget that these energy harvesting sources are, most of time, not "always available", in other words, average power that we can draw may be even smaller (See Table 12.1 for more details). On the other hand, if we choose a coin-cell battery, we have tens of mAh in the budget, which is still very limited.

Noise and Offset: IoT interface circuit may have near DC up to hundreds of kHz as its bandwidth. In this bandwidth, there are intruders that we refer to as "in-band noise". As shown in Fig. 12.1, these include *1/f* noise, 50/60 Hz noise, thermal noise and static/dynamic offset. In many cases, the noise well embeds the weak signal. Hence noise-aware design is of crucial.

1/f Noise: Flicker noise, often referred to as *1/f* noise, is unavoidable in all CMOS amplifiers, caused by charge carriers randomly trapped/ released between the gate oxide and the substrate.

J. Yoo (✉)
Electrical and Computer Engineering, National University of Singapore, Singapore 119077, Singapore
e-mail: jyoo@nus.edu.sg

© Springer International Publishing AG 2017
M. Alioto (ed.), *Enabling the Internet of Things*, DOI 10.1007/978-3-319-51482-6_12

Table 12.1 Estimated power output values per harvesting methods (Table from Belleville et al. (2009))

Source		Source characteristics	Physical efficiency	Harvested power
Photovoltaic				
Office		0.1 mW/cm^2	10–24%	10 μW/cm^2
Outdoor		100 mW/cm^2		10 mW/cm^2
Vibration/motion				
Human		0.5 m@1 Hz 1 m/s^2@50 Hz	Maximum power is source dependent	4 μW/cm^2
Industry		1 m@5 Hz 10 m/s^2@1 kHz		100 μW/cm^2
Thermal energy				
Human		20 mW/cm^2	0.10%	25 μW/cm^2
Industry		100 mW/cm^2	3%	1–10 mW/cm^2
RF				
GSM	900 MHz	0.3–0.03 μW/cm^2	50%	0.1 μW/cm^2
	1800 MHz	0.1–0.01 μW/cm^2		

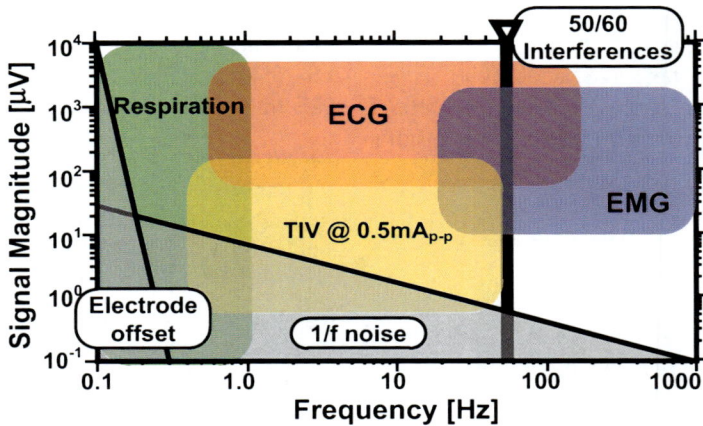

Fig. 12.1 Example IoT noise and signal bandwidth: case of wearable sensor applications (figure courtesy of Dr. Long Yan, Samsung Electronics)

The average power of *1/f* noise is given by Eq. (12.1):

$$\overline{V_n^2} = \frac{\kappa}{C_{ox}WL} \cdot \frac{1}{f} \qquad (12.1)$$

The existence of *1/f* noise is problematic because it overlaps with many IoT application's signal bandwidth; specifically, the *1/f* nature of the noise makes the dominant component near DC. As Eq. (12.1) shows, in order to decease the 1/f, we need to have larger device (with larger WL), or use dynamic offset cancellation technique, which will be discussed in detail in Sect. 12.3.

12.1.2 Design Requirement and Performance Metrics

System Resolution: Depending on what are the back-end and post-processing needs, we can decide the system resolution. For example, a wearable physiological sensor may require >10b signal accuracy, therefore the A/D Converter (ADC) should have >10b resolution.

ADC Reference Voltage and Interface Amplifier Gain: Once the ADC bit widths is decided, the Least-Significant Bit (LSB) system resolution can be calculated by the ADC reference voltage. For example, if 1.10 V is used as a ADC voltage reference, and the ADC is 10b, then the LSB resolution becomes 1.07 mV. At this point, if we want to have the minimal signal level that we can detect to be 1 μV, the interface amplifier need to have at least 60 dB gain; a 60 dB gain means we need a multi-stage amplifier. Note that the minimal signal level is related to the noise floor, where Signal-to-Noise Ratio (SNR) becomes 1.

Sampling Rate: ADC's sampling speed is strictly related to the signal processing throughput at the back-end. Also, for the AFE perspective, this impacts the anti-aliasing filter specification. Considering the stringent energy/power and form factor (area), we need to choose the right order (roll-off factor). Choosing too high order will increase the power and area consumption.

Common Mode Rejection Ratio (*CMRR*): The interface circuitry may be composed of off-chip devices such as capacitors and resistors in the signal path. In such cases, CMRR has two components: on-chip AFE ($CMRR_{AFE}$) and the off-chip interface ($CMRR_{IF}$), where the overall $CMRR_{SYS}$ is determined by Eq. (12.2) (Yoo 2014):

$$\frac{1}{CMRR_{SYS}} = \frac{1}{CMRR_{AFE}} + \frac{1}{CMRR_{IF}} \quad (12.2)$$

It should be noted that in general off-chip devices have limited matching, which makes it challenging to have $CMRR_{IF}$ of over 60 dB; consequently, as Eq. (12.2) shows, however high the $CMRR_{AFE}$ would be, the $CMRR_{IF}$ will determine the overall $CMRR_{SYS}$. The message here is clear: when it comes to IoT system, integrate as many components on-chip as possible.

Noise Efficiency Factor (*NEF*): One of the performance metric we can use for the IoT AFE is NEF. Due to variety of gains that different amplifiers have, it is not fair to compare the output noise directly. Instead, we divide the output noise with the gain of the amplifier, and this is called the Referred-to-Input (RTI) noise. Note that noise RTI exists regardless of technology node we use, and there is a theoretical lower limit on the power

consumption of the amplifier, dependent on the noise spectral density requirement.

To probe this further, we can start the noise analysis with a Bipolar Junction Transistors (BJTs), which has a lower noise spectral density RTI than CMOS transistors have for a given bias current. For a single common-emitter amplifier, a short-circuit voltage noise density is well known as in Eq. (12.3):

$$V_{ni,RMS} = \sqrt{\frac{\pi q V_t^2 BW}{I_C}} \quad (12.3)$$

Where $V_{ni,RMS}$ is the RMS value of the noise RTI integrated over a bandwidth of BW, I_C is the collector current, q is the charge on an electron, and V_t is the thermal voltage of kT/q. With this, we can now compare the performance of the different amplifiers to that of the theoretical limit of a single BJT; this is called the Noise Efficiency Factor (NEF) (Steyaert et al. 1987):

$$NEF = V_{ni,RMS} \sqrt{\frac{I_{overall}}{\pi V_t 4kT \, BW}} \quad (12.4)$$

The NEF shows the trade-off between the current consumption and the noise of an amplifier. Note that the NEF of a single BJT having only thermal noise equals to 1, where a differential BJT is $\sqrt{2} = 1.414$. Since we know that CMOS amplifier in general have worse noise spectral density for a given bias current, we can expect that NEF of a CMOS amplifier will be higher than that of a BJT.

12.2 Ultra Low Power/Energy Interface Design

We should keep in mind that in IoT applications, we have a very tight power and energy constraint. As probed earlier, we have largely two scenarios in power source: energy harvesting/scavenging, and battery. With this in mind, in this Sect. 12.2, we will take a closer look at the interface circuit design, in power and energy perspective. We will also cover design strategies for low-cost, energy-efficient filter.

12.2.1 Power Vs. Energy: Choosing the Right Goal

When an energy harvesting/scavenging is used for the IoT power supply, it is very important to design the circuitry not to exceed the instantaneous current driving capability of the power source. Table 12.1 shows the estimated output values of energy harvesting (for different sources). Depending on the IoT application, the harvested power amount varies significantly. We should not forget that these amount are optimistic values on average; for example, in PV, if there is an instantaneous blockage of the cell, the generated power will drop abruptly. Therefore, if the circuit needs continuous operation, the designer may want to introduce a secondary energy storage such as super capacitor or battery.

Nevertheless, when using the harvesting/scavenging as the main source, keeping the "peak power consumption" of the circuitry well below that of the generator's limit should be the key design target. This is aligned with what we saw from Sect. 1.3.1.

With this in mind, it is helpful to see what will be the consequences of designing lower power interface circuit. First of all, the amplifier will be impacted significantly. The bias current cannot exceed the limit set by the supply, therefore the thermal noise will increase. Available power level of µW order means the transistors will likely operate in weak inversion (or - sub-threshold), which will be prone to Voltage and Temperature variation. Also, the impact of dynamic offset will become more serious. Strategies to overcome these issues will be described in detail in Sect. 12.3.

Now let us consider the case where we use battery as the main power supply. A coin cell battery, or a small flexible battery, will have the capacity of tens to hundreds of millamp-hour (mAh). Compared to energy harvesting/scavenging case, we have more room for maximum allowable current. However, as we discussed in Sect. 1.3.1, now the energy consumption (or the

average power consumption) of the circuity will determine the system lifetime. This means, when the battery is the power supply, then minimizing the "energy consumption", or the average power consumption, should be the target goal. Systematic approach to achieving energy efficiency was already covered in Sects. 1.3 and 2.3. Also, in Sect. 12.3, we will see the design strategies for power- and energy-efficient analog interface circuit design.

12.2.2 Top-Down Approach for Power and Energy Efficiency

Especially for IoT system/sensor, it is very important we design the system in to-down approach to meet the power and energy requirement. This means, when designing such system, we should answer the following questions, from top to bottom:

- What is the application we are targeting for?
- What is the energy source? (Harvesting/Scavenging/Battery/Hybrid/etc.)
- What are the system components that will be included/excluded? (analog interface, filters, voltage reference, analog-to-digital converter, post processing, communication block, memory)
- What is the available power budget for each block?
- What is the performance requirement for each block?
- What are the constraints/bottleneck in achieving such performance under such environment?
- Defining the specification of each block, and start designing.

Without considering the system perspective, one might end up designing the lowest-power consuming analog interface circuit, but that block is only a small fraction of overall system power consumption; as a result, in such cases, the total system power consumption would not decrease much. This can be avoided by breaking down the system power budget from the beginning.

12.2.3 Small Form Factor, Efficient Filter Design

Because of the very strict power budget in IoT, often the filter design becomes an issue in interface circuits. Not only the power and energy, but also the area is of concern. Most of time we do not have luxury of using high order filters in such area- and power- constrained system. Therefore, approaching from a system perspective is very crucial.

As previously described in Sect. 12.1, this becomes particularly important in dynamic offset compensation circuit, where extensive filtering is needed. Here are some strategies we can take for an area- and energy-efficient filter design:

- Active filtering of ripple-induced noise by real-time sampling and subtracting (Fan et al. 2011): suitable for an area-constrained environment. The amplifier features a ripple-reduction loop (see Fig. 12.18) to sample the noise and subtract it in real time; by doing so, we can avoid using area-consuming high-order filters.
- Balance of analog interface and digital processing filtering (Altaf et al. 2015; Yoo et al. 2013): Choose the oversampling rate and chopping frequency to use only 2nd order filter in analog interface; then the digital processor also aids filtering.

12.3 Noise-Aware Interface Circuit Design

In this subsection, we will look into details of the interface circuit design for IoT. We will start with basic instrumentation amplifier (IA), compare the strengths and weaknesses. After that, we will see the two widely used dynamic offset compensation techniques: Auto-Zeroing and Chopper-Stabilization.

12.3.1 Instrumentation Amplifier Basics

Now let us look into the Instrumentation Amplifier (IA) basics. Understanding the IA basics is crucial for expanding the perimeter to IoT applications.

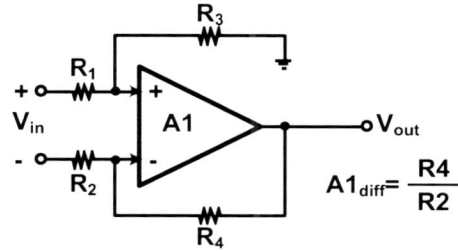

$$A1_{diff} = \frac{R4}{R2}$$

Fig. 12.2 Single-amp IA

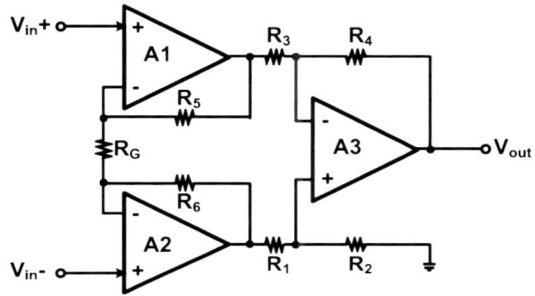

Fig. 12.3 Three-amp IA

Single-Amp IA: The simplest IA is a Single-amp IA shown in Fig. 12.2. The gain is defined by R4/R2 with a large input voltage range. However, this structure has a fundamental problem: low input impedance. The input impedance is determined by R_1 and R_2, but we cannot increase them too much, because otherwise the gain will be limited. Hence, loading effect may become an issue in IoT applications. More importantly, in IoT, with stringent low power requirement (low bias current), the thermal noise may be large enough to corrupt the input signal, and there is no mitigation technique of it.

Three-Amp IA: The classical 3-amp IA shown in Fig. 12.3 solves the low input impedance problem of a Single-amp IA. This is done by adding buffer amplifiers (A1 and A2) to each input of the differential amplifier (A3).

Here, the gain is defined as shown in Eq. (12.5):

$$Gain = \left(1 + \frac{R_5 + R_6}{R_G}\right)\left(\frac{R_4}{R_3}\right) \qquad (12.5)$$

Note that when R_G is removed, A1 and A2 will operate as a unity-gain buffer amplifier. By adding R_G, and additional gain is achieved. On top of

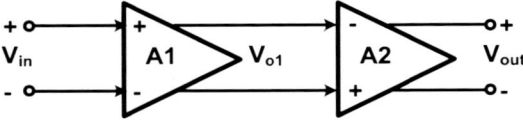

Fig. 12.4 CMRR of a multi-stage amplifier

that, R_G is free from matching issues, since only one resistor is added. This is why this structure is widely adopted to-date in IA domain.

However, for IoT, this is still not enough due to lack of dynamic offset removal—the offsets will be a problem on top of that, with such low power requirement, the bias current cannot be high, and subsequently noise is significant.

CMRR of a multi-stage IA: It is also worth remembering that in a multi-stage amplifier, CMRR of each stage will impact differently in overall CMRR performance. Let us assume that the 2-stage amplifier shown in Fig. 12.4 has gain of A1 and A2, with CMRR of each stage to be $CMRR_1$ and $CMRR_2$, respectively.

Now let us assume the input voltage of V_{in} $= V_{id} \pm \frac{V_{ic}}{CMRR_1}$ is applied, where V_{id} and V_{ic} are the differential and common mode component of the input, respectively. Then, in Fig. 12.4, V_{o1} can be expressed as $V_{o1} = A1 \cdot V_{in}$. Therefore, V_{out} will become:

$$V_{out} = A2 \left(V_1 \pm \frac{V_{ic}}{CMRR_2} \right)$$

$$= (A1 \cdot A2) \cdot V_{id}$$

$$\pm \left(\frac{A1 \cdot A2}{CMRR_1} + \frac{A2}{CMRR_2} \right) \cdot V_{ic} \qquad (12.6)$$

Observing Eq. (12.6), we find that the differential gain is $(A1 \cdot A2)$ and the common-mode gain becomes $\left(\frac{A1 \cdot A2}{CMRR_1} + \frac{A2}{CMRR_2} \right)$. Since CMRR is the ratio of differential gain to the common mode gain, now the CMRR of the overall amplifier becomes as in Eq. (12.7):

$$\frac{1}{CMRR_{amp,overall}} = \frac{1}{CMRR_1} + \frac{1}{A1 \cdot CMRR_2} \qquad (12.7)$$

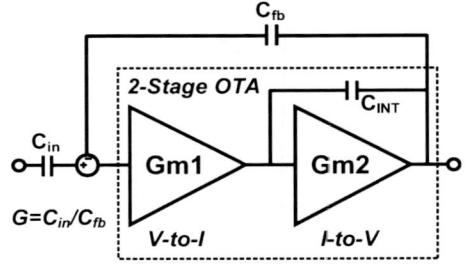

Fig. 12.5 IA with a capacitive gain element

This gives us an important message: in a multi-stage amplifier, the CMRR of the first stage will dominate the overall CMRR performance! Hence, in an IoT AFE/IA design, it is very important we give an attention on the low-noise yet high-CMRR performance of the first stage amplifier.

Capacitive Coupled IA (CCIA): Noting that in general capacitors have better matching than the resistor has, the amplifier shown in Fig. 12.5 utilizes capacitive gain element (Harrison and Charles 2003). Here, the gain is defined as C_{in}/C_{fb}. This type of an amplifier is called Capacitive Coupled IA (CCIA).

Using capacitive gain element has another very strong advantage: C_{in} blocks DC and therefore the amplifier accepts rail-to-rail input. Since then, this simple-but-powerful idea is widely adopted in IA domain.

Aforementioned amplifier types—single-amp, three-amp or CCIA—has a fundamental limit, when it comes to IoT applications: static and dynamic offset. None of above amplifiers have the offset mitigation scheme, and especially in low power (with low bias current) environment, the dynamic offset including $1/f$ noise becomes a series issue.

12.3.2 Auto-Zeroing

Aforementioned issues in IA—especially the dynamic offset—must be mitigated in IoT applications. One technique to overcome the issue is Auto-Zeroing (Enz and Temes 1996). The concept of the auto-zeroing is simple:

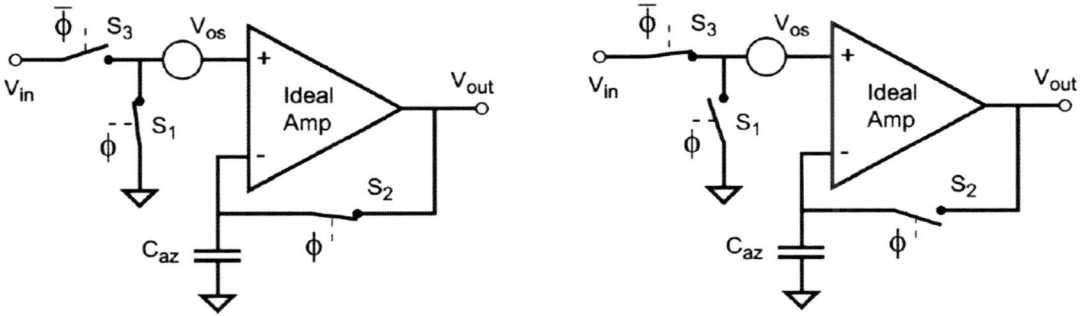

Fig. 12.6 Auto-zeroing concept: sampling phase (*left*), and amplification phase (*right*) (figure from Makinwa et al. (2007))

sample the offset (from either input or output), and then subtract it. This is illustrated in Fig. 12.6.

In auto-zeroing, there are two phases: Sampling Phase (shown in left), and Amplification Phase (right). Assuming an ideal amplifier in Fig. 12.6, during the sampling phase S1 and S2 are closed (forming a buffer amplifier), and the offset of the amplifier is sampled at the output and stored into the storage capacitance C_{az}. Alternatively, in other structures, intermediate or input offset can be sampled.

Now in the amplification phase, S1 and S2 are open, and S3 is closed. Since the C_{az} stores the offset value, and is connected to the negative input, the offset is removed and only the signal component will be amplified.

The auto-zeroing can be explained in sampled signal perspective as well. Let us consider the auto-zeroing system as a sample-and-hold (S&H) structure as shown in Fig. 12.7. Here, the $V_{n,az} z(f) = V_n(f) \cdot (1 - H(f))$, where $H(f)$ is the transfer function of the S&H block. S&H is gate function in time domain, and therefore $H(f) = sinc(\pi f / f_S)$. Noting that sinc function acts as a low-pass filter (LPF) with passband in DC, $1-H(f)$ is a high-pass filter (HPF). This is why the auto-zeroing mitigates the *1/f* noise (near DC) and the other offset as well.

Auto-zeroing also has one limitation and that is the noise folding into baseband. This is due to limited bandwidth of the S&H; during the sampling

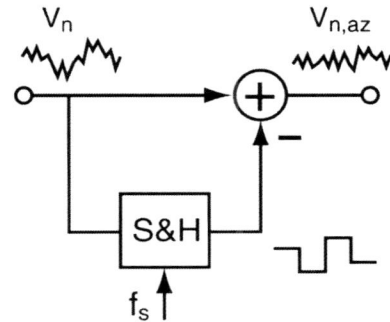

Fig. 12.7 Auto-zeroing in signal processing perspective (figure from Makinwa et al. (2007))

phase, the higher frequency component (mostly thermal noise) will be under sampled, and this will be folded into baseband during amplification.

Regardless of the limitation, the auto-zeroing can be widely used in IoT applications, especially those with sampled system or with applications having single-ended signals as their inputs.

12.3.3 Chopper Stabilization

Chopper-stabilization (Enz and Temes 1996) is another commonly used dynamic offset cancellation technique that suits well with IoT application. In a nutshell, chopper stabilization modulates the input signal into higher frequency before the 1/f and other dynamic offset corrupts the signal.

Figure 12.8 illustrates a chopper-stabilized amplifier in time domain. It is composed of a differential amplifier (Amp), a modulator/demodulator pair, and a low pass filter (LPF). The signal drawn in blue, denoted as V_{IN}, is differential. As the input chopper swaps V_{IN}^+ and V_{IN}^- with the chopping frequency f_{chop}, the input of the differential amplifier will be amplitude modulated (shown as a square wave in Fig. 12.8 at the amplifier input). After the amplifier, the modulated input signal is amplified, and the dynamic offset such as $1/f$ noise, denoted by red dotted line, is added to the output. Then the "modulated and amplified" input signal, plus the noise, are demodulated at the frequency f_{chop}. At this stage, the signal is demodulated (with an image at $2 \times f_{chop}$), and the noise is "up modulated" at f_{chop}. After filtering out the higher frequency component, clean amplified signal is obtained. This is just as in amplitude modulation/demodulation.

Figure 12.9 shows the same in frequency domain. Again, the signal component is denoted by blue (and the gray shade in spectrum), and the aggressors are in red. Since the demodulation leaves harmonics at higher frequencies, careful design of LPF is needed. f_{chop} should be higher than the $1/f$ corner frequency, in order to sufficiently mitigate it. One may think that choosing higher f_{chop} will ease the filter specification, because Fig. 12.9 implies having baseband as far away from f_{chop} will enable using lower order filters. However, doing so will also introduce more frequent chopping induced spikes due to charge injection, which in turn translates to a residual offset. Therefore, balancing the filtering specification and chopping frequency becomes a crucial design choice! It should be also noted that the choosing f_{chop} is also dependent on the $1/f$ corner frequency and the Gain Bandwidth (GBW) of the amplifier—the amplifier BW should be sufficiently higher than the f_{chop} to avoid gain errors.

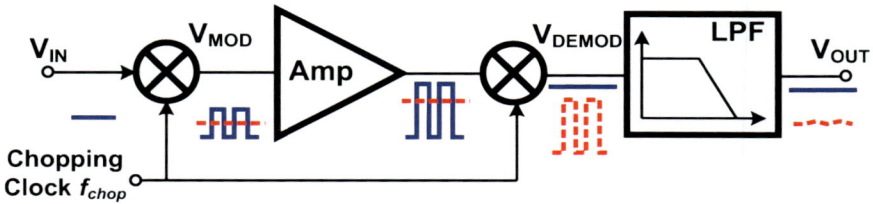

Fig. 12.8 chopper-stabilization (time domain)

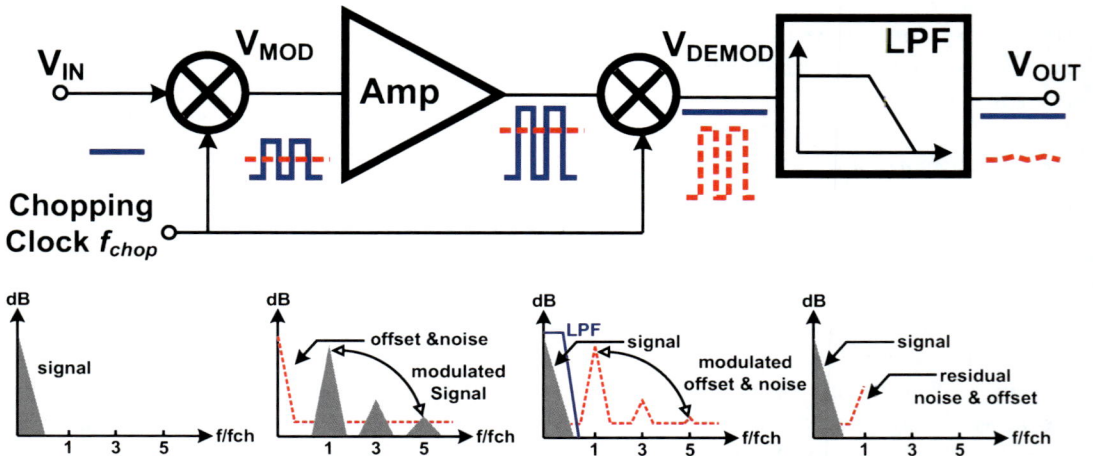

Fig. 12.9 Chopper-stabilization (frequency domain)

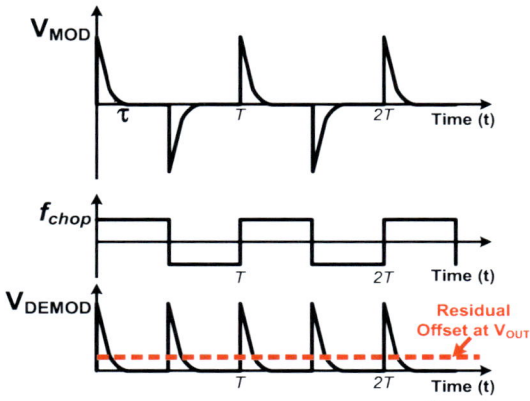

Fig. 12.10 Chopper-induced residual offset (figure from Yan et al. (2010))

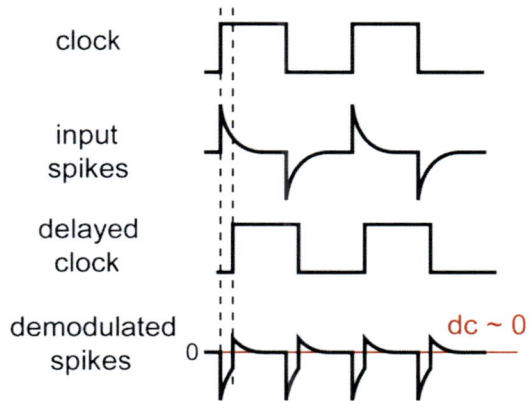

Fig. 12.12 Delayed sampling (figure from Makinwa et al. (2007))

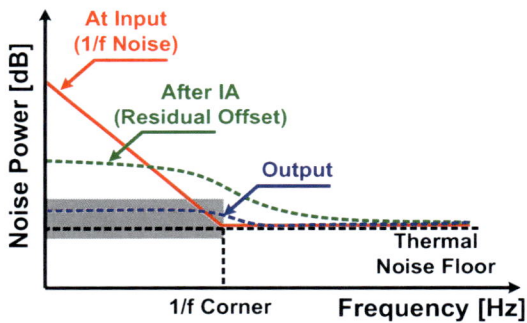

Fig. 12.11 Noise power in spectrum

Figures 12.10 and 12.11 denotes the chopper-induced spikes and its residual offset. An ideal chopper switch would not induce such issues; however, in real world, due to the parasitic capacitance between the gate and source/drain, clock feedthrough and the chopper induced spikes will show up at the chopper output, as in V_{MOD}. After the demodulation and the LPF, these spikes will show up as a DC offset—in this text, we denote this as the "residual DC offset". Of course, the higher the f_{chop}, the more frequent the spikes happen, and the more residual DC offset we will get.

The residual DC offset will incur the distortion and gain error, hence we need to mitigate it. First of all, we should remember the source of

such spikes are from parasitic capacitance between gate and drain/source; hence, using smaller switch will reduce the parasitic capacitance, consequently reduce the spike strength. However, using smaller switch means the on resistance of the switch becomes higher, which maybe an issue for a given application.

To mitigate the residual offset in more active way, delayed sampling (Menolfi et al. 1999) can be used. Shown in Fig. 12.12, a demodulator matched to a modulator is operated by a "delayed clock" with respect to the modulator clock. The amount of delay is controlled in a such way that the demodulated spikes, when averaged out, will have DC offset of nearly zero. Controlling the delayed timing is difficult, however, due to PVT variation and load-dependent spike amount.

Another technique is nested chopping (Bakker et al. 2000), as shown in Fig. 12.13. The inner chopper, operating at f_{CHOP_H}, performs the $1/f$ and offset mitigation. This will generate chopper-induced spikes. The outer chopper, operating at f_{CHOP_L}, embraces the spikes to alternate and swap the group of spikes, as shown in the figure. This will effectively average out the spikes, thereby reducing the residual DC offset. We should keep in mind, however, that the nested chopping will reduce the bandwidth considerably, because now the signal bandwidth is bound by f_{CHOP_L}, but not by f_{CHOP_H}.

Fig. 12.13 Nested chopping (figure from Yan et al. (2010))

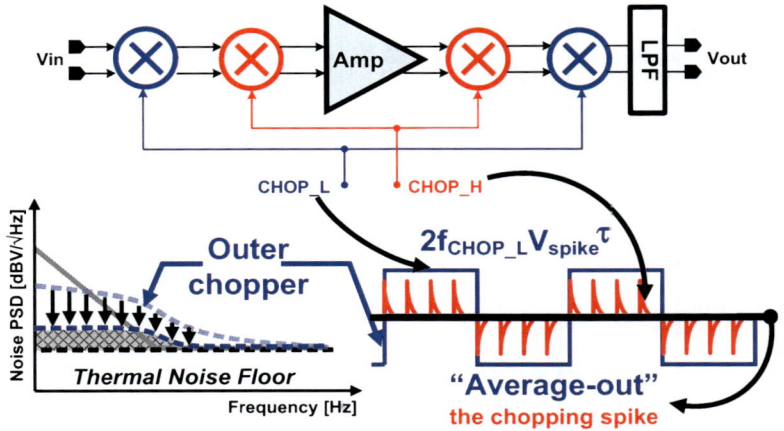

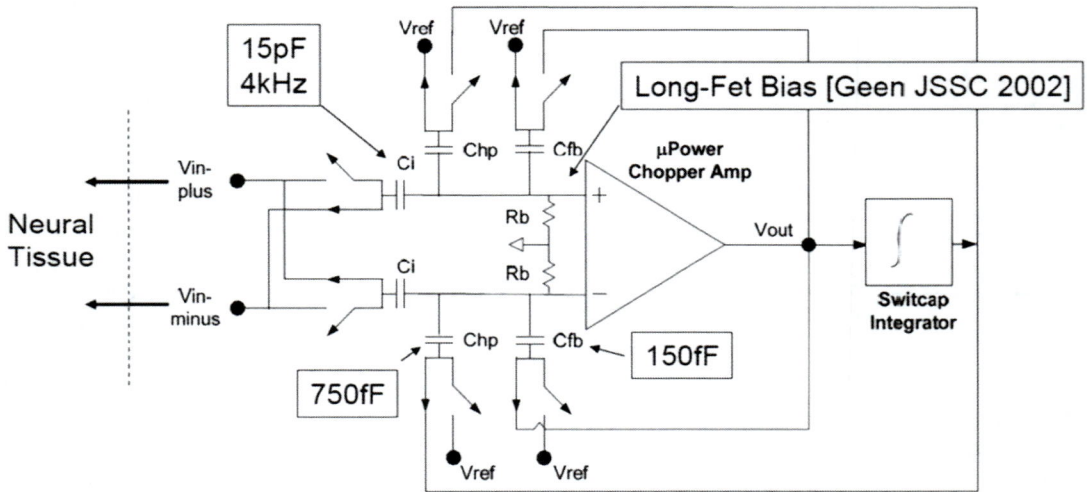

Fig. 12.14 Chopper-Stabilized Capacitive-Coupled IA (CS-CCIA) with chopper switch moved before the DC blocking capacitance (Denison et al. 2007)

12.3.4 State-of-the-Art IA Design

In this subsection, we will explorer several state-of-the-art IA design, adopting low-noise, low-power and low-energy topologies. Pros and cons of each design is also presented.

12.3.4.1 Chopper-Stabilized Capacitive-Coupled IA (CS-CCIA) with Chopper Ahead of Input Capacitor: High CMRR, Low Noise

The amplifier shown in Fig. 12.14 has the input chopper prior to DC the input capacitor (C_i)

(Denison et al. 2007). By doing so, any mismatch on C_i can be modulated and mitigated. With the capacitive-coupled structure with capacitive gain element, the amplifier has very high CMRR (100 dB), low integrated input referred noise (0.98 μV_{rms} for 0.5–100 Hz), and well-defined gain with excellent high-pass corner is obtained.

One limitation of the strcuture, though, is the exitance of chopper at the input stage drops the differential input impedance ($Z_{in,diff}$) of the amplifier; the parasitic capacitance within the chopper switch, combined with high frequency switching, degrades the $Z_{in,diff}$ significantly. As a result, the amplifier has only $Z_{in,diff}$ of 8 MΩ.

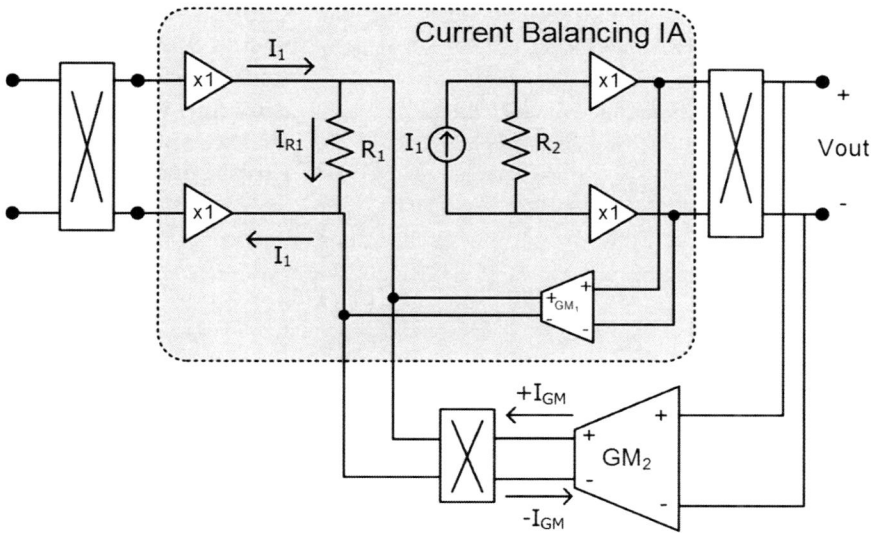

Fig. 12.15 Chopper-stabilized resistive IA with DC servo loop

This is perfectly okay in the application of the amplifier (Denison et al. 2007), which is an implantable device; however, for other IoT applications, this may not be suitable due to loading effect from input side.

12.3.4.2 Chopper-Stabilized Resistive IA: DC Offset Removal with a Servo Loop

Figure 12.15 shows a chopper-stabilized IA that uses resistive gain element (Yazicioglu et al. 2011). It introduces a IA offset reduction loop (GM1) to reduce the IA offset. Additionally, it samples the output DC level with a DC servo loop (GM2), which provides a negative feedback to stabilize and mitigate the DC offset. This smart approach mitigates the chopper induced offset as well as the DC offset caused by mismatch at the sensor interface (in this example, DC offset mainly due to electrode mismatch). As a result, the amplifier shows excellent low noise performance (noise floor of 60 nV/$\sqrt{\text{Hz}}$), high input impedance (>100 MΩ), well defined gain and a DC offset removal.

One downside of the structure is the limited DC headroom by the current-based DC servo loop; in some IoT applications, where rail-to-

rail input is required, this structure may impose major limitation.

12.3.4.3 Chopper-Stabilized Capacitive-Coupled IA (CS-CCIA) with Chopping at Virtual Ground: High Input Impedance

To improve the limited $Z_{in,diff}$ of a CS-CCIA, a groundbreaking structure was proposed in (Verma et al. 2010). Shown in Fig. 12.16, the amplifier moves the chopper to the input virtual ground, which boosts the $Z_{in,diff}$ to a very high value. Also note that now the $Z_{in,diff}$ will include the impedance of the capacitors C_{IN} as well. Therefore, this structure is free from loading effect, and ideal for many IoT applications. Also, using capacitive gain component means it has well defined gain of C_{IN}/C_{FB}. Additionally, the bias resistors are implemented by switched capacitors with high resistance (15 GΩ) to save power. The amplifier shows excellent low noise performance (noise floor of 60 nV/$\sqrt{\text{Hz}}$), high input impedance (>700 MΩ), well defined gain and a rail-to-rail input.

The only downside of the strcuture, though, is that the off-chip capacitor C_{IN}. Since C_{IN} are outside the chopper modulation, any mismatch in the devices will affect the CMRR; as we discussed in Sub-section 12.1 and in Eq. (12.2), the offchip device typically has worse matching than the on-chip device has. As a result, the CMRR of the overall system in (Verma et al. 2010) is limited at 60 dB.

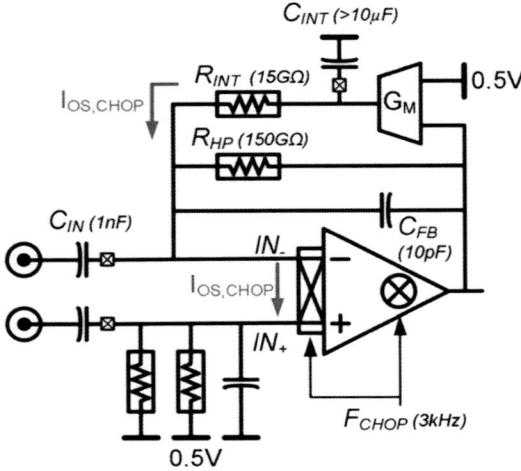

Fig. 12.16 chopping at virtual ground to improve input impedance

12.3.4.4 Chopper-Stabilized Capacitive-Coupled IA (CS-CCIA) with Active Filtering, Impedance Boosting Loop and a Ripple Reduction Loop: Low Noise, High Input Impedance and Small Area

In order to resolve the limited $Z_{in,diff}$ of a CS-CCIA with chopper at input stage, the CS-CCIA in Fig. 12.17 introduces a ground-breaking concept of using positive feedback impedance boosting loop [FSH14]. The amplifier samples the output and forms a positive feedback loop (PFL), which aids the input signal current. Hence, the amount of current needed to achieve desired output swing is dropped, and $Z_{in,diff}$ is effectively boosted. It inherits all the features of CS-CCIA, so it has extremely low-power (1.8 μW), low-noise (60 nV/√Hz) operation, and therefore ideal for many IoT applications.

To further suit for an area-constrained environment, a ripple-reduction loop can be used (Wu et al. 2009). As shown in Fig. 12.18, the amplifier samples the output of the CS-CCIA (with the chopper-induced ripple denoted as $V_{out,ripple}$). It is then passed through C_4, CH_6 then to $Gm6/C_{int}$ to a residual DC offset. This is

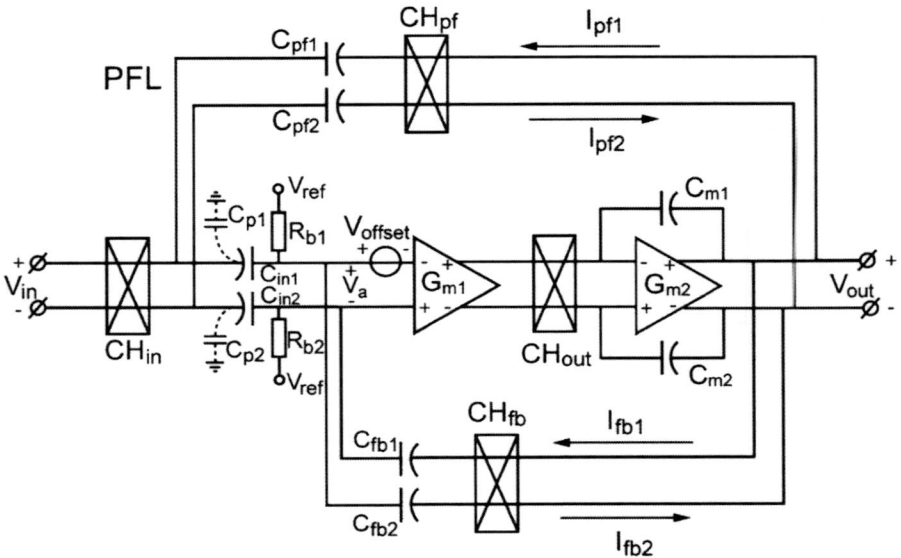

Fig. 12.17 Improving $Z_{in,diff}$ with a positive feedback impedance boosting loop

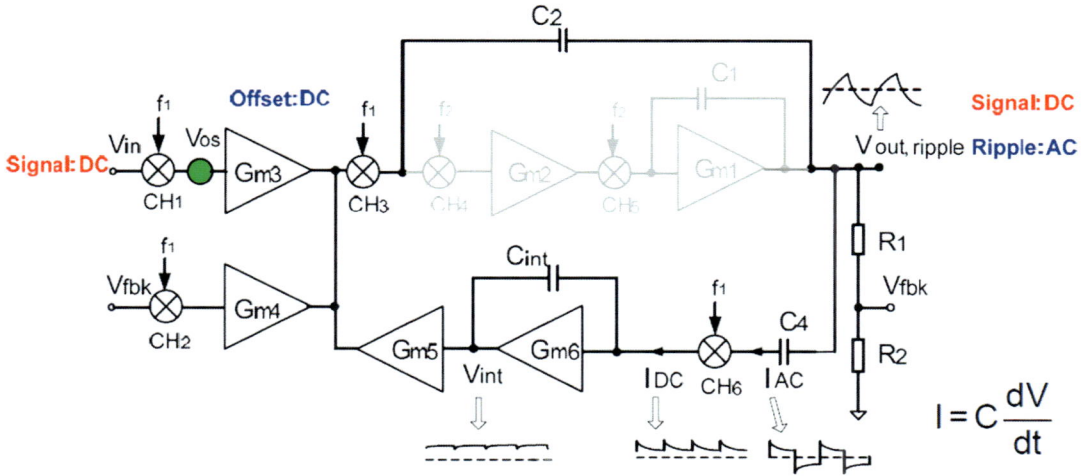

Fig. 12.18 Reducing the chopper-induced ripple with ripple-reduction loop, removes the area-consuming high-order filters

then given as a negative feedback at the input of the amplifier, which effectively removes the residual DC offset at the output stage. With this scheme, use of high-order passive filter can be removed, and significant area reduction can be achieved. This scheme is particularly helpful in IoT where stringent power and area filtering is needed.

12.3.4.5 Chopper-Stabilized Capacitive-Coupled IA (CS-CCIA) with An Additional Chopper at DC Servo Loop: Lower Noise, Small Area and High Input Impedance

As we have seen so far, using DC servo loop to remove DC offset is an effective means to remove DC offset (Fan et al. 2011; Verma et al. 2010; Yoo et al. 2013), but an issue with such structure is that the DC servo loop feedback is applied to an "up modulated" signal; this means, even after the demodulation, the DC servo loop-induced noise cannot be removed. To solve this issue, along with dynamic offset cancellation, the amplifier shown in Fig. 12.19 uses another chopper (operating at lower frequency f_L) around the DC servo loop to effectively remove the elevated noise (Altaf and Yoo 2016). Unlike the nested chopper case, this additional chopper does not affect the main amplifier signal bandwidth, since

the DC servo loop is primarily used to remove the DC component, and hence the passband of the loop itself is can be extremely low.

Additionally, the amplifier has a ripple reduction loop and an impedance boosting loop; therefore, the amplifier achieves extremely low power (1.1 μA), low noise (0.81 μVrms for 0.5–100 Hz) while having high input impedance (>500 MΩ).

12.4 Summary and Future Perspectives

So far we have explored the challenges, requirements and design techniques of analog interface circuit for IoT. In this section, we will look into future perspectives of the IA for IoT, and summarize what we have covered so far.

12.4.1 Future Trends on IoT Interface Circuits

With more and more IoT devices being used, recent research on analog interface circuits for IoT shows direction towards power and area efficiency with system-level design consideration; here, we will see two good examples.

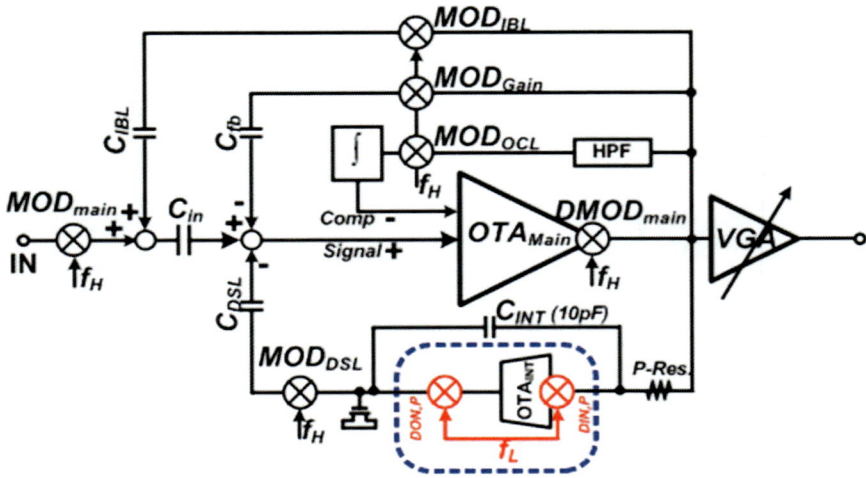

Fig. 12.19 CS-CCIA with ripple-reduction loop, impedance boosting loop and chopped DC servo loop

12.4.1.1 Bulk Switching

Chopper-stabilized IAs described in previous section may suffer from low input impedance, which causes loading effect with high-impedance signal source. We have seen the amplifiers with positive-feedback impedance boosting loop (Fan et al. 2011; Yoo et al. 2013) may mitigate the issue, but at the cost of additional control circuitry and careful design of stability test. To overcome this, a new concept of Bulk Switching was recently introduced (Han et al. 2015). Figure 12.20 shows the concept applied to a CS-CCIA.

The amplifier core (PMOS input pair) has its bulk terminal switched on and off with frequency f_s. As we can observe from the figure, the gate of the input pairs are free from chopper switch, so the input impedance of the amplifier is very high. With the bulk of the input pair switched on and off, the transistor is also turned on and off, thereby decreasing the $1/f$ noise. This smart approach solves the low input impedance issue of chopper-stabilized amplifiers with chopper switch at the input, without adding any complex circuitry. As a result, the measurement shows the amplifier has NEF of 2.2, input referred noise of 0.75 μV_{rms} over 1–200 Hz for an 100 kΩ source impedance,

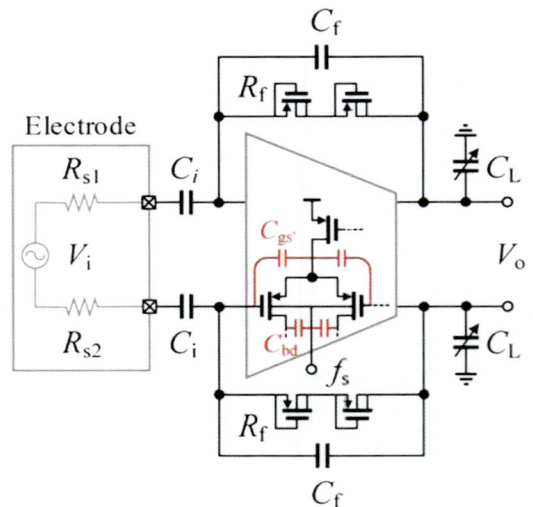

Fig. 12.20 Bulk switching CS-CCIA

while consuming only 3.96 μW from 1.2 V supply. The amplifier occupies 0.053 mm^2 in 65 nm CMOS, which is by far one of the smallest in IAs with this performance.

The bulk switching is a good example of the research directions in IA for IoT, where area- and power-efficiency are both an important design metric.

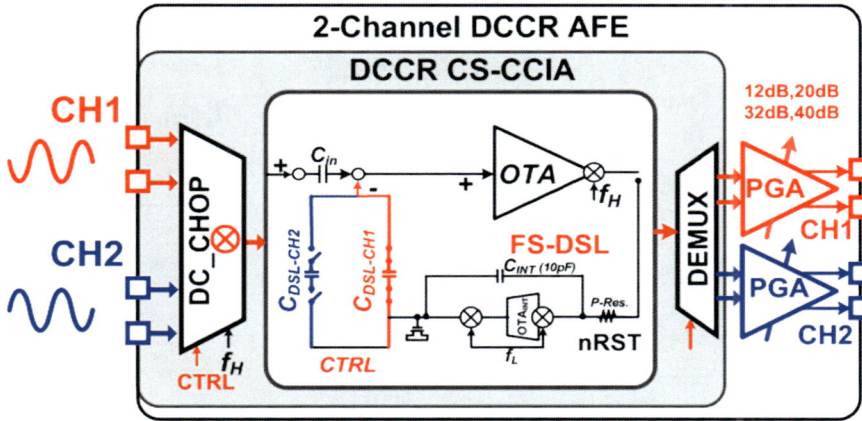

Fig. 12.21 Channel sharing: Dual Channel Charge Recycled (DCCR) CS-CCIA

12.4.1.2 Channel Sharing

An aggressive approach of saving power and area in an IA was recently proposed (Altaf et al. 2015). The Dual Channel Charge Recycled (DCCR) CS-CCIA, switches among two channel, as shown in Fig. 12.21. The idea is to have an IA shared and time-multiplexed among two separate channels.

To avoid bias settling issues, the amplifier core (OTA) samples and stores the bias point. That is, when C_{DSL_CH1} is in amplification phase, the C_{DSL_CH2} and the OTA internal bias storing capacitance hold its bias point; in the next phase, C_{DSL_CH1} holds its bias point and C_{DSL_CH2} resumes its amplification. This effectively "recycles" the bias current of OTA, thereby decreasing the power consumption significantly.

An advantage of such this method is that we do not need to have a separate circuitry to multiplex the channels. Rather, the Dual Channel Chopper (DC_CHOP) (Fig. 12.22) is employed. The DC_CHOP uses two gated clocks, which are originated from a single chopper clock, to perform chopping function and multiplexing function at once!

The implementation and measurement results (Fig. 12.23) show that there is minimal impact on integrated noise (10% increase) while amplifying two channels with a single amplifier. The input is 180-degrees out-of-phase between channels (one of the worst case for multiplexing). The improvement is very clear: 43% and 28% reduction in current consumption (per channel) and in area (per channel), respectively, when compared to using two separate IAs. As a result, it comes only 1.62 µW/channel, integrated input referred noise of 0.9 μV_{rms} (0.5–100 Hz), NEF of 3.29/channel, CMRR of 97 dB while occupying 0.716 mm^2 in 0.18 µm CMOS.

The channel sharing IA is a breakthrough concept in IA for IoT; this is another good example where system-level design consideration and ideas, balanced with the IA's circuit idea, improves the overall system performance.

12.4.2 Summary

As we have seen so far, designing an analog interface for IoT faces many challenges, especially towards the strict power, energy and area constraint. It is very important that system-level consideration is made prior to designing each component. Therefore, top-down approach is highly recommended.

IoT covers many different applications, from environmental sensing/monitoring to wearable healthcare applications, with sensing/processing/communication/storing functions. Each application has different needs, and it is crucial that the IA performance meets the proper target within the application. These metrics include system resolution, sampling rate, CMRR, power/energy consumption, noise performance, NEF and area.

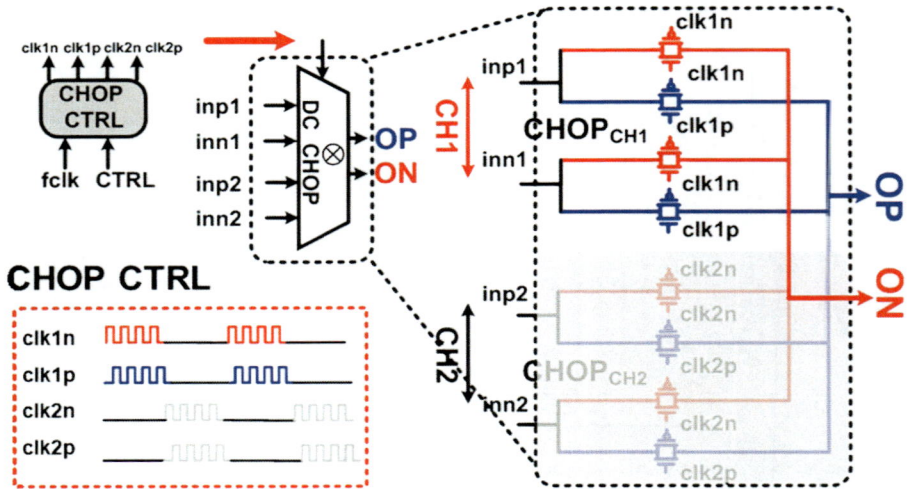

Fig. 12.22 Dual channel chopper switch

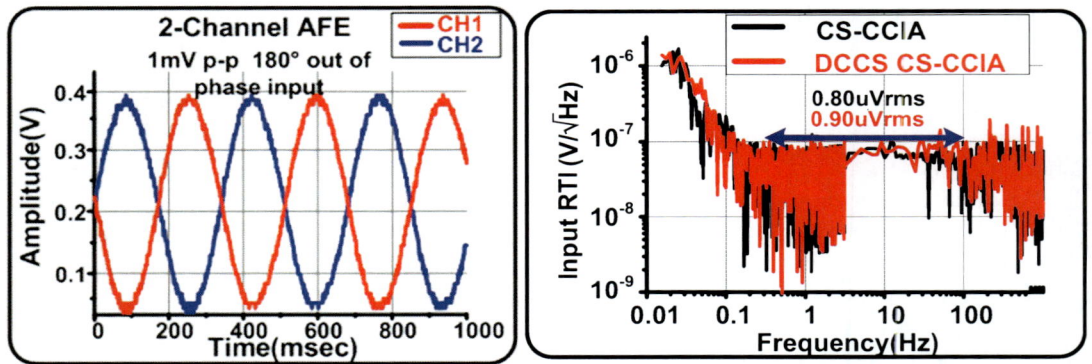

Fig. 12.23 Time-domain output of out-of-phase inputs, and the IA's input-referred noise

References

M.A.B. Altaf, J. Yoo, A 1.83µJ/classification, 8-channel, patient-specific epileptic seizure classification SoC using a non-linear support vector machine. IEEE Trans. Biomed. Circuits Syst. **10**(1), 49–60 (2016)

M.A.B. Altaf, C. Zhang, J. Yoo, A 16-channel patient-specific seizure onset and termination detection SoC with impedance-adaptive transcranial electrical stimulator. IEEE J. Solid State Circuits **50**(11), 2728–2740 (2015)

A. Bakker, K. Thiele, J.H. Huijsing, A CMOS nested chopper instrumentation amplifier with 100nV offset. IEEE J. Solid State Circuits **35**(12), 1877–1883 (2000)

M. Belleville, E. Cantatore, H. Fanet, P. Fiorini, P. Nicole, M. Pelgrom, C. Piguet, R. Hahn, C. Van Hoof, R. Vullers, M. Tartagni, Energy autonomous systems: future trends in devices, technology, and systems, CATRENE Report on Energy Autonomous Systems (2009)

T. Denison, K. Consoer, W. Santa, A.-T. Avestruz, J. Cooley, A. Kelly, A 2µW 100nV/√ Hz chopper-stabilized instrumentation amplifier for chronic measurement of neural field potentials. IEEE J. Solid State Circuits **42**(12), 2934–2945 (2007)

C.C. Enz, G.C. Temes, Circuit techniques for reducing the effects of op-amp imperfections: autozeroing, correlated double sampling, and chopper stabilization. Proc. IEEE **84**(11), 1584–1614 (1996)

Q. Fan, F. Sebastiano, J.H. Huijsing, K.A.A. Makinwa, A 1.8µW 60nV/sqrtHz capacitively-coupled chopper instrumentation amplifier in 65nm CMOS for wireless sensor nodes. IEEE J. Solid State Circuits **46**(7), 1534–1543 (2011)

M. Han, B. Ki, Y.-A. Chen, H. Lee, S.-H. Park, E. Cheong, J. Hong, G. Han, Y. Chae, Bulk switching instrumentation amplifier for a high-impedance source in neural signal recording. IEEE Trans. Circuits Syst. Exp. Briefs **62**(2), 194–198 (2015)

R.R. Harrison, C. Charles, A low-power low-noise CMOS amplifier for neural recording applications. IEEE J. Solid State Circuits **38**(6), 958–965 (2003)

K.A.A. Makinwa, Dynamic-offset cancellation techniques in CMOS, in *IEEE International Solid-State Circuits Conference*, Tutorial-04 (2007)

C. Menolfi, Q. Huang, A chopper modulated instrumentation amplifier with first-order low-pass filter and delayed modulation scheme, in *Proceedings of the IEEE European Solid-State Circuits Conference* (1999), pp. 54–57

M. Steyaert, W. Sansen, C. Zhongyuan, A micropower low-noise monolithic instrumentation amplifier for medical purposes. IEEE J. Solid State Circuits **SC-22**, 1163–1168 (1987)

N. Verma, A. Shoeb, J. Bohorquez, J. Dawson, J. Guttag, A.P. Chandrakasan, A micro-power EEG acquisition SoC with integrated feature extraction processor for a chronic seizure detection system. IEEE J. Solid State Circuits **45**(4), 804–816 (2010)

R. Wu, K.A.A. Makinwa, J.H. Huijsing, A chopper current-feedback instrumentation amplifier with a 1mHz $1/f$ noise corner and an AC-coupled ripple reduction loop. IEEE J. Solid State Circuits **44**(12), 3232–3243 (2009)

L. Yan, J. Yoo, B. Kim, H.-J. Yoo, A 0.5-μV_{rms} 12-μW wirelessly powered patch-type healthcare sensor for wearable body sensor network. IEEE J. Solid State Circuits **45**(11), 2356–2365 (2010)

R.F. Yazicioglu, S. Kim, T. Torfs, H. Kim, C. Van Hoof, A 30μW analog signal processor ASIC for portable biopotential signal monitoring. IEEE J. Solid State Circuits **46**(1), 209–223 (2011)

J. Yoo, L. Yan, D. El-Damak, M.A.B. Altaf, A.H. Shoeb, A.P. Chandrakasan, An 8-channel scalable EEG acquisition SoC with patient-specific seizure classification and recording processor. IEEE J. Solid State Circuits **48**(1), 214–228 (2013)

J. Yoo, Design strategies for wearable sensor interface circuits: from electrodes to signal processing, in *IEEE International Solid-State Circuits Conference*, Short Course on Biomedical and Sensor Interface Circuits, 13 Feb (2014)

Ultra-Low Power Analog-Digital Converters for IoT

13

Pieter Harpe

This chapter addresses ADCs for IoT nodes, which are needed to digitize sensor information before processing, storage or wireless transmission. ADCs are also required for the radio communication channel. This chapter focusses on successive approximation (SAR) ADCs, a popular architecture for IoT thanks to their high power-efficiency. After deriving requirements for IoT, the design basics of SAR ADCs are discussed, followed by various design examples to illustrate key enabling techniques.

13.1 ADC Requirements for IoT

This section first reviews basic ADC definitions, and then discusses ADC requirements in the context of IoT nodes. State-of-the-art ADCs are highlighted and the SAR architecture is selected for further elaboration in this chapter thanks to its suitability for IoT.

The function of an Analog-to-Digital Converter (ADC) is to convert analog information into the digital domain. The three most important performance metrics of ADCs are accuracy, speed and power consumption. While the ADC's accuracy can be expressed in different ways, a common way is to use SNDR (Signal-to-Noise-and-Distortion-Ratio), i.e., the ratio between signal power and power caused by all forms of noise and distortion (Pelgrom 2017). The SNDR can also be recalculated to ENOB (Effective-Number-Of-Bits) with equation (13.1). Note that an ideal Nyquist-rate ADC with N-bit output codes will have an ENOB equal to N. From this, the maximum theoretical SNDR of such an ADC can be determined as well. In reality, ENOB and SNDR will be lower than this theoretical upper bound.

$$SNDR = 6.02 \cdot ENOB + 1.76 \ [\text{dB}] \quad (13.1)$$

In terms of speed, two metrics are relevant: the sample rate f_s of the ADC and the signal bandwidth (BW) that can be converted. For a Nyquist rate ADC, f_s will be twice the value of BW to satisfy the Nyquist criterion. Oversampled ADCs have an f_s that is much higher than BW to allow for instance noise-shaping techniques. In this chapter, $f_{s,Nyq}$ denotes the equivalent Nyquist sample-rate of an ADC, which is thus twice BW. The last performance parameter of an ADC is its power consumption, here denoted with P. Combining the above three metrics, different Figure-Of-Merits can be calculated to evaluate the overall performance with a single number. Two commonly used FOMs are the FOMW and FOMS as shown in equations (13.2) and (13.3), respectively. Both FOMs show the trade-off between power, accuracy, and speed, albeit with different weighting functions.

P. Harpe (✉)
Eindhoven University of Technology, Eindhoven, Netherlands
e-mail: p.j.a.harpe@tue.nl

© Springer International Publishing AG 2017
M. Alioto (ed.), *Enabling the Internet of Things*, DOI 10.1007/978-3-319-51482-6_13

361

Fig. 13.1 ADCs in IoT nodes: wireless frontends and sensor interfaces

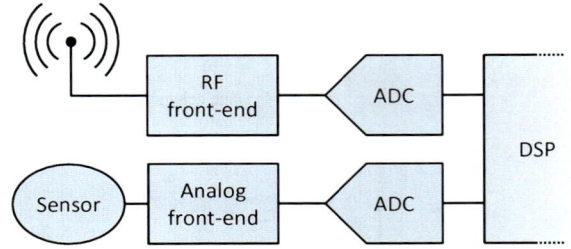

$$FOMW = \frac{P}{f_{s,Nyq} \cdot 2^{ENOB}} \quad [\text{J/conversion-step}]$$

$$(13.2)$$

$$FOMS = SNDR + 10 \cdot \log\left(\frac{f_{s,Nyq}}{2P}\right) [\text{dB}] \quad (13.3)$$

Inside IoT nodes, data converters are generally found in two locations (Fig. 13.1): inside the baseband of the wireless receiver and as the final stage of the sensor interface. Due to energy constraints, ADCs in early-generation IoT nodes are typically optimized for low power consumption by reducing accuracy and speed down to the minimum acceptable performance. For instance, 8/9-bit ADCs with up to a few MS/s sample rate are used in low-power proprietary standard ISM-band radios (Vidojkovic et al. 2011) or in standard compliant BLE (Bluetooth Low Energy) radios (Liu et al. 2015; Prummel et al. 2015; Sano et al. 2015). For sensor interfaces, traditional high-precision applications typically use ADCs with 12 to 16-bit resolution (Makinwa et al. 2015), but IoT systems have sacrificed precision to achieve power reduction. For instance, 7 to 10-bit performance was proposed in systems used for IoT and wearable sensing applications (Harpe et al. 2015; Yip et al. 2011). With the progress of technology, a trend towards higher resolutions like 10 to 12-bit can be seen more recently (Konijnenburg et al. 2016). The required speed of IoT ADCs strongly depends on the application. In wireless communication, this depends on the signal bandwidth which is typically in the kHz or low-MHz range, requiring up to a few MS/s of sample rate from

the ADC. For sensors, some applications require only quasi-static conversion, for instance in case of temperature, pressure, light, or humidity monitoring. Many other sensors have relatively low speed of operation and require ADCs up to a few kS/s, for instance in case of bio-potential recording, accelerometers, or gyroscopes. Lastly, some particular sensors, such as image sensors, could require a very high conversion bandwidth.

In summary, most IoT applications can be covered with ADCs with relatively low speed (DC up to a few MS/s) and moderate resolution (ENOB around 7 to 12-bit). With today's CMOS technology, such speed of operation is straightforward to achieve, and hence the focus in design can be mostly on power efficiency.

Figure 13.2 gives an overview of state-of-the-art ADCs (1997–2016, (Murmann 2016)) in terms of efficiency (energy consumption per sample) versus SNDR. It can be seen that for the resolutions of interest, Successive Approximation Register (SAR) ADCs are the most power efficient solution. Therefore, the remainder of this chapter focusses on SAR ADCs.

As stated before, IoT nodes have minimized ADC accuracy and speed to save power. However, this could for instance reduce the interference tolerance of the radio, or it may give suboptimal sensor performance. For those reasons, higher performance ADCs are demanded. Fortunately, technology scaling and design techniques enable this demand. As shown in Fig. 13.3, it can be observed that the power-efficiency of ADCs has been improving steadily over the last decades, reaching more than a factor 1000 of power reduction in 20 years. Thanks to

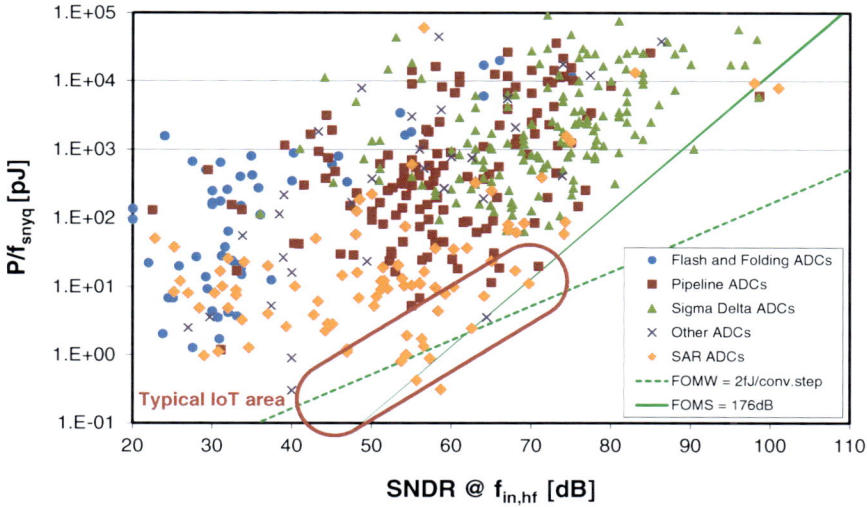

Fig. 13.2 State-of-the-art ADC power-efficiency versus SNDR, data from (Murmann 1997)

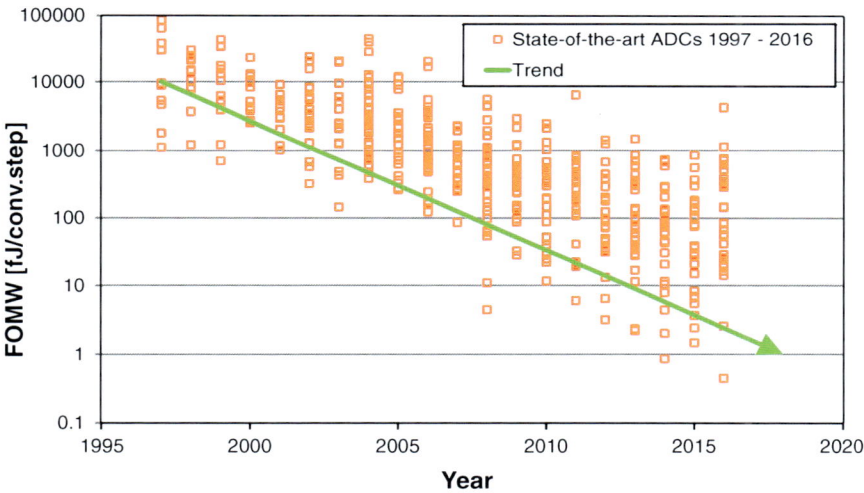

Fig. 13.3 Improvement of ADC power-efficiency over time, data from (Murmann 1997)

this trend, future IoT nodes will either benefit from lower power consumption or improved precision/speed.

Besides the critical trade-off between power, precision, and speed, that lead to the selection of the SAR architecture, there are several other aspects that are relevant in the context of IoT. First, IoT nodes are often operated on-demand, implying that the node is inactive for a long time, and only operational in short bursts. For this reason, ADCs with low standby consumption

and instantaneous operation are preferred. SAR ADCs suit well in this context, because they can be made without static bias currents, allowing automatic power-down in standby phases. Moreover, as they are Nyquist converters, a single conversion can be made instantaneously (as opposed to oversampled converters that could suffer from memory or start-up effects). Reconfigurability, i.e., the ability to operate efficiently at different speeds and resolutions, is another key parameter that is favorable for IoT.

For instance, this allows to digitize signals with a variable speed or resolution, dependent on the activity, application, or environmental conditions.

In the remainder of this chapter, Sect. 13.2 describes the general design aspects of low-power SAR ADCs. The main requirements of IoT are discussed by means of design examples, namely: low power consumption and duty cycling (Sect. 13.3), developments towards higher precision (Sect. 13.4), and reconfigurability (Sect. 13.5). Section 13.6 elaborates on voltage references that are needed for the ADC, and Sect. 13.7 discusses perspectives and trends in the field of ADCs for IoT.

13.2   Basics of SAR ADC Design

In this section, the basics of low-power SAR ADC design are discussed, addressing all major components in the system.

13.2.1   SAR Principle of Operation

Figure 13.4 shows the topology and timing diagram of a SAR ADC where an analog input voltage V_{in} is first sampled at the sample clock f_s using a Track & Hold (T&H). Next, the SAR logic performs a binary search algorithm to find the N-bit output code D_{out} that matches the sampled input voltage. This is done by comparing V_{DAC}, the voltage corresponding to the estimated output code, to the sampled voltage V_s. After N cycles of approximation, V_{DAC} will be very close to V_s, implying that D_{out} is the output code describing the sampled input V_{in}. As shown in the timing diagram, sampling takes place in a first clock cycle, after which there are N clock cycles to determine the N bits. After this, the final code is obtained and the operation repeats for the next input sample. In order to implement this ADC, a T&H, comparator, logic and DAC are needed. These components will be described in the following paragraphs.

13.2.2   Track&Hold

The basic T&H topology that is often used in SAR ADCs is shown in Fig. 13.5a. V_{in} is sampled on capacitor C_s at the sampling rate f_s. When the control signal is high, the output tracks in the input signal; when the control signal is low, the signal is held on the capacitor. Crucial for the T&H performance is the implementation of the switch. The most straightforward way is to use an NMOS (or PMOS) device as switch (Fig. 13.5b). However, assuming the control signal going to the gate of the NMOS is limited by the supply voltage, this NMOS device cannot conduct for rail-to-rail input signals: the NMOS only conducts when $V_{in} < VDD - V_{tn}$, where VDD is the supply voltage and V_{tn} the threshold voltage of the NMOS. Similarly, a PMOS only conducts when $V_{in} > |V_{tp}|$, where V_{tp} is the PMOS threshold voltage.

A conventional solution to this problem is to use a CMOS switch, where an NMOS and PMOS device are connected in parallel. In this way, as

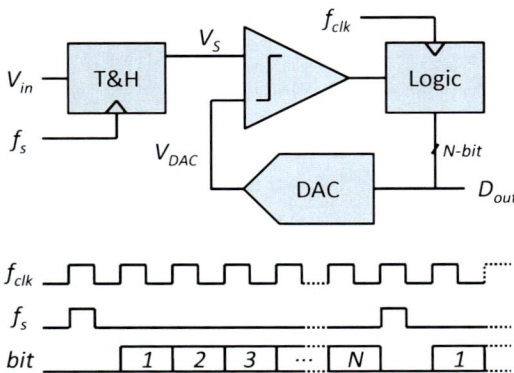

Fig. 13.4 Basic SAR ADC topology and timing diagram

(a)   **(b)**

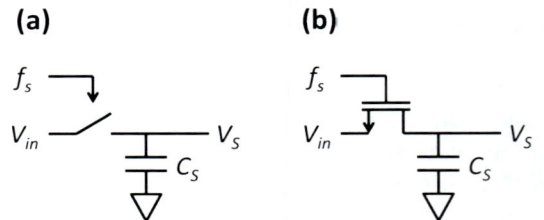

Fig. 13.5 Basic T&H topology (**a**) and T&H with NMOS switch (**b**)

long as $|V_{tp}| + V_{tn} <$ VDD, a rail-to-rail input swing can be supported. Unfortunately, VDD tends to scale down faster than the threshold voltages in advanced processes, making it more and more difficult to satisfy this requirement. Moreover, IoT nodes may prefer to use a reduced supply voltage to save power, making it even harder to reach this requirement.

Due to the above issues, techniques such as clock boosting (Cho and Gray 1995) and bootstrapping (Abo and Gray 1999) are often used. Both techniques essentially use the circuit of Fig. 13.5b with an NMOS only, but they control the gate with a voltage beyond VDD to allow a rail-to-rail input swing and a better switch conductivity. In clock boosting, a fixed voltage multiplier is used to increase the gate voltage from VDD to kVDD, where k is typically between 1.5 and 2x of gain. In bootstrapping, a level-shifter is used to lift the gate voltage to VDD + V_{in}, making the gate-source voltage always equal to VDD. A limitation of boosting and bootstrapped techniques is that they rely on a charge pump. For ADCs operating at extremely low speed or on-demand (which could be the case in IoT nodes), these charge pumps could fail to operate properly due to charge leakage. In such cases, a CMOS switch could be preferable.

Having discussed the basic topologies of the T&H, the most important imperfections are discussed next, namely: sampling noise, on-resistance, charge injection, leakage and capacitive coupling.

At the moment the switch samples the input voltage on capacitor C_s, the thermal noise produced by the transistor will also be sampled on the capacitor. This leads to an output noise power P_{nth} of kT/C_s, where k is the Boltzmann constant and T the temperature. For ADCs with higher SNDR, a lower P_{nth} can be tolerated, implying that they will need a larger C_s. Typical low-power SAR ADCs can have a C_s in the order of 0.3–1 pF (Harpe et al. 2013, 2015).

Two important imperfections causing distortion are the finite on-resistance and the charge injection of the sampling switch (Pelgrom 2017). However, as the speed and resolution requirements

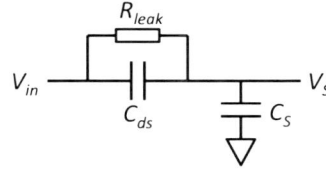

Fig. 13.6 T&H leakage and capacitive coupling in the hold mode

are relatively mild in IoT nodes, boosted or bootstrapped switches can usually achieve sufficient linearity. In some cases, a CMOS switch might also be adequate. Only if the supply voltage is scaled down too much, the above problems become a limiting factor. As example of what is feasible, the T&H in Harpe et al. (2014) achieves a linearity well above 80 dB for a signal bandwidth of 16 kHz using a 0.8 V supply, a 65 nm technology and a clock-boosted switch.

The last two problems, leakage and capacitive coupling, occur when the T&H is in the hold mode. Ideally there should an infinite impedance between the input and output node of the switch during hold mode. However, there might be drain-to-source leakage from the transistor (modeled by a resistance R_{leak}) and capacitive coupling from the layout (modeled by C_{ds}), as shown in Fig. 13.6. As a result, the output voltage is not completely isolated from the input signal during the conversion, which can cause distortion in the output code (Harpe et al. 2011). In particular in advanced CMOS nodes, it can be important to simulate these effects to verify the impact on the performance.

13.2.3 DAC

In this subsection, the design of the feedback DAC is discussed. While there are various possible implementations, the most typical solution is a capacitive charge-redistribution DAC, using voltage-mode operation. In the context of IoT, one important advantage of a switched-capacitor DAC is that its power consumption is fully dynamic. It thus scales intrinsically with the sampling rate and it allows on-demand operation and standby, simply by enabling/disabling the control

Fig. 13.7 3-bit charge-redistribution DAC (**a**) and switch simplification (**b**)

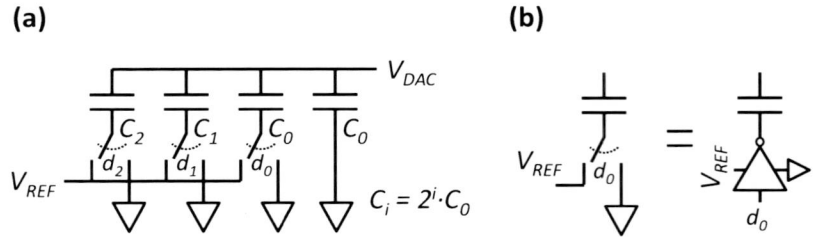

signals. Figure 13.7a shows a single-ended 3-bit example of such a DAC. It is composed of binary-scaled capacitors C_i, controlled by a digital binary code $d_{2:0}$. The digital control can switch the bottom plate of each capacitor between various reference voltages, in this example GND and V_{REF}. If code d_i switches capacitor C_i from GND to V_{REF}, an output voltage step ΔV_{DAC} of $C_i/C_S \cdot V_{REF}$ is induced, where C_s is the sum of all capacitors ($8C_i$ in this example). Based on superposition, the overall DAC voltage is hence proportional to the value of the binary code:

$$V_{DAC} = V_{REF} \frac{C_0}{C_S} \sum_{i=0}^{N-1} d_i 2^i [\text{V}] \qquad (13.4)$$

Note that the DAC is shown here as a stand-alone component. However, as will be explained in Sect. 13.3, the DAC often acts as sampling capacitance of the T&H as well, such that a dedicated C_S (as in Fig. 13.5) is not needed anymore.

The above circuit shows only one example of a 3-bit DAC that creates a transfer function as in equation (13.4). In reality, different topologies with different switching schemes can achieve the same transfer function while saving power or resources by using for instance multiple references, semi-differential switching schemes, split capacitors, or charge recycling schemes. Without being complete, a few examples are: (Ginsburg et al. 2006; Hariprasath et al. 2010; Liu et al. 2010; Liou et al. 2013; Liu et al. 2016b; Tai et al. 2014; Zhu et al. 2010). These schemes can save a substantial amount of power, and are thus very relevant to IoT nodes.

In terms of reference voltages, a trend is to use only voltages close (or equal) to the supply and

ground levels, as this simplifies the design of the switches in Fig. 13.7a. At reduced supplies it is particularly difficult to make a well-conducting switch for signal levels in the middle between supply and ground. On the other hand, the ground level can be easily switched with an NMOS while the supply level can be easily switched with a PMOS device. By doing so, the switching network simplifies to a digital-style inverter, where the control signal is applied to the input of the inverter, and the reference voltage V_{REF} (often equal to VDD) is the supply of the inverter (Fig. 13.7b).

The selection of the value of the unit capacitor (C_0), implying a total DAC capacitance C_S equal to $2^N \cdot C_0$, is the most critical decision in the DAC design process. In the following, the various considerations are discussed.

In terms of noise, the DAC has several contributions. First, when the DAC is reused as sampling capacitor, it will exhibit kT/C-noise as explained in Sect. 13.2.2. Furthermore, noise from the reference voltages and noise from the DAC switches contribute additional noise to the DAC's output. In terms of linearity, the ideal transfer function is given in equation (13.4), but this is only valid if the capacitors C_i are perfectly binary scaled. In reality, each capacitor element C_0 will experience a random mismatch σ_0, causing the function to become non-linear and thus introducing INL/DNL errors and loss of SNDR. Generally speaking, a larger value of C_S is required to suppress noise and mismatch.

As opposed to the above, power consumption and chip area benefit from using smaller capacitors. To first order, both power and chip area are linearly related to the total capacitance C_S, suggesting one should reduce its value as much as possible down to noise and linearity limitations. In practice, the smallest possible element can be dictated by the minimum element available from the foundry's library. A solution to that is to develop custom designed sub-fF capacitors (Harpe et al. 2011) which can save power, while having relatively good matching (given the capacitor value), and a small form-factor layout. However, the chip area will also be related to the number of interconnections and spacing requirements, which will be proportional to the number of control signals (N), or the number of elements (2^N). As such, for higher resolution ADCs, the area might be dictated by the large number of units (2^N). This can be reduced by using a split-capacitor array (Agnes et al. 2008) or by using multiple layout units (Harpe et al. 2014).

13.2.4   Comparator

In this subsection, the design of the comparator is discussed. Conceptually, in a single-ended case, the comparator compares the sampled input voltage V_S against the DAC voltage V_{DAC}, as illustrated in Fig. 13.8a. Comparators in SAR ADCs are usually dynamic, i.e., they perform a comparison and a reset for each clock period f_{cmp}. Similar to the DAC, this allows convenient power scaling dependent on the sample rate. In most practical implementations, the signals V_S and V_{DAC} are differential. To avoid a 4-input comparator, V_{DAC} is usually subtracted from V_S prior to the comparator. Then, a 2-input comparator can simply decide the sign of $V_S - V_{DAC}$ as shown in Fig 13.8b to obtain the required information. An advantage of doing so is that the input common mode of the comparator can

become signal-independent and remain around mid-supply.

The most important non-idealities of the comparator are its input-referred noise, offset and decision time. Comparator noise can lead to decision errors once the input signal magnitude is similar to the noise level. For that reason, comparator noise is as important as T&H noise and quantization noise. As the comparator is often critical for the overall ADC's power consumption, the power-efficiency is a particular point of attention. Fundamentally, each 6 dB noise improvement costs a factor of four in power consumption. However, several strategies can be employed to optimize the efficiency. For instance, the dynamic topology in van Elzakker et al. (2010) achieves relatively good efficiency, which is further improved by biasing the critical devices in sub-threshold. A further enhancement of the above circuit is made in Liu et al. (2016) where a factor of two in power is saved by using both phases for amplification rather than wasting one for reset. Another approach is to minimize the supply voltage, as the power of a dynamic circuit scales with the square of the supply. Besides circuit innovations, system-level solutions such as in Harpe et al. (2012, 2013) have helped to reduce the comparator power consumption by using adaptive performance during the conversion.

A second imperfection of the comparator is its - input-referred offset. Fortunately, the comparator's offset is equivalent to an ADC input-referred offset and does not introduce distortion. For IoT applications where the signal being converted is not containing DC information, this input-referred offset might be ignored. For applications where the offset is critical, for instance an offset calibration could be performed, or a system-level chopping

Fig. 13.8 Dynamic comparator in single-ended (**a**) and differential case (**b**)

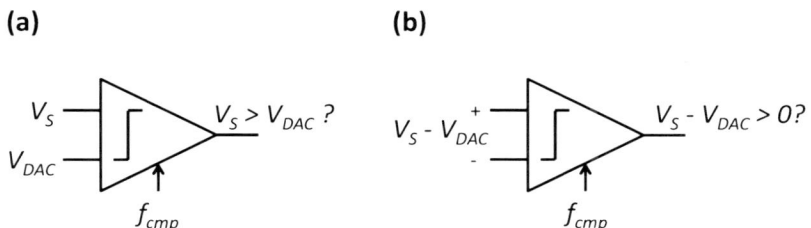

(a) $V_S > V_{DAC}$? f_{cmp}

(b) $V_S - V_{DAC} > 0$? f_{cmp}

technique could be applied to cancel the ADC's offset (Harpe et al. 2014).

The third consideration for the comparator is it speed. As soon as the comparator is triggered by the clock signal, it still takes a certain delay until the comparator reaches its decision. This delay depends on the input signal magnitude and increases for smaller input signals (Pelgrom 2017). The clock signal should thus allow sufficient time to reach a decision. Moreover, the comparator also needs a sufficiently long reset phase to reset the comparator to the initial condition. If this time is too short, the next decision can be affected by the previous one, causing signal dependency and hence non-linearity. While IoT applications are not very demanding in terms of speed, the comparator delay will increase substantially when lowering the supply voltage. Moreover, when operating in sub-threshold, the delay can also be severely affected over PVT variations, thus requiring sufficient design margin with respect to the timing. As discussed in the next paragraph, asynchronous timing can alleviate the comparator delay variation to some extent.

13.2.5 Logic

The logic in a SAR ADC is conventionally build around two strings of flip-flops (Fig. 13.9a).

A first string (or register) is acting as a thermometer counter to memorize in which phase of the conversion process the SAR ADC is. Assuming that the ADC uses $N + 1$ clock cycles (1 for tracking and N for the N-bit SAR conversion), the thermometer counter will count from 0 up to N. A second register contains the actual DAC code that will ultimately compose the ADC output code at the end of the conversion. Besides these two registers, additional combinational logic needs to be added to generate various internal signals to control the comparator, T&H and DAC, based on the state of the registers. For simplification this is not shown.

The basic operation of the registers is shown in the timing diagram of Fig. 13.9b. The external clock f_{clk}, at $N + 1$ times the sample rate f_s, drives the thermometer counter, thus creating $N + 1$ counter values from 0 to N. Combinational logic will combine f_{clk} with the counter value to generate a sampling clock f_s in the first counter cycle (counter $= 0$) and to generate N comparator clock cycles (f_{cmp}) in the other counter cycles. The incrementing counter also strobes the data register, addressing each bit one by one (starting with the MSB down to the LSB), to store the comparator output in the correct data bit. The data register drives the DAC, but some additional combinational logic might be needed, dependent on the switching scheme used in the DAC.

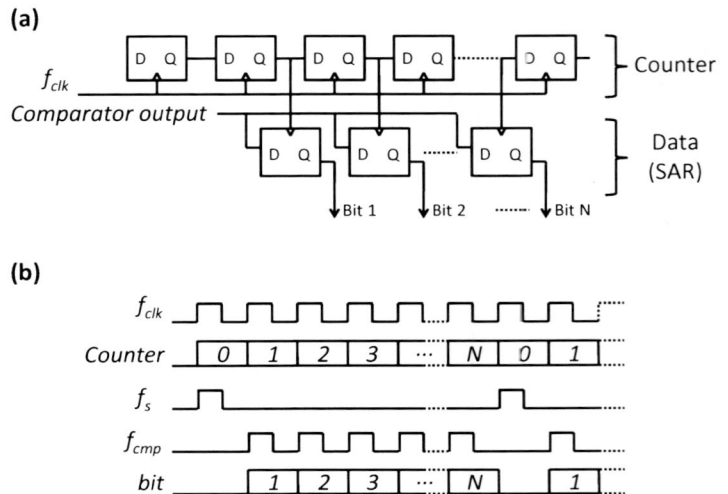

Fig. 13.9 SAR logic core (a) and timing diagram (b)

The above solution has two drawbacks. First, the logic requires an external clock which is N + 1 times faster than the actual sample rate. The higher speed could increase system-level power consumption, but it also complicates duty-cycled operation as one has to keep track of the number of cycles of f_{clk}. A second disadvantage is in terms of comparator metastability handling. As explained in Sect. 13.2.4, the comparator decision time depends on the signal magnitude applied to the comparator. Ultimately, when the signal becomes very small, the comparator is close to metastability where the decision can take a long time. During the N cycles of the SAR conversion, the input signal to the comparator will vary substantially. As a result, the comparator decision time varies significantly from cycle to cycle. However, with a fixed f_{clk} rate, the maximum f_{clk} is limited by the slowest decision time of the comparator. Since in most cases the comparator is much faster, this implies that time is wasted in the other cycles.

A solution to the above problems is to use asynchronous timing inside the SAR ADC. In this case, as sketched in Fig. 13.10, only an external clock at the sample rate f_s is provided. The comparator clock f_{cmp} is internally generated, usually by means of a feedback loop that automatically produces N comparator cycles. The actual cycle time can be made variable by waiting exactly until the comparator decision has been made. As a result, some cycles will be faster and others slower, to accommodate the timing variation of the comparator without wasting time. By doing so, the average clock cycle is now set by the average delay of the comparator rather than the worst-case delay, hence improving speed and improving the ability to deal with metastability issues. This also simplifies the system-level design, as a single clock pulse on f_s is now sufficient to perform a single (on-demand) conversion. Examples of asynchronous timing are for instance given in Chen and Brodersen (2006), Harpe et al. (2011, 2013), and detailed logic implementations are described in Harpe et al. (2011, 2012).

13.3   An Ultra-Low Power SAR ADC

This section discusses a design example of an ultra-low power SAR ADC with 10 bit resolution and a variable sampling rate from DC to 100kS/s, consuming down to 0.15 nW of power (Harpe et al. 2015, 2016). While originally developed for low-power bio-potential recording, the specifications are suitable for versatile sensing applications where especially the power consumption is critical. Thanks to nW-level operation, this ADC allows extremely small form-factor devices powered by energy harvesting or tiny batteries.

The topology of the ADC, which is rather standard, is shown in Fig. 13.11. To save power, the nominal supply is reduced to 0.6 V, well below the regular 1.2 V supply of the applied 65 nm CMOS technology. At the same time, the supply of 0.6 V is still sufficiently high such that conventional circuits can operate correctly. The sampling switches are clock-boosted to achieve sufficient linearity (Cho and Gray 1995). As mentioned earlier, the T&H has no explicit sampling capacitance, but the input signal is sampled directly on the DAC capacitor array instead. Because all circuits are dynamically biased, the power scales inherently with the sample frequency. To reduce the leakage power consumption, high threshold-voltage transistors are used.

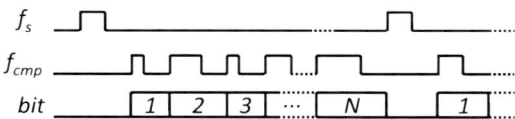

Fig. 13.10 Example of asynchronous timing

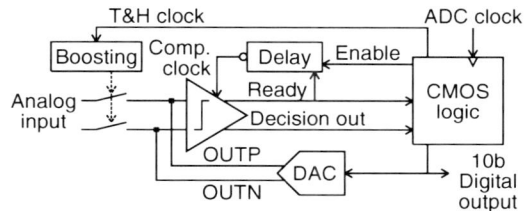

Fig. 13.11 10 bit asynchronous SAR ADC

Fig. 13.12 Asynchronous on-demand operation of the ADC

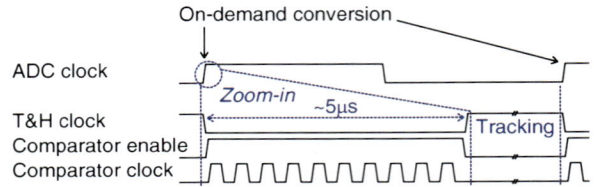

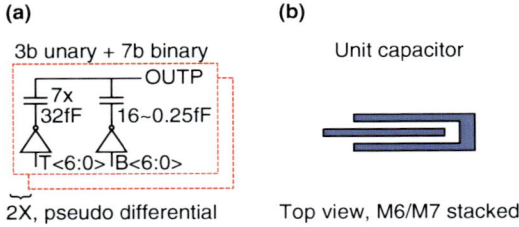

(a)

3b unary + 7b binary

2X, pseudo differential

(b)

Unit capacitor

Top view, M6/M7 stacked

Fig. 13.13 Implementation of the DAC (**a**) and the capacitor elements (**b**)

The ADC only requires a single external clock at the sample rate f_s. An internal loop around the comparator creates the clock that is needed for the comparator and logic. As shown in the timing diagram in Fig. 13.12, a rising edge of the external ADC clock initiates a complete conversion. The logic enables the feedback loop around the comparator (Fig. 13.11): this loop will start the comparator clock first. As soon as the comparator has resolved a decision, its Ready output will be engaged. By means of the inverted delay, this Ready signal will disable the comparator clock. As a result, the comparator is reset and its Ready will switch off again. As the Ready turns off, this will initiate a next comparison clock cycle. In this way, the consecutive bit cycles are generated until all 10 bits are resolved. At that point, the logic disables the feedback loop, produces an ADC output code, returns the DAC to tracking mode, and then places the ADC into standby. As the timing diagram shows, the entire conversion takes approximately 5 μs, after which the ADC returns to sleep. Since the ADC only requires a single rising clock-edge on its external clock to trigger a conversion, the ADC is very suitable for on-demand operation.

The implementation of the capacitive DAC is shown in Fig. 13.13a. The differential DAC is implemented by two identical single-ended

structures. Normally, a 10-bit binary scaled array would be needed. In this case, the 10 bit is segmented in 3 MSBs and 7 LSBs. The LSBs are binary coded and controlled by code $B_{<6:0>}$, the 3 MSBs are thermometer encoded and therefore control 7 identically sized capacitors, controlled by thermometer code $T_{<6:0>}$. Using thermometric coding has two advantages: it saves switching energy in the DAC and it reduces the maximum DNL error as the worst-case number of switching elements is reduced (Harpe et al. 2013).

To maximize the dynamic range and to simplify the ADC system integration, the DAC only uses ground and VDD as its reference voltages, as explained before in Fig. 13.7b. Because of that, the ADC requires only a single supply voltage that is used by all components.

To save power in the DAC, as well as in the (external) reference buffer and in the analog buffer driving the ADC, the capacitance of the DAC is minimized as much as possible by developing custom designed fringing capacitors. Similar to Harpe et al. (2011), this work uses capacitive elements as shown in Fig 13.13b with a unit value of 0.25fF. To improve capacitor density, two metal layers (6 and 7) are stacked. The lower layers (1 to 5) are not used, because they would increase the parasitic capacitance to the substrate, causing an attenuation of the signal range. By using these small capacitors, the total input capacitance of the DAC is only 0.3 pF, which leads to a kT/C-limited SNR of approximately 66 dB, which is still sufficient for a 10 bit ADC.

Figure 13.14 presents the die photo of the ADC, implemented in a 65 nm CMOS process and occupying 180 μm·80 μm. The chip area is mostly dominated by the capacitive DAC. However, a large part of the DAC area (~75 %) is due

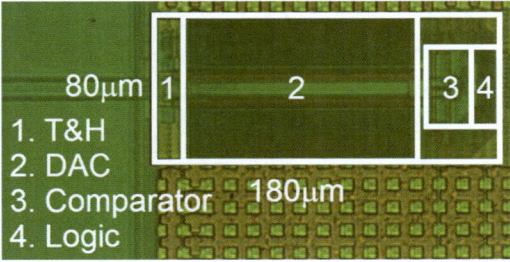

Fig. 13.14 Die photo of the ADC in 65 nm CMOS

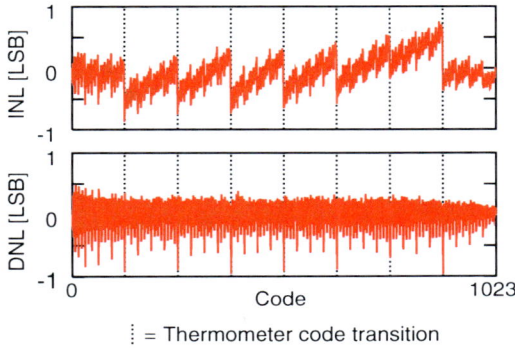

Fig. 13.16 Measured ENOB of the ADC

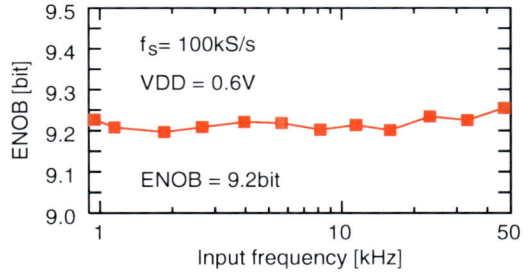

Fig. 13.15 Measured INL/DNL of the ADC

to interconnections that were not really optimized for chip area, while the capacitors only take <20 % of the total DAC area.

The precision of the ADC is verified by measuring the INL, DNL and ENOB. The INL and DNL are shown in Fig. 13.15. Both parameters remain within 1LSB. As expected, the largest DNL errors (and thus the largest discontinuities in the INL) happen at the thermometer code transitions.

The ENOB, measured as a function of the input signal frequency at a sample rate of 100 kS/s is given in Fig. 13.16. The performance is constant over the entire Nyquist zone, showing the ADC has sufficient bandwidth despite the reduced supply voltage.

Lastly, Fig. 13.17 displays the measured power consumption versus sampling rate together with the simulated power breakdown at 1 kS/s. It can be seen that the power scales proportional to the sample rate. Due to limitations of the measurement setup, the lowest frequency measured is around 0.2kS/s. A standby

leakage of 0.15 nW is measured by disabling the clock altogether and measuring the supply current. From the simulated breakdown (post-layout), it is clear that the comparator contributes most to the overall power. The DAC contribution is small thanks to the small unit capacitors, even though the DAC switching scheme was not optimized.

Table 13.1 shows a performance summary and comparison to prior-art. The efficiency is comparable to state-of-the-art, but not as good as (Tai et al. 2014). However, as can be seen from Fig. 13.2, this is still one of the few ADCs under the indicated FOMW trend line of 2 fJ/conversion-step. Another advantage of the proposed ADC is that it has the lowest leakage power, allowing to maintain power efficiency even when the sample rate is reduced to well below 1 kS/s.

This section described a 10-bit ADC with a versatile sampling rate. As shown, the architecture and implementation are relatively basic. Thanks to technology scaling, the simple architecture and circuits, the small unit capacitors, and a reduced supply voltage, this still allows to achieve state-of-the-art power efficiency. Moreover, by using only a single supply and a single clock-edge for triggering conversion, the ADC is simple to integrate and use in an IoT system.

13.4 A High-Precision SAR ADC

The previous section described a SAR ADC with 10 bit resolution. While this can be sufficient for basic sensing applications, other applications

Fig. 13.17 Measured power consumption and simulated power breakdown of the ADC

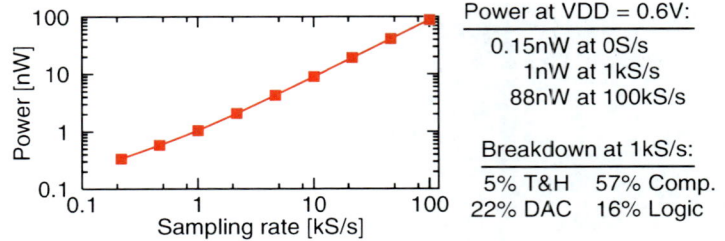

Power at VDD = 0.6V:

0.15nW at 0S/s
1nW at 1kS/s
88nW at 100kS/s

Breakdown at 1kS/s:

| 5% T&H | 57% Comp. |
| 22% DAC | 16% Logic |

Table 13.1 ADC performance summary and comparison

	This work, Harpe et al. (2015)	Zhang et al. (2012)	Harpe et al. (2013)	Tai et al. (2014)
Process [nm]	65	65	65	40
Supply [V]	0.6	0.7	0.6	0.45
Total power [nW]	88	3	72	84
Leakage power [nW]	0.15	0.67	0.4	N/A
Resolution [bit]	10	10	10	10
ENOB [bit]	9.2	9.1	9.4	8.95
Sample rate [kS/s]	100	1	40	200
FOMW [fJ/conv.step]	1.5	5.5	2.7	0.85

could demand higher precision. Therefore, this section describes a SAR ADC with a relatively high precision, selectable from 67.8 dB up to 79.1 dB of SNDR (Harpe et al. 2014).

In order to increase a SAR ADC's SNDR, the main challenge is to reduce noise and distortion contributions while maintaining power-efficiency. In this design, a combination of oversampling, chopping, dithering and data-driven noise-reduction is applied to achieve this. These techniques will be described one by one.

Oversampling is a known technique to improve the SNR (Signal-to-Noise-Ratio) in a signal band of interest by sampling faster than the Nyquist rate. Given that the sampling rate is increased with a certain oversampling ratio (OSR), the in-band noise power will be reduced by the same factor:

$$P_{n,inband} = \frac{P_{n,total}}{OSR} \qquad (13.5)$$

This implies that every factor 4 of oversampling reduces the in-band noise by 6 dB, thus improving the SNR by 6 dB at the cost of 4x higher speed and power. Oversampling allows to mitigate all random noise contributions

of a SAR ADC while maintaining a constant FOMS. However, correlated errors (such as distortion and 1/f noise) cannot be solved by oversampling.

The second technique being applied is system-level chopping. Chopping is a known technique to mitigate offset and 1/f noise problems of amplifiers, and can be applied likewise to an ADC. Figure 13.18 shows a simplified explanation of an ADC converting an analog input X to a digital code Y. As shown in the upper graph, the ADC might add signal-dependent harmonic distortion (HD(X)), offset (O) and 1/f noise (1/f). In the second graph, system-level chopping is applied to this ADC. The input signal is modulated with a chopping clock (f_c) before the conversion, and demodulated by the same clock at the output of the ADC. If f_c is set to half of f_s, this means that the system is transparent in the odd clock cycles, while the input and output signals are inverted in the even clock cycles. The equations in Fig. 13.18 show how the output codes in the two phases (Y_1 and Y_2) depend on the input and various imperfections.

Next, the harmonic distortion is separated in two components, namely the even-order and the

Fig. 13.18 System-level
chopping of an ADC

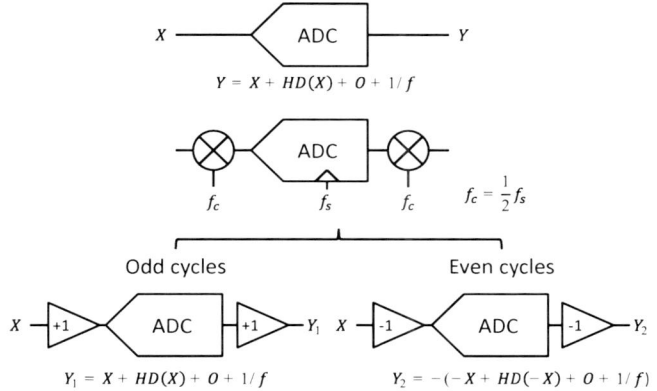

odd-order distortion components, denoted by $HD_e(X)$ and $HD_o(X)$, respectively. This means that the relations in equations (13.6) and (13.7) hold.

$$HD(X) = HD_e(X) + HD_o(X) \qquad (13.6)$$

$$\begin{cases} HD_e(-X) = HD_e(X) \\ HD_o(-X) = -HD_o(X) \end{cases} \qquad (13.7)$$

As a final step, the average value of Y_1 and Y_2 can be determined, which yields the result in equation (13.8). The result shows that chopping removes offset, 1/f noise and even-order distortion when looking to the average of codes (or in reality: when looking to the low-frequency part of the spectrum). In fact, these imperfections are not removed, but modulated to the chopping frequency. If oversampling is applied together with chopping, this implies that offset, 1/f noise and even-order distortion will be moved out of band, hence allowing to improve the SNDR of a SAR ADC.

$$\frac{Y_1 + Y_2}{2} = X + HD_o(X) \qquad (13.8)$$

The third technique used to improve linearity is dithering. It can be observed (for instance in Fig. 13.15) that the ADC distortion is often rather irregular due to capacitor mismatch, causing strong local variations. Dithering aims to smooth out these irregularities by adding an amount of dither to the input signal before conversion. In this way, the local irregularities can be averaged,

causing an improvement of local linearity. However, it should be noted that since the dither is limited in amplitude, it cannot solve global non-linearities.

Figure 13.19a shows the implementation of the ADC including the chopping and dithering techniques. It supports a native resolution of 12 or 14-bit. To integrate chopping, there are two sets of sampling switches, allowing to sample the input signal alternatingly in the normal way (φ_1) or with inverted polarity (φ_2). The second chopper, which is in the digital domain, simply needs to forward the output bits in one clock cycle, and invert the output bits in the next clock cycle.

The dithering technique is implemented inside the DAC, as illustrated in Fig. 13.19b. The actual circuit is differential, but only one half is shown here. Similar to the ADC design discussed in Sect. 13.3, the lower bits (9 down to 0) are binary encoded, while the upper 4 bits are unary encoded (requiring 15 identical-sized capacitors). The dither signal can be injected with a capacitive network connected to the DAC. In time, the input signal is first sampled on the top plates of all the capacitors. Then, the delayed sample clock will switch the dither capacitors, causing a dither value to be added to the sampled input signal. After that, the normal conversion starts, such that the ADC converts the dithered input signal. Additional details on this technique are explained in Harpe et al. (2014).

The implementation of the DAC capacitors leads to a practical challenge. Since the

(a)

(b)

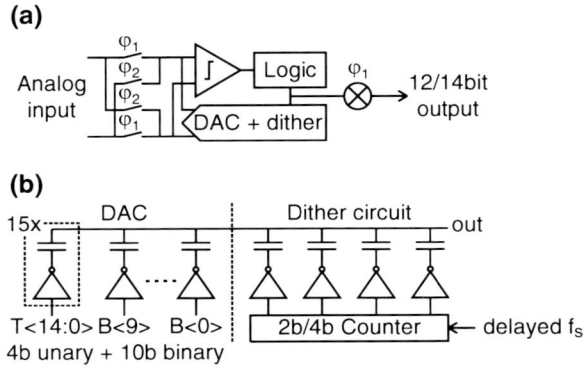

Fig. 13.19 Implementation of chopping technique (**a**) and DAC with dithering (**b**)

Table 13.2 DAC capacitor implementation

	Unit capacitor	Number of units
$T_{<14:0>}$	8.8fF	15 × 64 (960 in total)
$B_{<9:4>}$	8.8fF	32, ..., 1 (63 in total)
$B_{<3>}$	4.4fF	1
$B_{<2>}$	2.2fF	1
$B_{<1>}$	1.1fF	1
$B_{<0>}$	1.1fF	½
Total	9 pF	1027

ADC has 14-bit of resolution, it theoretically needs $2^{14} = 16,384$ capacitive elements, which is unpractical. A split capacitor array (Agnes et al. 2008) could reduce this, but may lead to additional non-linearities. In this case, the number of units is reduced by using multiple layout units, of respectively 8.8, 4.4, 2.2 and 1.1fF. Table 13.2 gives an overview of how the DAC capacitors are composed. The largest capacitors (for the thermometer bits and the binary bits 9 down to 4) are composed of 8.8fF layout elements to reduce the number of devices. The lower bits (bit 3, 2, and 1) use the down-scaled layout elements of 4.4, 2.2 and 1.1fF. Lastly, bit 0 effectively uses half a unit of 1.1fF. This is done by placing a capacitor of 1.1fF on the positive side of the DAC array, and no capacitor on the negative side of the differential topology. In this way, the effective value is 0.55fF when looking to the differential operation. The slight common-mode imbalance is irrelevant as this is very small. From Table 13.2 it is clear that the number of units is reduced to 1027 in this way,

saving almost a factor of 16 in the number of devices and thus making the layout more simple and compact. The total DAC capacitance is 9 pF, which leads to about 85 dB of kT/C-related SNR at 0.8 V supply.

The last technique applied in this design is Data-Driven Noise-Reduction (DDNR). Its goal is to improve the comparator noise level in a more efficient way than by simple analog circuit scaling which costs 4x in power for a 6 dB better SNR. This is critical in low-power ADCs, as the comparator can dominate the overall ADC power consumption (e.g., 57 % in Fig. 13.17). When looking to Fig. 13.20a, it can be observed that during the SAR conversion, the input signal magnitude applied to the comparator is often large, and only in a few cycles it will be small (in the order of an LSB). Comparator noise can cause decision errors, but this will only happen if the input signal is in the same order of magnitude as the noise. For those cycles where the input is very large, the comparator noise is in fact not critical at all. Thus, rather than using a comparator with a fixed noise level, DDNR saves power by modifying the comparator noise level on-the-fly for each cycle, dependent on the input signal magnitude: for large signals a high noise level is tolerated to save power, and for small signals the noise level is reduced to achieve better precision. In order to implement this concept, two components are needed: (1) it should be detected (within each cycle) whether the input signal magnitude is large or small; (2) the noise-level of the

(a)

Comparator input
during approximation

* = Small input magnitude

(b)

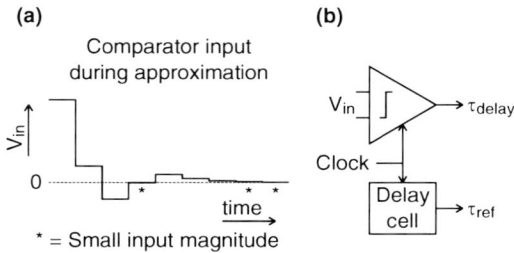

Fig. 13.20 Comparator input during conversion (**a**) and decision time monitor (**b**)

comparator should be tunable instantaneously during the conversion process.

To address the first problem, recall that the comparator decision time is related to the input signal magnitude: smaller signals lead to a longer decision time (Sect. 13.2.4). Thus, by observing the decision time, the input signal magnitude can be classified. As shown in Fig. 13.20b, a tunable delay cell is triggered together with the comparator. By comparing the delay of the comparator (τ_{delay}) against this reference delay (τ_{ref}), it can be decided whether the input signal was large or small. In practice, by tuning the reference delay cell by means of feedback, the reference delay τ_{ref} can be stabilized to a desired value regardless of PVT variations (Harpe et al. 2014).

The second problem is to tune the noise-level of the comparator dynamically. While this could be done by tuning the analog circuit, this is cumbersome and could induce new errors (such as offset variations). Therefore, a digitally-intensive solution is applied by majority voting on a repeated set of comparator decisions to enhance the effective noise level. For instance, when the same comparator decision is repeated five times, and the majority vote of those five decisions is used as final output, this effectively reduces the input-referred noise by 6 dB, as shown in the presentation of Harpe et al. (2013). In fact, this is a bit similar to oversampling, where a higher sample rate is used to reduce in-band noise.

Combining the above two components, the overall approach is as follows: each SAR cycle starts with a single comparison. If the decision time is faster than the reference delay, this decision is immediately forwarded to the SAR logic

and the SAR continues with the next bit cycle. However, if the decision was slower than the reference delay, four additional comparisons are performed on the same input signal. The majority vote on five decisions is taken and forwarded to the SAR logic. Only then will the SAR logic proceed to the next bit cycle.

A die photo of the implemented ADC is shown in Fig. 13.21. The largest portion of the indicated area is occupied by supply decoupling capacitors and the DAC. The ADC operates at a nominal supply of 0.8 V, and can work either in 12 bit or 14 bit Nyquist mode, or in oversampling modes. 12 bit resolution is simply implemented by skipping the last 2 conversion cycles and by disabling majority voting to save power. When oversampling is enabled, the chopping and dithering techniques can be turned on to improve the in-band SNDR.

Figure 13.22 shows an example of a measured output spectrum, in this case in 14 bit 16x oversampling mode with 128kS/s sampling rate. The in-band SNDR is 80.0 dB while the linearity reaches 87.5 dB and the ENOB is 13 bit. In this mode, the power breakdown (based on post-layout simulations) is as follows: DAC 51 %, comparator 40 %, logic 4 %, chopped T&H 3 %, dithering 2 %. Figure 13.23 shows the SNDR of this ADC in different modes of operation (12 bit/14 bit, Nyquist and oversampling), and an overall performance summary is given in Table 13.3. Compared to prior-art (Fig. 13.2), this work achieves state-of-the-art power-efficiency (FOMS from 173.8 to 176.8 dB) and also enables a relatively high SNDR up to 79.1 dB.

The SAR ADC presented in this section shows several simple features that enhance the SNDR while maintaining state-of-the-art power-efficiency. Also, the ADC can cover different performance settings (67.8 dB to 79.1 dB of SNDR) to allow a flexible trade-off between performance and power.

13.5 A Reconfigurable SAR ADC

The design examples of the previous sections are mostly relevant for sensor interfaces because of their sample rates in the kHz range. However,

Fig. 13.21 Die photo of
the ADC in 65 nm CMOS

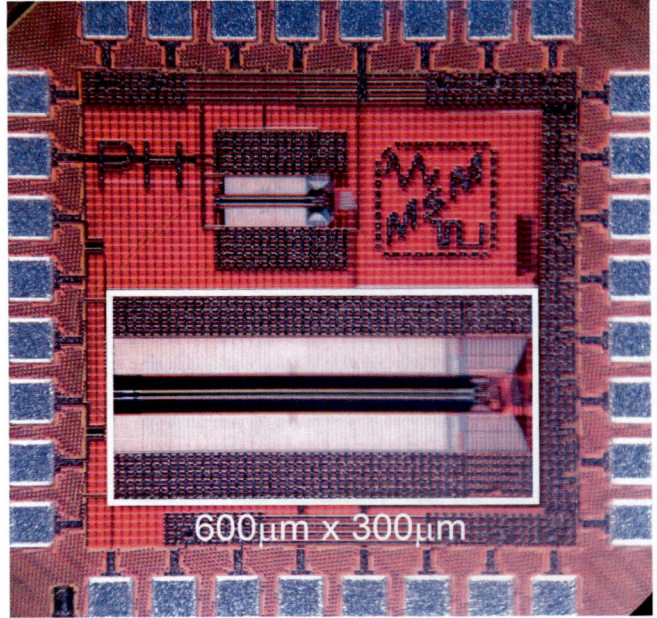

Fig. 13.22 Measured
output spectrum of the
ADC in 14 bit mode with
16x OSR

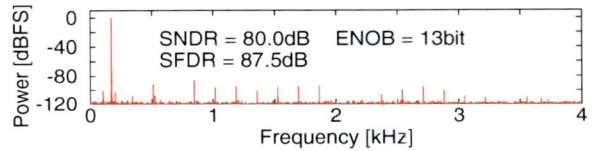

Fig. 13.23 Measured
SNDR of the ADC in
12 and 14 bit mode, with
and without oversampling

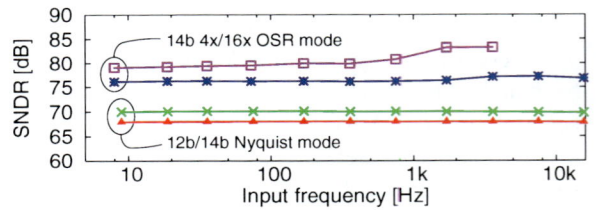

Table 13.3 ADC performance summary

Process [nm]	65			
Supply [V]	0.8			
Resolution [bit]	12	14		
Sample rate [kS/s]	32	32	128	128
OSR	–	–	4x	16x
Bandwidth [kHz]	16	16	16	4
SNDR [dB]	67.8	69.7	76.1	79.1
Power [µW]	.310	.352	1.367	1.370
FOMW [fJ/conv.step]	4.8	4.4	8.2	23.2
FOMS [dB]	174.9	176.3	176.8	173.8

similar SAR topologies can be used easily to
operate in the MHz range, allowing to re-use
the converter in low power radios. In this section,
a SAR ADC with flexible resolution (7, 8, 9, or
10 bit) and a variable speed (DC to 2 MS/s) is
described, allowing to re-use the design for either
low-power sensing or low-power narrow-band
communication (Harpe et al. 2012).

Figure 13.24 shows the architecture of the
reconfigurable SAR ADC. The basic architecture

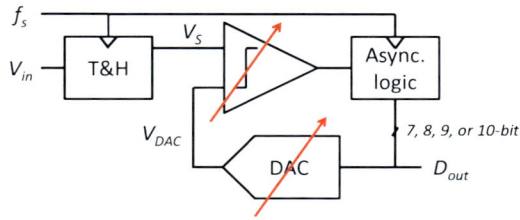

Fig. 13.24 Architecture of the reconfigurable SAR ADC

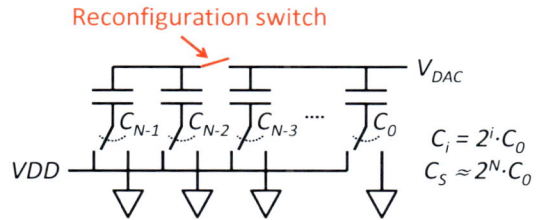

Fig. 13.25 Implementation of the resolution reconfigurable DAC

is similar to the previous design examples, using asynchronous clocking and a single supply for all components (including the DAC reference voltage). In order to make the sample rate reconfigurable, the ADC is implemented with dynamic circuits, similar to the examples discussed earlier. Therefore, the main attention in this section is on how to reconfigure the ADC resolution while maintaining power-efficiency.

For the digital logic inside a SAR ADC, the complexity and power consumption theoretically scale linear with the resolution N, as can be seen from the diagram in Fig. 13.9. To implement reconfigurable logic where the resolution can be adapted on-the-fly is straightforward: for instance, if 10 registers are implemented, up to 10 bit ADC resolution can be supported. If only 7 out of the 10 registers are activated, the ADC resolution is effectively reduced from 10 to 7 bit, while the power is scaled down proportionally as well, following the expected linear trend with N. Since the ideal power-scaling trend can be achieved with this reconfigurable implementation, the power efficiency is maintained for the logic.

As opposed to the power consumption of the digital part, the power consumption of the comparator and DAC should theoretically scale exponentially with N, because these blocks are limited by physical constraints such as noise and matching. To maintain the best possible ADC efficiency throughout a reconfigurable range of resolutions, these analog blocks need to be reconfigured in such a way that they achieve exponential power scaling with N.

Figure 13.25 shows the implementation of the reconfigurable DAC. First, it is assumed that the

reconfiguration switch is permanently connected. In this case, the DAC has a resolution of N bit, and the power consumption is proportional to the total capacitance C_S, which is equal to $2^N \cdot C_0$. To reduce the DAC's resolution, the reconfiguration switch can be disconnected. In this situation, the largest two capacitors (C_{N-1} and C_{N-2}) are permanently disconnected and do not contribute to the DAC's resolution or power consumption. Thus, the DAC has now N−2 bit of resolution, while the effective capacitance is reduced to $2^{N-2} \cdot C_0$. This implies that the power consumption is exponentially scaled down with a factor 2^2, following the exponential scaling requirement in N.

In theory, it would be possible to add 3 reconfiguration switches, such that DAC resolutions of 7, 8, 9, and 10 bit can be implemented with a relative power consumption of 100 %, 50 %, 25 % and 12.5 %, respectively. However, for sake of simplicity, the implementation is limited to have only 2 modes of operation as shown in the graph, with a factor of 4 in power scaling. The DAC is set to 8-bit mode to support ADC resolutions of 7 and 8 bit, and it is set to 10-bit mode to support ADC resolutions of 9 and 10 bit. The unit capacitance (C_0) is 0.6fF and implemented similar to Fig. 13.13b. Note that the proposed technique manages to scale the DAC power with 2^N, which achieves constant-FOMW scaling. However, the scaling is still not optimal, as the DAC's noise limit would theoretically allow scaling with 4^N.

For the comparator, the design is based on the dynamic two-stage topology proposed in van Elzakker et al. (2010). In that design, a first stage acts as a dynamic pre-amplifier, while the second stage implements a latch. As the first stage contributes most of the comparator noise, it also dominates the total power consumption. Hence, it is sufficient to reconfigure the dynamic pre-amplifier only. The pre-amplifier (Fig. 13.26) is enabled when the CLK signal is turned on: a current will start to flow through the tail transistor. Dependent on the differential input signal, the differential pair will generate a differential output current that will be integrated on the load capacitors, thus creating a dynamically amplified output voltage. As analyzed in van Elzakker et al. (2010), the effective input-referred noise voltage of this stage is inversely proportional to $\sqrt{C}$, while the power consumption is given by $2 \cdot C \cdot VDD^2$. As 1 bit additional resolution requires a 6 dB better SNR, this can be achieved by including a 4x larger C. In this way, the capacitor size and power consumption scale with 4^N, following the ideal noise-power trade-off. The implemented pre-amplifier has therefore a programmable load capacitance. This is done by adding a small array of capacitors and switches such that the value can be programmed digitally. Four different noise/power settings are supported, with which the power can be scaled almost by 4x while the noise voltage scales about 2x. A wider scaling range would be preferable, but is hard to achieve: further down-scaling of C

hardly helps, because by then the power is dominated by other components. Further up-scaling of C also has a limited impact, because by then the noise is dominated by other contributors.

So far, the reconfigurability of logic, DAC and comparator has been explained. A last feature integrated in this ADC is redundancy in order to save power. Rather than using a binary-scaled DAC which requires N cycles to find the N-bit output code, this ADC uses N + 1 cycles and a non-binary DAC. This redundancy allows to relax precision requirements in the early conversion cycles, as the redundancy can solve errors in the later conversion cycles. As shown in Giannini et al. (2008), this can be exploited to save power: Giannini et al. (2008) uses a noisy low-power comparator in the first SAR cycles to save power, and a precise higher-power comparator in the last few cycles to obtain the required precision. While (Giannini et al. 2008) required two separate comparators, this work can achieve the same result by simply reconfiguring the load capacitance of the comparator during the conversion process. More details can be found in Harpe et al. (2012).

The ADC was implemented in a 90 nm CMOS technology (Fig. 13.27) and operates

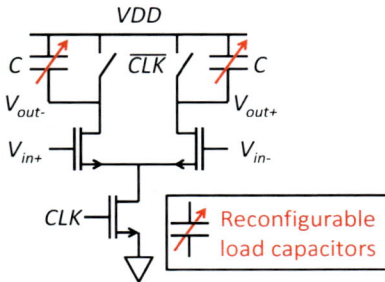

Fig. 13.26 Implementation of the resolution reconfigurable comparator

Fig. 13.27 Die photo of the ADC in 90 nm CMOS

Fig. 13.28 Measured ENOB of the ADC for different resolutions

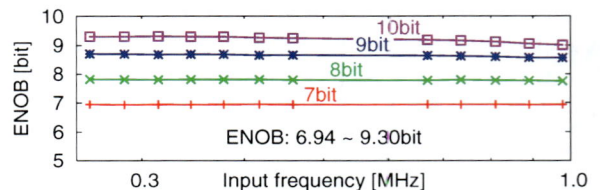

Fig. 13.29 Measured power consumption for different resolutions and sampling rates

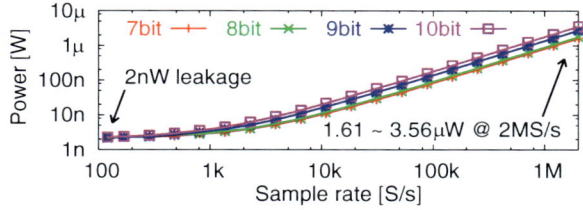

Table 13.4 ADC performance summary

Process [nm]	90			
Supply [V]	0.7			
Sample rate [MS/s]	2			
Resolution [bit]	7	8	9	10
ENOB [bit]	6.94	7.81	8.70	9.30
Power [μW]	1.61	1.77	2.72	3.56
FOMW [fJ/conv.step]	6.6	3.9	3.3	2.8

from a 0.7 V supply. Figure 13.28 shows the measured ENOB as function of the input frequency, while operating at 2 MS/s. The ENOB for the different resolutions (7, 8, 9, and 10 bit) varies from 6.94 to 9.30 bit.

Figure 13.29 shows the measured power consumption versus sampling rate for the different resolutions. The power scales linear with the sample rate. The minimum power is 2 nW, caused by leakage. The power also scales with resolution, but this is not very well visible due to the logarithmic axis in the graph. At 2 MS/s, the power scales from 1.61 up to 3.56 μW when the resolution is changed from 7 to 10 bit.

A summary of the performance is given in Table 13.4. The ADC achieves good power efficiency for all resolutions and covers a useful resolution and speed range for IoT. For instance, for biopotential or environmental monitoring, the ADC could operate at 10 bit resolution and 1kS/s, where it consumes only 4nW. For a low-power radio, it could work at 8 bit resolution and 2MS/s where it consumes 1.61 μW. To expand the application range, it would be interesting for future work to include higher resolutions for more precise sensing and higher speeds to support more advanced wireless communication standards.

13.6   On-Chip Voltage References

The previous sections described low-power ADCs. To use such ADCs, analog signal conditioning (discussed in Chap. 12) and references are also required. As seen from the basic topology (Sect. 13.2), the SAR ADC usually only requires a voltage reference for the DAC that sets the full-scale range of the ADC (equation (13.4)). Moreover, as can be observed from the design examples (Sects. 13.3–13.5), this DAC reference voltage is often equal to the VDD of the other circuit blocks. Therefore, this section focusses on Reference Voltage Generators (RVG) and their application to ADCs for IoT.

The reference voltage provided to the DAC can experience several different types of imperfections, here classified as random variations and systematic variations. Random variations happen for instance due to thermal noise, or due to incidental glitches/disturbances. Since these effects are directly seen by the DAC, they will immediately modulate with the signal being digitized. However, if the problem is truly random, it can be filtered out either by filtering the reference with a by-pass capacitor, or by filtering the digitized samples in the digital domain. Systematic variations of the reference voltage could happen due to gradual temperature, process or supply voltage variations. These errors might drift over time but are usually strongly correlated from sample to sample. As such, they cannot be filtered out. However, it depends on the application if these errors are a problem or not. For instance, ratiometric measurements can be insensitive to the precise reference voltage. Or, in case of a wireless link, the absolute amplitude of a received signal

Table 13.5 Examples of low-power voltage references

	Magnelli et al. (2011)	Seok et al. (2012)	Dong et al. (2016)	Liu et al. (2016)
Process [nm]	180	130	180	65
Minimum supply [V]	0.45	0.5	1.2	0.62
Power [nW]	2.6	0.0022	0.114	2.5–25
Line sensitivity [%/V]	0.44	0.033	0.38	0.07
PSRR @100Hz [dB]	≈ -45	-53	-42	-62
Temp. coefficient [ppm/°C]	165	231	124	108
Temp. range [°C]	0 to 125	-20 to 80	-40 to 85	-25 to 110

does not matter in itself. Communication can still be reliable if the reference is unprecise, as long as it is stable during the time of a transmission. However, for applications where the absolute voltage of the sensed signal is relevant, the reference voltage needs to be tightly controlled.

In case a reference voltage needs to be generated, several requirements of IoT should be noted. Low power consumption is mandatory. To save further power, low-voltage operation and the ability for duty-cycling are preferred as well. Lastly, to integrate the references together with the rest of the IoT system, technology portability to scaled CMOS nodes is also relevant. Table 13.5 shows several state-of-the-art examples of low-power voltage references. They all manage to achieve pW to nW levels of power consumption, albeit with different performance characteristics. The line sensitivity and PSRR describe how sensitive the generated reference is with respect to variations or disturbances of the input voltage. The temperature coefficient describes the sensitivity to temperature variations in the indicated temperature range. Most designs, except (Dong et al. 2016), can operate at sub-1 V supplies. Liu et al. (2016) is the only design in this list that is integrated in an advanced CMOS node, and the only one that includes duty-cycling ability. On the other hand, the power consumption of the other designs is so low that duty-cycling is not even necessary. In Magnelli et al. (2011), the generated reference is determined by the threshold voltage of a single MOS transistor. In Dong et al. (2016), Liu et al. (2016), Seok et al. (2012) the generated reference is dependent on the threshold voltage difference between two MOS transistors. However, their actual principles and circuit implementations

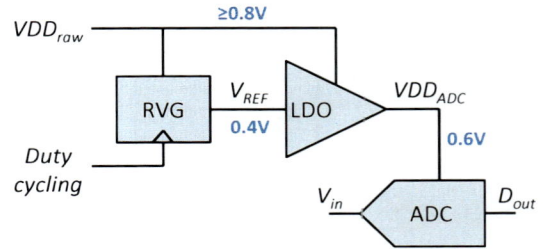

Fig. 13.30 Duty-cycled Voltage Reference Generator (RVG), LDO and ADC

are very different. Seok et al. (2012) uses the threshold difference between a native and a thick-oxide device, Dong et al. (2016) uses the difference caused by the body bias effect, and Liu et al. (2016) uses the difference based on different ion implant levels in two thin-oxide transistors. While these designs prove that low-power reference generation is feasible, these circuits have a very high output impedance and are thus not able to directly drive the DAC, as the DAC requires substantial power from the reference. For that reason, a buffer or low-dropout regulator (LDO) is still required to connect the RVG to the ADC.

A complete system example, published in Liu et al. (2016), is shown in Fig. 13.30. Here, the RVG provides a reference voltage to an LDO. The LDO multiplies the reference and powers the entire 10bit ADC. A raw supply of at least 0.8 V is needed to generate a reference voltage of 0.4 V and an ADC supply of 0.6 V. To save reference power, the RVG can be duty-cycled. A S&H at the RVG output is included so that the generated reference is continuously available for the LDO even when the RVG is powered down.

Table 13.6 ADC performance summary and comparison

	Harpe et al. (2015), Section 13.3	Liu et al. (2016), Section 13.6
Process [nm]	65	65
Supply [V]	0.6	0.8
Resolution [bit]	10	10
ENOB [bit]	9.2	9.1
Sample rate [kS/s]	100	80
Total power [nW]	88	106
FOMW [fJ/conv.step]	1.5	2.4
Including RVG and LDO	NO	YES

The overall system was measured while the RVG was duty-cycled at 10 %. Table 13.6 shows the measured performance and compares it against the very similar 10 bit ADC that was discussed in Sect. 13.3. The ENOB of both designs is similar, while the FOMW increased from 1.5 to 2.4fJ/conversion-step. The increase of FOMW is because this design includes the power of the RVG and LDO, and it uses a higher supply voltage. Nonetheless, this example shows that an ADC including reference generation can be power-efficient. It also confirms that the ADC is still dominant over the RVG and LDO in terms of power consumption.

13.7 Perspectives and Trends

In this chapter, ADCs for IoT nodes were discussed. The SAR ADC is a suitable solution in this context, allowing very low power consumption with suitable speed and precision for most applications. Several examples described techniques to achieve state-of-the-art in terms of efficiency and precision. As also shown, the SAR ADC can deal well with modern technologies, and allows operation at low supply levels. The dynamic power consumption of SAR ADCs and the Nyquist operation enable automatic power scaling with the sample rate and on-demand operation. Techniques to implement versatility in the ADC's resolution to expand the application range of a single design were also introduced.

In the future, the ongoing improvement of ADC power-efficiency (illustrated in Fig. 13.3) will enable further benefits for IoT. Either power could be saved, or advantage can be taken from improved speed and precision at existing power levels. However, due to technology scaling, leakage starts to dominate the overall power consumption for ADCs that are heavily duty-cycled. For instance for quasi-static monitoring, the leakage power will be higher than the active power, requiring leakage mitigation techniques to maintain efficiency.

Present state-of-the-art ADCs for IoT can be so power-efficient, that the bottleneck in terms of power is usually not inside the ADC anymore, but in the components that surround the ADC. For instance the reference voltage generation, the analog input buffer, or the anti-aliasing filter could consume similar or more power than the ADC itself. This is particularly true in on-demand sensing applications as the ADC can be duty-cycled easily but the other analog blocks often experience static consumption that does not scale down with the activity. Hence, research is needed in those circuits to take full advantage of low-power ADCs at the system level.

Lastly, while the simple SAR architecture suits well to IoT requirements, other topologies should not be discarded. Hybrid architectures are becoming more and more popular in recent years. For instance, Shu et al. (2016) combines a SAR structure with noise-shaping and mismatch-shaping techniques, allowing to reach a far greater SNDR (101 dB) than typical SAR ADCs with state-of-the-art power-efficiency (FOMS = 180 dB). The power-efficiency of Sigma-Delta Modulators is also improving, such as Billa et al. (2016) which achieves an SNDR of 98.5 dB for 24 kHz signal bandwidth with a FOMS of 177.8 dB. Hybrid and Sigma-Delta converters especially stand out for high-precision applications that cannot be covered easily with pure SAR ADCs.

References

A.M. Abo, P.R. Gray, A 1.5-V, 10-bit, 14.3-MS/s CMOS pipeline analog-to-digital converter. IEEE J. Solid-State Circuits **34**(5), 599–606 (1999)

A. Agnes, E. Bonizzoni, P. Malcovati, F. Maloberti, A 9.4-ENOB 1V 3.8 μW 100kS/s SAR ADC with time-domain comparator. ISSCC Dig. Tech. Papers (Feb. 2008), pp. 246–247

S. Billa, A. Sukumaran, S. Pavan, A 280 μW 24 kHz-BW 98.5 dB-SNDR chopped single-bit CT ΔΣM achieving <10Hz 1/f noise corner without chopping artifacts. ISSCC Dig. Tech. Papers (Feb. 2016), pp. 276–277

S.-W.M. Chen, R.W. Brodersen, A 6-bit 600-MS/s 5.3-mW asynchronous ADC in 0.13-μm CMOS. IEEE J. Solid-State Circuits **41**(12), 2669–2680 (2006)

T.B. Cho, P.R. Gray, A 10 b, 20 M sample/s, 35 mW pipeline A/D converter. IEEE J. Solid-State Circuits **30**(3), 166–172 (1995)

Q. Dong, K. Yang, D. Blaauw, D. Sylvester, A 114-pW PMOS-only, trim-free voltage reference with 0.26% within-wafer inaccuracy for nW systems, in *IEEE Symposium on VLSI Circuits* (June 2016)

V. Giannini, P. Nuzzo, V. Chironi et al., An 820μW 9b 40MS/s noise–tolerant dynamic-SAR ADC in 90 nm digital CMOS. ISSCC Dig. Tech. Papers (Feb. 2008), pp. 238–239

B.P. Ginsburg, A.P. Chandrakasan, A 500 MS/s 5b ADC in 65 nm CMOS. in *IEEE Symposium on VLSI Circuits* (June 2006), pp. 140–141

V. Hariprasath, J. Guerber, S.-H. Lee, U.-K. Moon, Merged capacitor switching based SAR ADC with highest switching energy-efficiency. IET Electron. Lett. **46**(9), 620–621 (2010)

P. Harpe, C. Zhou, Y. Bi et al., A 26 μW 8 bit 10 MS/s asynchronous SAR ADC for low energy radios. IEEE J. Solid-State Circuits **46**(7), 1585–1595 (2011)

P. Harpe, G. Dolmans, K. Philips, H. de Groot, A 0.7 V 7-to-10 bit 0-to-2 MS/s flexible SAR ADC for ultra low-power wireless sensor nodes. in *European Solid-State Circuits Conference* (Sep. 2012), pp. 373–376

P. Harpe, E. Cantatore, A. van Roermund, A 2.2/2.7fJ/conversion-step 10/12b 40kS/s SAR ADC with data-driven noise reduction. ISSCC Dig. Tech. Papers (Feb. 2013), pp. 270–271

P. Harpe, E. Cantatore, A. van Roermund, An oversampled 12/14b SAR ADC with noise reduction and linearity enhancements achieving up to 79.1dB SNDR. ISSCC Dig. Tech. Papers (Feb. 2014), pp. 194–195

P. Harpe, H. Gao, R. van Dommele, E. Cantatore, A. van Roermund, A 3 nW signal-acquisition IC integrating an amplifier with 2.1 NEF and a 1.5fJ/conv-step ADC. ISSCC Dig. Tech. Papers (Feb. 2015), pp. 382–383

P. Harpe, H. Gao, R. van Dommele, E. Cantatore, A.H.M. van Roermund, A 0.20 mm² 3 nW signal acquisition IC for miniature sensor nodes in 65 nm CMOS. IEEE J. Solid-State Circuits **51**(1), 240–248 (2016)

M. Konijnenburg, S. Stanzione, L. Yan et al., A battery-powered efficient multi-sensor acquisition system with simultaneous ECG, BIO-Z, GSR, and PPG. ISSCC Dig. Tech. Papers (Feb. 2016), pp. 480–481

C.-Y. Liou, C.-C. Hsieh, A 2.4-to-5.2fJ/conversion-step 10b 0.5-to-4MS/s SAR ADC with Charge-Average Switching DAC in 90 nm CMOS, in *ISSCC Dig. Tech. Papers* (Feb. 2013), pp. 280–281

C.-C. Liu, S.-J. Chang, G.-Y. Huang, Y.-Z. Lin, A 10-bit 50-MS/s SAR ADC with a monotonic capacitor switching procedure. IEEE J. Solid-State Circuits **45**(4), 731–740 (2010)

Y.-H. Liu, C. Bachmann, X. Wang et al., A 3.7 mW-RX 4.4 mW-TX fully integrated bluetooth-low-energy/IEEE802.15.4/proprietary SoC with an ADPLL-based fast frequency offset compensation in 40 nm CMOS. ISSCC Dig. Tech. Papers (Feb. 2015), pp. 236–237

M. Liu, K. Pelzers, R. van Dommele, A. van Roermund, P. Harpe, A 106nW 10 b 80 kS/s SAR ADC with duty-cycled reference generation in 65 nm CMOS. IEEE J. Solid State Circuits **51**(10), 2435–2445 (2016a)

M. Liu, A. van Roermund, P. Harpe, A 7.1fJ/conv.-step 88dB-SFDR 12b SAR ADC with energy-efficient swap-to-reset, in *European Solid-State Circuits Conference* (Sep. 2016b), pp. 409–412

L. Magnelli, F. Crupi, P. Corsonello, C. Pace, G. Iannaccone, A 2.6 nW, 0.45 V temperature-compensated subthreshold CMOS voltage reference. IEEE J. Solid-State Circuits **46**(2), 465–474 (2011)

K.A.A. Makinwa, A. Baschirotto, P. Harpe (eds.), *Efficient Sensor Interfaces, Advanced Amplifiers and Low Power RF Systems—Advances in Analog Circuit Design 2015* (Springer, Berlin, 2016), ISBN 978-3-319-21184-8

B. Murmann, ADC Performance Survey 1997–2016. http://web.stanford.edu/~murmann/adcsurvey.html

M. Pelgrom, *Analog-to-Digital Conversion* (Springer, Berlin, 2017) ISBN 978-3-319-44970-8

J. Prummel, M. Papamichail, M. Ancis, A 10mW bluetooth low-energy transceiver with on-chip matching, in *ISSCC Dig. Tech. Papers* (Feb. 2015), pp. 238–239

T. Sano, M. Mizokami, H. Matsui, A 6.3mW BLE transceiver embedded RX image-rejection filter and TX harmonic-suppression filter reusing on-chip matching network, in *ISSCC Dig. Tech. Papers* (Feb. 2015), pp. 240–241

M. Seok, G. Kim, D. Blaauw, D. Sylvester, A portable 2-transistor picowatt temperature-compensated voltage reference operating at 0.5 V. IEEE J. Solid-State Circuits **47**(10), 2534–2545 (2012)

Y.-S. Shu, L.-T. Kuo, T.-Y. Lo, An oversampling SAR ADC with DAC mismatch error shaping achieving 105dB SFDR and 101dB SNDR over 1 kHz BW in 55 nm CMOS, in *ISSCC Dig. Tech. Papers* (Feb. 2016), pp. 458–459

H.-Y. Tai, Y.-S. Hu, H.-W. Chen, H.-S. Chen, A 0.85fJ/conversion-step 10b 200kS/s subranging SAR ADC in 40 nm CMOS, in *ISSCC Dig. Tech. Papers* (Feb. 2014), pp. 196–197

M. van Elzakker, E. van Tuijl, P. Geraedts et al., A 10-bit charge-redistribution ADC consuming 1.9 μW at 1 MS/s. IEEE J. Solid-State Circuits **45**(5), 1007–1015 (2010)

M. Vidojkovic, X. Huang, P. Harpe et al., A 2.4 GHz ULP OOK single-chip transceiver for healthcare applications, in *ISSCC Dig. Tech. Papers* (Feb. 2011), pp. 458–459

M. Yip, A.P. Chandrakasan, A resolution-reconfigurable 5-to-10b 0.4-to-1V power scalable SAR ADC, in *ISSCC Dig. Tech. Papers* (Feb. 2011), pp. 190–191

D. Zhang, A. Alvandpour, A 3-nW 9.1-ENOB SAR ADC at 0.7 V and 1 kS/s, in *European Solid-State Circuits Conference* (Sep. 2012), pp. 369–372

Y. Zhu, C.-H. Chan, U.-F. Chio et al., A 10-bit 100-MS/s reference-free SAR ADC in 90 nm CMOS. IEEE J. Solid-State Circuits **45**(6), 1111–1121 (2010)

Circuit Techniques for IoT-Enabling Short-Range ULP Radios

14

Pui-In Mak, Zhicheng Lin, and Rui Paulo Martins

This chapter addresses the design of cost-aware ultra-low-power (ULP) radios for both 2.4-GHz and sub-GHz ISM bands. Starting from the system aspects that provide the essential insights, effective circuit techniques are presented to improve the radio performances and power efficiency, while minimizing the die area and number of external components.

14.1 ULP Wireless Nodes in the IoT Landscape

Smart cities, environmental monitoring, energy management and healthcare systems, just to name a few, are all inside the gigantic landscape of Internet of Things (IoT) (Stankovic 2014) or Internet of Everything (IoE). The estimated IoT market by 2020 will be close to hundreds of billion dollars (annually ~16 billions). To accelerate the proliferation of IoT products in different application sectors, it is opportune to develop *ultra-low-cost software-defined ULP radios* that are flexible to support different data rates (e.g., from kb/s to a few Mb/s), different standards [e.g., *ZigBee and Bluetooth Low Energy (BLE)*] and a wide range of frequency

P.-I. Mak (✉) • Z. Lin • R.P. Martins
University of Macau, Macau, China

Instituto Superior Técnico, Universidade de Lisboa, Lisbon, Portugal
e-mail: pimak@umac.mo

(e.g., sub-GHz and 2.4-GHz ISM bands), while occupying a small die area and entailing a minimum number of external components. These next-generation ULP radios will be decisive for a wide variety of products that have strong competition among cost, performance and time-to-market. Nevertheless, the tradeoff analysis between cost, size and power for an ULP wireless link can involve many parameters that must be co-designed, implying that deeper understanding of the system aspects and effective circuit techniques are both essential to reach an optimum solution.

14.2 System Aspects of Short-Range ULP Radios

Focusing on short-range connectivity with a RF link budget of ~80 to 90 dB, the physical (PHY) layer specifications of Zigbee and BLE are not particularly tough for modern RF skills. Yet, traditional textbook RF and analog techniques can unlikely help to bring down the radio's power by orders of magnitude, while allowing it to be universal enough to serve multiple bands without resorting from costly external components. The following sub-sections briefly discuss the PHY layer of Zigbee and BLE standards. The pros and cons of opting different frequency bands and supply voltages (VDD) are

Table 14.1 Key PHY specifications of ZigBee and BLE standards

	ZigBee	BLE
Frequency (GHz)	0.3, 0.4, 0.7, 0.9, 2.4	2.4
Bandwidth (MHz)	2	1
Channel Spacing (MHz)	5	2
Modulation	BPSK, OQPSK	GFSK
Range (m)	10 to 200	10 to 100
Data Rate (Mbps)	0.25	1
Network Topology	Star/Mesh	Star/P2P

also mentioned; all are correlated to the overall cost, size and power efficiency of the radios.

14.2.1 ZigBee and Bluetooth Low Energy (BLE) Standards

Both ZigBee and BLE standards are suitable for short-range ULP communication as they draw low peak and average power. Their key features are briefed next.

ZigBee was developed as a wireless personal-area network (WPAN) standard with the IEEE 802.15.4 to define the PHY and Media Access Control (MAC) layers. It can operate at a very low duty cycle (<1%) and is allowed in three different frequency bands. The first band (868 MHz) is for Europe only offering only a single channel. It supports a low bit rate of 20 kbps using binary phase-shift keying (BPSK) modulation. The second band (915 MHz) permits 10 channels and is widely adopted in North America, Australia, New Zealand, and some countries in South America. Each channel supports 40 kbps using BPSK modulation. The third band is 2.4 GHz available worldwide, and has a total of 16 channels with 250 kbps each. Unlike the sub-GHz bands, this third band exploits offset quadrature phase-shift keying (OQPSK) with half sine-wave shaping for its modulation. Beyond these three bands, the IEEE 802.15.4c/d study groups also considered to open 314 to 316 MHz, 430 to 434 MHz, and 779 to 787 MHz bands for use in China, and 950 to 956 MHz for use in Japan. Obviously, an international market will be opened if the ULP radio can be reconfigured to support multiple bands from sub-GHz to 2.4 GHz. The key PHY specifications of ZigBee and BLE standards are summarized in Table 14.1.

BLE is a prospective short-range wireless standard ratified in 2009. It supports 40 channels in the 2.4-GHz band, each of which is 2-MHz wide. It is based on Gaussian frequency-shift keying (GFSK) modulation with an index of 0.5. The state-of-the-art 2.4-GHz receiver (Liu et al. 2014) achieves an energy efficiency of ~1.2 nJ/b at 2 Mb/s. Unsurprisingly, >40% of the receiver power is dissipated by the forefront low-noise amplifier (LNA) and mixer to maximize the sensitivity (−92 dBm). Such a high sensitivity seems overkill, but it is indeed effective to reduce the power consumption of the transmitter which normally has a lower energy efficiency to fulfill an RF link budget of ~90 dB. Thus, it is highly desired to develop circuit techniques for better LNA, mixer and voltage-controlled oscillator (VCO) for a better overall energy efficiency. In fact, for the 2.4-GHz band, Zigbee and BLE share a similar PHY, and modern solutions can easily support both. For the sub-GHz bands, multi-band operation poses additional challenges. To achieve this without leveraging the cost, a fully-integrated RF-tunable ULP radio will be of great relevance.

14.2.2 Cost, Size and Power

Ultra-scaled CMOS technologies are still the best platform for full integration of ULP radios that have RF (transceiver), analog (sensor and power management) and digital (microcontroller and memory). Established technology nodes (e.g., 90 or 65 nm) are regaining lots of interest for low-cost fast-to-market IoT products, as they can leverage more reasonably between the manufacturing cost, development time and power consumption. Apparently, the system cost and size can be optimized by reducing the chip area, number of external components and battery volume that depends on the targeted lifetime of the system. Although using on-chip passives (inductors and transformers) can help to reduce the VDD and system power, we will describe later that recent cost-aware ULP RF and analog circuits can balance better the power, chip area and cost. For instance, a fully-integrated input matching network not only can reduce the cost and system form factor of an ULP receiver, but also enhance its power efficiency. Also, the matching network can offer passive pre-gain to enhance the sensitivity of the receiver. Other low-power techniques such as current-reuse and function-reuse are will be introduced later in this chapter.

14.2.3 Frequency Bands: 2.4 GHz vs. Sub-GHz

Most existing ULP radios were designed for the 2.4-GHz band as it is available worldwide and has a smaller antenna size suitable for integration. Yet, the sub-GHz ISM bands offer other advantages such as longer propagation distance and less interference that are worth to be considered when the power budget is the priority.

Communication range—In highly congested environments, the 2.4-GHz signal can weaken rapidly, which adversely affects the signal quality. To quantify the influence of frequency on path loss with respect to the wavelength λ, we can use the simplified Friis transmission equation,

$$L = 20\log_{10}\left(\frac{4\pi d}{\lambda}\right) \qquad (14.1)$$

Hence, it can be calculated that the path loss at 2.4 GHz is 8.5 dB higher than that at 900 MHz. This translates into a $2.67\times$ longer range for a 900-MHz radio. Since the range almost doubles with every 6 dB increment of power, a 2.4-GHz

radio will entail an increment of power budget (by 8.5 dB), in order to match the range of a 900-MHz radio. Besides, biological tissues absorb RF energy as a function of frequency. Lower frequencies can penetrate the body easily without being absorbed, meaning a better RF link for sub-GHz when compared to 2.4 GHz for body-area networks.

Interference—The 2.4-GHz band has a high chance to come across interferences due to the co-existence of other wireless standards, degrading the link reliability. For example, the IEEE 802.11 (WiFi) can transmit an output power 10x to 100x higher than the ZigBee. Signals from Bluetooth-enabled computer, cell phone peripherals and microwave ovens can also be considered as "jammers", which have a much lower output power. Sub-GHz ISM bands are mostly used for proprietary low-duty-cycle links and are not as likely to interfere with each other. A quieter spectrum means easier transmissions and fewer retries, which is more efficient to save the battery power. In fact, due to the limited power budget, it is hard for an ULP radio to tolerate large out-of-channel blockers.

Antenna size—One disadvantage of sub-GHz operation is the larger antenna size since most antenna types are designed to be resonant at their intended operation frequency. Since the antenna size is inversely proportional to the frequency, a small wireless node would prefer the 2.4-GHz band. Communication distance, low potential interference and low power consumption are the obvious advantages of the sub-GHz bands.

14.2.4 Supply Voltage (VDD)

To minimize the system size, short-range ULP radios should run preferably from a tiny battery, thus sub-2 V supply voltages are highly desired. Radios that work down to 1.2 V allow extra flexibility in sensors' design and reduce the power management constraints (Rajan 2012). Besides, low peak current and sub-1 V VDD also benefit wireless sensors that run from harvested energy sources which will enhance flexibility, lower the maintenance cost, and open up more applications. For example, on-chip solar cells only can provide an output voltage between 200 and 900 mV, while thermoelectric generators exhibit an even lower VDD (50–300 mV) (Bandyopadhyay et al. 2011). Although boost converters can be employed to boost up the output voltage, their efficiency is still quite limited (~75%). Besides, a low peak current consumption will ease the design of the power management. Furthermore, radio operating at higher VDD is only required when a higher output power is entailed. This is not the case for short-range communications, as the output power rarely exceeds 0 dBm. Thus, a low VDD is in general the simplest way to reduce the power consumption at the system level.

In a low VDD design, however, due to the limited dynamic range, for the given parameters such as third-order intercept point (IIP3), noise-figure (NF), gain etc., the current should be larger than that with a high VDD. For example, for the given NF requirement, the current-reuse P-type metal-oxide-semiconductor (PMOS) and N-type metal-oxide-semiconductor (NMOS) self-biased amplifier with a VDD of 1 V consumes half of the current of a single NMOS (or PMOS) without current-reuse and with a VDD of 0.5 V. This constraint is even tighter if a small chip area and/or no/limited external components are imposed for cost reduction. As an example, inductors/transformers can help to boost the operating frequency and bias the circuit with lower voltage headroom and noise. If inductors/transformers cannot be used due to the limited area budget, only resistors or transistors can be adopted instead. This imposes a hard trade-off with IIP3, NF and bandwidth. Thus, to balance the VDD, current, area and external components with the key performance metrics (NF and out-of-band (OB) IIP3), effective system-to-circuit-level co-design, RF and analog circuit techniques become highly important and correlated. The next two sections present the key circuit techniques applied into two state-of-the-art cost-aware ULP receivers: one for the 2.4-GHz band and one for the sub-GHz bands.

14.3 Current-Reuse ULP Receiver Techniques for the 2.4-GHz ISM Band

Nanoscale CMOS offers sufficiently high f_t and low V_t favoring the design of ULP receivers via stacking the RF-to-baseband (BB) functions in one cell, while sharing the smallest possible bias current. Also, the signals can be conveyed in the current domain to enhance the area efficiency (i.e., no AC-coupling capacitor), RF bandwidth and linearity at those inner nodes. The proposed Zigbee receiver (Lin et al. 2013, 2014a) is inspired by the above hypothesis, and its block diagram is depicted in Fig. 14.1.

The single-ended RF input (V_{RF}) is taken by a low-Q input-matching network before reaching the Balun-LNA-I/Q-Mixer (Blixer). Merging the Blixer with the hybrid filter not only saves power, but also reduces the voltage swing at internal nodes benefitting the linearity. The wide-band input-matching network is also responsible for the passive pre-gain to reduce the NF. Unlike the LMV cell that only can utilize single-balanced mixers (Tedeschi et al. 2010), here the balun-LNA featuring a differential output ($\pm i_{LNA}$) allows the use of double-balanced mixers (DBMs). Driven by a 4-phase 25% LO, the I/Q-DBMs with a large output resistance robustly correct the differential imbalances of $\pm i_{LNA}$. The balanced BB currents ($\pm i_{MIX,I}$ and $\pm i_{MIX,Q}$) are then filtered directly in the current domain by a current-mode Biquad stacked atop the DBM. The Biquad features in-band noise-shaping centered at the desired intermediate frequency (IF, 2 MHz). Only the filtered output currents ($\pm i_{rLPF,I}$ and $\pm i_{rLPF,Q}$) are returned as voltages ($\pm V_{o,I}$ and $\pm V_{o,Q}$) through the complex-pole load, which performs both image rejection and channel selection. Out of the current-reuse path there is a high-swing variable-gain amplifier (VGA). It essentially deals with the gain loss of its succeeding 3-stage passive RC-CR polyphase filter (PPF), which is responsible for large and robust image rejection over mismatches and process variations. The

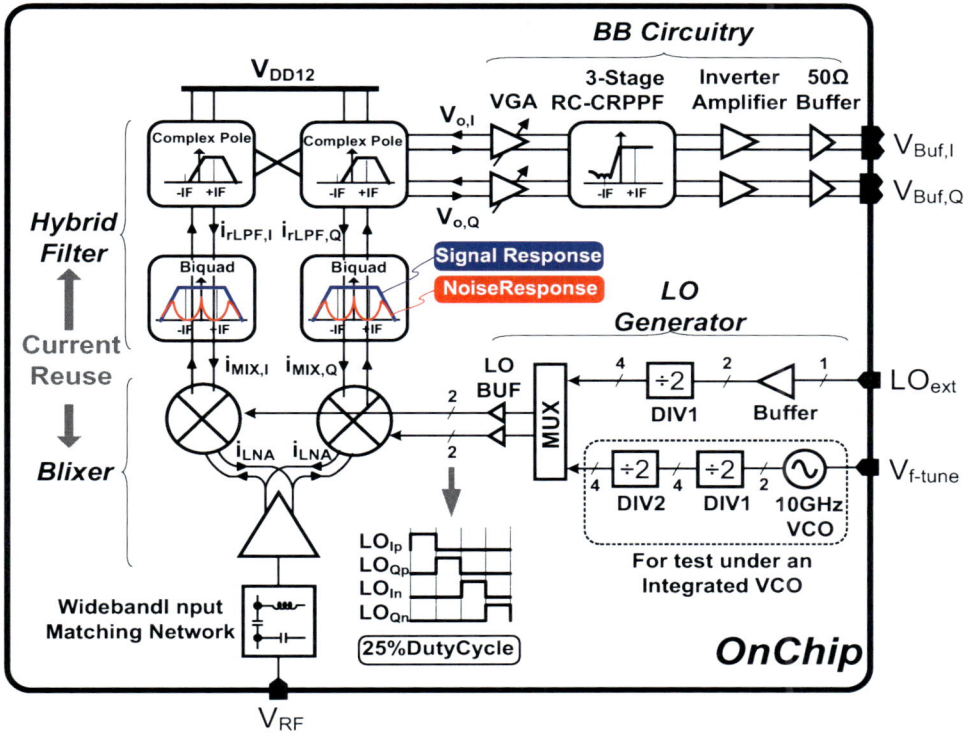

Fig. 14.1 Proposed RF-to-BB-current-reuse ULP 2.4-GHz receiver

final stage is an inverter amplifier before 50-Ω test buffering. The 4-phase 25% LO can be generated by an external 4.8-GHz reference (LO$_{ext}$) after a divide-by-2 (DIV1) that features 50%-input 25%-output, or from an integrated 10-GHz VCO after DIV1 and DIV2 (25%-input 25%-output) for additional testability.

14.3.1 Circuit Implementation

Wideband Input-Matching Network—As shown in Fig. 14.2a, a low-Q inductor (L_M) and 2 tapped capacitors (C_p and C_M) can be employed for impedance down-conversion resonant and passive pre-gain. A high-Q inductor is unnecessary since the Q of the LC matching is dominated by the low input resistance of the LNA. Thus, a low-Q inductor results in area savings, while averting the need of an external inductor for cost savings. L_M also serves as the bias inductor

for M_1. R_p is the parallel shunt resistance of L_M. C_p stands for the parasitic capacitance from the pad and ESD diodes. R_{in} and C_{in} are the equivalent resistance and capacitance at node V_{in}, respectively. R'_{in} is the downconversion resistance of R_{in}.

L_{BW} is the bondwire inductance and R_s is the source resistance. To simplify the analysis, we first omit L_{BW} and C_{in}, so that L_M, C_p, C_M, R_S and R_T ($= R_p//R_{in}$) together form a tapped capacitor facilitating the input matching. Generally, $S_{11} \leq -10$ dB is required and the desired value of R'_{in} is from 26 to 97 Ω over the frequency band of interest. Thus, given the R_T and C_M values, the tolerable C_p can be derived from $R'_{in} = R_T \left(\frac{C_M}{C_M + C_p} \right)^2$. The pre-gain value ($A_{pre, amp}$) from V_{RF} to V_{in} is derived from $\frac{V_{in}^2}{2R_T} = \frac{V_{RF}^2}{2R_S}$, which can be simplified as $A_{pre, amp} = \sqrt{\frac{R_T}{R_S}}$. The $-$3-dB bandwidth of $A_{pre,amp}$ is related to the

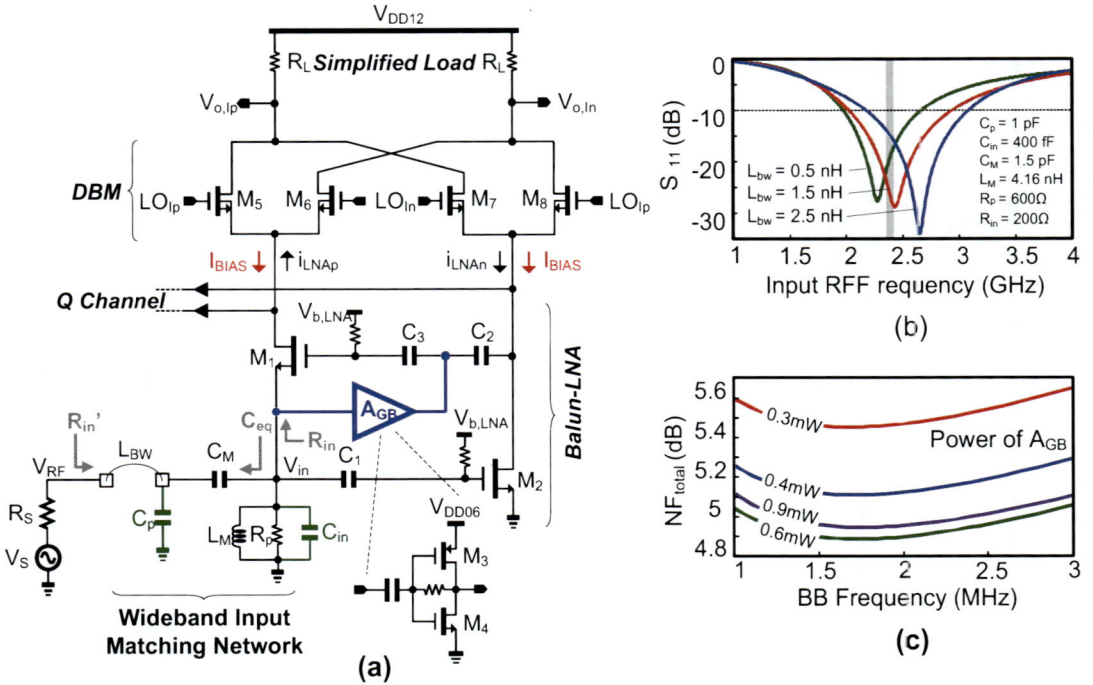

Fig. 14.2 (a) Proposed wideband input matching network, balun-LNA and I/Q-DBMs (Q channel is omitted and the load is simplified as R_L). (b) Variation of S_{11}-bandwidth with bondwire inductance L_{BW}. (c) Power of A_{GB} versus NF

network's quality factor (Q_n) as given by: $Q_n = \frac{R_T}{2\omega_0 L_M} = \frac{\omega_0}{\omega_{-3dB}}$, with $\omega_0 = \frac{1}{\sqrt{L_M C_{EQ}}}$ and $C_{EQ} = \frac{C_M C_p}{C_M + C_p}$.

In our design ($R_T = 150\ \Omega$, $C_M = 1.5$ pF, $L_M = 4.16$ nH, $R_p = 600\ \Omega$, $C_p = 1$ pF and $R_{in} = 200\ \Omega$), $A_{pre,amp}$ has a passband gain of ~4.7 dB over a 2.4-GHz bandwidth (at RF = 2.4 GHz) under a low Q_n of 1. Thus, the tolerable C_p is sufficiently wide (0.37 to 2.1 pF). The low-Q L_M is extremely compact (0.048 mm^2) in the layout and induces a small parasitic capacitance (~260 fF, part of C_{in}). Figure 14.2b demonstrates the robustness of S_{11}-bandwidth against L_{BW} from 0.5 to 2.5 nH. The variation of C_{in} to S_{11}-bandwidth was also studied. From simulations, the tolerable C_{in} is 300 to 500 f. at $L_{BW} = 1.5$ nH.

Balun-LNA—The common-gate (CG) common-source (CS) balun-LNA (Blaakmeer et al. 2008a) avoids the off-chip balun and achieves a low NF by noise canceling, but the asymmetric CG-CS transconductances and loads make the output balancing not wideband consistent. In Blaakmeer et al. (2008b), output balancing is achieved by scaling M_{5-8} with cross-connection at BB, but that is incompatible with this work that includes a hybrid filter. In Mak and Martins (2011), by introducing an AC-coupled CS branch and a differential current balancer (DCB), the same load is allowed for both CS and CG branches for wideband output balancing. Thus, the NF of such a balun-LNA can be optimized independently. This technique is transferred to this ULP design, but only with the I/Q-DBMs inherently serves as the DCB, avoiding a high VDD (Mak and Martins 2011). The detailed schematic is depicted in Fig. 14.2a. To maximize the voltage headroom, M_1 (with $g_{m,CG}$) and M_2 (with $g_{m,CS}$) were sized with non-minimum channel length (L = 0.18 μm) to lower their V_T. The AC-coupled gain stage is a self-biased inverter amplifier (A_{GB}) powered at 0.6-V (V_{DD06}) to enhance its transconductance ($g_{m,AGB}$)-to-current ratio. It gain-boosts the CS branch while creating a loop gain around M_1 to enhance its effective transconductance under less bias current (I_{BIAS}). This scheme also allows the same I_{BIAS} for both M_1 and M_2, requiring no scaling of load (i.e., only R_L). Furthermore, a small I_{BIAS} lowers the supply requirement, making a 1.2-V supply (V_{DD12}) still adequate for the Blixer and hybrid filter, while relaxing the required LO swing (LO_{Ip} and LO_{In}). C_{1-3} for biasing are typical metal-oxide-metal (MoM) capacitors to minimize the parasitics.

The balun-LNA features partial-noise cancelling. To simplify the study, we ignore the noise induced by DBM (M_5–M_8) and the effect of channel-length modulation. The noise transfer function (TF) of M_1's noise ($I_{n,CG}$) to the BB differential output ($V_{o,Ip} - V_{o,In}$) can be derived when LO_{Ip} is high, and the input impedance is matched,

$$TF_{I_{n,CG}} = -\frac{1}{2}(R_L - R_{in}G_{m,CS}R_L) \qquad (14.2)$$

where $G_{m,CS} = g_{m,CS} + g_{m,AGB}$. The noise of M_1 can be fully canceled if $R_{in}G_{m,CS} = 1$ is satisfied. However, $R_{in} \approx 200\ \Omega$ is desired for input matching at low power. Thus, $G_{m,CS}$ should be ≈ 5 mS, rendering the noises of $G_{m,CS}$ and R_L still significant. Thus, device sizings for *full* noise

cancellation of M_1 should not lead to the lowest total NF (NF_{total}). In fact, one can get a more optimized $G_{m,CS}$ (via $g_{m,AGB}$) for stronger reduction of noise from $G_{m,CS}$ and R_L, instead of that from M_1. Although this noise-canceling principle has been discussed in Bruccoleri et al. (2004) for its single-ended LNA, the output balancing was not a concern there. Here, the optimization process is alleviated since the output balancing and NF are decoupled. The simulated NF_{total} up to the $V_{o,Ip}$ and $V_{o,In}$ nodes against the power given to the A_{GB} is given in Fig. 14.2c. NF_{total} is reduced from 5.5 dB at 0.3 mW to 4.9 dB at 0.6 mW, but is back to 5 dB at 0.9 mW. Due to the use of passive pre-gain and a larger R_p that is ~3 times of R_{in}, the noise contribution of the inductor is <1% from simulations. The simulated NF at the outputs of the LNA and test buffer are 5.3 and 6.6 dB, respectively.

Double-Balanced Mixers Offering Output Balancing—The output balancing is inherently done by the I/Q-DBMs under a 4-phase 25% LO. For simplicity, this principle is described for the I channel only under a 2-phase 50% LO, as shown in Fig. 14.3, where the load is simplified as R_L. During the first-half LO cycle when LO_{Ip} is high, i_{LNAp} goes up and appears at $V_{o,Ip}$ while i_{LNAn} goes down and appears at $V_{o,In}$.

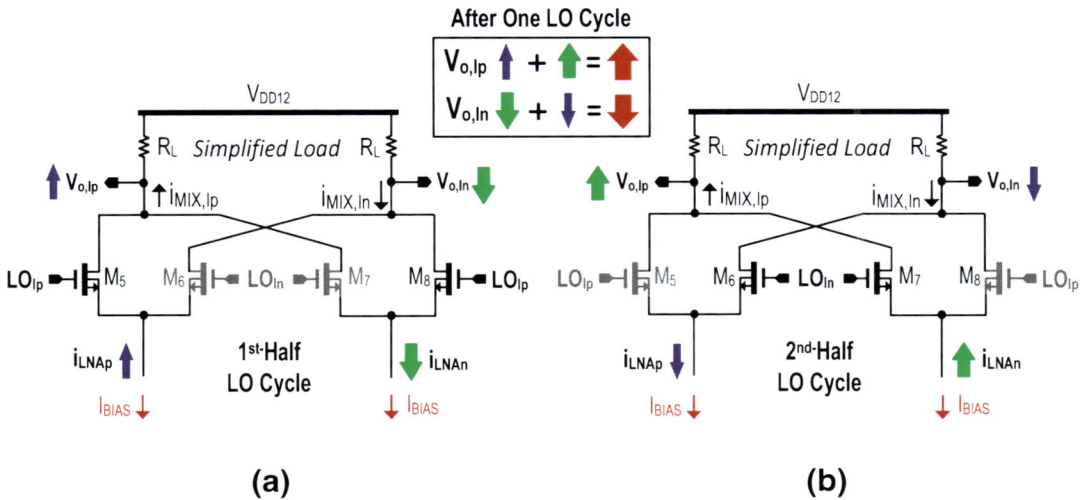

Fig. 14.3 Operation of the I-channel DBM. It inherently offers output balancing after averaging in one LO cycle as shown in their (**a**) 1st-half LO cycle and (**b**) 2nd-half LO cycle

In the second-half LO cycle, both of the currents' sign and current paths of i_{LNAp} and i_{LNAn} are flipped. Thus, when they are summed at the output during the whole LO cycle, the output balancing is robust, thanks to the large output resistance (9 kΩ) of M_5-M_8 enabled by the very small I_{BIAS} (85 μA). To analytically prove the principle, we let $i_{LNAp} = \alpha I_A \cos(\omega_s t + \varphi_1)$ and $i_{LNAn} = -I_A \cos(\omega_s t + \varphi_2)$, where I_A is the amplitude, ω_s is the input signal frequency, α is the unbalanced gain factor and φ_1 and φ_2 are their arbitrary initial phases. When there is sufficient filtering to remove the high-order terms, we can deduce the BB currents $i_{MIX,Ip}$ and $i_{MIX,In}$ as given by,

$$
\begin{aligned}
i_{MIX,Ip} &= \frac{2}{\pi}\alpha I_A \cos(\omega_s t + \varphi_1) \\
&\times \cos\omega_0 t + \frac{2}{\pi} I_A \cos(\omega_s t + \varphi_2) \times \cos\omega_0 t \\
&= \frac{\alpha I_A}{\pi}\cos(\omega_s t - \omega_0 t + \varphi_1) \\
&+ \frac{I_A}{\pi}\cos(\omega_0 t - \omega_s t + \varphi_2)
\end{aligned}
$$

(14.3)

$$
\begin{aligned}
i_{MIX,In} &= -\frac{I_A}{\pi}\cos(\omega_s t - \omega_0 t + \varphi_2) \\
&- \frac{\alpha I_A}{\pi}\cos(\omega_0 t - \omega_s t + \varphi_1) \\
&= -i_{MIX,Ip}
\end{aligned}
$$

(14.4)

and a consistent proof for I/Q-DBMs under a 4-phase 25% LO is obtained. Ideally, the DBM can correct perfectly the gain and phase errors from the balun-LNA, independent of its different output impedances from the CG and CS branches. In fact, even if the conversion gain of the 2 mixer pairs (M_5, M_8 and M_6, M_7) does not match (e.g., due to non-50% LO duty cycle), the double-balanced operation can still generate balanced outputs (confirmed by simulations). Of course, the output impedance of the DBM can be affected by that of the balun-LNA [Fig. 14.2a], but is highly desensitized due to the small size of R_L (i.e., the input impedance of the hybrid filter) originally aimed for current-mode operation. Thus, the intrinsic imbalance between $V_{o,Ip}$ and $V_{o,In}$ is negligibly small (confirmed by simulations).

For devices sizing, a longer channel length (L = 0.18 μm) is preferred for M_{5-8} to reduce their 1/f noise and V_t. Hard-switch mixing helps to desensitize the I/Q-DBMs to LO gain error, leaving the image rejection ratio (IRR) mainly determined by the LO phase error that is a tradeoff with the LO-path power. Here, the targeted LO phase error is relaxed to ~4°, as letting the BB circuitry (i.e., the complex-pole load and 3-stage RC-CR PPF) to handle the IRR is more power efficient.

Hybrid Filter 1st Half—Current-Mode Biquad with IF Noise-Shaping—The current-mode Biquad [Fig. 14.4a] proposed in Pirola et al. (2010) is an excellent candidate for current-reuse with the

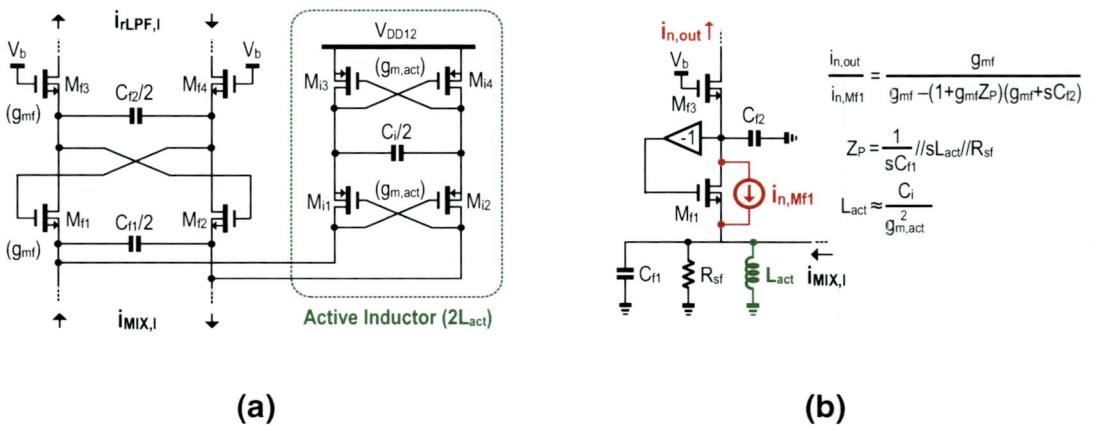

(a)

(b)

Fig. 14.4 (a) Proposed IF-noise-shaping Biquad and (b) its small-signal equivalent circuit showing the noise TF of M_{f1}

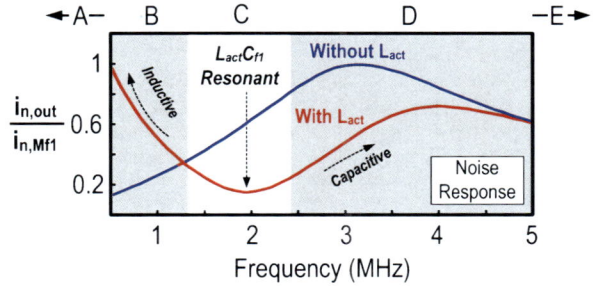

Fig. 14.5 (**a**) Equivalent impedance of Z_P versus ω_{or}, and (**b**) simulated noise TF of $\frac{i_{n,out}}{i_{n,Mf1}}$ with and without L_{act}

Blixer for channel selection. However, this Biquad only can generate a noise-shaping zero spanning from DC to $\sim 2\pi 0.1 Q_B \omega_{0B}$ MHz for M_{f1}–M_{f2}, where Q_B and ω_{0B} are the Biquad's quality factor and -3-dB cutoff frequency, respectively. This noise shaping is hence ineffective for our low-IF design having a passband from ω_1 to ω_2 ($= \omega_{0B}$), where $\omega_1 > 0.1 Q_B \omega_{0B}$. To address this issue, an active inductor (L_{act}) is added at the sources of M_{f1}–M_{f2}. The $L_{act}C_{f1}$ resonator shifts the noise-shaping zero to the desired IF. The cross-diode connection between M_{i1}–M_{i4} (all with $g_{m,act}$) emulate $L_{act} \approx C_i/g_{m,act}^2$ (Ler et al. 2008; Chen et al. 2012). The small-signal equivalent circuit to calculate the noise TF of $i_{n,Mf1}/i_{n,out}$ is shown in Fig. 14.4b. The approximated impedance of Z_P in different frequencies related to ω_{0r} is summarized in Fig. 14.5a, where $\omega_{0r} = \frac{\omega_1 + \omega_2}{2}$ is the resonant frequency of $L_{act}C_{f1}$ at IF. The simulated $i_{n,Mf1}/i_{n,out}$ is shown in Fig. 14.5b. At the low frequency range, Z_P behaves inductively, degenerating further $i_{n,Mf1}$ when the frequency is increased. At the resonant frequency, $Z_P = R_{sf}$, where R_{sf} is the parallel impedance of the active inductor's shunt resistance and DBM's output resistance. The latter is much higher when compared with R_L thereby suppressing $i_{n,Mf1}$. At the high frequency range, Z_P is more capacitive dominated by C_{f1}. It implies $i_{n,Mf1}$ can be leaked to the output via C_{f1}, penalizing the

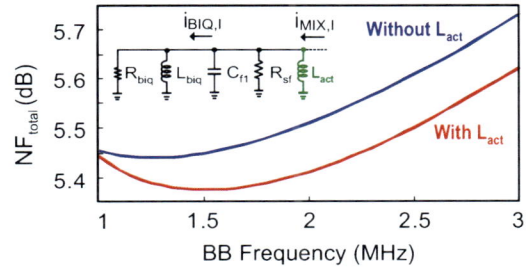

Fig. 14.6 Simulated NF_{Total} (at $V_{o,Ip}$ and $V_{o,In}$) with and without L_{act}

in-band noise. At even higher frequencies, the output noise decreases due to C_{f2}, being the same as its original form (Pirola et al. 2010).

The signal TF can be derived from Fig. 14.6. Here $R_L = \frac{1}{g_{mf}}$, $L_{biq} = \frac{C_{f2}}{g_{mf}^2}$. For an effective improvement of NF, $L_{act} \gg L_{biq}$ should be made. The simulated NF_{total} at $V_{o,Ip}$ and $V_{o,In}$ with and without the L_{act} is shown Fig. 14.6, showing about 0.1 dB improvement at the TT corner (reasonable contribution for a BB circuit). For the SS and FF corners, the NF improvement reduces to 0.04 and 0.05 dB, respectively. These results are expected due to the fact that at the FF corner, the noise contribution of the BB is less significant due to a larger bias current; while at the SS corner, the IF noise-shaping circuit will add more noise by itself, offsetting the NF improvement. Here M_{f1}–M_{f4} use isolated P-well

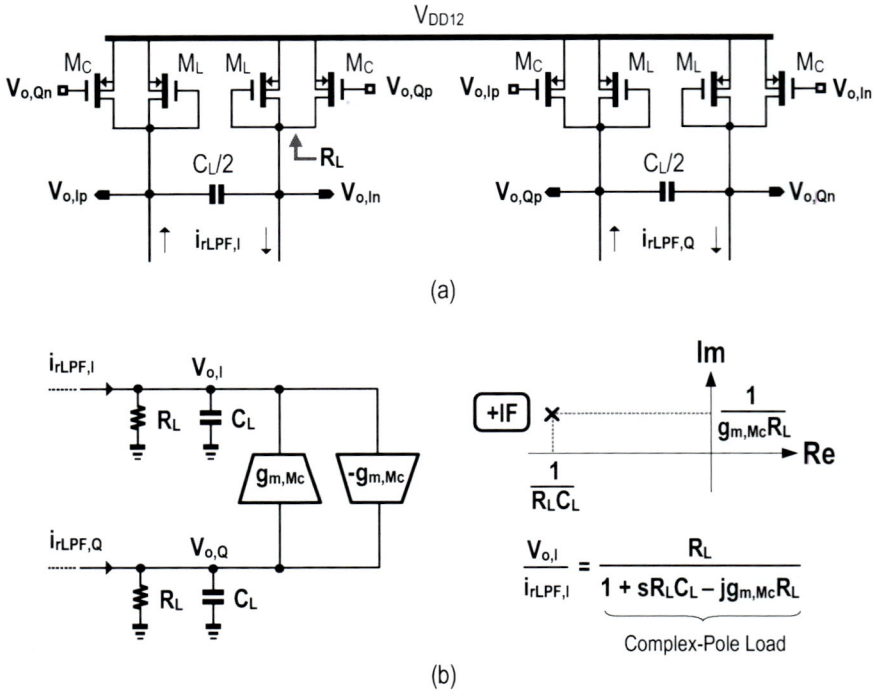

Fig. 14.7 (a) Proposed complex-pole load and (b) its small-signal equivalent circuit and pole plot

for bulk-source connection, avoiding the body effect while lowering their V_T.

Hybrid Filter 2nd Half—Complex-Pole Load—Unlike most active mixers or the original Blixer (Blaakmeer et al. 2008a, b) that only use a RC load, the proposed "load" synthesizes a 1st-order complex pole at the positive IF (+IF) for channel selection and image rejection. The circuit implementation and principle are shown in Fig. 14.7a and b, respectively. The real part (R_L) is obtained from the diode-connected M_L, whereas the imaginary part ($g_{m,Mc}$) is from the I/Q-cross-connected M_C. The entire hybrid filter offers 5.2-dB IRR, and 12-dB (29-dB) adjacent (alternate) channel rejection as shown in Fig. 14.8 (the channel spacing is 5 MHz). Similar to g_m-C filters the center frequency is defined by $g_{m,Mc}R_{L.}$. When sizing the −3-dB bandwidth, the output conductances of M_C and M_L should be taken into account.

Current-Mirror VGA and RC-CR PPF—Outside the current-reuse path, $V_{o,I}$ and $V_{o,Q}$ are AC-coupled to a high swing current-mirror

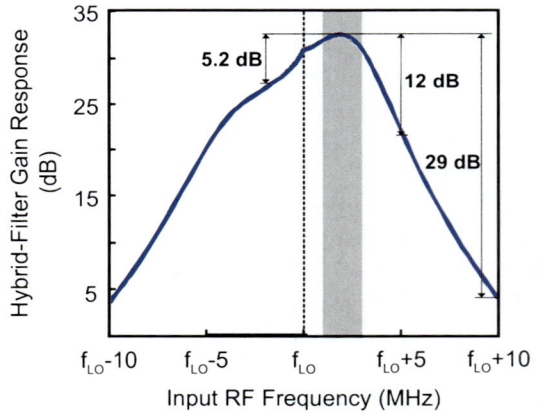

Fig. 14.8 Simulated hybrid-filter gain response

VGA formed with M_L [Fig. 14.7a] and a segmented M_{VGA} (Fig. 14.9), offering gain controls with a 6-dB step size. To enhance the gain precision, the bias current through M_{VGA} is kept constant, so as its output impedance. With the gain switching of M_{VGA}, the input-referred noise of M_{VGA} will vary. However, when the RF signal

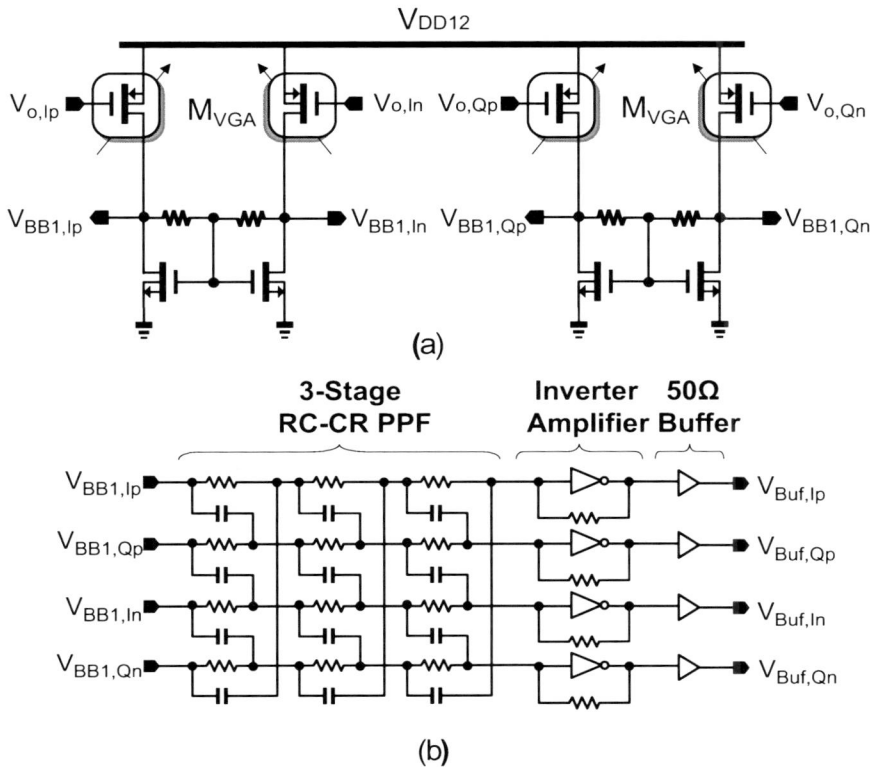

Fig. 14.9 Schematics of the BB (**a**) VGA, and (**b**) 3-stage RC-CR PPF, inverter amplifier and 50-Ω buffer

level is low the gain of the VGA should be high, rendering the gain switching not influencing the receiver's sensitivity. The VGA is responsible for compensating the gain loss (30 dB) of the 3-stage passive RC-CR PPF that provides robust image rejection of >50 dB (corner simulations). With the hybrid filter rejecting the out-band blockers the linearity of the VGA is further relaxed, so as its power budget (192 μW, limited by the noise and gain requirements).

A 3-stage RC-CR PPF can robustly meet the required IRR in the image band (i.e., the –IF), and cover the ratio of maximum to minimum signal frequencies (Kaykovuori et al. 2008; Behbahani et al. 2001). In our design, the expected IRR is 30 to 40 dB and the ratio of frequency of the image band is f_{max}/f_{min} (=3). However, counting the RC variations as large as $\pm 25\%$, the conservative $\Delta f_{eff} = f_{max\_eff}/f_{min\_eff}$ should be close to 5. The selected RC values are guided by Behbahani et al. (2001)

$$\frac{\sigma(\text{Image Out})}{\text{Desired Out}} = 0.25\sqrt{\left(\frac{\sigma_R}{R}\right)^2 + \left(\frac{\sigma_C}{C}\right)^2}$$

(14.5)

Accordingly, the matching of the resistors (σ_R) and capacitors (σ_C) can be relaxed to 0.9% (2.93%) for 40-dB (30-dB) IRR with a 3σ yield. Here, ~150-kΩ resistors are chosen to ease the layout with a single capacitor size (470 fF), balancing the noise, area and IRR. The simulated worst IRR is 36 dB without LO mismatch, and still over 27 dB at a 4° LO phase error checked by 100x Monte-Carlo simulations. Furthermore, if the 5-dB IRR offered by the complex-pole load is added the minimum IRR of the IF chain should be 32 dB. The final stage before 50-Ω output buffering is a self-biased inverter amplifier (power = 144 μW), which embeds one more real pole for filtering. The simulated overall IF gain response is shown in Fig. 14.10, where the notches at DC offered by the AC-coupling network, and around the −IF offered by the 3-stage RC-CR PPF, are visible. The IRR is about 57 dB [=52 dB (RC-CR PPF) + 5 dB (complex-pole load)] under an ideal 4-phase 25% LO for the image band from [f_{LO} − 3, f_{LO} − 1] MHz.

VCO and Dividers and LO Buffers—To fully benefit the speed and low-V_t advantages of fine linewidth CMOS, the entire LO path is powered at a lower supply of 0.6 V to reduce the dynamic power. For additional testability, an on-chip VCO is integrated. It is optimized at ~10 GHz to save area and allows division by 4 for I/Q

generation. The loss of its LC tank is compensated by complementary NMOS-PMOS negative transconductors.

The divider chain [Fig. 14.11a] cascades two types of div-by-2 circuits (DIV1 and DIV2) to generate the desired 4-phase 25% LO, from a 2-phase 50% output of the VCO. The two latches (D1 and D2) are employed to build DIV1 that can directly generate a 25% output from a 50% input (Razavi et al. 1995), resulting in power savings due to less internal logic operation (i.e., AND gates) and load capacitances. Each latch consists of two sense devices, a regenerative loop and two pull up devices. For 25%-input 25%-output division, DIV2 is proposed that it can be directly interfaced with DIV1. The 25% output of DIV1 are combined by M_{D1} to M_{D4} to generate a 50% clock signal for D3 and D4.

For testing under an external LO_{ext} source at 4.8 GHz, another set of D1 and D2 is adopted. The output of these two sets of clocks are combined by transmission gates and then selected. Although their transistor sizes can be reduced aggressively to save power, their drivability and robustness in process corners can be degraded. From simulations, the sizing can be properly optimized. The four buffers (Buf_{1-4}) serve to reshape the pulses from DIV2 and enhance the

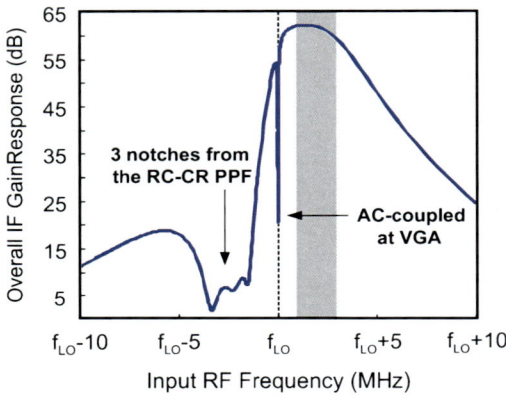

Fig. 14.10 Simulated overall IF gain response

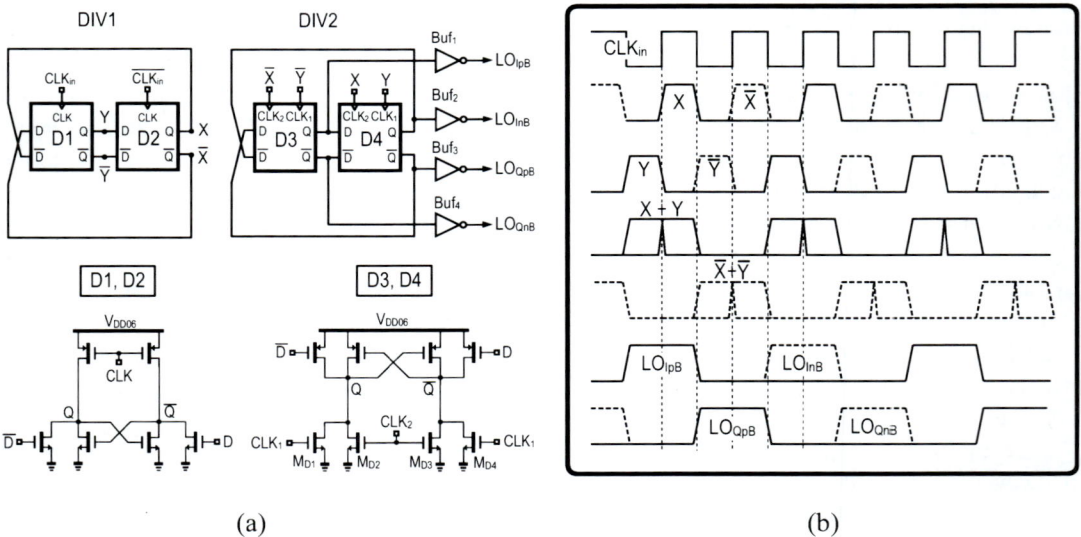

(a)   (b)

Fig. 14.11 (a) Schematics of DIV1 and DIV2, and (b) their timing diagrams

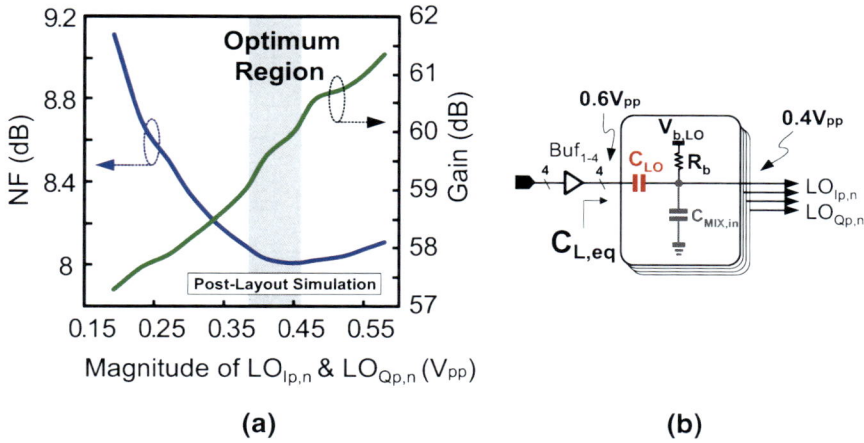

Fig. 14.12 (a) Post-layout simulation of NF and gain versus LO's amplitude, and (b) additional C_{LO} generates the optimum LO's amplitude

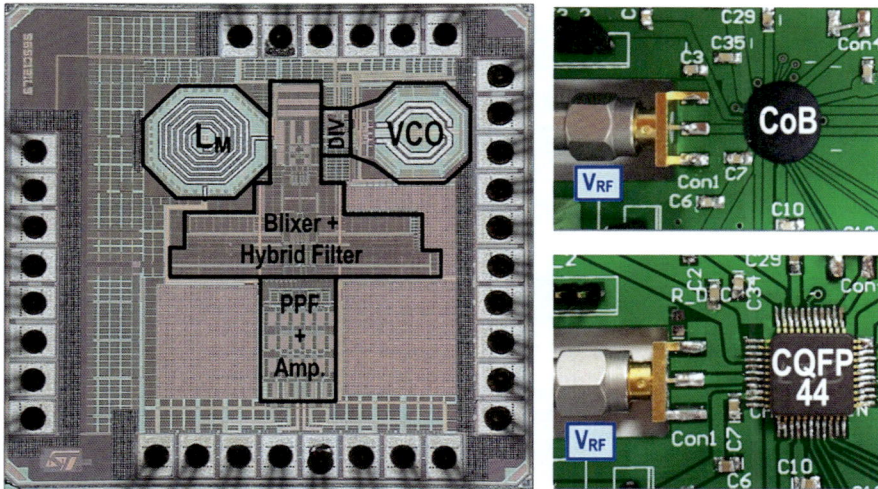

Fig. 14.13 Chip micrograph of the receiver. It was tested under CoB and CQFP44 packaging. No external component is entailed for input matching

drivability. The timing diagram is shown in Fig. 14.11b. Due to the very small I_{BIAS} for the I/Q-DBMs, a LO amplitude of around 0.4 V_{pp} is found to be more optimized in terms of NF and gain as simulated and shown in Fig. 14.12a. To gain benefits from it C_{LO} is added to realize a capacitor divider with $C_{MIX,in}$ (input capacitance of the mixer) as shown in Fig. 14.12b. This act brings down the equivalent load ($C_{L,eq}$) of Buf_{1-4} by ~33%.

14.3.2 Experimental Results

The ZigBee receiver was fabricated in 65-nm CMOS (Fig. 14.13) and optimized with dual supplies (1.2 V: Blixer + hybrid filter, 0.6 V: LO and BB circuitries). The die area is 0.24 mm^2 (0.3 mm^2) without (with) counting the LC-tank VCO. Since there is no frequency synthesizer integrated, the results in Fig. 14.14a–d were measured under LO_{ext} for accuracy and data

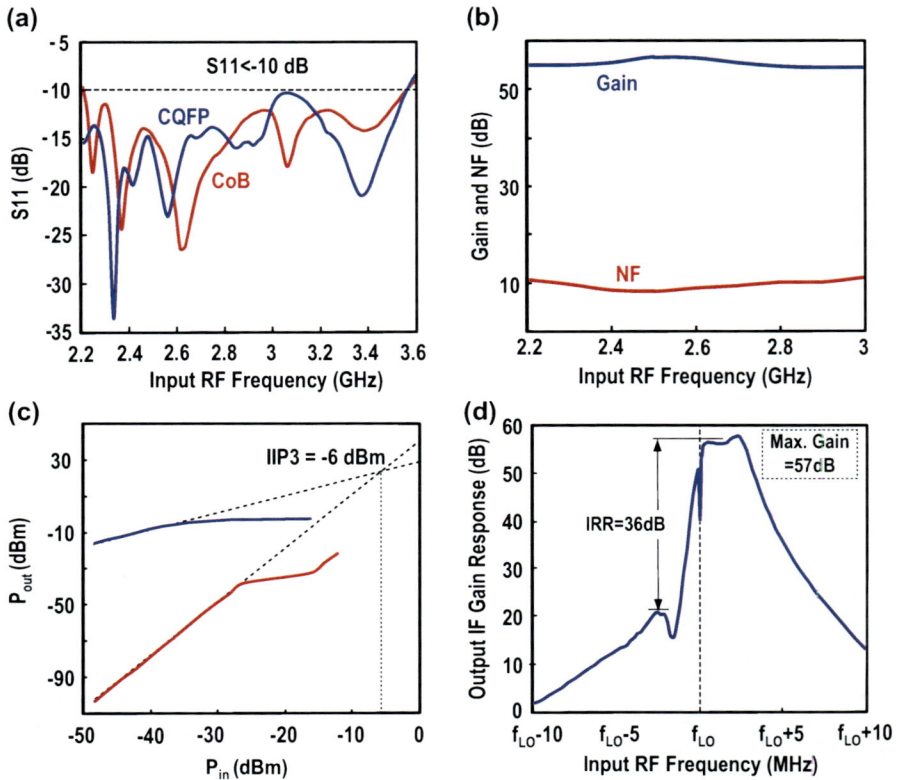

Fig. 14.14 Measured (**a**) S_{11}, (**b**) wide band gain and NF, (**c**) IIP3$_{out\text{-}band}$, and (**d**) low-IF filtering profile

repeatability. The S_{11}-BW (<-10 dB) is ~1.3 GHz for both chip-on-board (CoB) and CQFP-packaged tests [Fig. 14.14a], which proves its immunity to board parasitics and packaging variations. The gain (55–57 dB) and NF (8.3–11.3 dB) are also wideband consistent [Fig. 14.14b]. The gain peak at around 2.4–2.5 GHz is from the passive pre-gain. Following the linearity test profile of Tedeschi et al. (2010), two tones at [LO + 12 MHz, LO + 22 MHz] are applied, measuring an IIP3$_{out\text{-}band}$ of − 6 dBm [Fig. 14.14c] at the maximum gain of 57 dB (there is 24-dB gain loss in Fig. 14.14c associated with the test buffer and used 1:8 transformer). This high IIP3 is due to the direct current-mode filtering at the mixer's output before signal amplification. The asymmetric IF response [Fig. 14.14d] shows 22-dB (43-dB) rejection at the adjacent (alternate) channel, and 36-dB IRR. Differing from the simulated IF frequency response that has three notches at the

image band under an ideal LO, the measured notches are merged. Similar to Behbahani et al. (2001), this discrepancy is likely due to the LO gain and phase mismatches, and the matching and variations of the RC-CR networks. The layout design is similar to Behbahani et al. (2001) that uses dummy to balance the parasitic capacitances. The filtering rejection profile is ~80 dB/decade. The spurious free dynamic range (SFDR) is close to 60 dB according to Tedeschi et al. (2010),

$$\mathrm{SFDR} = \frac{2(\mathrm{P_{IIP3}} + 174\mathrm{dBm} - \mathrm{NF} - 10\log\mathrm{BW})}{3}$$
$$- \mathrm{SNR_{min}}$$

$$(14.6)$$

where SNR$_{min}$ = 4 dB is the minimum signal-to-noise ratio required by the application, and BW = 2 MHz is the channel bandwidth.

The receiver was further tested at lower voltage supplies as summarized in Table 14.2. Only the NF degrades more noticeably, the IIP3, IRR and BB gain are almost secured. The better IIP3 for 0.6-V/1-V operation is mainly due to the narrower −3-dB bandwidth of the hybrid filter. For the 0.5-V/1-V operation, the degradation of IIP3$_{out-band}$ is likely due to the distortion generated by A$_{GB}$. Both cases draw very low power down to 0.8 mW, being comparable with other ULP designs such as Herberg et al. (2011).

The LC-tank VCO was tested separately. Its power budget is related with its output swing and

is a tradeoff with the phase noise, which measures—114 dBc/Hz at 3.5 MHz that has an enough margin to the specifications (Liscidini et al. 2008) [Fig. 14.15a]. Porting it to the simulation results, it can be found that the corresponding VCO's output swing is 0.34 V$_{pp}$ and the total LO-path power is 1.7 mW (VCO + dividers + BUFs). Such an output swing is adequate to lock DIV1 as shown in its simulated sensitivity curve [Fig. 14.15b].

The chip summary and performance benchmarks are given in Table 14.3, where (Tedeschi et al. 2010) is a current-reuse architecture and (Zhang et al. 2013) is an ultra-low-voltage design. For this work, the results measured under a 10-GHz on-chip VCO are also included for completeness, but they are more sensitive to test uncertainties. The degraded NF and IRR are mainly due to the phase noise of the free-running VCO. In both cases, this work succeeds in advancing the IIP3$_{out-band}$, power and area efficiencies, while achieving a wideband S$_{11}$ with zero external components. Particularly,

Table 14.2 Key performances of the receiver at different supply voltages

Supply voltage (V)	0.6/1.2	0.6/1	0.5/1
Power (mW)	1.7	1.2	0.8
Gain (dB)	57	58	57.5
IIP3$_{out-band}$ (dBm)	−6	−4	−8
NF (dB)	8.5	11.3	12
IRR (dB)	36	38	35

Fig. 14.15 (a) The measured phase noise has enough margin to the specifications. From simulations, it can be shown that it is a tradeoff with the power budget according to the VCO's output swing. (b) Simulated sensitivity curve of DIV1 showing its small input-voltage requirement at ~10 GHz

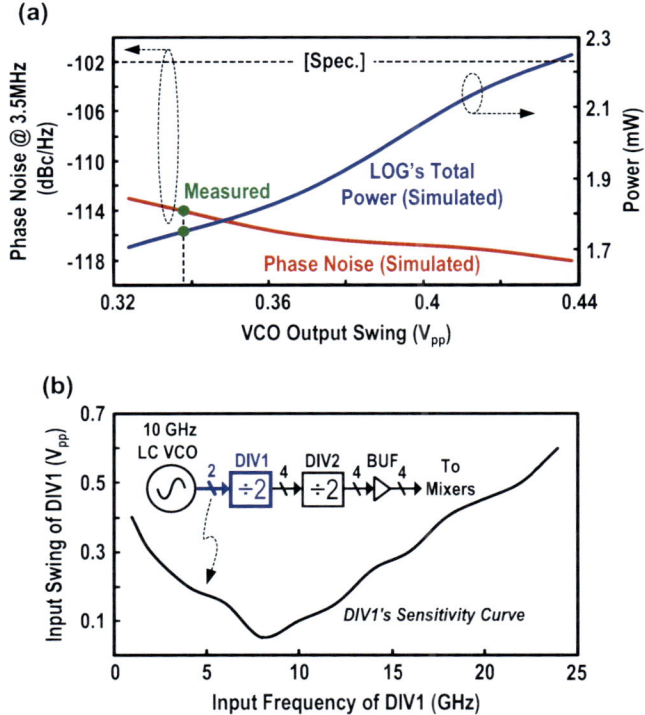

Table 14.3 Performance summary and benchmark with the state-of-the-art

	This work (Lin et al. 2013, 2014a)		JSSC '10 (Tedeschi et al. 2010)	ISSCC '13 (Zhang et al. 2013)
Application	ZigBee		ZigBee	Energy Harvesting
Architecture	Blixer + Hybrid-Filter + Passive RC-CR PPF		LMV Cell + Complex Filter	LNA + Mixer + Frequency-translated IF Filter
BB Filtering	*1 Biquad + 4 complex poles*		3 complex poles	2 real poles
External I/P matching components	*zero*		1 inductor, 1 capacitor	2 capacitors, 1 inductor
$S_{11} < -10$ dB Bandwidth (MHz)	*1300 (2.25 to 3.55 GHz)*		<300 (2.3 to 2.6 GHz)	>600 (<2 to 2.6 GHz)
Integrated VCO	No	Yes	Yes	Yes
Gain (dB)	57	55	75	83
Phase Noise (dBc/Hz)	NA	−115 @ 3.5 MHz	−116 @ 3.5 MHz	−112.8 @ 1 MHz
NF (dB)	8.5	9	9	6.1
IIP3$_{\text{out-band}}$ (dBm)	*−6*	*−6*	−12.5	−21.5
IRR (dB)	36 (worst of 5 chips)	28	35	N/A
SFDR (dB)	*60.3*	*60*	55.5	51.6
LO-to-RF Leak (dBm)	−61	−61	−60	N/A
Power (mW)	*1.7*[a]	*2.7*	3.6	1.6
Active Area (mm^2)	*0.24*	*0.3*	0.35	2.5
Supply Voltage (V)	0.6/1.2		1.2	0.3
Technology	65 nm CMOS		90 nm CMOS	65 nm CMOS

[a]Breakdown: 1 mW: Blixer + hybrid filter + BB circuitry, 0.7 mW: DIV1 + LO Buffers

when comparing with the recent work (Zhang et al. 2013), this work achieves 8x less area and 15.5 dBm higher IIP3, together with stronger BB channel selectivity.

14.4 Function-Reuse ULP Receiver Techniques for the Sub-GHz ISM Bands

Differing from the previous design that is for single band, the function-reuse receiver (Lin et al. 2014b, c) to be described here can flexibly support multiple bands (433/860/915/960 MHz) and can operate at a single low VDD. It features a gain-boosted N-path switched-capacitor (SC) network embedded into a function-reuse RF front-end, offering concurrent RF (common-mode) and BB (differential-mode) amplification, LO-defined RF filtering, and input impedance matching with zero external components. The details are presented next.

14.4.1 Receiver Architecture

Specifically, the gain-boosted 4-path SC network [Fig. 14.16a] separates the output of each gain stage G_m (G_m has a transconductance of g_{m3}, output resistance of $4R_L$, and feedback resistor of R_{F3}) with capacitor C_o that is an open circuit at BB. The I/Q BB signals at $V_{B1,I\pm}$ and $V_{B1,Q\pm}$ are further amplified along the Path C [Fig. 14.16b] by each G_m stage. With the memory effect of the capacitors, the functional view of the gain response is shown in Fig. 14.16c. In order to achieve current-reuse between the RF LNA and BB amplifiers without increasing the VDD, the circuit published in Han and Gharpurey (2008) with an active mixer has a similar function. However, the BB NF behavior and the RF filtering behavior are different from the N-path passive mixer applied here that is at the feedback path. For the BB amplifiers, it is one G_m with one R_{F3}, balancing the BB gain and OB-IIP3. After considering that the BB amplifiers have been

(a)

Common-Mode RF Signal ◀ (A)

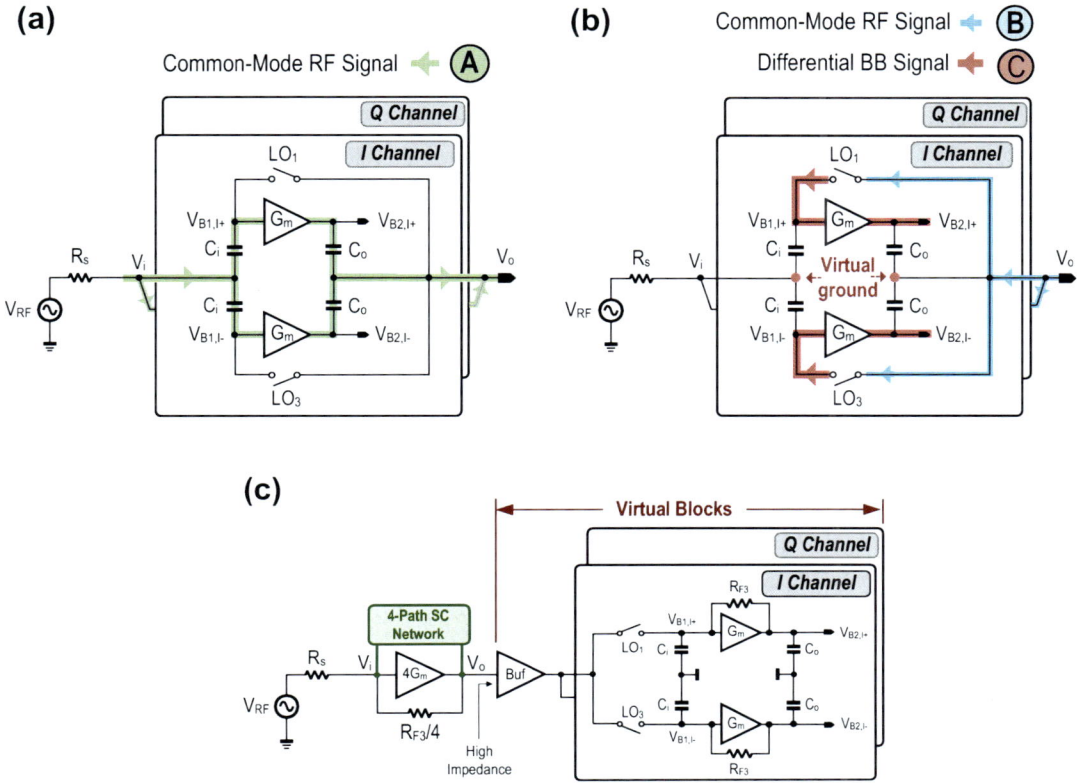

(b)

Common-Mode RF Signal ◀ (B)

Differential BB Signal ◀ (C)

(c)

Fig. 14.16 (a) Function-reuse receiver embedding a gain-boosted 4-path SC network and its operation for RF signal, (b) BB signals and (c) its functional view to model the gain response

absorbed in the LNA, the I/Q passive mixers and capacitors absorbed by the 4-path SC network, the blocks after the LNA can be assumed virtual. These virtual blocks reduce the power, area and NF. To validate the above viewpoint, the gain and noise performances under two sets of R_{F3} are simulated. Here, the virtual blocks in Fig. 14.16c are implemented with physical transistors and capacitors for the BB amplifiers and the mixers while the buffer is ideal. Thus, the power of the modeled receiver is at least 2x larger than the proposed receiver. For the IB BB gain at $V_{B2,I\pm}$ ($V_{B2,Q\pm}$) between the proposed function-reuse receiver and its functional view, the difference is only 1 dB at a large R_{F3} of 150 kΩ [Fig. 14.17a]. For a small R_{F3}, the gain error goes up to 2 dB [Fig. 14.17b], which is due to the gain difference between the model of the N-path tunable LNA [Fig. 14.16c] and the

implementation of the function-reuse receiver that has AC-coupling. For the NF difference (ΔNF), with a large (small) R_{F3}, it is ~0.8 dB (3.5 dB) as compared in Fig. 14.18a, b. This is due to the lower gain at the LNA's output, forcing the input-referred noise from the downconversion passive mixers and the BB amplifiers to increase with a small R_{F3}. Either with a small or large R_{F3}, it is noteworthy that the variation of BB NF is small (i.e., for $R_{F3} = 20$ kΩ it is 3.6 dB while for $R_{F3} = 150$ kΩ it is 3.4 dB), because the BB NTF has a weak relation with R_{F3}. It also indicates that the BB NTF is weakly related with the gain at the LNA's output, which is dissimilar to the usual receiver where the NF should be small when the LNA's gain is large. Similarly, the NF at the LNA's output (now shown) can be larger than that at BB due to the different NTFs.

Fig. 14.17 Simulated BB gain response of the function-reuse receiver and its functional view with (**a**) a large R_{F3} and (**b**) a small R_{F3}

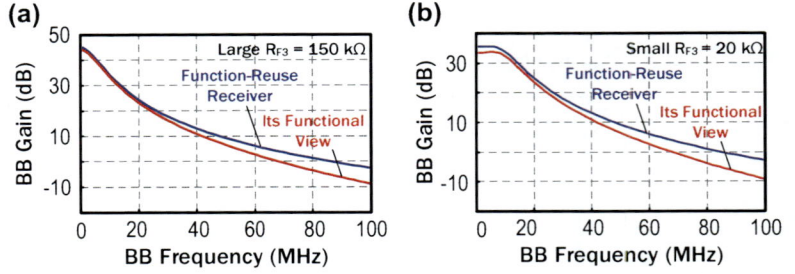

Fig. 14.18 Simulated BB NF of the function-reuse receiver and its functional view with (**a**) a large R_{F3} and (**b**) a small R_{F3}

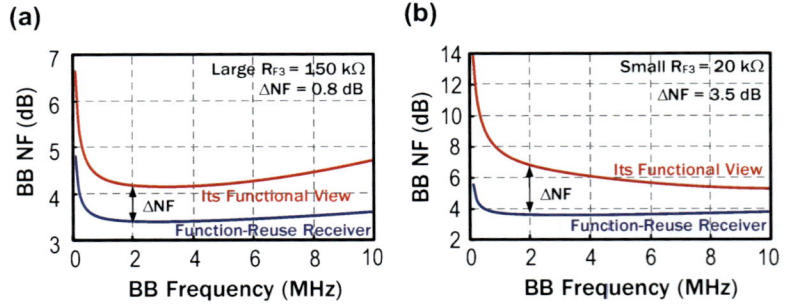

14.4.2 Low-Voltage Current-Reuse VCO-Filter

In order to further optimize the power, the VCO is designed to current-reuse with the BB complex low-IF filter (Fig. 14.19). The negative transconductor of the VCO is divided into multiple M_v cells. The aim is to distribute the bias current of the VCO to all BB gain stages (A_1, $A_2...A_{18}$) that implement the BB filter. For the VCO, M_V operates at the frequency of $2f_s$ or $4f_s$ for a div-by-2 or div-by-4 circuit. Thus, the VCO signal leaked to the source nodes of M_V ($V_{F1,I+}$, $V_{F1,I-}$) is pushed to very high frequencies ($4f_s$ or $8f_s$) and can be easily filtered by the BB capacitors. For the filter's gain stages such as A_1, M_b (g_{Mb}) is loaded by an impedance of ~1/ $2g_{Mv}$ when L_p can be considered as a short circuit at BB. Thus, A_1 has a ratio-based voltage gain of roughly g_{Mb}/g_{Mv}, or as given by $4Tg_{Mb}/G_{mT}$, where G_{mT} is the total trans-conductance for the VCO tank. The latter shows how the distribution factor T can enlarge the BB gain, but is a tradeoff with its input-referred noise and can add more layout parasitics to $V_{vcop,n}$ (i.e., narrower VCO's

tuning range). The –R cell using cross-coupled transistors is added at $V_{F1,I+}$ and $V_{F1,I-}$ to boost the BB gain without loss of voltage headroom. For the BB complex poles, $A_{2,5}$ and C_{f1} determine the real part while $A_{3,6}$ and C_{f1} yield the imaginary part. There are three similar stages cascaded for higher channel selectivity and image rejection ratio (IRR). R_{blk} and C_{blk} were added to avoid the large input capacitance of $A_{1,4}$ from degrading the gain of the front-end.

14.4.3 Experimental Results

Two versions of the receiver were fabricated in 65-nm CMOS (Fig. 14.20) and optimized with a single 0.5-V VDD. With (without) the LC tank for the VCO, the die area is 0.2 mm^2 (0.1 mm^2). Since the measurement results of both are similar, only those measured with VCO in Fig. 14.21a–d are reported here. From 433 to 960 MHz, the measured BB gain is 50 ± 2 dB. Two tones at [f_s + 12 MHz, f_s + 22 MHz] are applied, measuring an OB-IIP3 of –20.5 ± 1.5 dBm at the maximum gain. The IRR is

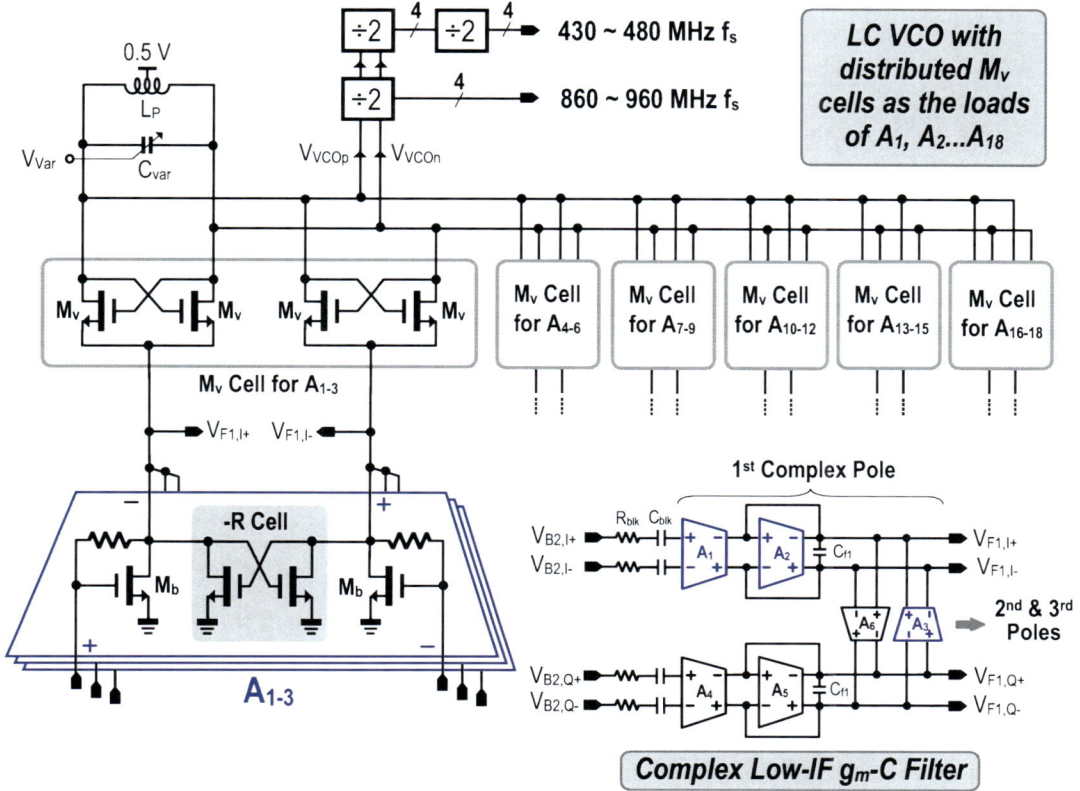

Fig. 14.19 Proposed low-voltage current-reuse VCO-filter

Fig. 14.20 Chip micrograph of the function-reuse receiver with a LC-tank for the VCO (*left*) and without it (*right*)

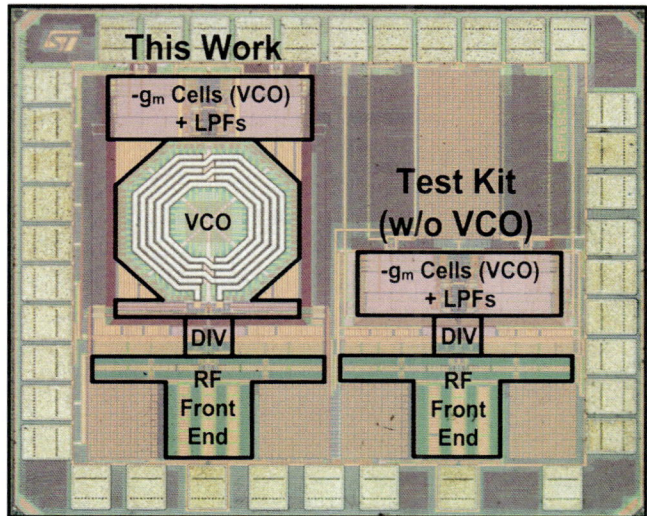

20.5 ± 0.5 dB due to the low-Q of the VCO-filter. The IIP3 is mainly limited by the VCO-filter. The measured NF is 8.1 ± 0.6 dB.

Since the VCO is current-reuse with the filter, it is interesting to study its phase noise with the BB signal amplitude. For negligible phase noise

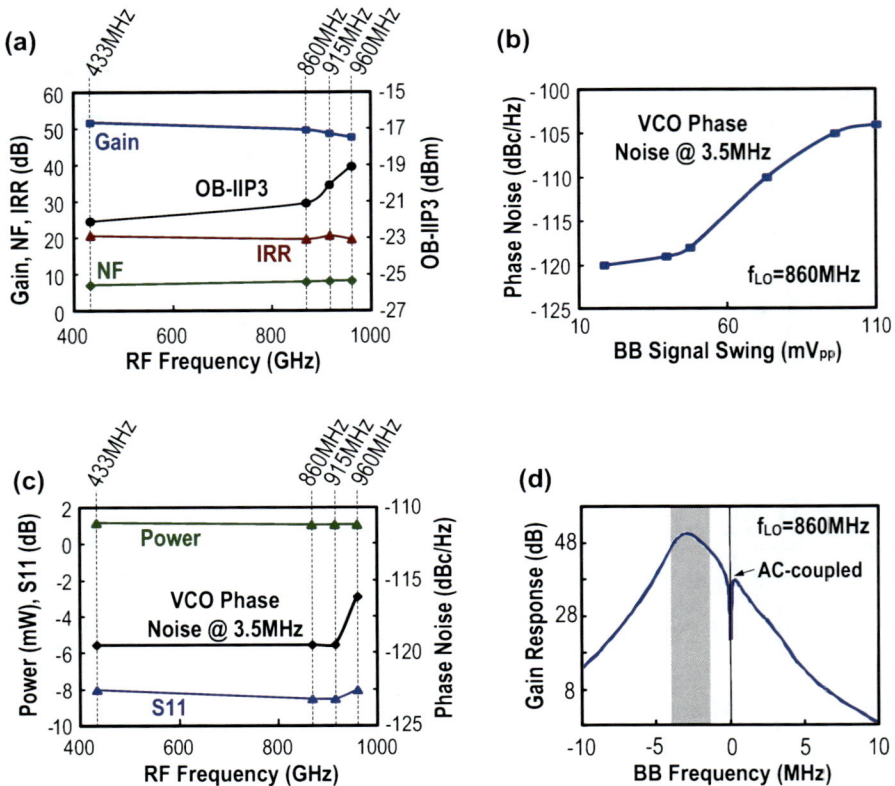

Fig. 14.21 Measured key performance metrics: (**a**) Gain, NF, IRR and OB-IIP3. (**b**) VCO phase noise versus BB signal swing. (**c**) S11, power and VCO phase @ 3.5-MHz offset. (**d**) BB complex gain response centered at −2-MHz IF

degradation, the BB signal swing should be <60 mV$_{pp}$, which can be managed by variable gain control. If a 60-mV$_{pp}$ BB signal is insufficient for demodulation, a simple gain stage (e.g., inverter amplifier) can be added after the filter to enlarge the gain and output swing. The total power of the receiver is 1.15 mW (0.3 mW for the LNA + BB amplifiers and 0.65 mW for VCO-filter and 0.2 mW for the divider), while the phase noise is −117.4 ± 1.7 dBc/Hz at 3.5-MHz frequency offset. The S$_{11}$ is below −8 dB across the whole band. The asymmetric IF response shows 24-dB (41-dB) rejection at the adjacent (alternate) channel.

To study the RF filtering behavior, the P$_{1dB}$ and blocker NF are measured. For the in-band signal, the P$_{1dB}$ is −55 dBm while with a frequency offset frequency of 20 MHz, it increases to −35 dBm, which is mainly due to the double-

RF filtering [Fig. 14.22a]. For an offset frequency of 60 MHz, the P$_{1dB}$ is −20 dBm, limited by the current-reuse VCO-filter. For the blocker NF, with a single tone at 50 MHz, the blocker NF is almost unchanged for the blocker <−35 dBm. With a blocker power of −20 dBm, the NF is increased to ~14 dB [Fig. 14.22b].

This work is compared with the prior art in Table 14.4, where (Lin et al. 2013) is the current-reuse architectures described previously, while (Zhang et al. 2013) is the cascade architecture with ULV supply for energy harvesting. For this work, the results measured under an external LO are also included for completeness. In both cases, this work succeeds in advancing the power and area efficiencies with multi-band convergence, while achieving tunable S$_{11}$ with zero external components. When comparing with the most recent ULV design (Zhang et al. 2013), this

Fig. 14.22 Measured (**a**) P_{1dB} versus input offset frequency and (**b**) blocker NF versus input power

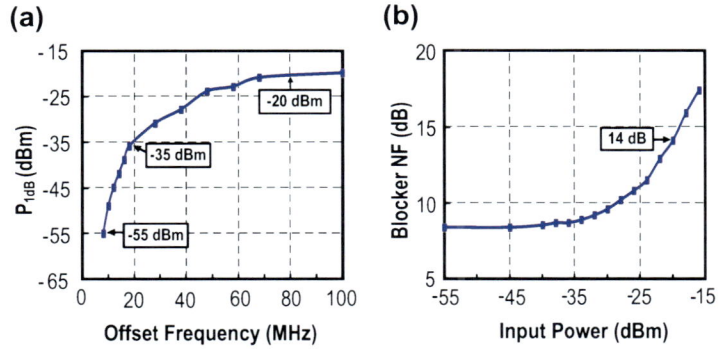

Table 14.4 Performance summary and benchmark with the state-of-the-art RXs

	This work (Lin et al. 2014b)	ISSCC '13 (Lin et al. 2013)	ISSCC '13 (Zhang et al. 2013)
Application	**433/860/915/960 MHz (ZigBee/ IEEE802.15.4c/d)**	2.4 GHz (ZigBee/IEEE 802.15.4)	2.4 GHz (Energy Harvesting)
Architecture	**Function-Reuse RF Front-End + N-path Tunable LNA + Current-Reuse VCO-Filter**	Blixer + Hybrid Filter + Passive RC-CR Filter + LC VCO	CG LNA + Passive Mixers + N-Path SC IF Filter + LC VCO
BB Filter	3 complex poles	1 Biq., 4 complex poles	2 real poles
Input matching technique	**On-chip N-path SC (tunable by LO, high Q)**	On-chip LC (fixed, low Q)	Off-chip LC (fixed, low Q)
External components	**zero**	zero	2 caps, 1 inductor
Input matching BW and tunability	**433 to 960 MHz (tunable by LO)**	2.25 to 3.55 GHz (fixed)	~2 to 2.6 GHz (fixed)
Active Area (mm²)	**0.2 (0.1[a])**	0.3	2.5
Power (mW) @V_{DD}	**1.15 ± 0.05 @ 0.5 V**	2.7 @ 0.6/1.2 V	1.6 @ 0.3 V
Gain (dB)	50 ± 2 (51 ± 3[a])	55	83
NF (dB)	8.1 ± 0.6 (8 ± 1[a])	9	6.1
OB-IIP3 (dBm)	−20.5 ± 1.5 (−23 ± 1[a])	−6	−21.5
IRR (dB)	20.5 ± 0.5 (21 ± 0.5[a])	28	N/A
VCO Phase Noise (dBc/Hz)	−117.4 ± 1.7 @ 3.5 MHz	−115 @ 3.5 MHz	−112 @ 1 MHz
Technology	65 nm CMOS	65 nm CMOS	65 nm CMOS

[a]Results measured from the test kit that has no VCO

work saves more than 10x of area while supporting multi-band operation with zero external components.

14.5 Sub-1 V ULP 2.4GHz Transmitter

This section briefly covers an ongoing-design of a sub-1 V ULP 2.4-GHz transmitter (TX) with scalable output power (P_{out}) and system efficiency for ZigBee and other Internet-of-Things wireless solutions.

To improve the system efficiency of a TX, which normally consists of a VCO (or DCO) and a power amplifier, it is worth to consider a current-reuse topology between the two blocks. The current-reuse VCO-PA (Li et al. 2015) has demonstrated good system efficiency (17.5%), but has a limited P_{out} (-1 dBm) after stacking.

Fig. 14.23 Class-F VCO (Babaie et al. 2013)

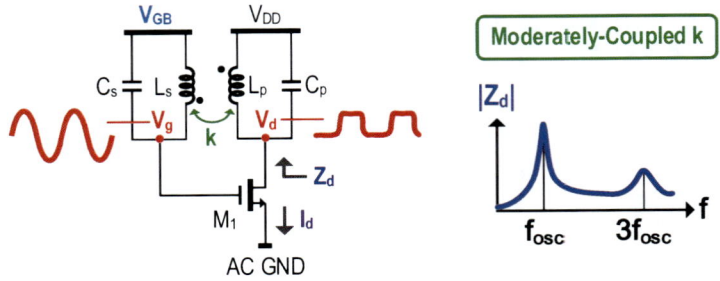

Fig. 14.24 Ongoing work on a Class-F VCO driving up the antenna (Peng et al. 2016)

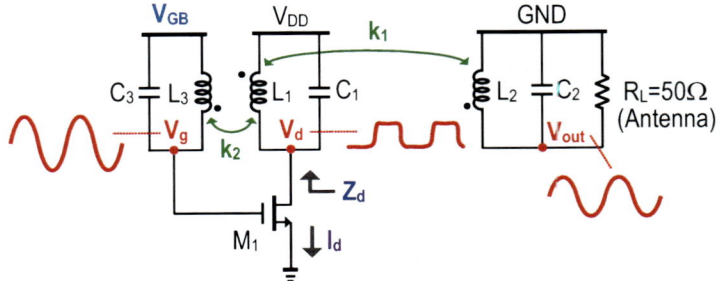

Table 14.5 Brief summary of this work with respect to the state-of-the-art TXs

Parameters	This work (Peng et al. 2017)	ISSCC '15 (Liu et al. 2015a)	ISSCC '15 (Prummel et al. 2015)
Applications	2.4 GHz ZigBee	2.4 GHz BLE/ZigBee/2 M Proprietary	2.4 GHz BLE
Architectures	Class-F VCO	Class-D PA + LC-DCO	Class-D PA + LC-VCO
On-chip inductor or transformer	1 (Shared by VCO, PA and O/P Matching)	2 (1 for LC-DCO, 1 for O/P Matching)	3 (1 for LC-VCO, 2 for O/P Matching)

Thus, our recent work (Peng et al. 2017) reported a function-reuse DCO-PA that not only shares bias current, but also upholds a full VDD for both the DCO and PA operation.

Here, the basic circuit is inspired by the class-F oscillator (Fig. 14.23), which is attractive for its high FoM of 192 dBc/Hz (Babaie et al. 2013). It features a resonant tank (L_pC_p, L_sC_s) with a moderately-coupling k to peak up the drain impedance (Z_d) at both 1st and 3rd harmonics, resulting in a pseudo-square drain voltage (V_d) to reduce the oscillator's impulse sensitivity function. Although V_d and the drain current (I_d) are alike a square-wave, the tank's resonant response recovers a sine V_g by suppressing the harmonic components, while offering a passive gain to V_g under step-up ratioing ($L_s > L_p$). These properties are preserved when it is modified as

a class-F VCO that can directly up the antenna (Fig. 14.24). This scheme reuses the amplifying device (M_1) to unify the DCO and PA functions for power and area savings (Table 14.5).

L_{1-3} form a transformer enabling self-oscillation and boosts up V_g by step-up ratioing ($L_3 > L_1$), and output-impedance matching to deliver adequate P_{out} (>0 dBm) to R_L. The coupling coefficients can be customized for optimum performance.

14.6 Conclusions and Future Perspectives

This chapter described the system aspects and circuit techniques of building cost-aware ultra-low-power (ULP) radios for both 2.4-GHz and sub-GHz ISM bands. We demonstrated that effective co-design between the system and circuit levels, as well as RF and analog block levels, are decisive to concurrently balance the cost and power budgets, while keeping up the radio performance without resorting from external components. Other state-of-the-art techniques are summarized in Lin et al. (2016) and Yu et al. (2017).

For future development, to cope with the fast market shift and many upcoming applications, multi-band multi-standard ULP radios with flexible data rate will become promising for the future IoT growth. In addition to the obvious goal of high energy efficiency during the active mode of the radios, they should also be designed for very low sleep/leakage power, preferably in the range of pW (Paidimarri et al. 2015), such that after heavily duty cycling, the average system power can be minimized. The technology choices also offer the flexibility of using low-V_t thin-oxide transistors for core circuits such that a lower VDD can be used to save power, whereas high-V_t or thick-oxide transistors can be used to suppress the sleep current. Mixed-VDD design can be a chip-level strategy for power savings (Mak and Martins 2012).

To save power, it is possible to avoid the RF PLL for channelization by using a temperature-compensated thin-film bulk-acoustic-wave resonator (FBAR) that assists the RF-to-IF downconversion. Thus, the channel selection can be delayed to lower frequency (Wang et al. 2014).

The average power is critical for a long battery lifetime, as it will dominate the maintenance cost of massive-scale wireless sensor networks. This fact urges the need of highly autonomous ULP radios that can survive with mainly/only energy harvesting. To this point, fully-integrated ULP power management units and multi-source energy harvesters will be of great importance (Masuch et al. 2013).

For the transceiver, although the sensitivity of a state-of-the-art ULP receiver is better than −90 dBm (Liu et al. 2014), their tolerability to large out-of-band blockers should have room to be further improved. The gain-boosted N-path

filtering technique (Lin et al. 2014c) can be a helpful technique to enhance the resilience of the receiver. For the state-of-the-art transmitters, their power efficiency is still not that high (Liu et al. 2015a) at a 0-dBm output power. Thus, it is worth to revisit the design of ULP PA and VCO as described in (Peng et al. 2017).

LO generation can consume significant power and area when approaching multi-band operation. For example, for a universal radio to cover the 2.4 GHz and sub-GHz ISM bands, the tuning range of the VCO should be 57% if a 2.4-GHz VCO is selected and followed by a divide-by-4 circuit. Such a wide tuning range should consume more power than the single-band design. In fact, from area and tuning range's viewpoint, a ring oscillator can be more attractive. However, to meet the required phase noise, ULP consumption is still challenging. Time-interleaved ring oscillator (Yin et al. 2016a, b) with effective phase noise reduction offers the potential to alleviate this tradeoff.

Acknowledgement This research is funded by the Macau Science and Technology Development Fund (FDCT)—SKL Fund, and the University of Macau—MYRG2015-00040-FST.

References

M. Babaie, R. Staszewski, Third-harmonic injection technique applied to a 5.87-to-7.56 GHz 65nm class-F oscillator with 192 dBc/Hz FoM. ISSCC Dig. Tech. Papers (Feb. 2013), pp. 348–349

S. Bandyopadhyay, A. Chandrakasan, Platform architecture for solar, thermal and vibration energy combining with MPPT and single inductor, in *Proceedings Symposium on VLSI Circuits* (2011), pp. 238–239

F. Behbahani, Y. Kishigami, J. Leete, A.A. Abidi, CMOS mixers and polyphase filters for large image rejection. IEEE J. Solid-State Circuits 36, 873–887 (2001)

S. Blaakmeer, E. Klumperink, D. Leenaerts, B. Nauta, Wideband balun-LNA with simultaneous output balancing, noise-canceling and distortion-canceling. IEEE J. Solid-State Circuits 43, 1341–1350 (2008a)

S. Blaakmeer, E. Klumperink, D. Leenaerts, B. Nauta, The blixer, a wideband Balun-LNA-I/Q-mixer topology. IEEE J. Solid-State Circuits 43, 2706–2715 (2008b)

F. Bruccoleri, E. Klumperink, B. Nauta, Wide-band CMOS low-noise amplifier exploiting thermal noise canceling. IEEE J. Solid-State Circuits **39**, 275–282 (2004)

Y. Chen, P.-I. Mak, L. Zhang, Y. Wang, A 0.07mm², 2mW, 75MHz-IF, 4th-order BPF using a source-follower-based resonator in 90 nm CMOS. IET Electron. Lett. **48**, 552–554 (2012)

J. Han, R. Gharpurey, Recursive receiver down-converters with multiband feedback and gain-reuse. IEEE J. Solid-State Circuits **43**, 1119–1131 (2008)

A.C. Herberg, T.W. Brown, T.S. Fiez, K. Mayaram, A250-mV, 352-μW GPS receiver RF front-end in 130-nm CMOS. IEEE J. Solid-State Circuits **46**(4), 938–949 (2011)

J. Kaykovuori, K. Stadius, J. Ryynanen, Analysis and design of passive polyphase filters. IEEE Trans. Circuits Syst. I, Reg. Papers **55**, 3023–3037 (2008)

C.L. Ler, A.K. A'ain, A.V. Kordesh, CMOS source degenerated differential active inductor. IET Electron. Lett. **44**, 196–197 (2008)

C. Li, A. Liscidini, A current re-use PA-VCO cell for low-power BLE transmitters, in *European Solid-State Circuits Conference*, Paper No. 1198 (Sept. 2015)

Z. Lin, P.-I. Mak, R.P. Martins, A 1.7 mW 0.22 mm² 2.4 GHz ZigBee RX exploiting a current-reuse Blixer + Hybrid filter topology in 65nm CMOS. ISSCC Dig. Tech. Papers (Feb. 2013), pp. 448–449

Z. Lin, P.-I. Mak, R.P. Martins, A 2.4-GHz ZigBee receiver exploiting an RF-to-BB-current-reuse Blixer + Hybrid filter topology in 65-nm CMOS. IEEE J. Solid-State Circuits **49**, 1333–1344 (2014a)

Z. Lin, P.-I. Mak, R. P. Martins, A 0.5V 1.15mW 0.2mm² Sub-GHz ZigBee receiver supporting 433/860/915/960MHz ISM bands with zero external components. ISSCC Dig. Tech. Papers (Feb. 2014b), pp. 164–165

Z. Lin, P.-I. Mak, R.P. Martins, A Sub-GHz multi-ISM-band ZigBee receiver using function-reuse and gain-boosted N-path techniques for IoT applications. IEEE J. Solid-State Circuits **49**, 2990–3004 (2014b)

Z. Lin, P.I. Mak, R.P. Martins, *Ultra-Low-Power and Ultra-Low-Cost Short-Range Wireless Receivers in Nanoscale CMOS*. Series of Analog Circuits and Signal Processing (ACSP) (Springer, Berlin, 2016)

A. Liscidini, M. Tedeschi, R. Castello, A 2.4GHz 3.6mW 0.35mm² quadrature front-end RX for ZigBee and WPAN applications. ISSCC Dig. Tech. Papers (Feb. 2008), pp. 370–371

Y.-H. Liu, A. Ba, J. van den Heuvel et al., A 1.2 nJ/bit 2.4 GHz receiver with a sliding-IF phase-to-digital converter for wireless personal/body area networks. IEEE ISSCC Dig. Tech Papers (Feb. 2014), pp. 166–167

Y.-H. Liu, C. Bachmann, X. Wang et al., A 3.7mW-RX 4.4mW-TX fully integrated bluetooth low-energy/IEEE802.15.4/proprietray SoC with an ADPLL-based fast frequency offset compensation in 40 nm CMOS. ISSCC Dig. Tech. Papers (Feb. 2015), pp. 236–237

W.-H. Yu, H. Yi, P.-I. Mak, J. Yin, R.P. Martins, A 0.18 V 382 μW bluetooth low-energy (BLE) receiver with 1.33 nW sleep power for energy-harvesting applications in 28 nm CMOS, in *IEEE ISSCC Dig., Tech Papers*, 2017, to appear

P.-I. Mak, R.P. Martins, A 0.46-mm² 4-dB NF unified receiver front-end for full-band mobile TV in 65-nm CMOS. IEEE J. Solid-State Circuits **46**, 1970–1984 (2011)

P.-I. Mak, R.P. Martins, *High-/Mixed-Voltage Analog and RF Circuit Techniques for Nanoscale CMOS*. Series of Analog Circuits and Signal Processing (ACSP) (Springer, Berlin, 2012)

J. Masuch, M. Delgado-Restituto, *Ultra Low Power Transceiver for Wireless Body Area Networks*, Series of Analog Circuits and Signal Processing (ACSP) (Springer, Berlin, 2013)

A. Paidimarri, N. Ickes, A. P. Chandrakasan, A +10 dBm 2.4 GHz transmitter with sub-400pW leakage and 43.7% system efficiency. ISSCC Dig. Tech. Papers (Feb. 2015), pp. 246–247

X. Peng, J. Yin, P.-I. Mak, W.-H. Yu, R.P. Martins, A 2.4-GHz ZigBee Transmitter using a function-reuse Class-F DCO-PA and an ADPLL Achieving 22.6% (14.5%) system efficiency at 6-dBm (0-dBm) Pout. IEEE J. Solid-State Circuits (2017), to appear

A. Pirola, A. Liscidini, R. Castello, Current-mode, WCDMA channel filter with in-band noise shaping. IEEE J. Solid-State Circuits **45**, 1770–1780 (2010)

J. Prummel, M. Papamichail, M. Ancis et al., A 10 mW bluetooth low-energy transceiver with on-chip matching. ISSCC Dig. Tech. Papers (Feb. 2015), pp. 238–239

R. Rajan, *Ultra-Low Power Short-Range Radio Transceiver* (Microsemi Corporation, Aliso Viejo, CA, 2012)

B. Razavi, K.F. Lee, R.H. Yan, Design of high-speed, low-power frequency dividers and phase-locked loops in deep submicron CMOS. IEEE J. Solid-State Circuits **30**, 101–109 (1995)

J.A. Stankovic, Research directions for the internet of things. IEEE Internet Things J. **1**(1), 3–9 (2014)

M. Tedeschi, A. Liscidini, R. Castello, Low-power quadrature receivers for ZigBee (IEEE 802.15.4) applications. IEEE J. Solid-State Circuits **45**, 1710–1719 (2010)

K. Wang, J. Koo, R. Ruby, B. Otis, A 1.8 mW PLL-free channelized 2.4 GHz ZigBee receiver utilizing fixed-LO temperature-compensated FBAR resonator. ISSCC Dig. Tech. Papers (Feb. 2014), pp. 21–22

J. Yin, P.-I. Mak, F. Maloberti, R.P. Martins, A time-interleaved ring-VCO with reduced 1/f3 phase noise corner, extended tuning range and inherent divided output. IEEE J. Solid-State Circuits **51**, 2979–2991 (2016)

F. Zhang, K. Wang, J. Koo, Y. Miyahara, B. Otis, A 1.6 mW 300 mV supply 2.4 GHz receiver with −94 dBm sensitivity for energy-harvesting applications. ISSCC Dig. Tech. Papers (Feb. 2013), pp. 456–457

Battery Technologies for IoT

15

Jeff Sather

Although IoT devices appear in myriad physical configurations and serve countless purposes, the battery requirements for any particular category of IoT devices can be evaluated by recognizing their physical, electrical, and functional elements as follows:

- physical size—typically miniature assemblies having a battery compartment, socket, or wiring harness for connection to the power source;
- operating environment—wide range of conditions, from indoor, outdoor, body-worn and implantable, to extreme environments;
- functionality—sensing, data collection, and radio communications;
- components—sensors, radios, microcontrollers/microprocessors, including human interfaces in some cases;
- purpose—remote monitoring/diagnostic, preventive maintenance, controls, gathering statistics, machine-to-machine communication.

The physical size of the IoT device might in turn limit the physical size of the battery that can be considered. The operating environment will dictate whether the battery options are limited to those having industrial, automotive, or commercial temperature ratings. Components and

J. Sather (✉)
Cymbet Corporation, Elk River, MN, USA
e-mail: jsather@cymbet.com

functionality will determine the energy and power requirements, as well as whether a primary or rechargeable battery is better suited for the application. Duty cycle of radio transmit/receive pulses, communication protocols, sensor activity, and activation of visual indicators are some of the most power consumptive factors influencing the battery capacity needed in order to operate for an acceptable period between replacement or recharge.

Using this broad framework of what an IoT node does and how it performs those functions, and with an understanding of the power requirements of the disparate components and system as a whole, a survey of power sources can be undertaken, comparing how each delivers in terms of performance, cost, size, and other parameters. Of course, not all IoT devices need a battery or other portable power source. However, a significant number of devices are deployed where batteries are the only viable power source. In some cases, rechargeable batteries are used as the main power source, where a recharging source is available and practical, or as a backup power source in the event of main power loss—where main power can be a larger battery, grid power, or a renewable supply such as photovoltaic cells. In many other embodiments, a non-rechargeable battery is the more useful power source, as there will be no practical means of replenishing the energy in a discharged battery.

© Springer International Publishing AG 2017
M. Alioto (ed.), *Enabling the Internet of Things*, DOI 10.1007/978-3-319-51482-6_15

Because IoT devices encompass any number of combinations of active and passive components, are configured in untold shapes and sizes, operate over a range of power consumption modes, and can have product operating lives of days to years, the battery specifications required to support these features and functions can vary almost as widely as the devices themselves. Whatever the IoT implementation, it is important to select the battery that meets minimum performance objectives under all possible operating conditions, will last the intended life of the product or, in the case where battery replacement is expected, can be replaced with minimal expense, difficulty, and in compliance with disposal regulations. This chapter presents the characteristics of batteries that, because of their size, voltage, power delivery, and cost profiles, are candidates as power sources in IoT devices. Special-use batteries such as those used specifically in implantable medical devices and aerospace applications; stationary batteries; and batteries having chemical compositions no longer in widespread use, will not be covered in this chapter. A much wider selection of batteries than those covered here are commercially available for every imaginable application. Aside from references to materials comprising the positive (cathode) and negative (anode) electrodes of different battery chemistries, fundamental battery principles and electrochemical reactions will not be discussed. Throughout this chapter, the terms "cell" and "battery" will be used interchangeably, although in the strictest sense they are distinct terms.

Battery designations as defined by active material composition—negative electrode, electrolyte, and positive electrode—are categorized by both ANSI and IEC standards, yet each of those standards uses its own nomenclature. Designations used in this chapter will refer to the letter codes in some cases, but also the more common parlance of, for example, "AA alkaline" or "lithium coin cell" when generalities will suffice. When making specific technical comparisons, the more precise nomenclature will be used.

15.1 An Overview of Primary and Secondary Battery Chemistries

Batteries are designed as either non-rechargeable or rechargeable cells—referred to as primary and secondary, cells, respectively. Within each of these categories, chemistries vary widely, from alkaline chemistry in the ubiquitous primary batteries such as AA, AAA, C, and D-cells, and coin variety including the commonly used "CR" family (e.g., CR2032), to secondary batteries like Li-ion cylindrical (e.g., 18650) and Li-polymer pouch cells. Each chemistry and construction method has its tradeoffs in terms of performance and cost, with lithium-based primary and secondary cells of various types having gained widespread use in recent years, owing to their relatively high energy density, higher (relative to alkaline cells) nominal voltage of 3.0–3.6 V in many of the cell types, and typically low self-discharge. Cells are available in consumer and industrial grade, offering a range of performance in parameters such as cycle life, self-discharge, pulse current delivery, and cost. An overview of chemistries in widespread use, their common embodiments, and noteworthy characteristics, is given in Table 15.1.

Storage capacities of the batteries listed range from a few tens of microampere-hours, to several ampere-hours.

15.2 Voltage, Power, and Battery Capacity Demands in IoT Applications

Given the range of features, functions, and applications in which IoT devices are deployed, each category of devices, and different models within a category, will impose different demands on the battery, not only in operating voltage and current, but in physical size constraints and product life.

Components deployed in IoT products operate from a range of supply voltages and present load

Table 15.1 A selection of battery types commonly used in portable applications and IoT devices

Designator/ common name	Common type	Anode (negative electrode)	Cathode (positive electrode)	Rechargeable?	Nominal volgate (V)	Standard continuous load rating (typical)	Shelf/operating life in years (room temp.)	Temp. range (discharge) (°C)
CR	Coin	Li	MnO_2	No	3	C/100	10	−30 to +60
BR	Coin	Li	CF_x	No	3	C/1000	10	−30 to +80
ML	Coin	Li-Al	Mn_xO_y	Yes	3	C/300	10	−20 to +60
AA alkaline (LR6)	Cylindrical	Zn	MnO_2	No	1.5	C/10	7–10	−20 to +55
AA lithium (FR6)	Cylindrical	Li	FeS_2	No	1.5	1C	20	−40 to +60
Li-ion (18650)	Cylindrical	C	$LiCoO_2$	Yes	3.6	1C	5	−20 to +60
Li thionyl chloride	Coin and cylindrical	Li	$SOCl_2$	No	3.6	C/10,000	20+	−55 to +85
Solid state thin film	Thin film/ planar	Li	$LiCoO_2$	Yes	3.8	2C	20+	−40 to +70

Fig. 15.1 Discharge capacity as a function of discharge rate for standard AA alkaline (LR6) batteries

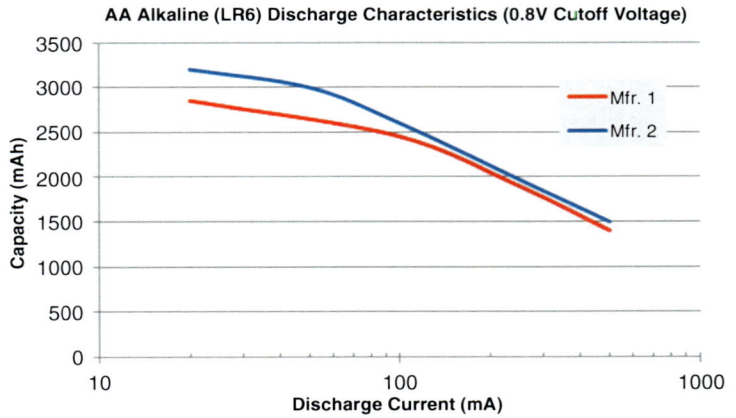

Fig. 15.2 Voltage drop as a function of discharge rate for typical AA alkaline (LR6) batteries

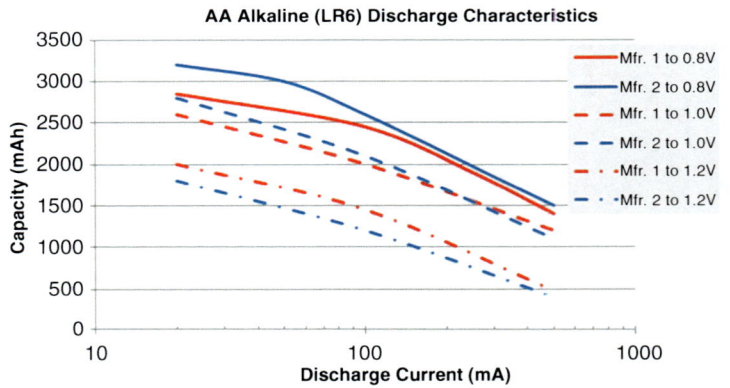

currents spanning a wide range—from a few tens of nanoamperes for sleep currents, memory retention, and low power clocks, to hundreds of milliamperes or in extreme cases, amperes needed to operate radios, actuators, and other mechanical devices. Pulsed loads can rapidly diminish the capacity of many battery chemistries. Figure 15.1 illustrates the diminished battery capacity as a function of load current for AA alkaline cells from two manufacturers (http://data.energizer.com/PDFs/EN91.pdf; http://na.industrial.panasonic.com/sites/default/pidsa/files/1.5vseries_datasheets_merged.pdf).

When considering battery lifetime for the IoT device, the average and peak loads to be placed on the battery must therefore be taken into account.

Not only do batteries deliver less total capacity when discharged above the rate at which the capacity is specified, but the internal resistance of the battery has an attendant voltage drop under high current loads that can cause the system to drop out of regulation or below the minimum operating voltage of some device components, resulting in a loss of device functionality—either momentarily or permanently (i.e., requiring a system reset). Note in Fig. 15.2 (http://data.energizer.com/PDFs/EN91.pdf; http://na.industrial.panasonic.com/sites/default/pidsa/files/1.5vseries_datasheets_merged.pdf) the capacity of the same AA alkaline cells from Fig. 15.1, for a given discharge current and cutoff voltage. In systems using alkaline or other lower voltage cells, a series combination of two will be used. Hence, the additive effects of the voltage drop will be proportional to the number of series-connected cells.

15.3   Attributes of Batteries Commonly Used in IoT Applications

Given there are literally thousands of combinations of battery shapes, sizes, capacities, chemistries, and other parameters to choose from when selecting the appropriate battery for an IoT device, it can be a daunting exercise to cull through all the options and select the battery that has just the right combination of specifications meeting the performance and cost objectives of the product designer, manufacturer, and end user alike. It is all too easy then to rely on a default selection process based on what has been used in similar designs, a battery type that is known and understood, or what is least expensive and just enough to do the job. With engineering time constraints always an implicitly or explicitly understood as being part of the design process, these approaches to battery selection are not necessarily unjustified. However, such approaches do not always lead to the optimum result when the interests of all parties are taken into account. The battery in the system might last well longer than the product life demands, sometimes resulting in a larger or more expensive battery being used that otherwise needed. Conversely, the battery might fail prematurely as a result of some mode or another of system operation demanding more from the battery than its intended use. As with most engineering exercises, there are tradeoffs between thoroughly investigating all options and getting the job done on time and on budget.

This section endeavors to provide guidance into the variety of batteries available to the designer, along with their respective advantages and limitations relative to one another in terms of performance, durability, cost, and other parameters important in IoT device design and operation in the field.

15.3.1   Battery Shapes and Sizes

Because IoT devices take on a wide range of physical embodiments, selection of a suitable battery often comes down to its physical size— that is, selecting the battery that fits within a pre-defined compartment, board footprint, or height constraint. Conversely, the size of the IoT device is in some cases determined largely by the physical size of the battery that has sufficient storage capacity and other attributes that serve the power and energy needs of the IoT device. Fortunately for product designers, batteries are available in sizes ranging from a cubic millimeter upwards, with operating voltages from <1 to >4 V for single cells and tens of volts for larger battery packs. Storage capacity ratings span more than five orders of magnitude, with discharge currents ranging from less than C/10,000 to 5C. The C-rate is the rate at which the battery is being discharged, notionally defined as the battery discharge current divided by the current draw under which the battery delivers its rated capacity in 1 h— although the rated capacity of most batteries today is at discharge rates much slower than 1 h. Moreover, a range of shapes and sizes can be found in both primary and secondary batteries. In this section, the emphasis will be on describing batteries that are physically no larger than ~25 cm^3. In addition to larger battery sizes being available, battery packs are also configured to serve applications requiring higher voltage, higher power, and/or higher storage capacity.

Table 15.2 lists the attributes of several commercially available battery types, in formats that can serve myriad IoT device sizes.

15.3.2   Comparisons of Energy and Power Densities

When selecting a battery for a particular system, designers usually want to get as much energy as possible in the smallest volume, other considerations—such as voltage, safety, cost, reliability—aside. Energy density is one of the more important specifications of a battery, yet it is not usually specified explicitly in data sheets. It is, rather, implicit in the rated capacity of a battery of a given physical size. Energy density is measured in volumetric and gravimetric terms. Volumetric energy density is measured

Table 15.2 Commonly used battery types and their general characteristics

Battery type	Common name	Depictions (not to scale)	Storage capacity range (mAh)	Typical discharge rate	Discharge voltage (V)
Coin/button	CR, ML, LR		1–1000	C/100	1.5–3.3
Cylindrical (Li-ion, alkaline, NiMH, NiCd)	18650, AAA		25–5000	C/10–C/2	1.2–3.7
Prismatic Li-ion	9 V		200–2000	C/10–C/2	3.6
Lithium polymer	Li-poly		5–500	C/10–1C	3.6
Solid state/thin film	EnerChip, EnFilm		0.01–20	2C	3.8

in units of watt-hours per liter (Wh/L), or its equivalent in milliwatt-hours per cubic centimeter (mWh/cm^3). Gravimetric energy density is a measure of the amount of energy stored per unit of mass and has units of watt-hours per kg (Wh/kg), or milliwatt-hours per gram (mWh/g). Gravimetric energy density is a more critical parameter in electric vehicles, where the battery power must propel its own weight as well as that of the vehicle itself, and in mobile devices, where weight presents a burden to humans. In stationary devices, gravimetric energy density is usually not critical. This section will therefore focus on volumetric energy density of various types of batteries.

Energy density can be a misleading parameter depending on the output voltage of the battery and the voltage required by the circuit to operate. That is, sometimes *charge* density is a more useful measure, especially if the battery voltage is applied directly to a circuit or circuit element

that could otherwise operate as efficiently at a lower voltage, or, if the battery voltage is being routed through a linear regulator, for example, that wastes energy in regulating to a lower voltage. In these cases, charge density is a truer metric of available capacity. Batteries are specified in amp-hours and not watt-hours.

Energy density varies widely, as a function of chemistry, physical form, packaging, and to some extent by manufacturer within a given battery type. Volumetric densities range from as low as 10 Wh/L, to over 1000 Wh/L. (Zn-air batteries, of the type commonly used in hearing aids, have energy densities exceeding 1200 Wh/L in many cases. But these cells are impractical in most IoT devices because the operational life is only a few days once the sealing tab is removed.) The highest energy densities are typically found in primary cells, with lithium thionyl chloride ($LiSOCl_2$) cells boasting energy densities >1200 Wh/L in some cell designs.

Of course, in a rechargeable cell, the rated capacity of a battery is multiplied n-fold, n being the number of charge-discharge cycles available to the user or system over the life of the products. In that sense, the energy density of rechargeable cells, when measured as lifecycle capacity (discussed later), is usually much higher than that of primary cells. Again, not all systems can take advantage of the rechargeable aspect of a battery. And this is true for many IoT devices.

Batteries nearing the physically smaller end of the spectrum tend to have lower energy density, as the packaging an inactive components of the cell tend to occupy a larger percentage of the total battery volume. In some cases, the size of the IoT device might be constrained in the footprint, but not so much in height. In such cases, the amount of energy per unit of board space is more important the total volume occupied by the battery.

Table 15.3 depicts a cross-section of primary and secondary coin and cylindrical cells and their respective energy densities and related metrics. Battery size and associated energy density assume an amount of area occupied by the battery on a circuit board. In other words, the coin and cylindrical cells are squared off in the x-y plane, as the region in proximity to the battery is typically unavailable for other components because of mounting tabs, sockets, or other keep-out provisions for manual assembly. Therefore, the actual energy density of the cells represented will be slightly greater than that shown. Figures are exclusive of battery sockets and holders needed to secure the battery, and connectors, harnesses, and cables used to make the electrical connection from battery to circuit.

The storage capacity and volumetric energy density for several battery types are depicted in Fig. 15.3. Generally, primary (non-rechargeable) cells have considerably higher volumetric energy density than the secondary (rechargeable) cells.

With rechargeable cells—provided some number of charge-discharge cycles can be practically exploited—the additional cycles are then effective multipliers to the energy density. Some secondary batteries, however, require more sophisticated charge control circuitry than others.

And while commercially available Li-ion and other battery charge/management integrated circuits are readily available to the designer, there is additional cost and component count associated with implementing the battery management circuit and not all IoT device applications will have a charging source that meets the requirements of constant current/constant voltage (CC/CV) battery charging.

Other secondary cells, such as ML/MT/NBL/VL coin cells, have less stringent charging requirements than do larger Li-ion cells. For example, 18650 Li-ion cells require CC/CV charge profiles, while ML coin cells, for example, accept constant voltage, requiring only a current limiting resistor and/or protection diode. Moreover, a restricted charging temperature range of Li-ion cylindrical cells, from 0 to 45 °C, can make them impractical in some applications.

In cases where a battery meets most system requirements, save the ability to deliver sufficient current to send short-duration radio transmissions, for example, a low ESR (Equivalent Series Resistance) capacitor can be deployed to deliver pulsed currents without forcing the designer to upsize the battery to deliver the power. When using a supporting capacitor to deliver pulse currents, it is important to recognize and account for the fact that the capacitor will have a leakage parameter associated with it, dependent on capacitor material, construction, ambient temperature, and other factors. The circuit designer must therefore consider whether the capacitor ought to be electrically isolated from the battery or other components when not actively delivering current.

In contrast to energy density, power density is the measure of how much current a battery sources relative to its volume or mass. As with energy density, power density varies by chemistry and construction. It is also highly dependent on temperature, increasing (decreasing) on the order of 2× per 10 °C increase (decrease) in temperature depending on the chemistry.

Some cells are designed to deliver relatively low currents (e.g., C/100, C/1000, C/10,000) for periods of months or years, while other cell types

Table 15.3 Parameters of commonly used batteries, including storage capacity, energy density, and related characteristics

Battery example	Typical capacity (mAh)	Rechargeable? (Y/N)	Nominal voltage (V)	Recharge cycles (to rated depth-of-discharge)	Effective capacity (mAh)	Dimensions (x, y, z; mm)[a]	Physical volume (mm^3)	Footprint (mm^2)	Energy density per unit area (mWh/cm^2)	Volumetric energy density (mAh/cm^3)	Volumetric charge density (mWh/cm^3)	Effective volumetric energy density (mWh/cm^3)[b]
BR2032	190	N	3	0	190	20 × 20 × 3.2	1280	400	142.50	445.3	148.44	445
CR2032	225	N	3	0	225	20 × 20 × 3.2	1280	400	168.75	527.3	175.78	527
CR2450	620	N	3	0	225	24 × 24 × 5.0	2880	576	322.92	645.8	215.28	234
VL621	1.5	Y	3	1000	150	6.8 × 6.8 × 2.15	99.4	46.2	9.73	45.3	15.09	4526
ML614	3.4	Y	3	1000	340	6 × 6 × 1.4	50.4	36	28.33	202.4	67.46	20,238
18650 Li-ion	3000	Y	3.6	500	1,200,000	18 × 65 × 18	21,060	11,700	92.31	512.8	142.45	205,128
AA thionyl chloride	2100	N	3.6	0	2100	14.5 × 50.5 × 14.5	10,618	7323	103.24	712.0	197.78	712
AA lithium	3000	N	1.5	0	3000	14.5 × 50.5 × 14.5	10,618	7323	61.45	423.8	282.55	424
AA alkaline (LR)	2800	N	1.5	0	2800	14.5 × 50.5 × 14.5	10,618	7323	57.36	395.6	263.71	396

[a]x–y dimensions represent occupied board space (rectangular rather than diametrical); energy density is calculated using those same dimensions
[b]Effective energy density is calculated from actual capcity, physical size, and number of recharge cycles available, at rated depth-of-discharge

Fig. 15.3 Illustration of
storage capacity and
volumetric energy density
of commonly used batteries

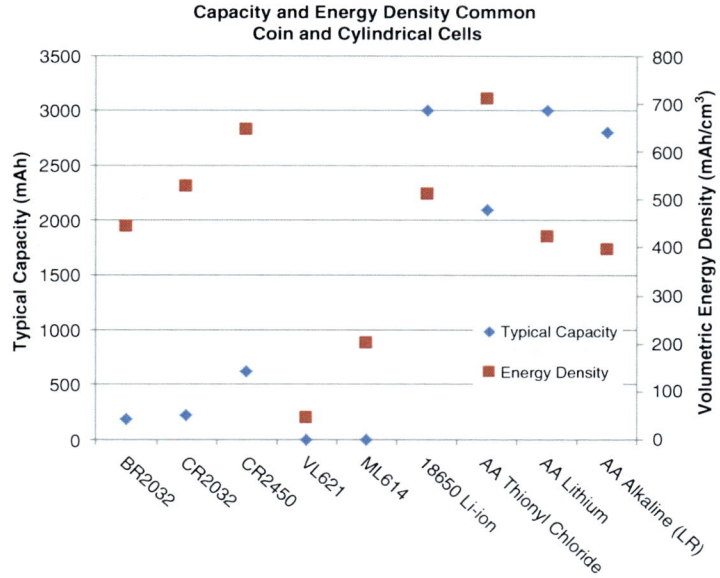

can source peak currents of 2C or more
without suffering significant capacity loss.
The profiles of Fig. 15.4 represent the discharge
rate performance at room temperature, to the
rated discharge voltage, for seven battery
types—four primary cell types and three
secondary:

- AA alkaline (LR6) (http://data.energizer.com/
  PDFs/EN91.pdf)
- AA lithium (FR6) (http://data.energizer.com/
  PDFs/l91.pdf)
- AA thionyl chloride (www.tadiranbat.com/
  pdf.php?id=TL-5104)
- 18650 Li-ion rechargeable cylindrical (http://
  industrial.panasonic.com/lecs/www-data/
  pdf2/ACI4000/ACI4000CE17.pdf)
- CR2032 primary coin (http://na.industrial.
  panasonic.com/sites/default/pidsa/files/
  crseries_datasheets_merged.pdf)
- ML1220 rechargeable coin (http://na.indus
  trial.panasonic.com/sites/default/pidsa/files/
  mlseries_datasheets_merged.pdf)
- Solid state thin film lithium micro-battery
  (http://www.cymbet.com/pdfs/DS-72-41.pdf)

Note that in some cases while the discharge
capacity might remain relatively constant over as
much as a $100\times$ range of discharge currents, the
total energy delivered is not the same from lowest discharge current to highest, as the cell voltage is markedly lower at the higher discharge
current. Depending on the series/parallel cell
combination, if any, this effect could force a
designer to add series- or parallel-connected
cells to sustain a minimum voltage at worst-
case current draw. Moreover, it is obvious that
not all AA cells are the same in this regard.
Lithium primary cells have relatively flat discharge profiles over most of the discharge
range, while alkaline cells sag measurably even
at relatively modest discharge currents. The
thionyl chloride battery is noted for its exception-
ally flat discharge profile—albeit at very low
C-rates. An adverse characteristic of the thionyl
chloride cell is the onset of a voltage recovery
after long-term storage due to passivation of the
lithium electrode.

The voltage decay, while often limiting the
ability to deliver the last portion of capacity at a
usable voltage, can be used to advantage as a low
voltage threshold detection level in circuits

Fig. 15.4 Effects of
discharge rate on battery
discharge capacity for
several cell types, shown as
a function of (**a**) C-rate and
(**b**) discharge current

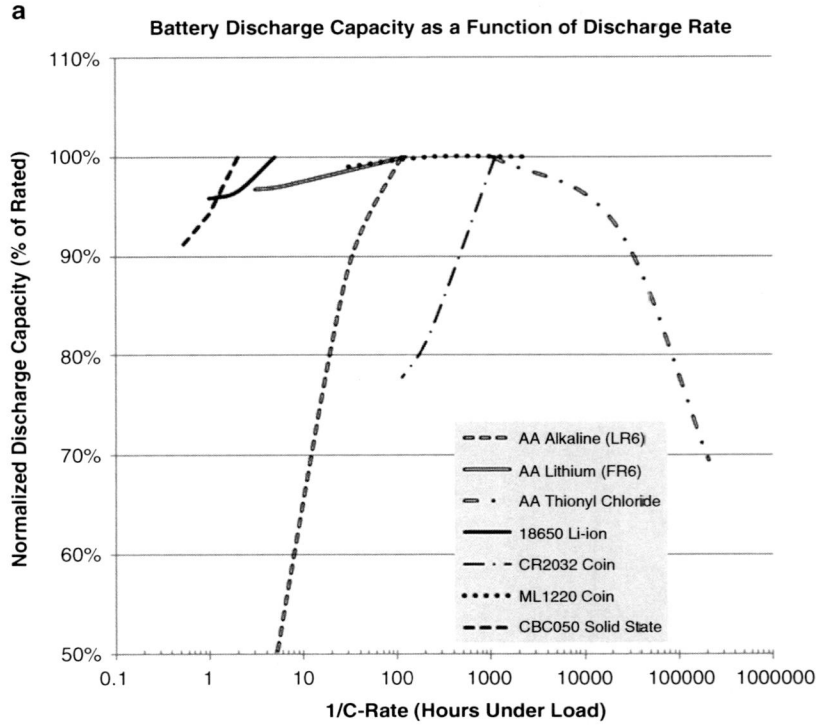

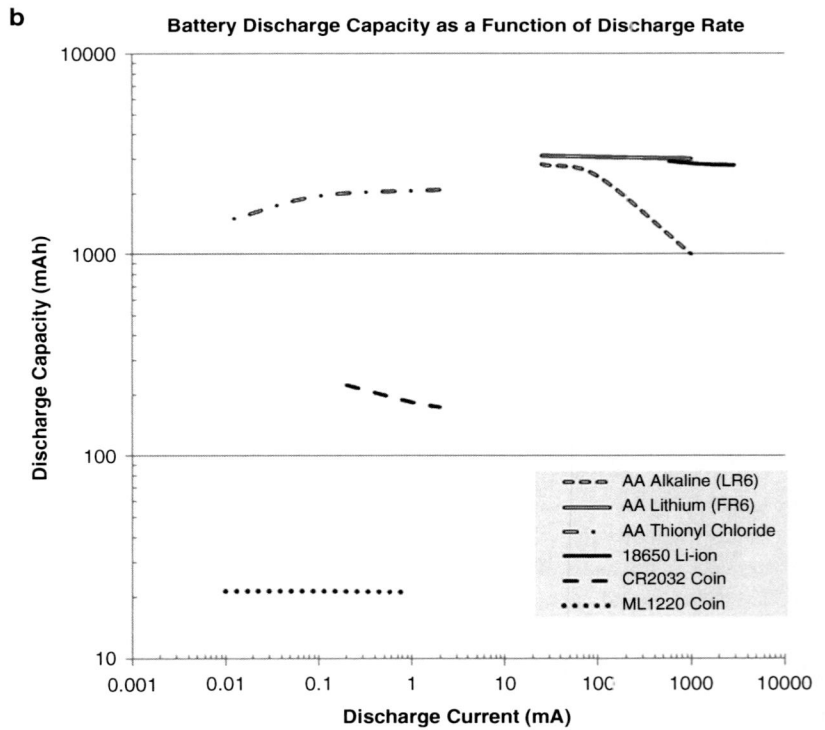

employing low battery indicators or for internal circuit housekeeping of system components that might otherwise undergo chaotic operation or catastrophic shutdowns if the battery were to fall below a critical voltage level without warning. Fuel gauges, while readily available for many types of batteries, can present a load to the battery that, in IoT applications operating from small batteries, is non-negligible. Because battery terminal voltages do not drop rapidly under light loads, in devices where it is important to monitor battery terminal voltage periodically to anticipate power loss, it is practical to use low power sampling circuits that measure the terminal voltage once every several seconds or minutes depending on operating conditions.

Power density is not only a function of battery chemistry and cell type, but also of operating temperature for a given battery. The discharge curves of Fig. 15.5 demonstrate the dependence of temperature on battery performance for a variety of cells:

- AA lithium (L91)
- AA alkaline (LR6) from two manufacturers
- CR2032 primary coin cell
- AA thionyl chloride (ER14505)
- 18650 Li-ion

At a given discharge rate, it is not unusual for there to be an order of magnitude or more disparity in discharge capacity over the rated operating temperature range. Some types of cells perform better than others in retaining capacity over the specified temperature range.

Note in some cases the competing effects of higher self-discharge and lower impedance at higher temperature, actually resulting in a diminished discharge capacity—relative to room temperature capacity—at the specified upper temperature limit over a wide range of discharge rates.

Manufacturers of certain types of batteries (e.g., alkaline cells) will often provide test results for their cells as tested in accordance with standardized IEC/ANSI test regimens. Such tests mimic applications in remote controls, radios, lighting, and others. Battery loads in these tests drain the battery over the course of several hours to several days. Although the standard suite of IEC/ANSI battery tests are performed under significantly higher loads than those encountered in most long operating life IoT devices, comparing standardized test results among different battery types, and from different manufacturers, can be helpful in selecting the best candidate from a particular battery category.

Fig. 15.5 Representative battery discharge characteristics as a function of operating temperature and discharge rate for various cell types

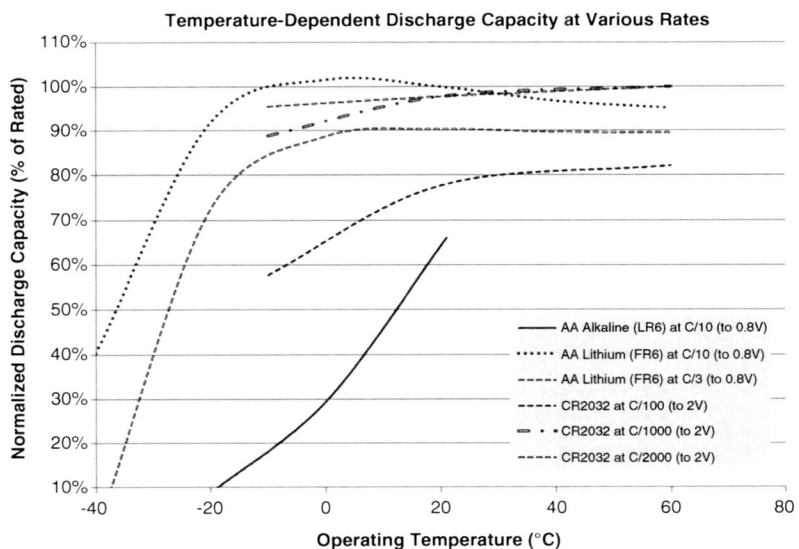

Temperature-Dependent Discharge Capacity at Various Rates

Legend:
- AA Alkaline (LR6) at C/10 (to 0.8V)
- AA Lithium (FR6) at C/10 (to 0.8V)
- AA Lithium (FR6) at C/3 (to 0.8V)
- CR2032 at C/100 (to 2V)
- CR2032 at C/1000 (to 2V)
- CR2032 at C/2000 (to 2V)

Y-axis: Normalized Discharge Capacity (% of Rated)
X-axis: Operating Temperature (°C)

Fig. 15.6 Battery voltage
as a function of delivered
capacity

Battery Voltage During Discharge - Various Chemistries

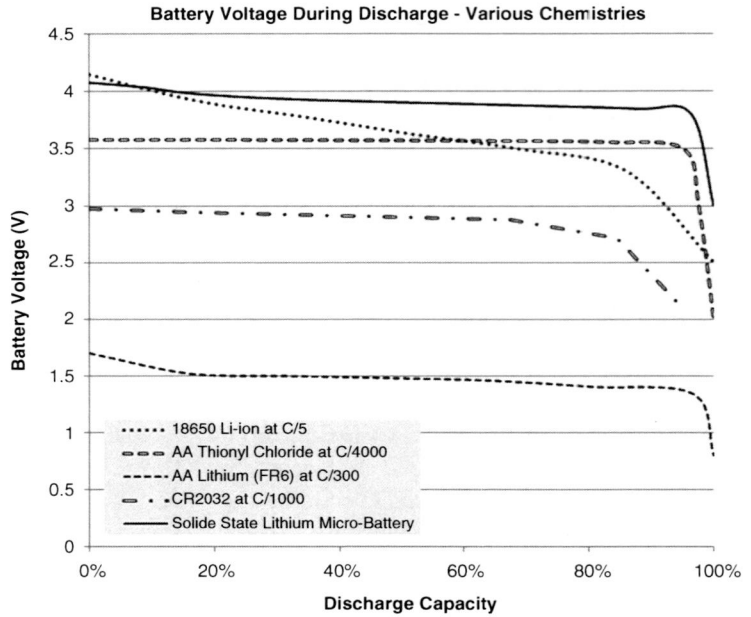

Legend:
...... 18650 Li-ion at C/5
= = = AA Thionyl Chloride at C/4000
- - - - AA Lithium (FR6) at C/300
— . —CR2032 at C/1000
——— Solide State Lithium Micro-Battery

15.3.3 Voltage Profiles and Load Requirements

Every battery type has a well characterized voltage discharge profile under the rated conditions of temperature and discharge rate. Typical voltage profiles for common battery types discharged at room temperature are shown in Fig. 15.6.

As described earlier, battery discharge profiles vary by chemistry, state-of-charge, operating temperature, and discharge rate, in addition to other factors. Active circuit elements require a minimum operating voltage to perform their one or several functions. Some functions can be performed at a lower voltage than others within the same integrated circuit, but all such elements still require a minimum voltage and therefore the power source must deliver adequate operating current at that voltage in order for that circuit to operate according to specification. The battery capacity available to the system is that capacity delivered at a voltage above the minimum operating voltage of the several components comprising the IoT device.

Pulse currents drawn from batteries in IoT devices can have unanticipated consequences to

Cell Resistance as a Function of State of Charge

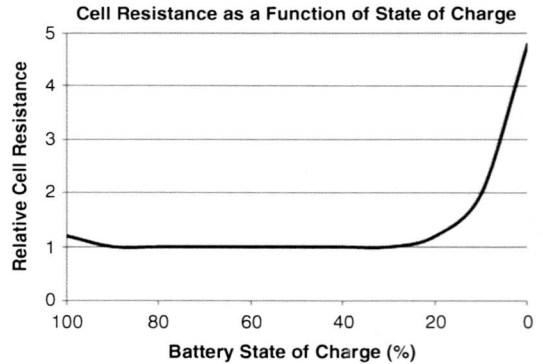

Fig. 15.7 Cell resistance as a function of state-of-charge

battery life and circuit performance in the case of momentary voltage drops if load current exceed the power rating of the battery.

From Fig. 15.7, a generic representation of cell resistance vs. state of charge, note how rapidly the cell resistance climbs as the battery reaches its final ~10% of stored charge (http://data.energizer.com/PDFs/l91.pdf). While the cell might be sustaining an adequate voltage level all the way to the end of life, should a high current load be presented during this high resistance phase of discharge, there could be a loss of power and circuit operation before any low voltage detection circuit can react and issue the necessary alerts to the system controller, or the controller circuit has time to perform any essential housekeeping operation prior to power loss.

Batteries also have a characteristic pulse response as a function of the electrochemical processes and parasitic elements within the cell. Therefore, it is sometimes necessary to support the battery with a low ESR capacitor to avoid effects of such behavior on circuit operation. Proper circuit design, with a view to worst-case circuit loads, will prevent such disruptions from ever occurring by properly sizing the battery—accounting for aging, state-of-charge, and other factors affecting battery impedance.

15.3.4 Operating Life, Wear-Out Mechanisms, and Factors Affecting Battery Aging

Battery performance includes more than simply delivering the necessary operating currents at a minimum voltage, having a meaningful number of recharge cycles (where applicable), and operating in a temperature range spanning the expected conditions of the IoT device. Battery longevity under these conditions is essential to not only the performance of the IoT device, but also its practicality in deployment. An otherwise perfectly useful battery will have limited applicability if its service life falls short of the expected operational life of the device it powers. This is especially true where replacement of the battery is difficult, expensive, or simply not possible. When selecting the proper battery for an IoT device, the system designer should consider how the typical and extreme operating conditions are likely to affect battery life. Important parameters determining battery longevity and long-term performance are:

- Temperature
- State-of-charge, overcharge, and over-discharge
- Magnitude of load currents
- Reliability of connectors, sockets, compartments, and holders

While cold temperature reduces power delivery, storage and operation at elevated temperature ages a battery as significantly as any other factor. Periodic excursions to high temperature, if kept to limited few in number and duration over the product life, can be accounted for in estimating battery life. Extended periods at temperatures nearing the upper specified temperature range of a battery can lead to premature failure. It is therefore important to understand the likely operating temperature of the IoT device and, where possible, thermally shield or isolate the battery from heat-generating sources in the device. Using a rule of thumb of a $2\times$ reduction in life per 10 °C increase in temperature, the designer might have the luxury of putting a larger capacity battery into the system to compensate for aging effects of elevated temperature on the battery.

The rate at which a battery is discharged also has affects the battery lifetime. Whether discharged under constant resistance, constant power, or constant current, the performance will vary accordingly. When testing a battery for a particular application, it is therefore advisable to test it under conditions—discharge rate, temperature, etc.—as similar as possible to the load expected during operation in the specific device and application.

Batteries are generally connected electrically and physical by one of four methods:

- Socket or holder soldered to the circuit board, with metal tabs making electrical connection to the battery terminals.
- Wire harness with a snap-on connection to the battery terminals, as commonly used with 9 V batteries, for example.
- Spring-load receptacle to house cylindrical batteries such as AA and AAA.
- Solder connection of the battery directly to the circuit board by way of through-hole tabs, or surface mount soldering of welded tabs on batter, to solder pads on the circuit board.

Each of these connection and mounting methods has associated reliability, cost, and physical size and weight advantages and disadvantages.

Corrosion and breakage of the socket, holder, and mounting tabs can occur in applications where mechanical shock and vibration occur, or in high humidity or other corrosive atmospheres. When selecting a proper mounting device or electrical connector, these environmental factors should be considered. Some connectors are specially designed for ruggedized applications.

Self-discharge of a cell is a function of material, construction, and varies by manufacturer for a given cell type. Self-discharge rates vary from as little as 0.5% per year for solid state batteries, to several percent per month for other chemistries. Primary batteries typically have lower self-discharge rates than secondary cells. The self-discharge of a battery becomes more important in long life applications, as the self-discharge contribution to total capacity loss competes increasingly with the load placed on the battery. In the case of secondary cells, self-discharge can become a significant contributor to capacity loss if long periods of time elapse between recharge events. Self-discharge increases with increasing temperature, although the battery will have higher current and capacity delivering capability at a higher temperature due to decreased internal impedance as the temperature increases.

15.3.5 Cost of Batteries, Holders, Connectors, Integration, and Replacement

As part of the battery selection process, cost must be weighed along with the technical specifications. In some cases, an otherwise satisfactory battery from a performance standpoint might be excluded from contention because of cost. The cost of the battery is not always so simple to determine, as the purchase price of the battery itself is usually not the only cost that comes into play.

Table 15.4 lists typical battery pricing as purchased through a component distributor or other volume outlet, made by reputable manufacturers, in modest volumes of 1000–5000 pieces (http://www.digikey.com/product-search/en?keywords=battery). Naturally, prices vary within any given battery subcategory, depending on manufacturer, cells adorned with or without mounting tabs, accompanying charging circuit (where applicable), temperature ratings, and other parameters.

Figure 15.8 depicts the data from the prior table, excluding the two most expensive (on a price/mAh measure) in order to expand the right-hand vertical axis to show differences among pricing of the other cells.

Note the wide range of absolute prices, as well as the wide range of prices per unit of energy. While selecting the lowest price battery type that meets the minimum energy, voltage, and power requirement of the system might at first be appealing, other factors should be considered. For example, note the large disparity in price between the BR2032 and CR2032 cells. Both have a nominal voltage of 3 V and storage capacity of ~200 mAh. But the chemistries differ and upon closer view of the respective data sheets, it is seen that the upper temperature limit of the BR series is higher than that of the CR series cells. This might be important in light of the application environment, for example if the IoT device

Table 15.4 Battery pricing in quantities of thousands of pieces, as might be purchased through component distributors

Battery example	Typical capacity (mAh)	Rechargeable? (Y/N)	Price ($US)	Price/mAh ($US/mAh)
BR2032	190	N	$0.60	$0.0032
CR2032	225	N	$0.15	$0.0007
CR2450	620	N	$0.60	$0.0010
VL621	1.5	Y	$2.35	$1.5667
ML614	3.4	Y	$1.10	$0.3235
18650 Li-ion	3000	Y	$2.00	$0.0007
AA thionyl chloride	2100	N	$3.50	$0.0017
AA lithium	3000	N	$1.80	$0.0006
AA alkaline (LR)	2800	N	$0.20	$0.0001

Fig. 15.8 Absolute price
and price per unit of energy
of commonly used batteries

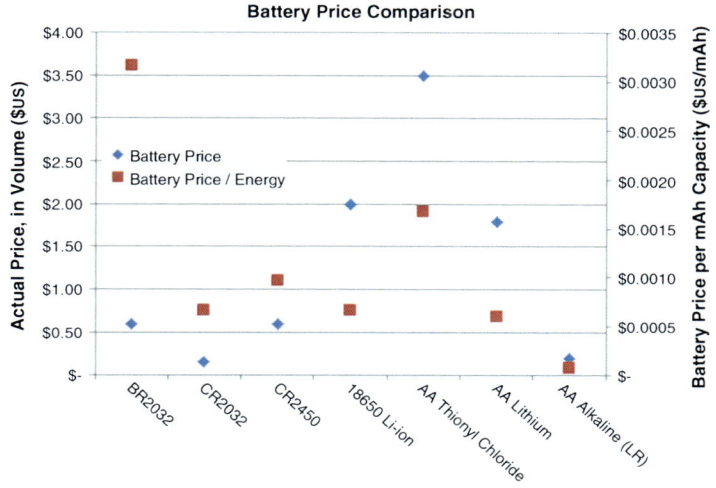

Fig. 15.9 Cycle life graph
of 18,650 cylindrical cell

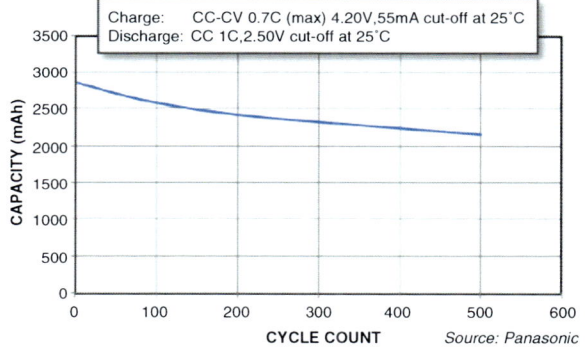

operates in an industrial setting where the device will be exposed to elevated temperature on a periodic or continuous basis. Another example is that of the AA lithium cells vs. the AA alkaline cell. The AA lithium cell is priced considerably higher, yet note that not only is the operating temperature range of that cell wider than that of the alkaline cell, but referring back to Fig. 15.4, the high current discharge performance of the lithium cell greatly exceeds that of the alkaline cell.

Again, those parametric differences might or might not be important. With some battery product data sheets, not all potentially critical parameters are routinely specified. Designers might be forced to turn to quality and reliability reports, user manuals, and other technical documents to find critical specifications of a battery from a specific manufacturer. This can be especially true when seeking longevity data, aging characteristics, and related data.

It is also interesting to see that the rechargeable 18650 Li-ion cell is cost competitive with the other cells when measuring price per unit of energy. And that is before factoring in the additional capacity that can be utilized with each of the several hundred recharge cycles available from the 18650 cell; that is, of course, if the IoT design and application can in fact take advantage of the rechargeable feature of the 18650 cell as depicted in Fig. 15.9.

When weighing the cost of a battery, it is necessary to factor in the cost of the compartment, holder, connector, or other component necessary for securing the battery and making electrical connection to the battery. In addition, assembly costs should be considered, as they vary as a function of the type of battery and connection scheme used.

If the battery will likely have to be replaced during product life, the replacement cost will depend on many factors, including accessibility to the device; labor cost in the case of industrial devices; possible equipment down time resulting from the IoT device being out of service, etc. Where batteries are soldered onto or into the circuit board, additional effort can be expended replacing such batteries. On the other hand, when batteries which can be soldered to the circuit board using automated assembly equipment are used, there can be an up-front cost savings in manufacturing.

In devices where rechargeable batteries are used, cost of battery management circuits and attendant board space must be considered.

Battery holders, connectors, compartments, and the need for waterproof enclosures in some devices, will all add component and assembly costs to the manufacture of the device.

Figure 15.10 illustrates nominal pricing of an assortment of battery holders, sockets, and connectors. Again, the cost of soldering wires, making through-hole solder connections, and inserting cells into holders or sockets must also

be considered when selecting the battery. Note that in some cases, the cost of the holder—assembly costs excluded—can be as much or more as the cost of the battery itself.

15.4 Battery and Power Management Topologies

IoT devices have many components, and while many integrated circuits have common operating voltage ranges based on the technology node in which they are designed, there are often different minimum/maximum voltage specifications among the disparate components of the system. Consequently, there is often a need for having multiple regulated supplies—1.8 V, 3.3 V, 5 V—on board, all of which might be sourced from a single battery or battery pack. Whenever DC-DC converters or other power converters are used in a system, their power conversion efficiencies under the range of load conditions must be accounted for when estimating the battery storage capacity needed for the product. This is true whether the system is operating from a rechargeable or primary battery, and whether or not energy harvesting circuits (discussed elsewhere in this chapter and more thoroughly in Chap. 11) are used.

Fig. 15.10 Assortment of battery holders, sockets, and cells with solderable tabs, with approximate prices for each

15.4.1 Battery Life Extension Using Efficient Power Management Architectures

Not only are the batteries themselves important considerations when identifying a power source for a system, but the power management architecture is equally critical to system performance. While many ICs operate over a wide range of voltages, it is often the case the battery can't directly supply power to all of the components in a system. Step-up or step-down converters are often needed, with attendant losses of energy in the conversion process.

Power management techniques including dynamic voltage scaling; extensive use of the various power savings modes resident within microcontrollers and microprocessors—such as hibernation/sleep modes and clock speed reduction—can have substantial impacts on battery life. Avoiding complete stack initializations and endlessly running loops in MCUs can also add appreciably to battery life.

Another approach to extending battery life is to use power manager integrated circuits that put components in deep sleep or hibernation modes whenever the functions those components perform are not needed. Optionally, the power manager, by way of a high side or low side switch, can completely disconnect those components from the supply or ground rail for a significant portion of time and reconnect them periodically or upon occurrence of an interrupt or other event. A schematic example of this concept is shown in Fig. 15.11 (http://www.cymbet.com/pdfs/AN-1059.pdf).

Typical microcontroller-based devices where energy conservation is an issue spend a large part of their time in a low-power "sleep" state. This is a state where the microcontroller is not running

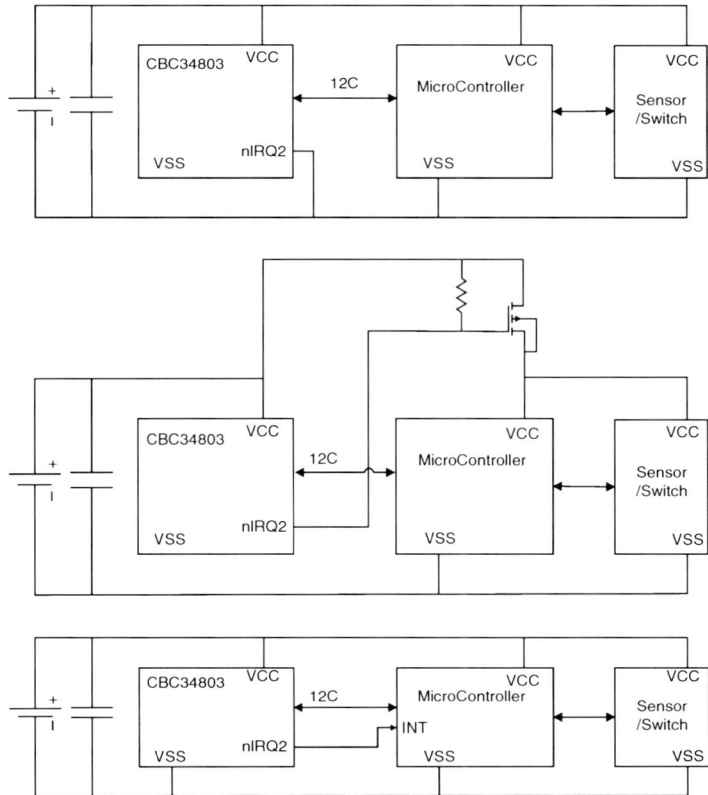

Fig. 15.11 Isolation and control of active system components from the power supply for the purpose of reducing power consumption and extending battery life using (*top*) switched VSS, (*middle*) switched VCC/VDD, and (*bottom*) interrupt configurations. (Courtesy of Cymbet Corporation)

but is waiting for an interrupt from either a sensor or a timer. When one of these interrupts is issued the microcontroller goes to a higher power active mode to process the event and then goes back to sleep. The active mode operation may include processing the sensor data and then may operate an actuator or send a message. Many times a message may be sent via a low-power radio protocol which requires significant processing cycles to correctly operate the protocol stack. The amount of processing depends greatly on the complexity of the protocol.

The average power consumption of the system is the sleep power times the percentage of time the system is asleep plus the active power times the percentage of time the system is active divided by 100.

$$P_{avg.} = \left(P_{sleep} \times \%time\ asleep + P_{active} \right.$$
$$\left. \times \%time\ active \right)/100$$

Minimizing the largest of these terms provides the greatest power savings. In some cases the active power term is much larger than the sleep power term either because the power per event is large or the active power events happen very often.

When the system that has a large sleep power compared to its active power is asleep there is an opportunity to reduce power by placing the microcontroller in its lowest power mode while using the standalone timer functions to provide periodic wakeups to the microcontroller and associated circuitry. This way the entire system is totally asleep for the majority of time and the microcontroller is only awakened for short periods to determine if it needs to service a sensor or switch.

Power savings are determined by accounting for and managing several variables, including active run times; sleep periods; active time/sleep time; and number of instructions executed by the microcontroller. In some applications, these techniques can extend battery life by 2× or much more, as given by the power savings ratios in Table 15.5 (http://www.cymbet.com/pdfs/AN-1059.pdf). The result can be the use of a much smaller, and/or less expensive battery than would otherwise be required to meet product performance and lifetime objectives.

The power savings technique described in the previous section works well with systems where the sleep power is a large or significant contributor to the average power. In systems where a radio is used the power for the radio and the power used by the microcontroller to run the software stack can be large. Many radio standards are complicated and use a sophisticated stack to manage the protocol. These stack implementations often have significant runtimes to initialize their internal memory structures. Often these stacks are supplied by the radio chip vendor and therefore are attractive since they save software development time. If the

Table 15.5 Power savings possible when utilizing a power manager to isolate a microcontrollers from the power supply (Courtesy of Cymbet Corporation)

Original sleep current in μA	μA/ MHz	MHz	Power savings with periodic interrupt					
			Active runtime in ms	Sleep period in ms	Active current when running (μA)	Active time/sleep time	Number of instructions	Power savings ratio
0.6	200	1	0.1	16	200.000	0.0063	100	0.47
0.6	200	1	0.1	33	200.000	0.0030	100	0.97
0.6	200	1	0.05	33	200.000	0.0015	50	1.89
0.6	200	1	0.1	100	200.000	0.0010	100	2.80
0.6	200	1	0.05	100	200.000	0.0005	50	5.26
0.6	200	1	0.1	250	200.000	0.0004	100	6.38
0.6	200	1	0.05	250	200.000	0.0002	50	11.11
0.6	200	1	0.2	250	200.000	0.0008	200	3.45
1.6	200	1	0.3	250	200.000	0.0012	300	6.30

stack needs a long time to initialize, then the previous power savings technique only makes sense if the radio operates very infrequently.

A typical radio application might require, say, 25 mA for 25 ms = 625 µA-s for each transmission. A simple stack may take essentially no time to initialize but a sophisticated one may take upwards of 400 ms of processing time at 5 mA, or 2000 µA-s to initialize. This highlights the importance of either a simple stack initialization or a microcontroller that has a low power mode to retain the initialization in RAM at low power.

Chapter 10 provides a more thorough analysis of this topic.

15.4.2 Using Primary and Secondary Batteries with Energy Harvesting (EH) Sources

In cases where an IoT device demands more from a battery than what is commercially available— or available at an acceptable cost—energy harvesting can be brought to bear as a supplementary source of power. There are, of course, attendant design and component costs when implementing energy harvesting circuits. The designer must consider whether the costs and complexities associated with energy harvesting circuits can be justified over the life of the IoT device. And the advantage of using energy harvesting might be apparent to one party in the supply chain that extends from IoT device designer, device manufacturer, to intermediate distribution or other sales outlet, to end user— and/or redound to their benefit—, but not to others involved in the transaction. Perhaps the end user would bear the expense of replacing a dead battery, but isn't willing to pay the additional cost of having energy harvesting circuits to minimize or eliminate the need for battery replacement. Or, to take another example, perhaps the IoT device is an intrusion sensor that is offered by a company with a service contract, who in turn can make or lose money with each battery replacement service call.

A simple first question to ask is: How much extra energy is needed throughout the life of the product, beyond which the preferred battery can store, retain for that product life, and deliver as needed? Once answered, the next two-part question is: Is that amount of energy harvestable under expected operating conditions and, if so, at what cost? The return on an energy harvesting investment is not always clearly measurable when designing a new IoT product. However, we can at least lay out the basic concomitant elements of energy harvesting architectures, their practical limits from a technical perspective, and rough costs of implementing such circuits.

Energy harvesting circuits can take many forms, from simple, widely deployed solar-powered devices charging a capacitor, to more elaborate power collection and energy management schemes utilizing peak power tracking circuits and algorithms relying on incoming power from transducers having characteristic impedance and power profiles according to their design, materials, and construction.

Beginning with a system-level view, a block diagram of a conceptual IoT node is illustrated in Fig. 15.12, including the basic elements of the energy harvesting module.

While core IoT functions vary by application, it is not uncommon to find basic sensor, MCU, and often, radio elements, as shown. Likewise, a device equipped with energy harvesting circuitry will normally include—in addition to the power transducer itself—a power conversion circuit and energy storage device, either a battery or adequately sized capacitor. When sizing the battery in an energy harvesting design, it is important to keep in mind that having a bulky, medium to high capacity energy storage cell is counterproductive if the transducer can never replenish it in practical operating environments. It is analogous to trying to fill a swimming pool one glass of water at a time. Similarly, if the power consumption of the device exceeds the average power production of the harvesting source and associated conversion circuitry, eventually the battery will become depleted and will be of little to no use in the long run.

IoT devices generally are comprised of several of the elements, or functions

Fig. 15.12 Block diagram of sensor node incorporating an energy harvesting module

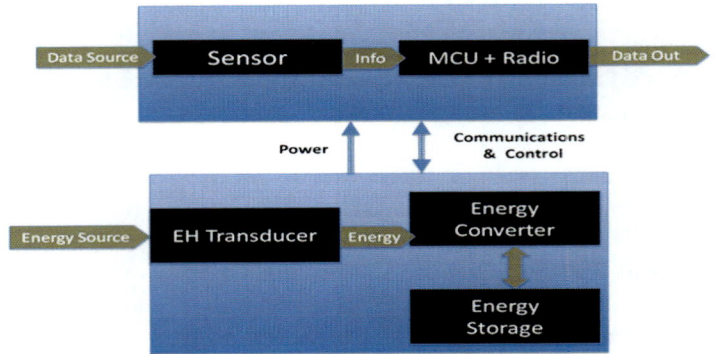

Table 15.6 Devices typically suitable to be powered by energy harvesting sources

<100 nW	<1 μW	<10 μW	<100 μW	<1 mW	<10 mW	<100 mW	<1 W	<10 W
Real-time clock	MCU standby	Watch/ calculator	RFID tag	Sensors/ remotes	Wireless sensors	Bluetooth transceiver	GPS	Mobile phone
	← Practical, indoor energy harvesting using low cost transducers →							

(e.g., microcontroller/microprocessor, sensor, radio), listed in Table 15.6, not all of which are operating at any given time, perhaps, but in some combination present a load to the battery and incoming power from the harvester. It is important to know not only the average power consumption in various IoT operating modes, but peak loads also, as the combination of the harvested power and battery must together support these loads. Depending on the battery choice, input power from the harvesting transducer under worst-case ambient conditions (e.g., times when minimum or no light is imparted to a solar cell), and other factors, the battery itself might be relied upon to support the peak loads. If the battery itself does not have sufficient power delivery under these worst-case conditions, a low ESR capacitor might be necessary as part of the power delivery chain.

By definition, energy harvesting requires a transducer to convert energy from one form to another, so it can be stored or delivered directly to the system load. Transducers come in a range of technologies, shapes, sizes, electrical characteristics, and suitability for any number of operating environments. However, not all transducers are practical in terms of cost, power generation, or in the physical size needed to

achieve intended results. A glimpse of the most conventional transducers, their limitations, challenges, and typical performance measures are listed in the following table. Such transducers have found their way into remote applications as a means of reducing or eliminating the need for batter replacement over the life of the product. In some manifestations, the energy harvesting circuit serves as way to extend the life of a primary battery by delivering power directly to the circuit when sufficient incoming power from light, temperature differential, vibration, or other sources is available from the environment. In these implementations, the energy harvesting circuit is connected as a logical 'OR' with the primary battery. In others, the energy harvesting circuit is used to regularly replenish the charge in a secondary battery when needed, and powers the circuit directly at times, when surplus power is available from the transducer and intermediate conversion circuitry—whether through a boost, buck, linear regulator, or other power conversion circuit.

With some exceptions, where the IoT device is also a wearable, energy harvesting is often not practical because of the difficulty in collecting ambient energy from body-worn transducers. One such exception is the use of a piezoelectric or inductive element buried in a shoe that generates energy with each impact. And of course, for many years low power watches have incorporated solar cells in the watch face.

Aside from technical limitations of generating and converting sufficient power on a predictable, regular basis, there are issues with commercial availability of many transducer types. Solar—or photovoltaic cells—are the most readily available in a range of configurations, conversion efficiencies, and prices. But of course ambient light is not always practical in the application setting. Battery charge time is important, especially if the rechargeable cell is completely depleted during the period while input power is not available to/from the transducer. If the cell takes many hours to be recharged—either because of its inherent characteristics, or because only small amounts of energy are available to trickle charge the cell—then not much stored energy will be available in the event it is discharged and is relied on again a short time after it has been depleted. Rechargeable coin cells can take hours for a full recharge, while some solid state batteries can be recharged to 80% capacity in a few minutes—provided sufficient input power is available to charge the cell.

The typical power transducers available to the energy harvesting circuit designer, along with their characteristics, are listed in Table 15.7. Transducers are available from a variety of manufacturers.

Conceptually, the approach to a power efficient energy harvesting circuit—at least the front-end portion of it—is to match the impedance of the transducer to the load in order to maximize power transfer. While easier idealized than implemented in some cases, the basic circuit structure is shown in Fig. 15.13.

The transducer characteristics shown in the preceding table must therefore be understood by the designer so the energy harvesting circuit elements can efficiently collect their output power. The circuit must also be adaptable as environmental conditions change and the transducer output in turn varies in its response. A critical piece of energy harvesting design that is often overlooked is the start-up phase of operation. Because the transducers are usually high impedance devices (compared with batteries), they cannot source much current. When sourcing low current loads and trickle charging batteries, this is not necessarily a problem. However, when starting from 0 V or being sourced by a slowly rising voltage delivered by some transducers, active components like microcontrollers and radios, if supplied by a current-starved source, might not make it through the power-up state and

Table 15.7 Conventional energy harvesting transducers, their characteristics, and challenges presented to designers

Energy source	Challenge	Typical impedance	Typical voltage	Typical power output	Normalized cost
Light	Conform to small surface area; wide input voltage range	Varies with light input—low $k\Omega$ to 10s of $k\Omega$	DC: 0.5–5 V [depends on number of cells in array]	10 µW–15 mW (outdoor: 0.15–15 mW) (indoor: <500 µW)	$0.50–$10.00
Vibrational	Variability of vibrational frequency	Constant impedance—10s of $k\Omega$ to 100 $k\Omega$	AC: 10s of volts	1 µW–20 mW	$2.50–$50.00
Thermal	Small thermal gradients: efficient heat sinking	Constant impedance—1 Ω to 100s of Ωs	DC: 10s of mV to 10 V	0.5–10 mW (20 °C gradient)	$1.00–$30.00
RF and inductive	Coupling and rectification	Constant impedance—low $k\Omega$s	AC: Varies with distance and power 0.5–5 V	Wide range	$0.50–$25.00

Fig. 15.13 Circuit model representing power conversion using matched impedances

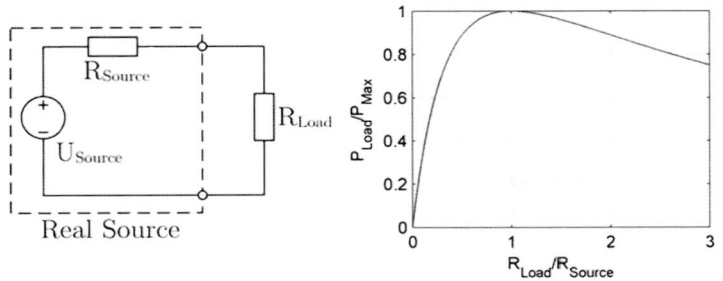

into normal operation. This is because there can be relatively high currents being drawn internally to these components as the various transistors and circuits are transitioning into their respective normal operating regimes. If the EH design does not account for these transition currents by way of sourcing current from a pre-charged capacitor—or using the on-board battery if available—the system can stall and never get successfully through the power-up phase. Use of a power supply supervisor integrated circuit can be used to switch on the input and output stages quickly once the pre-charged capacitor has been charged to a sufficient voltage to operate the MCU and other active components. Capacitance can be sized according to the amount of energy needed to power up the MCU, radio, or other component, or according to the amount of energy needed to send a radio transmission, for example, in the case of pre-charging a capacitor at the output stage.

While it is relatively straightforward to calculate a power budget and design a system to work within the constraints of the power and energy available, it is easy to overlook the power required to initialize the system to a known state and to complete the radio link with the host system or peer nodes in a mesh network. The initialization phase can sometimes take 2–3 times the power needed for steady state operation. Ideally, the hardware should be in a low power state when the system power-on reset is in its active state. If this is not possible, the microcontroller should place the hardware in a low power state as soon as possible. After this is done, the microcontroller should be put into a

sleep state long enough for the energy harvester to replenish the energy storage device. If the power budget is not exceeded during this phase, the system can continue with its initialization. Next, the main initialization of the system, radio links, analog circuits, and so forth, can begin. Care should be taken to ensure that the power consumed during the time the system is on during this phase does not exceed the power budget. Several sleep cycles might be needed to 'stair step' the system up to its main operational state.

In most system power budgets, the peak power drawn is not as critical as the length of time the power is required—in other words, the energy budget. Careful selection of the message protocol for the RF link can have a significant impact on the overall power budget. In many cases, using higher power analog circuits that settle quickly after being turned on, and which are turned off immediately thereafter, can decrease the overall energy consumed. Microcontroller clock frequency can also have a significant impact on the power budget. In some applications it might be advantageous to use a higher microcontroller clock frequency to reduce the time the microcontroller and peripheral circuits are active. Avoid using circuits that bias microcontroller digital inputs to mid-level voltages; this can cause significant amounts of parasitic currents to flow. Use large value (10–22 MΩ) pull-up/down resistors where possible. However, be aware that high circuit impedances coupled with parasitic capacitance can make for a slow rise/fall time that can place the voltage on the microcontroller inputs at mid-levels, resulting in parasitic current flow. One solution to the problem is to enable the internal pull-up/down resistor of the microcontroller input to force the input to a known state, then disable the resistor when it is time to check the state of the line. If using the microcontroller's internal pull-up/down resistors on the inputs to bias push-button switches in a polled system, leave the pull-up/down resistors disabled and enable the resistors only while checking the state of the input port. Alternatively, in an interrupt-driven system, disable the pull-up/down resistor within the first few instructions in the interrupt service routine. Enable the pull-up/down resistor only after checking that the switch has been opened. Microcontroller pull-up/down resistors are typically less than 100 kΩ and will present a meaningful load on the system if left on continuously while a switch is activated for any significant length of time. When using external pull-up/down resistors, bias the resistors not with the power rail, but with a microcontroller port instead.

As there are always cost tradeoffs or additional expenses in design complexity, versus simply using a larger battery or replacing a battery in coming years, it is important to assess these costs before accepting that energy harvesting is a necessary or desirable feature of an IoT design. The return on investment depends in part on the total energy budget for the system and whether a larger battery delivering to this energy budget can physically fit within the confines of the IoT device. The various costs are presented below, using cost per unit of energy for an energy harvesting design in comparison to that of the several battery options available and as discussed earlier. Considering the energy harvester as a variable capacity battery, the cost per unit of energy generated can be calculated and compared with that of a battery as discussed above. In the case of rechargeable batteries, the cost of the charging circuitry must be included. The cost of a battery is in the range of $0.001–$0.005 per mAh of capacity, including the cumulative capacity of multiple charge-discharge cycles in the case of secondary batteries.

As an example scenario, suppose a photovoltaic (i.e., solar) cell is the transducer, to which an off-the-shelf converter circuit is added, and the harvester is used as a battery life extender by way of powering the load when delivering sufficient power, and relying on the battery when ample harvested power is not available. The battery is not rechargeable in this case. Further, suppose the solar cell is a low cost amorphous silicon type, capable of generating 10 μW/cm^2 of cell area under modest indoor lighting conditions of 400 Lux. A cell 30 cm^2 in size will then produce 300 μW of power at, say, 3 V. This leaves 80 μA of current available to power the load, assuming 80% conversion efficiency through the DC-DC conversion. Let's further assume that 80 μA is sufficient average current to keep the system running continuously.

Over a 10-year life, producing power 12 h per day, the total energy (or charge) generated is 3.5 Ah. Now, what was the cost to generate that energy? The solar cell is on the order of $2.50 in modest volumes. The DC-DC converter and ancillary components is perhaps $1.50. Assume for the sake of simplicity that the incremental design effort is negligible because the cost is spread over a large number of units sold. Neglect also the cost of additional board space for the harvesting components, as well as the solar cell housing. Total cost of the harvesting components is then $4.00. That cost over 3.5 Ah is $1.14/Ah, or $0.0011/mAh. This is on par with the cost of alkaline and lithium primary cells, although in this simplistic analysis the cost of battery connectors, holders, compartments, and any other overhead associated with the battery is neglected, because in this scenario a battery is needed in addition to the EH circuit—it is just a matter of which size/capacity battery is going to be needed.

In a case where a rechargeable battery is used to receive harvested energy and deliver it to the load when energy is unavailable for harvesting, the battery lifecycle capacity can be calculated to determine whether its aggregate storage capacity over the rated number of charge-discharge cycles, at the rated depth-of-discharge, is adequate for the task. The lifecycle capacity is calculated as follows:

Lifecycle capacity = Rated battery capacity × Rated charge-discharge cycle life × Depth-of-discharge corresponding to rated cycle life

Example:

ML1220 rechargeable coin cell
Rated capacity = 17 mAh
Charge-discharge cycles = 1000 cycles
Depth-of-discharge = 10%
Lifecycle capacity = 17 mAh × 1000 cycles × 10%/cycle
=1.7 Ah.

An obvious problem encountered when analyzing an energy harvesting system that includes a rechargeable battery, is the difficulty in confining battery discharge to the recommended depth-of-discharge on each cycle with adding a monitor circuit that depletes significant battery capacity in the process. As described earlier, it is sometimes practical to make infrequent measurements of the battery voltage, sacrificing precision and real-time monitoring for the sake of energy conservation. And of course, cutting off the battery from the circuit after it is been depleted just 10% from a full charge means that only 1/10th the battery capacity is available to operate the system on any given charge-discharge cycle. Discharging the battery to deeper discharge depths will cut into the number of charge-discharge cycles available. Therefore, the designer must make a tradeoff between charge available per cycle and the number of cycles available over the life of the product. Flat discharge profiles make this measurement process more difficult to control, as charge removed does not correlate strongly with battery voltage over narrow segments of the discharge profile.

With the exception of some industrial grade cells, most rechargeable batteries are not rated to be recharged more than about 1000 cycles. As

such, a candidate rechargeable battery rated for 10- to 20-year life, recharged daily, is generally more expensive than is a consumer grade battery of similar capacity. Small solid state batteries, with recharge cycle life ratings to several thousand cycles, are now commercially available and suitable as supplemental power sources in IoT designs where energy harvesting is implemented.

Two energy harvesting scenarios will now be considered, the objective being to calculate the battery lifetime. Several commercially available rechargeable batteries are compared, with a range of IoT device power consumption. Rated capacity and cycle life at specified depth-of-discharge for each battery type are used to estimate operating life, assuming negligible self-discharge and ignoring irrecoverable charge loss mechanisms such as mechanical wear-out of gaskets (where applicable), degradation of electrolyte, high temperature aging effects, and other losses not directly induced by charge-discharge cycling of the cell. Such losses cannot be ignored, especially when the battery is being used over the course of many years and in harsh environments. However, the aging mechanisms differ widely by not only the type of battery, but by construction methods particular to a given manufacturer, materials that comprise the battery, and manufacturing process. Therefore, this exercise seeks to calculate the ideal lifetime of a battery under the two energy harvesting scenarios:

1. Recharge the battery during the 12-h period in which the IoT device is being powered by an energy harvesting transducer, and deliver power to the IoT device directly from the battery for 12 h per day, while the power transducer is not generating power.
2. Recharge the battery only when the battery is depleted to its rated depth-of-discharge. This assumes a coulomb counter or other fuel gauging method is used to measure battery capacity, to signal when the battery has reached its deepest rated depth-of-discharge and requires a recharge. A 1-h recharge time is assumed in this scenario. Recharging might be accomplished through a near-field

Fig. 15.14 Battery operating life in a system operating continuously on a 12-h charge, 12-h discharge cycle

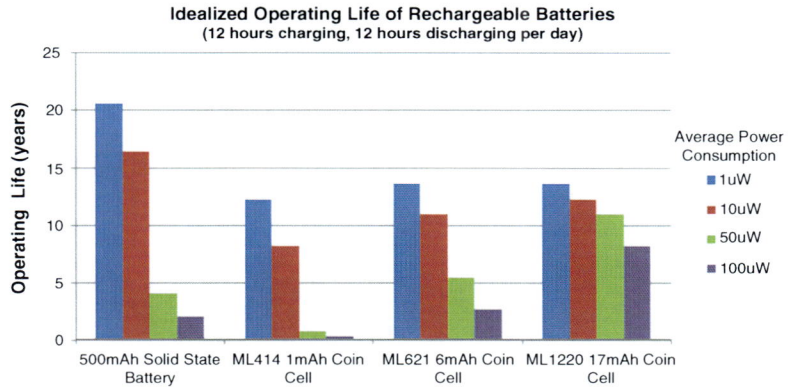

communication and inductive power transfer method when an IoT device is retrieved from the field for data transfer to a computer, for example; or, by way of a direct connection to a portable terminal periodically for the purpose of transferring information from the IoT device, at which time the device is recharged and redeployed to the field.

Battery lifetime depends on the number of rated charge-discharge cycles available—which varies as a function of depth-of-discharge—and the frequency at which the cell is being cycled. In scenario (1), the average power consumption will determine to what depth-of-discharge the battery is drained on each charge-discharge cycle, which in turn affects the number of charge-discharge cycles available. As rechargeable battery specifications do not declare cycle life as a function of depth-of-discharge at more than one or two depths, extrapolations and interpolations will be made for the sake of this exercise in order to calculate the operating life in scenario (1). Scenario (2) is more straightforward, as the depth-of-discharge used in this example will be that as explicitly defined in the respective product specifications. Another assumption is that the cell resistance of each battery considered in these examples is low enough to power the system that an additional energy storage element— such as a low ESR capacitor—is not needed to supply currents to IoT device elements that would otherwise consume higher power than the battery itself can deliver. Supplemental storage devices are accompanied by internal leakage currents that have to be accounted for in the average power consumption figures.

Figure 15.14 represents the operating life of various batteries under a range of IoT device load conditions, from 1 to 100 μW average power consumption, using the first scenario described above, whereby the battery is charged 12 h each day, and discharged 12 h each day. Results are calculated based on room temperature operation and typical battery specifications.

It is clear that, absent the permanent energy losses arising from aging mechanisms not related to battery cycling, an operating life greater than 10 years is possible with all batteries compared in this example.

Calculating battery life using the second scenario, whereby the battery is taken to a rated depth-of-discharge before being recharged, results as shown in Fig. 15.15 are obtained. Depending on the type of device and method used to recharge the battery, in the case of the smaller capacity batteries it might be impractical to recharge the battery as frequently as required by the higher operating power of 100 μW.

Again, this is an idealized situation that does not account for self-discharge and other aging mechanisms. The projected operating lifetimes are unrealistic in commercial batteries. Rechargeable batteries, with the exception of solid state batteries, are usually not expected to retain their energy for more than 10 years.

Note how the scenario two results differ vastly from the first scenario. This is primarily the result

Fig. 15.15 Battery operating life in a system in which the battery is recharged only upon depletion to rated depth-of-discharge (DoD)

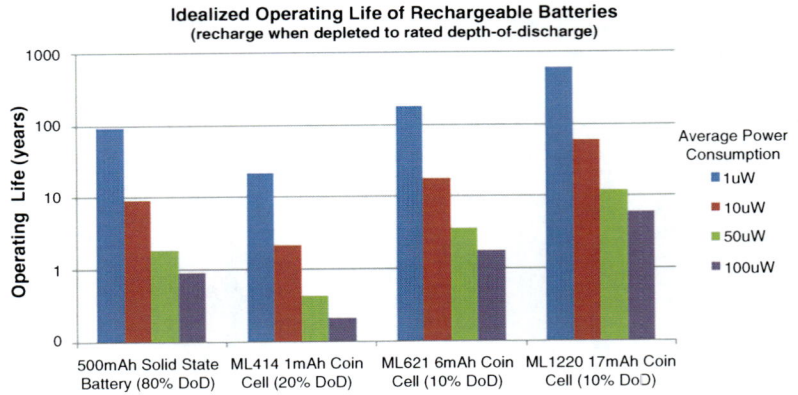

Idealized Operating Life of Rechargeable Batteries
(recharge when depleted to rated depth-of-discharge)

of the battery cycle life not being linear with depth-of-discharge, especially when the battery is under a charging voltage 50% of the time as in the first scenario. Continuously applying a charging voltage to the battery has the effect of reducing battery cycle life and retained capacity. It is better to remove the charging voltage from the cell when it is fully charged. Doing so might not be practical in some system configurations. In other words, keeping a battery at or close to a full state of charge has more deleterious effects to its service life than maintaining a lower state of charge.

By comparing the energy consumed over a 10-year period, for example, one can ascertain whether a primary battery can be found that has ample storage capacity, sufficiently low self-discharge, acceptably small physical size, low enough cost, and other attributes which could make it a suitable candidate for the application. A rule of thumb: every 1 µA of operating current results in 10 mAh of battery capacity consumed per year. This rule accounts for self-discharge of typical primary cells.

Therefore, other factors have to be considered when making the decisions of whether or not to implement energy harvesting, such as:

- Will a larger battery fit within the product size constraints? Is there an incremental cost of having a larger, or other, compartment, holder, or socket to hold the battery?

- If a battery fails, what are the consequences of not having the auxiliary power source the energy harvesting supply affords?
- How reliable and consistent is the source of ambient power to supply the energy harvesting transducer? How reliable is the transducer itself?
- How impractical or expensive is the effort of replacing and disposing of the battery if it dies before end of product life? Or does battery failure in fact determine the end of the product life?
- Can a lower cost battery type be used if energy harvesting is used to supplement the energy needs?
- Could use of energy harvesting in an IoT device lead to fewer batteries in the waste stream? If so, can a business case be made to justify on implementation of EH on this basis?
- Are there advantages in product marketing and possibly increased customer adoption through the use and promotion of a "green" design—whether actual or perceived?

With the thousands of IoT device formats, applications, and use cases to draw from, it is impossible to say for any given device whether it is a good candidate for energy harvesting without knowing more about the specifics of the application and system operation. However, there are general categories of devices that can be classified as likely or unlikely candidates for energy

Table 15.8 Sample IoT devices and their suitability for use with energy harvesting sources

Device/application	Features	EH transducer	Good fit for EH? Why? Why not?
Home thermostat	Wireless; bluetooth radio	Solar	Yes. Low power consumption and regular availability of ambient light
Indoor air quality monitor	CO_2, CO, humidity, temperature sensors	Solar	Perhaps. Some sensors draw more power than can be replenished with harvested power
Fitness band/monitor	Accelerometer; memory; processor	Thermoelectric or solar	No. High current draw from "always on" operation; inconsistent ambient power; on-body power transducers usually impractical
Smartwatch	CPU; real-time operating system; radios; MEMS, other sensors; display	Solar	No. Power consumption of the many functions overwhelm power production of small EH transducers
Indoor lighting controller	Wireless switch	Solar; electromagnetic	In some cases. Push-button switch or ambient light power the transmission
Medical patch	Temperature, pulse rate monitors; NFC communication	Solar; electromagnetic	Solar cell or near-field inductive (using NFC power) to recharge battery during use, device interrogation, or as needed
Water meter	Flow meter; wireless transmitter	Thermoelectric	Yes. Harvest energy from temperature differential of water pipe to ambient air
Outdoor sensor	Humidity, temperature sensors; wireless transmitter	Solar, electromagnetic (wind generator)	Yes, provided energy consumption is not too high to be stored in battery for operation in the dark
Animal monitoring and tracking	Various sensors; GPS	Solar	In some cases. GPS power consumption requires relatively large solar cell
Wireless parking meter	Vehicle detection sensors; radio communication	Solar	Yes. Daylight provides ample energy to monitor and communicate. Industrial battery needed for communication and operation in the dark
Smart door lock	Keypad; real-time clock; low power processor; data logger	Solar; possibly piezoelectric depending on latch mechanism	Yes. Low power consumption, except solenoid latch mechanism where used; ambient light is regularly available
Factory automation	Proximity, motion sensors; radio link	Solar, piezoelectric	Yes, if ambient energy sources are available
Machine monitor	Vibration, temperature, other sensors; radio link	Solar, thermoelectric, piezoelectric	Practical in some cases, depending on ambient power available for harvesting
Bluetooth smart beacons	Sensor, radio, low power MCU	Various	Yes. Low power wireless link and MCU modes are very energy efficient
Security monitor	Proximity sensor; wireless transmitter	Solar	Yes. Ambient light is available to recharge battery and supply energy during daytime operation

harvesting, at least from a technical standpoint, if not necessarily in cost-effectiveness. Table 15.8 lists several candidate devices/applications and whether and why they might or might not be suitable for harvesting ambient energy.

This concludes the brief exploration of the potential for energy harvesting in IoT devices. See Chap. 11 for a more thorough treatment of this topic.

15.5 Comparisons of Batteries to Other Energy Storage Devices: Performance and Cost

While batteries are often the power source of choice for wireless and mobile applications, other storage devices are occasionally used as the main energy storage device in the system, or, more commonly, as a supplemental energy storage device to complement the characteristics of the battery—such as delivering higher power form a capacitor in systems where pulsed loads are found. In other cases, supplemental power sources might be used with batteries to the opposite effect, whereby a fuel cell, for example, produces a continuous supply of energy until the fuel source is depleted, but might not have sufficient drive current capability to operate the circuit directly. The fuel cell might then be used to charge a secondary battery, which in turn delivers higher power to the load.

15.5.1 Super-Capacitors

Electric double-layer capacitors (EDLCs), commonly known as super-capacitors and ultra-capacitors, are typically low resistance devices used to deliver high power pulse currents. Their volumetric energy density is much lower than that of most batteries, but their power density is usually considerably higher than that of batteries. Capacitance of these devices ranges from several millifarads (several microampere-hours equivalent) to thousands of farads.

In cases where small amounts of energy are needed—such as in a backup power source used

to maintain real-time clock or memory in the event of main power loss—super-capacitors are suitable substitutes for batteries. More often, capacitors are used to augment a battery where higher current loads demand more current than a battery can deliver, or more than a battery can deliver without falling below a critical voltage or having its useful life diminished as a result of delivering current well in excess of its rated output. Another limitation of super-capacitors is the inherently linear discharge profile that results in a steadily decaying voltage as charge is removed. Consequently, not all charge within the capacitor can be accessed by the load at a useful voltage.

The self-discharge of super-capacitors is usually orders of magnitude higher than that of batteries. For that reason—in addition to having much lower energy density than batteries—super-capacitors are largely excluded as viable main power sources in IoT devices. There are hybrid battery/capacitor devices—lithium-ion capacitors—that have properties resembling that of lithium ion batteries at one electrode and electric double-layer capacitors at the other electrode. These devices achieve higher cell voltage, higher energy density, and lower self-discharge than pure electric double-layer capacitors, while retaining relatively good power density. Capacitances range from a few millifarads to hundreds of farads.

15.5.2 Fuel Cells

As fuel cell technology has migrated to the realm of portable applications, coupled with development of metal/air batteries, the distinction between fuel cells and batteries is sometimes not so clear. Depending on the amount of fuel available in the fuel cell, the operating life, power requirements, and whether the IoT device operates continuously, in special cases a fuel cell can serve as a practical power source. State-of-the-art technology in portable fuel cells achieves energy density comparable to common batteries. However, the cost of fuel cells is generally much higher than batteries having similar energy

Table 15.9 Standards and publications related to battery safety and transportation

Standards body	Publication	Title/purpose
American National Standards Institute	ANSI C18.1M	Portable primary cells and batteries with aqueous electrolyte—general and specifications
	ANSI C18.2M	Portable rechargeable cells and batteries—general and specifications
	ANSIC18.3M	Portable lithium primary cells and batteries—general and specifications
International Electrotechnical Commission	IEC 60086	Primary batteries—safety standard for lithium batteries
	IEC 61960	Secondary lithium cells and batteries for portable applications
	IEC61809	Product safety standard for sealed alkaline secondary cells and batteries
	IEC62133	Secondary cells and batteries containing alkaline or other non-acid electrolytes—safety requirements for portable sealed secondary cells, and for batteries made from them, for use in portable applications
Underwriters Laboratories	UL1642	Standard for household and commercial batteries (superseded by IEC 62133)
	UL2054	Standard for lithium batteries (cells)
Occupational Safety and Health Administration	Various OSHA-approved state plans	Safety and health in battery manufacturing

storage capacity as a small fuel cell cartridge, for example. Portable fuel cells generally have a narrow effective operating temperature range, thus limiting their practical use in many devices. Restrictions that might apply to the transport of fuel cells aboard commercial aircraft can limit their viability in mobile devices.

Portable—and micro—fuel cells having power output less than 5 W have been developed for applications in military systems, personal electronics, and toys. The cost of fuel cells is not yet competitive with commodity batteries, in part due to the expense of the components comprising the fuel cell, but also due in large part to the fact that fuel cells have generally not yet reached commercial production status (Linden and Reddy 2002).

15.5.3 Safety, Transportation, and Disposal of Batteries: Regulations and Trends

As batteries become increasingly energy dense and find greater use in an increasingly mobile world, standards, directives, and regulations pertaining to their use, transport, and disposal are being developed and adopted by commissions.

Some battery chemistries and types are inherently more hazardous than others in terms of being flammable, prone to exploding, toxic to humans, and harmful to the environment in disposed of improperly.

Table 15.9 is a partial list of bodies issuing safety standards, example publications, and the general purpose of each (Linden and Reddy 2002).

There are numerous independent, accredited laboratories providing services for testing to the various published standards. Because standards are updated regularly, it is important to always refer to the latest edition of a particular standard when testing a device or product for compliance.

In addition to the batteries being inherently more or less safe in their materials, chemical reactions, and construction, external charging circuits, protection devices, and enclosures are important aspects of product design as it pertains to batteries. Problems of thermal runaway, short-circuits, careless handling, external application of electrical supplies to the battery, and protection against heat sources are all important considerations for designers of battery-powered products.

For designers developing products to be used in explosive atmospheres, compliance with IEC standard 60079-11—pertaining to "intrinsically safe" equipment and the components making up that equipment—might be necessary. Batteries used in such equipment must therefore pass tests related to spark generation, leaking, venting, and generation of electrical and thermal energy.

Transportation of batteries on commercial carriers is coming under increasing scrutiny and regulation as safety concerns are elevated in the wake of several high profile incidents aboard airliners, in particular. Battery fires are highly reactive, with incredible amounts of energy liberated over a brief time, resulting in high temperatures and violent reactions. Consequently, restrictions are being proposed and enacted by government bodies, trade associations, and commercial enterprises to protect the safety of those in proximity to high capacity batteries of the most energetic chemistries.

Regulations have been developed that govern battery packaging to prevent batteries from becoming shorted across their terminals or from one cell to another during transport, and from leaking or exploding if dropped or otherwise mishandled. There are restrictions in the amount of lithium that can be transported aboard aircraft or other carriers without taking special precautions. Regulations of particular types or capacities (amount of lithium, for example) of cells subject them to restrictions or prohibitions on passenger-carrying aircraft. Various agencies have published guidelines or regulations, including U.S. Department of Transportation (DOT), Federal Aviation Administration (FAA), United Nations (UN) and International Air Transport Association (IATA). International transportation regulations can be found in publications UN 3480/3481 and UN 3090/3091.

Equivalent lithium content of a lithium-ion or lithium polymer battery is calculated based on the ampere-hour rating of the cell, battery, or battery pack, as follows:

Lithium content (in grams) = rated capacity (in ampere-hours) per cell × 0.3 × number of cells in the battery or pack.

Example: 18650 Li-ion cell

Capacity = 3 Ah → lithium content = 0.9 g lithium.

As batteries of all types become more widely used in stationary and especially portable equipment and devices, rules governing their disposal have been enacted and continue to be developed. For example, the EU Battery Directive is a comprehensive regulatory document that regulates the manufacture and disposal of batteries and accumulators in the European Union.

Enforcement of disposal regulations can be difficult, with billions of battery-operated devices in use throughout the world and compliance not always foremost in the minds of users when disposing of batteries specifically, or the devices containing them. Achieving battery recycling objectives relies on public cooperation and participation.

Cytotoxicity of batteries is also of ongoing concern as more and more battery-powered medical (implantable and non-implantable) devices are being designed and used in everyday life. Some types of solid state batteries have been tested and shown to be non-cytotoxic.

15.6 New Battery Technologies, Trends, and Opportunities in IoT Applications

Since the introduction of Li-ion batteries in particular, great strides have been made in energy density, the array of cell types available, rate capability, package sizes, and other important attributes. Yet, as with electronic components generally, ever-increasing demands are placed on batteries. Consumers want more, for less, as with most products. And while the trend in pricing per unit of energy has been steadily downward in recent years, improvements in energy density have not kept pace with the proliferation of device functions and capabilities of products such as IoT devices. While the power consumption—active, quiescent, leakage—for a particular function is generally much lower than what is was just a few years ago, the inexorable tendency, precipitated by user demand, is for designers to make the next generation of devices more feature rich than the preceding generation, and thus the demand on batteries tends to increase, not diminish.

In response, scores of organizations in government, academia, and private industry are investing large sums of money and scientific and engineering talent into developing new battery chemistries, materials, and manufacturing techniques in an effort to bring batteries with advanced capabilities to market for any number of applications. Efforts encompass every aspect of battery design, materials development, and construction. Advances range from anode development—such as replacing carbon-based anodes with silicon and lithium anodes in Li-ion batteries—to higher capacity and higher power cathode materials, to new formulations in electrolyte composition, including new polymers and solid state materials. Li-ion materials tolerant to higher charging voltages are also in development.

As described earlier, the energy density of advanced Li-ion rechargeable cells is on the order of 650 Wh/L, and just above that in some primary lithium cells. On the horizon, rechargeable cells exceeding 1000 Wh/L are now in development. New rechargeable chemistries—using zinc rather than lithium—are in development, and lithium polymer cells are now commercially available in myriad sizes and storage capacities.

Significant resources are being deployed in the development and commercialization of solid state batteries, promising higher energy density, a greater number of charge-discharge cycles, and improved safety relative to batteries employing liquid and polymeric materials in their construction. Such solid state batteries have already made their way to the market in the form of thin film micro-batteries—allowing for special battery shapes to be designed and integrated with portable electronic products including IoT devices.

Thin film and thick film printed batteries are finding applications in wearable devices and other devices incorporating wireless sensors, radios, and processing capabilities requiring just a few milliampere-hours of storage capacity in thin, usually flexible, formats.

Many challenges face the battery industry, including pricing pressures driven by production of commoditized cells from manufacturers in low cost regions; quality concerns over products from certain manufacturers; a finite number of elements on the periodic table from which to develop new chemistries and the aforementioned increasing demands in performance generated by the vast proliferation of electronic appliances of every sort.

Numerous trade associations and annual conferences are dedicated to advancing battery technology and making improvements in energy density, safety, manufacturing, and in incorporating materials more benign to the environment through the lifecycle of materials extraction, battery production, disposal, and recycling. Through their common objectives, these organizations serve a purpose in continuing to advance battery technology and products to meet the growing needs of consumers worldwide, while also continuing to recognize and address the concerns over depletion of resources, disposal hazards, and effects on the environment generally.

Forecasts for unit volumes of IoT devices deployed in the coming years vary widely, with some forecasters projecting upwards of 100 billion sensors to be made and sold each year by 2020. While not all of these sensors will be used in IoT devices per se, supporting a sizable fraction of these devices in battery-powered systems will require a significant increase in the number of batteries or other suitable energy storage devices to be manufactured each year. Should such markets and applications develop in line with such forecasts, and to the extent that those devices are "unplugged" from wall power, there will be opportunity and incentives for battery technologists and manufacturers alike to develop and introduce battery technologies of including more of the type in widespread use today, to nascent battery technologies currently in development.

References

Cymbet Corporation AN-1059: Extend Battery Life by Reducing System Power Using the EnerChip RTC, http://www.cymbet.com/pdfs/AN-1059.pdf. Accessed Dec 2015

Cymbet Corporation Solid State Rechargeable Battery, http://www.cymbet.com/pdfs/DS-72-41.pdf. Accesed Dec 2015

Digi-Key Online Catalog, http://www.digikey.com/product-search/en?keywords=battery. Accessed Nov 2015

Energizer AA Alkaline (EN91), http://data.energizer.com/PDFs/EN91.pdf. Accessed Nov 2015

Energizer AA Lithium (L91), http://data.energizer.com/PDFs/l91.pdf. Accessed Nov 2015

D. Linden, T. Reddy, *Handbook of Batteries*, 3rd edn. (McGraw-Hill, New York, 2002)

Panasonic AA Alkaline (LR6), http://na.industrial.panasonic.com/sites/default/pidsa/files/1.5vseries_datasheets_merged.pdf. Accessed Nov 2015

Panasonic CR2032 Manganese Dioxide Lithium Coin Battery, http://na.industrial.panasonic.com/sites/default/pidsa/files/crseries_datasheets_merged.pdf. Accesed Nov 2015

Panasonic Manganese Lithium Coin Battery, http://na.industrial.panasonic.com/sites/default/pidsa/files/mlseries_datasheets_merged.pdf. Retrieved Nov 2015

Panasonic NCR18650 Li-ion, http://industrial.panasonic.com/lecs/www-data/pdf2/ACI4000/ACI4000CE17.pdf. Accessed Nov 2015

Tadiran Model TL-5104 data sheet, www.tadiranbat.com/pdf.php?id=TL-5104. Accessed Nov 2015

System Packaging and Assembly in IoT Nodes

16

You Qian and Chengkuo Lee

The internet of things is the networks of physical "thing" embedded with sensors, integrated circuits and power source. Thus, packaging all these elements into one IoT nodes is essential for promoting this technology. In this chapter, some of the popular and commercially available packaging technologies such as wire bonding, flip-chip bonding, tape automated bonding, etc. are reviewed. Secondly, wafer level packaging is emphasized as it can be used as a low cost packaging technology, usually because the packaging cost is much higher than other cost associated with device manufacturing. Recent progress on packaging for wearable electronics, silicon Photonics and energy harvesters are also included. In the later part, infrared sensors have been used to demonstrate the packaging constraints imposed by device performance and application requirements, various packaging solutions available at an individual sensor level and complicated array level.

16.1 Introduction

The internet of Things (IoT) is the network majorly comprises of multiple miniaturized sensors, actuators, integrated circuit and power source. It creates a seamless connection between the physical world and computational systems, and resulting in improved accuracy and efficiency. Having a massive amount of sensors and other elements communicated in wireless manner has emerged as a disruptive wireless sensor network technology in data collection between every IoT node. Integration or packaging of the above mentioned components influences the overall system performance, cost and time-to-market.

From application point of view, the IoT allow users to view the phenomena of interest from multiple yet simultaneous vantage points, and with the vast aspect of applications. More specifically, such features enable users to simultaneously monitor the health of civil structures, e.g. from bridges, dams to tunnels and buildings. The IoT nodes can be used to localize and to measure the condition of transport goods, thereby optimizing the logistics cost and energy consumption. Besides, the same platform technology realizes wearable (or embedded) miniature sensors which promote higher-quality medical procedures and research, as well as next-generation preventative healthcare solutions. All these different applications require careful design of the packaging, to allow the sensors to monitor the environment effectively, and to provide the whole node with a reliable working condition.

To implement the above-mentioned applications, advances in hardware technology and engineering design have led to significant

Y. Qian • C. Lee (✉)
National University of Singapore, Singapore, Singapore
e-mail: elelc@nus.edu.sg

© Springer International Publishing AG 2017
M. Alioto (ed.), *Enabling the Internet of Things*, DOI 10.1007/978-3-319-51482-6_16

reductions in size, power consumption, and cost of all the basic components of the IoT nodes including sensors, digital circuitry, and wireless communications. In particular, microelectromechanical systems (MEMS) technology has enabled various kinds of sensors in micrometer scale, which are more sensitive, occupy significantly less area and requires very lower power than conventional sensors. For example, MEMS based gas sensors, pressure sensors and air flow sensors only demand 1–2 µW for operation (Guo et al. 2012; Lou et al. 2012; Zhang et al. 2012). While most of MEMS sensors are fabricated on silicon substrates with the potential of monolithic integration with integrated circuits (ICs), MEMS sensors make the distributed IoT become feasible.

Another major part of a IoT node is battery. However, due to use of large batteries as the power source, single packaged chip for entire sensor node is not possible. Compared with traditional batteries, MEMS energy harvesters seems to be a viable candidate, because of its compactness and unlimited lifetime. Additionally, Energy harvesters can be made with similar process like MEMS sensors and provide the entire IoT node as a single packaged chip.

Thus, various levels of system integration involve MEMS based sensor, chip for computation work and communication are likely assembled individually packaged components on a single board (Printed Circuit Board level packaging) or housing all components in a single package (System-in-Package or System-on-Chip approach) have been demonstrated. Packaging is inevitable; as it provides a significant improvement in device performance and ensures long term reliability of components in it nodes. Packaging of IoT nodes and/or its components are more challenging, because of the variety of components, like sensors, actuators, Integrated circuit (IC) controllers, etc., that make up such nodes, where such elements require different condition of the packaging. Numerous packaging solutions have also been explored for both component level and system level packaging of IoT nodes.

In view of great impact made by IoT, we will introduce the prevailing packaging technology available in microelectronics and MEMS industries first, followed by discussion on design, requirements and technology developed for MEMS sensors in case study manner. At the end of this chapter, the state-of-the-art MEMS packaging technology will be provided for understanding the packaging technology trend in the applications of IoT.

16.2 Packaging Technology

Electronics Packaging is the engineering of establishing interconnections with electrical components, managing heat dissipation, and protection of functional elements. The continuing scaling trend in microelectronics/integrated circuit (IC) technology has a significant impact on the different packaging technologies. Considering the microelectronics manufacturing flow, the first step is the fabrication processes involved at semiconductor wafer fabs, where ICs are completed with wafer-level fabrication, inspection and known-good-die test. Secondly each wafer is then mounted on a special sticky tape and sawed into individual dices. Then the back-end microelectronics manufacturing flow takes place at the packaging house. This manufacturing flow includes the next few steps: (1) one selected good die, i.e., the chip, is attached to a carrier, e.g. a ceramic package, a metal header, or a premolded plastic lead frame with an embedded metal lead frame; (2) electrical interconnects between the IC chip and carrier are made by wire bonding, flip chip, or another method; (3) either having plastic molded over the assembled chip on the plastic lead frame, or having a ceramic, metal, glass, or plastic cap sealed on the assembled ceramic package or metal header is used to provide the required reliability; (4) a final test and calibration is made to conclude the processes. The above packaging is widely considered as the first level of packaging. The modern first-level packages might not only contain one but several IC chips and are then referred to as multichip modules (MCM). Single-chip modules and MCM are assembled together with other components, such as resistors and capacitors, onto a printed circuit board (PCB) via soldering process, i.e., surface mount technology (SMT) process. This PCB package is considered as the second level of package. Moreover, the motherboard which is used in the personal computer (PC) providing sockets for various PCB cards to be plugged in; is considered as the third level package. Generally speaking, an electronic system, like a PC or a mobile phone, is a device with electronics in the third level package assembled in a single housing.

In order to realize sensor nodes which are truly ubiquitous, we really rely on the advanced packaging technology to fabricate sensor nodes in a very compact format and at a low cost way. A wireless sensor node typically consists of one or more sensors, an IC controller, a wireless transceiver, an antenna and a battery, or even better, an energy scavenger. To precisely report real-world variables, sensors are required to detect pressure, temperature, heat, flow, force, vibration, acceleration, shock, torque, humidity, and strain. In the applications such as security, surveillance and traffic control, video and images are taken by image sensors. Some of these sensors are commercialized in MEMS based solutions, whereas others are conventional solutions that have been available for decades, for example, ceramics and polymer sensors. Ranging from IC, passive elements (like capacitors, resistors and inductors), and power source like battery to various kinds of sensors, different fabrication technologies are required to make these components in rather different manufacturing lines. Thus the current cost-effective way of making a wireless sensor node is to assemble and integrate the above-mentioned ICs, various components and sensors in a PCB package, i.e., the second level package as shown in Fig. 16.1a. Heterogeneous integration is the technical term to describe the technology for integrating multifunctional elements and devices into a compact standalone system. While system-on-chip (SoC) technology represents the

integration effort of more and more functions on a single chip based on IC fabrication technology, some other elements and components made of non-silicon materials could be integrated together by system-in-a-package (SiP) solutions. Figure 16.1b shows a prototype of wireless sensor node under research based on a SiP approach. Significant size reduction shows that the SiP approach is a promising solution for heterogeneous integration in the fabrication of wireless sensor nodes.

16.2.1 Current Commercial Packaging Technology for Internet of Things

Packaging is one of the most essential and expensive part of the sensor nodes in the IoT system. The main challenge comes from the packaging controller IC with different MEMS sensor and battery. Even though, IC packaging is a well-developed technology, the MEMS packaging is relatively new and is more difficult owing to moving structural elements. Yet another complication for MEMS sensor packaging is that, most MEMS sensors should interact with the environment and transfer the coupled energy effectively and at the same time has to minimize noise and cross talk interference. Hence, not only electrical signals need to be routed through the package, but also the sensing energy has to be effectively coupled to MEMS devices. Different MEMS sensors demand for different packaging techniques which are depending on the physical domain signals to be sensed (Bloss 2012; Sparjs 2001). The unique packaging requirements for each individual MEMS sensor lead to additional complexity of designing a common packaging platform technology among different MEMS sensors at the first place. Furthermore, the mechanical requirements for a sensor package are typically much more stringent than the packaging standard of ICs, because most of the MEMS devices have fragile suspended structures, and so even a small stress or strain can influence the performance or the long term

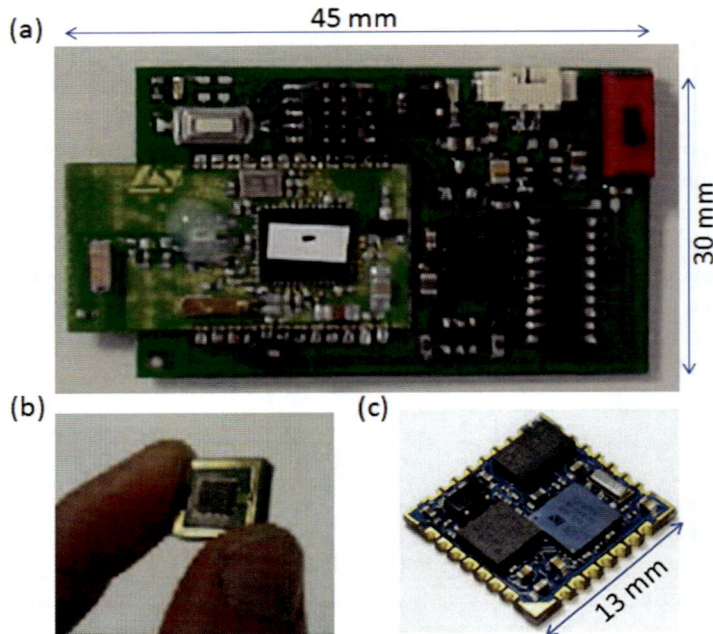

Fig. 16.1 (**a**) A wireless sensor node in production and realized in the second level package, i.e., the PCB package; (**b**) a prototype of miniaturized wireless sensor node in system-in-package (SiP) format from IMEC Research Center, Belgium; (**c**) a state-of-the-art inertial sensor module in a system-on-board (SoB) package format from STMicroelectronics. It integrates multiple sensors with a powerful computational core: a six-axis geomagnetic module, a three-axis gyroscope and a 32-bit microcontroller unit (MCU), i.e., a processor. This nine-degree-of-freedom (DoF) inertial system represents a fully integrated solution for various applications such as virtual reality, augmented reality, image stabilization, human machine interfaces, robotics and inertial body tracking [Courtesy: ST Microelectronics]

reliability of the device. In the worst case, it may also cause the device completely malfunction (Lau et al. 2010). MEMS sensors are designed for sensing various physical parameters, so in order to achieve certain device specifications, appropriate choice of materials is inevitable, which can introduce more constrains to packaging specifications. Secondly, as advancement in IoT, MEMS actuators are also integrated into the unified standalone system, along with the sensors and controller IC, to autonomously control the system by physically reacting to the decision signal from the controller. This feature of integrating actuators has greatly enhanced their capabilities and range of applications. This also means additional complexity in overall system packaging capabilities. These are inherent problems of packaging criterion demanded by different components in the IoT, thus post a major roadblock in the development of a common packaging platform technology that could provide higher reliability and lower cost devices.

The sensor packaging design is considered equivalently important as the design of the sensors itself. The sensor packaging majorly influences the performance of the device, especially with respect to factors such as long-term drift, stability and reliability. More importantly, the cost of the package and its development will often be significantly higher than that of the MEMS device fabrication. Hence packaging of integrated systems like IoT is extremely crucial for high performance, low cost and high reliable product. Researches on heterogeneous integration for building innovative standalone systems

Fig. 16.2 Block diagram showing the different levels of possible system integration (Reichl et al. 2009)

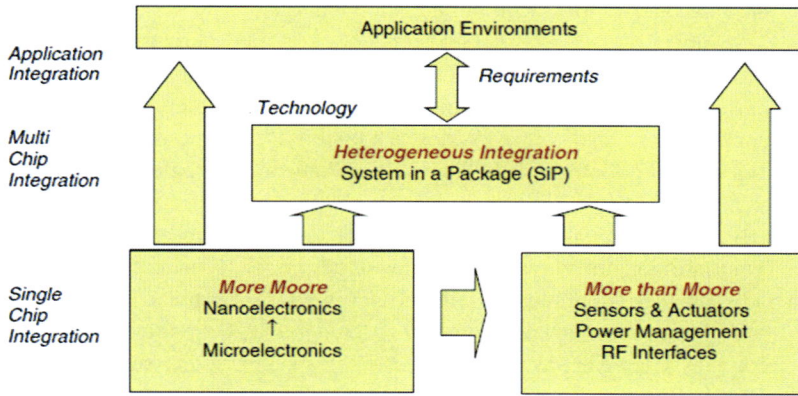

are gaining more interests globally. Various packaging techniques and procedures are successfully demonstrated for Internet of Things.

Packaging can be achieved at different levels depending upon the number of different functional chips that are packaged in a single housing as shown in Fig. 16.2. Heterogeneous integration can be achieved using different level of packaging as follows:

16.2.1.1 Integration/Assembly of Multiple Chips (or) PCB Level Packaging

The integration and assembly of individually packaged chips on to a dedicated Printed Circuit Board (PCB) is the most straight-forward means of integrated system packaging. Hence it is known as "PCB level packaging". The main advantage of this approach is that each component can be packaged with the best suitable technology possible, and it enables the best of individual functional blocks to get together. The complexity level for integrating and assembling the overall system is low. Additionally, each component can be bought from various vendors and integrated into the PCB later. This packaging approach is highly compatible with functionality upgrade and design changes, while the lead time to market is significantly minimal. However, the weakness of PCB level packaging is large footprint of the packaged systems. The different levels of integrated standalone sensor network packages as shown in Fig. 16.2 and mentioned in previous sections will be briefly discussed in next sections.

16.2.1.2 First Level Packaging

The first level packaging refers to tlhe encapsulation of discrete functional chips onto individual packages. The fabricated wafers from the semiconductor wafer fabrication facility are tested and passed the known-good-die inspection step, and then are diced into individual dies for further packaging. In first level packaging, the passed die is placed onto the package carrier holder and is electrically connected to the carrier pins. First level of packaging is also known as "Single Chip Packaging". Single chip packages can be categorized in terms of various aspects, such as package material (plastic, ceramic, or metal), interconnect arrangement (peripheral or area array interconnects), assembly technology (surface-mount technology, chip-stacking, multi-chip-module, or other package formats). Some of the main stream single chip packaging techniques are described briefly,

1. **Single in line packages (SIPs)**: SIPs have pins for through-hole assembly, only along one side of the package. SIPs are generally plastic packages and mostly used for memory modules packaging. They are also limited to application demanding fewer input/output (I/O) lead pins.

2. **Dual in line packages (DIPs)**: DIPs provides a rectangular housing with two parallel rows of electrical connecting pins. The package is a through-hole mounted package. They usually made of ceramics or plastics. The DIP is a commodity through-hole package with up to

84 pins and a most common lead pitch of 2.54 mm.

3. **Small outline packages (SOPs)**: A small-outline package is a lead frame-based plastic package with leads on two sides of the body, and is well-adopted in SMT packaging lines. SOP is 30–50% less area than an equivalent DIP, with a typical thickness that is 70% less. They are generally available in the same pin outs as their counterpart DIPs. SOP is the most common package type for low pin count ICs and MEMS pressure sensors.

4. **Quad flat packages (QFPs)**: Quad flat packages are SMT compatible packages with leads extending out from all four sides of the package. QFP is available as ceramic (CQFP) or lead frame-based molded plastic (PQFP) packages. The lead pins are placed close and are very thin to achieve maximum I/O pins.

5. **Chip carriers (CCs)**: Chip carriers are also surface-mountable packages with peripheral leads on all four sides like the QFP counterparts. However, in contrast to the QFP which has gull wing leads, chip carriers feature either J-leads or no leads at all. The leadless chip carriers (LLCCs) have a laminated ceramic substrate, leaded chip carriers (LDCCs) with J-leads are either lead frame based molded plastic packages or utilize ceramic substrates.

6. **Pin grid arrays (PGAs)**: PGAs are through-hole packages with an area array of interconnect pins. The package is square or roughly square, and the pins are arranged in a regular array on the backside of the package with either laminated ceramic (CPGA) or laminated plastic (PPGA) substrates. CPGA packages are available in two configurations—cavity-up or cavity-down. In the cavity up configuration, the IC is on the opposite side as the pin array, while the IC is placed on the same side as the pin array in the cavity down configuration. PGAs allow for more pins per IC than older packages such as DIP. Intel Corporation adopted PGA technology for several generations of its microprocessors.

7. **Ball grid arrays (BGAs)**: A BGAs are the surface-mountable versions of the PGA with an area array of interconnect using attached solder balls, which are popularly used for permanent mounting of devices like microprocessors. A BGA can provide more interconnection pins than the number can be put on a dual in-line or flat package. The whole bottom surface of the device can be used, instead of just the perimeter. The leads are also on average shorter than with a perimeter-only type, leading to better performance at high speeds. Soldering of BGA devices requires precise control and is usually done by automated processes. Plastic BGAs (PBGAs) based on laminated multilayer plastic substrates and ceramic BGAs (CBGAs) based on multilayer ceramic substrates are available. The chip interconnection is accomplished either by wire bonding or by flip chip bonding.

The advancement in the single-chip package is focused towards increased I/O counts and reduced package form factor for both ICs and MEMS packaging as shown in Fig. 16.3a, b respectively. These advancements have led to the high density, low cost and much efficient single chip packages. Even though the standard packaging evolved to provide more I/O pins with reduced footprint, the major technological improvement that aided this growth was the bonding techniques. Some of the well-established techniques are discussed below:

1. Wire bonding

   Wire bonding is the process of providing electrical connection between the chip and the external lead pins of the chip carrier by using fine bonding wires. The wire used in wire bonding is usually made either of gold (Au) or aluminum (Al), although copper (Cu) wire bonding is starting to gain a foothold in the semiconductor manufacturing industry.

   During gold ball wire bonding, a gold ball is first formed by melting the free end of the wire. The gold ball is then brought into contact with the bond pad. Once the contact is made, desired amounts of pressure, heat, and/or ultrasonic forces are then applied to

(a) General Packaging Trends

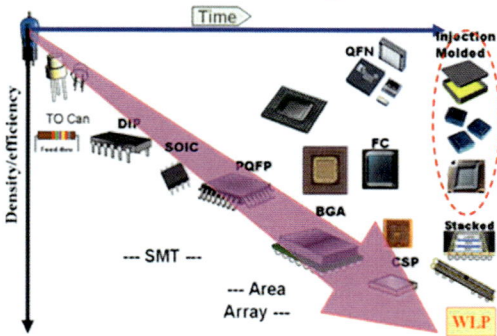

(b) MEMS Packaging

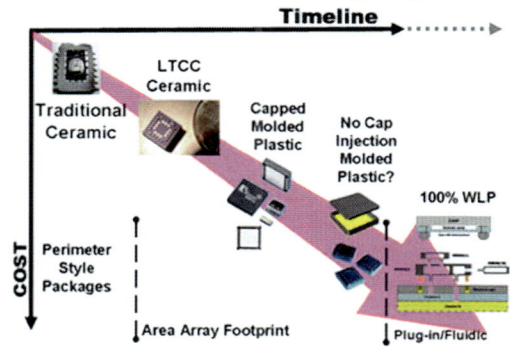

Fig. 16.3 Trend in packaging industry: (**a**) general packaging trend and (**b**) MEMS packaging trend [Courtesy: MEMS Journal]

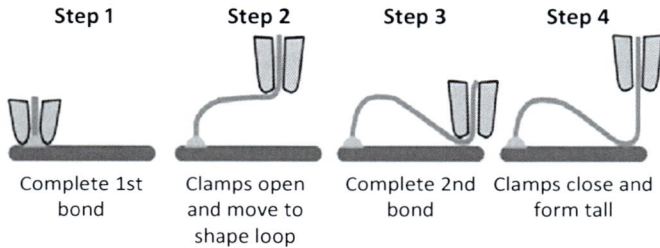

a

Step 1 — Complete 1st bond

Step 2 — Clamps open and move to shape loop

Step 3 — Complete 2nd bond

Step 4 — Clamps close and form tall

b

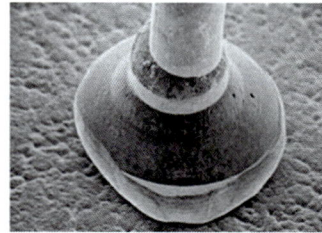

Fig. 16.4 (**a**) Schematic demonstration of gold ball wire bonding process [Courtesy: Freescale Semiconductor]; (**b**) SEM photo of ball bonded gold wire [Courtesy: Delft Institute of Microsystems and Nanoelectronics (DIMES) Technology Center, Netherlands]

the ball for a specific amount of time, and to form the weld between the ball and the bond pad. The wire is then run to the corresponding finger of the lead frame, forming a gradual arc or "loop" between the bond pad and the lead finger. Pressure and ultrasonic forces are applied to the wire to form the second bond this time with the lead finger. Finally, the wire bonding machine breaks the wire in preparation for the next wire bonding cycle (Fig. 16.4).

During aluminum wedge wire bonding, a clamped aluminum wire is brought in contact with the aluminum bond pad. Ultrasonic energy is then applied to the wire for a specific duration while being held down by a specific amount of force, forming the first wedge bond between the wire and the bond pad. The wire

is then run to the corresponding lead finger; against which it is again pressed. The second bond is again formed by applying ultrasonic energy to the wire. The wire is then broken off by clamping and movement of the wire (Fig. 16.5).

Among the aforementioned wire bonding techniques, gold ball bonding is much faster than aluminium wedge bonding, and so are extensively used in plastic packaging. Unfortunately, gold ball bonding on Al bond pads cannot be used in hermetic packages, primarily because the high sealing temperatures (400–450 °C) used for these packages will tremendously accelerate the formation of Au-Al intermetallics, leading to early life failures. Gold ball bonding on gold bond pads, however, may be employed in hermetic

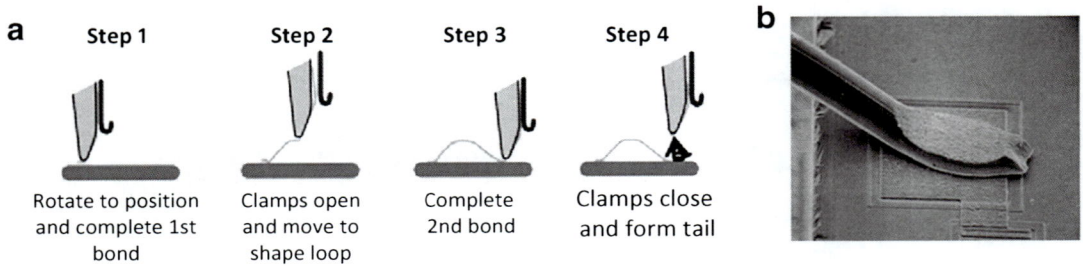

a

Step 1	Step 2	Step 3	Step 4

Rotate to position and complete 1st bond | Clamps open and move to shape loop | Complete 2nd bond | Clamps close and form tail

b

Fig. 16.5 (**a**) Schematic demonstration of Al wedge wire bonding process [Courtesy: Freescale Semiconductor]; (**b**) SEM photo of wedge bonded Al wires [Courtesy: Delft Institute of Microsystems and Nanoelectronics (DIMES) Technology Center, Netherlands]

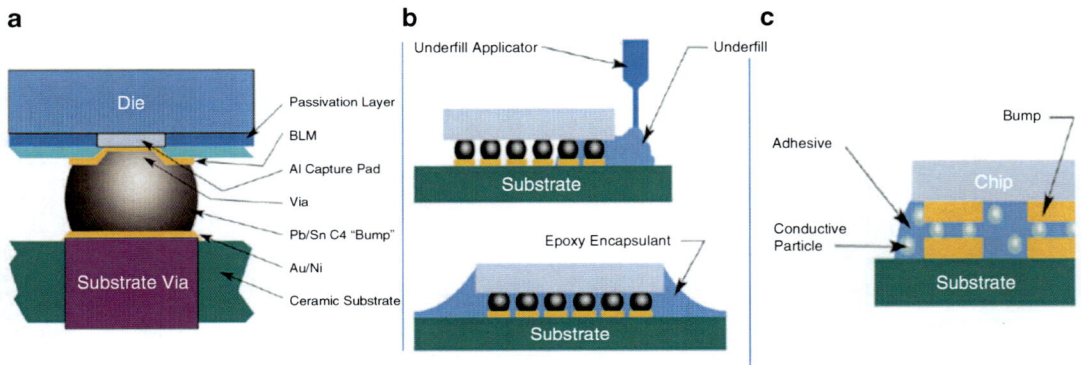

Fig. 16.6 Schematic demonstration of (**a**) flip chip bumping; (**b**) flip chip underfilling; (**c**) flip-chip assembly process from ACI technologies Inc

packages. Alternatively, the Al-Al ultrasonic wedge bonding does not require heat to facilitate the bonding process, unlike the Au-Al thermosonic ball bonding. The Al bond pad is also harder than the Au ball bond, making good bonding between them through purely ultrasonic means, thereby avoiding any thermal damage caused to wires, bond pads, or silicon substrates.

2. Flip-Chip Bonding

Flip-Chip bonding is the process of flipping the chip over and bonding it onto a substrate, board, or carrier in a 'face-down' manner. Prior to the bonding process, conductive bumps are precisely formed in the desired target positions on the chip surface, so while mounting the chip, it has to be faced down to enable electrical connection to the chip carrier. These bumps provide electrical connection, mechanical

stability and also enable heat transfer to substrate. Flip chip assembly is quite different from the traditional approach. Flip chip assembly consists of three major steps (Gianchandani et al. 2008; Lau 1996; Zhong et al. 2007):

(a) Flip-Chip Bumping—The process of forming bump on a bond pad of the die as shown in Fig. 16.6a. This can be done by various techniques like Under Bump Metallization (UBM), Plated Bumping, Stud Bumping, and Adhesive Bumping.

(b) Flipping and attaching the chip to the chip carrier.

(c) Flip-Chip Underfilling—It is a process of filling the open spaces between the chip and the substrate or board with a non-conductive but mechanically protective material. Needle dispensation followed by thermal curing to form

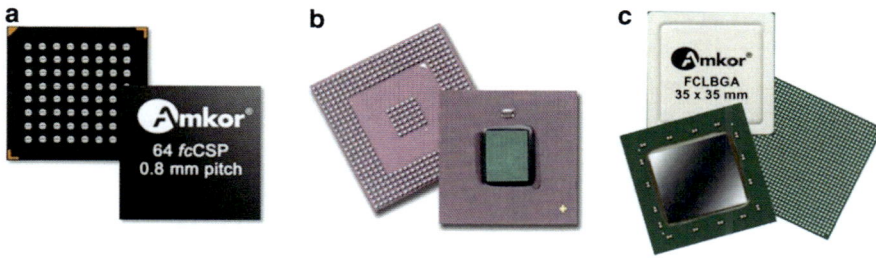

Fig. 16.7 Various commercially available flip-chip packages with different interconnect formats offered by Amkor Technology (**a**) fcCSP package, (**b**) FC ceramic package (**c**) SuperFCTM

permanent bond is used, as shown in Fig. 16.6b.

The major advantage of flip-chip bonding approach is that, their size is much smaller than their conventional counterparts, as the chips do not require wire bonds. Additionally, all problems related to inductance and capacitance that are associated with bond wires are eliminated and so provides improved performance. Flip-chip bonding is one the most preferred techniques in chip-scale packaging and is commercially available with lot of variation from number of interconnects to the substrate, package material, etc., as shown in Fig. 16.7.

3. Tape Automated Bonding

Tape Automated Bonding (TAB) is the process of mounting a die on a flexible tape made of polymer material. Initially conductive bumps or balls are formed on the targeted locations on the chip. Then the bumped chip is connected to fine conductors on the tape (known as inner lead bonds). The other side of tape (known as outer lead bonds) connects the tape to the external circuit. Sometimes the tape on which the die is bonded already contains the other chips (Fig. 16.8).

The main advantages of tape automated bonding include: it allows the use of smaller bond pads and finer bonding pitch; and the bond pads can be all over the die therefore increasing the possible I/O count of a given die size. The unique feature of TAB is that it allows circuits to be physically flexible. However, on the other hand, the tailoring of the tape and the entire process may demand higher cost and time. Thus, tape automated bonding is a better alternative to conventional wire bonding if very fine bond pitch, reduced die size, and higher chip density are desired. It is also the technique of choice when dealing with applications that need the circuits to be flexible.

Second Level Packaging

In the second level of packaging, the packaged chips with different functionalities are mounted on the printed circuit board, in which the electrical routing between discrete components are designed and printed beforehand. The mounted chips and other electronic components (i.e. resistors, capacitors, inductors, power source ports and other leads) are electrically connected through soldering process. Various PCB level integrated IoT has been reported by research groups all over the world. Some of the PCB assembled sensor nodes are shown in Fig. 16.9.

PCB level module integration has the advantage of providing best possible packaging demanded by individual chips and later integrating to realize the functionality of the entire system, which in this case is the IoT. The PCB technology and MCM integration at PCB level are well established, and such platform provides higher yield and flexibility to design change or modification in particular module.

However, the biggest disadvantage of this kind of PCB level integration is larger footprint of the final product, and increased capacitive and inductive interference from the routing wires. In order to reduce the overall footprint, more

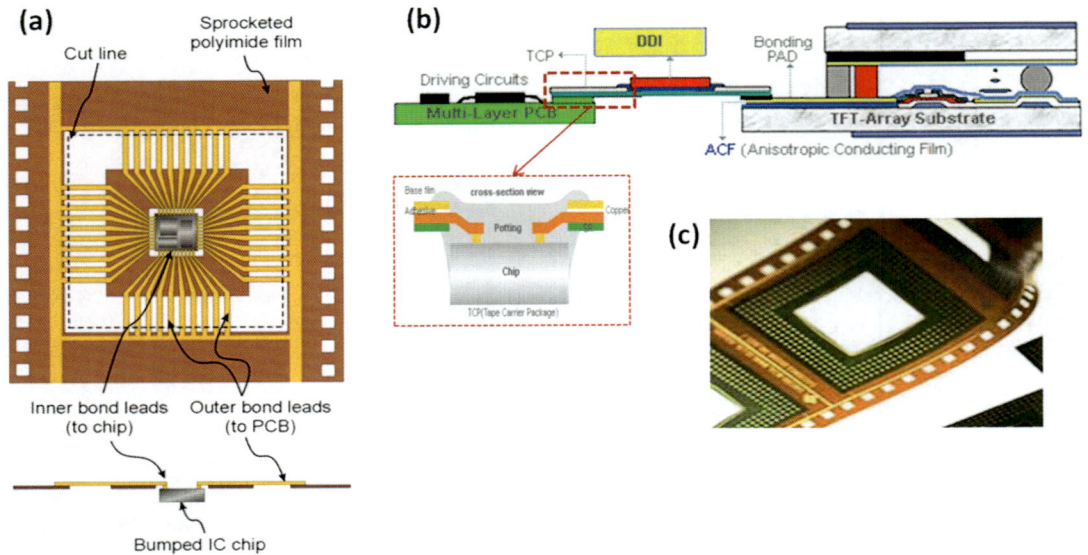

Fig. 16.8 (a) Schematic drawings of tape automated bonding (TAB) and its respective cross section; (b) schematic demonstration of using TAB for flexible interconnects based application; (c) shows real tape automated bonded devices for flexible application

Fig. 16.9 Various sensor nodes based on PCB level packaging—(a) in-plane packaging of discrete component of WSN, (b) effective 3D PCB level assembly demonstration for WSN (Reichl et al. 2009) and (c) compact PCB level packaged WSN [Courtesy: Wireless Sensornets Laboratory, Western Michigan University]

interestingly the individual chips can also be placed on either side of the PCB, which straight away reduces the footprint to half but it also increases the packaging complexity at the same time. Despite the above mentioned challenge, this 3D integration of ICs on a PCB has been embraced by a lot of companies and PCB level packaging remains as the leading technology for heterogeneous integration. For example, the Google NEXUS 7 tablet computer has the various functionality chips, assembled onto a PCB in 3D manner as shown in Fig. 16.10.

16.2.1.3 Multiple Chip Integration (or) System in Package

Even though PCB level integration is the popular choice of technology for system integration, various other technologies are also gaining interests and are explored actively worldwide. For Internet of Things, which employ synchronous functioning of multiple sensors, IC controller, discrete electrical components and power source, the most attractive scheme is the System-in-Package (SiP). As mentioned in the introduction of this chapter, different functional dies are

BACK   **FRONT**

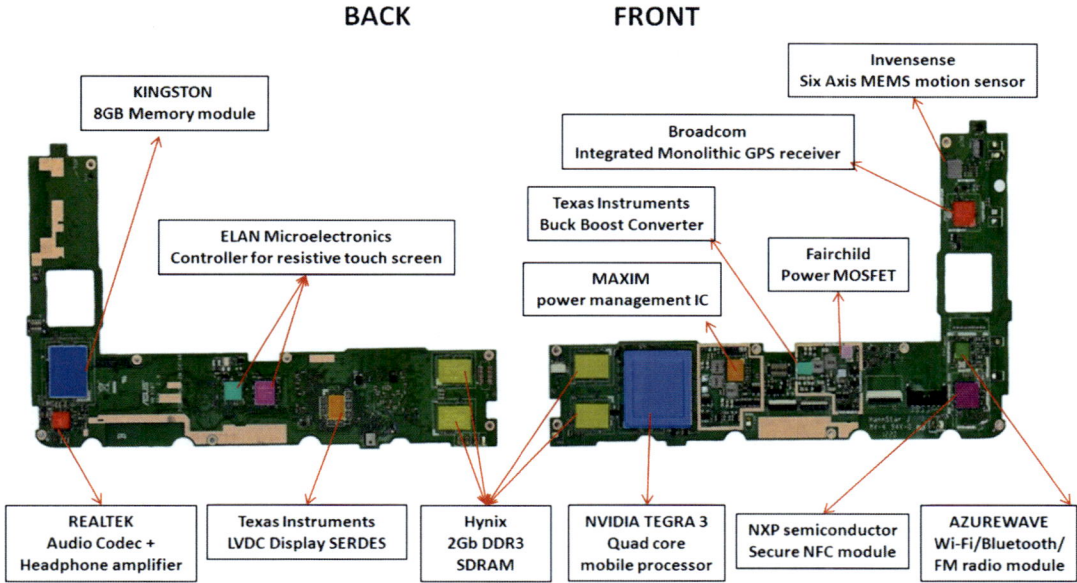

KINGSTON
8GB Memory module

ELAN Microelectronics
Controller for resistive touch screen

Invensense
Six Axis MEMS motion sensor

Broadcom
Integrated Monolithic GPS receiver

Texas Instruments
Buck Boost Converter

Fairchild
Power MOSFET

MAXIM
power management IC

REALTEK Audio Codec + Headphone amplifier	Texas Instruments LVDC Display SERDES	Hynix 2Gb DDR3 SDRAM	NVIDIA TEGRA 3 Quad core mobile processor	NXP semiconductor Secure NFC module	AZUREWAVE Wi-Fi/Bluetooth/ FM radio module

Fig. 16.10 The communication board of Google NEXUS 7 tablet computer has various individual chip assembled on either side of the PCB, which provides significant reduction in footprint [Courtesy: Google NEXUS 7]

manufactured independently, but are packaged in a single carrier. By this means, the footprint can be drastically reduced. Various techniques are available for packaging multiple functional dies on to a single package. There two major ways of achieving SiP—(1) Planar SiP packaging, where individual chips are placed alongside laterally to each other in a housing and (2) Stacked SiP packaging, where individual dies are placed one on top of another in a vertical fashion in a housing (Goldstein 2001).

In the case of 2D planar SiP approach, the individual functional dies are picked and placed onto a dedicated carrier one by one after the known-good-die testing. Once the dies are placed side by side onto the carrier, they are wire bonded to the bond pads of the lead pins of the package, which is followed by encapsulation of the entire system under a single package. The schematics and commercialized 2D planar SiP based pressure sensor from Infineon Technologies are shown in Fig. 16.11. The main advantage of 2D planar SiP technique is the reduced footprint as compared to PCB level packaging.

More aggressive scheme of high density integration is the stacked SiP approach, where individual dies are placed on top of one-another. This effectively reduces the overall footprint of the system. Different approaches are proposed to realize this concept. Some of them are:

1. Stacked Packages—Different packaged chips are placed on top of each other and are wire bonded to bond pads of the carrier substrate (see Fig. 16.12a).
2. Stacked Chip-Scale Package—More compact approach is a package that bare dies and are stacked and wire-bonded to bond pads that connect to the leads of solder balls on the substrate. This package is further connected to a PCB to form a complicated system via the solder balls based on SMT manufacturing lines (see Fig. 16.12b).
3. Folded Package—It is an advanced way of achieving smaller package, i.e., chips are initially placed on a flat polyimide tape, followed by folding the flexible tape to yield a stacked package of small footprint (see Fig. 16.12c).

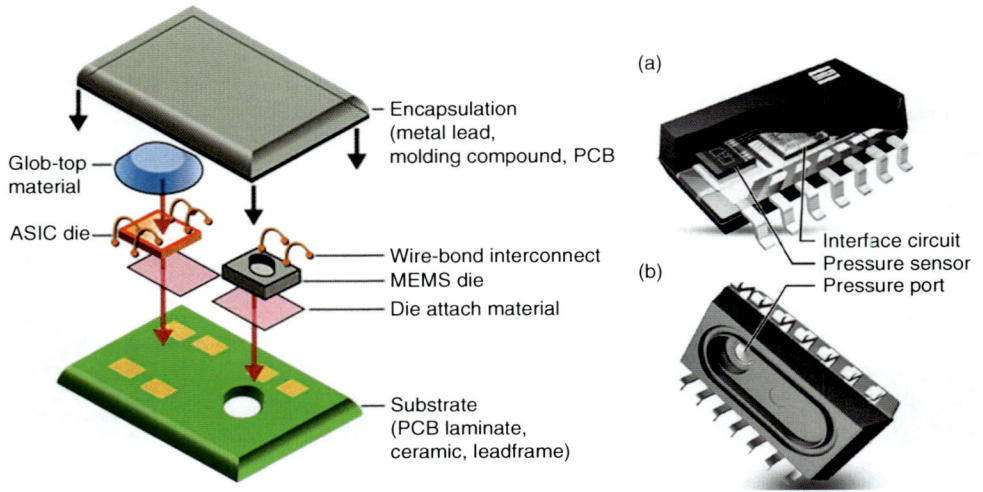

Fig. 16.11 Schematic demonstration of system in package (Yole Developpement 2012) and SiP assembled device [Courtesy: Infineon Technologies]

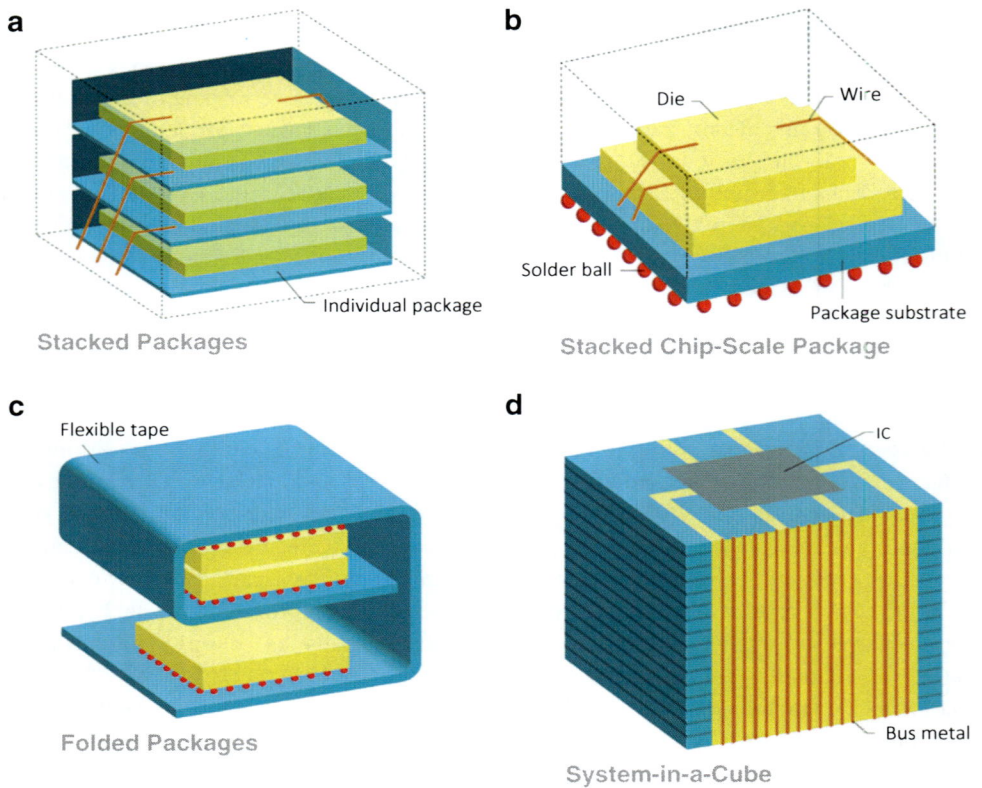

Fig. 16.12 Schematic demonstration various packaging approaches to achieve 3D integration of individual chips (**a**) stacked packages, (**b**) stacked chip level packages, (**c**) folded packages and (**d**) system-in-a-cube approach to provide electrical interconnects through vertical bus lines (Goldstein 2001)

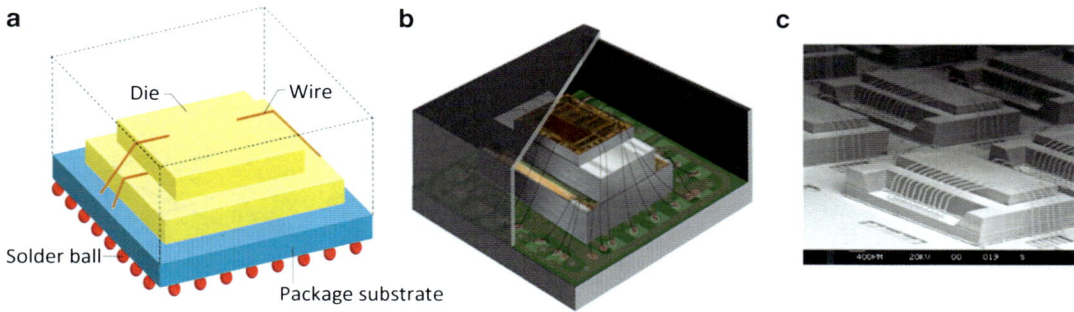

Fig. 16.13 (a) Conceptual drawing of 3D stacking of individual chips in a package (Goldstein 2001); (b) conceptual drawing of 3D chip level stacking in a package for an inertial sensor modfule [Courtesy: InvenSense Inc]; (c) SEM photo of fabricated ST microelectronics gyroscope demonstrating the concept of chip level stacking in a package [Courtesy: ST Microelectronics]

4. System-in-a-cube—The densest package so far relies on a stack of epoxy layers and embedded chips. Metal interconnects to the layers are made along the sides of the stack (see Fig. 16.12d).

Even though the technologies enabling 3D stacking looks fascinating in terms of foot print reduction, the sole purpose of integrating the multiple sensor chips with windows to interact with the environment will be limited or impossible to achieve in some case. Thus, for the packaging, the planar and stacked chip level system-in-package looks very attractive (Fig. 16.13).

Without comprising with the high cost and lead time associated with system-on-a-chip development, manufacturing SiP has the added benefit of compatibility with die design changes and integration of various die technologies (e.g., Si, GaAs, SiGe, SOI, MEMS). Thus, together with SoC technology, interconnect and packaging technologies are key enabling technologies for realizing smart SiP solutions.

16.2.2 Advanced Packaging Technology for Internet of Things

More advanced packaging involves packaging of dies at wafer level and is commonly known as "Wafer Level Packaging (WLP)" technology (Esashi 2008; Ko and Chen 2010; Lau et al.

2010). Wafer level packaging has the ability to enable true integration of wafer fabs, packaging, test, and burn-in at wafer level in order to streamline the manufacturing process undergone by a device from bare silicon substrate to customer shipment. Hence WLP can be considered as the extension of the wafer fabrication processes to include device interconnection and device protection processes. There are various methods of achieving WLP as shown in Fig. 16.14.

WLP also provides significant performance improvement over chip level packaging with almost zero increment in footprint, due to package. This enables WLP to revolutionize the next generation IC, MEMS and integrated system packaging and opened up whole new areas of applications as shown in Fig. 16.15.

The ultimate advantage of WLP is that it combines the possibly smallest implementation form of packaging through 3D hybrid integration of die stacking, with the economic benefits of wafer batch processing. The two most attractive WLP approaches for packaging Internet of Things are discussed briefly.

16.2.2.1 Wafer to Wafer (W2W) Bonding

Wafer to Wafer bonding is the process of firmly joining two wafers (or more) to create a stacked and bonded wafer. Wafer bonding is used in the MEMS field to build 3D microstructures and to manufacture silicon-on-insulator (SOI) substrates. Wafer bonding techniques are not only used to bond two silicon wafers to each

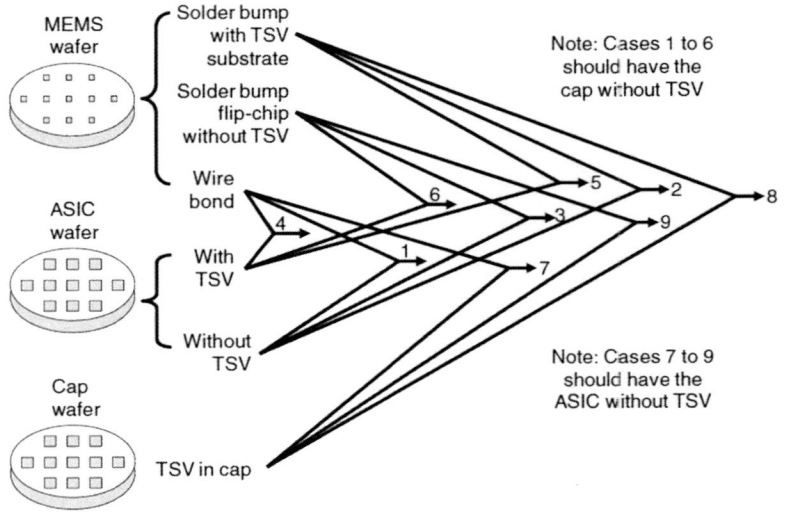

Fig. 16.14 Different designs of 3D MEMS packaging for the MEMS wafer, ASIC wader and cavity cap wafer (Esashi 2008)

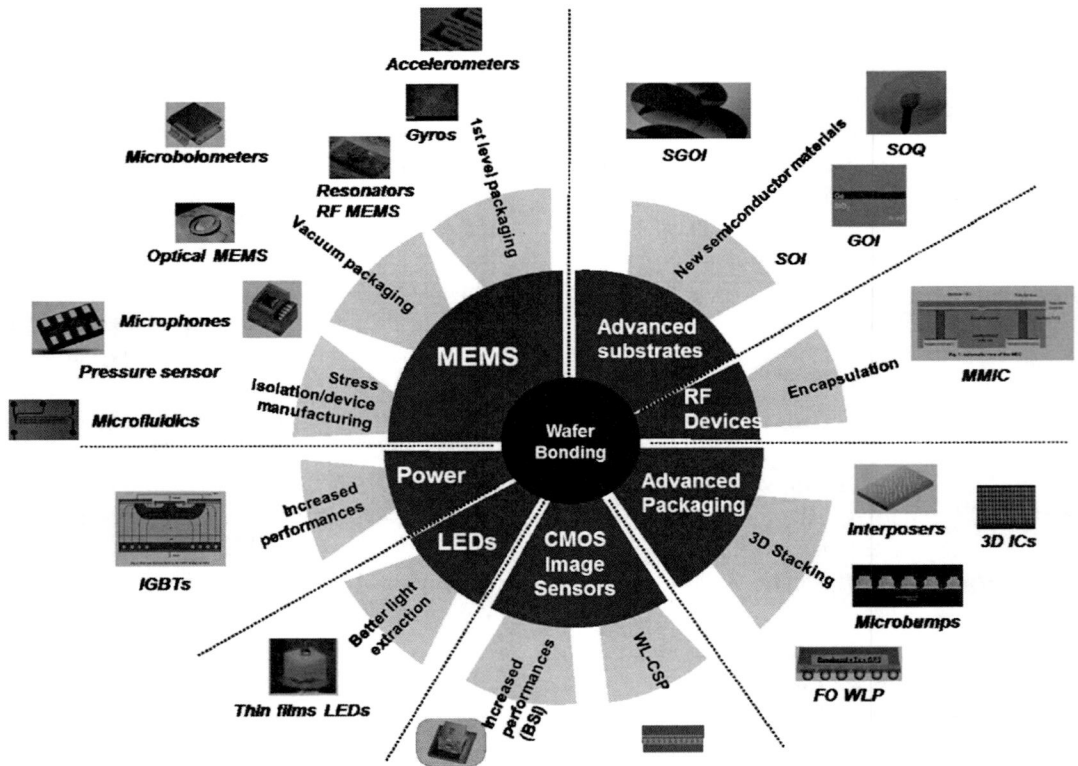

Fig. 16.15 Application scope of wafer level packaging technology (Yole Developpement 2012)

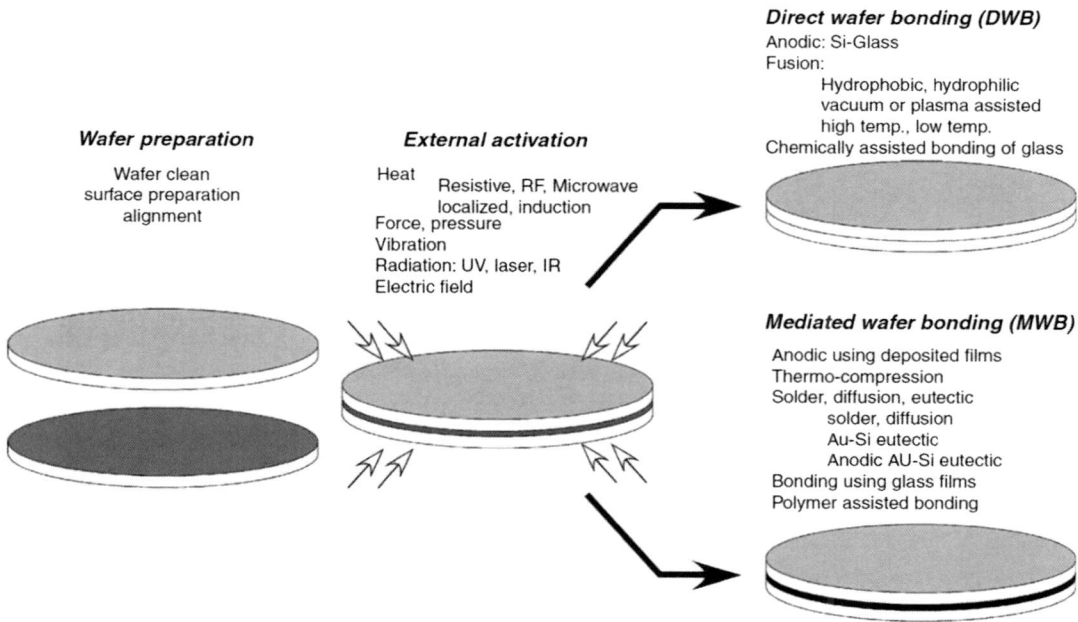

Fig. 16.16 Essential elements of wafer bonding: wafer preparation, external activation and wafer bonding (Gianchandani et al. 2008)

other, but can be employed for different substrates, including glass wafers or non-silicon semiconductor (e.g., GaAs) wafers as well. In general, two wafers are processed, one with functional devices and other with capping cavity. One wafer is flipped over and aligned onto the other wafer and bonded together through different bonding techniques as shown in Fig. 16.16.

There are two main kinds of wafer bonding processes—direct wafer bonding and intermediated wafer bonding. The most suitable choice of wafer bonding technology depends on the particular applications and the materials involved. The two broad categories of W2W bonding approaches explained briefly below:

1. **Direct wafer bonding (DWB)**: In direct wafer bonding approach, the two wafer surfaces are cleaned and polished to high degree of smoothness and are brought into intimate contact, thus allowing the atoms on the surfaces of these wafers to covalently bond together and form a permanent bond. The most popular subcategories of DWB

are—(a) Fusion bonding—where very high temperature is applied after intimate contact to form bonding process and is usually used for Si-Si wafer bonding and (b) Anodic or field assisted bonding where a high electrical bias is applied to form the bonding and is usually done is case of glass to conductive wafer bonding.

2. **Intermediated Wafer Bonding (IWB)**: In intermediated wafer bonding approach, an intermediate layer or glue is used in between two wafers to permit their bonding and permanent attachment. The intermediate layer can be metals, semiconductors, inorganic insulators (like glass, oxides, or nitrides), or organic insulators (like polymers, epoxies, and other adhesives). Some of the popular MWB techniques are—(a) Eutectic bonding (a thin gold layer as an adhesive layer to bond silicon wafers), (b) Glass-frit bonding (uses a glass frit seal at the interface between the wafers), (c) Solder-based thermocompression bonding (Bonding at room temperature by

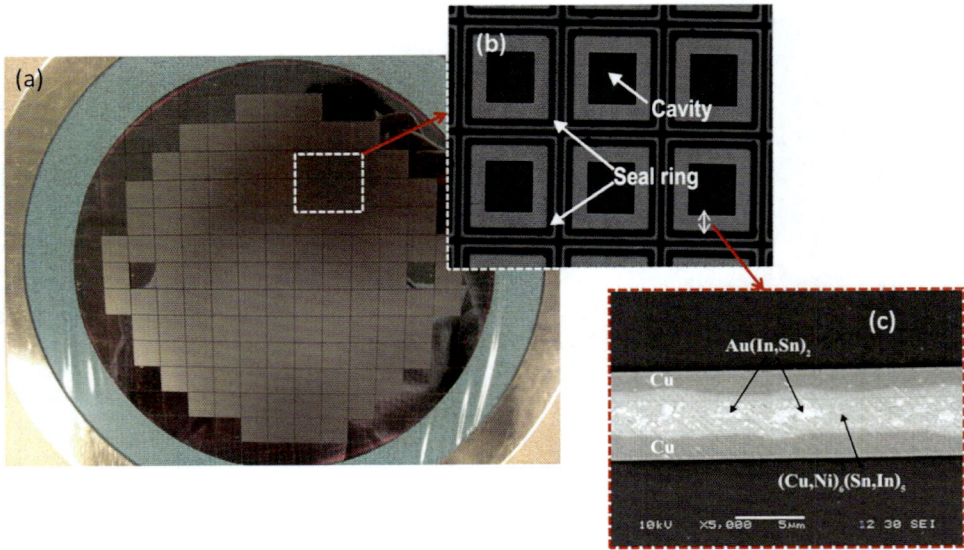

Fig. 16.17 (**a**) Photo of a successfully bonded wafer with 100% yield (Esashi 2008); (**b**) scanning acoustic micrograph (SAM) of bonded devices using Ti/Cu/Ni/ Au UBM with the bonding temperature of 180 °C (Yu et al. 2009); (**c**) interfacial microstructure of the joint bonding at 180 °C—center region of the joint (Yu et al. 2009)

using solder as interface layer and with application of very high pressure), and (d) Adhesive bonding (polymer-based adhesive bonding) (Lindroos et al. 2009; Niklaus et al. 2006).

It is very important to provide complementary metal oxide semiconductor (CMOS) compatible wafer bonding technology to drive the cost down. However, the traditional silicon fusion based direct bonding requires annealing temperatures around 1000 °C, and so it is not compatible with standard CMOS metallization. However different approaches are demonstrated using intermediated wafer bonding technology. The low temperature bonding approach that gained a lot of research attention was the formation of the bonding interface at low temperature with low-temperature solder (e.g., indium) for hermetically sealed packaging. The key advantage of this technology is that low bonding temperature is achieved because of low melting point (LMP) solder; however, the same package can withstand much higher temperature because of the high-melting-temperature intermetallics compounds (IMCs) formed at the interface during the

bonding process. Stable and high-temperature resistant bonding interfaces rely on the formation of IMCs from reaction between a low-melting-point (LMP) component, such as In or Sn, and a high-melting-point (HMP) component, such as Au, Ag, or Cu. In the material selection for the LMP component, In-Sn alloy is more preferred over single component solder of Sn or In, owing to the low eutectic temperature (118 °C) and better wettability with various commonly available substrates. For HMP component, Cu is more preferred as the use of Au would lead to a high cost for wafer bonding, and also Cu is used widely in modern IC packaging technology and in 12-in. CMOS process technology. More importantly it is much cheaper than Au. Hence In-Sn of eutectic composition for Cu-based metallization provides a low cost alternative (Fig. 16.17).

In order to achieve high yield of good dies through wafer bonding technology, issues of wafer warpage and topography differences along the bonding rings across a wafer can be overcome by a solder-reflow step when sufficient solder is placed in the bonding rings. However, the LMP solder has to be fully consumed by

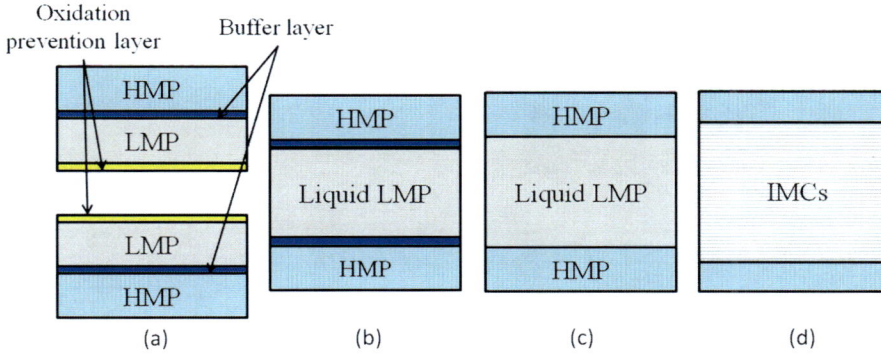

Fig. 16.18 Schematic drawing of the role of a buffer layer during bonding: (**a**) before bonding; (**b**) liquid solder formed at the beginning of wafer bonding; (**c**) buffer layer dissolved into solder; (**d**) high-temperature IMC joint finally formed during bonding. Redrawn from (Yu et al. 2009)

reacting with HMP metals during the bonding and annealing steps to form reliable and high temperature-resistant IMC phases at the bonding rings for hermetic sealing applications. Hence reasonable thickness of solder layers is critical. One of the major issues with this approach is the interdiffusion of material into other layers. Considering an example of microstructure of Sn/In/Au/Cu metallization, the diffusion rates between low-temperature solders and HMP metal substrates, e.g. Cu and Au, are very high. A portion of the as-deposited solders on the HMP metal would be consumed before the wafer-bonding step; thereby forming $Cu_6(Sn,In)_5$ ternary phase at the Cu substrate side due to interdiffusion. Similarly, AuIn phase is formed before the bonding step completed due to the fast diffusion of Au into In. These compounds may not be high temperature resistant and will adversely affect the long time reliability of packaged devices. Therefore, it is very important to prevent the interdiffusion between LMP and HMP components during the fabrication process and room-temperature storage. One way to prevent interdiffusion is to include a thin buffer layer between the Cu substrate and the solder layers such that interdiffusion is inhibited, whereas this thin buffer layer preferentially dissolves into the melted solder quickly at the beginning of soldering reaction. Then the diffusion between the liquid solder and the Cu starts, and finally, all solder was converted into IMCs.

Since the buffer layer saves the low-temperature materials for a successful solder-reflow step during bonding, the production yield of wafer bonding is improved significantly. Ni is well known as a barrier layer for Sn-based solder and Cu substrate, and at the same time, it can easily diffuse into Cu-based IMCs. For In/Sn/Cu systems after eutectic bonding, all low-temperature phase will be converted to high-temperature IMCs. According to the ternary-phase diagram of Cu-In-Sn, the calculated thickness ratio of Cu to In-Sn needs to be larger than 0.5 to form $Cu_6(Sn,In)_5$ compounds. To protect In-Sn from oxidization, a thin Au layer is deposited on top of the solder layers. Since In will form Au-In IMCs with Au, there will be two kinds of IMCs in the final seal joint—$Cu_6(Sn,In)_5$ and Au-In. To get a robust joint, the Au layer should be as thin as possible to reduce the volume of the Au-In IMC (Fig. 16.18).

Through Silicon Vias (TSVs)

Another key technology that essentially made wafer bonding technology achievable was the through silicon vias (TSV), by which electrical connections can be made between two bonded wafers without the need for wire bonding or bump based bonding.

For 3D IC packaging, TSVs are the most important enabling technology. TSVs provide advanced SiP solutions such as C2C (Chip-to-Chip), C2W (Chip-to-wafer), and W2W

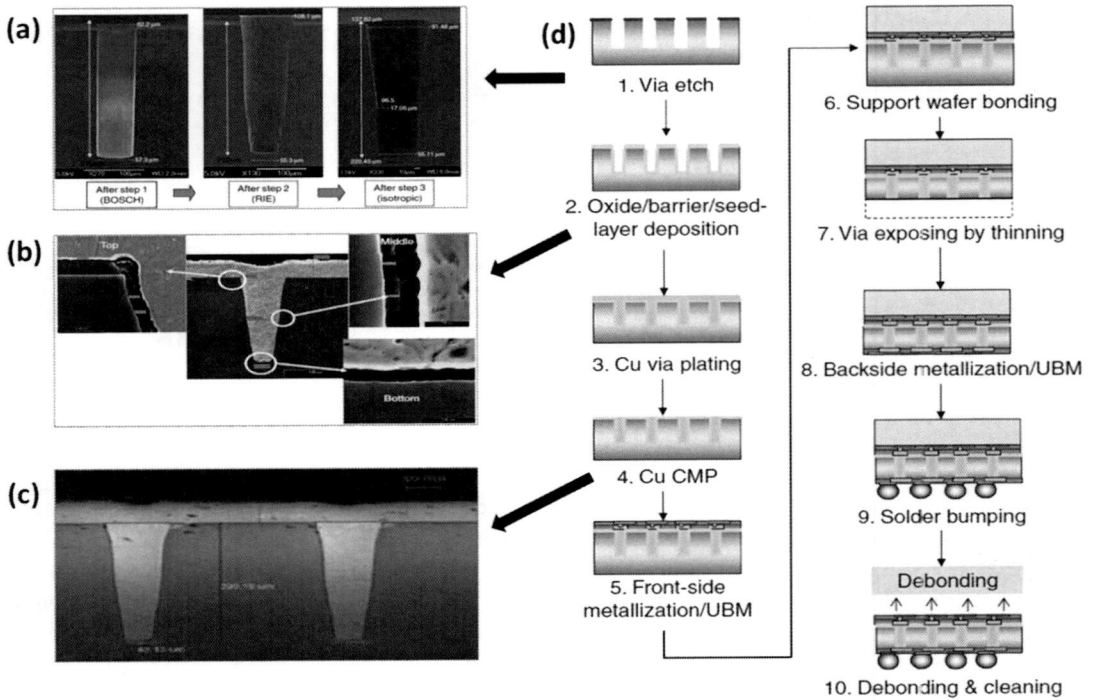

Fig. 16.19 (a) SEM image showing the via cavity formation by DRIE process, (b) SEM image of oxide/barrier layer deposition, (c) SEM image showing via filled with Cu, followed by CMP planarization and (d) process flow for TSV formation, Cu filling followed by metallization through TSV (Esashi 2008)

stacking, WLP; with the shortest electrical path (vertical electrical feedthrough) between two sides of a silicon chip. The most critical step in TSV technology is via formation, because it demands very high aspect ratio via cavity to achieve maximum number of interconnects between the two chips. Via cavity in silicon wafer can be formed by either wet or dry etching process. At times, laser drilling is also used to achieve high aspect ratio via cavity. However, the most popular technology to form via cavities on wafer level is by deep reactive-ion etching (DRIE), which is a highly anisotropic etching process. The silicon vias are formed in inductively-coupled plasma (ICP)-based DRIE system based on process comprising specially designed switched steps of etch and passivation processes, also known as the BOSCH process.

The high-aspect-ratio tapered silicon vias are formed by three independently controlled process steps: (1) Straight vertical via formation step by the BOSCH etch process, (2) Via tapering step by a controlled isotropic etch process, and (3) Corner-rounding step by a global isotropic etch process as shown in Fig. 16.19a. Once the vias are formed, the thermal oxidation is used to provide the electrical isolation layer, followed by deposition of Ti barrier and copper seed layer as shown in Fig. 16.19b. Now the via cavity is filled through copper electroplating process, followed by Copper (Cu) planarization through Chemical Mechanical Polishing (CMP) (see Fig. 16.19c). Once the Cu vias are formed, the wafer is attached to a supporting wafer using a temporary adhesive material, followed back-grinding (thinning) and backside processing to expose the vias. The bonded wafers then are processed for back-end metallization/UBM/solder bumps. The backside UBM is electroplated copper. After all these processes, the bonded wafers are debonded by sliding the wafers apart at higher temperature (200 °C). Now the TSV

Fig. 16.20 (**a**) Process
flow to achieve vertical
feedthrough on either side
of the Si wafer without via
cavity filling and (**b**)
optical microscope image
of the highly conformal
metal vertical feedthrough
without filling via cavity
(Kuhmann et al. 1999)

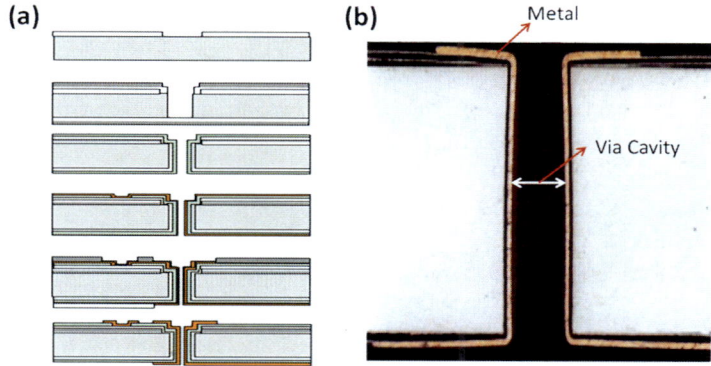

ASIC (Application specific Integrated circuit) wafer is ready for the MEMS device (C2W) bonding or the MEMS wafer (W2W) bonding (Chen et al. 2008; Lau et al. 2010; Lindroos et al. 2009).

Alternatively, vertical feedthrough is also possible by conformal electroplating of thin metal, rather than filling up the entire via cavity. The process of achieving the above mentioned vertical feedthrough is shown in Fig. 16.20a. The Si wafer is etched using DRIE process with a hard mask. Then a conformal coating of thin metal is achieved through electroplating, thus providing electrical path between either sides of the Si wafer. This approach looks attractive, however the current carrying capability through these vias are limited as compared to entirely metal filled via.

Another means of vertical feedthrough is the Si V-groove vias. At first, the via cavity is formed through anisotropic wet-etched Si V-grooves followed by dielectric material coating and metal layer deposition (Fig. 16.21). The process of achieving V-grooves vias by utilizing anisotropic wet etch is shown in Fig. 16.22d. The footprint of through-wafer V-groove vias is relatively larger, owing to the larger angle of the V-grooves as shown in Fig. 16.22a. The size of the through-wafer V-groove via is determined by the thickness of wafer, hence using a thin device layer of a silicon-on-insulator (SOI) wafer, the V-groove vias can be shrunk to a very small area as shown in Fig. 16.22b, c.

TSV is the key enabling technology to successfully implement W2W bonding technology.

TSV enables high performance devices, compared to its alternative 3D packaging technology such as package-on-package or stacked chip approach, because of the higher density and shorter length of the vertical electrical feedthrough. Additional TSV technology also enables zero footprint increment due to chip stacking and perfectly flat package with minimal step height. TSVs could complement the W2W bonding technology to provide highly efficient and low cost product.

16.2.2.2 Thin Film Encapsulation

The emerging MEMS technology in the recent years has open up advanced packaging technique that can potentially enable better performance and lower manufacturing cost. One such essential prerequisite is MEMS encapsulation which can be imperative to the device's capability and reliability. For example, increased reliability and higher Q-factor performance are reported by using vacuum based encapsulation on MEMS resonator (Candler et al. 2003; Pourkamali et al. 2007; Rais-Zadeh et al. 2007) and energy harvester (Xie et al. 2010). In today's context, established manufacturing technologies in MEMS encapsulation are mostly performed at chip level back-end assembly, which means each device has its own dedicated packaging where the housings or encapsulations are mostly consists of polymers and molding material. Consequently, this method leads to inefficient production and higher manufacturing cost. In contrast, some other wafer level packaging technique, for example, glass frit is utilized in wafer

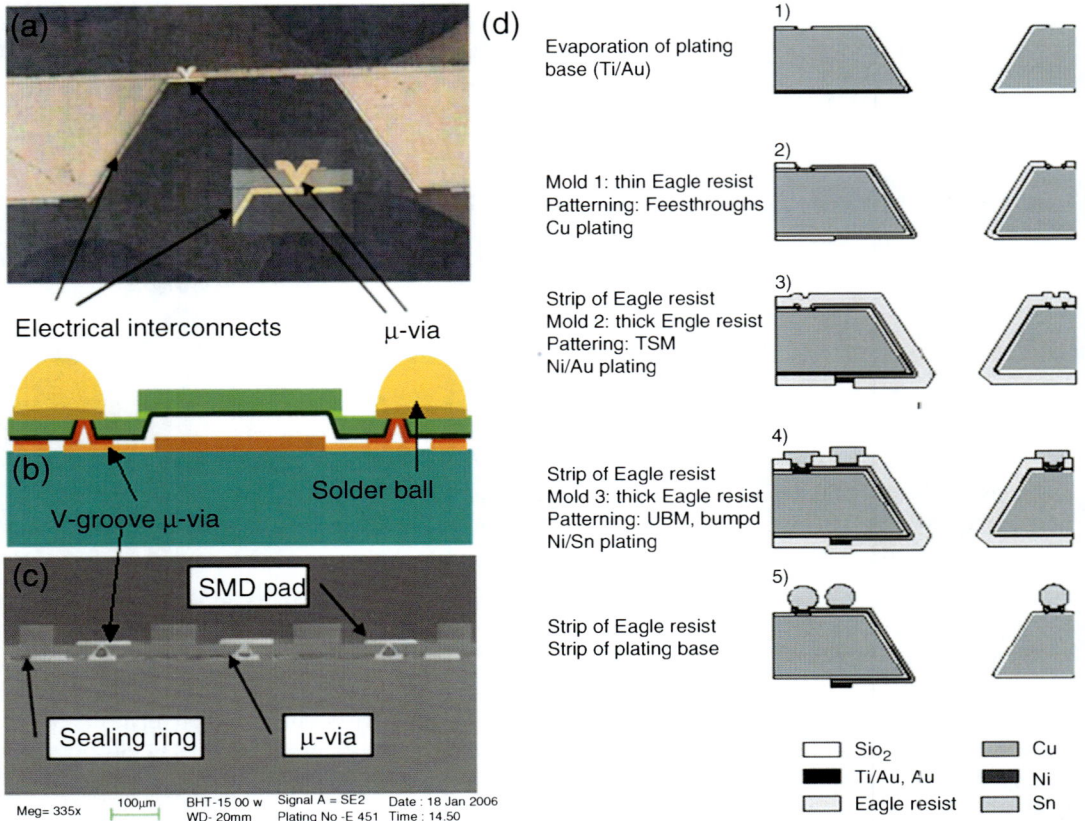

(d)

1) Evaporation of plating base (Ti/Au)

2) Mold 1: thin Eagle resist
Patterning: Feesthroughs
Cu plating

3) Strip of Eagle resist
Mold 2: thick Engle resist
Pattering: TSM
Ni/Au plating

4) Strip of Eagle resist
Mold 3: thick Eagle resist
Patterning: UBM, bumpd
Ni/Sn plating

5) Strip of Eagle resist
Strip of plating base

☐ Sio₂	☐ Cu
■ Ti/Au, Au	■ Ni
☐ Eagle resist	☐ Sn

Fig. 16.21 (**a**) Optical microscope image of the electrical interconnections and the μ-via [Hymite Corporation]; (**b**) schematics of V groove μ-via (Shiv et al. 2006); (**c**) SEM of sowing the μ-via (Shiv et al. 2006); (**d**) process flow for V groove feedthrough with electroplated photoresist and metal (Kuhmann et al. 1999)

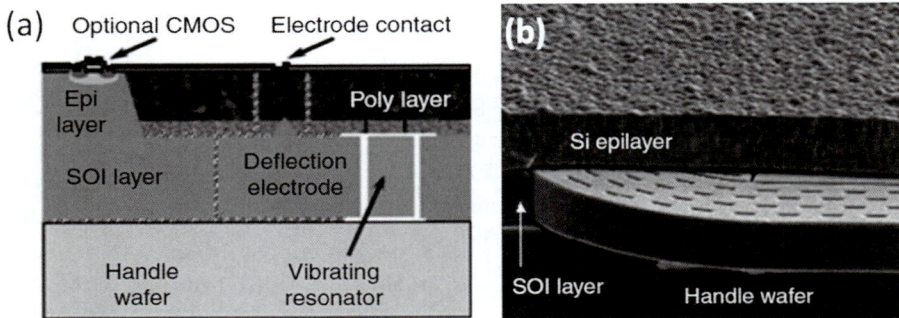

Fig. 16.22 Wafer level packaging approach to form thin film encapsulation. (**a**, **b**) High temperature WLE device [Courtesy: SiTime Corporation]

level spin coating and encapsulation using bonding scheme as reported by Knechtel et al. (Knechtel et al. 2006) may be a viable solution as the process is low cost and does not involved complicated machinery. However, this technique is not vacuum based and often creates excessive particle count resulting in difficulty to keep the wafer in pristine condition. In fact, vacuum based

Fig. 16.23 Fabrication process flow of an epi-seal devices. (**a**) Si deep trench etching. (**b**) SiO$_2$ overfill to seal trench. (**c**) Via definition with device release. (**d**) Si epitaxy sealing and electrode isolation. (**e**) Passivation opening. (**f**) Al metallization (Soon et al. 2014)

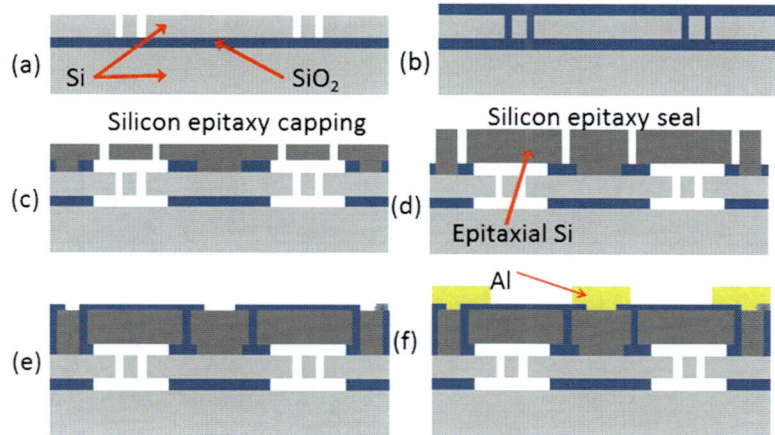

sealing is one of the most significant features in wafer level encapsulation (WLE).

Thin Film Encapsulation (TFE) consists of using a sacrificial material to cover the MEMS device and a deposited thin film as capping layer to encapsulate the MEMS device, followed by removal of the sacrificial layer by wet etching or vapor-phase etching through the release holes on the capping layer and then sealing up the release holes by deposition of a thin film sealing layer. The deposition of the sealing layer is usually carried out in a reaction chamber, which is under vacuum and this ensures that the sealed cavity of the MEMS device can be preserved at low vacuum.

One of the pioneer works come from the epi-seal process. It involves deposition of a 20 μm layer of epi-polysilicon over unreleased devices to act as a sealing cap, and a final seal of the parts in 7 mbar (700 Pa) vacuum as shown in Fig. 16.22a, b. A more detailed example of process can be found in Fig. 16.23. Initial silicon on insulator (SOI) wafer of 40 μm thick device layer with 1 μm buried oxide is used, where the active parts are defined by deep reactive ion etching (DRIE) in this silicon device layer. After that, sacrificial SiO$_2$ is overfill the trenches. It is then etched to define where the electrode should be interfaced, at the same time the opening area serves as the anchor area for the following Si epitaxy capping. The release hole is patterned and etches in the Si capping layer to provide

access to the SiO$_2$ underneath. Then, isotropic vapor hydrofluoric acid is used to release the active parts. Subsequently the release hole is sealed by a second layer of Si epitaxy. During the sealing, the wafer is exposed to high temperature with hydrogen gas content and this process removes most impurities and polymers as reported (Hong et al. 2013). Subsequently, top Si is isolated with SiO$_2$ and finally the contact pads are open and metalized with aluminum. The finish chip has flat topography and only the metal pads are visible.

This high temperature process (1080 °C for the epitaxy) provides a MEMS-first front-end wafer-level packaging, which completely defines, releases, and seals the microstructures in the front-end processing steps. Thus, it provides reliable wafer-level vacuum packaging at low cost. One advantage of using a high-temperature sealing process (i.e., Epitaxial Si deposition) is that the absorbed molecules inside the cavity will be removed during the high-temperature sealing step, just as in the hot baking treatment. Thus, it enables the high vacuum of the sealed cavity. The derived products exhibit excellent performance in terms of ultra-low long-term drift of device features and good stability against temperature variation.

The effect of this epi-seal process to the reliability of MEMS in vacuum have been investigated by several groups (Jin et al. 2003; Shea et al. 2006). Below we show a report of

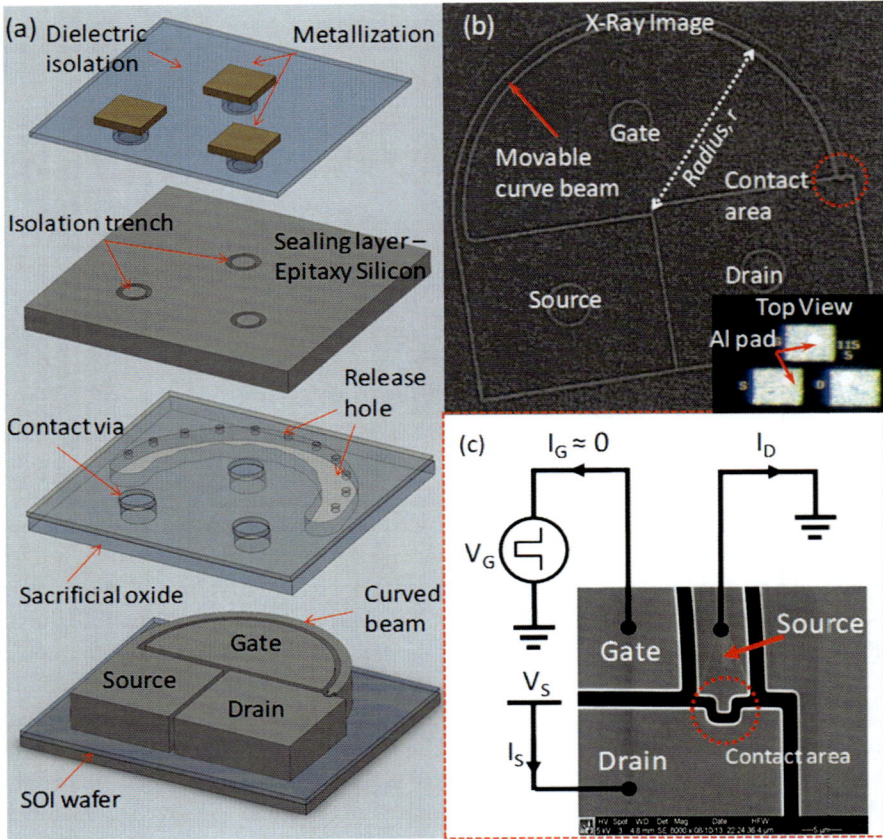

Fig. 16.24 (a) Exploded view of the Si-to-Si MEMS switch with encapsulation layers. (b) X-ray image showing the encapsulated three terminal device consisting of a movable curve beam (source), control terminal (gate) and contact terminal (drain). (*Inset*) top view of the finished chip with metalized aluminum pad. The encapsulated device is flat with minimal topology (Soon et al. 2015). (c) Scanning electron micrograph (SEM) of a curved beam Si-to-Si MEMS switch. The operation of the switch is shown where V_G is used to control the actuation with gate terminal and V_S is used to detect signal from source to drain

increased MEMS switch reliability under epi-sealed vacuum environment. Si based MEMS switch working under air ambient normally shows the reliability of only ~10 cycles. With epi-seal process, the reliability of the MEMS switch is dramatically improved as the device is working under vacuum environment with prevention of oxidation at the contacts.

The epi-sealed MEMS switch with the entire moving structure and all electrodes are encapsulated in a vacuum environment with electrical interconnects built through the encapsulation to the chip surface and metalized by aluminum (Fig. 16.24). These terminals consist of source (movable beam, fixed at one end), gate (fixed) and drain (fixed) defined on silicon-on-insulator wafer. The movable beam (source terminal) can be electrostatically pulled in to contact the drain terminal by applying a gate voltage between the gate and source terminal. Subsequently, electrical signals can pass from source to drain. The reliability of the MEMS switch is characterized by repeatedly turn on and off the gate terminal, while monitoring the contact resistance between source and drain terminal.

The measurement shows even under accelerated conditions, while the device operates under 400 °C, Fig. 16.25 there is no degradation after 10^6 cycles. The epi-seal process, not only elevate the reliability of Si-based MEMS switch from 10 cycles to 10^6 cycles, it also makes the MEMS switch technology become feasible and practical for future development.

However, the high-temperature epitaxial Si sealing process limits material selection for MEMS devices. Some metals and low-temperature oxides cannot sustain their material characteristics under such high temperatures. As shown in Fig. 16.26, a low-temperature process has been proposed to provide a low temperature alternative WLE process (Soon et al. 2012). First,

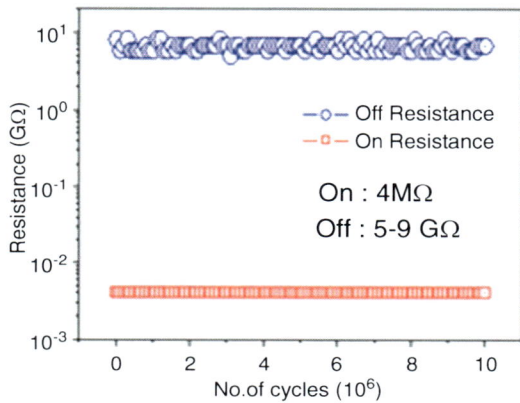

an insulation layer of silicon Si_3N_4 is plasma enhanced chemical vapor deposition (PECVD) deposited, followed by 4 μm of low stress silicon oxide (SiO_2) sacrificial layer in the same tool. The insulation layer of Si_3N_4 acts as an etch stopper during the sacrificial layer etching at the same time it is also the dielectric layer which is very commonly used in MEMS devices, in order to prevent leakage into the bulks Si. The sacrificial SiO_2 then defines the entire area where the MEMS devices should be encapsulated. Next, a thinner layer of 1 μm creates a spacer area larger than then first sacrificial layer. A dual layer deposition of 1 μm Si_3N_4 and 0.2 μm of amorphous Si is deposited to complete the top housing and defines the structure for the encapsulation. This layer is patterned with release holes. After that, the wafer is dipped into buffered oxide etcher (BOE) wet etching to remove the sacrificial oxide. After the cavity is formed, Al is sputtered to seal the etch hole. The vacuum level is around 0.01 mTorr during the deposition. In other mean, the encapsulation itself is assumed to be around the same order of vacuum level as the chamber.

Another approach of low temperature WLE using vapor hydrofluoric acid to release the devices instead of BOE is shown in Fig. 16.27.

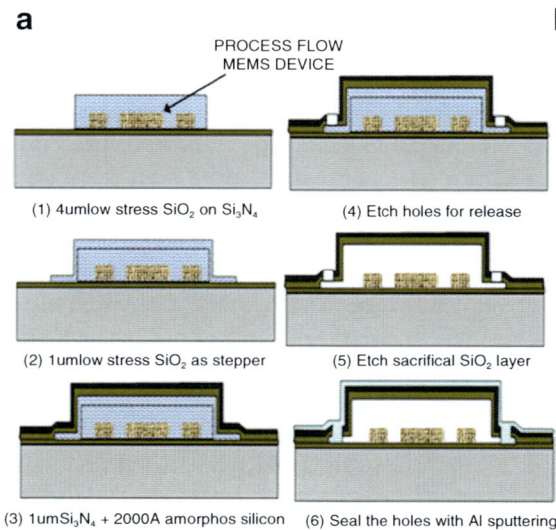

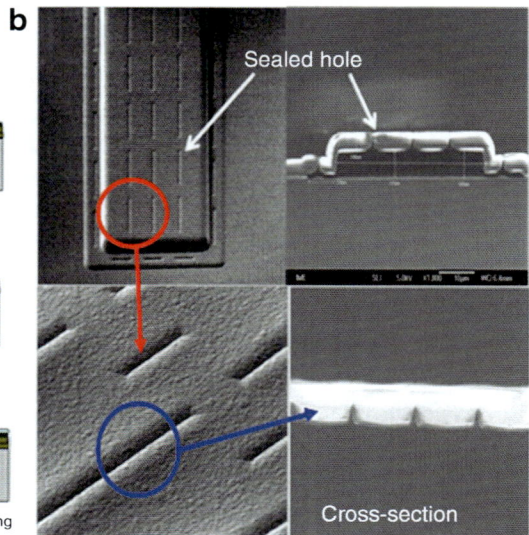

Fig. 16.25 Contact resistance of on and off state for one million cycles

Fig. 16.26 (**a**) Schematics of process flow for low temperature wafer level MEMS encapsulation. (**b**) Encapsulation using Al with sealed opening and cross section of encapsulation (Soon et al. 2012)

a

b

Si   SiO₂   Mo   AlN   AJ

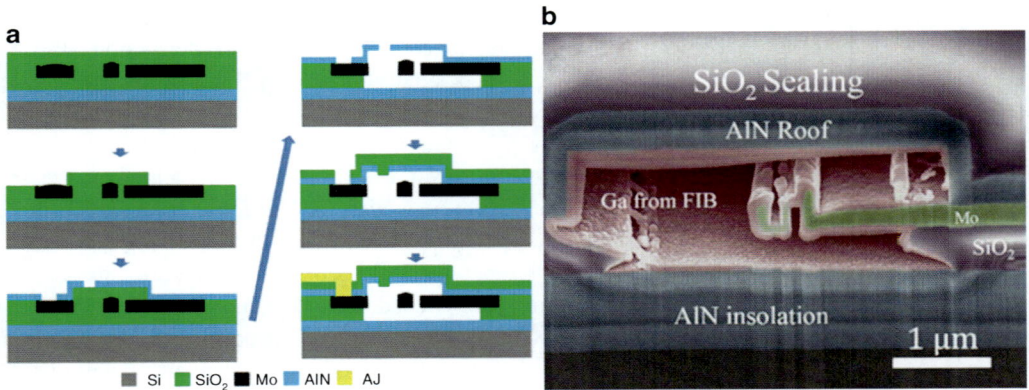

Fig. 16.27 (**a**) Process flow for wafer level encapsulated metallic device (**b**) cross-section of an encapsulated metallic device (Qian and Development of nano/ microelectromechanical system (N/MEMS) switches 2015)

Vapor hydrofluoric acid release is less likely to have stiction. Commonly used metal in CMOS process like Al cannot survive in BOE but it does not react with Vapor hydrofluoric acid at all. Figure 16.27a shows the process flow, after formation of the active devices, the whole wafer is covered with SiO_2, followed by etching to define the active area. A 500 nm AlN layer is deposited by sputtering and pattern for releasing hole. The devices are then released by VHF. The sealing is conducted by 1 μm SiO_2 deposition. Lastly, the metal pad is opened for Al interconnection.

There are several advantages of low temperature wafer level encapsulation technology, such as (1) Low-temperature processing—the technology is suitable for packaging of MEMS devices that are sensitive to high temperatures and thermally induced residual stress; (2) Low cost and simple packaging. The process does not require cap-to-wafer alignment or a wafer-bonding step; and (3) High-yield process.

16.2.2.3 Packaging for Wearable, Flexible Electronics

Recent advances and developments in the field of wearable sensors with emphasis on devices that are able to perform highly-sensitive analysis. Novel fabrication methodologies have resulted in the demonstration of sensors able multiple physical measurements (i.e. heart rate, EEG, ECG, etc.). Thereby providing added dimensions of rich, analytical information to the personnel in a timely manner. The field of wearable sensors has developed substantially over the past decade, with most solutions targeting healthcare applications via the transduction of physical parameters such as heart rate (Chiarugi et al. 2008; Paradiso et al. 2005), respiration rate (Chiarugi et al. 2008; Merritt et al. 2009; Rovira et al. 2011), oxygenation of the blood (Rothmaier et al. 2008), skin temperature (Jung et al. 2006), bodily motion (Hasegawa et al. 2008; Lorussi et al. 2004; Lorussi et al. 2005), brain activity (Devot et al. 2007), and blood pressure (Espina et al. 2006). As a component of biotelemetry systems, these sensors leverage the wireless infrastructure to form a IoT node capable of relaying the physiological or environmental information to the personnel in real-time. Novel fabrication concepts have been developed whereby robust sensors and biosensors are implemented directly on flexible substrate such that, sensors are able to withstand the rigors of field-based use and exhibit uncompromising resiliency to various forms of mechanical strain and deformation that are typically encountered during routine use.

Flexible substrates such as Polyimide (Kapton), polyethylene naphthalate (PEN), polyethylene terephthalate (PET), polytetrafluoroethylene (Teflon), among others have long been employed in the printed electronics industry. Owing to their intrinsic plasticity, hydrophobicity, excellent dielectric and insulate

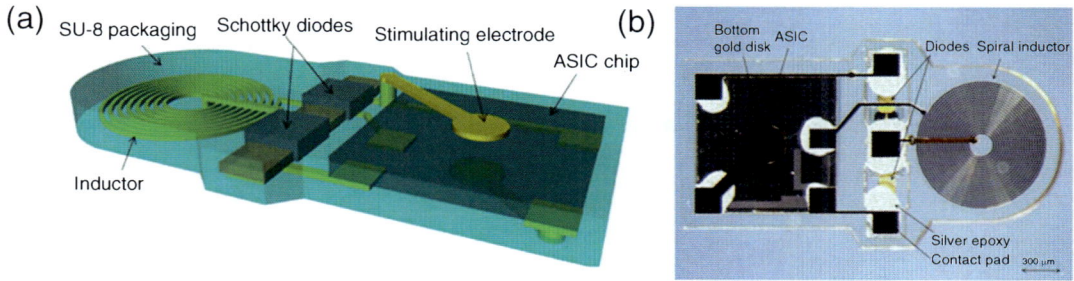

Fig. 16.28 (**a**) Schematic diagrams of SU-8-packaged wireless neurostimulator showing a receiver coil, two rectifying Schottky diodes, a stimulating ASIC chip, and biphasic stimulating electrodes; (**b**) Photomicrographs of the fabricated wireless neurostimulator

properties, thermal stability and a low coefficient of thermal expansion, structural resiliency against repeated deformities forces, and compatibility with roll-to-roll fabrication processes, flexible substrates have served as the platform of choice in various demonstrations of the wearable electronics.

Advances in microelectromechanical systems (MEMS) now allow nearly seamless integration of an array of sensors to collect information. Microcontrollers have shrunk to the point that they can be hidden away neatly and consume very little power. Wireless communication and low power standards allow for a network of independently operating components to easily integrate with each other. Low-power LEDs and flexible display technology are close to bringing science fiction into reality. With very limited board space available, the packaging of the wearable electronics is heavily relying on advanced technology like wafer level packaging (WLP) or System-in-Package (SiP).

In the meantime, many concepts of how to integrate electronics are still new and very much in the prototyping phase. Form factor plays a major role in wearable technology—not just for the processor, but also the entire system. Chip-scale and wafer-based packaging is naturally found a home in wearables with early adopters. Package size and integration of common functions are also important to wearables. Here, three cases are introduced to the advanced packaging technology for wearable, or flexible electronics application. The bonding between ICs and flexible substrate are accomplished by conductive epoxy, solder paste, and anisotropic conductive film (ACF) respectively.

First, we introduce a fully integrated minimally invasive compact polymer based wireless neurostimulator (Fig. 16.28). These microprobes on neurostimulator have been designed to be implanted to act as an interface between neurons and electronics for the stable detection of neural spike signals or for efficient stimulation of neurons. To avoid these issues, wireless operation of the implanted neural stimulation devices is highly desirable. In order to realize such a wireless neural stimulator, wireless power transmission from the outside of the body of the implanted devices in the body is used.

The designed wireless neurostimulator was fabricated by SU-8-based surface micromachining technique (Fig. 16.29). Where the spiral inductor for wireless power transmit is embedded inside the Su-8 layer, only leave via on the Top Su-8 layer. The via is filled with 15-µm-thick electroplated gold. Then A 156-µm-deep SU-8 socket structure was then patterned in order to make sure that both the diodes and the ASIC are completely enclosed by SU-8 socket. For bonding the diodes and ASIC board, a small amount of biocompatible conductive epoxy was applied to the contact pads of the sockets. The ASIC chip and diodes were slid into the sockets with their contact pads side down. Because solvent inside the biocompatible conductive epoxy can affect the subsequent SU-8 process, it should be completely evaporated. Thus, the sample was cured in a 95 °C convection oven for 5 h to

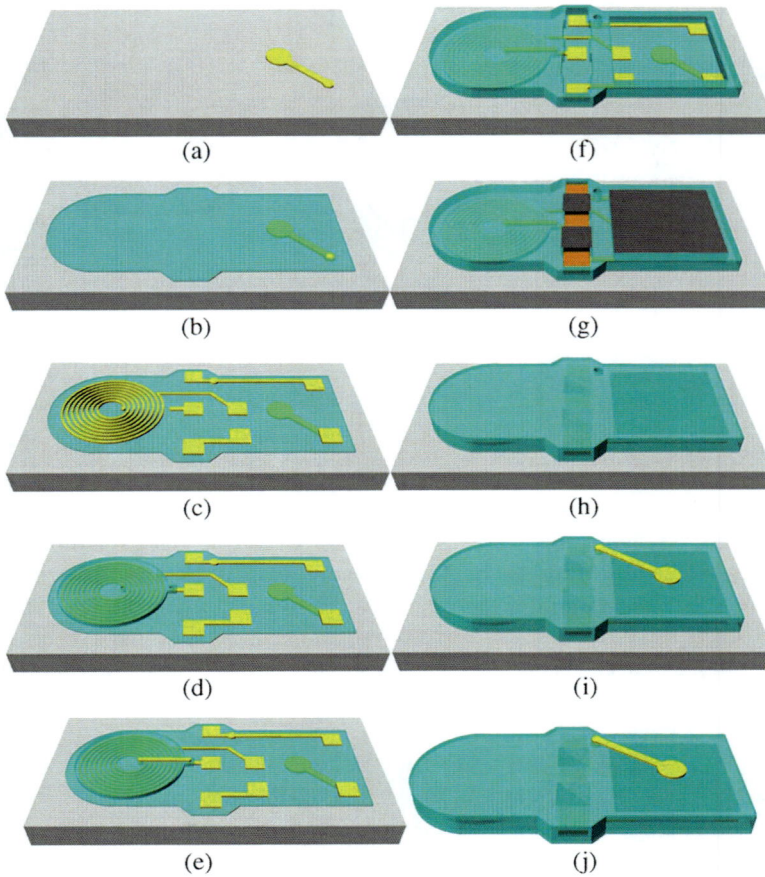

Fig. 16.29 Fabrication process flow of the wireless neurostimulator: (**a**) patterning of the bottom gold electrode; (**b**) formation of the base layer of the SU-8 socket platform; (**c**) conductor traces of the spiral inductor and the contact pads; (**d**) SU-8 insulation on top of the spiral inductor with via holes; (**e**) conductor bridge over the spiral inductor; (**f**) 156-μm-thick SU-8 sockets; (**g**) diodes and ASIC chip packaging into the sockets; (**h**) 100-μm-thick SU-8 protection layer with the output via hole; (**i**) patterning of the top gold stimulating electrode; and (**j**) device completely released from the substrate

thoroughly evaporate the solvent of the conductive epoxy. Subsequently, the device was thoroughly sealed by 100-μm thick SU-8, and one of the via holes were created through the SU-8 sealing for the purpose of the output pad connection.

The second case is an adhesive radio-frequency identification (RFID) sensor bandage, which can be made completely intimate with human skin, a distinct advantage for chronological monitoring of biomarkers in sweat (Fig. 16.30). In this demonstration, a commercial RFID chip is adapted with minimum components to allow potentiometric sensing of solutes in sweat, and surface temperature, as read by an Android smartphone app. All circuitry is solder-reflow integrated on a standard Cu/Polyimide flexible-electronic layer including an antenna, but while also allowing electroplating for simple integration of exotic metals for sensing electrodes.

Fabrication begins at the flexible electronic circuit board with a sheet of Dupont Pyralux AC (18-μm-thick Cu foil clad to 12 μm Kapton) for the Cu wire. The electrodes of the sensor are fabricated by electrodeposition of Pd and Ag on the previously defined bare copper electrodes.

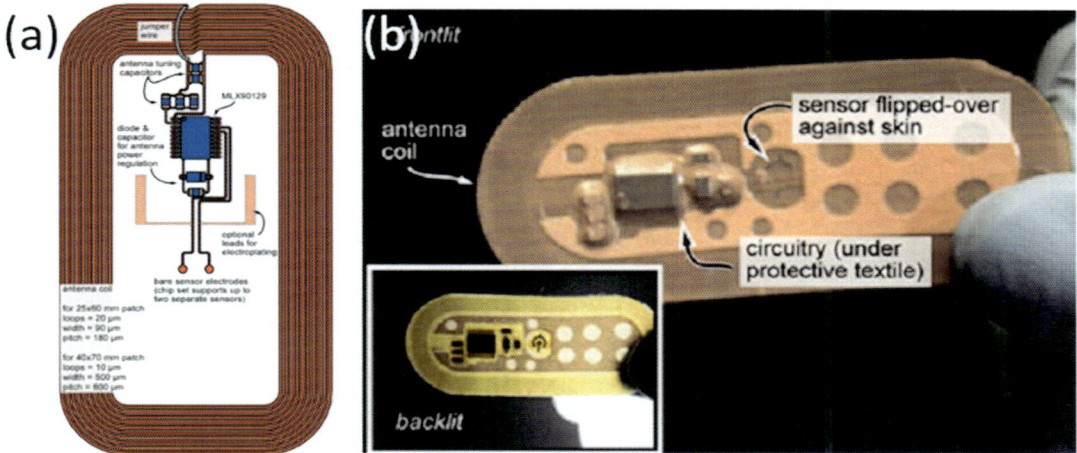

Fig. 16.30 (**a**) Circuit layer layout detailing antenna coil geometry and electronics placement. (**b**) Device photographs to illustrate electronic and sensor features

An alloy of Sn/Ag (96.5/3.5%) no-clean solder paste (Superior Flux & Mfg. Co. P/N: 3033-85) is applied to each pad using an air-metered syringe dispenser, or higher volume fabrication by stencil printing. The chips are then placed, the flexible circuit board is then placed in a programmable, convection reflow oven that applies a heating profile according to the solder supplier specification. The reflow oven is connected to a pressurized N_2 gas supply, which is used to purge the sample chamber and prevent oxidation of the Cu surfaces.

In applications where fluid contact could occur, such as sweat sensing, the electronics must be hermetically sealed. This is easily achieved by masking the sensor electrodes with Kapton Tape (silicone adhesive), and conformal coating the flexible circuit and electronics with 10 μm of Parylene. Ideally, this is performed before sensor functionalization, to ensure that the Kapton masking tape does not damage or contaminate the final sensor surfaces.

Another commercially available method for the packaging of flexible electronics is using anisotropic conductive film (ACF). The principles of the ACF are that the electrical connections are established through conductive particles and the mechanical interconnections are maintained by the cured adhesive. For example, Fig. 16.31 shows a typical process based on ACF's thermo-compress interconnection techniques. First, an ACF is laminated to a substrate. Pressure and temperature are applied during the lamination process to ensure positioning accuracy, uniformity, etc. Then, the bumps on integrated circuits (ICs) are aligned with the electrode prepared on the flexible substrates. Finally, the IC chip is pressed onto the substrate at a specified high temperature and pressure. The conductive particles are trapped between the bumps and tracks, while the adhesive resin is squeezed out. The interconnections are established by the compressive force between the electrodes due to the shrinkage of the adhesive after curing. Consequently, the electrical conduction is restricted to the z-direction and the electrical isolation is maintained in the x–y plane. Additionally, it is necessary to know the deformation of conductive particles during die mounting process. Generally, the conductive particles consist of a polymer core material coated with a thin metal layer (nickel and/or gold), and then insulated again with a polymer material on its exterior surface. The conductive particles in an ACF are of ball shape before die bonding. When the IC chip is pressed onto the glass in a high temperature and pressure, the particles are entrapped and deformed between the bumps of the chip and the glass substrate, and then the external insulation layer of the particle is damaged and the interior metal layer is exposed. Thus, the

Fig. 16.31 Anisotropic conductive film interconnection process

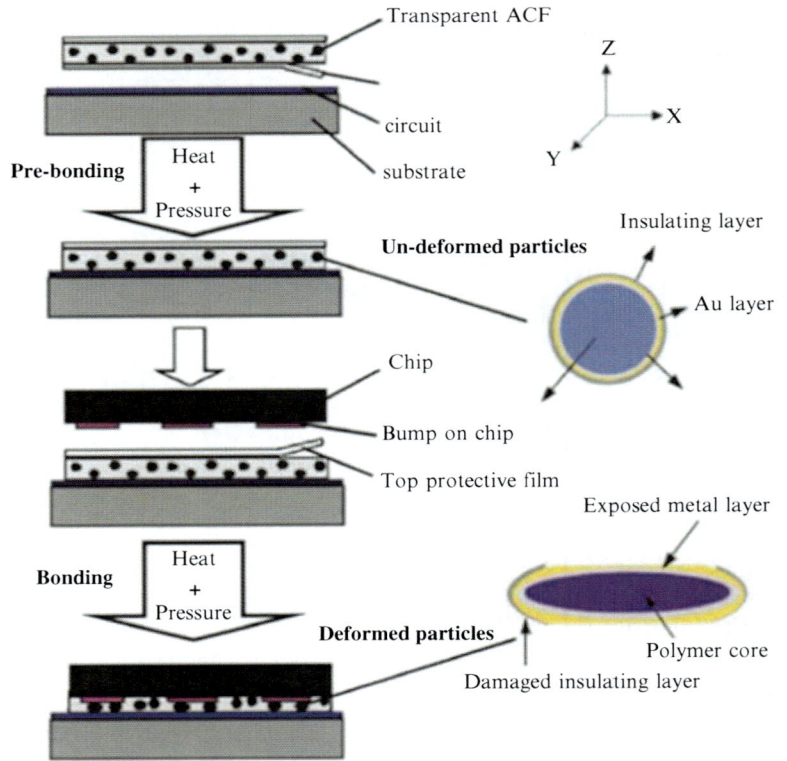

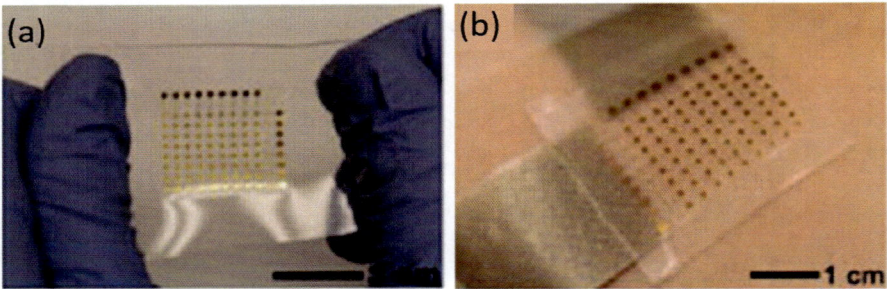

(a)

(b)

1 cm

Fig. 16.32 (**a**) A fabricated device under a stretching deformation. (**b**) Images of use of an epidermal hydration sensor (Huang et al. 2013)

electrical connections are established through conductive particles.

The ACF is capable of bonding ICs on a flexible substrate, it is also possible to connect two flexible electrodes. As Fig. 16.32 presents a hydration monitor that uses ultrathin, stretchable sheets with arrays of embedded impedance sensors for precise measurement and spatially

multiplexed mapping. The devices contain miniaturized capacitive electrodes arranged in a matrix format. The flexible electronics are a sandwiched structure Polyimide (PI; 1 μm thick), Au (400 nm) and another layer of PI (1 μm). Selected regions of PI are removed by reactive ion etching (RIE) to form via contacts. To obtain the signal from the capacitive electrodes, a flexible wire is attached to the functional electronics with ACF.

16.2.2.4 Packaging with Silicon Photonics

It's no secret that this IoT benefits photonics product manufacturers; specifically, those companies in the optical fiber communications business, RFID and bar-code reader technology suppliers, imaging and sensor companies, and hosts of other high-technology equipment suppliers. Exponentially increasing Internet traffic along with the IoT will place a huge burden on next generation, cloud resident data centers. The new requirements include: higher system performance, coping with higher power consumption via more effective cooling concepts, faster interconnect speeds (between components, modules, cards, and racks). The challenge for designers is to provide faster compute/storage/networking systems with more effective bandwidth and performance.

Silicon photonics is a new technology that should at least enable electronics and optics to be integrated on the same optoelectronic circuit chip, leading to the production of low-cost devices on silicon wafers by using standard processes from the microelectronics industry. For more than 30 years, the optical interconnect solutions have been gradually implemented from very long to short distances due to the continually increasing bandwidth demand. As a result, optical fiber networks have been progressively deployed, in long links first, down to the enterprise local area network, metropolitan, and access networks. Compared with the electrical interconnect solutions, optical links exhibit several demonstrated advantages, such as low signal attenuation, lower dispersion, and crosstalk leading to superior bandwidth by distance products. Another significant advantage of the optical solutions is the immunity of the signal transmission to electromagnetic interference, which makes them very well suited to wireless systems in IoT. However, optical links require more research effort. Silicon photonics, using highly confined optical modes in a silicon waveguide, appears as a unique opportunity to cope with this integration challenge. It will allow manufacturers to build optical components using

the same semiconductor equipment and processes. As a result, the next generation of optical components needs to be low cost and compatible with high-volume manufacturing.

An obvious way to achieve this requirement is to increase the integration of optical devices. In spite of many technology developments since the beginning of optical communications, one should admit that the integration level has remained low compared to the microelectronic technology, due to various reasons, particularly the size of integrated optics components and the heterogeneity of photonic materials.

Consider a packaging between photonics and ICs, the 3-D technology is attractive since it allows two or more different process technologies to be stacked and interconnected. 3-D integration can be realized in various technologies. The 3-D technology allows two or more different process technologies to be stacked and interconnected. In an example of advanced mixed material short-wave infrared focal plane arrays Error! Reference source not found (Keast et al. 2008), the 3-D technology enables complex focal architectures with ideal fill factors and a broad range of spectral sensitivities. 3-D integration technology has been used to fabricate four different focal planes, including: a two-tier 64×64 imager with fully parallel per-pixel A/D conversion; a three-tier 640×480 imager consisting of an imaging tier, an A/D conversion tier, and a digital-signal-processing tier; a two-tier 1024×1024 pixel, imaging module for tiling large mosaic focal planes 3; and a three-tier Geiger-mode avalanche photodiode (APD) 3-D light detection and ranging (LIDAR) array (Fig. 16.33).

In IoT application, system complexity and functionality are increasing. Everything from front end through signal processing and storage to sensors and actuators, are housing with integrated energy sources. This can be satisfied by heterogeneous integration (see Fig. 16.34) of different technologies, leading to the best compromise in systems functionality and cost of ownership for higher multifunctional converging systems. Heterogeneous integration (Reichl et al. 2005) requires a set of technologies like thinning

Fig. 16.33 Cross-sectional SEM of a single pixel from a functional three-tier Geiger-mode avalanche photodiode (APD) 3-D LIDAR array (Keast et al. 2008)

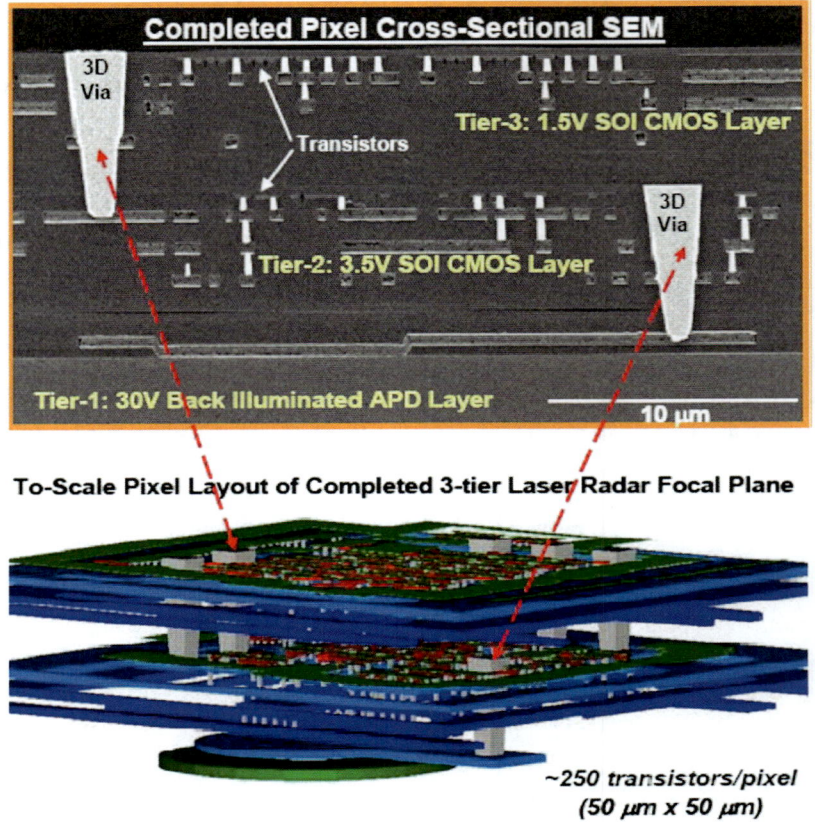

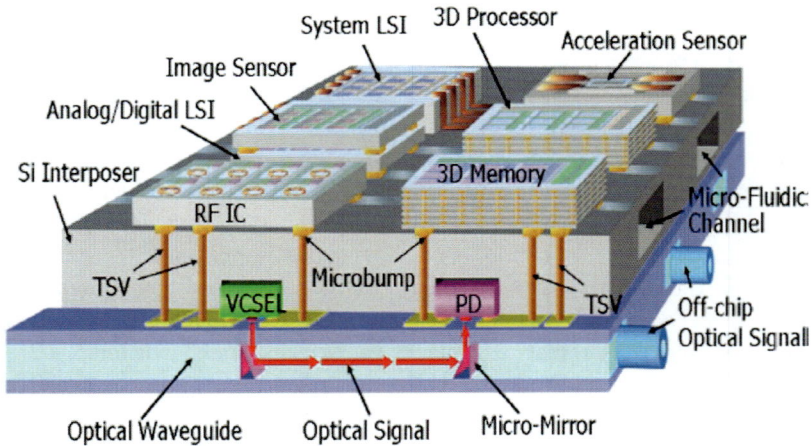

Fig. 16.34 Conceptual structure of 3-D heterogeneous optoelectronic integrated system-on-silicon for intelligent vehicle system with variable signal-processing function depending on moving speed (Lee et al. 2009b)

of wafers and components, vertical system integration, functional layer technologies, assembly of thin components, thin interconnect technologies, wafer molding, bumping/ball placing, and dicing. All these technologies need to be optimized and adapted into a modular integrated process flow.

System-on-silicon for intelligent vehicle system with variable signal-processing functions depending on moving speed is shown in Fig. 16.34. The proposed system including image sensor stacked with Analog-to-digital converter is for high-performance image processing. MEMS sensor, optical sensor, and RF IC are for high-sensitivity sensing of moving speed. Three-dimensional memory and 3-D processor are for high-performance computing. Optical interconnection is for high-speed information networking. Microfluidic channels are for heat sinks for high power consumption. Such heterogeneous integration of LSI, MEMS, and optoelectronic devices is very difficult due to incompatible processes.

16.2.2.5 Packaging for Energy Harvesters

The advancement in packaging of semiconductor chips has been keenly focused reducing on the overall product size by integrating more functional units into a single package. In case of packaging of IoT, which employs multiple functional units, the packaging trend has been shown from the PCB level system packaging to more complex multichip packaging like SiP solutions to highly complicated SoC approach. This integration of multiple functional devices into a single package as in the case of SiP or into a single chip like SoC also provides means of achieving high performance devices with improved reliability and at a lower cost. Currently approaches like SiP look very attractive for IoT as it can incorporate most of the functional units like controller IC, MEMS sensors and actuators into a single package. However due to use of large batteries as the power source of a sensor network to operate autonomously, single packaged chip for entire sensor node is not possible. Compared with traditional batteries, MEMS energy

harvesters seems to be a viable candidate as a power source for WSN application, because of its compactness, environment friendliness, unlimited lifetime, and autonomy (Xie et al. 2009). Additionally, Energy harvesters can be made with similar process like MEMS sensors and hence enables technologies like SiP to provide the entire IoT node as a single packaged chip. Energy harvesters can ultimately open up the room for providing SoC level of integration and WLP of the SoC based IoT module.

The power unit of a wireless sensor node may be supported by a power scavenging unit such as batteries or solar cells. In some application scenarios, replenishment of power resources might be impossible. Therefore, sensor nodes are implemented in a "deploy and forget" scenario since the cost of battery replacement will be prohibitive and impractical (Mathúna et al. 2008). Sensor node lifetime, therefore, shows a strong dependence on battery lifetime. Not to mention that the sensors and electronics of a wireless sensor node will be far smaller than 1 cm^3, in this case, the battery would dominate the system volume. Therefore, the development of alternative power sources for wireless sensor nodes is acute. Significant research is ongoing to deliver power from the environment using energy harvesting technology such as solar energy, vibration/motion energy, thermal gradient, et al. (Niyato et al. 2007; Raghunathan et al. 2006), which can deliver energy directly to a wireless sensor load or to a storage element such as a rechargeable battery or capacitor. Some of the major benefits of energy harvesting technology for WSNs are stated as (Thomas et al. 2006). Firstly, energy harvesting solution can reduce the dependency on battery power and provide long-term solutions. With the advancement of microelectronics, the power consumptions of sensor nodes have been reduced significantly. Hence harvested ambient environmental energy may be sufficient to replace battery completely. Secondly, energy harvesting solution would reduce installation and maintenance cost. Self-powered sensor nodes do not require power cables wiring and conduits, hence they are very easy to install. The heavy installation cost can be reduced greatly

(Kheng et al. 2010). Clearly, it can be deduced that energy harvesting technology is a promising solution to power WSNs for extended operation with the supplement of the energy storage devices.

There are various sources of energy available for energy harvesting, and indeed, many works have been presented on generating electrical energy from solar energy (Lee et al. 1994; Randall 2003), temperature gradients (Böttner et al. 2002; Stordeur et al. 1997; Xie et al. 2010), ambient radio frequency (RF) (Hudak and Amatucci 2008; Rabaey et al. 2000), vibrations (Mitcheson et al. 2004; Roundy et al. 2002), and human motions (Shenck and Paradiso 2001). Thermoelectric power generators (TPGs) scavenge energy from the body heat or any heat sources available in environment, where there is a temperature gradient to generate electricity. Ideally, all devices lose power as heat and hence this heat can be used collectively to generate electricity in turn. Researchers have demonstrated that a TPG chip of 1 cm × 1 cm size at 5 K temperature difference across two sides of device provides an open-circuit voltage of 16.7 V, and equivalent output power of 1.3 μW under matched load resistance (Xie et al. 2009; Xie et al. 2011). Such TPGs are quite attractive to be considered as a promising power source in the WSN nodes.

Ambient vibrations are present in many environments, such as air and water flows, structures (building, towers, bridges, etc.), industrial machines, equipment and transportations (washing machines, machines tools, automobiles, air planes, etc.), movement of human bodies (walking, running, pulse beats, and blood flow) and so on. Vibration-based energy harvesting techniques convert energy in the form of mechanical vibrations in the ambient environment into electricity energy by using piezoelectric, electromagnetic, and electrostatic mechanisms (Beeby et al. 2006; Liu et al. 2012a; Liu et al. 2012b; Liu et al. 2012c; Lee et al. 2009a; Liu et al. 2011a; Liu et al. 2011b; Liu et al. 2012d; Yang and Lee 2009; Yang and Lee 2010; Yang et al. 2012; Yang et al. 2010a; Yang et al. 2010b).

An example of a wideband energy harvesting system (Liu et al. 2012a) is illustrated in Fig. 16.35a. It contains a low-resonant-frequency cantilever (denoted as LRF cantilever) and a high-resonant-frequency cantilever (denoted as HRF cantilever). The LRF cantilever as shown in Fig. 16.35b comprises of a silicon supporting beam deposited with a PZT layer and silicon inertial mass. The PZT layer consists of ten PZT stripes electrically isolated from one another. The HRF cantilever (Fig. 16.35c) has similar dimensions and configuration with that of the LRF cantilever, but without an inertial mass. The resonant frequencies of the LRF and HRF cantilevers are 36 Hz and 618 Hz, respectively. The LRF and HRF cantilevers are assembled with a pre-determined stop-spacing. The vibration amplitude of the LRF cantilever is larger than the stop-spacing and it is able to respond to ambient vibrations nearby its resonant frequency and further extend over a wider frequency bandwidth by the HRF cantilever which acts as a frequency up-converter (FUC) stopper. In the meantime, the HRF cantilever is triggered by the inertial mass of the LRF cantilever into a high frequency self-oscillation. Due to the piezoelectric effect, the electric current is generated once the PZT layer on both LRF and HRF cantilevers deform. The advantages of this assembly strategy in employing the HRF cantilever as a FUC stopper are that it protects the excitation oscillator from damage during vibration and broadens the operation range. The system also converts ambient low frequency vibration to a high frequency self-oscillation. Using MEMS technology, the proposed system is applied to a piezoelectric-based energy harvesting system by the use of cantilever beams deposited with piezoelectric PZT material.

On the other hand, a demonstration of wafer level packaging using thin film encapsulation for thermoelectric power generator, shows the ease of process integration of WLE technology for packaging with the MEMS process technology. The main process steps and images are illustrated in Fig. 16.36. The 0.7 μm-thick poly-Si is deposited at 580 °C in furnace by LPCVD as the thermoelectric layer, followed by dry etching of the poly-Si layer to form thermoelectric leg

Fig. 16.35 Piezoelectric based wideband energy harvesting system (Liu et al. 2012a)

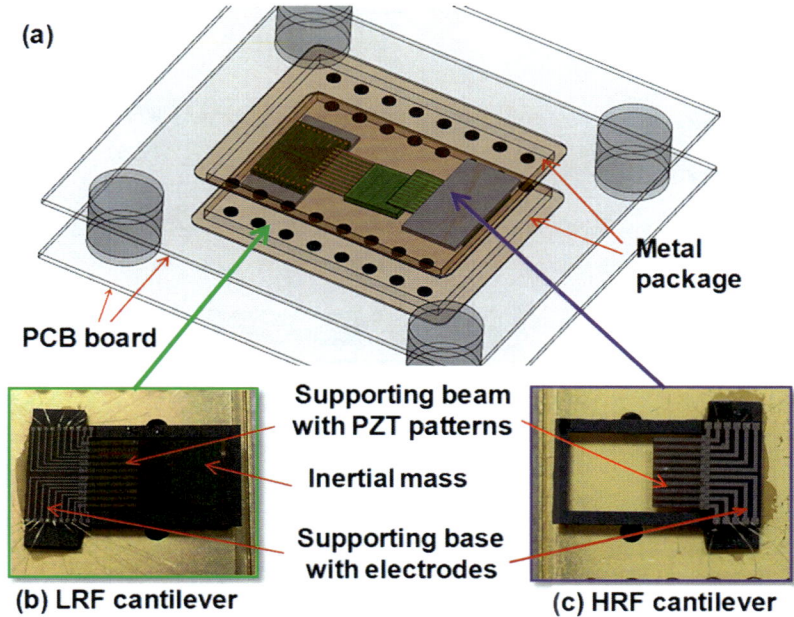

(a)

Metal package

PCB board

Supporting beam with PZT patterns

Inertial mass

Supporting base with electrodes

(b) LRF cantilever

(c) HRF cantilever

patterns with 5 μm in width and 16 μm in length (see Fig. 16.36). Afterwards, n-type and p-type thermoelectric legs are formed by implanting one half of poly-Si with phosphorous ions at 180 keV energy and other half with boron ions using 80 keV in the other half to form the p-type thermoelectric legs respectively. The doping dose of 10^{16} cm^{-2} is used in both the cases. The doped poly-Si is annealed in a furnace at 1000 °C for 30 min to activate the implanted dopants. SiO$_2$ insulating layer deposition is carried out, following which contact via opening, aluminum layer deposition and aluminum etching is performed to form electrical interconnects between the p-type and n-type thermoelectric legs in series (see Fig. 16.36b). The bottom cavity and top cavity are created with the pre-described process of wafer-level sealing and encapsulation—first perforating the holes, followed by DRIE, and isotropic etching silicon (see Fig. 16.36c), sealing bottom cavities (see Fig. 16.36d), patterning USG sacrificial layer, opening etching holes (see Fig. 16.36e), and finally removing the USG sacrificial layer and sealing the top cavities (see Fig. 16.36f). In the

last step, 0.7 μm-thick aluminum is deposited to cover the whole device area as the heat sink layer (see Fig. 16.36g). The cross section of the fabricated device in (see Fig. 16.36j) shows that the thermal legs are embedded into the top cavities and the bottom cavities. The sealed bottom cavity is about 30 μm deep and 42 μm wide, while the encapsulated top cavity is 2 μm high and 35 μm wide. The lengths of both bottom and top cavities are about 670 μm.

Thus from the above example of thermoelectric power generator, the flexibility of adopting WLE onto the device fabrication process is quite straight forward and does not demand any more complications. This WLE technology can be readily used for encapsulation of IR sensors as well. Hence from the presented case study example, it can be seen that the advancement in MEMS fabrication and packaging technology has improved the possibility of providing many functional ICs along with the MEMS sensors and actuators to be fabricated onto a single wafer and diced. After dicing, the chips can be effectively packaged into a single package to provide extremely small and efficient IoT node at relative low cost.

Fig. 16.36 Process flow to
fabricate thermoelectric
power generator and
corresponding device
images (OM or SEM)
are shown in right
(Xie et al. 2011)

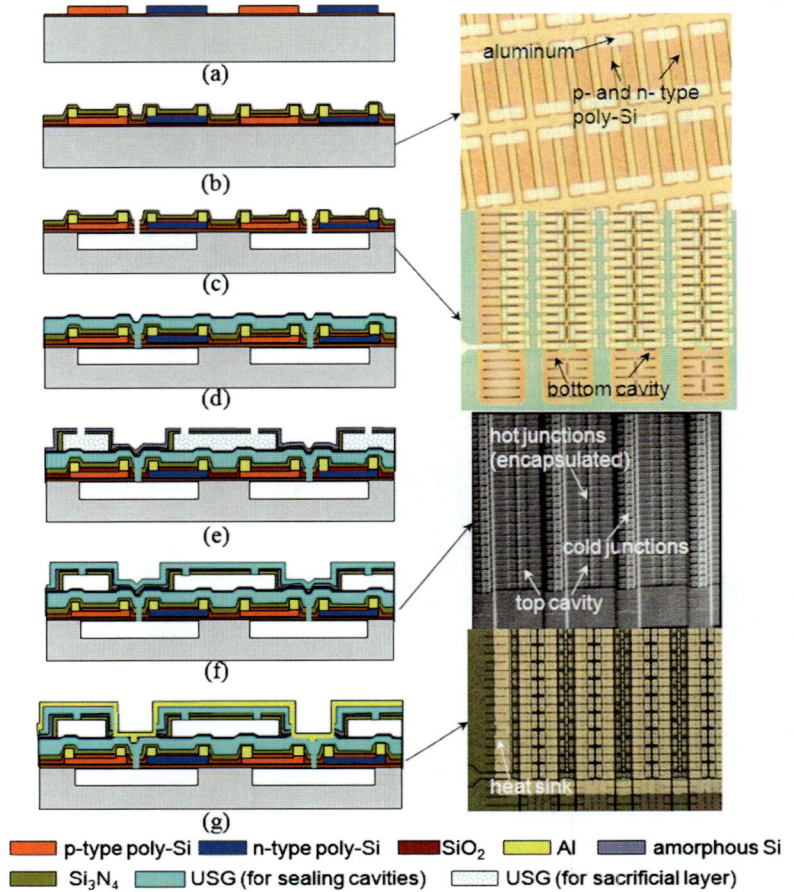

(a)
(b)
(c)
(d)
(e)
(f)
(g)

aluminum
p- and n- type
poly-Si
bottom cavity
hot junctions
(encapsulated)
cold junctions
top cavity
heat sink

▬ p-type poly-Si ▬ n-type poly-Si ▬ SiO₂ ▬ Al ▬ amorphous Si
▬ Si₃N₄ ▬ USG (for sealing cavities) ▭ USG (for sacrificial layer)

16.3 Case Study of Infrared Sensor Packaging

In this section, we will consider the packaging of
infrared (IR) sensors as the case to understand the
basic functionality of a MEMS sensor and its
packaging requirements, followed by technological improvements and finally state-of-art packaging technology.

The infrared spectrum is part of the electromagnetic spectrum, with wavelengths ranging
from about 1 to 1000 μm. The most attractive
application lies especially in the specific wavelength regimes ranging from 1 to 12 μm, and are
categorized as (1) short-wave infrared (SWIR)
from 1 to 3 μm; (2) mid wave infrared (MWIR)

from 3 to 5 μm and (3) long-wave infrared
(LWIR) from 8 to 12 μm. Infrared sensors have
widespread applications including security, surveillance, search and rescue, firefighting, night
vision, traffic systems, law enforcement, process
control, and predictive maintenance. Basically
IR detectors are classified into two major classes—photon detectors and thermal detectors. The
photon detectors utilize quantum well structures
band gap designed for particular IR wavelength,
and has very sensitivity and shorter response
time, but demands constant cooling hence
making the whole system heavy and costly. On
the other hand, the thermal detectors work on the
principle of change of temperature dependent
material and physical characteristics (like resistance, ferroelectricity, pyroelectricity, coefficient

a

b

c

d

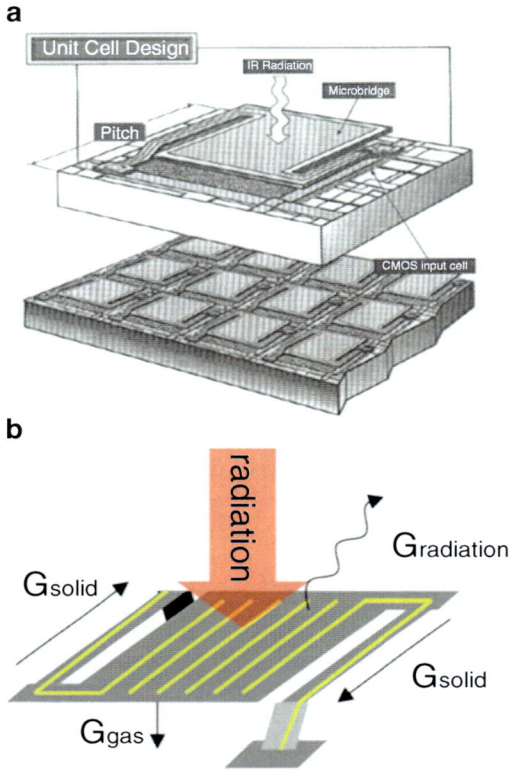

Fig. 16.37 (**a**) Schematics of IR sensor array and a single unit cell design (Kohin et al. 2004); (**b**) schematics showing the incoming radiation and the various thermal loss path in a single IR sensor pixel structure; (**c**) shows the dependence of thermal loss due to solid conduction, air convection, radiation and overall thermal conductance with respect to pressure level in the sensor package (Chou et al. 1995); (**d**) shows the dependence of IR pixel responsivity with respect to pressure level in the package with different gases (Liddiard 1984)

of thermal expansion, resonant frequency, etc.) of the devices and have lower sensitivity and longer response time, but doesn't demand cooling and so are more cost effective. With the advancement of MEMS technology, thermal IR detectors are made with very high sensitivity and lower response time. Most of the IR detectors consist of the following parts, (1) absorber membrane to absorb IR radiation, (2) sensor structure to convert IR radiation to different read out parameter, (3) long mechanical legs to provide high thermal isolation from substrate and mechanical stability at the same time. One of the most important considerations for IR detectors is the thermal losses. As shown in Fig. 16.37b, there are three possible thermal

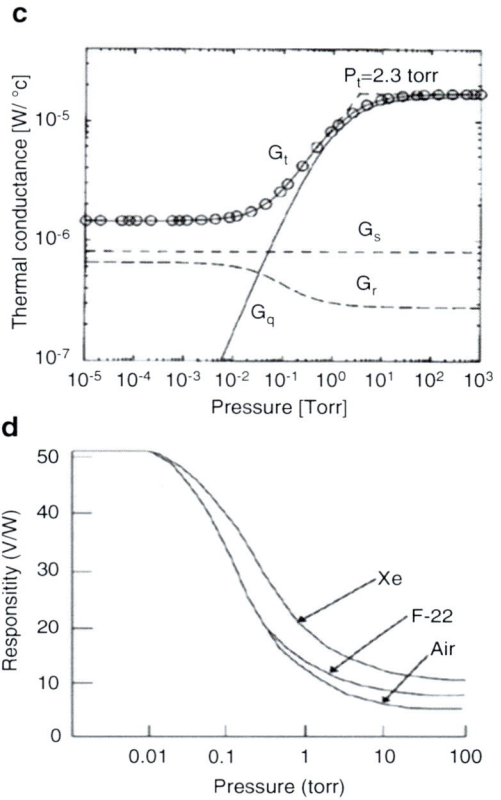

loss path—(1) radiation loss due to the heated membrane and defines the fundamental limit of the thermal loss, (2) convection loss due to the gas present in between the active sensor membrane to outside environment or substrate and (3) conduction loss due to the mechanical leg to support the sensor active membrane. It can be seen from Fig. 16.37c, d, that air convection loss is the dominating factor for the overall loss for a pressure level of more than 0.1 Torr. It is highly desirable to operate the IR detectors in low pressure; thus making vacuum packaging inevitable for IR sensors.

The commercially available IR detector usually involves four levels of packaging and assembly: (1) Pixel level packaging—vacuum packaging of

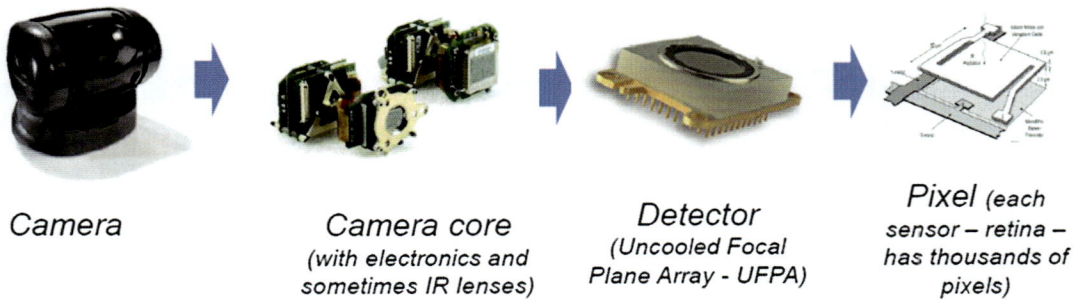

Camera   Camera core
(with electronics and sometimes IR lenses)

Detector
(Uncooled Focal Plane Array - UFPA)

Pixel *(each sensor – retina – has thousands of pixels)*

Fig. 16.38 Various levels of device assembly and integration for IR sensor product (Yole Developpement 2012)

individual pixel is achieved through micromachining or wafer bonding techniques; (2) Array level packaging—vacuum packaging of an array device is also possible through wafer bonding technique and generic chip level packaging technologies; (3) Integration of vacuum packaged IR detector array with electronic circuitry on PCB level to perform various readout operations and pixel to pixel compensation schemes and may also include IR lenses for focusing; (4) Assembly of integrated PCBs on to a compact casing to deliver the final product that can be easily handled. The four levels of packaging and assembling for IR detectors is shown in Fig. 16.38.

16.3.1 Conventional Packaging for IR Sensors

The conventional packaging at chip level involves packaging individual die in a chip carrier using either a silicon or germanium window. In this approach, the die are mounted in a chip carrier, usually with a solder perform attachment. The package and lid are placed in a vacuum chamber and vacuum baked at elevated temperature to outgas the package and lid, prior to sealing the package. Once outgassing is complete, the lid is sealed to the package using a solder seal. Depending upon the application, different packaging approaches are considered like metal can packaging or ceramic packaging.

Metal packaging is usually used in applications like automotives, where good heat conduction and durability at higher temperature are desired. In the

metal can packaging technique, the IR sensor array is mounted on a metal plate with a metal can crimped on top of it. The metal package can be attached to a heat sink, there by dissipating, the excess heat from the sensor array, which is highly significant for high sensitive IR sensors. Thermal compounds are also used to improve heat transfer between the device case and the heat sink. Since the device case is one of the electrical connections, an insulator may be required to electrically isolate the component from the heat sink. Insulating washers made of mica or other materials with good thermal conductivity are often used (Fig. 16.39).

The most attractive feature of metal can packaging is that almost all the surface of the package are metal, hence provides better heat dissipation and durability than most of other available packaging technologies. Additionally, Metal-glass seal is hermetic to protect the semiconductor from liquids and gases. However, on the other hand, metal can packaging is more expensive than the counterpart solutions using plastic or ceramics. Also, achieving high level vacuum within the package is quite challenging and so the gas convection in the metal can package limits the detector's performance.

Due to the limitations of metal can packaging, ceramic and metal based flip chip packaging became extremely popular as higher levels of vacuum could be achieved with these technologies (Hunter et al. 2008). The flip-chip packaging provided smaller packaging, and also enabled much more components to be integrated into a single package as shown in Fig. 16.40. Some of the key components are IR window

(a)

Component-level packaging

(b)

(c)

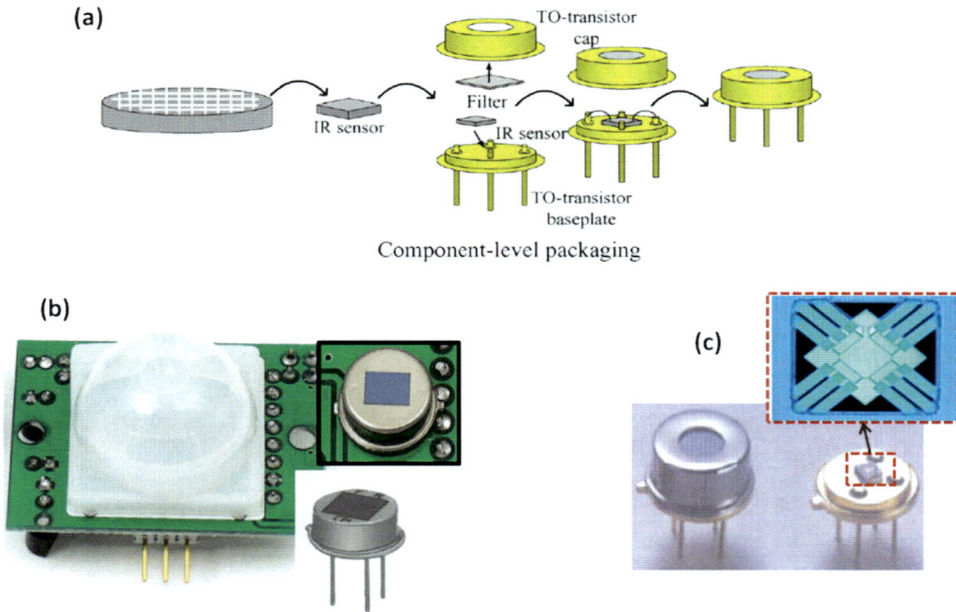

Fig. 16.39 (**a**) Schematic demonstration of metal can packaging process for IR sensor array (Xu et al. 2010), (**b**) metal can packaged IR sensor by OMRON technologies and (**c**) shows metal can packaged thermopile based IR sensor array (Du and Lee 2002)

with antireflection coating to allow IR radiation to reach the IR sensor array without getting absorbed or reflected. Another important component is the thermistor, which enables to determine the temperature of the sensor array dynamically and regulate the temperature in the package by providing feedback input to the thermoelectric cooler. The thermoelectric cooler (or TE stabilizer) is one of the key components in photon IR detectors, as smallest increase in temperature can elevate the thermal noise in the system thereby making device performance deteriorate significantly. As IR sensors need high level vacuum packaging, the key challenge was with the outgassing of the deposited material layers after vacuum packaging. Hence electrically activated getters were used to restore back the vacuum level inside the package even after sealing it. The getter is a material, which went provided with certain amount of energy, reacts with the gas molecules in the package, thereby providing a higher degree of vacuum.

16.3.2   Wafer Level Packaging for IR Sensors

Wafer level packaging of IR sensors involves addition of extra process technology to the sensor fabrication process in order to achieve packaging of IR sensor in array or pixel level. The most exciting advantage provided by WLP for IR sensor is that very high vacuum levels can be achieved without the use of external components like getters. Additional WLP reduces the cost and time to market by a large margin. Certain application demands the use of particular wavelength IR signal, in which case the optical filter wafer can be directly bonded with the IR sensor to miniaturize the overall sensor module. Recently, low-temperature wafer-level bonding technologies have also been proposed for optical device fabrication and packaging.

In the wafer bonding technology, the active IR sensor array is fabricated on a silicon MEMS wafer and corresponding companion silicon lid

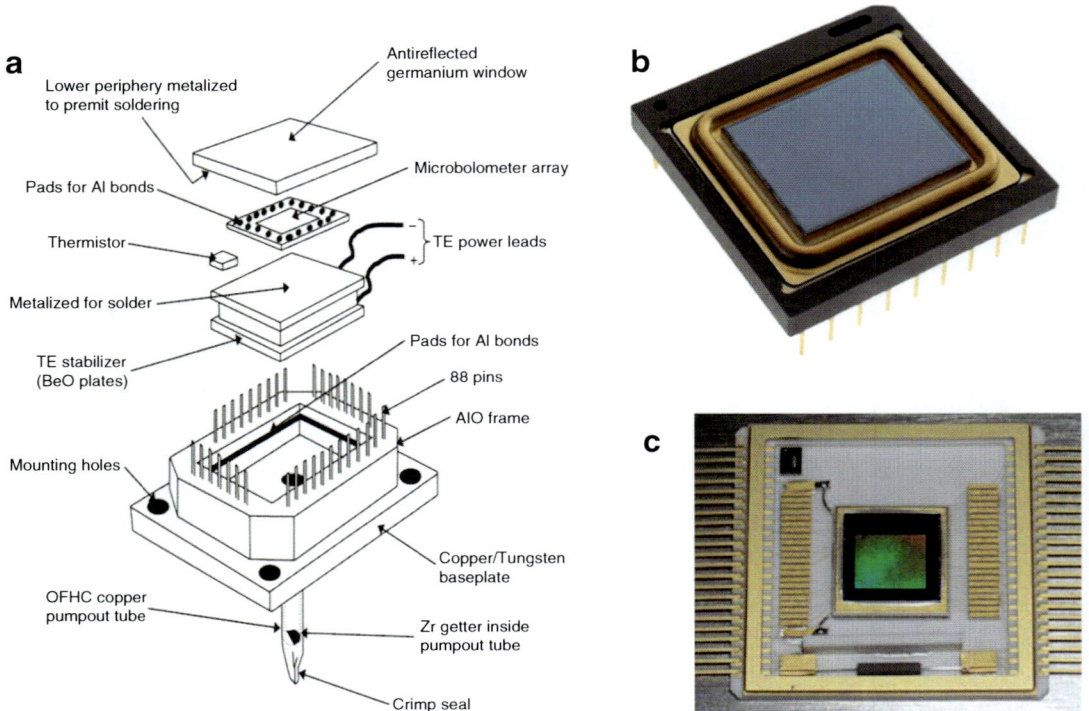

Fig. 16.40 (a) Vacuum package for vanadium oxide based microbolometer uncooled infrared focal plane array (IRFPA) (Capper and Elliott 2013), (b) completely packaged IR imager [Courtesy: ULIS-IR] and (c) metal packaged IR imaging module from Multispectral Imaging Inc (Hunter et al. 2008)

wafer is fabricated with the seal ring to cover the entire sensor array in the first wafer. The lid wafer provides the IR window for the sensor array. Vacuum packaging is carried out by placing the MEMS wafer and lid wafer in a vacuum chamber, vacuum-baking them at elevated temperature to outgas, and finally sealing the wafers using either a solder seal or a thermocompression bond seal as in Fig. 16.41 and 16.42.

16.4 Summary

In summary, the importance of packaging of individual chips or as a system is explained. Various conventional and more advanced WLP technologies for autonomous sensor net-work has been discussed. Following which, the various packaging technologies were explained as a case study of packaging MEMS based IR sensor. Finally, the more futuristic approach of using energy harvesting as power source and ultimately integrating the whole IoT system within a single package has been explained. Thus it is can be confidently stated that the increasing efforts with keen focus of semiconductor manufacturing and packaging industries in "More than Moore" technological roadmap; by integrating multiple functional chips either through packaging (SiP approach) or in process level (SoC approach) will enable in achieving extremely small IoT modules at lower cost, thereby making it a disruptive technology for a wide range of applications.

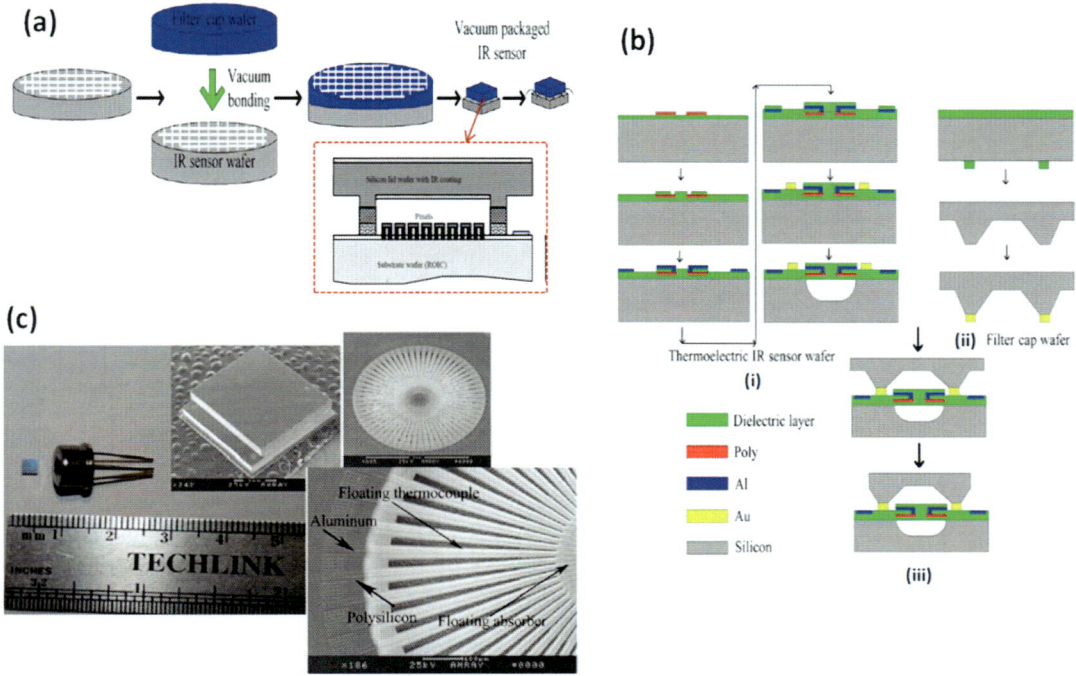

Fig. 16.41 (**a**) Schematic demonstration of wafer level packaging of IR sensor using wafer bonding approach (*inset*: schematic cross section showing the vacuum package IR sensor array) (Xu et al. 2010) (**b**) (i) micromachined IR sensor fabrication process, (ii) filter cap wafer fabrication process, (iii) wafer-level packaging of the sensor by bonding the sensor wafer with the filter cap wafer using Au-Au thermocompression bonding (Xu et al. 2010) and (**c**) wafer level, vacuum packaged thermopile based IR sensor later assembled into a metal can package (*inset*: wafer level packaged single unit array of IR sensor with the close up view of thermopile structure) (Xu et al. 2011)

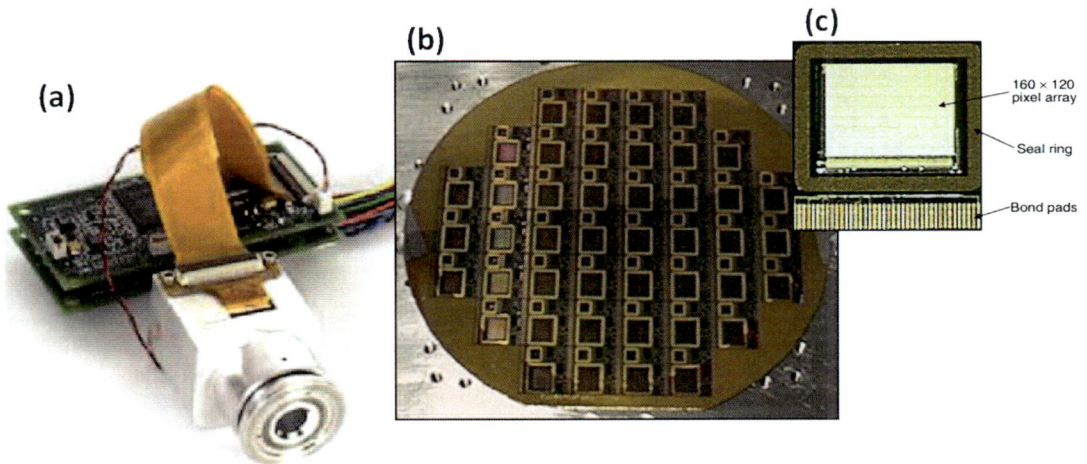

Fig. 16.42 (**a**) The final product of wafer level packaged IR sensor array [Courtesy: Raytheon Company]; (**b**) wafer level packaged IR sensors in a 6 in. wafer and (**c**) close up view of wafer level packaged single IR sensor array (Xu et al. 2010)

References

S.P. Beeby, M.J. Tudor, N. White, Energy harvesting vibration sources for microsystems applications. Meas. Sci. Technol. **17**(12), R175 (2006)

R. Bloss, Numerous recent shows highlight the latest in sensor innovations. Sens. Rev. **32**(2), 101–105 (2012)

H. Böttner, Thermoelectric micro devices: current state, recent developments and future aspects for technological progress and applications, in *Twenty-First International Conference on Thermoelectrics, 2002. Proceedings ICT'02* (2002), pp. 511–518

R.N. Candler, W.-T. Park, H. Li, G. Yama, A. Partridge, M. Lutz, T.W. Kenny, Single wafer encapsulation of MEMS devices. IEEE Trans. Adv. Packag. **26**(3), 227–232 (2003)

P. Capper, C. Elliott, *Infrared Detectors and Emitters: Materials and Devices* (Springer US, New York, 2013)

K. Chen, C. Premachandran, K. Choi, C. Ong, X. Ling, A. Ratmin, M. Pa, J. Lau, C2W low temperature bonding method for MEMS applications, in *IEEE Proceedings of Electronics Packaging Technology Conference, Singapore, 2008* (2008), pp. 1–7

F. Chiarugi, I. Karatzanis, G. Zacharioudakis, P. Meriggi, F. Rizzo, M. Stratakis, S. Louloudakis, C. Biniaris, M. Valentini, M. Di Rienzo, Measurement of heart rate and respiratory rate using a textile-based wearable device in heart failure patients. Comput. Cardiol. **35**, 901–904 (2008)

B. Chou, J.-S. Shic, Y.-M. Chen, A highly sensitive Pirani vacuum gauge, in *The 8th International Conference on Solid-State Sensors and Actuators, 1995 and Eurosensors IX. Transducers' 95* (1995), pp. 167–170

S. Devot, A.M. Bianchi, E. Naujokat, M.O. Mendez, A. Brauers, S. Cerutti, Sleep monitoring through a textile recording system, in *29th Annual International Conference of the IEEE Engineering in Medicine and Biology Society. EMBS 2007* (2007), pp. 2560–2563

C.-H. Du, C. Lee, Characterization of thermopile based on complementary metal-oxide-semiconductor (CMOS) materials and post CMOS micromachining. Jpn. J. Appl. Phys. **41**(6S), 4340 (2002)

M. Esashi, Wafer level packaging of MEMS. J. Micromech. Microeng. **18**(7), 073001 (2008)

J. Espina, T. Falck, J. Muehlsteff, X. Aubert, Wireless body sensor network for continuous cuff-less blood pressure monitoring, in *3rd IEEE/EMBS International Summer School on Medical Devices and Biosensors, 2006* (2006), pp. 11–15

Y.B. Gianchandani, O. Tabata, H.P. Zappe, *Comprehensive Microsystems* (Elsevier, Amsterdam, 2008)

H. Goldstein, Packages go vertical. IEEE Spectr. **38**(8), 46–51 (2001)

H. Guo, L. Lou, X. Chen, C. Lee, PDMS-coated piezoresistive NEMS diaphragm for chloroform vapor detection. IEEE Electron Device Lett. **33**(7), 1078–1080 (2012)

Y. Hasegawa, M. Shikida, D. Ogura, Y. Suzuki, K. Sato, Fabrication of a wearable fabric tactile sensor produced by artificial hollow fiber. J. Micromech. Microeng. **18**(8), 085014 (2008)

V. Hong, B. Lee, D.L. Christensen, D. Heinz, E.J. Ng, C.H. Ahn, Y. Yang, T.W. Kenny, XY and Z-axis capacitive accelerometers packaged in an ultra-clean hermetic environment, in *2013 Transducers and Eurosensors XXVII: The 17th International Conference on Solid-State Sensors, Actuators and Microsystems (TRANSDUCERS & EUROSENSORS XXVII)* (2013), pp. 618–621

X. Huang, H. Cheng, K. Chen, Y. Zhang, Y. Zhang, Y. Liu, C. Zhu, S.-C. Ouyang, G.-W. Kong, C. Yu, Epidermal impedance sensing sheets for precision hydration assessment and spatial mapping. IEEE Trans. Biomed. Eng. **60**(10), 2848–2857 (2013)

N.S. Hudak, G.G. Amatucci, Small-scale energy harvesting through thermoelectric, vibration, and radiofrequency power conversion. J. Appl. Phys. **103** (10), 101301 (2008)

S.R. Hunter, G.S. Maurer, G. Simelgor, S. Radhakrishnan, J. Gray, K. Bachir, T. Pennell, M. Bauer, U. Jagadish, Development and optimization of microcantilever based IR imaging arrays, in *SPIE Defense and Security Symposium* (2008), pp. 694013–694013-12

Y. Jin, Z. Wang, P. Lim, D. Pan, J. Wei, J. Wong, MEMS vacuum packaging technology and applications, in *2003 5th Conference (EPTC 2003) Electronics Packaging Technology* (2003), pp. 301–306

S. Jung, T. Ji, V.K. Varadan, Point-of-care temperature and respiration monitoring sensors for smart fabric applications. Smart Mater. Struct. **15**(6), 1872 (2006)

C.L. Keast, B. Aull, J. Burns, C. Chen, J. Knecht, B. Tyrrell, K. Warner, B. Wheeler, V. Suntharaligam, P. Wyatt, Three-dimensional integration technology for advanced focal planes, in *MRS Proceedings* (2008), pp. 1112-E01-02

T.Y. Kheng, Analysis, design and implementation of energy harvesting systems for wireless sensor nodes, PhD Thesis, National University of Singapore, Singapore (2010)

R. Knechtel, M. Wiemer, J. Frömel, Wafer level encapsulation of microsystems using glass frit bonding. Microsyst. Technol. **12**(5), 468–472 (2006)

C.-T. Ko, K.-N. Chen, Wafer-level bonding/stacking technology for 3D integration. Microelectron. Reliab. **50**(4), 481–488 (2010)

M. Kohin, N.R. Butler, Performance limits of uncooled VO_x microbolometer focal plane arrays, in *Defense and Security* (2004), pp. 447–453

J.F. Kuhmann, M. Heschel, S. Bouwstra, F. Baleras, C. Massit, Through wafer interconnects and flip-chip bonding: a toolbox for advanced hybrid technologies for MEMS, in *13th European Conference on Solid-State Transducers, 1999* (1999), pp. 265–272

J.H. Lau, *Flip Chip Technologies* (McGraw-Hill, New York, 1996)

J.H. Lau, C. Lee, C.S. Premachandran, Y. Aibin, *Advanced MEMS Packaging* (McGraw-Hill Professional, 2010)

J.B. Lee, Z. Chen, M.G. Allen, A. Rchatgi, R. Ayra, A high voltage solar cell array as an electrostatic MEMS power supply, in *MEMS'94, Proceedings, IEEE*

Workshop on Micro Electro Mechanical Systems (1994), pp. 331–336

C. Lee, Y.M. Lim, B. Yang, R.K. Kotlanka, C.-H. Heng, J.H. He, M. Tang, J. Xie, H. Feng, Theoretical comparison of the energy harvesting capability among various electrostatic mechanisms from structure aspect. Sens. Actuators, A **156**(1), 208–216 (2009a)

K.-W. Lee, A. Noriki, K. Kiyoyama, S. Kanno, R. Kobayashi, W.-C. Jeong, J.-C. Bea, T. Fukushima, T. Tanaka, M. Koyanagi, 3D heterogeneous opto-electronic integration technology for system-on-silicon (SOS), in *2009 I.E. International Electron Devices Meeting (IEDM)* (2009) pp. 1–4

K. Liddiard, Thin-film resistance bolometer IR detectors. Infrared Phys. **24**(1), 57–64 (1984)

V. Lindroos, S. Franssila, M. Tilli, M. Paulasto-Krockel, A. Lehto, T. Motooka, V.-M. Airaksinen, *Handbook of Silicon Based MEMS Materials and Technologies* (Elsevier, Amsterdam, 2009)

H. Liu, C.J. Tay, C. Quan, T. Kobayashi, C. Lee, Piezoelectric MEMS energy harvester for low-frequency vibrations with wideband operation range and steadily increased output power. J. Microelectromech. Syst. **20** (5), 1131–1142 (2011a)

H. Liu, C.J. Tay, C. Quan, T. Kobayashi, C. Lee, A scrape-through piezoelectric MEMS energy harvester with frequency broadband and up-conversion behaviors. Microsyst. Technol. **17**(12), 1747–1754 (2011b)

H. Liu, C. Lee, T. Kobayashi, C.J. Tay, C. Quan, Investigation of a MEMS piezoelectric energy harvester system with a frequency-widened-bandwidth mechanism introduced by mechanical stoppers. Smart Mater. Struct. **21**(3), 035005 (2012a)

H. Liu, C. Lee, T. Kobayashi, C.J. Tay, C. Quan, A new S-shaped MEMS PZT cantilever for energy harvesting from low frequency vibrations below 30 Hz. Microsyst. Technol. **18**(4), 497–506 (2012b)

H. Liu, C. Lee, T. Kobayashi, C.J. Tay, C. Quan, Piezoelectric MEMS-based wideband energy harvesting systems using a frequency-up-conversion cantilever stopper. Sens. Actuators, A **186**, 242–248 (2012c)

H. Liu, S. Zhang, R. Kathiresan, T. Kobayashi, C. Lee, Development of piezoelectric microcantilever flow sensor with wind-driven energy harvesting capability. Appl. Phys. Lett. **100**(22), 223905 (2012d)

F. Lorussi, W. Rocchia, E.P. Scilingo, A. Tognetti, D. De Rossi, Wearable, redundant fabric-based sensor arrays for reconstruction of body segment posture. IEEE Sensors J. **4**(6), 807–818 (2004)

F. Lorussi, E.P. Scilingo, M. Tesconi, A. Tognetti, D.D. Rossi, Strain sensing fabric for hand posture and gesture monitoring. IEEE Trans. Inf. Technol. Biomed. **9**(3), 372–381 (2005)

L. Lou, S. Zhang, W.-T. Park, J. Tsai, D.-L. Kwong, C. Lee, Optimization of NEMS pressure sensors with a multilayered diaphragm using silicon nanowires as piezoresistive sensing elements. J. Micromech. Microeng. **22**(5), 055012 (2012)

M.P. Market, Technology trend report, Yole Développement, 2012

C.Ó. Mathúna, T. O'Donnell, R.V. Martinez-Catala, J. Rohan, B. O'Flynn, Energy scavenging for long-term deployable wireless sensor networks. Talanta **75** (3), 613–623 (2008)

C.R. Merritt, H.T. Nagle, E. Grant, Textile-based capacitive sensors for respiration monitoring. IEEE Sensors J. **9**(1), 71–78 (2009)

P.D. Mitcheson, T.C. Green, E.M. Yeatman, A.S. Holmes, Architectures for vibration-driven micropower generators. J. Microelectromech. Syst. **13**(3), 429–440 (2004)

F. Niklaus, G. Stemme, J.-Q. Lu, R. Gutmann, Adhesive wafer bonding. J. Appl. Phys. **99**(3), 031101 (2006)

D. Niyato, E. Hossain, M.M. Rashid, V.K. Bhargava, Wireless sensor networks with energy harvesting technologies: a game-theoretic approach to optimal energy management. IEEE Wirel. Commun. **14**(4), 90–96 (2007)

R. Paradiso, G. Loriga, N. Taccini, A wearable health care system based on knitted integrated sensors. IEEE Trans. Inf. Technol. Biomed. **9**(3), 337–344 (2005)

S. Pourkamali, F. Ayazi, Wafer-level encapsulation and sealing of electrostatic HARPSS transducers, in *Sensors, 2007 IEEE* (2007), pp. 49–52

Y. Qian, Development of nano/microelectromechanical system (N/MEMS) switches, PhD Thesis, National University of Singapore, Singapore (2015)

J.M. Rabaey, M.J. Ammer, J.L. da Silva, D. Patel, S. Roundy, PicoRadio supports ad hoc ultra-low power wireless networking. Computer **33**(7), 42–48 (2000)

V. Raghunathan, S. Ganeriwal, M. Srivastava, Emerging techniques for long lived wireless sensor networks. IEEE Commun. Mag. **44**(4), 108–114 (2006)

M. Rais-Zadeh, H.M. Lavasani, F. Ayazi, CMOS-compatible encapsulated silver bandpass filters, in *IEEE Microwave Symposium (IMS'07)* (2007), pp. 1301–1304

J.F. Randall, On the use of photovoltaic ambient energy sources for powering indoor electronic devices. Thesis, EPFL, Switzerland (2003)

H. Reichl, R. Aschenbrenner, H. Potter, *Hetero System Integration-Enabling Technology for Future Products* (Productions, Frankfurt/Main, 2005) pp. 45–49

H. Reichl, R. Aschenbrenner, M. Töpper, H. Pötter, Heterogeneous integration: building the foundation for innovative products, in *More than Moore* (Springer, New York, 2009) pp. 279–303

M. Rothmaier, B. Selm, S. Spichtig, D. Haensse, M. Wolf, Photonic textiles for pulse oximetry. Opt. Express **16** (17), 12973–12986 (2008)

S. Roundy, P.K. Wright, K.S. Pister, Micro-electrostatic vibration-to-electricity converters, in *ASME 2002 International Mechanical Engineering Congress and Exposition* (2002), pp. 487–496

C. Rovira, S. Coyle, B. Corcoran, D. Diamond, F. Stroiescu, K. Daly, Integration of textile-based sensors and Shimmer for breathing rate and volume measurement, in *2011 5th International Conference on Pervasive Computing Technologies for Healthcare (PervasiveHealth)* (2011), pp. 238–241

H.R. Shea, Reliability of MEMS for space applications, in *Moems-mems 2006 micro and nanofabrication* (2006), pp. 61110A–61110A-10

N.S. Shenck, J.A. Paradiso, Energy scavenging with shoe-mounted piezoelectrics. IEEE Micro **21**(3), 30–42 (2001)

L. Shiv, M. Heschel, H. Korth, S. Weichel, R. Hauffe, A. Kilian, B. Semak, M. Houlberg, P. Egginton, A. Hase, Ultra thin hermetic wafer level, chip scale package, in *2006 Proceedings of 56th Electronic Components and Technology Conference* (2006), 7 pp

B.W. Soon, N. Singh, J.M. Tsai, C.-K. Lee, Vacuum based wafer level encapsulation (WLE) of MEMS using physical vapor deposition (PVD), in *2012 I.E. 14th Electronics Packaging Technology Conference (EPTC)* (2012), pp. 342–345

B.W. Soon, E.J. Ng, V. Hong, Y. Yang, C.H. Ahn, Y. Qian, T.W. Kenny, C. Lee, Fabrication and characterization of a vacuum encapsulated curved beam switch for harsh environment application. J. Microelectromech. Syst. **23**(5), 1121–1130 (2014)

B.W. Soon, Y. Qian, E.J. Ng, V. Hong, Y. Yang, C.H. Ahn, T.W. Kenny, C. Lee, Investigation of a vacuum encapsulated Si-to-Si contact microswitch operated from −60 °C to 400 °C. J. Microelectromech. Syst. **24**(6), 1906–1915 (2015)

D. Sparjs, Packaging of microsystems for harsh environments. IEEE Instrum. Meas. Mag. **4**(3), 30–33 (2001)

M. Stordeur, I. Stark, Low power thermoelectric generator-self-sufficient energy supply for micro systems, in *Proceedings ICT'97. XVI International Conference on Thermoelectrics, 1997* (1997), pp. 575–577

J.P. Thomas, M.A. Qidwai, J.C. Kellogg, Energy scavenging for small-scale unmanned systems. J. Power Sources **159**(2), 1494–1509 (2006)

J. Xie, C. Lee, M.-F. Wang, Y. Liu, H. Feng, Characterization of heavily doped polysilicon films for CMOS-MEMS thermoelectric power generators. J. Micromech. Microeng. **19**(12), 125029 (2009)

J. Xie, C. Lee, H. Feng, Design, fabrication, and characterization of CMOS MEMS-based thermoelectric power generators. J. Microelectromech. Syst. **19**(2), 317–324 (2010)

J. Xie, C. Lee, M.-F. Wang, H. Feng, Seal and encapsulate cavities for complementary metal-oxide-semiconductor microelectromechanical system thermoelectric power generators. J. Vac. Sci. Technol. B **29**(2), 021401 (2011)

D. Xu, E. Jing, B. Xiong, Y. Wang, Wafer-level vacuum packaging of micromachined thermoelectric IR sensors. IEEE Trans. Adv. Packag. **33**(4), 904–911 (2010)

D. Xu, B. Xiong, Y. Wang, Micromachined thermopile IR detector module with high performance. IEEE Photon. Technol. Lett. **23**(3), 149–151 (2011)

B. Yang, C.K. Lee, A wideband electromagnetic energy harvester for random vibration sources. Adv. Mater. Res. **74**, 165–168 (2009)

B. Yang, C. Lee, Non-resonant electromagnetic wideband energy harvesting mechanism for low frequency vibrations. Microsyst. Technol. **16**(6), 961–966 (2010)

B. Yang, C. Lee, W.L. Kee, S.P. Lim, Hybrid energy harvester based on piezoelectric and electromagnetic mechanisms. J. Micro/Nanolithogr. MEMS MOEMS, **9**(2): 023002–023002-10 (2010)

B. Yang, C. Lee, R.K. Kotlanka, J. Xie, S.P. Lim, A MEMS rotary comb mechanism for harvesting the kinetic energy of planar vibrations. J. Micromech. Microeng. **20**(6), 065017 (2010b)

B. Yang, C. Lee, G.W. Ho, W.L. Ong, J. Liu, C. Yang, Modeling and experimental study of a low-frequency-vibration-based power generator using ZnO nanowire arrays. J. Microelectromech. Syst. **21**(4), 776–778 (2012)

D.-Q. Yu, C. Lee, L.L. Yan, W.K. Choi, A. Yu, J.H. Lau, The role of Ni buffer layer on high yield low temperature hermetic wafer bonding using In/Sn/Cu metallization. Appl. Phys. Lett. **94**(3), 034105 (2009)

S. Zhang, L. Lou, C. Lee, Piezoresistive silicon nanowire based nanoelectromechanical system cantilever air flow sensor. Appl. Phys. Lett. **100**(2), 023111 (2012)

Z. Zhong, T. Tee, J. Luan, Recent advances in wire bonding, flip chip and lead-free solder for advanced microelectronics packaging. Microelectron. Int. **24**(3), 18–26 (2007)

An IPv6 Energy-Harvested WSN Demonstrator Compatible with Indoor Applications

17

Pascal Urard, Liviu Varga, Mališa Vučinić, and Roberto Guizzetti

In this chapter, we present the results of a research project that was developed in ST with key partners over several years. It is an energy-harvested, self-healing, secured IPv6 Wireless Sensor & Actuator Network (WSAN), designed for indoor environments. Pairing and installation have been taken into account to propose a complete system solution. GreenNet has demonstrated that it is possible to achieve autonomy with upcoming radio generations. However, this is possible only if certain cross-layer optimizations are performed throughout the software stack. Details and results are provided along the chapter.

17.1 GreenNet at a Glance

17.1.1 Introduction

The system solution named GreenNet project is a prototype developed by STMicroelectronics in partnership with the Laboratoire d'Informatique de Grenoble (LIG) and few other partners. It has not been commercialized; however, the level of performance and efficiency reached by this demonstrator are still among the best published. GreenNet demonstrates today what could be

P. Urard (✉) • L. Varga • M. Vučinić • R. Guizzetti
STMicroelectronics, Crolles, France
e-mail: urard.pascal@gmail.com; liviu.varga@gmail.com; malishav@gmail.com; roberto.guizzetti@st.com

achieved with commercial, secured-IPv6 networks in a few years. The idea of a self-powered WSN has been identified in 2007 during a boosters-of-innovation session. In 2010, the RF power consumption reached such a low level that we started the project R&D. It has been held in ST in 2014, and closed at the end of 2015 in the university labs. The project enabled a vision of the upcoming IoT low-power nodes and the upcoming low-power roadmaps to serve this market.

GreenNet is a Wireless Sensor and Actuator Network enabling true End-to-End IPv6 connectivity, where some network nodes can be energy-harvested and capable of a fully autonomous operation. This is done without sacrificing security features required for most of the applications running on WSANs. Figure 17.1. depicts vision of the final application where autonomous GreenNet nodes communicate through a gateway with hosts located anywhere in the Internet.

17.1.2 Overview of GreenNet System Features

Bi-directional wireless network. GreenNet is a wireless network of sensors and actuators operating in 2.4 GHz ISM band. It primarily targets low-cost networks, such as home or building automation, inventory management,

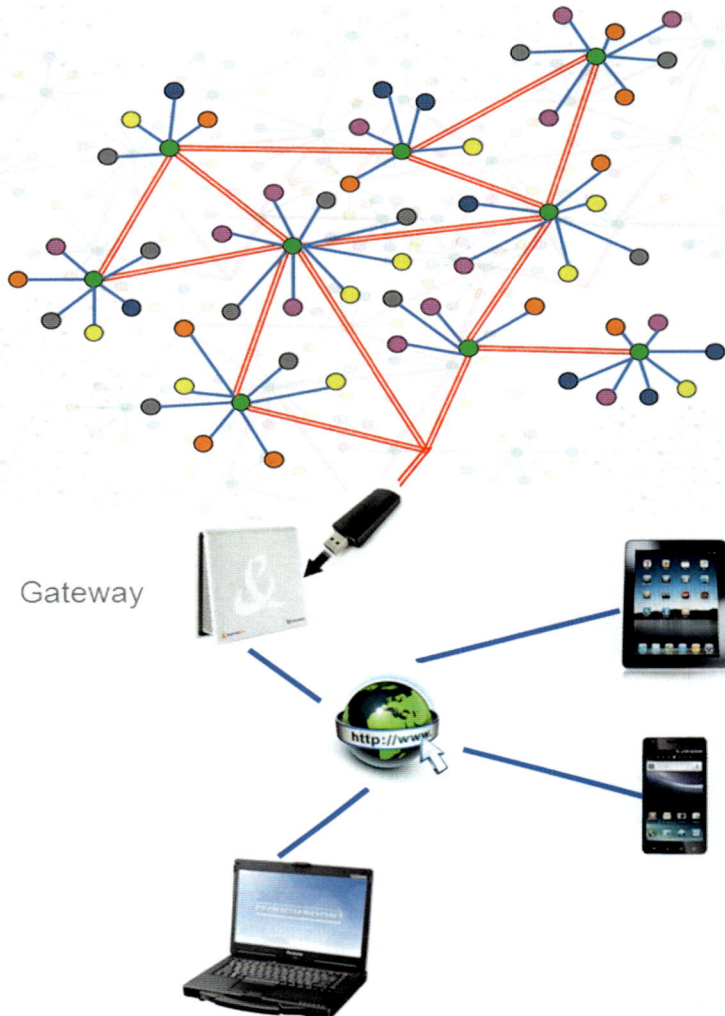

Fig. 17.1 Vision of a possible final application based on a GreenNet network

infrastructure monitoring, where wired solutions are neither cost effective, nor flexible. GreenNet ensures true bidirectional communication among nodes or between a node and end-user application (e.g. running on PC, smartphone or tablet), allowing remote sensor or actuator control.

IoT ready. GreenNet is designed to support true End-to-End connectivity over IPv6. This feature is foreseen as a key market shaper enabling on one hand, massive development of interoperable applications that are agnostic of wireless

technology, and on the other hand, various low-power hardware wireless solutions.

User-friendly experience. Based on IEEE 802.15.4 PHY+MAC standard, GreenNet targets self-organizing, scalable, self-healing networks making final user experience quite simple from installation or maintenance perspectives.

Security. GreenNet protects against network threats by implementing security features that guarantee data confidentiality, data integrity, host and data-origin authentication and network availability, both in the presence of external and internal attackers.

Based on open standards. The goal of GreenNet is to counteract the current market status quo where many proprietary solutions already take place, being themselves a de-facto limiting factors of market development. For that reason, GreenNet develops a fully optimized network stack based on open-source implementations.

Sustainable energy. GreenNet is designed to target operation of energy-harvested nodes. The goal is achieved by combining both advanced hardware design techniques (namely on the radio chip and energy charger chip level) and relevant standard and communication protocol options. Altogether, this yields a system where a temperature sensor node can, for instance, send a reading with a 5 min period, just by being exposed to 100Lux Compact Fluorescent Light (CFL) during 6 h a day. This amount of light corresponds to conditions typically found bellow an office desk. With more energy (i.e: more light), a node can be sustainable for years. Experiment has also demonstrated that such a node is able to operate during more than 75 days in the dark from a fully charged battery.

17.1.3 Network Topology at a Glance

Network description. GreenNet network is based on a cluster-tree topology depicted in Fig. 17.2. Nodes in the network correspond to different types of devices. The Network Coordinator (dark blue) controls the entire network and provides external connectivity. Network Coordinator is also called *Edge Router* (ER) or Sink. Leaf nodes (yellow) are either sensors or actuators. Leaf nodes can attach to the Network Coordinator or to Hop Routers (light blue) if the Network Coordinator is outside of the wireless range. Hop Routers (light blue) ensure connectivity between *Leaves* and the Network Coordinator allowing physical network expansion.

Self-organizing network. GreenNet is a Self-Organizing Network (SON), which means that

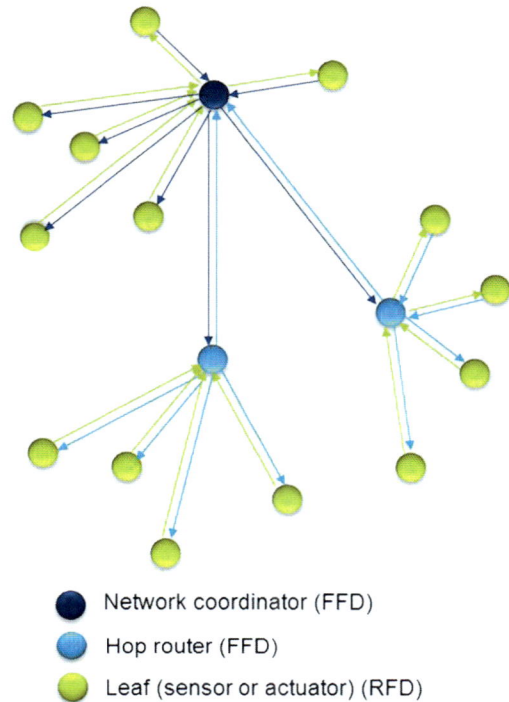

● Network coordinator (FFD)

● Hop router (FFD)

● Leaf (sensor or actuator) (RFD)

Fig. 17.2 GreenNet Network Topology

the network is built automatically, without any programming needed from the user. Each node in the network selects the best available parent based on the advertisements in its radio vicinity.

17.2 GreenNet Hardware

17.2.1 Introduction

We showed the layout of GreenNet hardware demonstrator boards in Fig. 2.5 in Chap. 2. The three core components of the board are: an STM32L1 (STMicroelectronics) microcontroller; an advanced low-power radio transceiver compliant with IEEE 802.15.4(/e) standard; and a low leakage battery charger. Apart from the three core chips, an STM32F101 microcontroller provides USB interface during software development. GreenNet board is powered by a low capacity Li-Ion cell, which is recharged from a

"GreenNet" node side connectors
(debug / USB for Edge Router)

"GreenNet" v2.1 chipset

"GreenNet" node internal PCB
with the rechargeable Li-Ion battery

"GreenNet" node with its photovoltaïc cell
and control LEDs

Fig. 17.3 GreenNet hardware demonstrator

small photovoltaic panel. Three LEDs are used for control and debug.

For demonstration purposes, the board includes various ST sensors: accelerometer, magnetometer, gyroscope, pressure, temperature, and light sensors. An NFC EEPROM (M24LR64E-R datasheet) is added to demonstrate user-friendly bootstrapping operations. GreenNet can support additional sensors through the use of a dedicated port extension. The same hardware board (see Fig. 17.3) can be programmed as a leaf node, hop router or network coordinator node (also called *Edge router*). The same board can as well be programmed with several applications using different MEMs.

17.2.2   Functional Description

17.2.2.1   Microcontroller

GreenNet boards use the STM32L1 microcontroller to control and communicate with the radio transceiver, sensors and actuators. STM32L1 is an ultra-low-power 32-bit microcontroller based on ARM Cortex-M3 core. We chose the version with 256 KB Flash memory and 32 KB RAM for data. This part was selected for its low power properties.

17.2.2.2   Radio

GreenNet boards communicate wirelessly through a low-power radio transceiver, compliant with IEEE 802.15.4 standard and its "/e" amendment. The radio chip implements the analog interface operating in 2.4 GHz band as well as a dedicated hardware in charge of the MAC operations as defined by the IEEE 802.15.4 standard. Advanced MAC features implemented in hardware lighten the load of the microcontroller. In addition to the nominal 250 Kbits/s data rate proposed by the IEEE 802.15.4 standard, the chip also provides a proprietary extension to increase the data rate up to 2 Mbits/s.

17.2.2.3   Power Management

Power Management (PM) delivers power supply to the rest of the system: main MCU (STM32L1) & MEMs, 2.4 GHz radio. It also plays the role of battery charger: it insures solar-cell management, battery charging and battery safety (Todeschini and Dimensionnement energetique de reseaux de capteurs ultra-compacts autonomes en energie. PhD thesis, Supelec 2014). The battery is thus connected directly to this chip (see Fig. 17.4). PM chip limits the charging current to a maximum of 2 mA to preserve the

Fig. 17.4 GreenNet node
power management
architecture

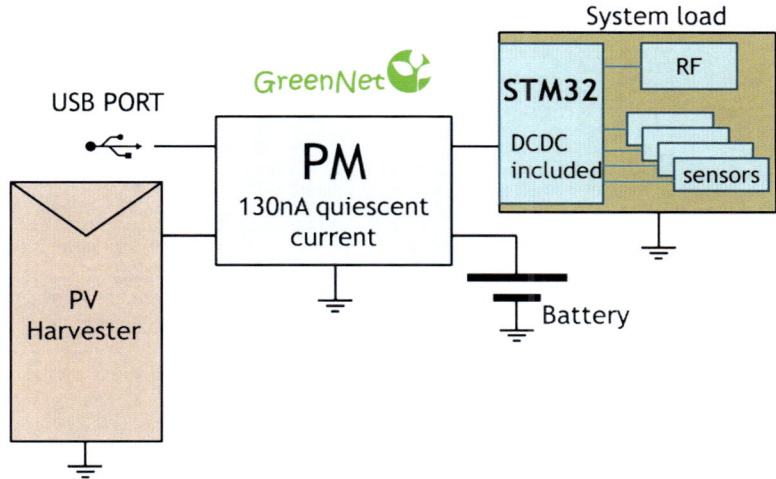

Fig. 17.4 GreenNet node power management architecture

battery. It stops charging when battery reaches 3.2 V (high cut-off) and stops the full system if battery goes lower than 2.35 V (low cut-off). The PM core area is 0.4 mm^2 in ST M10ULP (90 nm embedded NVM CMOS process) to allow a possible future integration in a SoC. PM quiescent current in dark conditions is 130 nA. The energy harvester is a photovoltaic (PV) cell: a standard amorphous cell from a PV cell provider. Its active area is 16 cm^2, and its open circuit voltage is $V_{oc} = 4.4$ V at 200Lux. It has been co-designed with our PM to provide the best possible energy efficiency in the critical zone: between 70 and 200 Lux CFL light (indoor conditions). The maximum energy efficiency (MPP) is obtained around 200Lux. Charging voltage varies from 2.4 V under 20Lux, to 3 V under 200Lux, demonstrating a global efficiency of the full system (PV + PM + battery) above 86.4% at 200Lux (Todeschini et al. 2013). Given the good results of this architecture, the added value of a more complex MPP algorithm has not been demonstrated. PM chip is able to perform battery charging at a minimum current of 290 nA, obtained with only a few Lux. This corresponds to ~1 μW depending on the voltage of the battery. This is not sufficient to operate in a sustainable way but it enables to extend the

lifetime of an operating node placed in harsh conditions during a long period.

When a node is connected to USB for debug purpose, only STM32F1 microcontroller is powered by USB (default mode) unless a connector allows the PM to charge (optional mode).

17.2.2.4 USB Interface

The USB interface is built from a physical USB connector and an ARM-based 32-bit microcontroller (STM32F101) that implements the bridge towards the main STM32L1. This interface supports the following functions:

– Energy-supply of the node during development and debug phase.
– Bidirectional communication with PC for on-chip and serial port debugging, bootstrapping configuration, and flashing of STM32L1 program memory with firmware image.
– Connection to the gateway or a PC, in the case the node is programmed as an Edge Router.

17.2.2.5 Sensors

GreenNet node includes six types of sensors with key metrics summarized in Table 17.1.

Table 17.1 Summary of the embedded sensors

Type	Reference	Digital/analog output	Dynamic ranges	Accuracy
3D Gyroscope	L3GD20	Digital	±250 dps ±500 dps ±2000 dps	16-bit-rate value data output
Temperature sensor	STLM20	Analog	−40 to 85 °C	±1.5 °C at NTC = 25 °C <±2.5 °C at ETC
3D Accelerometer 3D Magnetometer	LSM303DLHC	Digital	Acceleration ±2 g/±4 g / ±8 g/±16 g Magnetic field ±1.3 Gs ±1.9 Gs ±2.5 Gs ±4.0 Gs ±4.7 Gs ±5.6 Gs ±8.1 Gs	16 bit data output
Pressure sensor	LP331AP	Analog	260–1260 mbar	0.020 mbar rms
Light sensor	Custom design	Analog	20 lux to 1 sun	±15% (between 20 lux & 7000 lux, indoor)

17.2.2.6 NFC

An **NFC** EEPROM (M24LR64E-R datasheet) enables user-friendly read/write access to the main memory. The NFC transceiver is based on ISO-15693 communication standard. Network pre-shared (security) keys, radio channels and similar parameters can be configured before the deployment over NFC. For instance, a user can copy the parameters from the *Edge Router* with a smartphone and write them to each node that is about to join the network.

17.3 Energy-Sustainable Operation: Intersection of Network Protocols and Low-Power Hardware

17.3.1 Duty-Cycled Operations

Low-power hardware design is a necessary but not a sufficient condition to reach sustainable operation of a network with energy-harvested nodes. Optimization of average current consumption takes place both at the hardware and network level. Romaniello's PhD thesis (Romaniello et al. 2015) describes explorations

and tradeoffs for energy efficient protocols enabling energy-harvested Wireless Sensor Networks.

For communication, GreenNet nodes use the IEEE 802.15.4 standard in its beacon-enabled mode that enables efficient **duty cycling.** They sleep most of the time and wake up only to perform their application task, such as sensing of a physical quantity and the radio transmission of the measured value. Sleep periods correspond to the low current draw, mainly caused by leakage currents while the active periods require more current due to the involvement of the microprocessor and radio transceiver. The average current consumption depends on the duty cycle—the more time node spends sleeping, less current it consumes on average. We depict this operation for two types of network devices in Fig. 17.5.

Duty cycles can typically vary between 10% (typical value for routers) down to 0.01% for leaf nodes.

In order to reach the balance between consumed and harvested energy, the total energy consumption of the node while sleeping must be in the range of µAs. To reach this goal, not only the radio and all the peripherals including sensors must be put in the lowest possible

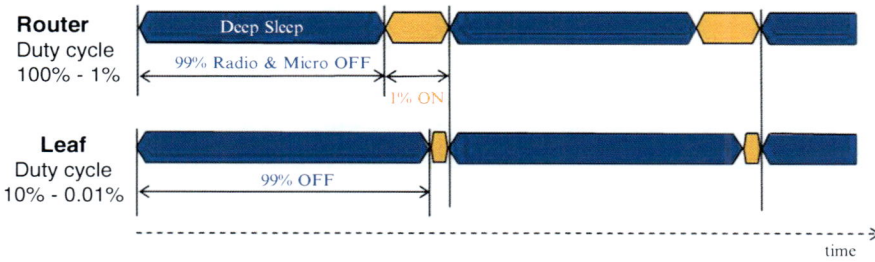

Fig. 17.5 Duty-cycled network operation

Fig. 17.6 Duty-cycle/
luminosity dependency
(simulation)

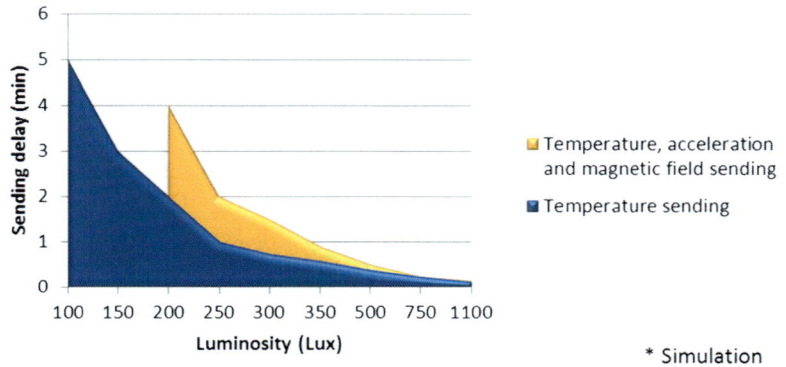

consumption mode, but also the microcontroller. This has an important impact on the network stack, as the hardware delays of these operations can be significant. We always need to maintain positive (among a certain duration) the energy balance between harvested energy while sleeping versus the energy needed for an active operation, as depicted in Fig. 2.4 in Chap. 2.

17.3.2 Relationship Between Light Conditions and Duty Cycle

Light conditions impact the allowed duty cycle. The less light there is, the longer the node has to sleep to recharge its battery. To enable autonomy, power consumption has to be balanced by harvested power on a daily or week basis, depending on the use case. If a node has to send some measurement at a regular time interval, it has to stay awake some period to sense and transmit. Typical value would be around 30 ms but it varies with different sensors and application requirements. When the node goes back to

sleep, the duration of the sleep period must be long enough to recharge the battery for the subsequent operation. The interval depends on light conditions and the type of measurement, as estimated in Fig. 17.6.

Even though GreenNet targets indoor applications with light intensity as low as 100 Lux, there could be cases where energy provided by PV panel at this intensity is not enough for a particular application. In those scenarios, nodes can either be connected to an external power source or to a larger battery. GreenNet communication stack allows the networks to be comprised of both duty-cycled and always-on nodes.

17.4 GreenNet Software Communication Stack

17.4.1 Introduction

In order to reach true end-to-end IPv6 connectivity and still operate in energy-constrained conditions, GreenNet uses a protocol stack

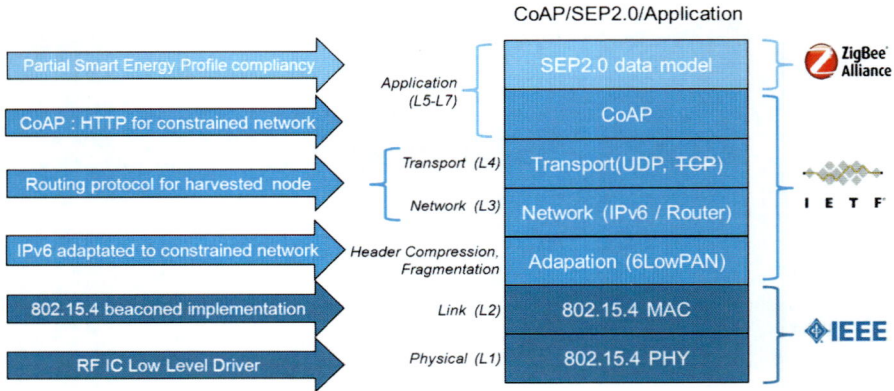

Fig. 17.7 GreenNet protocol stack based on Open Standards

built upon **6LoWPAN** (Montenegro et al. 2007) that optimizes IPv6 signaling overhead. Figure 17.7 summarizes the GreenNet protocol stack and the main points addressed in its implementation.

The implementation of this communication stack was developed under Contiki environment (Swedish ICT; Dunkels et al. 2004). The following sections provide a brief overview of each abstraction layer and how it relates to GreenNet.

17.4.2   I.E. 802.15.4 PHY Layer (L1)

IEEE 802.15.4 is arguably the most prominent standard (IEEE 2011) in low-power technology and the one that was chosen for GreenNet project, as outlined in Chap. 3 of Vučinić's PhD thesis (Vučinić et al. 2015c) and detailed hereafter. IEEE 802.15.4 standard specifies multiple physical layers that can be used in different parts of the world, depending on local regulations. The physical layer in the ISM band at 2.4 GHz guarantees worldwide use free of any licensing requirements, which is in practice the most widely deployed and naturally supported with GreenNet's radio transceiver.

The 2.405–2.480 GHz band is split into 16 frequency channels that are 5 MHz apart. Each channel is only 2 MHz wide, with the remaining band used as a guard against adjacent-channel interference. The band at 868 MHz with better propagation characteristics is also popular but is

only available in Europe and has a single frequency channel available.

IEEE 802.15.4 physical layer at 2.4 GHz uses DSSS technique for robustness: each 4 bits of data are encoded as 32 chips (physical bits) (Palattella et al. 2013). This helps recover from errors caused by narrow band interference. O-QPSK modulation is then used and results in physical rate of 2 Mchips/s and effective data rate of 250 kb/s.

Before the upper-layer information can be exchanged, transmitter starts by sending a preamble, a pre-defined sequence of ones and zeros that allows the receiver to synchronize. Transmission of the preamble lasts 128 us, and is followed by start frame delimiter (SFD), another pre-defined sequence. SFD signals that the subsequent byte corresponds to the physical-layer payload. First byte of the payload indicates the length of the encapsulated radio frame. IEEE 802.15.4 specifies that the maximum length frame, i.e. link-layer Maximum Transmission Unit (MTU), is 127 bytes.

17.4.3   *IEEE* 802.15.4 Beacon-Enabled MAC Layer L2

17.4.3.1   Superframe Concept

Harvested nodes have very stringent constraints in terms of available energy. Enabling IPv6 communication in this case requires the selection of an ultra-low-power MAC protocol. This point

Fig. 17.8 Time-beaconed superframe as described in IEEE 802.15.4-2006

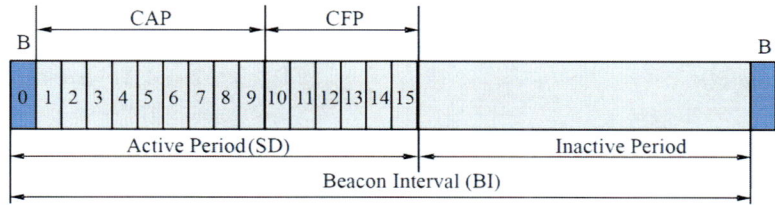

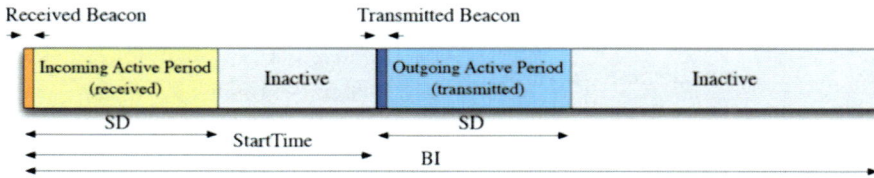

Fig. 17.9 Incoming and outgoing superframe structure in IEEE 802.15.4 standard

together with the main requirement of using only standardized solutions at all levels brought to the choice of using the CSMA/CA slotted option part of IEEE 802.15.4 beacon-enabled standard (IEEE 2011). At MAC (L2) level GreenNet stack is therefore using Superframe structure as specified in the IEEE specification. See Fig. 17.8.

The standard divides time in Superframes with regular beacons sent at a Beacon Interval (BI). Superframe is itself divided in

- an Active Period (SD) during which communication takes place with CSMA/CA mechanism for collision avoidance during
  - Contention Active Period (CAP)
  - Contention Free Period (CFP) which remains unused in GreenNet
- an Inactive Period during which the node may sleep.

Beacon Intervals (BI) and *Active Period* (SD) durations can be changed according to the IEEE-802.15.4 specifications between 15.36 ms up to ~4.2 min[1] allowing the use of various duty-cycles (ON/ON + SLEEP ratio) which best adapt to available energy.

---

[1] Minimum Active Period = 60 Symbols × 16 μs × 16 slots = 15.36 ms
Maximum Beacon Interval = 2^{14} Symbols × 16 μs × 16 slots = ~4.2 min.

17.4.3.2 Network Association and Disassociation

Also part of the MAC layer is the association and disassociation procedure (how a device joins the network) which is specified by IEEE 802.15.4. The network association implemented in GreenNet stack respects this standard.

17.4.3.3 Incoming and Outgoing Active Period Allocation

Each *Network Coordinator* or *Router* sends its own beacons to its sons further down in the tree (see also Sect. 17.1.3), and can decide where to place its active period into the *Beacon Interval* (BI). While a *Leaf node* only receives incoming beacons from its parent router, a router first receives incoming beacons from its parent (*Network Coordinator* or other *Router*) and then sends beacons to its children (*Leaf nodes* or other *Routers*), see Fig. 17.9.

IEEE 802.15.4 does not specify how the active period can be allocated; this is left to the implementation choice (Active Period Manager). In order to guarantee by construction that there is no overlap between active periods of each *Router* in the network, an algorithm has been implemented. It is based on Zigbee *Distributed Address Allocation Algorithm* (**DAAA**). *Active Period* start time for each potential router is a function of hereafter $Cskip(d)$ computed according to the following formula

(*Cm*: max number of children, *Lm*: max depth, *Rm*: max number of routers, *d*: actual depth of the router):

$$Cskip(d) = \frac{1 - Rm^{Lm-d}}{1 - Rm} \, assuming \; Cm$$
$$= Rm \; with \; Rm > 1 \quad for \quad 1 < d$$
$$\leq Lm$$

The above algorithm is a centralized mechanism, which does not take into account, the radio range of the nodes. More flexible active period allocation algorithms can be implemented.

17.4.3.4 Data Exchange Between Devices

It should be noted that, based on IEEE 802.15.4 beaconed standard, *Network Coordinator* or *Router* never initiate a transmission, only reply to solicitations from children nodes. They notify children nodes by signaling in beacon frames their corresponding address, together with a flag that they have a packet pending. Once the children node wakes up and parses the beacon, it can solicit the data packet by transmitting a *data request*. Such a mode of operation allows the children devices to skip beacons and sleep extensive periods of time. In case of traffic going upwards (from children towards the sink), the nodes simply transmit the packet within the corresponding CAP, using the CSMA/CA algorithm.

17.4.4 Routing Layer L3

17.4.4.1 Introduction

As this was introduced in Sect. 17.1, the GreenNet nodes can be of the following types (see Fig. 17.10):

- *Edge Router* or *Network Coordinator* is the central node (a.k.a sink). It is connected to the gateway of the network (this is a FFD)
- *Routers* act as a relay for the other nodes, to propagate the data in the whole network. They can also themselves sense, capture and send their own data (this is a FFD)
- *Leaf nodes* are at the bottom of the tree, sensing and sending their own data (this is a RFD)

Routing consists in finding the best route (according to some objective function) from any network node to the edge router. Chapter 7 of Varga's PhD thesis (Varga et al. 2015b) describes the innovative routing solution in GreenNet, which is particularly suited for harvested nodes. The solution will be described hereafter.

17.4.4.2 Limitations of RPL Routing Protocol for Harvested Nodes

IETF Routing Over Low power and Lossy networks (ROLL) Working Group specified RPL (Winter et al. 2012) routing protocol.

Fig. 17.10 IP based GreenNet network example

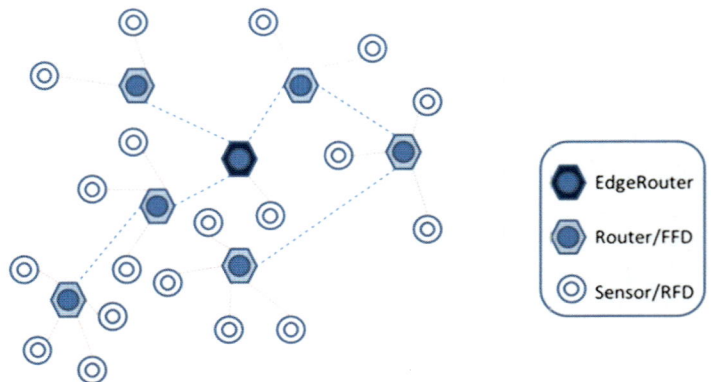

Unfortunately, RPL is not well suited for energy-harvested nodes having very stringent energy constraints.

Major drawbacks of RPL are (see (Clausen et al. 2011) for further details):

– DIO (DODAG[2] Information Object) are *descending* messages sent periodically to create upward routes. With RPL these messages may be fragmented
– RPL is engineered with preamble sampling in mind i.e. DIO are sent with a Trickle timer to avoid collisions and lower their global sending rate, which solves the problem only partially and makes the protocol unnecessarily complex
– In storing mode[3] problem of memory (routing table size) and DAO[4] message footprint
– Complexity (executable RAM size: ~50 Kbytes[5])
– Routers MUST be always-on (i.e. you cannot use duty-cycled routers) to keep track of the radio environment to ensure proper trickle functioning.

17.4.4.3 Optimized Solution for Routing of Harvested Nodes

– An optimized solution was found by implementing a cross layer (**L2/L3**) construction of the Cluster Tree (La et al. 2013), exploiting the best of both **RPL** (Winter et al. 2012) and **LOADng** (Clausen et al. 2015) routing protocols. This innovative implemented solution
– Keeps the network structure built by **RPL** (Collection Tree)
– Uses very compact code with few parameters

– Sends routing info in beacon payloads (piggyback)
– Keeps *Routers* by construction duty-cycled
– Makes *route request* initiated ONLY by the *Edge Router*
– Avoids B*roadcast* messages issue
– The following sub-sections give more details on how this innovative routing solution is working.

17.4.4.4 Upward Route Request

In LOADng, when a node does not know the route to a destination, it initiates a route request (RREQ), causing a flooding in the network with broadcast messages (as discussed above). In GreenNet stack, only the Edge Router (ER) can initiate a route request.

As can be seen from Fig. 17.11 below, if a node A does not know a destination address B then:

– It sends data packet up to its coordinator until it reaches ER (action 1 in Fig. 17.11)
– If the ER does not know the destination either, it will send down a broadcast RREQ which is recursively broadcasted by each router of interest[6] (action 2 in Fig. 17.11)

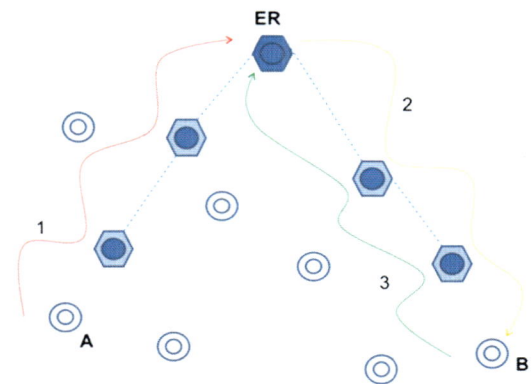

Fig. 17.11 Route request customization in GreenNet stack

---

[2] DODAG : Destination Object Directed Assisted Graph.

[3] A node uses routing tables for forwarding packets to leafs.

[4] DAO (Destination Advertisement Object): upward messages sent to create downward routes (see Winter et al. 2012).

[5] We did not succeed in implementing RPL and ContikiMAC (Dunkels et al. 2011) within a typical 8 KB RAM low End MCU.

---

[6] During this flooding, each router creates also a reverse path inserting in its routing table the next hop to ER (actually next-hop in upward direction is always the parent).

Fig. 17.12 For broadcast
(D-1) router behaves as a
proxy for its children
leaves

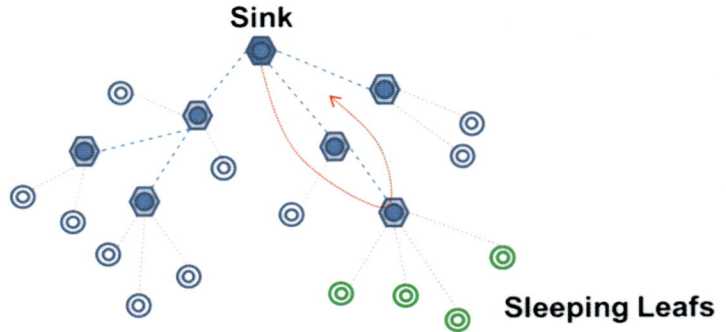

- When destination node receives it, it replies with a route replay (RREP) that will go up to the ER through the same path[7] (action 3 in Fig. 17.11)
- Once the RREP arrives to the ER the path has been properly created (on all nodes of interest)
- From this time on, each time A wants to send a packet to B the same path (through ER) as created by this RREQ/RREP, mechanism will be used: this avoids the overhead caused by any routers broadcasting RREQ to anyone (as required in LOADng).

17.4.4.5 Downward Routes Toward Sleeping Nodes (Broadcast and Unicast)

Wireless Sensor Networks are typically built to sense, collect and transmit information to a *Sink* (i.e. *Edge Router*) which is then possibly connected to a gateway. The inherent tree of WSN is used to easily collect information up to the sink.

It is more complex to send data down into the tree (i.e. configuration request to a leaf and/or control an actuator) especially when a leaf is most of the time sleeping.

In the downward routes, we have to further categorize in:

- Broadcast like the RREQ of previous section

- Unicast i.e. when we want to send a data (or a configuration) to a single node.

For Broadcast (see Fig. 17.12), GreenNet routing solution is such that the router at depth (D-1) in the tree (i.e. the latest one before the sleeping *Leaf*), behaves as a proxy, i.e. *replies on behalf of its sleeping children*, to the RREQ.

For Unicast, implemented solution tends to avoid the drawback described hereafter: when packets have to be sent to a sleeping leaf (i.e. a low duty cycle leaf programmed for skipping beacons), only the first packet (i.e. that one received immediately after the beacon) is likely to arrive to destination. The next ones have a high probability to be lost (even though a buffer mechanism could be applied which anyway would be limited by the RAM size).

GreenNet implemented solution tends to avoid this limitation (see Fig. 17.13): when a packet arrives to a sleeping leaf, RDC layer stops skipping beacons for a certain (programmable) number of times (assuming that other messages could arrive after the first one just received).

17.4.5 Application Layers (L5–L7)

17.4.5.1 CoAP

IETF CoRE working group standardized CoAP (Shelby et al. 2014) to answer the needs of IoT devices and replace the more energy hungry HTTP. CoAP runs over User Datagram Protocol (UDP) rather than over Transmission Control

---

[7] During this phase each router creates also the downward route for the destination of interest.

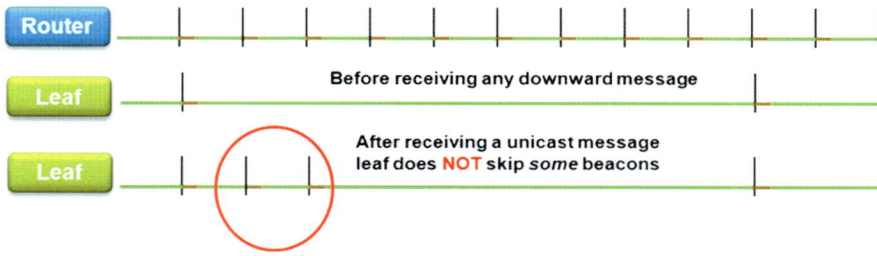

Fig. 17.13 After receiving Unicast message, some beacons are NOT skipped

Table 17.2 Comparison of communication overhead between HTTP and CoAP

	HTTP/TCP	CoAP/UDP
# Messages	9	2
# Bytes	681	111
RTT (ms)	7.2	4.4

Protocol (TCP) that is used in the traditional Internet. We show an example in Table 17.2 how CoAP/UDP is more efficient than HTTP/TCP, in terms of the number of messages, total byte overhead and round trip time (RTT) for the same application-level transaction.

CoAP is not a blind compression of HTTP, although it is commonly referred to as such due to the same RESTful design and purpose. Instead, CoAP implements a subset of HTTP features relevant to IoT devices and further specializes for constrained environments by leveraging typical traffic patterns and by minimizing overhead.

CoAP leverages the traditional client-server architecture where a client interested in a given resource makes a request towards the server, hosting that resource. The server prepares a response that potentially encapsulates a resource representation corresponding to the requested resource. For example, a request to resource "temperature" would result in response containing temperature measurement in the payload. There could also be a request to change the value of some resource on the server, in which case the response typically acknowledges the change. The separation of client and server roles in IoT environments is not as straightforward as in the traditional Internet. The same device often executes both roles, depending on the application

semantics. The term "server" should by no means be associated with powerful machines hosting HTTP servers in the Internet, as in our context CoAP "server" corresponds to a GreenNet board exposing its resources to CoAP clients (e.g. gateways, smart phones, or other GreenNet nodes).

GreenNet uses two open source implementations of CoAP. **Erbium** ("C") is used for GreenNet nodes while **Californium** ('Java') is used for the gateway.

17.4.5.2 ZigBee Smart Energy Profile (SEP2.0)

Zigbee Alliance specified SEP2.0 (ZigBee Alliance), which is a software layer standardizing, at the application level, data exchanges between the network nodes so that these exchanges remain independent of the lower-layer protocols. Different profiles were standardized for various market segments. A prototype subset of the SEP2.0 *Function Sets* has been implemented under Contiki for GreenNet nodes, while a more exhaustive version is used for the gateway, running in Java.

When a new node joins the network, it triggers a SEP2.0 *Node Registration* phase towards the gateway ("End Device" message). Then, each node sends upwards few status parameters, one of them being its parent address. The information from each node will be used to update the graph in the gateway GUI and corresponds to the *Status Information Exchange Discovery* phase. This is followed by *Function Sets Discovery* phase, *Resource Discovery* phase and *Data Exchange* phase.

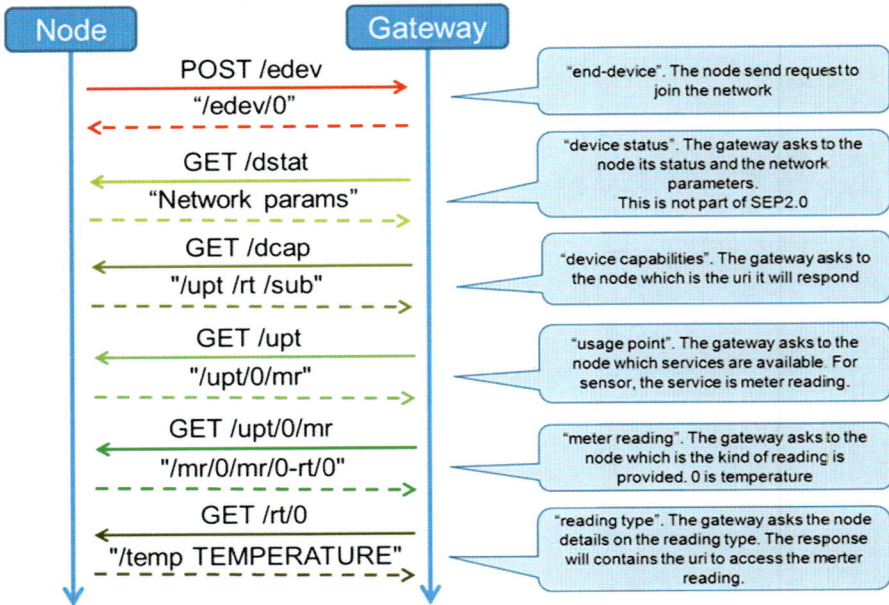

Fig. 17.14 SEP registration procedure example

An example of SEP registration phase is described in Fig. 17.14.

17.5 Security

As a wireless network, GreenNet is prone to both passive (e.g. eavesdropping) and active (e.g. packet injection, replay) attacks. Network security protocols and mechanisms use cryptographic techniques to protect against such threats, for example by encrypting and/or authenticating packets. By protecting information at different abstraction layers and by controlling access to cryptographic keys, it is possible to selectively place trust on different nodes in the network.

For example, security at L2 protects the communication over the radio channel and provides the first level of protection against *external* attackers. Since GreenNet nodes typically send data to the gateway, that means that every node on the multihop path has to decrypt and encrypt the packet as it travels towards its destination.

For that reason, we refer to security at L2 as hop-by-hop security. An attacker can physically compromise one of the nodes in the network and therefore obtain access to the information originating at other nodes through the compromised node. We refer to such an attacker as an *insider*. In addition, an insider can easily tamper with existing information flows or inject false readings to make damage to the system.

Defense against *inside* attackers is end-to-end security that protects information travelling end-to-end, such that no intermediary node on the path has access to it. In practice, it means that parts of the packet that correspond to the application are encrypted and authenticated with cryptographic keys different from those used at L2. Figure 17.15 depicts how security at different layers protects information flows in GreenNet networks.

Additionally, GreenNet may leverage an advanced security scheme OSCAR that protects application data, while it is stored at intermediaries, such as untrusted Cloud servers.

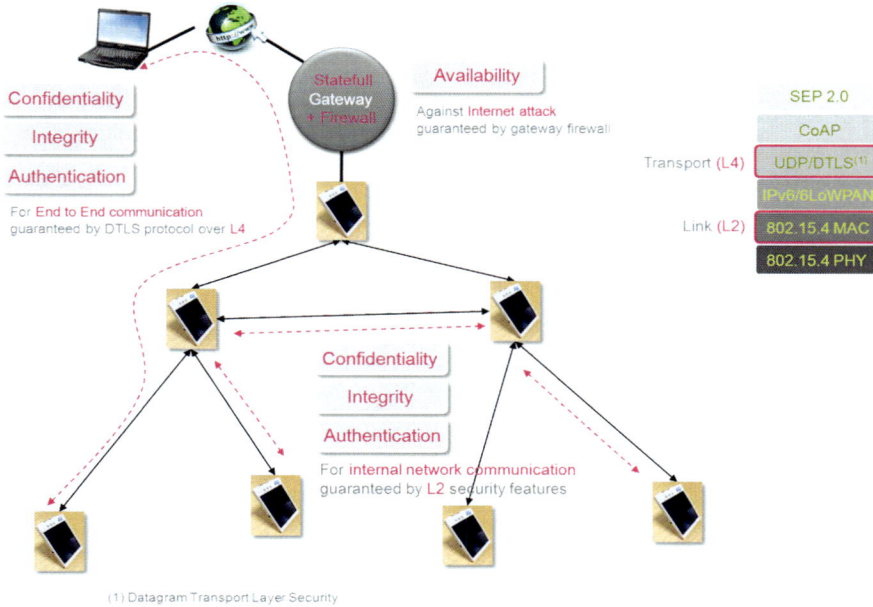

Fig. 17.15 Security embedded in L2 (hop-by-hop) and L4 (end-to-end) layers

17.5.1 Hop-by-Hop Security (L2)

Security mechanisms at L2 are tightly bound to the radio technology, which in GreenNet case corresponds to IEEE 802.15.4 standard. The standard completely relies on symmetric cryptographic primitives to provide confidentiality, data origin authentication and protection from replay attacks on a per frame basis. The standard delegates host authentication and key exchange functions to upper layers of the network stack. Since hop-by-hop security is only the first layer of protection, GreenNet uses group keys: all nodes in the network share the keys used for protection of radio frames.

The basic cryptographic primitive of IEEE 802.15.4 standard is Advanced Encryption Standard (AES), a block cipher that encrypts blocks of 16 bytes. To use AES in practice, it is necessary to apply different modes of operation for protection of messages longer than 16 bytes. IEEE 802.15.4 standard uses the CCM mode (AES-CCM) to encrypt and authenticate radio frames in the same time.

CCM is a particular mode of operation that uses CTR mode for encryption and CBC-MAC for authentication. It encrypts and/or authenticates a sequence of plaintext bytes.[8] When applied to a radio frame, this means that CCM can encrypt the MAC payload while keeping the MAC header intact (Sciancalepore et al. 2016). When used to authenticate a radio frame, CCM calculates a Message Integrity Code (MIC) over the complete frame. This MIC is truncated to the desired length (4, 8 or 16 bytes), and appended at the end of the radio frame.

Each frame secured with a given key must use a different nonce. Encrypting two frames with the same nonce has severe consequences on security: plaintext of both frames may be easily recovered. By constructing the nonce with a monotonic counter, it is possible to ensure replay protection for two communicating nodes.

---

[8] While the upper bound on message length for CCM does exist, for practical purposes in IEEE 802.15.4 WSN it can be considered arbitrarily large.

Fig. 17.16 Message exchanges during DTLS handshake to agree on a common key

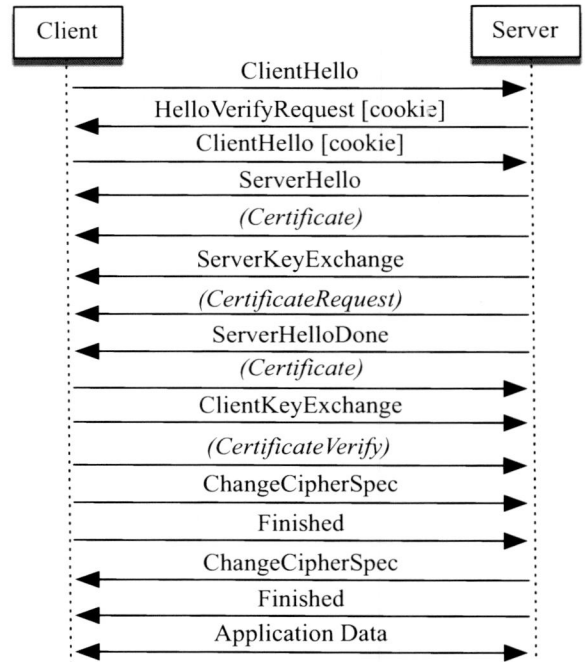

17.5.2 End-to-End Security

Securing end-to-end traffic is often achieved by transferring data over a secure logical channel between the two communicating end-points. The de-facto security standard for the Internet application traffic is SSL and its IETF successor TLS (Dierks and Rescorla 2008). TLS was designed for client-server applications that operate over a reliable transport. To establish a secure channel, a client and a server first perform the TLS handshake during which they authenticate each other and derive the symmetric keys to use during the session.

DTLS (Rescorla et al. 2012) is an extension of TLS for datagram transport and runs over UDP rather than TCP. Like TLS, DTLS protects the payload with encryption and authentication. Figure 17.16 illustrates the exchanges during a DTLS handshake.

Our evaluations of DTLS in the context of GreenNet and other MAC protocols (Vučinić et al. 2015b), however, suggest that its usage is acceptable for applications for which a DTLS

handshake is performed a limited number of times during the node lifetime due to the significant performance issues.

17.6 Protecting Application Data

In addition to end-to-end security, many use cases require access control and authorization (Seitz et al. 2016), but sensor nodes are not access control decision makers for the data they generate—access control is rather a policy of the network owner (Varga et al. 2015a). Recognizing significant performance issues of the DTLS handshake in radio duty-cycled networks, we have jointly addressed the problems of end-to-end security and authorization by proposing OSCAR, a stateless content-centric security architecture (Vučinić et al. 2015a). The main idea behind OSCAR is to protect the data rather than the communication channel, which allows the secured data to be cached at intermediaries without the need for decryption (Vučinić et al.

2015c). OSCAR relies on Authorization Servers that provide clients with Access Secrets required to request resources from constrained nodes. Nodes reply with the requested resources that are signed and encrypted. The scheme intrinsically supports multicast and asynchronous traffic. The architecture achieves end-to-end authenticity even in the presence of application-level gateways that do not allow direct communication from the outside of a sensor network, for security reasons. OSCAR leverages secure channels established by means of DTLS and stays compatible with DTLS-only networks, but uses it in a way that does not require the repetition of expensive handshakes (Varga et al. 2015a).

17.7   Technical Evaluation

17.7.1   Total Power Measurement

We have seen in Sect. 17.3 that the GreenNet node is able to charge its embedded battery starting from 290 nA—corresponding to 20lux. However, the capability to operate harvesting in harsh conditions is not sufficient to ensure sustainability. It has to be combined with *cross-the-board* optimizations:

– optimization of the hardware (Sect. 17.2)
– optimization of the software stack described in Sect. 17.4.

The system total power consumption can be measured in real time, thanks to some board debugging capability (Fig. 17.3).

Figure 17.17 shows a typical active part of a typical super-frame for a self-powered temperature sensor node. We can detail wake-up (A), sensor measurement (B) with in parallel radio calibration, auto-synchronization through beacon reception (C) with in parallel IPv6 frame preparation and security encryption, channel sensing using CSMA-CA in 802.15.4 (D) and 802.15.4 frame Tx (E), frame acknowledgement by the network (F), next wake-up preparation and back to sleep (G). The (C) section is longer than beacon reception to anticipate low oscillator errors. Despite the length of the active period, it only represents an integrated charge of 192 µC, so around 576 µJ

Fig. 17.17 802.15.4-beaconned active part details

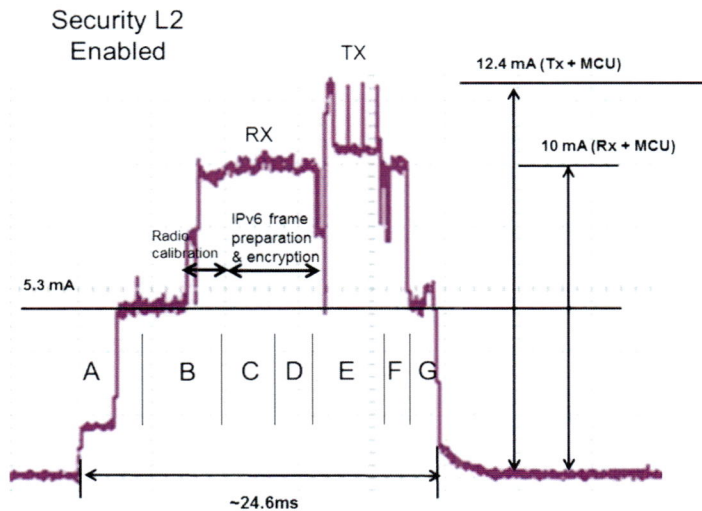

17.7.2 Contributors to Energy Consumption and the Cost of Hop-by-Hop Security

We evaluate the main contributors to the overall energy consumption in steady state over a 6-h period, using software estimation techniques (Dunkels et al. 2007). We focus on the energy spent during four main phases:

1. The energy spent during sleep mode (leakages of the MCU and the rest of the board)—Low Power Mode (LPM);
2. The energy spent during active computations (CPU);
3. The energy spent while the radio transceiver is in receive mode (RX);
4. The energy spent while the radio transceiver is in transmit mode (TX).

Then we compare the cost of security in different security modes (ENC-MIC-xx). The higher the ENC-MIC value, the higher the security level and lower the probability of blind guessing the MIC value. All evaluated modes include both encryption and authentication.

We note in Table 17.3 that as much as 69% of energy is spent during sleep mode. We see that the percentage of energy spent on radio communication increases with the security level. A higher security level corresponds to a longer MIC and a larger signaling overhead, which directly affects the radio transmission/reception cost. Cryptographic processing causes a notable increase in CPU usage from No Security to ENC-MIC-32 modes. The LPM drop observed with increasing security level is due to the overhead of time (and so overhead of energy consumption) during other phases.

Figure 17.18 presents the average power consumption measured over the experiment duration of 6 h. It is interesting to note in Fig. 17.18 the amount of power drawn while the radio transceiver is in receive mode. Although the application executing during the experiments induced typical convergecast traffic, with no packets going in the downward direction to the nodes, power drawn while receiving is three times higher compared to power drawn while transmitting. Our node sleeps most of the time, but just before transmitting the sensor reading, it needs to resynchronize to the rest of the network by receiving a beacon frame at a precise interval. Delays caused by hardware imperfections, such as crystal oscillator warm up time, radio

Table 17.3 Steady-state contributors to energy consumption of GreenNet temperature sensor

Security level	LPM (%)	CPU (%)	RX (%)	TX (%)
No security	69.77	15.77	10.92	3.54
ENC-MIC-32	68.57	16.35	11.30	3.78
ENC-MIC-64	67.62	16.63	11.69	4.06
ENC-MIC-128	67.11	16.80	11.81	4.28

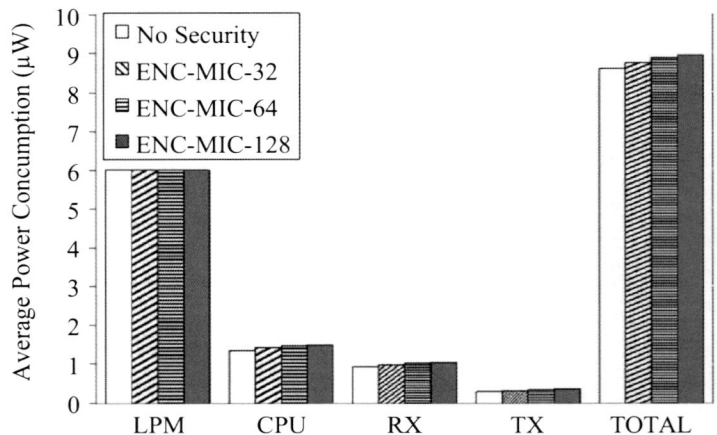

Fig. 17.18 Average power consumption of GreenNet temperature sensor

Fig. 17.19 Energetic cost of hop-by-hop (IEEE 802.15.4) security for GreenNet temperature sensor

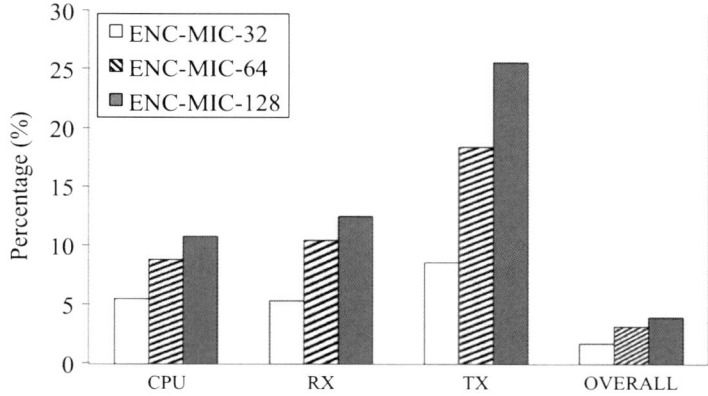

calibration, clock drifts, and similar, induce margins that are an important component in the overall power consumption. Roughly, only 15% of the receive mode consumption comes from the actual physical reception of a beacon frame, whose size increases with the security level, while the rest is due to idle listening.

We summarize in Fig. 17.19 the overall energetic cost of security for a GreenNet temperature sensor. Overall cost of IEEE 802.15.4 hop-by-hop security in our scenario ranges from 1.75% to 3.96%, depending on the security level. For GreenNet, this result directly corresponds to the requirement that 1.75–3.96% extra energy needs to be harvested from the environment.

Glossary

AES-CCM Advanced Encryption Standard (Counter with CBC MAC)
BI Beacon Interval
CoAP Constrained Application Protocol
CoRE Constrained RESTfull Environments
CSMA/CA Carrier Sense Mulitple Access with Collision Avoidance
DAAA Distributed Address Allocation Algorithm
DAO Destination Advertisement Object
DIO DODAG Information Object
DODAG Destination Object Directed Assisted Graph
DTLS Datagram Transport Layer Security
ER Edge Router
FFD Full Functionality Device
GUI Graphical User Interface
IETF Internet Engineering Task Force
IoT Internet of Things
LLN Low power and Lossy Network
Lux SI unit for luminous emittance
LOADng 6LoWPAN Ad hoc On-Demand distance vector routiNG
MAC Medium Access Control
NDEF NFC Data Exchange Format
NFC Near Field Communication
PAN Personal Area Network
PHY PHYsical link (ie: the HW radio)
PV Photo Voltaic cell
RDC Radio Duty Cycle
REST Representational State Transfer
RFD Reduced Functionality Device
ROLL Routing Over Low power and Lossy networks
RPL IPv6 Routing Protocol for LLNs
RREQ Route Request
RREP Route Replay
RTT Round Trip Time
SD Superfame duration

SON Self Organizing Network
TLS Transport Security Layer
WSAN Wireless Sensor & Actuator Network
6LoWPAN IPv6 over Low power Wireless Personal Area Networks

References

M24LR64E-R datasheet: 64-Kbit EEPROM with password protection, dual interface & energy harvesting: 400 kHz I^2C bus & ISO 15693 RF protocol at 13.56 MHz. http://www.st.com/internet/com/TECHNICAL_RESOURCES/TECHNICAL_LITERATURE/DATASHEET/DM00047008.pdf

T. Clausen, U. Herberg, M. Philipp, A critical evaluation of the ipv6 routing protocol for low power and lossy networks (rpl), in *Proceedings of the 2011 I.E. 7th International Conference on Wireless and Mobile Computing, Networking and Communications (WiMob)* (IEEE, 2011), pp. 365–372

T.H. Clausen, A.C. de Verdiere, J. Yi, A. Niktash, Y. Igarashi, H. Satoh, U. Herberg, C. Lavenu, T. Lys, J. Dean, *The Lightweight On-demand Ad hoc Distance-Vector Routing Protocol—Next Generation (LOADng)*. Draft-clausen-lln-loadng-13. Work in progress (2015)

T. Dierks, E. Rescorla, The Transport Layer Security (TLS) Protocol Version 1.2. RFC 5246 (2008)

A. Dunkels, B. Gronvall, T. Voigt, Contiki—a lightweight and flexible operating system for tiny networked sensors, in *Proceedings of the 29th Annual IEEE International Conference on Local Computer Networks, 2004* (2004), pp. 455–462

A. Dunkels, F. Osterlind, N. Tsiftes, Z. He, Software-based on-line energy estimation for sensor nodes, in *Proceedings of the 4th Workshop on Embedded Networked Sensors, EmNets '07* (ACM, New York, NY, 2007), pp. 28–32

A. Dunkels, The ContikiMAC radio duty cycling protocol. Technical report, Swedish Institute of Computer Science (SICS) (2011)

IEEE. 802.15.4-2011: IEEE standard for local and metropolitan area networks—Part 15.4: Low-Rate Wireless Personal Area Networks (LR-WPANs) (2011)

C.-A. La, M. Heusse, A. Duda, Link reversal and reactive routing in low power and lossy networks, in *Proceedings of the 2013 I.E. 24th International Symposium on Personal Indoor and Mobile Radio Communications (PIMRC)* (2013), pp. 3386–3390.

G. Montenegro, N. Kushalnagar, J. Hui, D. Culler, Transmission of IPv6 packets over IEEE 802.15.4 networks. RFC 4944 (2007)

M.R. Palattella, N. Accettura, X. Vilajosana, T. Watteyne, L.A. Grieco, G. Boggia, M. Dohler, Standardized protocol stack for the internet of (important) things. Commun. Surv. Tutorials IEEE **15**(3), 1389–1406 (2013)

E. Rescorla, N. Modadugu, Datagram Transport Layer Security Version 1.2. RFC 6347 (2012)

G. Romaniello, Energy Efficient Protocols for Harvested Wireless Sensor Networks. PhD thesis, Grenoble Alps University, March 2015

S. Sciancalepore, M. Vucinic, G. Piro, G. Boggia, T. Watteyne, Link-layer security in TSCH networks: effect on slot duration. Trans. Emerg. Telecommun. Technol. (2016)

L. Seitz, S. Gerdes, G. Selander, M. Mani, S. Kumar, *Use Cases for Authentication and Authorization in Constrained Environments*. RFC 7744 (2016)

Z. Shelby, K. Hartke, C. Bormann, Constrained Application Protocol (CoAP). RFC 7252 (2014)

STMicroelectronics. STM32L1 Series. http://www.st.com/stm32l1

Swedish ICT. The Contiki OS. http://www.contiki-os.org/

F. Todeschini, Dimensionnement energetique de reseaux de capteurs ultra-compacts autonomes en energie. PhD thesis, Supelec, 2014

F. Todeschini, C. Planat, P. Milazzo, S. Tricomi, P. Urard, P. Benabes, A nano quiescent current power management for autonomous wireless sensor network, in *Proceedings of the IEEE 20th International Conference on Electronics, Circuits, and Systems (ICECS) December 2013* (2013), pp. 815–818

L.-O. Varga, G. Romaniello, M. Vučinić, M. Favre, A. Banciu, R. Guizzetti, C. Planat, P. Urard, M. Heusse, F. Rousseau, O. Alphand, E. Duble, A. Duda, GreenNet: an energy harvesting IP-enabled wireless sensor network. IEEE Internet Things J. (2015)

L.-O. Varga, Multi-hop Energy Harvesting Wireless Sensor Networks: Routing and Low Duty-cycle Link Layer. PhD thesis, Grenoble Alps University, December 2015

M. Vučinić, B. Tourancheau, F. Rousseau, A. Duda, L. Damon, R. Guizzetti, OSCAR: Object security architecture for the Internet of Things. Elsevier Ad Hoc Netw **32**, 3–16 (2015a)

M. Vučinić, B. Tourancheau, T. Watteyne, F. Rousseau, A. Duda, R. Guizzetti, L. Damon, DTLS Performance in duty-cycled networks, in *Proceedings of the IEEE 26th Annual International Symposium on Personal, Indoor, and Mobile Radio Communication (PIMRC), August 2015* (2015), pp. 1–6

M. Vučinić, Architectures and Protocols for Secure and Energy-Efficient Integration of Wireless Sensor Networks with the Internet of Things. PhD thesis, Grenoble Alps University, November 2015

T. Winter, P. Thubert, A. Brandt, J.W. Hui, R. Kelsey, P. Levis, K.S.J. Pister, R. Struik, J. Philippe Vasseur, R.K. Alexander, RPL: IPv6 Routing Protocol for Low-Power and Lossy Networks. RFC 6550, March 2012

ZigBee Alliance. ZigBee Smart Energy Profile 2.0 http://www.zigbee.org/zigbee-for-developers/application standards/zigbeesmartenergy/

Ferro-Electric RAM Based Microcontrollers: Ultra-Low Power Intelligence for the Internet of Things

18

Sudhanshu Khanna, Mark Jung, Michael Zwerg, and Steven Bartling

Microcontrollers (MCUs) serve a central role in the design of IoT leaf nodes. A typical microcontroller comprises of a processing core, program and data memory, serial communication interfaces, general purpose IO (GPIOs) ports, comparators and ADCs, clock generation and power regulators. Microcontrollers have modest clock frequencies and memory capacities, keeping the IC cost low. High level of integration helps improves performance, lowers power and helps achieve a small form factor. All of these are critical in the design of ICs for IoT applications. This chapter focuses on the design of low power microcontrollers using two Texas Instruments (TI) microcontrollers as examples. Both microcontrollers feature embedded Ferro-electric RAM (FRAM) as a low power non-volatile unified data and program memory. Low write power non-volatile memory with high write endurance (number of write cycles) is critical in IoT applications that feature data logging. FRAM has low power writes at 1.5 V, practically unlimited write endurance ($>10^{15}$), and high yields with millions of shipped parts. In the following sections we describe analog, digital and system design

techniques that help achieve the low power metrics in TI microcontrollers.

18.1 MSP430FR5xx/6xx: 130 nm FRAM Based Microcontrollers Optimized for IoT Applications

This section discusses the design of the TI FRAM Microcontroller series MSP430FR5xx/6xx. We discuss a cross-hierarchy optimization process starting from technology node selection, digital and analog circuit techniques, optimized digital design methodologies, and analog design styles. The biggest knob in power optimization is system design: we see how smart peripherals and low power modes help achieve the ultra-low power (ULP) goals in this MCU. FRAM is a key enabler for IoT applications that need data logging. In this section we describe the technical reasons and discuss FRAM operation, advantages and limitations.

The MSP430FR5xx/6xx family of microcontrollers is designed for applications that demand ultra-low active and standby energy. Hence, it is tailored for applications that require low average and peak currents. With active currents in a range of 100 µA/MHz and standby currents of 450 nA including supply-voltage supervision and active real-time clock (RTC), it serves the requirements of typical IoT applications. The FR5xx provides seven low power modes that address different

S. Khanna (✉) • M. Zwerg • S. Bartling
Texas Instruments, Dallas
e-mail: skhanna@ti.com; m-zwerg@ti.com; bartling@ti.com

M. Jung
Texas Instruments, Deutschland GmbH
e-mail: m-jung@ti.com

© Springer International Publishing AG 2017
M. Alioto (ed.), *Enabling the Internet of Things*, DOI 10.1007/978-3-319-51482-6_18

Table 18.1 Typical ULP applications

Application	Duty cycle		Period	Avg. IDD	IDD dominated by:
	Active (%)	LPM (%)			
Sensor hub	10	90	1 s	89.6 µA	Active
Fitness tracker	1	99	1 ms	52.9 µA	Wakeup frequency
Blood glucose meter	0.4	99.6	3×/day	4.7 µA	LPM
Heat cost allocator	0.01	99.99	4 s	5.3 µA	LPM at high temp

scenarios: They serve for both applications with frequent sleep/active transitions in the order of hundreds of microseconds with fast wakeup requirements, and applications with long sleep periods in the order of seconds or more.

Power consumption in IoT applications varies due to several factors. Table 18.1 lists some typical ULP applications. The duty cycle (active to low power mode ratio) amongst these applications varies by orders of magnitude. The longer an application stays in low power mode, the lower the average current (IDD) should be. Nevertheless, the average current consumption is not a direct function of the duty cycle. The sensor hub takes sensor measurements and does further processing on the sensor data. In this case the average current is clearly dominated by the active mode current, with a high 10% duty cycle. However, while the fitness tracker application has a much lower duty cycle than the sensor hub (1% vs. 10%), the reduction in IDD is not proportional. The small wakeup period in the fitness tracker (1 ms) prevents the MCU from going into the lowest low power mode. This shows that the time and energy cost of going into a low power mode is also important. Other ULP applications like blood glucose metering remain in the deepest low power mode for a majority of their lifetime (low duty cycle and long time periods) and are dominated by the leakage current of the MCU. Blood glucose metering devices turn on a few times a day, but stay active for several minutes. Here the MCU can make best use of the deepest low power mode. The heat cost allocator is a special case, where the application is exposed to high temperatures for most of its lifetime. Even though it has the lowest duty cycle, the low power current is dominated by the high

temperature leakage of the process, causing IDD to be higher than that of blood glucose metering. This shows how temperature is also a factor in application IDD. Thus an IoT MCU has to be designed considering not only the duty cycle range across applications, but also the application time period, and temperature range.

A main differentiator of the FR5xx is the use of FRAM which has a significantly lower write access time and energy compared to flash memory. In typical IoT applications this shortens the data logging phase and hence the active mode time. Due to the 1.5 V write voltage of the FRAM, no charge pump is required.

The power supply architecture of the FR5xx is optimized for the requirements of the low-power modes: Switchable low drop-out regulators (LDO) supply dedicated FRAM, debug, real-time-clock and core domains on demand. The core domain is subdivided into further switchable power gated domains to reduce leakage currents during low-power modes. The predictive core LDO quickly adapts to the load requirements of the digital system which optimizes the bias current of the LDO depending on the current load scenario.

In addition the digital multi-VT standard cells enable optimum balance between leakage and active mode performance. Along with the variety of provided slow 9 kHz up to fast 16 MHz clocks and the extensive use of common low power techniques like clock or bus gating, the design is optimized for low energy through all levels.

By means of the Texas Instruments MSP430FR59xx which is part of the FR5xx platform, it will be demonstrated how the ambitious low power requirements of an IoT application impacts the architecture and design of a microcontroller unit. The FR59xx family is developed

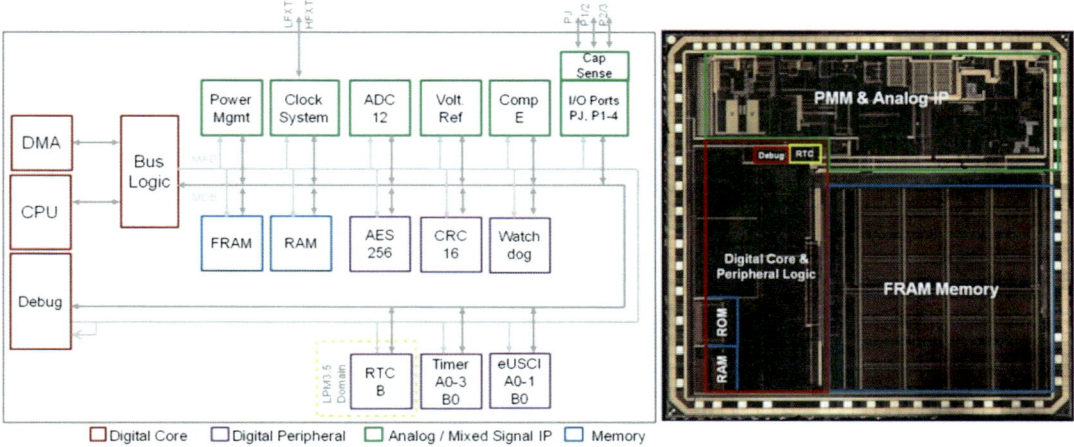

Fig. 18.1 FR59xx mixed-signal microcontroller block diagram (MSP430 FR59xx Mixed Signal Microcontrollers) and chip photo (Baumann et al. 2013)

in a 130 nm CMOS process which fulfills the requirements of a highly integrated ultra-low leakage process. Figure 18.1 represents the high-level block diagram of the FR59xx devices and a chip photo of an example die of the FR59xx family.

FR59xx consists of:

- The digital MSP430 core and debug unit,
- the analog core and peripheral components like the power management system and clock oscillators,
- the digital peripherals and accelerators, and
- the FRAM and SRAM memories for code and data storage.

All peripherals are optimized for ultra-low energy operation required by IoT applications. Accelerators like the AES or the CRC reduce the encoding time and therefore enable a reduction of active mode time. For example, AES hardware encryption is 32 times faster compared to a software implementation on an MSP430 (AES for Texas Instruments MSP430 Microcontroller; MSP430 FR59xx Mixed Signal Microcontrollers). Mixed-signal peripherals like the ADC12 consume only 145 μA and are able to run during standby mode without the need for waking up the CPU core.

18.1.1 Low Power Modes

The MSP430FR5xx microcontrollers provide seven different low-power modes (LPM) which all target different application scenarios. Each of the modes is a tradeoff between low bias and leakage currents versus quick wakeup times for the system and its components. Table 18.2 summarizes the main characteristics of the low-power modes. For instance, LPM0 has all high-frequency peripherals available and keeps the core LDO active. This results in minimum wakeup times. However, it comes along with relatively high bias currents e.g. of the power management system or the active oscillators. In contrast, LPM3 only allows for low frequency peripherals up to 50 kHz which can bring down typical overall current consumption to only 400 nA. Power gating with full data retention results in low digital leakage even at high temperatures. LPM3.5 and LPM4.5 provide lowest possible system currents, as the entire digital core is powered down. The penalty is a significant wakeup time increase, as the entire system has to reboot after the low power mode wakeup. In most instances LPM0 and LPM1 are best suited for short sleep phases, while LPM3-LPM4.5 show strength in case of long inactive periods.

For a holistic comparison of the LPM current consumption, the wakeup energy must be

Table 18.2 Power mode characteristics of the MSP430FR59xx devices (MSP430 FR59xx Mixed Signal Microcontrollers)

	Active	LPM0/LPM1	LPM2	LPM3/LPM4	LPM3.5/LPM4.5
Description		CPU Off	Standby	Standby/off	RTC/Shutdown
Max. frequency	16 MHz	16 MHz	50 kHz	50 kHz/0 kHz	50 kHz/0 kHz
Typical currents	~100 μA/MHz	70 μA@1 MHz/ 35 μA@1 MHz	Downto 0.7 μA	Downto 0.4 μA/0.3 μA	Downto 0.25 μA/0.02 μA
Typical wakeup		Instant/6 μs	6 μs	7 μs	250 μs 1000 μs
CPU	On	Off	Off	Off	Off
FRAM	On	Off or standby/off	Off	Off	Off
HF peripherals	Yes	Yes	Off	Off	Off
LF peripherals	Yes	Yes	Yes	Yes/off	Yes/off
Dig. power gating	No	No	No	Yes, full retention	Yes, no retention

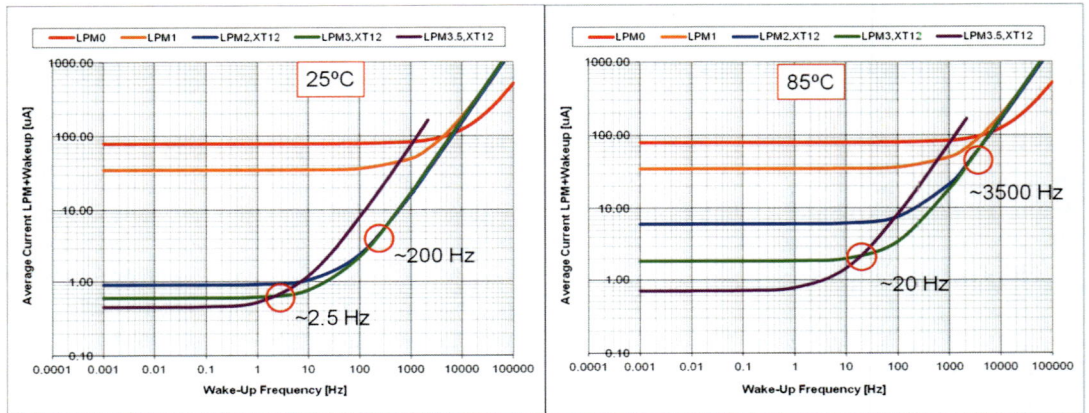

Fig. 18.2 FR59xx mixed-signal microcontroller block diagram (MSP430 FR59xx Mixed Signal Microcontrollers)

considered. Figure 18.2 shows that each of the power modes has a range in which it provides optimum average current consumption. The intersection of two curves means that the two low power modes show same current consumption for the given frequency.

18.1.2 Oscillators

Clock sources are critical in ULP applications. The MSP430FR5xx microcontroller platform provides different clock sources with varying frequency, power and startup time tradeoffs as shown in Table 18.3 (*the REFO is not present in MSP430FR59xx but is present in MSP430FR2xx

(MSP430FR2xx Mixed Signal Microcontrollers) and MSP430F5xx (MSP430 F5xx Mixed Signal Microcontrollers) as a crystal replacement. We mention it here to supplement the discussion on tradeoffs in oscillator choices).

The VLO is the slowest clock but also lowest power oscillator in the system. The VLO is always on and is mainly used by the power management for internal low frequency housekeeping. For example in low power mode the bandgap reference can be operated in a pulsed sample and hold scheme which drastically reduces the average power consumption of the voltage reference. In the FR57xx devices the VLO helps to reduce overall system power by providing a clock for the sample and hold circuit.

Table 18.3 Types of oscillators in MSP430FR5xx

	VLO	XTAL	REFO*	MODO	DCO
Frequency	9 kHz	32 kHz	32 kHz	4.8 MHz	1–24 MHz
IDD	100 nA	200 nA	3 μA	25 μA	(~15 μA @ 8 MHz) (~23 μA @ 16 MHz)
Temp drift	0.5%/°C	0.03 ppm/°C	0.01%/°C	0.08%/°C	0.01%/°C
Supply drift	4%/V	None	1%/V	1.4%/V	2%/V
Startup time	Always on	1 s	50 μs	100 ns	5 μs

The VLO can also be beneficial for other modules in the power modes LPM3 and LPM4. For example it can be used as the clock source for the watch dog timer or LCD module. Both are low frequency peripherals that can easily tolerate the wider temperature and supply drift. Since the VLO is always on in the FR57xx family, it is also considered a failsafe clock in this family. For example it is used to detect whether the crystal oscillator has stopped working and in case of a failure situation it will switch to the VLO clock so the system can avoid total failure. In this way modules will always see a clock source and continue working.

The MODO is used as a fixed mid-range frequency oscillator on demand. It has fast start up time and thus can be activated for just for a few clock cycles to save power. For example, it is used for active ADC sampling or for noise filtering on the external reset pin. The high active power of this oscillator is the tradeoff for the short on-time. Since the frequency is fixed and aligned across device families, peripheral modules can make use of this clock without the need to adjust clock dividers to match the DCO selected system frequency. This is important because the user can change the DCO frequency by software according to the application requirement. But this should not impact modules that require a constant frequency.

The DCO serves the purpose of a high frequency clock with configurable frequency. The DCO is optimized for a wide tuning range with a fairly low active current at the cost of startup time.

The DCO is optimized for a wide operating frequency range to support different kind of applications. With a current limited energy source it might be better to avoid high peak currents by choosing a low frequency continuous operation rather than bursting the system using high frequency activity bursts. For minimum energy it is better to run the oscillator as slow as possible, rather than sub-dividing a higher, fixed frequency oscillator clock. From power and startup time it fits between the REFO and MODO. It has a faster startup time than the REFO, but not as quickly as the MODO. On the other hand its current consumption is better as the MODO for the same frequency and therefore it is better suited for the active mode system clock.

The REFO is used as an on-chip replacement for the 32 kHz crystal oscillator on MSP430F5xx and MSP430FR2xx. It is not as accurate as an external crystal but can be used for clocking peripherals in low power modes, for example in timers for accurate milli second or second count. The UART can use the REFO for low baudrate communication (max 9600 baud). The main advantage of the REFO in comparison to the crystal oscillator is the overall system cost. In many low cost applications the crystal is an expensive component and on the other hand the accuracy of the crystal is not required. Therefore the REFO is a cost effective replacement for a 32 kHz crystal, but cannot match an external crystal in low power and accuracy. In MSP430FR59xx, accuracy is prioritized over cost and hence the XTAL is used instead of the REFO. In MSP430FR59xx another clock, which is a divided by 128 version of MODOSC, replaces the REFO.

18.1.3 Power Supply Architecture

The power architecture of the FR59xx is designed to consume minimum possible power in each of

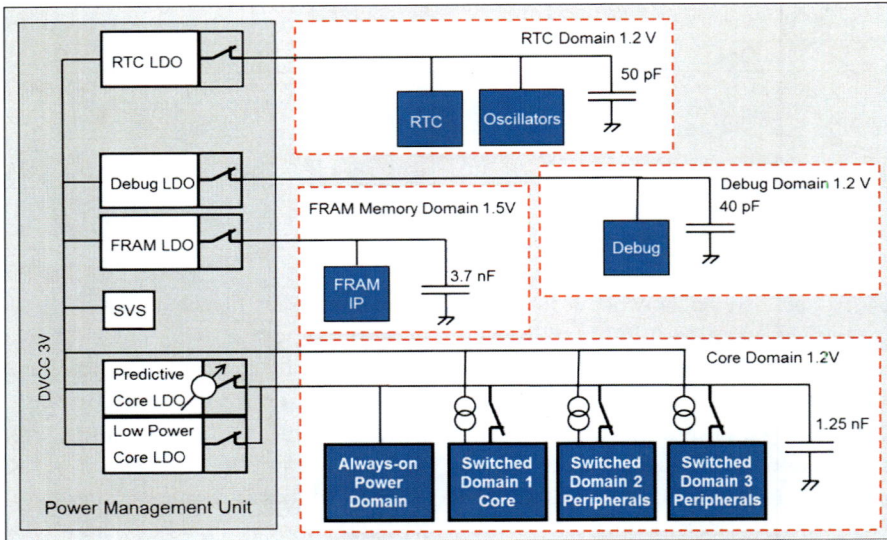

Fig. 18.3 Power Supply Architecture of the MSP430FR59xx devices (Baumann et al. 2013)

the active and low power states. It is illustrated in Fig. 18.3. The power management unit (left block of Fig. 18.3) is one of the central components for the ULP design of the microcontroller. It contains five low drop-out regulators (LDOs) which are enabled and configured depending on the current active or low power mode.

All of the LDOs can be disabled to reduce analog bias and digital leakage currents during defined device states: For instance, the supply for the digital RTC logic is turned off during LPM4.5, the supply for the FRAM from LPM0 onwards, and the debug LDO during user application mode. In LPM3.5 the RTC LDO is the only active regulator which supplies the RTC and the 32 kHz oscillator which is required for the RTC operation. All LDO output supplies use internal capacitors which reduces the number of required supply pins and external components. In addition, external voltage fail mechanisms have to be considered only for the main VDD supply.

During active mode, LPM0 and LPM1, the digital core and peripheral components are supplied by the Predictive Core LDO. The supported LDO load and hence the inherent LDO current is adapted depending on the actual system activity, which is determined by the number and frequency of the active clocks (Lueders et al. 2014).

The digital core logic is supplied by the Low Power Core LDO during standby and off modes. At these power modes, the voltage and current biasing circuits rely on a sampled mode concept (Fig. 18.4). This is based on the idea that the charges of a capacitor are frequently refreshed to ensure robust bias and reference conditions. This short refresh cycle is reflected by the "On" state in the diagram of Fig. 18.4. In sampled mode, the reference and supply-voltage supervisor consume less than 200 nA. This stands in contrast to the static full accuracy mode used in active mode which requires currents in the range of few μA.

18.1.4 Digital Core System

The digital core logic of the FR59xx devices, shown in the core domain box in Fig. 18.1, is partitioned into an always-on domain, which is always powered if one of the core LDOs is active, and several power gated domains that can be switched off during LPM. While the core domain is always switched off in the higher LPMs, the peripheral domains can remain active in case they are requested by one of the modules. In order to enable faster wakeup from low power

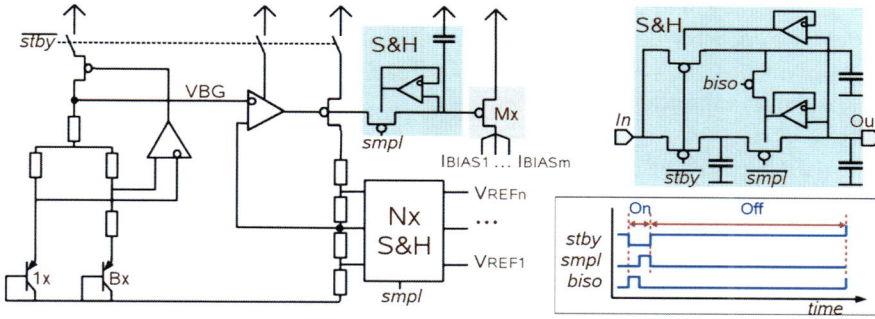

Fig. 18.4 Switched mode reference generation of the FR59xx (Baumann et al. 2013)

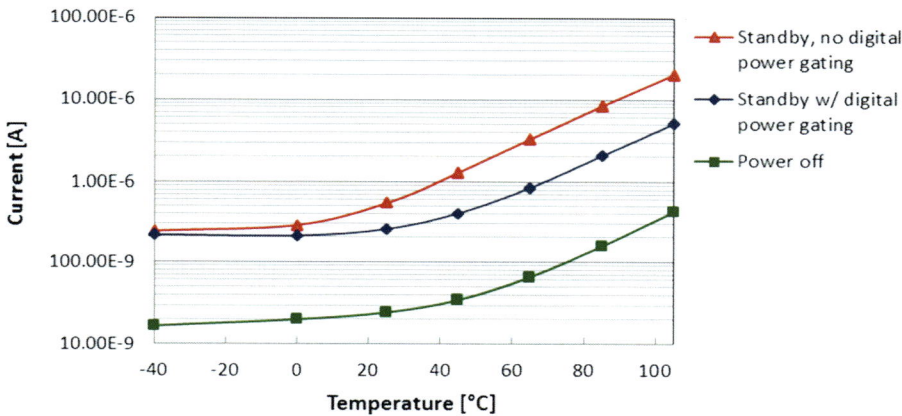

Fig. 18.5 Low power mode current trend for the described architecture

mode, current sources (shown in the bottom part of Fig. 18.3) recharge the switchable power domains in parallel to the LDO. Once the domain supply hits a defined voltage, the current sources are disabled again to keep quiescent currents low.

Digital currents are dominated by leakage during standby modes, especially when the device operates at higher temperatures. Figure 18.5 shows the typical currents trend over increasing temperatures (Baumann et al. 2013).

The red curve stands for a system in standby mode with no power gating, so all digital logic in the core domain is powered. At low temperature the total microcontroller current is just slightly higher compared to the same system in standby mode with full power gating of all switchable domains. This is due to the fact that the system's

current is dominated by the quiescent currents consumed by the analog reference and supervision circuits. The advantage of power gating is becomes obvious with increasing temperatures, when the overall system currents are dominated by digital leakage. For reference, the green curve presents the power-off mode of a device with active brownout, which only retains the wake-up capability via external I/O. Reference system and LDOs are switched off.

Digital power gating becomes even more attractive in mixed-VT designs, which make use of standard cells from different Vt classes. Cells with higher Vt (HVt) transistors are used to significantly reduce leakage currents, however they tend to be slower compared to cells with standard Vt (SVt) transistors. Hence, SVt cells are usually used in timing critical logic. If all the timing critical logic can be moved to the switched core domain which is turned off during the leakage sensitive standby modes, the system leakage can be reduced without impacting system performance in active mode.

However, digital power gating comes along with additional area requirements with for the power switches, logic retention signal isolation and additional floorplanning constraints. For logic retention, the data in the sequential cells like flops or latches has to be powered even when the rest of the domain is unpowered. Isolation cells are placed at the boundary of an unpowered and a powered domain to avoid floating inputs into the powered domain when the switched domain is shut off. The floorplan usually segregates the domains, which introduces placement and logic optimization complexities. These additional requirements and limitations are strongly dependent on the design architecture, e.g. number of domains, number of ports between domains, number of retainable cells or maximum domain currents. A typical area penalty for the described system can be in between 10 and 20%.

18.1.5 Ferro-Electric RAM

In this section we describe the operation, advantages and limitations of Ferroelectric RAM (FRAM).

Data is stored in FRAM by electrically switching ions between stable positions in a Lead Zirconium Titanate (PZT) crystal as shown in Fig. 18.6. Movement of the Ti/Zr ion is under the influence of an applied electric field and once the field is removed, the ion retains its position. Thus, based on the applied electric field each PZT molecule has a stored polarization state. Similar to magnetic domains in a magnet, domains in FRAM material align when an electric field is applied. When the field is removed, some domains get depolarized, but most domains retain polarization state, resulting in a net spontaneous polarization. Thus the FRAM material behaves as a programmable capacitor with hysteretic behavior similar to ferroelectric magnetic

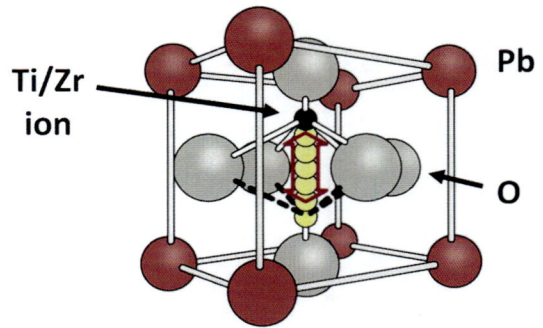

Fig. 18.6 Structure of the PZT crystal

materials, and thus the name ferroelectric RAM. However, unlike ferroelectric magnetic materials, FRAM is immune to magnetic fields and alpha particles. Figure 18.7 shows the hysteretic behavior of a ferroelectric (FE) capacitor. The materials to manufacture FRAM are CMOS compatible, and add only two extra masks to the baseline CMOS process. This makes FRAM processing much lower cost as compared to flash memory, which requires around 12 additional masks. TI has been mass producing standalone and embedded FRAM products for more than 10 years with millions of shipped parts demonstrating high yields and robust operation.

Fecaps (ferroelectric capacitors) can be written fast (<10 ns), at low voltage (1.5 V), and without high write currents. This makes FRAM much more like SRAM than flash in its write characteristics. Since the write mechanism involves only atomic movements, there is no wear and tear involved, resulting in high write endurance (>10^{15} cycles). Fast, low voltage, low current writes, and near-infinite write endurance makes FRAM ideal for IoT applications that need data logging. This is in contrast with flash memory which requires multiple high voltages, large write currents, and is also limited to ~10 K write cycles. The high write energy in flash memory is a fundamental limitation in IoT applications.

Figure 18.7 shows the circuit diagram of the simplest FRAM bitcell. It consists of an access transistor and the Fecap. This is very similar to a DRAM bitcell. The difference is the Plateline

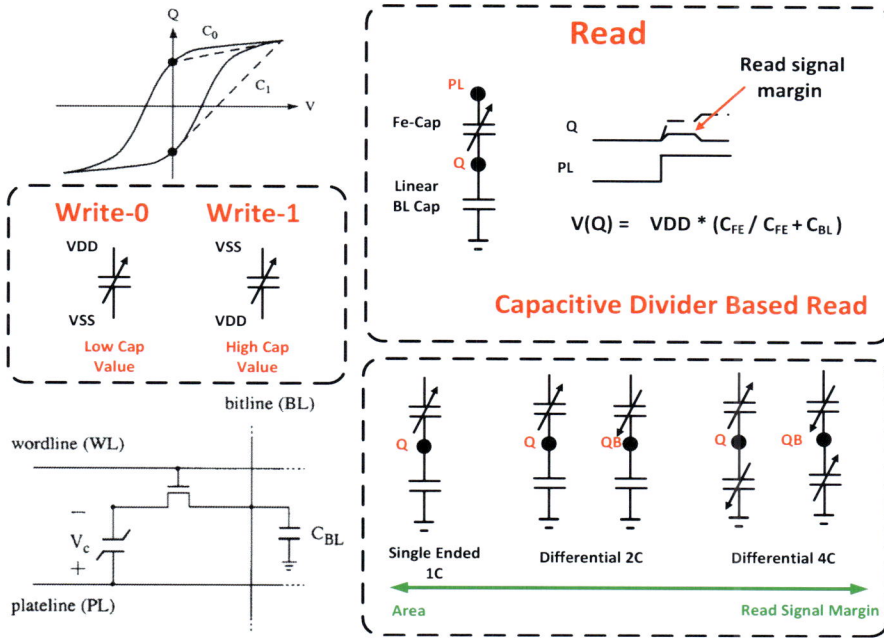

Fig. 18.7 Hysteretic behavior of fecaps (*top left*), writing (*top left*) and reading (*top right*) a fecap. Structure of a FRAM bitcell (*bottom left*)

(PL) terminal connected to the bottom plate of the Fecap. To write a Fecap to a state of "1" a voltage of 1.5 V is applied at bit-line (BL) and 0 V at PL. To write a "0" 0 V is applied at the BL and 1.5 V at the PL. For a NMOS pass gate, the word-line (WL) signal must be boosted for a short period of time (few ns) to let the 1.5 V signal pass while writing a "1". Applying these voltages stores a high or low capacitance in the Fecaps. To read the bitcell, the BL is floated and PL is toggled from 0 V to 1.5 V as shown in Fig. 18.7. This forms a capacitor divider between the Fecap and the parasitic BL capacitor. In differential FRAM schemes, each bitcell consists of two copies of the circuit and when the differential bitcell is read, BL and BLB develop voltages near VDD/2 when PL is toggled low to high. Since the two sides of the bitcell have Fecaps in low and high states, the voltage on BL and BLB differ slightly, and a sense amplifier is used to read the signal. Toggling the BL and PL is the most energy intensive operations since Fecaps are dense capacitors.

Read and write speed and energy are almost the same for FRAM. TI FRAM in 130 nm works at 8 MHz for both read and write operations. The big enabler for IoT applications is that FRAM write energy per bit is ~250× lower than flash. Read energy is ~2× lower as well. With low power and fast reads and writes, FRAM can be used as a unified memory for both program code and data logging. Data retention in FRAM is also resistant to electric/magnetic fields as well as radiation. Since FRAM state is not stored as a charge, alpha particles are not likely to cause bits to flip and the FRAM Soft Error Rate (SER) is below detectable limits.

18.2 Compute Through Power Loss Using Non-Volatile Logic

IoT applications are duty cycled and use small batteries or energy harvesting energy sources. Thus, lowering energy consumption (dynamic, leakage and mode transition) is critical in achieving the power and lifetime goals. As described in the previous sections, ULP MCUs use process technologies that are specially optimized for low leakage. Voltage is lowered to the minimum possible to meet latency. Non-volatile FRAM is used instead of SRAM so that in standby mode intermediate data and sensor data can be retained with zero energy consumption. HVT logic and power gates are used extensively to selectively reduce leakage currents with lesser impact to drive strengths. Clock gating is used to only enable switching in required sections of logic. Sampled mode analog blocks are used to reduce quiescent currents. In addition, MCUs implement multiple low power modes that tradeoff standby power for wakeup time by incrementally turning off different parts of the SoC.

While dynamic and leakage power is critical metrics in an IoT MCU, energy and time consumed in mode transitions and MCU wakeup are important metrics too. MCUs use thousands of flip-flops (FFs) as pipeline registers, general purpose registers and status flags. FFs are also used to save system configuration (available memory, peripheral count) and modes, analog calibration and trim settings. Unlike non-volatile memory like FRAM, flip-flops are volatile in nature, meaning that must have a constant power supply to retain their logic state. Thus, when a MCU (with volatile or even non-volatile FRAM) faces a power interruption or is put in a power gated low power mode, all flip-flops and latches in a system are reset. This means that after each wakeup event or mode transition from a power gated state to active mode, the MCU must re-boot. Bootup not only consumes precious energy resources, it can also be limiting in real time applications where the system must respond quickly to external events.

Table 18.4 shows how the boot-up process takes 10,000s of cycles, 10s of milliseconds and 100s of nJ of energy. Boot-up is costly for catalog MCUs because they are built for a range of applications and cost points and are thus configured and trimmed at boot. This "factory boot" is then followed by the customer's boot-up code or C-initialization with the variables and data structures used in the program are set to their default values. For ULP applications the parasitic wakeup time and energy can be prohibitive. For simple applications like a temperature sensor, boot-up is the major component of the overall application active energy and time cost. Further, real time applications like circuit breakers that need fast response times cannot use such a low power mode at all due to the long boot-up time. The mode transition can thus be a limiter in energy harvesting applications where the energy source is erratic and energy or latency is critical (Zwerg et. al. 2011).

To prevent having to boot-up upon every wakeup, FF state retention is used. Here, a portion of each flip-flop is powered by a separate "always-on" supply which is not power gated even in standby mode. This helps decrease wakeup time and energy by eliminating need for boot-up but it increases the standby power due to leakage in "always-on" portion of retention flip-flops and the LDO generating the "always-on" supply. In this manner, existing low power modes trade off standby power, wakeup time, and design volatility, but no single technology provides all of the above benefits together. Also, in energy harvesting applications, there may not be an always on supply available, so retention may not be possible at all (without using supercapacitors or re-chargeable batteries).

In this section, we describe non-volatile logic (NVL). NVL helps achieve standby power levels close to a fully power gated mode while still having fast wakeup similar to case where retention is used. We describe a custom MCU that can seamlessly "compute through power loss" by backing up all flip-slops and latches in the MCU before a power loss. This also helps achieve lower application level power because the MCU can now be power gated with lower energy and time overheads. We discuss different architectures to enable compute through power loss and describe our implementation in detail along with silicon measurements.

The NVL MCU achieves zero low power mode current while also having an ultrafast sub-400 ns wakeup time. Non-volatile FRAM mini-arrays distributed throughout the logic domain of the MCU snapshot the state of all

Table 18.4 Wakeup times of typical MSP430 applications

Power Management
- Brownout
- BG & REF
- LDO

Boot Code
- Trim & Calibration
- Access Protection
- Debug Support

C-Init User-Init
- Variable init
- Queue/Buffer init
- Peripheral init

Application
- Work Load Task

Application [1]	C-init [2]	User-Init [3]
DC/DC FET	50us	23us
USB FET BIOS	29ms	11ms
ULP Bench	2.9ms	18us
TI-RTOS	44ms	1ms
Graphics Lib 400x240 pix.	101ms	298us

(1) data based on real MSP430 applications
(2) C-init executes at default frequency (MSP430: 1MHz)
(3) User-init executes after clock config (here 16Mhz)

sequential elements before the MCU goes into a low power mode. A high bandwidth parallel connection between the flip-flops and mini-arrays helps achieve fast wakeup time. The MCU also has ferroelectric RAM (FRAM) as a unified data and program memory that retains data in the absence of a supply voltage. Thus during a low power mode, the MCU can be fully power gated, and no power is required to maintain system state. When the MCU wakes up from a low power mode, data from NVL arrays is transferred back to the flip-flops and latches, and the MCU can resume operation from where it left off. Upon wakeup no bootup, C-initialization or user-initialization is required.

Figure 18.8 shows the system block diagram of the 1.5 V 8 MHz 75 µA/MHz NVL MCU with zero standby power and 384 ns wakeup time. The MCU is built using a High VT (HVT) 130 nm CMOS process. The MCU contains a 32 bit core, 64 KB FRAM (unified program and data memory), 8 KB SRAM (for test only), 10 KB ROM, UART and SPI peripherals, and a serial wire debug interface. Note that there is no power management and analog blocks in this MCU. So, low power mode current and wakeup time contributions from these blocks would need to be added to get a representative number for a full MCU SoC.

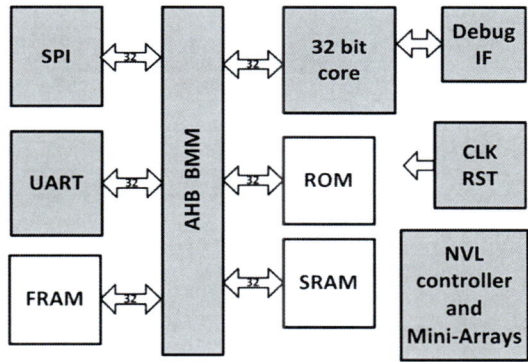

Fig. 18.8 System architecture showing the 32b core, peripherals, bus interface and the NVL controller. NVL arrays are dispersed in the chip

Section 18.2.1 introduces the NVL mini-array architecture and compares it with existing NVL architectures. Section 18.2.2 describes the NVL bitcell and array periphery and summarizes the performance and power results from the NVL MCU.

18.2.1 NVL Mini-Array Architecture

Figure 18.9 shows the functional diagram of an NVL MCU.

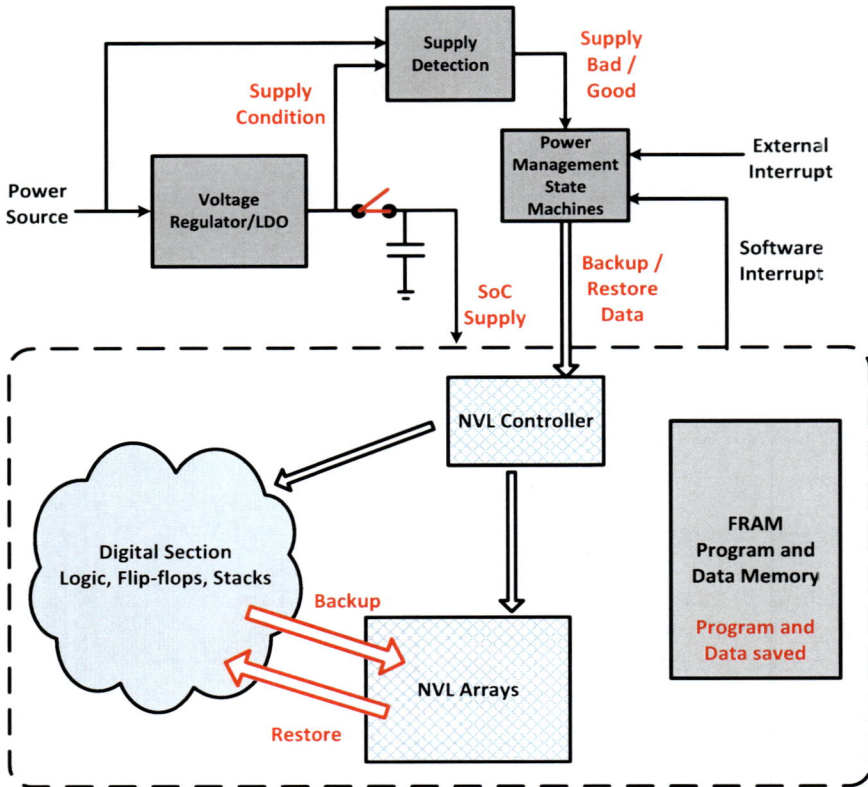

Fig. 18.9 NVL SoC functional diagram. In this version of the NVL SoC, voltage regulation, supply detection and PM state machines are off-chip

When the supply detection (SD) circuit detects a supply droop, the switch at the output of the LDO is turned off to prevent spurious voltages from reaching the MCU core. The supply capacitor holds enough charge to support the state backup. Then the SD activates the power management (PM) state machines with a supply bad signal. Following this, data from all flip-flops and latches in the MCU is backed up into NVL arrays. Since the program and data memory is already non-volatile (FRAM), no action needs to be taken for the memory backup/retention. In this way the entire system state is saved before power loss. In the case of software based power interrupt, the same sequence is followed, except that the supply switch is turned off last. The NVL controller orchestrates data backup and restore, and controls communication between the system flip-flops and the NVL arrays. When power is

resumed, following a similar procedure, data from NVL arrays is restored back into the system flip-flops, and system operation resumes from where it left off. No boot-up is required. We rely on off-chip components to regulate the power supply for this version of the chip.

NVL implementations reported so far (Wang et al. 2012; Qazi et al. 2013; Sakimura et al. 2014; Liu et al. 2016) propose adding non-volatile elements in flip-flops (FF). Implementing NVL at the FF level is a yield risk because non-volatile elements (like ferroelectric capacitors) are not qualified to operate in the irregular environment of standard cell layout. Adding non-volatile elements to each FF also makes the FF large because read/write/test circuits must be duplicated in each FF and cannot be shared across multiple bits. Further, test related reference voltages and a high voltage

Fig. 18.10 (**a**) NVL mini-array architecture. Arrays dispersed throughout the SoC serve 256 neighboring FFs. A central NVL controller synchronizes communications between the arrays and the FFs. (**b**) NVL FF with an additional data and enable port as compared to a regular FF (**c**) 256FF are connected to a NVL array using a 32b wide 8–1 mux

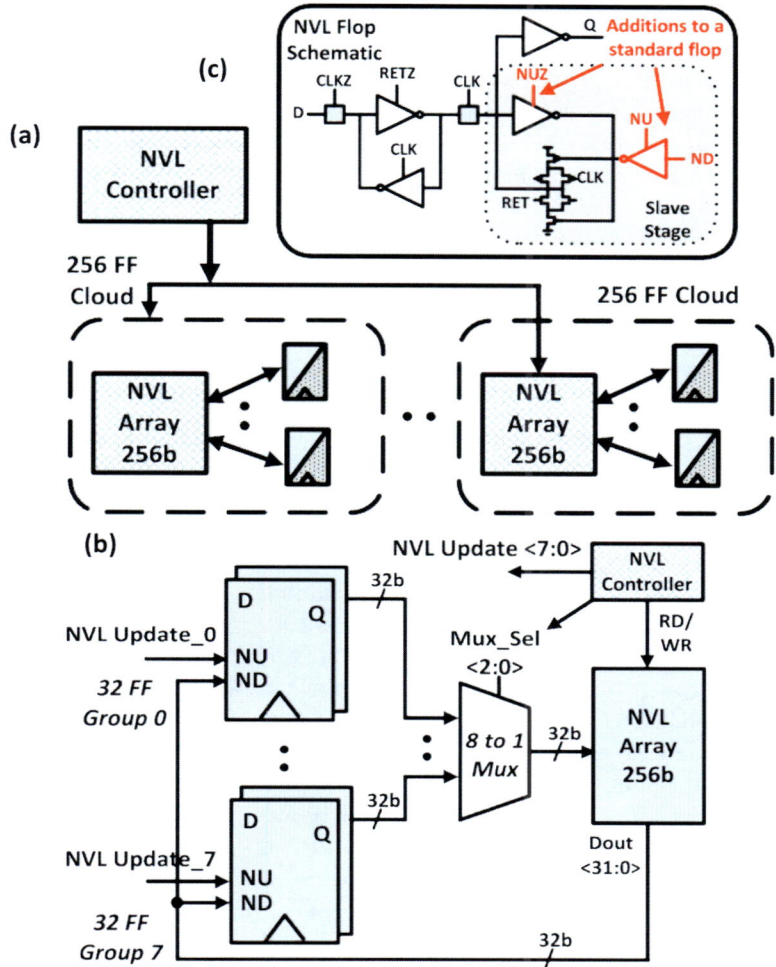

supply have to be routed to every FF in the design. Finally, spreading non-volatile elements makes test complex and results in high test time. A centralized NVL implementation helps overcome the above yield, density and test issues. However, a centralized array results in limited bandwidth and a slow wakeup time. For example, the 64 KB FRAM we use in our MCU can read/write 64 bits of data at 8 MHz. This translates to a theoretical minimum 5 μs wakeup time for our 32-bit core (which has 2537 FFs). If a more complex core like ARM Cortex M3 or MSP430 is used, the FF count and wakeup time would be 3–4× the above number. Routing is also an issue in the centralized approach since

signals to and from all FFs in the system have to be routed to a central location. Limited bandwidth and unscalable nature of a centralized NVL architecture makes it less attractive for the most energy and latency critical applications.

We use a novel NVL implementation called the NVL Mini-Array Architecture (Khanna et al. 2014) (shown in Fig. 18.10a). Custom 256b FRAM mini-arrays are used to backup FF data. Multiple such arrays are dispersed throughout the digital area with each array serving 256 neighboring FFs. This approach allows us to amortize fecap related read, write and test circuitry over 256 bits in the mini-array. Testability and bit level screening features available in a regular

FRAM array can be similarly implemented for such a mini-array. This is critical for making a viable product. Being in a regular array layout environment also drives fecap yield higher. Since each mini-array has a 32b interface and all mini-arrays can be accessed in parallel, system backup and restore is fast and wakeup time remains independent of the total FF count. Thus, the mini-array architecture provides speed similar to a FF level implementation with the yield, density, and test advantages of a centralized NVL implementation.

Figure 18.10b shows how NVL arrays connect to FFs. An NVL array is organized into 8 rows of 32 bits each. During a data backup, the NVL controller walks through all the Mux_Sel combinations transferring data from 256 FFs to 8 rows of an NVL array. To enable transferring data from NVL array to FFs we add an additional data port to a regular FF as shown in Fig. 18.10c. Upon wakeup, the NVL controller reads the NVL arrays row by row while simultaneously pulsing appropriate NU signals, thereby transferring data from the arrays to the FFs. The system CLK is kept low during backup and can be in any state during restore. RET is kept high during restore. This direct parallel connection between FFs and NVL arrays helps achieve a fast wakeup time. As MCU FF count increases, array count also increases, keeping bandwidth and wakeup time independent of the FF count. Also, since the MCU core is not involved in any data transfer, this NVL implementation is design agnostic and easy to add in any existing core or ASIC.

Conventional flip-flops have a single data input port. To insert data from NVL arrays back into a flip-flop, we added a second data input port (ND) and an enable port (NU) to a regular FF. As shown in Fig. 18.10b, the 32 bit data output bus of the NVL array is connected to all 8 FF groups. When row 0 of the NVL is being read, NU ports of the FF group 0 are enabled. In this fashion, all eight rows are sequentially read. Again, the NVL controller generates the NU signals and synchronizes data transfer between the NVL array and FFs. We call the modified flip-flop an NVL flip-flop.

Figure 18.10c shows the schematic of the NVL flip-flop. The NVL FF is a modified retention FF. The difference between a regular FF and a retention FF is that the slave latch (shaded) of a retention FF is powered by a separate supply. This allows the primary logic supply to be turned off during sleep mode to save leakage power. At the same time, state of the FF is retained by only keeping the minimum sized slave latch powered. The NVL FF modifies a retention FF by adding two new ports ND and NU to the slave stage as shown in red. During regular operation, NU = L and ND = X. Here, the NVL FF behaves like a regular retention FF. After coming back from a sleep mode, data must be transferred from the NVL array into the NVL FF. To accomplish this, NU is pulsed high for a cycle while ND has valid data. RET is kept high at this time to allow ND to propagate to QB (CLK maybe X because the logic domain may be off while restoring data). When NU goes low, the latch closes and ND is latched as QB. Since inverted version of ND is available at Q, an inverter is placed in the path between the NVL array and the NVL FF. We carefully add circuits only to the slave latch so that they are not on the critical CLK to Q path of the FF and are minimum sized. The NVL FF is 19% larger, 2% slower and has 7% higher power than a retention FF. At the system level performance and power impact is not perceptible. Moreover, these numbers are for F1 FF. For higher drive strengths the percent increase is lower.

18.2.2   NVL Bitcell Design

In this section we describe the NVL bitcell. Signal margin is a critical metric for evaluating robustness of a ferroelectric capacitor (FeCap) based bitcell. Large non-volatile memory arrays implement Error Correction Codes (ECC) to detect and correct bit errors after a read operation. However, ECC is impractical to implement in a small 256b mini-array. In absence of ECC the bitcell inherently must have a high signal margin. Thus we use a custom bitcell rather

Fig. 18.11 An FRAM bitcell with two FeCaps using BL capacitance as load. An NVL bitcell uses four active FeCaps resulting in 2X the signal margin. Local SA helps avoid charge sharing between storage nodes and the bitlines. Bitcell read waveforms are shown with modifications for signal margin measurement in dotted lines. During a normal read, Q and QB develop a nominal signal of ~0.5 V and ~1 V, respectively. Pre-charging Q to 0.3 V reduces signal differential from 0.5 V to 0.2 V

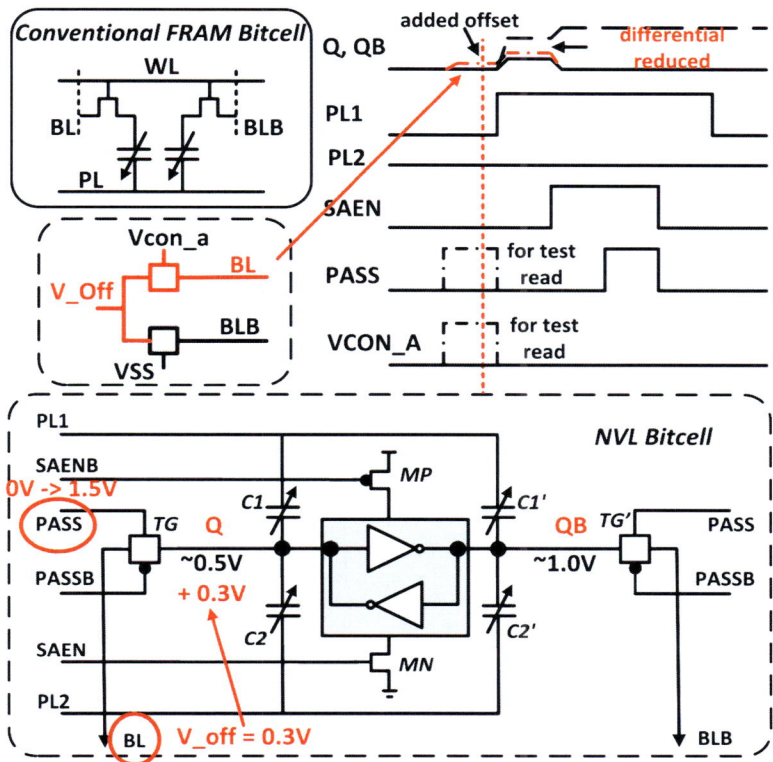

than a conventional FRAM bitcell. Figure 18.11 shows a conventional FRAM bitcell, the NVL bitcell and NVL read and signal margin measurement waveforms.

A ferroelectric capacitor is a programmable hysteretic capacitor whose value is set during a write based on the voltage polarity applied across the FeCap. In this way, a FeCap can be set to a low or high capacitance state. During a read, the FRAM bitcell thus behaves like a capacitive divider with the BL capacitor as a load. When PL is toggled, a signal develops on BL and BLB based on the value stored in the FeCaps. To get higher signal margin, we use an active FeCap load rather than the constant BL load. Data is stored in all four FeCaps rather than just two, resulting in almost double the signal margin as compared to an FRAM bitcell. Figure 18.12 shows the array periphery. The NVL array consists of the bitcell array, row drivers, IO circuits and a control and timing block.

The NVL bitcell is read using the waveform sequence shown in Fig. 18.11. NVL control signals are generated at clock edges of a high frequency NVL clock. Q and QB are initially discharged. PL1 is toggled high, which causes a signal to develop on Q (~0.5 V nominal) and QB (~1 V nominal). This differential (of 0.5 V) is then converted to a full swing signal by the sense amplifier (SA). Each bitcell has a local SA to avoid loss of signal due to charge sharing with bitlines. Loss due to charge sharing is small in nominal conditions, but can be significant for tail bits. Signal margin is measured by pre-charging the side storing a "0" to a disturb voltage. In the example shown in the figure, Q stores a "0". To measure signal margin, it is pre-charged to a disturb voltage (0.3 V here). Then, when PL1 is toggled, the signal that develops on Q is the sum of the disturb voltage and the signal that would have developed during a normal read (0.3 V + 0.5 V). The signal on the other side

Write Drivers   **RD latch and Scan flop**

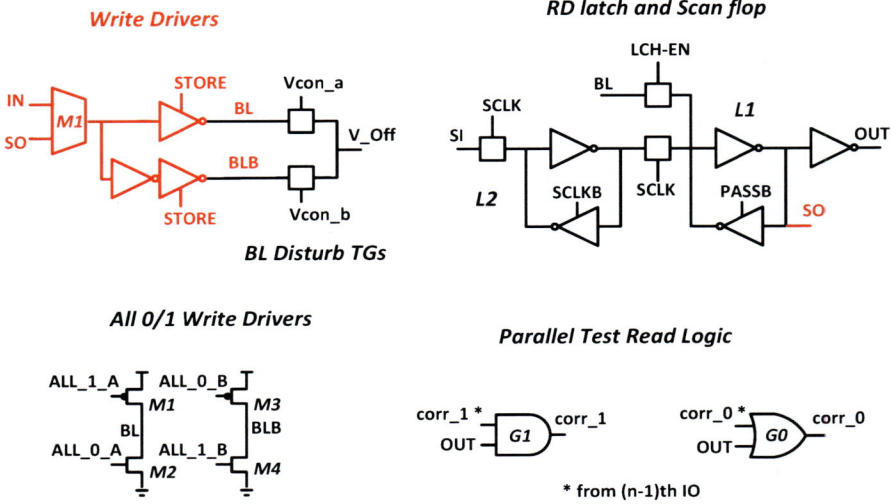

BL Disturb TGs

All 0/1 Write Drivers   **Parallel Test Read Logic**

Fig. 18.12 IO circuits used in the NVL array periphery. Read/write circuits are amortized over eight rows. Scan flop helps provide for a dedicated NVL scan chain. BL disturb TGs, and All 0/1 write and read circuits help implement fast on-chip NVL test. Implementing all these circuits in every FF is prohibitive, making the mini-array architecture the most appropriate for a practical NVL implementation

remains the same at 1.0 V. Hence, we are able to compress the differential available to the SA (from 0.5 V to 0.2 V). Signal margin for a bitcell is defined as the highest disturb voltage for a correct read.

During a write operation, store drivers in the array periphery are enabled and drive BL and BLB to the correct logic. In the waveform sequence, first PL1 and PL2 are pulsed high. Q and QB couple high due to capacitance of the FeCaps. Next, the pass gates are enabled and Q and QB are driven to the correct logic values. Then the SA is enabled which assists the store driver in holding Q and QB. At this time, the FeCaps on "0" side get written. Then the PL1 and PL2 and driven back low. At this time, the FeCaps in the "1" side get written. This marks the end of the write cycle.

NVL insertion into a design starts by addition of the NVL controller, NVL array and mux instantiations and an NVL controller interface on the power management module (PMM) RTL of the design. No changes are required to the core or peripherals. Note that at this point, FFs are not connected to NVL arrays. NVL FFs (which have two data inputs as opposed to one, as show in Fig. 18.10b) are not an RTL construct and are added to the design as a post synthesis step. This is very similar to adding scan to the design: only a scan controller and scan related IOs and interfaces are added to the RTL, but the actual scan chains and scan FFs are added only after synthesis. We have designed custom scripts that swap regular FFs with NVL FFs and connect them to NVL arrays in a physical aware manner during place and route.

Table 18.5 shows measured results from the 1.5 V HVT NVL MCU. The MCU runs at a maximum frequency of 8 MHz (at the slow corner) and consumes 75 µA/MHz and 170 µA/MHz while running code from SRAM and FRAM, respectively. NVL operates on a separate clock that can run up to 125 MHz resulting in an ultrafast 384 ns wakeup time. In comparison, conventional MCUs take 100s of microseconds to bootup from a power gated state. NVL has a one-time energy cost of backing up and restoring FF data. However, without NVL the sleep mode leakage due to the powered retention FF is 280 nW at 85C and 1740 nW at 125C. This results in a breakeven time of 34 ms and 5.5 ms, respectively. This is much smaller than

Table 18.5 NVL MCU simulation and measurement results

Regular SoC operation	
Technology node	130 nm, HVT
Clock frequency	8 MHz @ 1.5 V
Energy consumption (code in SRAM/FRAM)	SRAM: 75 µW/MHz FRAM: 170 µW/MHz
NVL sleep mode Lkg power	0.000 nW
Retention mode Lkg power (only slave stage powered)	280 nW @ 85C 1740 nW @ 125C
NVL operation	
Clock frequency	125 MHz
Chip backup cost	7.2 nJ, 320 ns
Chip restore cost	2.4 nJ, 384 ns
Breakeven time (based on leakage above)	34 ms @ 85C 5.45 ms @ 125C

Fig. 18.13 NVL MCU die photo

typical sleep times in most ULP applications. Moreover, we have not taken into account potential saving made possible by turning off the on-chip LDO that is usually part of an MCU SoC. Figure 18.13 shows the die photo. Since our MCU only has no power regulation, we achieve zero standby power. In a complete SoC the RTC, and some power management may be always on, but even then switching off the digital helps reduce standby power by about 40% according to our system level estimates. Area overhead at the SoC level is estimated to be ~10%.

18.3 Conclusion

Leakage power, dynamic power and mode transition time and energy are all critical in an MCU targeted for IoT applications. IoT applications are diverse and can vary in duty cycle, time period of usage and operating temperature range. An IoT MCU design must be optimized for all such usage scenarios. MCUs are highly integrated and thus a flexible but efficient system design is key to minimizing energy and latencies. In addition, innovations and optimizations are needed at all design hierarchies to get to the lowest energy and power targets. TI 430FR5XX devices feature multiple low power modes to accommodate applications with varying duty cycles. A range of LDOs and oscillators with different performance vs. power tradeoffs help achieve optimal performance and power number for each mode without compromising the other. Smart peripherals like the ADC, AES operate in low power mode and can alert the MCU core to wake up if needed. Technology node selection, transistor optimization, extensive clock gating, power gating, mixed Vt implementation, and retention flip-flops are used to minimize digital logic dynamic and leakage power. Some analog blocks in the chip are on even in the lowest power mode. Sampled analog techniques are used to reduce quiescent current in these blocks. This shows the range of optimizations needed to achieve low power targets in a ULP MCU.

On-chip Ferro-electric RAM (FRAM) offers non-volatility and low power, fast, random access write and read operation. Write energy per bit is 250× lower than flash memory, enabling data logging in energy constrained IoT applications. TI FRAM MCUs offer the combined benefits of an optimized ULP implementation and FRAM flexibility.

Mode transition time and energy restrict the use of the lowest power modes. Power gating the entire chip helps lower leakage, but requires a costly bootup process. Non-volatile logic helps lower leakage by transferring the contents of flip-flops into local NVL arrays before a power interruption. Upon wakeup, the data is transferred back, and no bootup is required. In this way, NVL helps lower the energy cost and breakeven time of using the lowest lower mode in an MCU. Further, the mini-array architecture helps provide a practical implantation of NVL as compared to other solutions that have yield, test time and area penalties.

References

AES for Texas Instruments MSP430 Microcontroller. http://jcewww.iaik.tu-graz.ac.at/index.php/sic/Products/Crypto_Software_for_Microcontrollers/AES_for_Texas_Instruments_MSP430_Microcontrollers

A. Baumann et al., A MCU platform with embedded FRAM achieving 350 nA current consumption in real-time clock mode with full state retention and 6.5 μs system wakeup time, in *VLSI Circuits (VLSIC), 2013 Symposium on, Kyoto, 2013* (2013), pp. C202–C203

M. Zwerg et al., An 82 uA/MHz Microcontroller with embedded FeRAM for energy harvesting applications, in *International Solid-State Circuit Conference (ISSCC)*, San Francisco, CA, 2011, pp. 334–335

S. Khanna, S.C. Bartling, M. Clinton, S. Summerfelt, J.A. Rodriguez, H.P. McAdams, An FRAM-based nonvolatile logic MCU SoC exhibiting 100% digital state retention at 0 V achieving zero leakage With 400-ns wakeup time for ULP applications. IEEE J. Solid-State Circ. **49**(1), 95–106 (2014)

Y. Liu et al., 4.7 A 65 nm ReRAM-enabled nonvolatile processor with 6× reduction in restore time and 4× higher clock frequency using adaptive data retention and self-write-termination nonvolatile logic, in *2016 I. E. International Solid-State Circuits Conference (ISSCC), San Francisco, CA, 2016* (2016), pp. 84–86

M. Lueders et al., Architectural and circuit design techniques for power management of ultra-low-power MCU systems. IEEE Trans. Very Large Scale Integr. Syst. **22**(11), 2287–2296 (2014)

MSP430 FR2xx Mixed Signal Microcontrollers: http://www.ti.com/lit/ds/symlink/msp430fr2533.pdf

MSP430 FR54xx Mixed Signal Microcontrollers: http://www.ti.com/lit/ds/symlink/msp430f5438a.pdf

MSP430 FR59xx Mixed Signal Microcontrollers: http://www.ti.com/lit/ds/symlink/msp430fr5969.pdf and http://www.ti.com/lit/ug/slau367g/slau367g.pdf

M. Qazi, A. Amerasekera, A.P. Chandrakasan, A 3.4pJ FeRAM-enabled D flip-flop in 0.13 μm CMOS for nonvolatile processing in digital systems, in *Solid-State Circuits Conference Digest of Technical Papers (ISSCC), 2013 I.E. International, 17–21 Feb. 2013* (2013), pp. 192–193

N. Sakimura et al., 10.5 A 90 nm 20 MHz fully nonvolatile microcontroller for standby-power-critical applications, in *Solid-State Circuits Conference Digest of Technical Papers (ISSCC), 2014 I.E. International, San Francisco, CA, 2014* (2014), pp. 184–185

Y. Wang, Y. Liu, S. Li, D. Zhang, B. Zhao, M.F. Chiang, Y. Yan, B. Sai, H. Yang, A 3us wake-up time nonvolatile processor based on ferroelectric flip-flops, in *Proceedings of the ESSCIRC (ESSCIRC), 17–21 Sept. 2012* (2012), pp. 149–152